Graduate Texts in Physics

Graduate Texts in Physics publishes core learning/teaching material for graduate- and advanced-level undergraduate courses on topics of current and emerging fields within physics, both pure and applied. These textbooks serve students at the MS- or PhD-level and their instructors as comprehensive sources of principles, definitions, derivations, experiments and applications (as relevant) for their mastery and teaching, respectively. International in scope and relevance, the textbooks correspond to course syllabi sufficiently to serve as required reading. Their didactic style, comprehensiveness and coverage of fundamental material also make them suitable as introductions or references for scientists entering, or requiring timely knowledge of, a research field.

Krishna Thyagarajan

Integrated Quantum Photonics

Springer

Krishna Thyagarajan
Former Professor
Indian Institute of Technology Delhi
New Delhi, Delhi, India

ISSN 1868-4513 ISSN 1868-4521 (electronic)
Graduate Texts in Physics
ISBN 978-3-031-85727-0 ISBN 978-3-031-85728-7 (eBook)
https://doi.org/10.1007/978-3-031-85728-7

This Springer imprint is published by the registered company Springer Nature Switzerland AG
The registered company address is: Gewerbestrasse 11, 6330 Cham, Switzerland

Preface

Quantum science and technology has assumed great importance in view of its applications in quantum computing, quantum key distribution, quantum sensing, quantum teleportation, and quantum information sciences. Quantum computing is expected to give us exponential increase in computing capacity, quantum key distribution will help us achieve secure information transmission, and quantum sensors will have sensitivities much beyond current capabilities. All these are based on the fundamental principles of quantum mechanics describing atoms and light through quantum description. Most advanced countries, including India, are significantly investing in this domain, as it is expected to become ubiquitous in future societies. In fact, it will be 100 years since the initial development of quantum mechanics, and recognizing this fact, the United Nations has proclaimed the year 2025 as the International Year of Quantum Science and Technology (IYQ).

Electromagnetic waves in the form of light waves with their quantum description in terms of photons are playing a vital role in this research. Photons are used in quantum key distribution for transfer of quantum keys for encryption and teleportation, quantum sensors based on photons are providing extreme sensitivities, and quantum computing with photons propagating through optical devices or interacting with atomic systems is helping us achieve increased complexity and speed of information processing. Today, advancement in the understanding and technology has provided us with sources emitting single photons, entangled photon pairs, squeezed photon states, single-photon detectors, etc., all of which are going to play a very vital role in the future of quantum science and engineering.

For widespread application and commercialization of quantum devices, it is imperative that the sources, detectors, and quantum information processing devices employing single photons have small footprint; consume low power; are reconfigurable, integrable into compact pre-aligned components, and scalable; and also provide novel configurations. Just as integrated microelectronic circuits transformed the world through integrated electronic circuits, the field of integrated quantum photonics is expected to help speed up the implementation of quantum devices to practical applications.

Integrated quantum photonics (IQP) deals with integration of photonic components operating at the quantum level (single or multiple photons) on a single chip, much like integrated electronic circuits and electrons. This enables all the advantages of integration in the form of small/low size, weight, and power (SWaP), novel functionalities, and ability to make future-proof technologies that are CMOS compatible. With its large-scale adoption, we can envision the ushering in of widespread application of quantum photonics in our daily lives. IQP also provides us with novel configurations of optical components to realize operations on photons that may not be feasible in bulk form. Increased research efforts in the field of integrated quantum photonics have resulted in chips with increased complexity and functionality with demonstrated applications.

With the increased importance of quantum information science and engineering, many educational institutions in the world have begun courses on the fundamentals of quantum photonics and its applications.

The proposed book aims to provide a comprehensive coverage of the fundamentals of integrated quantum photonics and some of its applications. The subject matter covered has been taught by the author at both undergraduate and graduate levels to students in top institutions in India and has also been presented in the form of short tutorials at international conferences in photonics.

Starting from a short recall of electromagnetic theory, the book covers the concepts of optical waveguides and their properties, various important basic waveguide devices such as directional couplers, electro-optic modulators and switches, waveguide arrays, ring resonators, and arrayed waveguide multiplexers, which form the basic components in integrated quantum photonic devices. The principles of nonlinear optics are used in the generation of novel nonclassical states of light; chapters on waveguide nonlinear optics provide the foundation of the field, and basic concepts are explained in detail. This is followed by the fundamentals of quantization of electromagnetic field and the concept of photons, which forms the foundation of quantum photonics. Single-mode photon states, multimode photon states, squeezed states, and entangled photon states are discussed from the point of view of clarifying the basic concepts. Integrated quantum photonic devices based on nonlinear optics provide us with novel ways of generating different nonclassical states of light, and different forms of entangled states are presented. Finally, some interesting and important counterintuitive experiments and brief discussion on some applications of integrated quantum photonics in quantum key distribution, quantum computing, and quantum teleportation are presented.

Since the focus of the book is toward understanding of the fundamentals, the various concepts presented throughout the book are derived from basic principles, and each chapter is provided with a number of solved exercises scattered throughout the text and also problems at the end. These have come out of the courses taught by the author and thus cover some basic concepts and applications of the subject matter discussed in the chapter. Some of the problems should also be of help to students to further their understanding of the field.

The book can be used as a text for a course on integrated quantum photonics with the aim of providing the basics of integrated optics, nonlinear optics, and a fundamental understanding and appreciation of the concepts of the quantum nature of light as well as the different nonclassical states of light. As the derivations and discussions are given in sufficient detail, the book can also be used for self-study to get introduced to this important topic to enable them to launch their research in this expanding field. The material provided in the book should help the student in his or her research or academic pursuits or in numerous applications and also in generating new ideas in this fertile field.

Bengaluru, India Krishna Thyagarajan

Acknowledgments

This book is a culmination of more than two decades of my collaborative research and teaching in the field of integrated quantum photonics and quantum electronics across the world. During this period, I have had the great fortune of interacting with some outstanding scientists and technologists and some bright students who attended my courses and worked with me toward their doctoral degree. I have thoroughly enjoyed the numerous discussions and interactions with them, many of which have helped me pursue significantly interesting scientific challenges and also develop a clearer understanding of the subject matter.

I would like to thank Dan Ostrowsky, Jean Sebastian Tanzilli, and Olivier Alibart of the Université Côte d'Azur, for our long-standing collaboration and insightful exchanges in the field of waveguide quantum photonics during my various visits to Nice, France, under an Indo-French collaborative program. These extended visits and stays were very helpful in clarifying many fundamental concepts while providing impetus for satisfying research. My visit to the University of Warsaw and interactions with Konrad Banaszek on multiple occasions helped me further in cementing my deeper understanding of many theoretical concepts in the area of quantum photonics. My deepest thanks to Professor Ajoy Ghatak, under whose mentorship I joined IIT Delhi for my doctoral research. He was instrumental in laying down my foundations in optical waveguides and their applications during our more than five decades of interactions thereafter. One has to be fortunate to have dedicated students to guide. Long discussions with my graduate students Jasleen Lugani, Divya Bharadwaj, and Pragati Aashna on their research topics of nonlinear and quantum photonics were enriching. On a similar vein, I would also like to thank the many students who attended my courses and posed challenging questions that led me further to a better understanding of the subject.

Additionally, this book covers a diverse range of sub-topics, and I would especially like to thank Sebastian Tanzilli, Olivier Alibart, and Konrad Banaszek who very patiently went through some of the sections in the book and gave me very valuable suggestions.

Thanks to Sam Harrison and the rest of the editorial team for their very prompt response to my queries and their inputs in making this work more fruitful.

I would like to thank Krishnan for his help in various facets of the manuscript and for his invaluable inputs. Since writing of the book took away much time that could have been spent with the family, I would like to express my gratitude to Abhinandan, Kalyani, Krishnan, Maitreya, and Raji for all their patience and understanding.

Finally, I hope that as a reader, you find reading this book as enjoyable as I did in writing it.

Competing Interests The author has no competing interests to declare that are relevant to the content of this manuscript.

Contents

About the Author

Krishna Thyagarajan a multicultural educator with more than four decades of experience as a Professor of Physics at the Indian Institute of Technology, Delhi (IITD), and then at Bennett University (Times Media Group), has played multiple roles across teaching, research, and administration, as a Professor, Head of the Department, and Officiating Director (IITD). Prof. Thyagarajan has germinated novel ideas and demonstrated high-impact scientific results in the fields of fiber optics, optical telecommunication, quantum optics, and integrated optoelectronics, in projects funded by both federal (Asia, Europe, USA) and commercial sources. During his career, he has held numerous prestigious visiting scientist positions across the globe including in the USA, France, Canada, Japan, Hong Kong, and Poland and has been decorated by multiple governments (India, France, USA) and industrial organizations for his outstanding contributions to science and education. He has held key positions in national and international federal committees overseeing funding allocations and scientific work in the areas of quantum optics, optical telecommunication, and integrated photonics. With more than 175 research papers, several patents, authorship of more than 10 internationally acclaimed textbooks, and creation of 73 nationally acclaimed physics education videos (quantum electronics, electromagnetics) on advanced topics to his name, Prof. Thyagarajan strives to bridge cutting-edge scientific research and STEM education to motivate future generations of scientists to pursue their passion in science and technology.

Chapter 1
Introduction

Abstract This chapter introduces the field of integrated quantum photonics and its importance in the field of quantum information science and technology. Brief descriptions of the topics covered in various chapters of the entire book are given.

The first quantum revolution occurred during the beginning of the last century and profoundly influenced our fundamental concepts of physics. The mathematical developments in quantum mechanics brought about during the revolution led to better understanding and helped in achieving many technological breakthroughs. These include the very important semiconductor technology, lasers, and many more fundamental components and devices with ubiquitous applications in science, technology, and medicine.

Today, we are witnessing the second quantum revolution (Dowling and Milburn 2003) through the field of quantum photonics, opening up immense opportunities in the domain of information and communication technologies. This is also enabling the creation of new avenues for our further understanding of the fundamental workings of nature. In this revolution, the primary objective is to harness the immense power and unique features of quantum physics to encode, process, and transmit information with the aim to dramatically improve and innovate in the fields of communication, computation, sensing, and measurement. This includes the ability to achieve unbreakable quantum keys for secure communication, to realize computational architectures with immense power not realizable with classical computing systems, and to be able to sense and measure various physical parameters at unprecedented levels of sensitivity.

In this adventure, optics is already playing a very significant role in advancing the necessary technology through innovations in the design and realization of photonic components such as single photon sources, squeezed sources of light, entangled photon sources, and low noise optical amplifiers, with the ability to be able to manipulate optical signals at the single-photon level. As with the electronics revolution where scalability and widespread adoption needed miniaturization and integration of heterogeneous electronic components on a chip, the field of quantum photonics will immensely benefit from the integration of disparate photonic components to process quantum information using single quanta of light—photons—on a

K. Thyagarajan, *Integrated Quantum Photonics*, Graduate Texts in Physics,
https://doi.org/10.1007/978-3-031-85728-7_1

single chip. This has led to the emergence of the new field of *integrated quantum photonics*. In bulk photonics, quantum information processing is carried out using bulk photonic components such as table-top lasers, beam splitters, and polarization components, detectors, requiring large optical tables. This inherently leads to challenges of alignment, power consumption, and overall size of the entire platform, with the intrinsic issue of their scalability for commercial applications. In contrast, integrated quantum photonics aims to integrate miniature photonic components on a single integrated quantum photonic chip which can process single or multiple photons at the quantum level. The field of integrated quantum photonics is developing rapidly with many demonstrations of complex quantum photonic circuitry in integrated form with hundreds of photonic components on a single chip. These include squeezed sources of light, generation of entangled pairs of photons, quantum processing of information using quantum gates, and quantum sensors with extreme sensitivity. With increased effort worldwide in the development of integrated quantum photonic circuits, it is evident that this will lead to increased utilization of quantum photonic applications in multiple areas, just as was the case with integrated electronic circuits (see, e.g., Gisin and Thew 2007; Politi et al. 2009; Aspuru-Guzik and Walther 2012; Pirandola et al. 2018; Slussarenko and Pryde 2019; Wang et al. 2019; Saravi et al. 2021; Moody et al. 2022; Chen et al. 2022; Pelucchi et al. 2022; Walmsley 2023; Labonté et al. 2024; Piergentili et al. 2024; Hu et al. 2023; Xie et al. 2024).

Figure 1.1 gives some integrated photonic components that are found in many integrated quantum photonic circuits. Many of these components also play a significant role in the field of light wave communication and sensing.

It is interesting to point out that the term integrated optics was coined by S E Miller from Bell Telephone Laboratories in 1969 (Miller 1969) with the aim to integrate optical components such as waveguides (to guide light efficiently), light

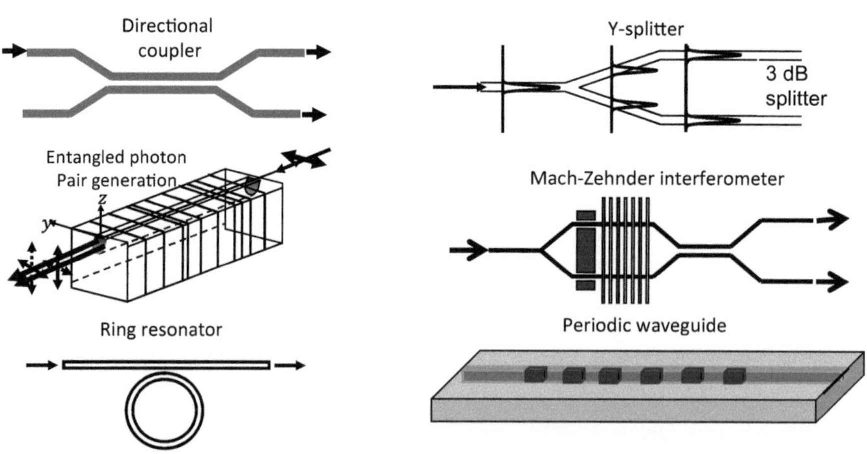

Fig. 1.1 Some integrated photonic components used in integrated quantum photonic devices

sources, detectors, modulators, switches, and polarization components, on a single optical chip. The primary objective was to build miniature optical devices which were of small size and compact footprint, realized through standard optical lithography techniques (like those used for integrated electronic circuits). This would enable the chips to perform optical functions efficiently and with low power consumption for applications in communication and optical signal processing. After the proposal, research and developmental efforts in the field progressed rapidly with the development of high-speed optical modulators, fast optical switches, wavelength division multiplexers and demultiplexers, polarization-based devices, and optical amplifiers. Today, many of these devices find widespread applications in high-speed optical fiber communication systems carrying hundreds of terabits of information traffic worldwide. New developments in communication techniques are leading to further developments of novel configurations of integrated optic devices with expanding applications (see, e.g., Shekhar et al. 2023; Li et al. 2023; Xie et al. 2024; Chen et al. 2024; Zhang et al. 2024a, b).

The field of integrated quantum photonics expands further the domain of operation of integrated optics (Walmsley 2015) and aims to realize photonic chips capable of processing quantum information. Since many integrated quantum photonic devices have different requirements, various suitable material platforms have been developed in parallel. Among these the important ones are based on lithium niobate, silicon, semiconductors (such as GaAs, InP), glass, silica, and hybrid combinations of some of these. Each one of these has its own advantages and applications and current research and development are exploring all these material systems for various applications. (see, e.g., Luo et al. 2023; Bogdanov et al. 2017; Wang et al. 2019; Elshaari et al. 2020; Baboux et al. 2023).

Lithium niobate is an extremely versatile material and has been one of the workhorses of the field of integrated optics providing low loss waveguides, high-speed modulators, switches, and many other devices for high-speed optical fiber communication systems and sensors. This has been primarily due to its excellent characteristics of transparency over a large band of wavelengths (0.4–5 μm), strong electro-optic coefficients leading to reduced voltage requirements for modulators and switches, strong acousto-optic coefficient for various applications such as acousto-optic spectrum analyzers, strong nonlinear optical coefficients for generation of new frequencies of light, and the ability to fabricate low loss waveguides on single crystal substrates. These important characteristics are also exploited in integrated quantum photonic devices based on lithium niobate (see, e.g., Tanzilli et al. 2012; Luo et al. 2019; Boes et al. 2023). A significant research effort is currently underway to use thin films of lithium niobate bonded on insulating substrate such as silica (referred to as lithium niobate on insulator-LNOI or thin film lithium niobate-TFLN) which is a very promising platform as it provides waveguides with extremely tight light confinement. This results in reduced power requirement for nonlinear optical applications and also reduced size of the fabricated devices (see, e.g., Zheng et al. 2023; Saravi et al. 2021).

Integrated quantum photonic devices based on silicon include silica-on-silicon, silica on insulator, and silicon nitride on silicon. Since silicon has been the workhorse of integrated electronic components, the technology for device design, fabrication, and characterization is extremely well developed. Unlike in the field of integrated electronics based on silicon, in the domain of integrated quantum photonics, the device needs to operate with photons instead of electrons. This implies that the appropriate technology to realize low loss optical waveguides, ensure proper coupling of light into the device, and realize waveguide optical components such as modulators, and multiplexers needs to be developed—something that has been progressing rapidly. In view of the large refractive index contrast between silicon and silica, it is possible to realize single mode optical waveguides with extremely small cross-sectional areas and small bend radii to achieve high component density in the device. In contrast to the second-order nonlinearity that is used in lithium niobate, in silicon the third-order nonlinearity is used for frequency conversion, etc. Some of the most advanced integrated quantum photonic devices have indeed been realized using this platform (see, e.g., Politi et al. 2008; Silverstone et al. 2016; Harris et al. 2014).

Realizing the importance of quantum information science and technology, most countries in the world are currently funding extensive research and development in the field of quantum technology with a view to develop practical systems for quantum computation, quantum key distribution, quantum teleportation, quantum sensing, and quantum metrology. In this spirit, integrated quantum photonics is expected to play a significant role in the commercialization and application of the field of quantum science and technology. This is leading to the initiation of educational programs in multiple universities across the globe, both at the undergraduate and post graduate levels, and of short courses and workshops in international conferences. With this perspective, there is thus a need for a textbook that can introduce the reader to the fundamental concepts related to integrated quantum photonics and help foster the skillsets and intellectual knowhow that will become necessary soon to power the industry. This textbook is an effort in that direction.

The book starts with two chapters, one on electromagnetics and the other on quantum mechanics which is more like a recall of the two subjects and covers some basic concepts. For more details on these two subject areas, readers are encouraged to read other excellent texts on these subjects.

The next part consists of five chapters related to the field of optical waveguides.

Chapter 4 discusses optical waveguides which guide light and introduces the concept of propagating modes and their characteristics in both planar and channel waveguides. Such waveguides form the basic unit of integrated quantum photonic devices and the characteristics of the modes determine the performance of the device. Since many materials such as lithium niobate, lithium tantalate, and potassium titanyl phosphate are anisotropic media, this chapter also discusses propagation characteristics of anisotropic waveguides. These results will be needed for discussions in later chapters when devices based on these waveguides are analyzed.

Chapter 5 discusses the very important concept of directional couplers which consist of a pair of closely placed optical waveguides that interact through evanescent coupling of power among the propagating modes. Two waveguide directional couplers act like bulk beam splitters and are a part of most integrated quantum photonic devices today. They exhibit very interesting quantum behavior and form the basic component in many integrated quantum photonic device architectures. Directional coupler devices with multiple interacting waveguides are used in many applications such as quantum random walk, and the Boson sampling problem, and are also discussed in the chapter.

Chapter 6 discusses the electro optic effect or the Pockels effect, which relates the changes in the optical property such as the refractive index and principal axes directions of a medium brought about by an applied external electric field. Waveguide devices based on the electro optic effect form the basic building blocks of high-speed optical modulators and high-speed optical switches. They are also extensively used in high-speed optical fiber communication systems and for achieving dynamic behavior of integrated quantum photonic devices.

Chapter 7 introduces two very important components, namely the Y-splitter and the Mach-Zehnder interferometer. In a Y-splitter, as the name implies, a waveguide branches into two waveguides across a Y-junction. Depending on the geometry of the Y-splitter, the incident power can divide equally among the two output waveguides, unequally among the waveguides, demultiplex two spatial modes, or separate two orthogonal polarizations. Such a Y-splitter is an extremely useful device in many integrated quantum photonic implementations. The Mach-Zehnder interferometer is the waveguide version of a standard bulk Mach-Zehnder interferometer and again forms an important building block of many integrated quantum photonic devices. Such a device is used in realizing high-speed optical modulators for optical fiber communication systems and for sensing applications and in photonic quantum computing applications.

Chapter 8 discusses ring resonators which utilize multiple interference due to the mode propagating along a waveguide in the form of a ring, to achieve very unique transmission characteristics. They are deployed in many implementations, for wavelength filtering, frequency comb generation, and entangled photon pair generation. The basic working principle of arrayed waveguide devices is presented. Arrayed waveguide multiplexers/demultiplexers form very important components in multi-wavelength integrated quantum photonic architectures.

Chapter 9 discusses how a periodic modulation in the property such as the refractive index, width or depth of the waveguide, etc. leads to coupling of power among different orthogonal modes in the waveguide. The periodic modulation acts like a perturbation and when certain phase matching conditions are satisfied, it can lead to very efficient power transfer from one spatial mode to another or result in a reflection of the propagating mode like a mirror. At the heart of distributed feedback lasers and distributed Bragg reflector lasers is periodic modulation of the waveguide characteristic. Periodic modulation of the linear optical property is used in coupling of power among different modes at the same frequency. Finally, periodic modulation of the nonlinear property of the waveguide finds wide applications in coupling of power among modes at different frequency and forms the heart of integrated quantum photonics. This is discussed in the next four chapters.

The next set of four chapters discuss nonlinear effects in optical waveguides.

Chapter 10 introduces the concept of nonlinear electric polarization that leads to the generation of new frequencies when light propagates through the waveguides. Depending on the medium, its nonlinear optical property and the propagating modes, it is possible to generate light at frequencies different from those of the incident waves. In this book, we restrict ourselves to second-order nonlinearity only, as it is one of the most important properties that are used in frequency conversion in waveguides. Using optical waveguides in frequency conversion devices provides us with additional degrees of freedom in the optimization of the conversion efficiency or new configurations that are not possible in bulk.

Chapter 11 discusses second harmonic generation in optical waveguides, in which light propagating with a frequency ω interacts with the nonlinear medium and generates light at the second harmonic frequency 2ω. Efficient generation involves satisfaction of certain phase matching conditions which in current practical implementations are resolved through quasi phase matching (QPM) technique in which the nonlinear property of the medium is periodically modulated along the propagation direction. This QPM technique allows one to achieve high efficiency conversion among frequencies. This is an extremely useful effect for conversion from a lower frequency to a higher frequency and is used in many quantum photonic devices.

Chapter 12 discusses sum frequency generation and difference frequency generation. In these processes, when light at two different frequencies is input into the waveguide, the nonlinear interaction within the medium leads to the generation of light at a frequency which is the sum of the two input frequencies or the difference among the two input frequencies. This is an extremely important technique to generate coherent light at different frequencies or to shift the frequency of a signal from one region such as infrared to a more convenient region such as the visible.

An interesting aspect of difference frequency generation is the possibility of amplification of one of the input frequencies. This is referred to as parametric amplification which is discussed in Chap. 13. Unlike conventional optical amplifiers such as erbium-doped fiber amplifiers, parametric amplifiers do not use the energy levels of the atomic system and hence are capable of amplifying signals over the band of transparency of the medium. The amplification process is sensitive to the phase of the input waves. Such phase-sensitive amplifiers have much better noise characteristics than conventional amplifiers and are being investigated for applications in long distance fiber optic communication systems, and the generation of squeezed light for applications in sensing.

The following Chaps. 14–18 present quantum aspects of light.

Chapter 14 discusses the basics of quantization of electromagnetic fields that leads to the concept of photons. It is shown that each mode of the electromagnetic field is equivalent to a harmonic oscillator and thus the mode is quantized in its energy. This naturally leads to the concept of a photon as a quantum of excitation of the electromagnetic field.

Chapter 15 discusses the important class of single mode photon states, number states, vacuum state, superposition states, and coherent states. Many of these states form the basis of applications of photonics in quantum information processing.

Chapter 16 discusses multimode photon states and introduces a single photon temporal wave packet.

Chapter 17 presents the basics of entangled states which form an extremely important aspect of quantum science and technology and have led to the development of applications in quantum key distribution, quantum teleportation, quantum computing, and quantum sensing, to name a few. Entangled states are very counterintuitive and heavily influence current applications of quantum photonics. Photons could be entangled in different degrees of freedom such as polarization, spatial mode, and frequency.

Chapter 18 discusses squeezed states which are quantum states that exhibit noise that is below that of a standard state coming out of a laser or a vacuum state. Such states are finding applications in improving the sensitivity of optical sensors and are currently already helping the laser interferometer gravitational wave (LIGO) detectors to improve their sensitivity in detecting gravitation waves, significantly.

Chapter 19 discusses the quantum aspect of beam splitters which form very important components in most quantum photonic applications. The well-known Hong-Ou-Mandel dip that appears when two identical photons are incident on a beam splitter is also presented. The quantum analysis of the Mach-Zehnder interferometer is also given.

Chapter 20 discusses the basics of optical homodyning technique that is used extensively in quantum photonics. This technique forms the basis of many quantum photonic measurement setups for characterizing the quantum behavior of the incident radiation.

Chapter 21 briefly discusses Schrödinger, Heisenberg, and interaction pictures of quantum mechanics.

Chapter 22 presents the quantum treatment of parametric processes. The process of spontaneous parametric down conversion (SPDC) which is one of the most used techniques for the generation of entangled photon pairs, for heralded single photons, and is a purely quantum mechanical process is discussed in detail. While the SPDC process produces entangled photon pairs, in many applications pure single photons that are not entangled to other photons are required. The process of generating photons that are unentangled is also discussed. This chapter also discusses the quantum treatment of phase-sensitive parametric amplifiers and their improved noise performance.

Chapter 23 presents different techniques for the generation of entangled photon pairs that are entangled in different degrees of freedom such as polarization, spatial mode of the waveguide, frequency, and energy-time. Time-bin entangled states that are extremely important for their applications in quantum key distribution are also presented.

Chapter 24 discusses some intriguing experiments in quantum photonics and presents a brief mathematical analysis of the results showing that quantum mechanical analysis can explain all the results of the experiments. These are some experiments that are counterintuitive and expose us to the beauty of the mathematical and philosophical foundations of quantum mechanics. They show how using the mathematical basis, it is possible to explain the experimentally observed behavior of photons.

Chapter 25 discusses briefly the basics of three very important applications of quantum photonics, namely: (a) quantum key distribution that is employed for secret exchange of information without any possibility of eavesdropping, (b) photonic quantum computing that uses the quantum properties of photons and integrated quantum photonic devices for photonic quantum information processing, and (c) quantum teleportation that involves teleportation of quantum information.

A very interesting application of integrated quantum photonic systems is the computational problem referred to as Boson sampling which consists in sampling from the output distribution of n indistinguishable bosons that interfere during their evolution through a linear network and which has been shown to be an intractable problem for classical computation. This problem was very nicely solved using multiple photons propagating through a passive linear optical interferometer consisting of an array of interacting waveguides (see, e.g., Crespi et al. 2013; Brod et al. 2019).

Although many of the examples discussed in the book are based on lithium niobate waveguides and substrates, the basic principles of operation of most of the devices discussed in the book such as propagation characteristics of optical waveguides, directional couplers, interferometers, ring resonators, and periodic interaction are also applicable to other material platforms. Lithium niobate possesses second-order nonlinearity and thus allows for three-wave interaction processes. Being extensively already used in the industry, lithium niobite also lends itself to an easier understanding for the more familiar reader. On the other hand, in waveguides based on silica and silicon one has to employ the third-order nonlinearity that allows for four wave interactions for which the readers are referred to other reference material.

Each chapter has many solved exercises scattered through the text and unsolved problems at the end of each chapter. These exercises and problems should help the reader to better understand the subject and the application of the concepts developed in the chapter. Most of these problems have been used in various courses taught by the author.

The field of integrated quantum photonics and its applications in various fields of information science and technology are expanding rapidly. Recent demonstrations of combining up to 650 optical and electronic components on a single chip that are capable of processing signals at the quantum level show the tremendous progress in the field pointing to the possibilities that are expected in the coming decades. The growth is expected to continue in the near future, promising more advanced processing and fabrication techniques, new designs, novel components, increased component density on chips, and much more. For more information on the roadmap of the future of quantum photonics, readers are referred to the following two excellent reviews written by leading scientists and technologists: Pelucchi et al. (2022) and Moody et al. (2022).

References

Aspuru-Guzik, A., and P. Walther. 2012. Photonic quantum simulators. *Nat Phys* 8:285–291.

Baboux, F., G. Moody, and S. Ducci. 2023. Nonlinear integrated quantum photonics with AlGaAs. *Optica* 10:917–931.

Boes, A., L. Chang, C. Langrock, M. Yu, M. Zhang, Q. Lin, M. Loncar, M. Fejer, J. Bowers, and A. Mitchell. 2023. Lithium niobate photonics: unlocking the electromagnetic spectrum. *Science* 379:eabj4396.

Bogdanov, S., M. Y. Shalaginov, A. Boltasseva, and V. M. Shalaev. 2017. Material platforms for integrated quantum photonics. *Opt Mater Express* 7:111–132.

Brod, D. J., E. F. Galvão, A. Crespi, R. Osellame, N. Spagnolo, and F. Sciarrino. 2019. Photonic implementation of boson sampling: a review. *Adv Photonics* 1:034001.

Chen, Y., L. Hong, and L. Chen. 2022. Quantum interferometric metrology with entangled photons. *Front Phys* 10:892519. https://doi.org/10.3389/fphy.2022.892519.

Chen, C., Y. Chen, Z. Fang, R. Ge, J. Wu, and X. Chen. 2024. Hybrid material integration for active photonic applications. *APL Photonics* 9:030903.

Crespi, A., R. Osellame, R. Ramponi, D. J. Brod, E. F. Galvão, N. Spagnolo, C. Vitelli, E. Maiorino, P. Mataloni, and F. Sciarrino. 2013. Integrated multimode interferometers with arbitrary designs for photonic boson sampling. *Nat Photonics* 7:545–549.

Dowling, J. P., and G. J. Milburn. 2003. Quantum technology: the second quantum revolution. *Phil Trans Roy Soc London A* 361:1655–1674.

Elshaari, A. W., W. Pernice, K. Srinivasan, O. Benson, and V. Zwiller. 2020. Hybrid integrated quantum photonic circuits. *Nat Photonics* 14:285–298.

Gisin, N., and R. Thew. 2007. Quantum communication. *Nat Photonics* 1:165–171.

Harris, N. C., D. Grassani, A. Simbula, M. Pant, M. Galli, T. Baehr-Jones, M. Hochberg, D. Englund, D. Bajoni, and C. Galland. 2014. Integrated source of spectrally filtered correlated photons for large scale quantum photonic systems. *Phys Rev X* 4:041047.

Hu, X.-M., Y. Guo, B.-H. Liu, C.-F. Li, and G.-C. Guo. 2023. Progress in quantum teleportation. *Nat Rev Phys* 5:339–353.

Labonté, L., O. Alibart, V. D'Auria, F. Doutre, J. Etesse, G. Sauder, A. Martin, É. Picholle, and S. Tanzilli. 2024. Integrated photonics for quantum communications and metrology. *Phys Rev X Quan* 5:010101.

Li, Z., R. N. Wang, G. Lihachev, J. Zhang, Z. Tan, M. Churaev, N. Kuznetsov, A. Siddharth, M. J. Bereyhi, J. Riemensberger, and T. J. Kippenberg. 2023. High density lithium niobate photonic integrated circuits. *Nat Commun* 14:4856. https://doi.org/10.1038/s41467-023-40502-8.

Luo, K.-H., S. Brauner, C. Eigner, P. R. Sharapova, R. Ricken, T. Meier, H. Herrmann, and C. Silberhorn. 2019. Nonlinear integrated quantum electro-optic circuits. *Sci Adv* 5:eaat1451.

Luo, W., L. Cao, Y. Shi, L. Wan, H. Zhang, S. Li, G. Chen, Y. Li, S. Li, Y. Wang, S. Sun, M. F. Karim, H. Cai, L. C. Kwek, and A. Q. Liu. 2023. Recent progress in quantum photonic chips for quantum communication and internet. *Light Sci Appl* 12:175.

Miller, S. E. 1969. Integrated optics: An introduction. *Bell Syst Tech J* 48:2059–2069.

Moody, G., V. J. Sorger, D. J. Blumenthal, P. W. Juodawlkis, W. Loh, C. Sorace-Agaskar, A. E. Jones, K. C. Balram, J. C. F. Matthews, A. Laing, M. Davanco, L. Chang, J. E. Bowers, N. Quack, C. Galland, I. Aharonovich, M. A. Wolff, C. Schuck, N. Sinclair, M. Lončar, T. Komljenovic, D. Weld, S. Mookherjea, S. Buckley, M. Radulaski, S. Reitzenstein, B. Pingault, B. Machielse, D. Mukhopadhyay, A. Akimov, A. Zheltikov, G. S. Agarwal, K. Srinivasan, J. Lu, H. X. Tang, W. Jiang, T. P. McKenna, A. H. Safavi-Naeini, S. Steinhauer, A. W. Elshaari, V. Zwiller, P. S. Davids, N. Martinez, M. Gehl, J. Chiaverini, K. K. Mehta, J. Romero, N. B. Lingaraju, A. M. Weiner, D. Peace, R. Cernansky, M. Lobino, E. Diamanti, L. T. Vidarte, and R. M. Camacho. 2022. Roadmap on integrated quantum photonics. *J Phys Photonics* 4:012501.

Pelucchi, E., G. Fagas, I. Aharonovich, D. Englund, E. Figueroa, Q. Gong, H. Hannes, J. Liu, C.-Y. Lu, N. Matsuda, J.-W. Pan, F. Schreck, F. Sciarrino, C. Silberhorn, J. Wang, and K. D. Jöns.

2022. The potential and global outlook of integrated photonics for quantum technologies. *Nature* 4:194–208.

Piergentili, P., F. Amanti, G. Andrini, F. Armani, V. Bellani, V. Bonaiuto, S. Cammarata, M. Campostrini, S. Cornia, T. H. Dao, F. De Matteis, V. Demontis, G. Di Giuseppe, S. D. Tchernij, S. Donati, A. Fontana, J. Forneris, R. Francini, L. Frontini, R. Gunnella, S. Iadanza, A. E. Kaplan, C. Lacava, V. Liberali, F. Marzioni, E. N. Hernández, E. Pedreschi, D. Prete, P. Prosposito, V. Rigato, C. Roncolato, F. Rossella, A. Salamon, M. Salvato, F. Sargeni, J. Shojaii, F. Spinella, A. Stabile, A. Toncelli, G. Trucco, and V. Vitali. 2024. Quantum information with integrated photonics. *Appl Sci* 14:387. https://doi.org/10.3390/app14010387.

Pirandola, S., B. R. Bardhan, T. Gehring, C. Weedbrook, and S. Lloyd. 2018. Advances in quantum sensing. *Nat Photonics* 12:724–733.

Politi, A., M. J. Cryan, J. Rarity, S. Yu, and J. L. O'Brien. 2008. Silica on silicon waveguide quantum circuits. *Science* 320:646–649.

Politi, A., J. C. F. Mathews, M. G. Thompson, and J. L. O'Brien. 2009. Integrated quantum photonics. *IEEE J Selected Topics Quant Electron* 15:1673–1684.

Saravi, S., T. Pertsch, and F. Setzpfandt. 2021. Lithium niobate on insulator: An emerging platform for integrated quantum photonics. *Adv Opt Mater* 9:2100789. https://doi.org/10.1002/adom.202100789.

Shekhar, S., W. Bogaerts, L. Chrostowski, J. E. Bowers, M. Hochberg, R. Soref, and B. J. Shastri. 2023. Roadmapping the next generation of silicon photonics. *Nat Commun* 15:75. https://doi.org/10.1038/s41467-024-44750-0.

Silverstone, J. W., D. Bonneau, J. L. O'Brien, and M. G. Thompson. 2016. Silicon quantum photonics. *IEEE J Select Topics Quant Electron* 22:6700113.

Slussarenko, S., and G. J. Pryde. 2019. Photonic quantum information processing: A concise review. *Appl Phys Rev* 6:041303.

Tanzilli, S., A. Martin, F. Kaiser, M. P. DeMicheli, O. Alibart, and D. B. Ostrowsky. 2012. On the genesis and evolution of integrated quantum optics. *Laser Photonics Rev* 6:115–143.

Walmsley, I. A. 2015. Quantum optics: Science and technology in a new light. *Science* 348:525–530.

Walmsley, I. 2023. Light in quantum computing and simulation: perspective. *Opt Quant* 1:35–40.

Wang, H., J. Qin, X. Ding, M.-C. Chen, S. Chen, X. You, Y.-M. He, X. Jiang, L. You, Z. Wang, C. Schneider, J. J. Renema, S. Höfling, C.-Y. Lu, and J.-W. Pan. 2019. Boson sampling with 20 input photons and a 60-mode interferometer in a 10^{14} dimensional Hilbert space. *Phys Rev Lett* 123:250503.

Xie, Z., F. Bo, J. Lin, H. Hu, X. Cai, X.-H. Tian, Z. Fang, J. Chen, M. Wang, F. Chen, Y. Cheng, J. Xu, and S. Zhu. 2024. Recent developments in integrated lithium niobate photonics. *Adv Phys X* 9:1–44.

Zhang, C., M. A. Tran, Z. Zhang, A. E. Dorche, Y. Shen, B. Shen, K. Asawa, G. Kim, N. Kim, F. Levinson, J. E. Bowers, and T. Komljenovic. 2024a. Integrated photonics beyond communications. *Appl Phys Lett* 123:230501.

Zhang, Z., C. You, O. S. Magaña-Loaiza, R. Fickler, R. de J. León-Montiel, J. P. Torres, T. S. Humble, S. Liu, Y. Xia, and Q. Zhuang. 2024b. Entanglement-based quantum information technology: a tutorial. *Adv Opt Photonics* 16:60–162.

Zheng, Y., H. Zhong, H. Zhang, L. Song, J. Liu, Y. Liang, Z. Liu, J. Chen, J. Zhou, Z. Fang, M. Wang, L. Li, R. Wu, and Y. Cheng. 2023. Electro-optically programmable photonic circuits enabled by wafer scale integration on thin-film lithium niobate. *Phys Rev Res* 5:033206.

Chapter 2
Quick Recall of Basic Electromagnetics

Abstract This chapter gives a brief recall of basics of electromagnetics starting form Maxwell's equations followed by the wave equation. After a discussion on plane waves and states of polarization using Jones' vector formulation, plane wave propagation through anisotropic media is presented.

2.1 Introduction

In this chapter, we will very briefly review the basics of electromagnetics and introduce concepts that will be required in later chapters in the book. Since anisotropic media play a very important role in integrated quantum photonics as many integrated optic devices are fabricated in anisotropic substrates, a brief discussion on propagation through anisotropic media is also given. For a more rigorous coverage of the topics, readers are referred to excellent literature on electromagnetics and optics (see, e.g., Born and Wolf 1980; Ghatak and Thyagarajan 1989).

2.2 Maxwell's Equations

In the absence of free charges ($\rho_f = 0$) and free currents $\left(\vec{J}_f = 0\right)$, the following four equations constitute the four fundamental Maxwell's equations which govern all electromagnetic phenomena:

$$\nabla \times \vec{E} = -\frac{\partial \vec{B}}{\partial t} \tag{2.1}$$

$$\nabla \times \vec{H} = \frac{\partial \vec{D}}{\partial t} \tag{2.2}$$

K. Thyagarajan, *Integrated Quantum Photonics*, Graduate Texts in Physics,
https://doi.org/10.1007/978-3-031-85728-7_2

$$\nabla.\vec{D} = 0 \tag{2.3}$$

$$\nabla.\vec{B} = 0 \tag{2.4}$$

Here $\vec{E}$ and $\vec{B}$ represent the electric and magnetic fields, respectively, $\vec{D}$ and $\vec{H}$ are related to the electric and magnetic fields through the following defining equations:

$$\vec{D} = \epsilon_0 \vec{E} + \vec{P} \tag{2.5}$$

$$\vec{H} = \frac{\vec{B}}{\mu_0} - \vec{M} \tag{2.6}$$

where $\vec{P}$ is the electric polarization and is the electric dipole moment per unit volume in the medium and $\vec{M}$ is the magnetization which is the magnetic dipole moment per unit volume. ϵ_0 and μ_0 are the electric permittivity and magnetic permeability of free space and

$$\mu_0 = 4\pi \times 10^{-7} \mathrm{N/A^2} \tag{2.7}$$

with

$$\epsilon_0 \mu_0 = \frac{1}{c^2} \tag{2.8}$$

with $c = 2.99792458 \times 10^8$ m/s being the speed of light in free space which is usually approximated to $c \approx 3 \times 10^8$ m/s. Using this value of c in Eq. (2.8), we obtain an approximate value of ϵ_0 as

$$\epsilon_0 \approx 8.854 \times 10^{-12} \mathrm{C^2/N.m^2} \tag{2.9}$$

In linear and isotropic media, we have the following relations between $\vec{P}$ and $\vec{E}$ and between $\vec{M}$ and $\vec{H}$:

$$\vec{P} = \epsilon_0 \chi \vec{E} \tag{2.10}$$

$$\vec{M} = \chi_m \vec{H} \tag{2.11}$$

where χ and χ_m represent the electric susceptibility and magnetic susceptibility, respectively, of the medium. Substituting in Eqs. (2.5) and (2.6), we obtain

$$\vec{D} = \epsilon_0 \vec{E} + \epsilon_0 \chi \vec{E} = \epsilon \vec{E} \tag{2.12}$$

$$\vec{B} = \mu_0\left(\vec{H} + \chi_m\vec{H}\right) = \mu\vec{H} \tag{2.13}$$

where

$$\epsilon = \epsilon_0(1+\chi) \tag{2.14}$$

and

$$\mu = \mu_0(1+\chi_m) \tag{2.15}$$

are the electric permittivity and magnetic permeability of the medium.

In this book, we will consider only media with $\mu \approx \mu_0$ and hence we will use

$$\vec{B} = \mu_0\vec{H} \tag{2.16}$$

The dielectric constant of the medium is related to ϵ through

$$K = \frac{\epsilon}{\epsilon_0} = 1+\chi \tag{2.17}$$

In Chap. 10, we will go beyond the linear approximation given by Eq. (2.10) and introduce nonlinear terms in the equation. These nonlinear terms lead to the rich field of nonlinear optics which has a significant role to play in integrated quantum photonics by providing us with a source of different quantum states of light, noiseless parametric amplifiers as well as sources of coherent light at frequencies where lasers may not be available. In this chapter, we restrict ourselves to the linear relation given by Eq. (2.12).

2.3 Wave Equation

Taking curl of Eq. (2.1) and using Eq. (2.2), we have

$$\nabla \times \nabla \times \vec{E} = -\frac{\partial(\nabla \times \vec{B})}{\partial t} = -\mu_0\frac{\partial(\nabla \times \vec{H})}{\partial t} = -\mu_0\frac{\partial^2\vec{D}}{\partial t^2} = -\mu_0\epsilon\frac{\partial^2\vec{E}}{\partial t^2}$$

Expanding the left hand side, we have

$$\nabla(\nabla.\vec{E}) - \nabla^2\vec{E} = -\mu_0\epsilon\frac{\partial^2\vec{E}}{\partial t^2}$$

Equation (2.3) gives us (in isotropic and homogeneous media where the permittivity ϵ is a scalar and independent of position) $\nabla.\vec{E}=0$; and thus we get the following wave equation satisfied by the electric field:

$$\nabla^2 \vec{E} = \mu_0 \epsilon \frac{\partial^2 \vec{E}}{\partial t^2} \tag{2.18}$$

We can similarly derive the following wave equation satisfied by the magnetic field:

$$\nabla^2 \vec{B} = \mu_0 \epsilon \frac{\partial^2 \vec{B}}{\partial t^2} \tag{2.19}$$

From the two wave equations, we obtain the speed of the electromagnetic wave in the medium as

$$v = \frac{1}{\sqrt{\mu_0 \epsilon}} \tag{2.20}$$

In free space $\epsilon = \epsilon_0$ and $v = c$. The refractive index of the medium is defined by

$$n = \frac{c}{v} = \sqrt{\frac{\epsilon}{\epsilon_0}} = \sqrt{K} \tag{2.21}$$

where K is the dielectric constant of the medium.

2.4 Energy Density and Intensity

The instantaneous power flow is described by the Poynting vector $\vec{S}$ defined below:

$$\vec{S} = \vec{E} \times \vec{H} \tag{2.22}$$

The magnitude of $\vec{S}$ gives the instantaneous rate of energy flow (transported by the electromagnetic waves) per unit time per unit area of a surface with its normal parallel to $\vec{S}$. Thus, the energy flow is perpendicular to both the electric and magnetic fields and is in the direction of their cross product. The direction of the Poynting vector $\vec{S}$ represents the direction of propagation of energy in the wave and its magnitude is a measure of the intensity of light.

The intensity I of the electromagnetic wave is the magnitude of the time averaged Poynting vector and represents the energy crossing per unit time per unit area (taken perpendicular to the propagation direction) and is given by the following:

$$I = \left| \left\langle \vec{E} \times \vec{H} \right\rangle \right| \tag{2.23}$$

where angular brackets imply time averaging. If we use complex notation for the fields, then we have

$$I = \frac{1}{2} \left| Re \left\langle \vec{E} \times \vec{H}^* \right\rangle \right| \tag{2.24}$$

The energy density of the electromagnetic wave is given by

$$U = \frac{1}{2} \left(\vec{E} . \vec{D} + \vec{B} . \vec{H} \right) = \frac{1}{2} \left(\epsilon E^2 + \mu_0 H^2 \right) \tag{2.25}$$

2.5 Plane Waves

We can solve Eq. (2.18) using the standard separation of variables technique in Cartesian coordinates and obtain solution corresponding to monochromatic plane waves which are described by the following electric field variation:

$$\vec{E} = \vec{E}_0 e^{i\left(\vec{k}.\vec{r} - \omega t\right)} = \vec{E}_0 e^{i\left(k_x x + k_y y + k_z z - \omega t\right)} \tag{2.26}$$

where ω is the angular frequency of the wave and

$$\vec{k} = k_x \hat{x} + k_y \hat{y} + k_z \hat{z} \tag{2.27}$$

is referred to as the propagation vector of the wave and defines the direction of propagation of the plane wave and $\vec{E}_0$ is a constant vector. The magnitude of $\vec{k}$ is given by

$$k = \left| \vec{k} \right| = \sqrt{k_x^2 + k_y^2 + k_z^2} = \frac{\omega}{c} n \tag{2.28}$$

The surfaces of constant phase of this wave are planes and are given by the equation

$$\vec{k}.\vec{r} = constant$$

Thus, $\vec{k}$ points in the direction perpendicular to phase fronts which are planes. For a plane wave propagating along the $z-$ direction, $\vec{k} = k\widehat{z}$ and the surfaces of constant phase are parallel to the $x - y$ plane.

Substituting in the Maxwell's equations, we obtain the corresponding magnetic field as

$$\vec{B} = \vec{B}_0 e^{i(\vec{k}.\vec{r} - \omega t)} = \frac{1}{v}(\widehat{\kappa} \times \vec{E}_0) e^{i(\vec{k}.\vec{r} - \omega t)} = \frac{1}{v}(\widehat{\kappa} \times \vec{E}) \tag{2.29}$$

where $\widehat{\kappa}$ is the unit vector along $\vec{k}$.

Note that $\vec{E}$, $\vec{B}$, and $\vec{k}$ form a right handed coordinate system (see Fig. 2.1).

Energy flow is defined by the Poynting vector and we obtain for intensity of the wave:

$$I = \frac{1}{2}\left| Re \langle \vec{E} \times \vec{H}^* \rangle \right| = \frac{1}{2\mu_0 v} |E_0|^2 = \frac{n}{2c\mu_0} |E_0|^2 = \frac{nc\epsilon_0}{2} |E_0|^2 \tag{2.30}$$

The intensity is hence proportional to the square of the amplitude of the electric field of the electromagnetic wave.

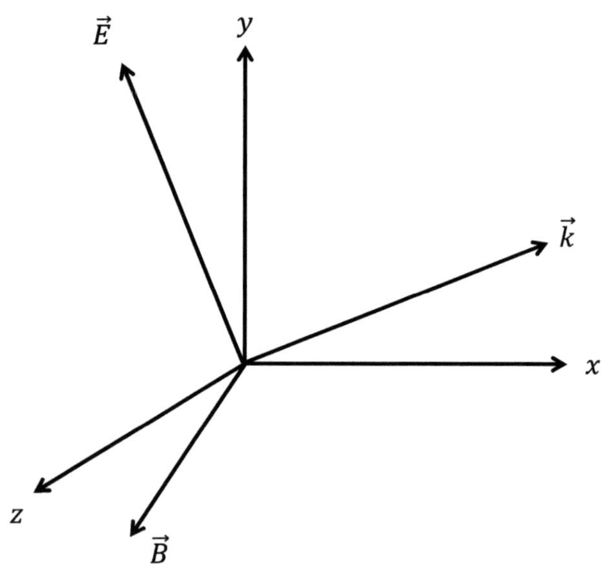

Fig. 2.1 The electric field, magnetic field and the propagation vector $\vec{E}$, $\vec{B}$, and $\vec{k}$ form a right handed coordinate system

Laser beams are characterized by a Gaussian intensity profile across the cross-section. We can write the corresponding transverse intensity profile as

$$I(x,y) = I_0 e^{-2(x^2+y^2)/w_0^2} \tag{2.31}$$

where I_0 is the intensity on the axis and w_0 is the spot size, which is the distance from the axis where the intensity drops by a factor of $\frac{1}{e^2}$ from that on the axis. It must be mentioned here that the Gaussian beam as it propagates will diffract and the spot size will increase with propagation distance (see, e.g., Ghatak and Thyagarajan 1978).

The power carried by such a beam would be

$$P = \iint I(x,y)dxdy = \frac{\pi w_0^2}{2} I_0 \tag{2.32}$$

As an example if we consider a Gaussian beam with a spot size $w_0 = 1$ mm carrying a power of 100 mW, the intensity on the axis would be approximately $\frac{2}{\pi}10^5$ W/m^2. The corresponding electric field on the axis will be $|E| \approx 6.9 \times 10^3$ V/m.

Exercise 2.1 What is the direction of propagation of a wave described by

$$\vec{E} = \vec{E}_0 e^{i\left[k\left(\frac{1}{2}x+\frac{\sqrt{3}}{2}z\right)-\omega t\right]}$$

Solution The propagation vector of the wave is given by $\vec{k} = k\left(\frac{1}{2}\hat{x} + \frac{\sqrt{3}}{2}\hat{z}\right)$. Hence, the wave is propagating in the $x - z$ plane making an angle $\theta = \cos^{-1}\left(\frac{k_x}{k}\right) = \frac{1}{2}$ with the x–axis giving $\theta = 60°$.

Exercise 2.2 Consider a laser beam at a wavelength of 800 nm and having a power of 1 W with a spot size $w_0 = 5$ mm propagating in air. What will be the intensity on the axis of the beam and what will be the corresponding electric field?

Solution The laser intensity on the axis would be $I_0 = \frac{2P}{\pi w_0^2} = \frac{8}{\pi} \times 10^4$ W/m^2. Using Eq. (2.30), the electric field comes out to be approximately 4400 V/m.

Exercise 2.3 If the same laser beam is focused into an spot with $w_0 = 10$ μm, what will be the corresponding electric field?

Solution The intensity would be $\frac{2}{\pi}10^{10}$ W/m^2 and the corresponding electric field would be about 2.2×10^6 V/m. Note the huge electric fields that get produced by focused lasers. The breakdown strength of air is 3×10^6 V/m. Never look straight into any laser beam as it can damage one's eye.

We will now consider some polarization states of light.

2.6 Polarization States

Light waves are transverse electromagnetic waves and for a given direction of propagation, there will be two possible independent orthogonally polarized states which will be the eigenmodes of propagation.

For plane wave propagating along the z–direction, an x–polarized wave will have its $\vec{E}$ along $\widehat{x}$ direction and the corresponding $\vec{B}$ will be along the $\widehat{y}$ direction. Thus, we have

$$\vec{E} = E_0\widehat{x}e^{i(kz-\omega t)} \tag{2.33a}$$

$$\vec{B} = B_0\widehat{y}e^{i(kz-\omega t)} \tag{2.33b}$$

Such a wave is termed linearly polarized along the x – direction.

Similarly a y–polarized wave will be characterized by

$$\vec{E} = E_0\widehat{y}e^{i(kz-\omega t)} \tag{2.34a}$$

$$\vec{B} = -B_0\widehat{x}e^{i(kz-\omega t)} \tag{2.34b}$$

Note that $\vec{E} \times \vec{B}$ is along the propagation direction z.

Using the Jones vector formalism, we represent these two polarization states in terms of column vectors with the two elements given by the x– and y–components of the electric field:

$$\begin{pmatrix} E_x \\ E_y \end{pmatrix} = E_0\begin{pmatrix} E_x/E_0 \\ E_y/E_0 \end{pmatrix} \tag{2.35}$$

A plane wave polarized linearly along the x–direction can be represented by the following normalized unit vector:

$$\begin{pmatrix} 1 \\ 0 \end{pmatrix}$$

Similarly a plane wave polarized along the y–direction can be represented by the following normalized unit vector:

$$\begin{pmatrix} 0 \\ 1 \end{pmatrix}$$

A plane wave with an arbitrary state of polarization can be expressed in terms of the two linearly polarized components as

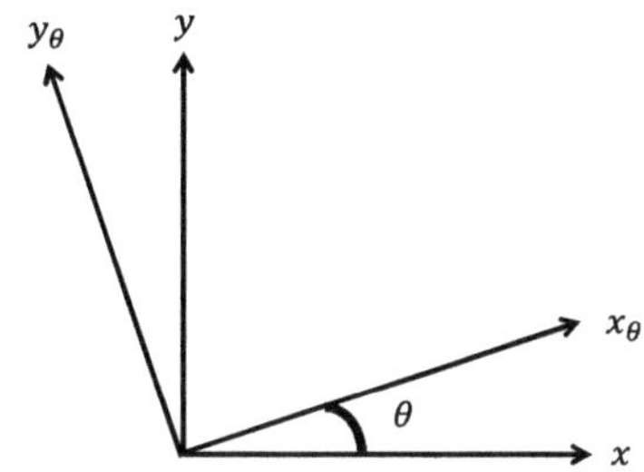

Fig. 2.2 Rotated coordinate system to express any input polarization state

$$\begin{pmatrix} E_x \\ E_y \end{pmatrix} = E_x \begin{pmatrix} 1 \\ 0 \end{pmatrix} + E_y \begin{pmatrix} 0 \\ 1 \end{pmatrix} \tag{2.36}$$

In this expansion, linearly polarized waves along the x– and y–directions are used as the basis set for the expansion. Here, the terms E_x and E_y are in general complex to account for different states of polarization.

For example if the wave is polarized linearly and oriented at 45° to the x–axis, then $E_x = E_y$. Similarly if the wave is circularly polarized, then the two Cartesian components of the electric field will have a phase difference of $\frac{\pi}{2}$ and we will have $E_y = \pm iE_x$ where the $\pm$ sign determines whether the wave is right circularly or left circularly polarized. In case of elliptically polarized wave with its two principal axes along the x– and y–directions, the magnitudes of E_x and E_y will not be equal and they will have a phase difference of $\frac{\pi}{2}$.

Instead of using the x– and y–axes as basis set, we may also use as the basis set the directions (x', y') which are oriented at 45° to the (x, y) axes. The corresponding unit vectors written in the (x, y) basis would be

$$\frac{1}{\sqrt{2}}\begin{pmatrix} 1 \\ 1 \end{pmatrix}; \qquad \frac{1}{\sqrt{2}}\begin{pmatrix} -1 \\ 1 \end{pmatrix}$$

which are again orthogonal to each other. Any input polarization can again be written as a superposition of these two components:

$$\begin{pmatrix} E_x \\ E_y \end{pmatrix} = E_{x'} \frac{1}{\sqrt{2}} \begin{pmatrix} 1 \\ 1 \end{pmatrix} + E_{y'} \frac{1}{\sqrt{2}} \begin{pmatrix} -1 \\ 1 \end{pmatrix} \tag{2.37}$$

where the amplitudes $E_{x'}$ and $E_{y'}$ are related to E_x and E_y and are in general complex.

If we wish to express the polarization in terms of a set of axes (x_θ, y_θ) with the x_θ axis making an angle θ with the x axis (see Fig. 2.2), then the corresponding unit vectors written in the (x, y) basis are

$$\begin{pmatrix} \cos\theta \\ \sin\theta \end{pmatrix}; \begin{pmatrix} -\sin\theta \\ \cos\theta \end{pmatrix}$$

and we can write

$$\begin{pmatrix} E_x \\ E_y \end{pmatrix} = E_{x\theta} \begin{pmatrix} \cos\theta \\ \sin\theta \end{pmatrix} + E_{y\theta} \begin{pmatrix} -\sin\theta \\ \cos\theta \end{pmatrix} \tag{2.38}$$

with components of E_{x_θ} and E_{y_θ} along the (x_θ, y_θ) directions, respectively.

Sometimes it is convenient to express the input polarization in terms of right and left circularly polarized waves rather than linearly polarized waves. Unit vectors for right and left circularly polarized waves are, respectively, given by

$$\frac{1}{\sqrt{2}} \begin{pmatrix} 1 \\ i \end{pmatrix}; \quad \frac{1}{\sqrt{2}} \begin{pmatrix} 1 \\ -i \end{pmatrix}$$

Note that in both cases the amplitudes of the x– and y–components are equal and they have a phase difference of $\frac{\pi}{2}$. The two vectors are orthogonal to each other.

Any state of polarization can now be written as a linear combination of the right and left circularly polarized waves:

$$\begin{pmatrix} E_x \\ E_y \end{pmatrix} = E_r \frac{1}{\sqrt{2}} \begin{pmatrix} 1 \\ i \end{pmatrix} + E_l \frac{1}{\sqrt{2}} \begin{pmatrix} 1 \\ -i \end{pmatrix} \tag{2.39}$$

where E_r and E_l are the components of the incident light in the right and left circular polarization states.

An $\hat{x}$ polarized light (with unit amplitude) will give

$$\begin{pmatrix} 1 \\ 0 \end{pmatrix} = E_r \frac{1}{\sqrt{2}} \begin{pmatrix} 1 \\ i \end{pmatrix} + E_l \frac{1}{\sqrt{2}} \begin{pmatrix} 1 \\ -i \end{pmatrix} \tag{2.40}$$

which gives us $E_r = E_l = \frac{1}{\sqrt{2}}$. Thus, adding a right circularly polarized and a left circularly polarized light in phase gives us an $\hat{x}$ polarized light. Similarly a $\hat{y}$ polarized light can be obtained if $E_l = -E_r = \frac{1}{\sqrt{2}}$ (apart from an overall phase factor).

Similarly a linearly polarized light wave making an angle of 45° with the x–axis can be represented as

$$\frac{1}{\sqrt{2}} \begin{pmatrix} 1 \\ 1 \end{pmatrix} = E_r \frac{1}{\sqrt{2}} \begin{pmatrix} 1 \\ i \end{pmatrix} + E_l \frac{1}{\sqrt{2}} \begin{pmatrix} 1 \\ -i \end{pmatrix} \tag{2.41}$$

with $E_r = \frac{(1-i)}{2} = e^{-i\pi/4}$ and $E_l = \frac{(1+i)}{2} = e^{i\pi/4}$.

Depending on the specific problem it is convenient to use different basis sets to describe polarization states of light. In anisotropic media that we will discuss in Sect. 2.8, the modes of propagation are linearly polarized. Hence, it is convenient to use linear polarization basis. In other situations like in Faraday effect in dielectrics which results in a rotation of the plane of polarization of an incident linearly polarized wave, the eigenpolarization states are circularly polarized and hence in this situation it may be more convenient to use circular polarization basis.

More discussion on polarization of light and using the operator formalism to treat problems dealing with polarization is given in Chap. 3. For detailed discussion on the polarization of light and applications, readers are referred to Ghatak and Kumar (2012).

Exercise 2.4 An electromagnetic wave propagating in a medium is described by the following equation $E(x, y, z, t) = Ae^{i[\pi \times 10^6 (2x+2y+z)) - 6\pi \times 10^{14} t]}$. What is the refractive index of the medium?

Solution The magnitude of the propagation constant of the wave is given by $k = \sqrt{4+4+1}\pi \times 10^6 = 3\pi \times 10^6\ \text{m}^{-1}$. The angular frequency of the wave is $\omega = 6\pi \times 10^{14}$ rad/s. Hence, the refractive index of the medium is $n = \frac{ck}{\omega} = 1.5$.

Exercise 2.5 Consider two light waves having same intensity but one of them being linearly polarized while the other circularly polarized. What will be the difference in the peak electric fields produced in the two waves?

Solution The electric field of a linearly polarized wave polarized along the x–direction and propagating along the z– direction is given by $\vec{E} = \widehat{x} E_0 e^{i(kz - \omega t)}$. The corresponding intensity would be $\frac{n}{2c\mu_0} E_0^2$. The electric field of a circularly polarized wave propagating along the z–direction would be given by $\vec{E} = E_c(\widehat{x} + i\widehat{y}) e^{i(kz - \omega t)}$. The corresponding intensity would be $\frac{n}{2c\mu_0} 2E_c^2$. If both the waves have the same intensity, then we must have $E_0 = \sqrt{2} E_c$. Hence, the peak electric field of the circularly polarized wave having the same intensity would be smaller by a factor $1/\sqrt{2}$.

2.7 Boundary Conditions

When electromagnetic fields cross boundaries between two different media, they need to satisfy certain boundary conditions in order to satisfy the Maxwell's equations. Here we give the boundary conditions on the various fields in the absence of free charges and free currents:

1. Tangential component of $\vec{E}$ must be continuous across the boundary.
2. Normal component of $\vec{D}$ must be continuous across the boundary.
3. Tangential component of $\vec{H}$ must be continuous across the boundary.
4. Normal component of $\vec{B}$ must be continuous across the boundary.

These boundary conditions help us to solve problems in electromagnetics in the presence of multiple regions with different dielectric and magnetic characteristics such as reflection and transmission coefficients across interfaces. These boundary conditions will be used in Chap. 4 to obtain the modes of propagation in optical waveguides.

2.8 Temporal Pulse and the Frequency Spectrum

In Sect. 2.4, we discussed about monochromatic waves having a single frequency. Perfectly monochromatic source is an idealization and realistic sources are characterized by a frequency spectrum.

Here we consider a light pulse in the form of a plane electromagnetic wave propagating along the z–direction which can be approximately described by the following Gaussian function at $z = 0$:

$$E(t) = Ae^{-t^2/2\tau_0^2}e^{-i\omega_0 t} \tag{2.42}$$

Here, ω_0 is the central frequency of the pulse, τ_0 characterizes the temporal duration of the pulse, and A represents the peak amplitude of the pulse. The peak of the pulse is at $t = 0$ and the amplitude of the wave drops by a factor of $1/\sqrt{e}$ of the maximum at $t = \tau_0$.

The pulse given by Eq. (2.42) is composed of many frequencies which can be obtained by using the Fourier transform operation. The frequency spectrum of the pulse is given by

$$F(\omega) = \frac{1}{\sqrt{2\pi}}\int_{-\infty}^{+\infty} E(t)e^{i\omega t}dt \tag{2.43}$$

Substituting for $E(t)$ from Eq. (2.42), we obtain

$$F(\omega) = \frac{A}{\sqrt{2\pi}}\int_{-\infty}^{+\infty} e^{-t^2/2\tau_0^2}e^{i(\omega-\omega_0)t}dt \tag{2.44}$$

which using the following standard integral

$$\int_{-\infty}^{+\infty} e^{-px^2+qx}dx = \sqrt{\frac{\pi}{p}}e^{q^2/4p} \tag{2.45}$$

gives us

$$F(\omega) = A\tau_0 e^{-\frac{(\omega-\omega_0)^2\tau_0^2}{2}} \tag{2.46}$$

Thus, for the Gaussian pulse given by Eq. (2.42), the frequency spectrum of the pulse is also Gaussian centered at $\omega = \omega_0$. The spectral amplitude is maximum at $\omega = \omega_0$ and reduces by a factor $1/\sqrt{e}$ at $\omega = \omega_0 \pm \frac{1}{\tau_0}$.

Narrower the pulse in the time domain, smaller is τ_0 and broader is the frequency spectrum. Similarly, broader the pulse in time domain, larger is τ_0 and narrower is the frequency spectrum. This is nothing but a manifestation of the uncertainty principle between time and frequency domains.

In free space, all the frequency components contained in the pulse travel at the same speed thus experiencing the same phase shift in propagating through a certain distance. Due to this the temporal shape of the pulse remains the same as it propagates. However, in media, the speed of propagation of electromagnetic waves depends on the frequency due to the phenomenon of dispersion and thus the phase shift suffered by different frequency components is different leading to a change in the temporal shape and duration of the pulse as it propagates through any medium. This phenomenon of pulse dispersion can play a significant role in pulses propagating over long distances such as in optical fiber communication through optical fibers or when the pulse is of ultrashort duration (hundreds of femtoseconds or narrower) with an extremely wide spectrum.

Pulse propagation is characterized by phase velocity and group velocity. The former specifies the velocity with which a particular phase of the wave propagates, while the latter defines the velocity at which the pulse propagates.

The phase velocity is given by

$$v_{ph} = \frac{\omega}{k} \tag{2.47}$$

and the group velocity is defined by

$$v_{gr} = \frac{d\omega}{dk} = \left(\frac{dk}{d\omega}\right)^{-1} \tag{2.48}$$

Exercise 2.6 Express phase and group velocities in terms of refractive index of the medium.

Solution The propagation constant of an electromagnetic wave propagating in a medium of refractive index n is given by

$$k = \frac{\omega}{c} n(\omega)$$

where an explicit dependence of refractive index on ω is shown.

Hence, we have

$$v_{ph} = \frac{\omega}{k} = \frac{c}{n(\omega)} \tag{2.49}$$

and

$$v_{gr} = \left(\frac{dk}{d\omega}\right)^{-1} = \frac{c}{n + \omega\frac{dn}{d\omega}} = \frac{c}{N(\omega)} \tag{2.50}$$

where

$$N(\omega) = n(\omega) + \omega\frac{dn}{d\omega} \tag{2.51}$$

represents the group refractive index of the medium.

As an example, in lithium niobate which is an anisotropic medium (see Sect. 2.8) at a wavelength of 1500 nm, the ordinary refractive index is $n_o(\omega) \approx 2.139$ and the ordinary group index is $N_o(\omega) = 2.1836$.

2.9 Anisotropic Media

Most devices based on the electro optic effect (see Chap. 6) or nonlinear optical effects (see Chap. 10) use anisotropic media. These effects will be discussed in greater detail in later chapters in the book.

In anisotropic media for plane wave propagating along any direction, there are two normal modes which are linearly polarized and which in general have different speeds of propagation. Normal modes of polarization for propagation are special states of polarization which do not change their state of polarization as the wave propagates; only the phase of the wave changes with propagation. This property leads to their applications in various polarization devices such as half wave and quarter wave plates, polarization compensators, and phase matching in nonlinear interactions. In this section, we will briefly discuss some of the properties of anisotropic media.

In contrast to isotropic media, anisotropic media are characterized by the fact that the electric field $\vec{E}$ and the electric polarization vector $\vec{P}$ are not, in general, parallel to each other. This is described mathematically through the following equation:

$$\vec{P} = \epsilon_0 \overline{\overline{\chi}} \vec{E} \tag{2.52}$$

where $\overline{\overline{\chi}}$ represents the electric permittivity tensor. Equation (2.52) shows that the electric polarization is linearly related to the applied electric field. We can also express Eq. (2.52) in component form as

$$P_i = \epsilon_0 \sum_{j=1}^{3} \chi_{ij} E_j; i = 1, 2, 3 \tag{2.53}$$

In Chap. 10, we will introduce additional nonlinear terms in Eq. (2.53) which leads to the rich field of nonlinear optics.

In view of Eq. (2.52), the displacement vector $\vec{D}$ is given by

$$\vec{D} = \epsilon_0 \vec{E} + \vec{P} = \epsilon_0 \left(\overline{\overline{I}} + \overline{\overline{\chi}} \right) \vec{E} = \overline{\overline{\epsilon}} \vec{E} \tag{2.54}$$

where $\overline{\overline{I}}$ is a unit tensor and $\overline{\overline{\epsilon}}$ is the dielectric permittivity tensor:

$$\overline{\overline{\epsilon}} = \epsilon_0 \left(\overline{\overline{I}} + \overline{\overline{\chi}} \right) \tag{2.55}$$

Equation (2.54) shows that in general the vectors $\vec{D}$ and $\vec{E}$ are not parallel to each other.

In component form, Eq. (2.54) has the form

$$D_i = \sum_{j=1}^{3} \epsilon_{ij} E_j = \epsilon_{ij} E_j; i = 1, 2, 3; \; j = 1, 2, 3 \tag{2.56}$$

where in the last step we have used Einstein's convention of summation over repeated indices.

We can write the above equation in matrix form as

$$\begin{pmatrix} D_x \\ D_y \\ D_z \end{pmatrix} = \begin{pmatrix} \epsilon_{xx} & \epsilon_{xy} & \epsilon_{xz} \\ \epsilon_{yx} & \epsilon_{yy} & \epsilon_{yz} \\ \epsilon_{zx} & \epsilon_{zy} & \epsilon_{zz} \end{pmatrix} \begin{pmatrix} E_x \\ E_y \\ E_z \end{pmatrix} \tag{2.57}$$

where ϵ_{ij}; $i, j = 1, 2, 3$ are the elements of the permittivity tensor.

Although, in general, the dielectric permittivity tensor has off diagonal elements (see Exercise 2.7), we can express the dielectric constant tensor in a specific coordinate system referred to as the *principal axis system* of the medium in which the tensor has only diagonal terms:

$$\overline{\overline{\epsilon}} = \begin{pmatrix} \epsilon_x & 0 & 0 \\ 0 & \epsilon_y & 0 \\ 0 & 0 & \epsilon_z \end{pmatrix} \tag{2.58}$$

where ϵ_x, ϵ_y, and ϵ_z are referred to as the principal dielectric permittivities of the medium.

In isotropic media

$$\epsilon_x = \epsilon_y = \epsilon_z = \epsilon \tag{2.59}$$

In uniaxial media

$$\epsilon_x = \epsilon_y \neq \epsilon_z \tag{2.60}$$

and in biaxial media

$$\epsilon_x \neq \epsilon_y \neq \epsilon_z \tag{2.61}$$

The dielectric permittivities are related to the principal refractive indices through the following equations:

In uniaxial media

$$\epsilon_x = \epsilon_y = \epsilon_0 n_o^2; \quad \epsilon_z = \epsilon_0 n_e^2 \tag{2.62}$$

where n_o and n_e represent the ordinary and extraordinary refractive indices of the medium. Media with $n_o < n_e$ are referred to as positive uniaxial media, while media with $n_o > n_e$ are referred to as negative uniaxial media.

In biaxial media

$$\epsilon_x = \epsilon_0 n_x^2; \quad \epsilon_y = \epsilon_0 n_y^2; \quad \epsilon_z = \epsilon_0 n_z^2 \tag{2.63}$$

with n_x, n_y and n_z representing the three principal refractive indices of the medium. By convention we choose the axes so that $n_x < n_y < n_z$.

Lithium niobate (LN) is a negative uniaxial crystal which is extensively used in integrated quantum photonics. The typical refractive index of lithium niobate at 633 nm is $n_o \approx 2.288$, $n_e \approx 2.189$. Lithium tantalate (LT) is a positive uniaxial crystal having refractive index values at 800 nm of $n_o \approx 2.1538$, $n_e \approx 2.1578$. Potassium titanyl phosphate (KTP) is a biaxial crystal which is also used in integrated quantum photonics. At 1064 nm, the refractive index values of KTP are $n_x \approx 1.7379$, $n_y \approx 1.7455$, and $n_z \approx 1.8297$.

Exercise 2.7 In the principal axis system, the dielectric permittivity tensor is diagonal. Obtain the dielectric permittivity tensor in a coordinate system that is rotated by an angle θ about the $y-$ axis.

Solution We call the rotated coordinate system rotated about the $y-$ axis by (x', y', z'). The rotated coordinates are related to the coordinates before rotation (x, y, z) by

$$x' = x\cos\theta + z\sin\theta$$

$$y' = y$$
$$z' = -x \sin\theta + z \cos\theta$$

The permittivity $\overline{\overline{\epsilon}}'$ in the rotated coordinate system would be given by

$$\overline{\overline{\epsilon}}' = \overline{\overline{R}}\,\overline{\overline{\epsilon}}\,\overline{\overline{R}}^{-1}$$

where $\overline{\overline{R}}$ is the rotation matrix given by

$$\overline{\overline{R}} = \begin{pmatrix} \cos\theta & 0 & \sin\theta \\ 0 & 1 & 0 \\ -\sin\theta & 0 & \cos\theta \end{pmatrix}$$

We also have

$$\overline{\overline{R}}^{-1} = \begin{pmatrix} \cos\theta & 0 & -\sin\theta \\ 0 & 1 & 0 \\ \sin\theta & 0 & \cos\theta \end{pmatrix}$$

Thus

$$\overline{\overline{\epsilon}}' = \begin{pmatrix} \cos\theta & 0 & \sin\theta \\ 0 & 1 & 0 \\ -\sin\theta & 0 & \cos\theta \end{pmatrix} \begin{pmatrix} \epsilon_x & 0 & 0 \\ 0 & \epsilon_y & 0 \\ 0 & 0 & \epsilon_z \end{pmatrix} \begin{pmatrix} \cos\theta & 0 & -\sin\theta \\ 0 & 1 & 0 \\ \sin\theta & 0 & \cos\theta \end{pmatrix}$$

Simplifying, we obtain

$$\overline{\overline{\epsilon}}' = \begin{pmatrix} \epsilon_x \cos^2\theta + \epsilon_z \sin^2\theta & 0 & (\epsilon_z - \epsilon_x)\sin\theta\cos\theta \\ 0 & \epsilon_y & 0 \\ (\epsilon_z - \epsilon_x)\sin\theta\cos\theta & 0 & \epsilon_x \sin^2\theta + \epsilon_z \cos^2\theta \end{pmatrix}$$

We see that in a nonprincipal axis system, the dielectric permittivity tensor has off diagonal elements.

2.9.1 Plane Waves in Uniaxial Media

In this section, we will briefly discuss plane wave propagation in *uniaxial media.*

Consider a plane wave propagating along some arbitrary direction in a uniaxial medium. It is possible to show from Maxwell's equations that along any direction, there are two eigenpolarization states which have linear polarization and which propagate unchanged in their polarization state with, in general different speeds.

Any arbitrary polarization state propagating along this direction will change its polarization state, in general, since its components along the eigenpolarization states will propagate with different speeds (see, e.g., Ghatak and Thyagarajan 1989).

The optic axis of the uniaxial medium is by convention chosen to be along the z–axis and is a special direction along which the two eigenpolarization states propagate with the same speed and given by $\frac{c}{n_o}$. Thus, propagation along this direction does not lead to any change in the polarization state of the light since both the normal polarization states propagate with the same speed.

If we consider plane wave propagation in an anisotropic medium and write for the plane wave

$$\vec{E} = \vec{E}_0 e^{i(\vec{k}.\vec{r} - \omega t)} \tag{2.64}$$

where $\vec{k}$ is the propagation vector. The corresponding $\vec{D}$ is given by Eq. (2.54).

Substituting in the Maxwell's equation given by Eq. (2.3), $\nabla.\vec{D} = 0$, we obtain

$$\vec{k}.\vec{D} = 0 \tag{2.65}$$

Thus, in anisotropic media $\vec{D}$ is always perpendicular to the propagation vector $\vec{k}$ and the state of polarization is defined with respect to the direction of $\vec{D}$ rather than $\vec{E}$ which is not, in general, perpendicular to the propagation vector.

The corresponding $\vec{H}$ is given by

$$\vec{H} = \vec{H}_0 e^{i(\vec{k}.\vec{r} - \omega t)} \tag{2.66}$$

Substituting Eqs. (2.64) and (2.66) in the two curl equations (Eqs. 2.1 and 2.2), we have

$$\vec{k} \times \vec{E} = \omega \mu_0 \vec{H} \tag{2.67}$$

$$\vec{k} \times \vec{H} = -\omega \vec{D} \tag{2.68}$$

From the above two equations, we may conclude the following for plane electromagnetic waves propagating through anisotropic media:

1. $\vec{D}$ is perpendicular to $\vec{k}$ and $\vec{H}$
2. $\vec{H}$ is perpendicular to $\vec{k}$ and $\vec{E}$
3. Thus, $\vec{k}$, $\vec{E}$, and $\vec{H}$ form a right handed orthogonal coordinate system.
4. Since in general $\vec{D}$ and $\vec{E}$ are not parallel to each other, $\vec{E}$ is in general not perpendicular to $\vec{k}$.

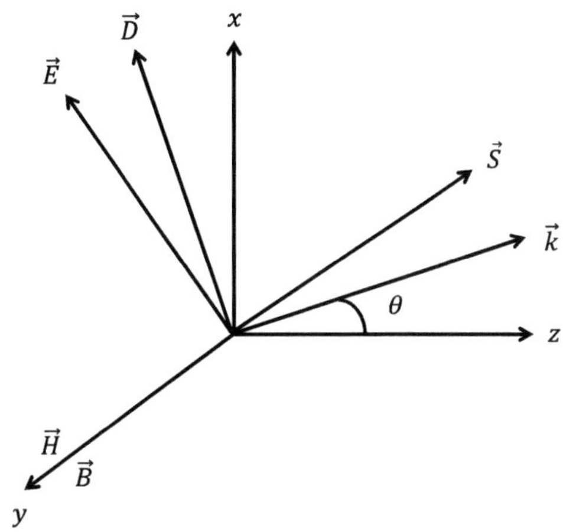

Fig. 2.3 The relative orientations of the various fields of a plane electromagnetic wave propagating in an anisotropic medium

5. The Poynting vector $\vec{S} = \vec{E} \times \vec{H}$, which defines the direction of energy propagation, is thus, in general, not parallel to the propagation vector $\vec{k}$. Thus, in anisotropic media the direction of energy flow and the propagation of the wavefront are not, in general parallel to each other.
6. The vectors $\vec{k}$, $\vec{S}$, $\vec{D}$, and $\vec{E}$ lie in a plane perpendicular to $\vec{H}$.

Fig. 2.3 shows the relative orientations of the various vectors in anisotropic media.

For propagation along any general direction, the two eigenpolarization states propagate with different speeds. For propagation along any one of the principal axis, the two eigenpolarizations have their $\vec{D}$ and $\vec{E}$ parallel to the other two principal axes. Thus, for propagation along the *x*– direction, the two eigenpolarization states will be the following (a) $\vec{D}$ and $\vec{E}$ parallel to *y*–direction with the wave having a phase velocity given by $\frac{c}{n_y}$ and (b) $\vec{D}$ and $\vec{E}$ parallel to *z*–direction with the wave having a phase velocity given by $\frac{c}{n_z}$.

Note that since in uniaxial media $n_x = n_y$, for propagation along the *z*–axis, the two phase velocities become equal. This direction is referred to as the optic axis. In uniaxial media, there is only one such direction where both eigenpolarizations propagate at the same velocity and hence the name. In biaxial media, there are two such directions and hence such media are referred to as biaxial.

For a general direction of propagation, Maxwell's equations lead to the following conclusions for *uniaxial media* (see, e.g., Ghatak and Thyagarajan 1989):

1. One of the eigenpolarization state referred to as the ordinary polarization has its $\vec{D}$ perpendicular to the optic axis (*z*–axis) and to the propagation vector $\vec{k}$. The speed of propagation of this wave is always $\frac{c}{n_o}$ irrespective of the direction of $\vec{k}$. Thus, this wave behaves very similar to wave propagation in an isotropic medium of refractive index n_o. This is referred to as the ordinary wave.

2. The other eigenpolarization referred to as the extra ordinary wave has its $\vec{D}$ perpendicular to the $\vec{k}$ and to the $\vec{D}$ of the ordinary polarization. The $\vec{D}$ of this wave lies in the plane containing the $\vec{k}$ and the optic axis or the z–axis. The speed of propagation of this wave depends on the angle subtended by the $\vec{k}$ of the wave with the optic axis. If the angle between the $\vec{k}$ of the wave and z–axis is ψ, then the speed of propagation of this wave is $\frac{c}{n_e(\psi)}$ where

$$\frac{1}{n_e^2(\psi)} = \frac{\cos^2\psi}{n_o^2} + \frac{\sin^2\psi}{n_e^2} \tag{2.69}$$

If $\psi = 0$, then the propagation direction is along the optic axis and the speed of the wave is equal to that of the ordinary wave and equal to $\frac{c}{n_o}$. In this case since both the eigenpolarizations propagate at the same speed and any input polarization state will continue to propagate without any change in the state of polarization.

On the other hand if the propagation direction is perpendicular to the optic axis $\psi = 90°$, then the speed of propagation of the extraordinary wave is $\frac{c}{n_e}$.

If $\vec{k}$ is along the x–axis, then the ordinary polarization will be along the y– axis and the extraordinary polarization will be oriented along the z – axis.

For propagation along an arbitrary direction, the speed of the ordinary wave will remain the same as $\frac{c}{n_o}$ independent of the angle ψ while that of the extraordinary polarization will lie between $\frac{c}{n_o}$ and $\frac{c}{n_e}$.

Thus, the maximum difference in speed of propagation between the ordinary and extraordinary waves corresponds to propagation perpendicular to the optic axis. Most polarization devices such as half wave plate ($\frac{\lambda}{2}$ plate), quarter wave plate ($\frac{\lambda}{4}$ plate), and polarization compensators are based on anisotropic media with the propagation direction chosen to be perpendicular to the optic axis. Also for propagation along any of the principal axes, the direction of energy flow is parallel to the propagation direction of the wave, that is, $\vec{S}$ and $\vec{k}$ are along the same direction.

Exercise 2.8 A plane wave with its $\vec{D}$ in the $x - z$ plane propagates along a direction given by $\hat{\kappa} = \frac{\sqrt{3}}{2}\hat{x} + \frac{1}{2}\hat{z}$ in lithium niobate which is a uniaxial medium with $n_o = 2.2111$ and $n_e = 2.1376$. What is the angle ϕ made by the $\vec{E}$ with the optic axis?

Solution Since the $\vec{D}$ of the wave lies in the plane containing the propagation direction and the optic axis, it must be an extraordinary wave. The $\vec{D}$ of the wave is perpendicular to the propagation direction and hence

$$\frac{D_x}{D_z} = -\frac{1}{\sqrt{3}}$$

Since $D_x = \epsilon_0 n_o^2 E_x$ and $D_z = \epsilon_0 n_e^2 E_z$, we have

$$\frac{E_x}{E_z} = \frac{n_e^2 D_x}{n_o^2 D_z} = -0.5396 = -\tan\phi$$

Thus, $\phi \approx 28.4°$. Note that the angle made by the $\vec{D}$ with the optic axis is 30°. Thus, $\vec{D}$ and $\vec{E}$ are not parallel to each other, and the angle between the two vectors is $\approx$1.6°. This is also the angle between the vectors $\vec{S}$ and $\vec{k}$.

If we consider propagation along a direction perpendicular to the optic axis, say along the y– axis, then from the discussion above, the ordinary polarization will be oriented along the x– axis and the extraordinary polarization will be oriented along the z– axis. Thus, if a light wave propagates through this medium along the y– axis, due to different speeds of propagation of the two eigenpolarizations which are along the x and z directions, a phase difference due to propagation will develop between the two components. If the length of propagation is l, then the phase difference introduced by the medium between the ordinary and extraordinary polarizations will be given by

$$\Delta\phi = \frac{2\pi}{\lambda_0}|(n_o - n_e)|l \tag{2.70}$$

where $\lambda_0 = \frac{2\pi c}{\omega}$ is the free space wavelength of the wave. This difference in phase between the two components of an incident wave will result in the change of the state of polarization of the wave. Depending on the medium, the wavelength of the incident wave and the length of propagation, we can introduce different phase differences between the components and thus generate different states of polarization.

There are two important wave plates namely the half wave plate and the quarter wave plate that are used extensively in quantum photonics to achieve desired polarization states of the wave.

If we consider a wave plate which introduces a phase difference of π between the two principal axes (referred to as a half wave plate) oriented with its principal axis making an angle ϕ with the x–axis, then it can be shown that such a plate can be represented by the following 2 × 2 matrix (see Exercise 2.12):

$$\widehat{O}_{HWP} = \begin{pmatrix} \cos 2\phi & \sin 2\phi \\ \sin 2\phi & -\cos 2\phi \end{pmatrix} \tag{2.71}$$

Similarly a wave plate which introduces a phase difference of $\pi/2$ between the two principal axes (referred to as a quarter wave plate) oriented with its principal axis making an angle ϕ with the x–axis is represented by (see Exercise 2.12)

$$\widehat{O}_{QWP} = \begin{pmatrix} \cos^2\phi + i\sin^2\phi & (1-i)\sin\phi\cos\phi \\ (1-i)\sin\phi\cos\phi & \sin^2\phi + i\cos^2\phi \end{pmatrix} \tag{2.72}$$

Consider an input state given by the following vector (written in the laboratory coordinate system with propagation along the z–direction):

$$\begin{pmatrix} E_x \\ E_y \end{pmatrix}$$

When this wave passes through the wave plate, the output will be given by

$$\widehat{O}\begin{pmatrix} E_x \\ E_y \end{pmatrix}$$

where $\widehat{O}$ represents the operator corresponding to the wave plate.

2.9.1.1 Half Wave Plate

Since a half wave plate introduces a phase difference of π, the required thickness can be calculated using Eq. (2.70) and comes out to be

$$l_h = \frac{\lambda_0}{2|(n_o - n_e)|} \tag{2.73}$$

If we consider a half wave plate with its principal axes oriented along the x and y axes (assuming propagation direction to be along the z axis), then $\phi = 0$ and Eq. (2.71) gives us the following transfer matrix:

$$\widehat{O}_{HWP} = \begin{pmatrix} 1 & 0 \\ 0 & -1 \end{pmatrix} \tag{2.74}$$

Thus, if the input polarization is linear and makes an angle θ with the x–axis, then the input state is given by

$$\begin{pmatrix} E_x \\ E_y \end{pmatrix} = E_0 \begin{pmatrix} \cos\theta \\ \sin\theta \end{pmatrix} \tag{2.75}$$

and the output after propagating through the half wave plate will be

$$\begin{pmatrix} 1 & 0 \\ 0 & -1 \end{pmatrix} E_0 \begin{pmatrix} \cos\theta \\ \sin\theta \end{pmatrix} = E_0 \begin{pmatrix} \cos\theta \\ -\sin\theta \end{pmatrix} \tag{2.76}$$

which is a linear polarization state making an angle of $-\theta$ with respect to the x– axis.

2.9.1.2 Quarter Wave Plate

Since a quarter wave plate introduces a phase difference of $\pi/2$, the required thickness can be calculated using Eq. (2.70) and comes out to be

$$l_q = \frac{\lambda_0}{4|(n_o - n_e)|} \tag{2.77}$$

If we consider a quarter wave plate with its principal axes oriented along the x and y axes (assuming propagation direction to be along the z axis), then $\phi = 0$ and Eq. (2.72) gives us the following transfer matrix:

$$\widehat{O}_{QWP} = \begin{pmatrix} 1 & 0 \\ 0 & i \end{pmatrix} \tag{2.78}$$

Thus, if the input polarization is linear and makes an angle θ with the x–axis, then the input state is given by Eq. (2.75) and the output after propagating through the quarter wave plate will be

$$\begin{pmatrix} 1 & 0 \\ 0 & i \end{pmatrix} E_0 \begin{pmatrix} \cos\theta \\ \sin\theta \end{pmatrix} = E_0 \begin{pmatrix} \cos\theta \\ i\sin\theta \end{pmatrix} \tag{2.79}$$

which corresponds to elliptic polarization state. If $\theta = \frac{\pi}{4}$, then the output will correspond to a circular polarization state.

It should be noted that the phase difference $\Delta\phi$ introduced by the wave plate depends on the wavelength of the propagating wave (see Eq. 2.70). There is an explicit dependence as well as an implicit dependence due to the wavelength dependence of the refractive index of the medium.

Both half wave and quarter wave plates are used extensively in integrated quantum photonic devices dealing with polarization degree of freedom.

Exercise 2.9 An anisotropic medium is characterized by $n_o = 2.62$, $n_e = 2.92$. What is the refractive index of an extraordinary wave propagating at 60° to the optic axis?

Solution Substituting the values in Eq. (2.69) we find the refractive index to be 2.835.

Exercise 2.10 A plane wave propagates in a uniaxial medium, with its $\vec{k}$ along a direction given by the following unit vector: $\hat{\kappa} = \frac{1}{2}\hat{x} + \frac{1}{\sqrt{2}}\hat{y} + \frac{1}{2}\hat{z}$. Obtain the unit vectors along the $\vec{D}$ of the ordinary and extraordinary waves.

Solution As discussed, the $\vec{D}$ of the ordinary wave has to be perpendicular to the optic axis and to the direction of propagation. Since the optic axis is along the z– direction, for the ordinary wave the unit vector along the direction of the $\vec{D}$ is given by $\hat{D}_o = \frac{2}{\sqrt{3}}\left(\frac{1}{\sqrt{2}}\hat{x} - \frac{1}{2}\hat{y}\right)$. The $\vec{D}$ of the extraordinary wave is perpendicular to the direction of propagation and to the direction of the $\vec{D}$ of the ordinary wave. This gives us the unit vector direction as $\hat{D}_e = \frac{1}{\sqrt{12}}\hat{x} + \frac{1}{\sqrt{6}}\hat{y} - \frac{3}{\sqrt{12}}\hat{z}$. It is left as an exercise to obtain the unit vectors along the $\vec{E}$ corresponding to the ordinary wave and the extraordinary wave.

Exercise 2.11 A quarter wave plate can be used to convert a linearly polarized wave to a circularly polarized wave. Consider a uniaxial medium with $n_o \approx 2.288$, $n_e \approx 2.189$ at 633 nm with which we need to realize a quarter wave plate having minimum thickness. (a) What direction of propagation should be chosen? (b) What is the minimum thickness of a quarter wave plate? (c) What should be the orientation of the incident linearly polarized light for the output to be circularly polarized?

Solution (a) To achieve maximum difference between the speed of the ordinary and extraordinary wave, the propagation direction should be perpendicular to the optic axis. Hence, the propagation direction can either be along x–axis or along y–axis. (b) The thickness should be so as to introduce a phase difference of $\frac{\pi}{2}$ between the ordinary and extraordinary components giving a thickness of $l = \frac{\lambda_0}{4(n_o - n_e)} \approx 1.6\ \mu\text{m}$. (c) Since circularly polarized light consists of equal amplitude components along the two orthogonal directions, the input polarization should be at 45° to the optic axis.

Exercise 2.12 Obtain Eq. (2.71).

Solution The electric field vector of a wave propagating along the z– direction is given by (apart from the time dependence)

$$\vec{E}_{in} = \widehat{x} E_x + \widehat{y} E_y \tag{2.80}$$

Let us assume that the wave passes through a half wave plate with its principal axis oriented at an angle ϕ with respect to the x–axis. Let $x' - y'$ be the axes corresponding to the principal axes of the wave plate. In this case, we have

$$\widehat{x} = \widehat{x}' \cos\phi - \widehat{y}' \sin\phi \tag{2.81}$$

$$\widehat{y} = \widehat{x}' \sin\phi + \widehat{y}' \cos\phi \tag{2.82}$$

Substituting these in Eq. (2.80) and simplifying we obtain

$$\vec{E}_{in} = \widehat{x}' \left(\cos\phi E_x + \sin\phi E_y\right) + \widehat{y}' \left(-\sin\phi E_x + \cos\phi E_y\right) \tag{2.83}$$

Since the half wave plate introduces a phase difference of π between the $\widehat{x}'$ and $\widehat{y}'$ components, the output from the half wave plate will be given by (assuming that the component along $\widehat{y}'$ has a phase change of π with respect to the component along $\widehat{x}'$)

$$\vec{E}_{out} = \widehat{x}' \left(\cos\phi E_x + \sin\phi E_y\right) - \widehat{y}' \left(-\sin\phi E_x + \cos\phi E_y\right) \tag{2.84}$$

Inverting Eqs. (2.81) and (2.82) and replacing $\widehat{x}'$ and $\widehat{y}'$ in terms of $\widehat{x}$ and $\widehat{y}$ we obtain the following equation:

$$\vec{E}_{out} = \widehat{x} \left(\cos 2\phi E_x + \sin 2\phi E_y\right) + \widehat{y} \left(\sin 2\phi E_x - \cos 2\phi E_y\right) \tag{2.85}$$

Thus, if we represent the input by the column vector $\begin{pmatrix} E_{x,in} \\ E_{y,in} \end{pmatrix}$ and the output by the column vector $\begin{pmatrix} E_{x,out} \\ E_{y,out} \end{pmatrix}$, then we have

$$\begin{pmatrix} E_{x,out} \\ E_{y,out} \end{pmatrix} = \begin{pmatrix} \cos 2\phi & \sin 2\phi \\ \sin 2\phi & -\cos 2\phi \end{pmatrix} \begin{pmatrix} E_{x,in} \\ E_{y,in} \end{pmatrix} \tag{2.86}$$

which gives Eq. (2.71) as the operator representing the half wave plate.

In the case of a quarter wave plate, which introduces a phase difference of $\frac{\pi}{2}$ between the $\widehat{x}'$ and $\widehat{y}'$ components, Eq. (2.84) becomes

$$\vec{E}_{out} = \hat{x}'(\cos\phi E_x + \sin\phi E_y) + i\hat{y}'(-\sin\phi E_x + \cos\phi E_y) \tag{2.87}$$

Replacing $\hat{x}'$ and $\hat{y}'$ in terms of $\hat{x}$ and $\hat{y}$ we obtain the operator for a quarter wave plate as given by Eq. (2.72).

Equation (2.69) can be rewritten in the case of biaxial media provided the $\vec{k}$ of the propagating plane wave lies in one of the principal planes, namely $(x - y)$ or $(y - z)$ or $(x - z)$.

For example if the $\vec{k}$ lies in the $(x - y)$ plane, making an angle of ψ with respect to the x–axis, then the two eigenpolarizations are given by

$\vec{D}$ parallel to $\hat{z}$ and the velocity equal to $v_1 = \frac{c}{n_z}$ and

$\vec{D}$ in the $(x - y)$ plane and lying perpendicular to $\vec{k}$ with a velocity given by $v_2 = \frac{c}{n_2(\psi)}$ with

$$\frac{1}{n_2^2(\psi)} = \frac{\sin^2\psi}{n_x^2} + \frac{\cos^2\psi}{n_y^2} \tag{2.88}$$

It can be seen that in this case if the wave propagates along the x–direction, then $\psi = 0$ and the second eigenmode has a velocity corresponding to $n_2 = n_y$. Similarly if the wave propagates along the y–direction, then $\psi = \frac{\pi}{2}$ and the velocity of the second eigenmode would be corresponding to $n_2 = n_x$.

For propagation in other principal planes, equation similar to Eq. (2.88) can be written by appropriate replacement of the principal refractive indices.

It is worth pointing out that when plane waves cross an interface between two anisotropic media or between an isotropic medium and an anisotropic medium, the boundary conditions of the electric and magnetic fields at the interface impose the condition of continuity of the tangential component of the propagation vector $\vec{k}$ in both the media. Since Snell's law is a consequence of this boundary condition, Snell's law is still valid for refraction between two anisotropic media as long as the angles correspond to the directions of $\vec{k}$. Since the energy flow is in general not parallel to the propagation vector, the energy flow directions will, in general, not satisfy Snell's law. Since knowing the direction of the propagation vector and the orientation of the optic axis is sufficient to determine the direction of the energy flow, it is quite straight forward to analyze refraction at such interfaces. Using Snell's law as valid for propagation vector directions, we first determine the direction of the propagation vector in the medium after refraction. Then knowing the direction of the propagation vector and the optic axis, we can determine the direction of energy flow which also defines the direction of ray propagation in the medium.

As an example we consider the normal incidence of a plane wave at an interface between an isotropic and a uniaxial medium. Irrespective of the orientation of the optic axis in the uniaxial medium, the propagation vector of both the ordinary and extraordinary waves will be perpendicular to the interface after refraction. For the ordinary wave since the propagation vector and the energy flow are in the same direction, the energy associated with the ordinary wave (which represents the ordinary ray) will travel perpendicular to the interface. If the optic axis is the uniaxial medium is either normal to the interface or parallel to the interface, then the energy flow of the extraordinary wave (representing the extraordinary ray) will also be perpendicular to the interface and both the energy of the ordinary and extraordinary waves (both ordinary and extraordinary rays) will propagate parallel to each other and perpendicular to the interface.

On the other hand, if the optic axis in the uniaxial medium is neither perpendicular nor parallel to the interface, then the energy associated with the extraordinary wave (extraordinary ray) will not propagate perpendicular to the interface and will bend to a different direction depending on the orientation of the optic axis and the ordinary and extra ordinary refractive indices. This is the mechanism that leads to double refraction in uniaxial media.

Exercise 2.13 Potassium titanyl phosphate (KTP) is a biaxial medium and at 800 nm, the principal refractive indices are given by $n_x = 1.7483$, $n_y = 1.7565$ and $n_z = 1.8446$. A linearly polarized plane wave polarized at 45° to the x–axis and with a wavelength of 800 nm propagates along the y–axis. After how much distance will the wave be again linearly polarized?

Solution For a wave propagating along the y–direction, the two eigenmodes of propagation are oriented along the x–and z– directions which propagate with speeds given by c/n_x and c/n_z, respectively. The minimum distance of propagation at which the polarization state will be again linearly polarized will correspond to a phase difference of π between the two components. This distance would be $l = \frac{\lambda_0}{2(n_z - n_x)} \approx 4.15\ \mu\text{m}$.

2.9.2 *Index Surfaces*

Figure 2.4 shows an important diagram representing the ordinary and extraordinary index surfaces. The figure is a plot of the variation in ordinary refractive index $n_o(\psi)$ and extra ordinary refractive index $n_e(\psi)$ versus ψ leading, respectively, to the ordinary index surface and the extraordinary index surface. Since the ordinary wave has the same refractive index for all propagation directions, the ordinary index surface is a sphere. The extraordinary refractive index $n_e(\psi)$ satisfies the equation given by Eq. (2.69) and which represents an ellipsoid (an ellipse in the $x - z$ plane) with the two principal axes as n_o and n_e oriented parallel and perpendicular to the optic axis. The extraordinary index surface and the ordinary index surface touch each other along the optic axis and have the maximum difference along the direction perpendicular to the optic axis.

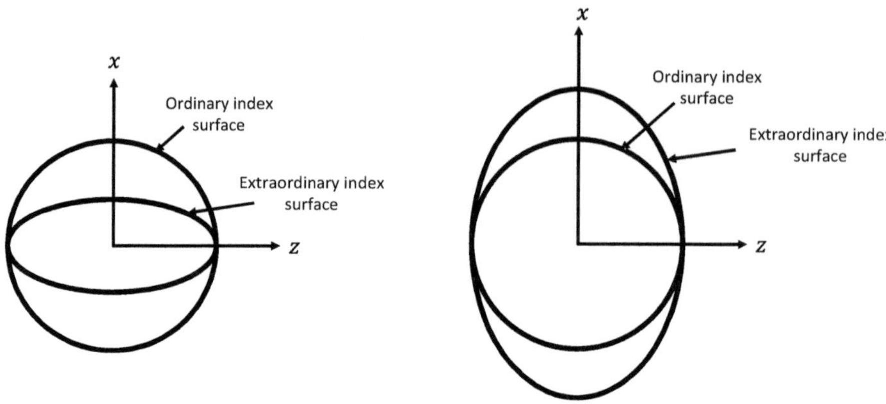

Fig. 2.4 The ordinary and extraordinary index surfaces in (**a**) negative and (**b**) positive uniaxial media with $n_o > n_e$ and $n_o < n_e$, respectively

If $n_o > n_e$ then the medium is referred to as negative uniaxial medium (see Fig. 2.4a) and if $n_o < n_e$, then the medium is referred to as positive uniaxial medium (see Fig. 2.4b).

Note that since the refractive indices change with wavelength/frequency, the index surfaces are different at different wavelengths.

The index surfaces are extremely useful in determining the propagation direction and orientation of the polarization of the interacting waves in birefringence phase matching that is used to maximize the nonlinear conversion of frequencies (see, e.g., Yariv 1991a, b).

2.9.3 *Index Ellipsoid*

Anisotropic media are also represented by the index ellipsoid which is defined by the following equation:

$$\frac{x^2}{n_x^2} + \frac{y^2}{n_y^2} + \frac{z^2}{n_z^2} = 1 \tag{2.89}$$

where n_x, n_y, and n_z represent the three principal refractive indices of the medium. The coordinate axes are oriented along the principal axes of the medium. The index ellipsoid will be a sphere for isotropic media, ellipsoid of revolution for uniaxial media, and ellipsoid for biaxial media. Figure 2.5 shows the index ellipsoid.

The index ellipsoid is a very useful construction for determining the eigenpolarization directions and the corresponding refractive indices for plane waves propagating along arbitrary directions in anisotropic media. By using the following procedure, one can determine the two eigenpolarizations:

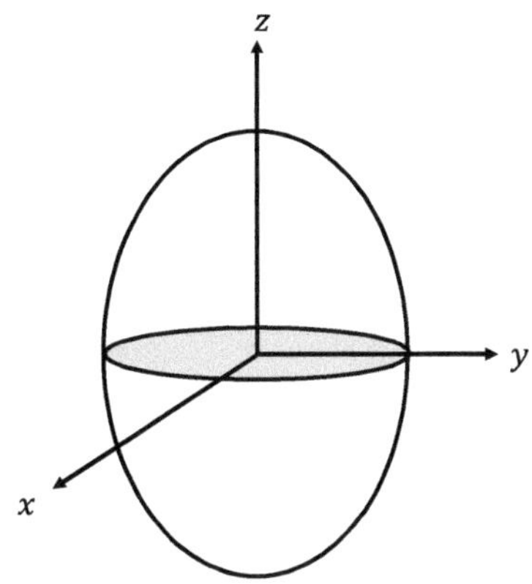

Fig. 2.5 The index ellipsoid

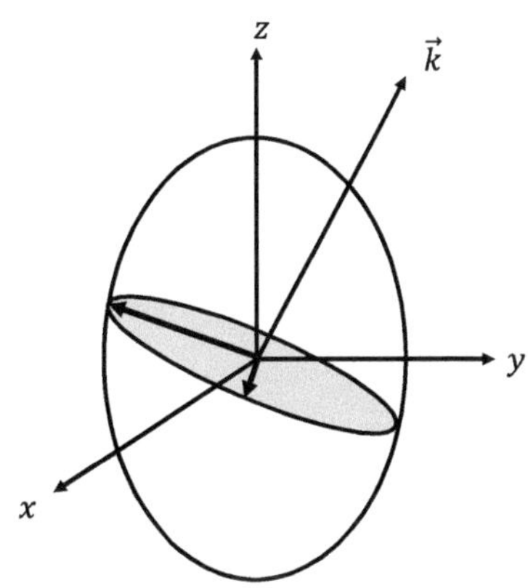

Fig. 2.6 Eigenpolarization directions (the two arrows perpendicular to $\vec{k}$) for a wave propagating along a direction given by $\vec{k}$. The lengths of the two arrows represent the refractive indices of the corresponding eigenpolarizations

- For a given direction of propagation of the plane wave, we draw a plane perpendicular to the direction of propagation and passing through the origin.
- The plane will intersect the index ellipsoid in an ellipse.
- The orientations of the principal axes of the ellipse give the two eigenpolarization directions for plane wave propagation along the chosen direction of propagation
- The lengths of the two principal axes give the corresponding refractive indices of the two eigenpolarization directions.

Figure 2.6 shows the orientations of the vectors along the eigenpolarizations for an arbitrary orientation of the $\vec{k}$.

As an example, we see that for a uniaxial medium the index ellipsoid is an ellipsoid of revolution around the z–axis or the optic axis. Hence, for propagation along the z–axis the plane to be considered is the $x - y$ plane passing through the origin which intersects the ellipsoid in a circle of radius n_o. Thus, all polarization states propagating along the optic axis will propagate without any change in the polarization state.

Similarly if we consider propagation along the x–axis, then the plane to be considered is the $y - z$ plane which intersects the ellipsoid in an ellipse with its principal axes along the y– and z– directions. Thus, the eigenpolarization directions are oriented along the y– and z– axes. The corresponding lengths are n_o and n_e. Thus, the refractive index of the eigenpolarization oriented along the y– direction is n_o while that of the eigenpolarization oriented along the z– direction is n_e.

In Chap. 6, we will show that the effect of an applied external electric field can be described mathematically through changes in the index ellipsoid equation. This effect is termed as electro optic effect and plays a very important role in integrated quantum photonics. In general, the applied electric field can change the orientations of the principal axes and also change the lengths of the principal axes. Since the linear optical properties of the medium are characterized by the index ellipsoid, the application of the electric field changes the optical properties of the medium.

Exercise 2.14 Given the following modified equation of index ellipsoid of a uniaxial medium which has cross terms, use rotation of coordinates to determine the new principal axis directions and also the new principal refractive indices:

$$\frac{x^2}{n_o^2}+\frac{y^2}{n_o^2}+\frac{z^2}{n_e^2}+2\alpha xy=1 \tag{2.90}$$

Solution Since the cross term involves the coordinates x and y only, the rotation needs to be about the z– axis. We define new axes $x' - y'$ rotated at an angle θ with respect to x–axis. Thus,

$$x=x'\cos\theta-y'\sin\theta \tag{2.91}$$

$$y=x'\sin\theta+y'\cos\theta \tag{2.92}$$

Substituting these in Eq. (2.90), we obtain

$$x'^2\left(\frac{1}{n_o^2}+2\alpha\cos\theta\sin\theta\right)+y'^2\left(\frac{1}{n_o^2}-2\alpha\cos\theta\sin\theta\right)+\frac{z^2}{n_e^2}+2\alpha x'y'\cos 2\theta=1 \tag{2.93}$$

We can get rid of the cross term by choosing $\theta=\frac{\pi}{4}$ which then gives us the following equation of the index ellipsoid written in the rotated coordinate system:

$$x'^2\left(\frac{1}{n_o^2}+\alpha\right)+y'^2\left(\frac{1}{n_o^2}-\alpha\right)+\frac{z^2}{n_e^2}=1 \tag{2.94}$$

which shows that the presence of the cross term in Eq. (2.90) has made the medium to be biaxial. The new principal axes are (x', y', z) now rotated with respect to the older axes (x, y, z) by an angle $\theta = \frac{\pi}{4}$ and two of the principal refractive indices have also changed. Note that the orientation of the new principal axes is independent of α.

Exercise 2.15 Equation (2.89) is written in the principal axis system. Write down the equation of the index ellipsoid in a coordinate system rotated about the x–axis by an angle θ.

Solution The coordinates in the rotated coordinate system and the original system are related through the following equations:

$$y = y' \cos\theta - z' \sin\theta \tag{2.95}$$

$$z = y' \sin\theta + z' \cos\theta \tag{2.96}$$

Substituting in Eq. (2.89) we obtain

$$\frac{x^2}{n_x^2} + y'^2\left(\frac{\cos^2\theta}{n_y^2} + \frac{\sin^2\theta}{n_z^2}\right) + z'^2\left(\frac{\sin^2\theta}{n_y^2} + \frac{\cos^2\theta}{n_z^2}\right) + 2y'z'\cos\theta\sin\theta\left(\frac{1}{n_z^2} - \frac{1}{n_y^2}\right) = 1 \tag{2.97}$$

The presence of the cross term shows that the equation is not written in the principal axis system of the ellipsoid.

Problems

2.1. Express the group refractive index $N(\omega)$ in terms of wavelength.

2.2. A plane electromagnetic wave with peak electric field amplitude of 100 V/m and wavelength 600 nm is travelling with its $\vec{k}$ in the $x - z$ plane making an angle of 45° with the x–axis and having $\vec{E}$ in the $x - z$ plane.

(a) Write complete expressions for the electric and magnetic field vectors of the wave

(b) Show that $\vec{E}.\vec{B} = 0$ and $\vec{E} \times \vec{B}$ points in the $\vec{k}$ direction.

2.3. Consider a plane electromagnetic wave propagating in a medium and described by an electric field

$$\vec{E} = \hat{z}E_0 e^{i\left[k\left(\frac{\sqrt{3}}{2}x - \frac{1}{2}y\right) - \omega t\right]}.$$

(a) Calculate the time averaged Poynting vector of the wave.

(b) If the refractive index of the medium is 1.5 and the wavelength is 1.5 μm, what is the value of E_0 for a wave with an intensity of 1 mW/mm^2.

2.4. A right circularly polarized wave and a left circularly polarized wave with equal amplitude are added with the following different phases (a) zero, (b) $\frac{\pi}{2}$ (c) π. What are the resultant states of polarization?

2.5. A plane wave propagates along the y–direction in a uniaxial medium with $n_o = 2.26$ and $n_e = 2.20$. Write down expressions for the $\vec{D}$ and $\vec{E}$ of the ordinary and extraordinary waves. Which of these waves travels faster in the medium?

2.6. Potassium titanyl phosphate (KTP) is a biaxial medium. At a wavelength of 800 nm the principal refractive indices are $n_x = 1.7483$, $n_y = 1.7565$ and $n_z = 1.8446$. A circularly polarized plane wave propagates in this medium with its propagation vector in the $x - z$ plane. At what angle with the z–axis should the wave propagate so that its polarization state does not change as it propagates?

2.7. Consider a medium with $n_o = 2.26$ and $n_e = 2.20$. A circularly polarized light wave at a wavelength of 800 nm propagates with its $\vec{k}$ lying in the $x - y$ plane and making an angle of 30° with the x-axis. After how much distance of propagation would the wave become linearly polarized?

2.8. The dielectric permittivity tensor for a uniaxial medium expressed in the principal axis system is given by

$$\epsilon = \epsilon_0 \begin{pmatrix} n_o^2 & 0 & 0 \\ 0 & n_o^2 & 0 \\ 0 & 0 & n_e^2 \end{pmatrix}$$

Obtain expressions for the permittivity tensor in coordinate systems (a) rotated about the z–axis by an angle θ and (b) rotated about the x–axis by an angle θ.

2.9. A lithium niobate plate of thickness 100 μm and with its optic axis parallel to the surface is placed between two crossed polarizers with their pass axes at ±45° to the optic axis. Calculate the wavelengths in the spectral region 750 nm to 950 nm which will be absent in the output. Given $n_o = 2.2553$, $n_e = 2.1755$. Neglect wavelength dependence of n_o and n_e.

2.10. A linearly polarized wave polarized at 30° to the x–direction propagates along the z–direction in a uniaxial medium. Here (x, y, z) represents the laboratory coordinate system. I wish to convert this to a circularly polarized wave.

(a) What wave plate would you use to achieve this?
(b) What should be the orientation of the axes of the wave plate with respect to the x–direction to achieve this?

2.11. Calculate the thickness of the thinnest half wave plate for operation at 0.8 μm that can be designed with a medium with $n_x = 1.748$, $n_y = 1.757$ and $n_z = 1.845$.

2.12. A right circularly polarized light wave at a wavelength of 680 nm is incident normally on a uniaxial medium of thickness 4 μm with $n_o = 1.66$ and $n_e = 1.49$. What is the state of polarization of the output wave?

2.13. Consider a medium with $n_x = 1.748$, $n_y = 1.757$ and $n_z = 1.845$. A light wave given by

$$\vec{E} = \left(\frac{1}{2}\hat{x} + \frac{\sqrt{3}}{2}\hat{y}\right)E_0 e^{i(kz-\omega t)}$$

propagates in the medium. What is the direction of the $\vec{D}$?

2.14. Consider a plane wave given by the following equation propagating in an anisotropic medium:

$$\vec{D} = \hat{n} D_0 e^{i\left(k\left[\frac{\sqrt{3}}{2}x - \frac{1}{2}y\right] - \omega t\right)}$$

What are the possible orientations of the unit vector $\hat{n}$?

2.15. In the presence of a magnetic field along z, an isotropic medium with refractive index n_0 becomes anisotropic with circularly polarized eigenmodes (Faraday effect) with principal refractive indices $n_0 + \delta$ and $n_0 - \delta$ where δ is a constant and depends on the externally applied magnetic field.

(a) Show that when a linearly polarized light wave propagates through the medium along z–direction, it remains linearly polarized except for a rotation in the plane of polarization.

(b) Obtain an expression for the angle of rotation per unit length of the medium.

References

Born, M., and E. Wolf. 1980. *Principles of optics*. Cambridge: Cambridge University Press.

Ghatak, A., and A. Kumar. 2012. *Polarization of light with applications in optical fibers*. New Delhi: Tata McGraw Hill.

Ghatak, A. K., and K. Thyagarajan. 1978. *Contemporary optics*. New York: Plenum Publishing Corporation.

Ghatak, A., and K. Thyagarajan. 1989. *Optical electronics*. New York: Plenum Publishing Corporation.

Yariv, A. 1991a. *Optical electronics*. Oxford: Oxford University Press.

———. 1991b. *Quantum electronics*. Hoboken: Wiley.

Chapter 3
Quantum Mechanics: A Short Overview

Abstract An overview of the basic concepts of quantum mechanics is provided in the chapter including the quantized states of linear harmonic oscillator. This chapter also presents the connection between the abstract bra and ket vectors and the wave function of quantum mechanics and a brief discussion on matrix representation of the states and operators. A brief discussion of time independent perturbation theory is also given.

3.1 Introduction

In this chapter, we will give an overview of the basic concepts of quantum mechanics and its operation. The discussion will primarily be using Dirac's formulation in terms of abstract vectors. For more detailed discussions on many of the topics given here and much more, readers are referred to excellent texts on quantum mechanics (Dirac 1982; Cohen-Tannoudji et al. 2019; Ghatak and Lokanathan 2004; Griffiths 1995; see also Zeilinger 1999).

Classical electromagnetic theory based on Maxwell's equations explains many phenomena and semi-classical treatments in which atoms are analyzed using the principles of quantum mechanics while light is still treated as a classical electromagnetic wave works in most situations. However, there are many instances such as spontaneous emission from atoms, spontaneous parametric down conversion from nonlinear crystals, states of light such as squeezed and entangled states have no classical explanation and we have to treat them using quantum mechanical principles. Many of these so-called non-classical states of light form very important aspects of quantum information science and technology.

In Chap. 14, we will quantize electromagnetic waves and introduce the concept of photons. We will show that each mode of the electromagnetic field *behaves* like a linear harmonic oscillator. Thus, quantization of electromagnetic fields is very closely linked with quantization of harmonic oscillator. Studying properties of mechanical harmonic oscillator can help us better understand quantized electromagnetic fields. In view of this, in this chapter, we will give a very brief basic formulation of quantum mechanics and introduce the Dirac bra-ket notation. We

K. Thyagarajan, *Integrated Quantum Photonics*, Graduate Texts in Physics,
https://doi.org/10.1007/978-3-031-85728-7_3

will then quantize a mechanical linear harmonic oscillator and obtain some of the properties which will be extensively used in Chap. 14. Readers who are familiar with the Dirac formulation of quantum mechanics may skip this chapter.

3.2 Basic Postulates

Quantum mechanics is based on a few fundamental postulates which are summarized below.

The state of any system is described by a state vector in an abstract space and is denoted by a complex vector $|\psi\rangle$, referred to as "ket ψ." All information about the system is contained in $|\psi\rangle$. Later we will describe how $|\psi\rangle$ is related to the wavefunction $\psi(\vec{r})$ that is used in standard quantum mechanics based on Schrödinger wave equation.

To every $|\psi\rangle$ we associate a vector $\langle\psi|$ called "bra ψ." Just as $|\psi\rangle$ describes the quantum mechanical system, so does $\langle\psi|$. This is used to describe the scalar product of the abstract vectors.

We define the scalar product between a vector $|\psi\rangle$ and a vector $|\phi\rangle$ through the following relations:

$$\alpha = \langle\psi|\phi\rangle \tag{3.1}$$

or

$$\alpha' = \langle\phi|\psi\rangle \tag{3.2}$$

Both α and α' are, in general, complex numbers and we define the connection between them through the following relation:

$$\alpha' = \alpha^* \tag{3.3}$$

where * represents complex conjugate. Thus, we have

$$\langle\psi|\phi\rangle = \langle\phi|\psi\rangle^* \tag{3.4}$$

We may note here that if we choose $|\phi\rangle = |\psi\rangle$, then we have

$$\langle\psi|\psi\rangle = \langle\psi|\psi\rangle^* \tag{3.5}$$

That is, the scalar product $\langle\psi|\psi\rangle$ is a real number. It is also imposed that $\langle\psi|\psi\rangle \geq 0$. The quantity $\langle\psi|\psi\rangle$ represents the length of the vector $|\psi\rangle$.

Two vectors $|\psi\rangle$ and $|\phi\rangle$ are said to be orthogonal if

$$\langle\psi|\phi\rangle = 0 \tag{3.6}$$

Observables like position, momentum, and energy are represented by linear operators which operate on the state vector of the system. Thus, an operator $\widehat{A}$ (operators will be represented by a hat on the symbol) can act on a ket and create another ket:

$$\widehat{A}|\psi\rangle = |\chi\rangle \tag{3.7}$$

Similarly the operator $\widehat{A}$ can operate on a bra $\langle\psi|$ and generate another bra:

$$\langle\psi|\widehat{A} = \langle\xi| \tag{3.8}$$

We must note that, in general the operator $\widehat{A}$ acting on $\langle\psi|$, does not give $\langle\chi|$, that is, in general, $\langle\psi|\widehat{A} \neq \langle\chi|$.

In order to obtain $\langle\chi|$ from $\langle\psi|$, we define another operator $\widehat{A}^\dagger$, referred to as the adjoint of the operator $\widehat{A}$ through the following equation:

$$\langle\psi|\widehat{A}^\dagger = \langle\chi| \tag{3.9}$$

An operator is referred to as self-adjoint or Hermitian if

$$\widehat{A}^\dagger = \widehat{A} \tag{3.10}$$

Now, if $\widehat{A}|\psi\rangle = |\chi\rangle$, then

$$\langle\xi|\widehat{A}|\psi\rangle = \langle\xi|\chi\rangle \tag{3.11}$$

Also

$$\langle\chi|\xi\rangle = \langle\psi|\widehat{A}^\dagger|\xi\rangle = \langle\xi|\chi\rangle^* = \langle\xi|\widehat{A}|\psi\rangle^* \tag{3.12}$$

Thus, the adjoint $\widehat{A}^\dagger$ of an operator $\widehat{A}$ is defined through the following relationship:

$$\langle\psi|\widehat{A}^\dagger|\xi\rangle = \langle\xi|\widehat{A}|\psi\rangle^* \tag{3.13}$$

If we choose $|\xi\rangle = |\psi\rangle$, then we obtain

$$\langle\psi|\widehat{A}^\dagger|\psi\rangle = \langle\psi|\widehat{A}|\psi\rangle^*$$

If $\widehat{A}$ is Hermitian then $\widehat{A}^\dagger = \widehat{A}$ and we have

$$\langle\psi|\widehat{A}|\psi\rangle = \langle\psi|\widehat{A}|\psi\rangle^* \tag{3.14}$$

implying that the quantity $\langle\psi|\widehat{A}|\psi\rangle$ is a real number if the operator $\widehat{A}$ is Hermitian.

Following are some properties of operators and state vectors:

- $\left(\widehat{A} + \widehat{B}\right)|\psi\rangle = \widehat{A}|\psi\rangle + \widehat{B}|\psi\rangle$
- $\widehat{A}(|\psi\rangle + |\chi\rangle) = \widehat{A}|\psi\rangle + \widehat{A}|\chi\rangle$
- $\widehat{A}^{\dagger\dagger} = \widehat{A}$ (adjoint of an adjoint of an operator is the operator itself)
- $\left(\widehat{A}\widehat{B}\right)^\dagger = \widehat{B}^\dagger\widehat{A}^\dagger$
- If $\widehat{A}$ is a complex number represented by c, then $\langle\psi|c|\xi\rangle = c\langle\psi|\xi\rangle$ and $\langle\psi|c|\xi\rangle^* = c^*\langle\psi|\xi\rangle^* = c^*\langle\xi|\psi\rangle = \langle\xi|c^*|\psi\rangle$. Thus, for a complex number, the adjoint is nothing but the complex conjugate.
- The bra corresponding to a ket $c|\psi\rangle$ is $c^*\langle\psi|$

3.2.1 *Eigenvectors and Eigenvalues*

The eigenvalue equation satisfied by an operator $\widehat{A}$ is defined through the following equation:

$$\widehat{A}|a_n\rangle = a_n|a_n\rangle \tag{3.15}$$

where a_n is the eigenvalue and $|a_n\rangle$ is the corresponding eigenvector. Since an operator may have multiple eigenvalues and eigenvectors, we are using a subscript on a to represent one of the eigenvectors and the corresponding eigenvalue. For convenience, we are using the same symbol for the eigenvalue and the eigenvector. Equation (3.15) implies that when the operator $\widehat{A}$ acts on the ket vector $|a_n\rangle$, the resulting ket vector is a number a_n multiplied by the same ket $|a_n\rangle$.

Now we also have

$$\langle a_n|\widehat{A}|a_n\rangle = \langle a_n|a_n|a_n\rangle = a_n\langle a_n|a_n\rangle \tag{3.16}$$

If $\widehat{A}$ is a Hermitian operator then $\langle a_n|\widehat{A}|a_n\rangle^* = \langle a_n|\widehat{A}|a_n\rangle$ and since $\langle a_n|a_n\rangle$ is also real, the eigenvalue a_n is also a real quantity. Thus, the eigenvalues of a Hermitian operators are real. All observables are represented by Hermitian operators and have real eigenvalues.

Exercise 3.1 Show that eigenvectors of a Hermitian operator are orthogonal to each other.

Solution Let $|a_n\rangle$ and $|a_m\rangle$ be two different eigenvectors of an operator $\widehat{A}$ with two different eigenvalues a_n and a_m, respectively. Hence

$$\widehat{A}|a_n\rangle = a_n|a_n\rangle$$

which gives us

$$\langle a_m|\widehat{A}|a_n\rangle = a_n\langle a_m|a_n\rangle$$

We also have

$$\widehat{A}|a_m\rangle = a_m|a_m\rangle$$

Now, if $\widehat{A}$ is Hermitian, then we have

$$\langle a_m|\widehat{A}^\dagger = \langle a_m|\widehat{A} = a_m^*\langle a_m| = a_m\langle a_m|$$

Thus,

$$\langle a_m|\widehat{A}|a_n\rangle = a_m\langle a_m|a_n\rangle$$

From above, we obtain

$$(a_m - a_n)\langle a_m|a_n\rangle = 0$$

Since $a_m \neq a_n$ (we are considering the state to be nondegenerate), we have

$$\langle a_m|a_n\rangle = 0; m \neq n$$

Implying that the eigenvectors are orthogonal. Hence, we may write

$$\langle a_m|a_n\rangle = \delta_{mn}$$

where δ_{mn} represents the Kronecker delta function with $\delta_{mn} = 0$; $m \neq n$ and $\delta_{nn} = 1$.

Eigenvectors of a Hermitian operator form a complete orthonormal set of vectors. Hence any ket $|\psi\rangle$ can be expanded in terms of the eigenkets:

$$|\psi\rangle = \sum c_n |a_n\rangle \tag{3.17}$$

In order to obtain the expansion coefficients c_n, we multiply Eq. (3.17) by $\langle a_m|$ and obtain

$$\langle a_m|\psi\rangle = \sum c_n \langle a_m|a_n\rangle = \sum c_n \delta_{mn} = c_m$$

where we have used the orthogonality of the eigenvectors. Thus,

$$|\psi\rangle = \sum c_n|a_n\rangle = \sum \langle a_n|\psi\rangle |a_n\rangle = \sum |a_n\rangle\langle a_n|\psi\rangle \tag{3.18}$$

Since the above equation is valid for any $|\psi\rangle$, we have the following relation satisfied by the eigenvectors:

$$\sum |a_n\rangle\langle a_n| = \widehat{I} \tag{3.19}$$

where $\widehat{I}$ is the identity operator. Equation (3.19) is referred to as the completeness condition satisfied by the eigenvectors.

For states which have a continuum of eigenvalues, if we represent the state by $|\xi\rangle$, instead of Eq. (3.17), we have the following equation

$$|\psi\rangle = \int \phi(\xi)|\xi\rangle d\xi \tag{3.20}$$

The corresponding normalization condition is given by

$$\langle \xi'|\xi\rangle = \delta(\xi - \xi') \tag{3.21}$$

where $\delta(x)$ represents the Dirac delta function. We also have

$$\phi(\xi) = \langle \xi|\psi\rangle \tag{3.22}$$

the completeness condition is given by

$$\int |\xi\rangle\langle\xi| d\xi = \widehat{I} \tag{3.23}$$

Equation (3.17) implies that the quantity c_m represents the projection of the quantum state $|\psi\rangle$ on to the eigenstate $|a_m\rangle$.

Let us consider the Hamiltonian operator represented by $\widehat{H}$. The eigenvalue equation satisfied by the Hamiltonian operator is written as

$$\widehat{H}|E_n\rangle = E_n|E_n\rangle \tag{3.24}$$

where $|E_n\rangle$ represents the eigenket and E_n represents the corresponding eigenvalue. For simplicity we are considering discrete eigenstates.

Any state can be written as

$$|\psi\rangle = \sum c_n|E_n\rangle \tag{3.25}$$

with

$$c_n = \langle E_n|\psi\rangle \tag{3.26}$$

When a measurement of energy is carried out on the system in thc quantum state $|\psi\rangle$, then according to the postulate of quantum mechanics, the state immediately jumps to one of the energy eigenstates $|E_n\rangle$ with a probability given by $|c_n|^2 = |\langle E_n|\psi\rangle|^2$.

Let us consider the quantity $\langle\psi|\widehat{H}|\psi\rangle$. Using Eq. (3.25), we can write

$$\langle\psi|\widehat{H}|\psi\rangle = \sum c_n^*\langle E_n|\widehat{H}\sum c_m|E_m\rangle$$

which using Eq. (3.24) gives us

$$\begin{aligned}\langle\psi|\widehat{H}|\psi\rangle &= \sum c_n^*\langle E_n|\sum c_m E_m|E_m\rangle = \sum\sum c_n^* c_m E_m\langle E_n|E_m\rangle = \sum\sum c_n^* c_m E_m\delta_{mn}\\ &= \sum |c_n|^2 E_n\end{aligned} \tag{3.27}$$

which is the expectation value of energy in the state represented by $|\psi\rangle$ with $|c_n|^2$ representing the probability of obtaining an energy value E_n during the measurement.

Note that since $|\psi\rangle$ has to be normalized, we have

$$\langle\psi|\psi\rangle = 1 \tag{3.28}$$

which using Eq. (3.25) can be written as

$$\sum c_m^*\langle E_m|\sum c_n|E_n\rangle = 1$$

Since the eigenfunctions are orthogonal to each other, we have

$$\langle E_m|E_n\rangle = \delta_{mn}$$

Hence, we have

$$\sum |c_n|^2 = 1 \tag{3.29}$$

All physical observables are represented by Hermitian operators and every precise measurement of the observable will result in one of the eigenvalues of the observable and the system will collapse into the corresponding eigenstate. Thus, if the state is described by $|\psi\rangle$ as given in Eq. (3.25), then a measurement of the energy of the state would result in one of the eigenvalues E_n with a probability $|c_n|^2$. Note that Eq. (3.29) implies that the total probability adds up to unity as it indeed should.

The quantity $\langle\psi|\widehat{A}|\psi\rangle$ represents the ensemble average of a number of measurements carried out on identically prepared ensemble of quantum systems described by the state $|\psi\rangle$.

3.2.2 Commutator Brackets

The commutator bracket between two operators $\widehat{A}$ and $\widehat{B}$ is given by

$$\left[\widehat{A}, \widehat{B}\right] = \widehat{A}\widehat{B} - \widehat{B}\widehat{A} = i\widehat{C} \tag{3.30}$$

showing that operators, in general, do not commute. Here $\widehat{C}$ is also an operator. In quantum mechanics, the observables corresponding to noncommuting observables cannot be measured simultaneously with arbitrary precision. This is represented by the Heisenberg uncertainty relation:

$$\Delta A.\Delta B \geq \frac{1}{2}\left|\left\langle\widehat{C}\right\rangle\right| \tag{3.31}$$

where ΔA and ΔB represent the uncertainties in the measurement of the two corresponding observables and are defined through the following equation:

$$\Delta A = \left(\left\langle\widehat{A}^2\right\rangle - \left\langle\widehat{A}\right\rangle^2\right)^{1/2} \tag{3.32}$$

with angular brackets representing expectation values.

One of the most important commutation relation is

$$[\widehat{x}, \widehat{p}] = i\hbar \tag{3.33}$$

where $\hbar = \frac{h}{2\pi}$ and h represents the Planck's constant, $\widehat{x}$ and $\widehat{p}$ represent the position and momentum operators. Heisenberg uncertainty principle implies that

$$\Delta x.\Delta p \geq \frac{\hbar}{2} \tag{3.34}$$

It is possible to show that a set of operators which have simultaneous eigenfunctions commute with each other and hence it is possible to simultaneously obtain precise values of the observables corresponding to the two commuting operators.

3.3 Matrix Representation of States and Operators: A Simple Example from Optics

In this section, we will consider the simple example of the polarization property of light (see Chap. 2) and use the formalism developed above to describe polarization. We consider plane electromagnetic waves propagating in free space along the z-axis. In such a case, there are two orthogonal eigenstates of propagation perpendicular to the z-axis. One of the possible basis states is in terms of linear polarization states namely x-polarized and y-polarized and any arbitrary state of polarization of a wave propagating along the z-axis can be represented by a linear superposition of the two orthogonal polarization states.

The electric field of a wave is described by

$$\vec{E} = \hat{x}E_x + \hat{y}E_y \tag{3.35}$$

Here $\hat{x}$ and $\hat{y}$ represent unit vectors along the x- and y-directions, respectively (not to confuse with operators which are also represented by a hat) with amplitudes E_x and E_y.

We use the bra-ket formalism and represent the vectors corresponding to x- and y-polarizations through the following kets related to column matrices (see also Chap. 2):

$$|x\rangle = \begin{pmatrix} 1 \\ 0 \end{pmatrix} \tag{3.36a}$$

$$|y\rangle = \begin{pmatrix} 0 \\ 1 \end{pmatrix} \tag{3.36b}$$

The corresponding bras are the row matrices

$$\langle x| = \begin{pmatrix} 1 & 0 \end{pmatrix} \tag{3.37a}$$

$$\langle y| = \begin{pmatrix} 0 & 1 \end{pmatrix} \tag{3.37b}$$

The kets $|x\rangle$ and $|y\rangle$ form the eigenvectors of polarization and form a complete orthonormal set implying that any polarization state can be described in terms of these two basis kets.

The basis kets satisfy the following orthogonality relation:

$$\langle x|y\rangle = \begin{pmatrix} 1 & 0 \end{pmatrix}\begin{pmatrix} 0 \\ 1 \end{pmatrix} = 0 \tag{3.38}$$

the following normalization conditions

$$\langle x|x\rangle = \begin{pmatrix} 1 & 0 \end{pmatrix}\begin{pmatrix} 1 \\ 0 \end{pmatrix} = 1 \tag{3.39}$$

$$\langle y|y\rangle = \begin{pmatrix} 0 & 1 \end{pmatrix}\begin{pmatrix} 0 \\ 1 \end{pmatrix} = 1 \tag{3.40}$$

and the following completeness condition

$$|x\rangle\langle x| + |y\rangle\langle y| = \begin{pmatrix} 1 \\ 0 \end{pmatrix}\begin{pmatrix} 1 & 0 \end{pmatrix} + \begin{pmatrix} 0 \\ 1 \end{pmatrix}\begin{pmatrix} 0 & 1 \end{pmatrix} = \begin{pmatrix} 1 & 0 \\ 0 & 1 \end{pmatrix} = \widehat{I} \tag{3.41}$$

Since the two eigenvectors form a complete set, an arbitrary polarization state can be represented as a superposition of the two eigenvectors. Thus, an electric field given by Eq. (3.35) can be represented as

$$|E\rangle = \begin{pmatrix} E_x \\ E_y \end{pmatrix} = E_x\begin{pmatrix} 1 \\ 0 \end{pmatrix} + E_y\begin{pmatrix} 0 \\ 1 \end{pmatrix} = E_x|x\rangle + E_y|y\rangle \tag{3.42}$$

Note that $E_x = \langle x|E\rangle$ and $E_y = \langle y|E\rangle$ (see Eq. 3.26).

Devices such as polarizers, half wave plates, and quarter wave plates acting on the polarization state of the light wave are represented by operators.

A polarizer with its pass axis making an angle θ with the x-axis is represented by the following matrix:

$$\widehat{P} = \begin{pmatrix} \cos^2\theta & \cos\theta\sin\theta \\ \cos\theta\sin\theta & \sin^2\theta \end{pmatrix} \tag{3.43}$$

An input state $|E_{in}\rangle$ passing through the polarizer is represented by operation of the polarizer operator $\widehat{P}$ acting on the input state. Thus, the output state will be given by

$$|E_{out}\rangle = \widehat{P}|E_{in}\rangle \tag{3.44}$$

For example if the initial state is an x-polarized wave with unit amplitude then $|E_{in}\rangle = |x\rangle$ and the output would be

$$|E_{out}\rangle = \widehat{P}|E_{in}\rangle = \widehat{P}|x\rangle = \begin{pmatrix} \cos^2\theta & \cos\theta\sin\theta \\ \cos\theta\sin\theta & \sin^2\theta \end{pmatrix}\begin{pmatrix} 1 \\ 0 \end{pmatrix} = \begin{pmatrix} \cos^2\theta \\ \cos\theta\sin\theta \end{pmatrix}$$

which is

$$|E_{out}\rangle = \cos^2\theta|x\rangle + \cos\theta\sin\theta|y\rangle \tag{3.45}$$

Thus, the operator operating on the input ket results in another ket. It should be noted that the output polarization state is written in terms of the laboratory coordinate system and not the system attached to the polarizer.

We can write the following representations of the different polarization states and the polarization devices in terms of ket vectors and operators, respectively:

- Linearly polarized light oriented at +45°: $|+45\,°\rangle = \frac{1}{\sqrt{2}}\begin{pmatrix} 1 \\ 1 \end{pmatrix} = \frac{1}{\sqrt{2}}|x\rangle + \frac{1}{\sqrt{2}}|y\rangle$
- Linearly polarized light oriented at −45°: $|-45\,°\rangle = \frac{1}{\sqrt{2}}\begin{pmatrix} 1 \\ -1 \end{pmatrix} = \frac{1}{\sqrt{2}}|x\rangle - \frac{1}{\sqrt{2}}|y\rangle$
- Right circularly polarized light: $|RCP\rangle = \frac{1}{\sqrt{2}}\begin{pmatrix} 1 \\ i \end{pmatrix} = \frac{1}{\sqrt{2}}|x\rangle + \frac{i}{\sqrt{2}}|y\rangle$
- Left circularly polarized light: $|LCP\rangle = \frac{1}{\sqrt{2}}\begin{pmatrix} 1 \\ -i \end{pmatrix} = \frac{1}{\sqrt{2}}|x\rangle - \frac{i}{\sqrt{2}}|y\rangle$
- Quarter wave plate with fast axis at +45° (see Chap. 2): $\widehat{Q} = \frac{1}{\sqrt{2}}\begin{pmatrix} 1 & -i \\ -i & 1 \end{pmatrix}$
- Half wave plate with fast axis at +45° (see Chap. 2): $\widehat{H} = \begin{pmatrix} 0 & 1 \\ 1 & 0 \end{pmatrix}$

Exercise 3.2 Express $|x\rangle$ and $|y\rangle$ in terms of $|RCP\rangle$ and $|LCP\rangle$.

Solution We have

$$|x\rangle = \begin{pmatrix} 1 \\ 0 \end{pmatrix} = \frac{1}{\sqrt{2}}\left(\frac{1}{\sqrt{2}}\begin{pmatrix} 1 \\ i \end{pmatrix} + \frac{1}{\sqrt{2}}\begin{pmatrix} 1 \\ -i \end{pmatrix}\right) = \frac{1}{\sqrt{2}}(|RCP\rangle + |LCP\rangle)$$

$$|y\rangle = \begin{pmatrix} 0 \\ 1 \end{pmatrix} = \frac{1}{\sqrt{2}}\left(\frac{1}{\sqrt{2}}\begin{pmatrix} 1 \\ i \end{pmatrix} - \frac{1}{\sqrt{2}}\begin{pmatrix} 1 \\ -i \end{pmatrix}\right) = \frac{i}{\sqrt{2}}(|RCP\rangle - |LCP\rangle)$$

3.4 Time Evolution

The time evolution of a quantum state is described by the Schrödinger equation:

$$i\hbar\frac{\partial}{\partial t}|\psi\rangle = \widehat{H}|\psi\rangle \tag{3.46}$$

where $\widehat{H}$ is the Hamiltonian operator and is the total energy of the system. Given the state of the system at any instant of time say $t = 0$, that is, given $|\psi(t = 0)\rangle$ solving Eq. (3.46) gives us the time evolution of the quantum system. From $|\psi(t)\rangle$ it is possible to obtain the time evolution of the expectation values of the various observables.

If $\widehat{H}$ is independent of time, a formal solution to Eq. (3.46) is

$$|\psi(t)\rangle = e^{-\frac{i\hat{H}t}{\hbar}}|\psi(t=0)\rangle \tag{3.47}$$

The exponential of an operator is defined through a Taylor series expansion:

$$e^{-i\hat{A}} = \sum_{n=0}^{\infty}\frac{\left(-i\widehat{A}\right)^n}{n!} = 1 - i\widehat{A} - \frac{\widehat{A}\widehat{A}}{2} - \ldots. \tag{3.48}$$

3.5 Linear Harmonic Oscillator

In this section, we will consider the quantum mechanical analysis of a one-dimensional linear harmonic oscillator of mass m oscillating with a frequency ω which is described by a potential energy function given by

$$V(x) = \frac{1}{2}m\omega^2x^2 \tag{3.49}$$

The Hamiltonian operator corresponding to the oscillator is given by

$$\widehat{H} = \frac{\widehat{p}^2}{2m} + \frac{1}{2}m\omega^2\,\widehat{x}^2 \tag{3.50}$$

where $\widehat{x}$ and $\widehat{p}$ represent the position and linear momentum operators, respectively. We may note that $\widehat{H}$, $\widehat{p}$, and $\widehat{x}$ correspond to observables and are Hermitian operators with real eigenvalues.

The operators $\widehat{x}$ and $\widehat{p}$ satisfy the following commutation relation:

$$[\widehat{x}, \widehat{p}] = i\hbar \tag{3.51}$$

Imposing this quantization condition leads to quantization of the harmonic oscillator.

For the analysis, we introduce the non-Hermitian operator $\widehat{a}$:

$$\widehat{a} = \frac{1}{\sqrt{2m\hbar\omega}}(m\omega\widehat{x} + i\widehat{p}) \tag{3.52a}$$

We then have

$$\hat{a}^\dagger = \frac{1}{\sqrt{2m\hbar\omega}}(m\omega\hat{x} - i\hat{p}) \tag{3.52b}$$

In obtaining Eq. (3.52b) we have used the fact that both $\hat{x}$ and $\hat{p}$ are Hermitian operators and hence $\hat{x}^\dagger = \hat{x}$ and $\hat{p}^\dagger = \hat{p}$.

We can invert Eqs. (3.52a) and (3.52b) and obtain

$$\hat{x} = \sqrt{\frac{\hbar}{2m\omega}}\left(\hat{a}^\dagger + \hat{a}\right) \tag{3.53a}$$

and

$$\hat{p} = i\sqrt{\frac{m\hbar\omega}{2}}\left(\hat{a}^\dagger - \hat{a}\right) \tag{3.53b}$$

Using the commutation relation satisfied by $\hat{x}$ and $\hat{p}$, it is easy to show that

$$\left[\hat{a}, \hat{a}^\dagger\right] = \hat{a}\,\hat{a}^\dagger - \hat{a}^\dagger\hat{a} = 1 \tag{3.54}$$

Using Eqs. (3.51), (3.52a), and (3.52b), we can express the Hamiltonian operator $\hat{H}$ (Eq. 3.50) in terms of the operators $\hat{a}$ and $\hat{a}^\dagger$ to obtain

$$\hat{H} = \frac{\hbar\omega}{2}\left(\hat{a}^\dagger\hat{a} + \hat{a}\,\hat{a}^\dagger\right) = \hbar\omega\left(\hat{a}^\dagger\hat{a} + \frac{1}{2}\right) \tag{3.55}$$

It is interesting to note that the factor of $\frac{1}{2}$ in Eq. (3.55) appears due to the noncommutation of $\hat{x}$ and $\hat{p}$ or equivalently the noncommutation of $\hat{a}$ and $\hat{a}^\dagger$.

Using the commutation relation given by Eq. (3.54), we can show that

$$\left[\hat{a}, \hat{a}^\dagger\hat{a}\right] = \hat{a} \tag{3.56a}$$

and

$$\left[\hat{a}^\dagger, \hat{a}^\dagger\hat{a}\right] = -\hat{a}^\dagger \tag{3.56b}$$

Using Eq. (3.55), we have

$$\left[\hat{a}, \hat{H}\right] = \hbar\omega\hat{a} \tag{3.57}$$

3.5.1 Eigenvalues of the Hamiltonian Operator

In order to calculate the eigenvalues of the energy of the harmonic oscillator, we start with the eigenvalue equation for the Hamiltonian operator:

$$\widehat{H}|E\rangle = E|E\rangle \tag{3.58}$$

where $|E\rangle$ is the eigenket and E is the corresponding eigenvalue. Multiplying Eq. (3.58) by $\widehat{a}$ on both sides, we obtain

$$\widehat{a}\,\widehat{H}|E\rangle = E\widehat{a}\,|E\rangle \tag{3.59}$$

Using Eq. (3.57), Eq. (3.59) gives us

$$\left(\widehat{H}\widehat{a} + \hbar\omega\widehat{a}\right)|E\rangle = E\widehat{a}\,|E\rangle$$

Thus,

$$\widehat{H}\widehat{a}\,|E\rangle = (E - \hbar\omega)\widehat{a}\,|E\rangle \tag{3.60}$$

implying that if $|E\rangle$ is an eigenket of $\widehat{H}$ with an eigenvalue E, then $\widehat{a}\,|E\rangle$ is an eigenket of $\widehat{H}$ with an eigenvalue $(E - \hbar\omega)$. This implies that operating with the operator $\widehat{a}$ on an eigenket of the Hamiltonian operator creates another eigenket of the Hamiltonian operator with an eigenvalue reduced by $\hbar\omega$.

Repeating in this fashion, we find that $\widehat{a}^n|E\rangle$ is also an eigenket of $\widehat{H}$ with the eigenvalue $(E - n\hbar\omega)$. Hence starting from any eigenvalue of $\widehat{H}$, we can keep on applying the operator $\widehat{a}$ on the eigenket and keep reducing the eigenvalue. We will then reach a stage when $(E - n\hbar\omega)$ will become negative for some value of n.

The question that arises is whether the operator $\widehat{H}$ can have negative eigenvalues. In order to find this out we substitute for $\widehat{H}$ from Eq. (3.55) in Eq. (3.58) and obtain

$$\hbar\omega\left(\widehat{a}^{\dagger}\widehat{a} + \frac{1}{2}\right)|E\rangle = E|E\rangle \tag{3.61}$$

We multiply the above equation by $\langle E|$ from the left and obtain

$$\hbar\omega\langle E|\,\widehat{a}^{\dagger}\widehat{a}|E\rangle + \frac{1}{2}\hbar\omega\langle E|E\rangle = E\langle E|E\rangle \tag{3.62}$$

We note that both $\langle E|E\rangle$ and $\langle E|\,\widehat{a}^{\dagger}\widehat{a}|E\rangle$ are positive quantities. ($\langle E|\,\widehat{a}^{\dagger}$ is the Hermitian conjugate of $\widehat{a}|E\rangle$ and as per our earlier discussion the quantity $\langle E|\,\widehat{a}^{\dagger}\widehat{a}|E\rangle$ will be positive.) This implies that E cannot be negative. The reduction of eigenvalue by operation with $\widehat{a}$ will get terminated if we find an eigenket $|E_0\rangle$ which satisfies the following equation:

$$\widehat{a}|E_0\rangle = 0 \tag{3.63}$$

Since any more operation with $\widehat{a}$ will not lead to any further eigenkets. The eigenket $|E_0\rangle$ will then be lowest eigenstate of $\widehat{H}$.

Writing down Eq. (3.62) for the ground state eigenket $|E_0\rangle$, we have

$$\hbar\omega\langle E_0|\,\widehat{a}^\dagger\widehat{a}|E_0\rangle + \frac{1}{2}\hbar\omega\langle E_0|E_0\rangle = E_0\langle E_0|E_0\rangle$$

which using Eq. (3.63) gives us the eigenvalue of the lowest eigenstate $|E_0\rangle$ as

$$E_0 = \frac{1}{2}\hbar\omega \tag{3.64}$$

We also note from Eq. (3.58) that

$$\widehat{a}^\dagger\widehat{H}|E\rangle = E\,\widehat{a}^\dagger|E\rangle$$

which using Eq. (3.56b) becomes

$$\widehat{H}\,\widehat{a}^\dagger|E\rangle = (E + \hbar\omega)\,\widehat{a}^\dagger|E\rangle \tag{3.65}$$

Thus, if $|E\rangle$ is an eigenket of $\widehat{H}$ with the eigenvalue E, then $\widehat{a}^\dagger|E\rangle$ is an eigenket of $\widehat{H}$ with an eigenvalue (E + $\hbar\omega$). Continuing in this fashion, we will find that $\widehat{a}^{\dagger^n}|E\rangle$ will be an eigenket of $\widehat{H}$ with eigenvalue (E + $n\hbar\omega$). In this fashion, we can generate all the eigenkets and eigenvalues of $\widehat{H}$ which will be

$$E_n = \left(n + \frac{1}{2}\right)\hbar\omega; \quad n = 0, 1, 2, \ldots.. \tag{3.66}$$

and the corresponding eigenket will be $\widehat{a}^{\dagger^n}|E_0\rangle$. Hence the energy of the linear harmonic oscillator is quantized and can increase or decrease in quanta of energy $\hbar\omega$.

Assuming the ground state $|E_0\rangle$ to be normalized, that is, $\langle E_0|E_0\rangle = 1$, then using the commutation relation between $\widehat{a}$ and $\widehat{a}^\dagger$ it is possible to obtain the following normalized eigenkets of the Hamiltonian operator $\widehat{H}$:

$$|E_n\rangle = \frac{1}{\sqrt{n!}}\,\widehat{a}^{\dagger^n}|E_0\rangle \tag{3.67}$$

It is straightforward to show that

$$\widehat{a}|E_n\rangle = \sqrt{n}|E_{n-1}\rangle \tag{3.68}$$

and

$$\widehat{a}^{\dagger}|E_n\rangle = \sqrt{n+1}|E_{n+1}\rangle \tag{3.69}$$

Since operation by $\widehat{a}$ reduces the eigenvalue by the quantum $\hbar\omega$, it is referred to as annihilation operator or destruction operator. Similarly since operation by $\widehat{a}^{\dagger}$ increases the eigenvalue by $\hbar\omega$, it is referred to as the creation operator.

Since the allowed energy eigenvalues of energy are $\left(n+\frac{1}{2}\right)\hbar\omega$, it is also customary to represent the eigenstate $|E_n\rangle$ as $|n\rangle$ with $n = 0, 1, 2, 3\ldots$. The ket $|n\rangle$ is also referred to as the number state since this state is characterized by a precise value of energy of $\left(n+\frac{1}{2}\right)\hbar\omega$. The lowest energy state or the ground state of the harmonic oscillator is represented by $|0\rangle$.

We also have the following eigenvalue equation of the operator $\widehat{a}^{\dagger}\widehat{a}$:

$$\widehat{a}^{\dagger}\widehat{a}|n\rangle = \widehat{a}^{\dagger}\sqrt{n}|n-1\rangle = \sqrt{n}\,\widehat{a}^{\dagger}|n-1\rangle = n|n\rangle \tag{3.70}$$

Thus, the eigenvalues of the operator $\widehat{a}^{\dagger}\widehat{a}$ are integers and give us the eigenvalue corresponding to the excitation state $|n\rangle$ of the harmonic oscillator. In view of this, we define the number operator $\widehat{N}$ by

$$\widehat{N} = \widehat{a}^{\dagger}\widehat{a} \tag{3.71}$$

The energy eigenstates of the harmonic oscillator form an orthonormalized set and we have

$$\langle n|m\rangle = \delta_{mn} \tag{3.72}$$

3.5.2 Expectation Values

It is now possible to obtain the expectation values of position and momentum of a harmonic oscillator in one of the eigenstates.

The expectation value of position when the oscillator is in an eigenstate $|n\rangle$ is given by

$$\langle\widehat{x}\rangle = \langle n|\widehat{x}|n\rangle = \langle n|\sqrt{\frac{\hbar}{2m\omega}}\left(\widehat{a}^{\dagger}+\widehat{a}\right)|n\rangle = \sqrt{\frac{\hbar}{2m\omega}}\left(\langle n|\widehat{a}^{\dagger}|n\rangle + \langle n|\widehat{a}|n\rangle\right) = 0 \tag{3.73}$$

using Eqs. (3.68) and (3.69). Similarly we can show that

$$\langle\widehat{p}\rangle = \langle n|\widehat{p}|n\rangle = 0 \tag{3.74}$$

It is interesting to note that although the harmonic oscillator is in one of the excited states, the expectation values of position and momentum are zero.

We also have

$$\left\langle\widehat{x}^2\right\rangle = \langle n|\widehat{x}^2|n\rangle = \frac{\hbar}{2m\omega}\langle n|\left(\widehat{a}^\dagger + \widehat{a}\right)^2|n\rangle$$

$$= \frac{\hbar}{2m\omega}\langle n|\left(\widehat{a}^{\dagger 2} + \widehat{a}^2 + \widehat{a}^\dagger\widehat{a} + \widehat{a}\widehat{a}^\dagger\right)|n\rangle = \frac{\hbar}{2m\omega}\langle n|\left(\widehat{a}^{\dagger 2} + \widehat{a}^2 + 2\widehat{a}^\dagger\widehat{a} + 1\right)|n\rangle$$

which gives us

$$\left\langle\widehat{x}^2\right\rangle = \frac{\hbar}{2m\omega}(2n+1) \tag{3.75}$$

Similarly

$$\left\langle\widehat{p}^2\right\rangle = \langle n|\widehat{p}^2|n\rangle = -\frac{m\hbar\omega}{2}\langle n|\left(\widehat{a}^\dagger - \widehat{a}\right)^2|n\rangle = \frac{m\hbar\omega}{2}(2n+1) \tag{3.76}$$

Thus, the uncertainties in position and momentum of a harmonic oscillator when it is in an eigenstate $|n\rangle$ are given by

$$\Delta x = \sqrt{\left\langle\widehat{x}^2\right\rangle - \langle\widehat{x}\rangle^2} = \sqrt{\frac{\hbar}{2m\omega}}(2n+1)^{1/2} \tag{3.77}$$

and

$$\Delta p = \sqrt{\left\langle\widehat{p}^2\right\rangle - \langle\widehat{p}\rangle^2} = \sqrt{\frac{m\hbar\omega}{2}}(2n+1)^{1/2} \tag{3.78}$$

Thus, we obtain the following uncertainty product

$$\Delta x.\Delta p = \hbar\left(n + \frac{1}{2}\right) \tag{3.79}$$

The ground state $|0\rangle$ has the lowest possible value $\frac{\hbar}{2}$ of the uncertainty product as dictated by Heisenberg uncertainty principle. Even when the harmonic oscillator is in its ground state, there is uncertainty in both position and momentum of the oscillator. Higher energy states have larger values of uncertainty.

The Schrödinger equation which gives us the time evolution of any quantum state is given by Eq. (3.46). We can use this equation to determine the time variation of the energy eigenstates. Thus, if we choose $|\psi\rangle = |n\rangle$ then we have

$$i\hbar \frac{\partial}{\partial t}|n\rangle = \widehat{H}|n\rangle = \hbar\omega\left(\widehat{a}^{\dagger}\widehat{a} + \frac{1}{2}\right)|n\rangle = \hbar\omega\left(n + \frac{1}{2}\right)|n\rangle \tag{3.80}$$

where we have used Eq. (3.70). We can formally integrate Eq. (3.80) and obtain the time variation of the number state as

$$|n(t)\rangle = |n(t=0)\rangle e^{-i\left(n+\frac{1}{2}\right)\omega t} \tag{3.81}$$

Exercise 3.3 Evaluate the time variation of the expectation value of position when the harmonic oscillator is in its energy eigenstate.

Solution We have (using Eq. 3.81)

$$\langle\widehat{x}(t)\rangle = \langle n(t)|\widehat{x}|n(t)\rangle = \langle n(t=0)|\widehat{x}|n(t=0)\rangle = 0 \tag{3.82}$$

Thus, in the energy eigenstate, the expectation value of the position of the harmonic oscillator is always zero.

3.6 Coherent State

The energy eigenstates $|n\rangle$ represent states of precise energy and we saw in Sect. 3.4 that in this state the expectation value $\langle\widehat{x}(t)\rangle$ is zero and independent of time. We are all familiar with the classical harmonic oscillator with its position oscillating with time in a sinusoidal fashion about the origin. A quantum mechanical analysis should give us an expectation value of the position namely $\langle\widehat{x}(t)\rangle$ to be varying sinusoidally with time. However, the energy eigenstates which are eigenstates of the operator $\widehat{H}$ and the number operator $\widehat{N}$ do not show this behavior. Hence they cannot represent a classical harmonic oscillator.

We will now discuss coherent states which are linear superpositions of the energy eigenstates which exhibit features closest to a classical harmonic oscillator.

We define coherent states as the eigenstates of the annihilation operator $\widehat{a}$:

$$\widehat{a}|\alpha\rangle = \alpha|\alpha\rangle \tag{3.83}$$

where $|\alpha\rangle$ represents the normalized ket corresponding to the coherent state with an eigenvalue α. Note that since $\widehat{a}$ is not a Hermitian operator, the eigenvalues α are, in general complex.

Let us first evaluate the expectation value of the position coordinate of the oscillator in a coherent state. Thus, we have

$$\langle\hat{x}\rangle = \langle\alpha|\hat{x}|\alpha\rangle = \langle\alpha|\sqrt{\frac{\hbar}{2m\omega}}\left(\hat{a}^{\dagger} + \hat{a}\right)|\alpha\rangle = \sqrt{\frac{\hbar}{2m\omega}}\left(\langle\alpha|\hat{a}^{\dagger}|\alpha\rangle + \langle\alpha|\hat{a}|\alpha\rangle\right) \quad (3.84)$$

For simplicity, we assume that the eigenvalue α is real. In this case Eq. (3.83) also implies

$$\langle\alpha|\hat{a}^{\dagger} = \langle\alpha|\alpha^{*} = \alpha\langle\alpha| \quad (3.85)$$

Hence, we have from Eq. (3.84)

$$\langle\hat{x}\rangle = \sqrt{\frac{\hbar}{2m\omega}}2\alpha\langle\alpha|\alpha\rangle = \sqrt{\frac{2\hbar}{m\omega}}\alpha \quad (3.86)$$

Similarly the expectation value of the momentum operator $\hat{p}$ is

$$\langle\hat{p}\rangle = \langle\alpha|\hat{p}|\alpha\rangle = \langle\alpha|i\sqrt{\frac{m\hbar\omega}{2}}\left(\hat{a}^{\dagger} - \hat{a}\right)|\alpha\rangle = i\sqrt{\frac{m\hbar\omega}{2}}\left(\langle\alpha|\hat{a}^{\dagger}|\alpha\rangle - \langle\alpha|\hat{a}|\alpha\rangle\right) = 0 \quad (3.87)$$

One can also evaluate the expectation values $\left\langle\hat{x}^2\right\rangle$ and $\left\langle\hat{p}^2\right\rangle$ and obtain

$$\left\langle\hat{x}^2\right\rangle = \frac{\hbar}{2m\omega}\left(4\alpha^2 + 1\right) \quad (3.88)$$

and

$$\left\langle\hat{p}^2\right\rangle = \frac{m\hbar\omega}{2} \quad (3.89)$$

Hence, we have

$$\Delta x = \sqrt{\left\langle\hat{x}^2\right\rangle - \langle\hat{x}\rangle^2} = \sqrt{\frac{\hbar}{2m\omega}} \quad (3.90)$$

and

$$\Delta p = \sqrt{\left\langle\hat{p}^2\right\rangle - \langle\hat{p}\rangle^2} = \sqrt{\frac{m\hbar\omega}{2}} \quad (3.91)$$

Hence, we have for the coherent state

$$\Delta x.\Delta p = \frac{\hbar}{2} \tag{3.92}$$

which is the minimum value as per Heisenberg uncertainty principle. Hence a coherent state is also a minimum uncertainty state (MUS).

The ratio of the uncertainty in position of the oscillator to the expectation value of the position is given by

$$\frac{\Delta x}{\langle \hat{x} \rangle} = \frac{1}{2\alpha} \tag{3.93}$$

which shows that as α increases, the fractional uncertainty in the position decreases and the oscillator behaves more like a classical oscillator. In Sect. 3.5.2, we will show that the time variation of the expectation values of position and momentum follows classical equations of motion.

The coherent state is a specific linear combination of number states of the harmonic oscillator (see Sect. 3.5.1). Here we obtain the expectation value of the number operator $\hat{N} = \hat{a}^{\dagger}\hat{a}$ in this state:

$$\left\langle \hat{N} \right\rangle = \langle \alpha | \hat{N} | \alpha \rangle = \langle \alpha | \, \hat{a}^{\dagger}\hat{a} | \alpha \rangle = |\alpha|^2 \tag{3.94}$$

Note that the Hamiltonian operator $\hat{H}$ is related to the number operator by $\hat{H} = \hbar\omega\left(\hat{N} + \frac{1}{2}\right)$. Thus, the expectation values and uncertainties in $\hat{N}$ are directly related to the expectation value and uncertainty in energy of the oscillator. In fact the expectation value of energy of the oscillator in the coherent state is $\left\langle \hat{H} \right\rangle = \hbar\omega\left(\left\langle \hat{a}^{\dagger}\hat{a} \right\rangle + \frac{1}{2}\right) = \hbar\omega\left(|\alpha|^2 + \frac{1}{2}\right)$.

We can also evaluate the expectation value of $\hat{N}^2$:

$$\langle \hat{N}^2 \rangle = \langle \alpha | \hat{N}^2 | \alpha \rangle = \langle \alpha | \, \hat{a}^{\dagger}\hat{a}\,\hat{a}^{\dagger}\hat{a} | \alpha \rangle = \langle \alpha | \, \hat{a}^{\dagger}\left(\hat{a}^{\dagger}\hat{a} + 1 \right)\hat{a} | \alpha \rangle = |\alpha|^4 + |\alpha|^2 \tag{3.95}$$

Hence the variance in the excitation state of the harmonic oscillator in a coherent state is given by

$$\Delta N = \sqrt{\left\langle \hat{N}^2 \right\rangle - \left\langle \hat{N} \right\rangle^2} = |\alpha| \tag{3.96}$$

which gives us the following ratio of uncertainty to the expectation value of

$$\frac{\Delta N}{\left\langle \hat{N} \right\rangle} = \frac{1}{|\alpha|} \tag{3.97}$$

Thus, as the value of $|\alpha|$ increases the fractional uncertainty in energy of the oscillator decreases and the state behaves more and more like a classical oscillator.

3.6.1 Expansion in Terms of Number States

Since number states form a complete set, we can expand the coherent state in terms of the number state and write

$$|\alpha\rangle = \sum c_n|n\rangle \tag{3.98}$$

The coefficients of expansion c_n can be evaluated and we obtain for the following expansion (see Exercise 15.2)

$$|\alpha\rangle = e^{-\frac{|\alpha|^2}{2}}\sum\frac{\alpha^n}{\sqrt{n!}}|n\rangle \tag{3.99}$$

We can use Eq. (3.81) to obtain the time evolution of the coherent state. Thus, if $|\psi(t=0)\rangle = |\alpha\rangle$, then

$$|\psi(t)\rangle = e^{-\frac{|\alpha|^2}{2}}\sum\frac{\alpha^n}{\sqrt{n!}}|n\rangle e^{-i\left(n+\frac{1}{2}\right)\omega t} \tag{3.100}$$

where we have used the fact that the time evolution of the number state is given by $|n\rangle e^{-i\left(n+\frac{1}{2}\right)\omega t}$. Equation (3.100) can be written as

$$|\psi(t)\rangle = e^{-\frac{|\alpha|^2}{2}}\sum\frac{\left(\alpha e^{-i\omega t}\right)^n}{\sqrt{n!}}|n\rangle e^{-\frac{i\omega t}{2}} = e^{-\frac{i\omega t}{2}}\left|\alpha e^{-i\omega t}\right\rangle \tag{3.101}$$

The state $|\alpha e^{-i\omega t}\rangle$ is also a coherent state with an eigenvalue $\alpha e^{-i\omega t}$. Thus, apart from a phase factor of $e^{-i\omega t/2}$, the coherent state remains as a coherent state and an eigenvector of $\hat{a}$ at all times as it evolves with time with the eigenvalue changing as $\alpha e^{-i\omega t}$.

3.6.2 Time Evolution of the Expectation Value

The expectation value of an operator $\hat{A}$ of a quantum system defined by the ket $|\psi\rangle$ is defined by

$$\langle \widehat{A} \rangle = \langle \psi | \widehat{A} | \psi \rangle \tag{3.102}$$

Differentiating Eq. (3.102) with respect to time, we have

$$\frac{d\langle \widehat{A} \rangle}{dt} = \frac{d\langle \psi |}{dt} \widehat{A} | \psi \rangle + \langle \psi | \widehat{A} \frac{d|\psi\rangle}{dt} \tag{3.103}$$

where we have assumed the operator $\widehat{A}$ to have no explicit time dependence. Using the Schrödinger equation given by Eq. (3.46) and its Hermitian conjugate, we have

$$\frac{d\langle \widehat{A} \rangle}{dt} = -\frac{1}{i\hbar} \langle \psi | \widehat{H}\widehat{A} | \psi \rangle + \frac{1}{i\hbar} \langle \psi | \widehat{A}\widehat{H} | \psi \rangle = \frac{1}{i\hbar} \langle \psi | \left(\widehat{A}\widehat{H} - \widehat{H}\widehat{A} \right) | \psi \rangle = \frac{1}{i\hbar} \langle \psi | \left[\widehat{A}, \widehat{H} \right] | \psi \rangle$$

Hence, we have

$$\frac{d\langle \widehat{A} \rangle}{dt} = \frac{1}{i\hbar} \langle \psi | \left[\widehat{A}, \widehat{H} \right] | \psi \rangle \tag{3.104}$$

We will now use Eq. (3.104) to obtain the time variation of the expectation value of $\widehat{a}$. Thus,

$$\frac{d\langle \widehat{a} \rangle}{dt} = \frac{1}{i\hbar} \langle \psi | \left[\widehat{a}, \widehat{H} \right] | \psi \rangle = \frac{1}{i\hbar} \langle \psi | \widehat{a} \hbar \omega | \psi \rangle = -i\omega \langle \widehat{a} \rangle$$

where we have used Eq. (3.57). The solution of the above equation is

$$\langle \widehat{a}(t) \rangle = \langle \widehat{a}(0) \rangle e^{-i\omega t} \tag{3.105}$$

Similarly we have

$$\langle \widehat{a}^\dagger(t) \rangle = \langle \widehat{a}^\dagger(0) \rangle e^{+i\omega t} \tag{3.106}$$

Now we can evaluate the time variation of the expectation value of the position and momentum in the coherent state. Thus, using Eqs. (3.105) and (3.106) we obtain (assuming α to be real)

$$\langle \widehat{x}(t) \rangle = \sqrt{\frac{\hbar}{2m\omega}} \left[\langle \widehat{a}^\dagger(t) \rangle + \langle \widehat{a}(t) \rangle \right] = \sqrt{\frac{\hbar}{2m\omega}} \left[\langle \widehat{a}^\dagger(0) \rangle e^{+i\omega t} + \langle \widehat{a}(0) \rangle e^{-i\omega t} \right]$$

which gives us

$$\langle \widehat{x}(t) \rangle = \sqrt{\frac{2\hbar}{m\omega}} \alpha \cos \omega t \tag{3.107}$$

Similarly,

$$\langle \widehat{p}(t) \rangle = -\sqrt{2m\hbar\omega}\, \alpha \sin \omega t \tag{3.108}$$

Notice that the expectation value of position varies with a time dependence given by $\cos\omega t$ exactly similar to a classical oscillator oscillating at a frequency ω. Also the expectation value of momentum varies as $\sin\omega t$. At $t = 0$, the quantum oscillator is in its extreme position where the expectation value of momentum is zero. The amplitude of the oscillation depends on the value of α.

It is interesting to note that a harmonic oscillator with $\langle \widehat{x}(t) \rangle$ given by Eq. (3.107) will, from a classical perspective have an energy of $\frac{1}{2}m\omega^2\left(\sqrt{\frac{2\hbar}{m\omega}}\alpha\right)^2 = \hbar\omega\alpha^2$. In contrast, from the quantum mechanical picture, the energy comes out to be $\hbar\omega\left(\alpha^2 + \frac{1}{2}\right)$; the factor $\frac{1}{2}\hbar\omega$ is the zero point energy of the oscillator.

Note that

$$\frac{d\langle \widehat{x}(t) \rangle}{dt} = \frac{\langle \widehat{p}(t) \rangle}{m}$$

This implies that the expectation values follow the classical equation of motion. This is referred to as Ehrenfest's theorem.

In Chaps. 14 and 15, we will discuss in detail coherent states of the electromagnetic field in which each mode of the field behaves like a harmonic oscillator.

3.7 Schrödinger Representation

The formulation discussed above uses abstract vectors and abstract operators. In this section, we will connect the abstract formulation to the standard quantum mechanical formulation in terms of wave function $\psi(\vec{r})$ which is referred to as Schrödinger representation.

To keep the analysis simple, we will discuss the one-dimensional case.

We first note that the position operator $\widehat{x}$ satisfies the following eigenvalue equation:

$$\widehat{x}|x'\rangle = x'|x'\rangle \tag{3.109}$$

where x' is the eigenvalue of the eigenket $|x'\rangle$. Note that the position eigenvalues and eigenvectors vary continuously as $x'$is varied. We now define the projection of the abstract vector $|\psi\rangle$ into position space through the following equation:

$$\psi(x') = \langle x'|\psi\rangle \tag{3.110}$$

Hence the value of $\psi(x')$ at any point x' is nothing but a component of the ket $|\psi\rangle$ on the basis vector $|x'\rangle$. Since x' is a continuous variable, $\psi(x')$ varies continuously as x' varies and thus, $\psi(x')$ represents a function of x'. This function is the wave function of Schrödinger wave equation.

We postulate the Schrödinger representation of the momentum operator through the following equation:

$$\langle x'|\hat{p}|\psi\rangle = -i\hbar\frac{\partial}{\partial x'}\langle x'|\psi\rangle = -i\hbar\frac{\partial\psi(x')}{\partial x'} \tag{3.111}$$

The two equations, Eqs. (3.110) and (3.111), are consistent with the commutation relation between $\hat{x}$ and $\hat{p}$.

Since $|x\rangle$ forms a continuum of eigenstates, the completeness condition is given by

$$\int |x\rangle\langle x|dx = \hat{I} \tag{3.112}$$

Thus, we have, using Eq. (3.112)

$$\langle\phi|\psi\rangle = \langle\phi|\hat{I}|\psi\rangle = \int \langle\phi|x\rangle\langle x|\psi\rangle dx = \int \phi^*(x)\psi(x)dx \tag{3.113}$$

where $\psi(x) = \langle x|\psi\rangle$ and $\phi^*(x) = \langle\phi|x\rangle$. Equation (3.113) is the Schrödinger representation of $\langle\phi|\psi\rangle$ in terms of wave functions.

From the above discussion, we find that the orthogonality of two states which is represented by $\langle\phi|\psi\rangle = 0$ in Dirac notation is given by $\int\phi^*(x)\psi(x)dx = 0$ in the Schrödinger representation.

Similarly considering the expectation value of an operator $\hat{A}(\hat{x},\hat{p})$, we have

$$\langle\psi|\hat{A}|\psi\rangle = \int \psi^*(x)\hat{A}\left(x, -i\hbar\frac{\partial}{\partial x}\right)\psi(x)dx \tag{3.114}$$

where $\hat{A}\left(x, -i\hbar\frac{\partial}{\partial x}\right)$ is the operator in which $\hat{x}$ is replaced by x and $\hat{p}$ is replaced by, $-i\hbar\frac{\partial}{\partial x}$. Thus, if the operator was $\hat{A}(\hat{x},\hat{p}) = \hat{x}\hat{p}$, then $\hat{A}\left(x, -i\hbar\frac{\partial}{\partial x}\right)\psi(x) = -i\hbar x\frac{\partial\psi(x)}{\partial x}$. Similarly if the operator is $\hat{A}(\hat{x},\hat{p}) = \hat{p}\hat{x}$, then $\hat{A}\left(x, -i\hbar\frac{\partial}{\partial x}\right)\psi(x) = -i\hbar\frac{\partial}{\partial x}(x\psi(x))$. Note that we need to pay attention to the order of the operators and not change them.

Using Eq. (3.114), we have

$$\langle\psi|\hat{p}|\psi\rangle = \int \psi^*(x)\left(-i\hbar\frac{\partial}{\partial x}\right)\psi(x)dx \tag{3.115}$$

which represents the expectation value of the momentum in the state $|\psi\rangle$ or in the Schrödinger representation, the state $\psi(x)$.

We can now proceed to obtain the Schrödinger representation of the coherent state $|\alpha\rangle$. From Eqs. (3.83) and (3.52a), we obtain

$$\langle x|\widehat{a}|\alpha\rangle = \langle x|\frac{1}{\sqrt{2m\hbar\omega}}(m\omega\widehat{x} + i\widehat{p})|\alpha\rangle = \alpha\langle x|\alpha\rangle$$

or

$$\frac{1}{\sqrt{2m\hbar\omega}}(m\omega\langle x|\widehat{x}|\alpha\rangle + i\langle x|\widehat{p}|\alpha\rangle) = \alpha\langle x|\alpha\rangle \tag{3.116}$$

Using the definitions given above, Eq. (3.116) can be written as

$$\frac{1}{\sqrt{2m\hbar\omega}}\left(m\omega x\psi_\alpha(x) + i(-i\hbar)\frac{\partial\psi_\alpha(x)}{\partial x}\right) = \alpha\psi_\alpha(x)$$

where $\psi_\alpha(x) = \langle x|\alpha\rangle$ (see Eq. (3.110)). The above equation can be simplified to

$$\frac{\partial\psi_\alpha(x)}{\partial x} + \gamma^2 x\psi_\alpha(x) - \sqrt{2}\gamma\alpha\psi_\alpha(x) = 0 \tag{3.117}$$

where

$$\gamma^2 = \frac{m\omega}{\hbar} \tag{3.118}$$

The solution of Eq. (3.117) is given by

$$\psi_\alpha(x) = Ae^{-\frac{1}{2}(\xi-\xi_0)^2} \tag{3.119}$$

where A is a normalization constant and

$$\xi = \gamma x; \quad \xi_0 = \sqrt{2}\alpha \tag{3.120}$$

The coherent state is thus a spatially shifted Gaussian wave packet having a spatial width related to $\frac{1}{\gamma}$ and shifted by a distance $x = \frac{\sqrt{2}\alpha}{\gamma}$. The Gaussian has the same spatial width as the ground state wavefunction of the harmonic oscillator (see Exercise 3.4).

Exercise 3.4 Obtain the Schrödinger representation of the ground state of the harmonic oscillator.

Solution The ground state of the harmonic oscillator is defined by (see Eq. 3.63):

$$\widehat{a}|E_0\rangle = \frac{1}{\sqrt{2m\hbar\omega}}(m\omega\widehat{x} + i\widehat{p})|E_0\rangle = 0$$

Thus,

$$m\omega\langle x|\widehat{x}|E_0\rangle + i\langle x|\widehat{p}|E_0\rangle = 0$$

which gives us

$$m\omega x\psi_0(x) + i(-i\hbar)\frac{\partial\psi_0(x)}{\partial x} = 0$$

where $\psi_0(x) = \langle x|E_0\rangle$ is the Schrödinger wave function of the ground state of the harmonic oscillator. Thus, we have

$$\frac{\partial\psi_0(x)}{\partial x} = -\frac{m\omega}{\hbar}x\psi_0(x)$$

whose solution is given by

$$\psi_0(x) = Ae^{-\gamma^2x^2/2} \tag{3.121}$$

where γ is defined in Eq. (3.118) and A is the normalization constant.

Exercise 3.5 Starting from Eq. (3.13) defining the adjoint of an operator, obtain the corresponding definition of adjoint in Schrödinger representation.

Solution We recall the definition of adjoint of an operator $\widehat{A}$ (see Eq. 3.13)

$$\langle\psi|\widehat{A}^\dagger|\phi\rangle = \langle\phi|\widehat{A}|\psi\rangle^*$$

Using the identity operator given in Eq. (3.112) we write

$$\langle\phi|\widehat{A}|\psi\rangle = \langle\phi|\widehat{I}\widehat{A}|\psi\rangle = \int\langle\phi|x\rangle\langle x|\widehat{A}|\psi\rangle dx$$

Now $\langle\phi|x\rangle = \phi^*(x)$ and $\langle x|\widehat{A}|\psi\rangle = \widehat{A}(x, -i\hbar\frac{\partial}{\partial x})\psi(x)$. Thus, we have

$$\langle\phi|\widehat{A}|\psi\rangle = \int \phi^*(x)\widehat{A}\left(x, -i\hbar\frac{\partial}{\partial x}\right)\psi(x)dx$$

Hence

$$\langle\phi|\widehat{A}|\psi\rangle^* = \int \left[\widehat{A}\left(x, -i\hbar\frac{\partial}{\partial x}\right)\psi(x)\right]^* \phi(x)dx$$

Similarly we have

$$\langle\psi|\widehat{A}^\dagger|\phi\rangle = \langle\psi|\widehat{IA}^\dagger|\phi\rangle = \int \langle\psi|x\rangle\langle x|\widehat{A}^\dagger|\phi\rangle dx$$

Using $\langle\psi|x\rangle = \psi^*(x)$ and $\langle x|\widehat{A}^\dagger|\phi\rangle = \widehat{A}^\dagger\left(x, -i\hbar\frac{\partial}{\partial x}\right)\phi(x)$, we have

$$\langle\psi|\widehat{A}^\dagger|\phi\rangle = \int \psi^*(x)\,\widehat{A}^\dagger\left(x, -i\hbar\frac{\partial}{\partial x}\right)\phi(x)dx$$

Thus, using the definition of adjoint, we see that in the Schrödinger representation, the adjoint is defined by

$$\int \psi^*(x)\,\widehat{A}^\dagger\left(x, -i\hbar\frac{\partial}{\partial x}\right)\phi(x)dx = \int \left[\widehat{A}\left(x, -i\hbar\frac{\partial}{\partial x}\right)\psi(x)\right]^* \phi(x)dx$$

or dropping the dependence of $\widehat{A}$ and $\widehat{A}^\dagger$, we have

$$\int \psi^*(x)\left[\widehat{A}^\dagger\phi(x)\right]dx = \int \left[\widehat{A}\psi(x)\right]^* \phi(x)dx \tag{3.122}$$

Replacing $\widehat{A}$ by $\widehat{B}^\dagger$ and $\widehat{A}^\dagger$ by $\widehat{B}$, we can rewrite Eq. (3.122) as

$$\int \left[\widehat{B}^\dagger\psi(x)\right]^* \phi(x)dx = \int \psi^*(x)\left[\widehat{B}\phi(x)\right]dx \tag{3.123}$$

which defines the adjoint of an operator in the Schrödinger representation.

Exercise 3.6 Using Eq. (3.123) shows that the operator $\widehat{p} = -i\hbar\frac{\partial}{\partial x}$ is a Hermitian operator, that is, $\widehat{p}^\dagger = \widehat{p}$.

Solution Let us consider the following

$$\int \psi^*(x)\widehat{p}\,\phi(x)dx = \int \psi^*(x)\left(-i\hbar\frac{\partial\phi(x)}{\partial x}\right)dx$$

Integrating by parts the right hand side, we have

$$\int \psi^*(x)\left(-i\hbar\frac{\partial\phi(x)}{\partial x}\right)dx = -i\hbar\left[\psi^*(x)\phi(x)|_{-\infty}^{+\infty} - \int \phi(x)\frac{\partial\psi^*(x)}{\partial x}dx\right] = i\hbar\int \phi(x)\frac{\partial\psi^*(x)}{\partial x}dx$$

where we have used the fact that the wave functions go to zero at infinity. The right-hand side can be written as

$$i\hbar\int \phi(x)\frac{\partial\psi^*(x)}{\partial x}dx = \int\left[-i\hbar\frac{\partial\psi(x)}{\partial x}\right]^*\phi(x)dx = \int [\widehat{p}\psi(x)]^*\phi(x)dx$$

From the definition of adjoint of an operator, we find that $\widehat{p}^\dagger = \widehat{p}$.

Exercise 3.7 Equation (3.46) gives the Schrödinger equation in the Dirac notation. Obtain the corresponding Schrödinger equation for a particle in a potential $\widehat{V}(x)$ in the Schrödinger representation.

Solution The Hamiltonian is given by

$$\widehat{H} = \frac{\widehat{p}^2}{2m} + \widehat{V}(x)$$

and the Schrödinger equation given by Eq. (3.46) becomes

$$i\hbar\frac{\partial}{\partial t}|\psi\rangle = \widehat{H}|\psi\rangle = \left(\frac{\widehat{p}^2}{2m} + \widehat{V}(x)\right)|\psi\rangle$$

Multiplying from the left side by $\langle x|$, we obtain

$$i\hbar\frac{\partial}{\partial t}\langle x|\psi\rangle = \langle x|\widehat{H}|\psi\rangle = \langle x|\frac{\widehat{p}^2}{2m}|\psi\rangle + \langle x|\widehat{V}(x)|\psi\rangle$$

Using Eqs. (3.110) and (3.111), we obtain

$$i\hbar \frac{\partial}{\partial t}\psi(x) = \frac{1}{2m}\left(-i\hbar \frac{\partial}{\partial x}\right)^2 \psi(x) + V(x)\psi(x)$$

which gives us the Schrödinger equation in the Schrödinger representation:

$$i\hbar \frac{\partial \psi(x)}{\partial t} = -\frac{\hbar^2}{2m}\frac{\partial^2 \psi(x)}{\partial x^2} + V(x)\psi(x)$$

3.8 Time-Independent Perturbation Theory

There are not many potential distributions for which analytical solutions of the Schrödinger equation are available. Approximation techniques are hence required to obtain the quantum characteristics of these systems. Perturbation theory, variational technique, and WKB approximation are some of the more important approximate methods developed to analyze such problems. In this section, we will develop one of the most important approximate technique, namely, the time-independent perturbation theory which is used to solve for stationary states of perturbed systems and also to obtain modes of optical waveguides for which analytical solutions do not exist. In Chap. 4, we will present the variational method as applied to optical waveguide problems.

We now discuss the time-independent perturbation technique wherein potential distributions which deviate only slightly from other exactly soluble problems can be approximately solved and the corresponding stationary state energy eigenvalues and the corresponding eigenfunctions can be obtained.

Let us consider a Hamiltonian of the form

$$\widehat{H} = \widehat{H}_0 + \widehat{W} \tag{3.124}$$

where $\widehat{H}_0$ is the unperturbed Hamiltonian for which exact solutions exist and $\widehat{W}$ represents a small time independent perturbation. For example if we consider a potential distribution given by

$$V(x) = \frac{1}{2}m\omega^2\widehat{x}^2 + k\widehat{x}^4 \tag{3.125}$$

with the term $k\widehat{x}^4$ being a small perturbation to the harmonic oscillator potential, then the effect of the perturbation on the energy eigenvalues of the harmonic oscillator can be obtained by using the perturbation method.

We introduce a parameter λ and write Eq. (3.124) as

$$\widehat{H} = \widehat{H}_0 + \lambda\widehat{W} \tag{3.126}$$

so that we can expand the perturbed eigenfunction and eigenvalue in terms of a power series in λ and then equate equal powers of λ to get the approximate solution to different orders. At the end we will set λ to unity. We will restrict our discussion to approximation up to first order only.

We begin by representing the eigenfunction and eigenvalue of the unperturbed Hamiltonian $\widehat{H}_0$ corresponding to the nth eigenstate as $\left|E_n^{(0)}\right\rangle$ and $E_n^{(0)}$, respectively:

$$\widehat{H}_0\left|E_n^{(0)}\right\rangle = E_n^{(0)}\left|E_n^{(0)}\right\rangle \tag{3.127}$$

The eigenfunction and eigenvalue of the perturbed Hamiltonian is represented by $|E_n\rangle$ and E_n, respectively:

$$\widehat{H}|E_n\rangle = E_n|E_n\rangle \tag{3.128}$$

We expand the eigenfunction $|E_n\rangle$ and eigenvalue E_n and write

$$|E_n\rangle = \left|E_n^{(0)}\right\rangle + \lambda\left|E_n^{(1)}\right\rangle + \lambda^2\left|E_n^{(2)}\right\rangle + \ldots \tag{3.129}$$

and

$$E_n = E_n^{(0)} + \lambda E_n^{(1)} + \lambda^2 E_n^{(2)} + \ldots \tag{3.130}$$

Here $E_n^{(m)}$ represents the mth order correction to the eigenvalue and $\left|E_n^{(m)}\right\rangle$ the mth order correction to the eigenfunction due to the perturbation. In the absence of perturbation $\lambda = 0$ and $|E_n\rangle = \left|E_n^{(0)}\right\rangle$ and $E_n = E_n^{(0)}$.

Substituting in Eq. (3.128) we have

$$\left(\widehat{H}_0 + \lambda\widehat{W}\right)\left(\left|E_n^{(0)}\right\rangle + \lambda\left|E_n^{(1)}\right\rangle + \lambda^2\left|E_n^{(2)}\right\rangle + \ldots\right) = \left(E_n^{(0)} + \lambda E_n^{(1)} + \lambda^2 E_n^{(2)} + \ldots\right)$$
$$\times\left(\left|E_n^{(0)}\right\rangle + \lambda\left|E_n^{(1)}\right\rangle + \lambda^2\left|E_n^{(2)}\right\rangle + \ldots\right)$$

Equating coefficients of equal powers of λ on both sides, we obtain

$$\widehat{H}_0\left|E_n^{(0)}\right\rangle = E_n^{(0)}\left|E_n^{(0)}\right\rangle \tag{3.131}$$

$$\widehat{H}_0\left|E_n^{(1)}\right\rangle + \widehat{W}\left|E_n^{(0)}\right\rangle = E_n^{(0)}\left|E_n^{(1)}\right\rangle + E_n^{(1)}\left|E_n^{(0)}\right\rangle \tag{3.132}$$

Equation (3.131) is nothing but the eigenvalue equation of the unperturbed Hamiltonian.

Equation (3.132) is rewritten as

$$\left(\widehat{H}_0 - E_n^{(0)}\right)\left|E_n^{(1)}\right\rangle = \left(E_n^{(1)} - \widehat{W}\right)\left|E_n^{(0)}\right\rangle \tag{3.133}$$

Since the eigenfunctions of the unperturbed Hamiltonian form a complete set of functions, we can expand $|E_n^{(1)}\rangle$ in terms of $|E_n^{(0)}\rangle; n=0,1,2,\ldots$ as

$$|E_n^{(1)}\rangle = \sum C_k |E_k^{(0)}\rangle \tag{3.134}$$

where C_k's are the expansion coefficients.

We note that we can add an arbitrary multiple of $|E_n^{(0)}\rangle$ to $|E_n^{(1)}\rangle$ without affecting the left-hand side of Eq. (3.132) which helps us to ensure that

$$\langle E_n^{(0)}|E_n^{(1)}\rangle = 0 \tag{3.135}$$

Substituting from Eq. (3.134) in Eq. (3.133), we obtain

$$\left(\widehat{H}_0 - E_n^{(0)}\right)\sum C_k|E_k^{(0)}\rangle = \left(E_n^{(1)} - \widehat{W}\right)|E_n^{(0)}\rangle$$

or

$$\sum C_k E_k^{(0)}|E_k^{(0)}\rangle - E_n^{(0)}\sum C_k|E_k^{(0)}\rangle = \left(E_n^{(1)} - \widehat{W}\right)|E_n^{(0)}\rangle \tag{3.136}$$

Multiplying by $\langle E_m^{(0)}|$ from the left, we obtain

$$\sum C_k E_k^{(0)}\langle E_m^{(0)}|E_k^{(0)}\rangle - E_n^{(0)}\sum C_k\langle E_m^{(0)}|E_k^{(0)}\rangle = E_n^{(1)}\langle E_m^{(0)}|E_n^{(0)}\rangle - \langle E_m^{(0)}|\widehat{W}|E_n^{(0)}\rangle$$

Using the orthonormality condition $\langle E_m^{(0)}|E_k^{(0)}\rangle = \delta_{mk}$ between the different eigenfunctions of $\widehat{H}_0$, we have

$$\sum C_k E_k^{(0)}\delta_{mk} - E_n^{(0)}\sum C_k\delta_{mk} = E_n^{(1)}\delta_{mn} - \langle E_m^{(0)}|\widehat{W}|E_n^{(0)}\rangle$$

or

$$C_m E_m^{(0)} - E_n^{(0)} C_m = E_n^{(1)}\delta_{mn} - \langle E_m^{(0)}|\widehat{W}|E_n^{(0)}\rangle \tag{3.137}$$

We choose $m = n$ and obtain

$$E_n^{(1)} = \langle E_n^{(0)}|\widehat{W}|E_n^{(0)}\rangle \tag{3.138}$$

Choosing $m \neq n$, we have

$$C_m = \frac{\langle E_m^{(0)}|\widehat{W}|E_n^{(0)}\rangle}{\left(E_n^{(0)} - E_m^{(0)}\right)}; m \neq n \tag{3.139}$$

We also note that (using Eq. 3.135)

$$\langle E_n^{(0)}|E_n^{(1)}\rangle = \sum C_k \langle E_n^{(0)}|E_k^{(0)}\rangle = C_n = 0 \tag{3.140}$$

Equations (3.138), (3.139), and Eq. (3.134) give us the first-order approximation of the eigenvalue and eigenfunction in the presence of the perturbation as

$$E_n = E_n^{(0)} + \langle E_n^{(0)}|\widehat{W}|E_n^{(0)}\rangle \tag{3.141}$$

and

$$|E_n^{(1)}\rangle = \sum_{k\neq n} \frac{\langle E_k^{(0)}|\widehat{W}|E_n^{(0)}\rangle}{\left(E_n^{(0)} - E_k^{(0)}\right)} |E_k^{(0)}\rangle \tag{3.142}$$

We can express Eq. (3.141) in the Schrödinger representation by the following equation:

$$E_n = E_n^{(0)} + \int \psi_n^{(0)*} \widehat{W} \psi_n^{(0)} dx \tag{3.143}$$

We will use the results of the perturbation theory in Chap. 4 when we discuss optical waveguides.

Exercise 3.8 Consider a linear harmonic oscillator and assume a perturbation given by

$$\widehat{W} = \frac{1}{2} m\rho\omega^2 \widehat{x}^2 \tag{3.144}$$

with $\rho \ll 1$. Obtain the perturbed eigenvalues of the perturbed Hamiltonian.

Solution For this case, we have

$$|E_k^{(0)}\rangle = |n\rangle$$

Hence

$$\langle E_n^{(0)}|\widehat{W}|E_n^{(0)}\rangle = \frac{1}{2} m\rho\omega^2 \langle n|\widehat{x}^2|n\rangle$$

Using the expression for $\widehat{x}$ in terms of $\widehat{a}$ and $\widehat{a}^\dagger$ (see Eq. 3.53a), we obtain

$$\langle E_n^{(0)}|\widehat{W}|E_n^{(0)}\rangle = \frac{1}{2}m\rho\omega^2\frac{\hbar}{2m\omega}\langle n|(\widehat{a}+\widehat{a}^\dagger)^2|n\rangle = \frac{\rho\hbar\omega}{2}\left(n+\frac{1}{2}\right)$$

Hence

$$E_n = \left(n+\frac{1}{2}\right)\hbar\omega + \frac{\rho\hbar\omega}{2}\left(n+\frac{1}{2}\right) = \hbar\omega\left(1+\frac{\rho}{2}\right)\left(n+\frac{1}{2}\right) \quad (3.145)$$

This expression for the perturbed eigenvalue of the nth eigenstate of the harmonic oscillator is accurate to first order (see also Problem 3.16).

Problems

3.1. Consider two Hermitian operators $\widehat{A}$ and $\widehat{B}$. Under what conditions will the product of these operators be also Hermitian?

3.2. Show that an appropriately oriented half wave plate operating on $|x\rangle$ gives $|y\rangle$ and an appropriately oriented quarter wave plate operating on $|x\rangle$ gives $|RCP\rangle$.

3.3. Consider a harmonic oscillator in a quantum state given by $|\psi\rangle = \frac{1}{2}(i|0\rangle + \beta|1\rangle)$. Normalize $|\psi\rangle$ and obtain the value of β.

3.4. Show that for a linear harmonic oscillator in the energy eigenstate $|n\rangle$, the expectation values of potential energy and kinetic energy are equal to each other.

3.5. Consider a linear harmonic oscillator state given by

$$|\psi\rangle = \frac{1}{2}|10\rangle + \frac{\sqrt{3}}{2}|11\rangle$$

(a) Obtain the expectation values $\langle x\rangle$ and $\langle p\rangle$.
(b) Calculate the expectation value of energy in this state.

3.6. Consider a linear harmonic oscillator in the following state:

$$|\psi\rangle = \frac{1}{\sqrt{2}}|2\rangle + \frac{1}{\sqrt{2}}|4\rangle$$

(a) Obtain the expectation value of energy in this state.
(b) Calculate the state $|\chi\rangle = \widehat{a}|\psi\rangle$ and obtain the expectation value of energy of the state $|\chi\rangle$. Since $\widehat{a}$ is the annihilation operator, check whether the expectation value of the energy has decreased by $\hbar\omega$.

3.7. Consider the following superposition state of a harmonic oscillator given at $t = 0$:

$$|\psi(t=0)\rangle = \frac{1}{2}|n\rangle + \frac{\sqrt{3}}{2}|n+1\rangle$$

(a) Is this an eigenstate of the number operator $\widehat{N} = \widehat{a}^{\dagger}\widehat{a}$?
(b) Obtain the expectation value of the energy of the oscillator in this state.
(c) What would be $|\psi(t)\rangle$?

3.8. Consider a linear harmonic oscillator in the following state:

$$|\Psi\rangle = |n\rangle$$

Check whether the above state corresponds to an eigenstate of any of the following operators; $\widehat{a}$, $\widehat{N}$ and $\widehat{H}$.

3.9. Consider a simple harmonic oscillator in a number state $|\psi\rangle = |n\rangle$ with $n = 10$ and a coherent state $|\phi\rangle = |\alpha\rangle$ with $\alpha = 10$.

(a) Which of these is an eigenstate of the number operator $\widehat{N} = \widehat{a}^{\dagger}\widehat{a}$?
(b) Obtain the expectation value of $\widehat{N}$ and $\widehat{N}^2$ in both these states.

3.10. Consider a harmonic oscillator in the coherent state $|\alpha\rangle$ with (a) $\alpha = |\alpha|$ and (b) $\alpha = |\alpha|e^{i\pi/4}$. Obtain the expectation values of the potential energy and kinetic energy of the oscillator in the two states and compare.

3.11. Consider an electron oscillating as a linear harmonic oscillator with angular frequency $\omega = 10^3$ rad/s in the number state $|n\rangle = |100\rangle$. For the oscillator, obtain the value of x of the turning point of the oscillator.

3.12. A squeezed state $|\beta\rangle$ is defined as the eigenstate of the operator $\widehat{b} = \cosh\sigma\widehat{a} + \sinh\sigma\widehat{a}^{\dagger}$. Obtain $\langle\widehat{x}\rangle$ and $\langle\widehat{x}^2\rangle$ in this state. Show that the uncertainty Δx in this state is less than that for a coherent state. Assume σ and the eigenvalue β to be real. Chapter 18 discusses squeezed states of light in greater detail.

3.13. Show that if a state is in an eigenstate of an operator $\widehat{F}$, then the uncertainty in the observable corresponding to the operator is zero, that is, $\Delta F = 0$.

3.14. Measurements of energy carried out on an ensemble of identically prepared LHO gives energy values of $\frac{5}{2}\hbar\omega$ with probability 0.2 and energy of $\frac{11}{2}\hbar\omega$ with a probability of 0.8. Write down two possible kets (at $t = 0$) that can lead to these measurements.

3.15. Referring to Exercise 3.6, we see that the perturbed potential energy is given by

$$V(x) = \frac{1}{2}m\omega^2\widehat{x}^2 + \frac{1}{2}m\rho\omega^2\widehat{x}^2 = \frac{1}{2}m(1+\rho)\omega^2\widehat{x}^2$$

which can be solved exactly. Obtain the corresponding eigenvalues and show that to first order the eigenvalue matches with that obtained using the first order perturbation theory.

3.16. Consider a harmonic oscillator with a perturbation given by $\widehat{W} = -Qe\widehat{x}$. Use the first order perturbation theory and obtain the first order correction to the eigenvalues due to the perturbation.

3.17. Using the equation $|E_1\rangle = \widehat{a}^\dagger |E_0\rangle$, obtain the Schrödinger representation of the first excited state of the harmonic oscillator.

References

Cohen-Tannoudji, C., B. Diu, and F. Laloë. 2019. *Quantum mechanics*. Vol. I, II and III. Hoboken: Wiley.

Dirac, P. A. M. 1982. *The principles of quantum mechanics*. Oxford: Clarendon Press.

Ghatak, A., and S. Lokanathan. 2004. *Quantum mechanics: Theory and applications*. Netherlands: Springer.

Griffiths, D. J. 1995. *Introduction to quantum mechanics*. Upper Saddle River: Prentice Hall.

Zeilinger, A. 1999. Experiment and the foundations of quantum physics. *Rev Mod Phys* 71:S288–S297.

Chapter 4
Modes in Optical Waveguides

Abstract This chapter discusses in detail the concept of modes in planar and channel dielectric optical waveguides which are the fundamental building blocks of integrated quantum photonic devices. Types of waveguides discussed include planar and channel waveguides, step index and graded index waveguides, metal clad waveguides, and anisotropic waveguides.

4.1 Introduction

An optical waveguide is a dielectric structure that can guide light and is a very important component in all integrated optical and integrated quantum photonic devices. It consists of a high refractive index region surrounded by lower refractive index regions. Figure 4.1 shows two primary types of optical waveguides, namely, planar waveguides and channel waveguides. In planar waveguides the light guidance takes place only along one transverse direction. For example for the geometry shown in Fig. 4.1a the wave is constrained along the x-direction while it can diffract in the $y - z$ plane. In contrast to planar waveguides which guide light only along one transverse direction, channel waveguides are surrounded in both directions by lower refractive index regions and thus the wave gets guided along both transverse directions (see Fig. 4.1b).

Step index planar waveguides consist of a homogeneous film of refractive index n_f and thickness d surrounded by homogeneous media of refractive indices n_s ($<n_f$) referred to as the substrate and n_c ($<n_s$ $<n_f$) referred to as the cover. In the geometrical optics picture light guidance by such a film may be considered due to total internal reflection at the film-substrate and film-cover boundaries. In the case of step index channel waveguides, a medium with constant refractive index n_f having a transverse cross-section of dimensions $a \times b$ is surrounded in all directions by lower refractive index media. We will mostly deal with waveguides with dimensions comparable to the wavelength of light and in such a case one has to employ the wave picture of light and obtain the characteristics by solving Maxwell's equations with appropriate boundary conditions.

K. Thyagarajan, *Integrated Quantum Photonics*, Graduate Texts in Physics,
https://doi.org/10.1007/978-3-031-85728-7_4

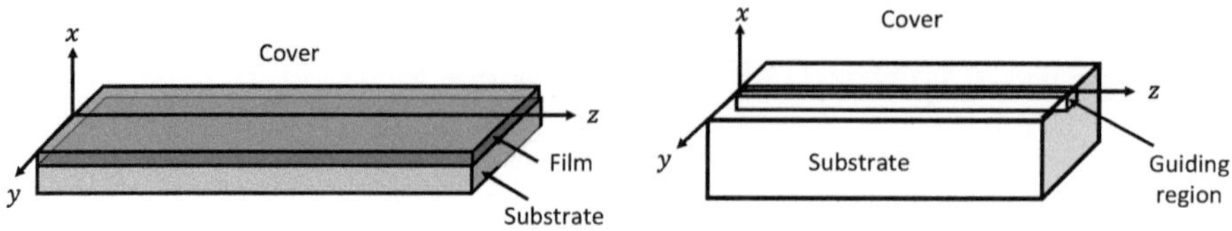

Fig. 4.1 Step index (**a**) planar waveguide guiding along one dimension and (**b**) a channel waveguide guiding along both the transverse coordinates

Graded index optical waveguides are characterized by a guiding region in which the refractive index is not a constant but graded and this graded index region is surrounded by media of lower refractive indices thus guiding the light wave.

Both planar and channel waveguides with refractive index profile independent of the propagation direction support normal modes of propagation which propagate with specific transverse field profiles.

Waveguides based on many different material systems are used in integrated quantum photonics. This includes anisotropic media such as lithium niobate, lithium tantalate, potassium titanyl phosphate, semiconducting substrates such as gallium arsenide, etc. and isotropic media such as silicon, silica, silicon nitride, glass, etc. Different fabrication techniques have been developed over many years to achieve low loss waveguides and waveguide devices using these substrates. Depending on the method of fabrication of the channel waveguide, the guiding region may have a constant refractive index or may be characterized by a graded refractive index profile which arises when ion diffusion techniques are used. Lithium niobate on insulator (LNOI) or thin film lithium niobate (TFLN) is another very interesting and important platform that is evolving and expected to play a very important role in integrated quantum photonic applications (see, e.g., Boes et al. 2018; Lin et al. 2020; Vazimali and Fathpour 2022; Zhu et al. 2021).

In this chapter, we will discuss the guided modes of planar and channel optical waveguides. Some of the waveguide geometries can be analyzed analytically leading to transcendental type equations that can be solved using simple techniques. Other waveguides need approximate analytical methods for their solutions. It is also possible to employ different numerical techniques to obtain the propagation characteristics of waveguides. In this chapter, we will only discuss waveguide geometries that can be solved using simple techniques. For more details readers may refer to many excellent texts on integrated optical waveguides (see, e.g., Marcuse and Liao 2012; Okamoto 2021; Nishihara et al. 1989; Suhara and Fujimura 2003).

4.2 Analysis of Planar Waveguides

We will first start with planar optical waveguides wherein the refractive index profile depends only on one coordinate, chosen here to be the x-coordinate (Fig. 4.1a). The propagation direction is assumed to be along the z-direction. We will first derive the wave equation that the fields have to satisfy. We will show that in such waveguides, the modes split into two separate independent groups namely the transverse electric (TE) modes and the transverse magnetic (TM) modes.

In order to determine the propagation characteristics of such a waveguide, we start from Maxwell's equations with no free charges and free currents

$$\nabla \times \vec{E} = -\frac{\partial \vec{B}}{\partial t} = -\mu_0 \frac{\partial \vec{H}}{\partial t} \tag{4.1}$$

$$\nabla \times \vec{H} = \frac{\partial \vec{D}}{\partial t} \tag{4.2}$$

$$\nabla . \vec{D} = 0 \tag{4.3}$$

$$\nabla . \vec{B} = 0 \tag{4.4}$$

where we have assumed the medium to have a permeability equal to that of free space and hence $\vec{B} = \mu_0 \vec{H}$. We also have

$$\vec{D} = \epsilon(x) \vec{E} = \epsilon_0 n^2(x) \vec{E} \tag{4.5}$$

where $\epsilon(x)$ is the x-dependent permittivity, ϵ_0 is the free space permittivity, and $n(x)$ represents the transverse refractive index profile of the planar waveguide.

We obtain solutions for monochromatic waves and assume a time dependence of the form $e^{-i\omega t}$. Thus, Eqs. (4.1) and (4.2) reduce to

$$\nabla \times \vec{E} = i\omega\mu_0 \vec{H} \tag{4.6}$$

$$\nabla \times \vec{H} = -i\omega\epsilon_0 n^2(x) \vec{E} \tag{4.7}$$

Taking curl of Eq. (4.6) and using Eq. (4.7), we obtain,

$$\nabla \times (\nabla \times \vec{E}) = i\omega\mu_0 \nabla \times \vec{H} = \frac{\omega^2}{c^2} n^2(x) \vec{E}$$

Using the vector identity

$$\nabla \times (\nabla \times \vec{E}) = \nabla(\nabla . \vec{E}) - \nabla^2 \vec{E} \approx -\nabla^2 \vec{E}$$

where we have assumed $\nabla . \vec{E} = 0$ which is valid in each of the constant refractive index regions of the step index waveguide. Thus, we get the following wave equation satisfied by the electric field $\vec{E}$:

$$\nabla^2 \vec{E} + \frac{\omega^2}{c^2} n^2(x) \vec{E} = 0 \tag{4.8}$$

A propagating mode is a specific transverse electric field distribution that propagates through the waveguide without any change in the transverse field distribution; only the phase changes as the mode propagates. These are eigensolutions to the propagation and the modes form a complete set of orthogonal solutions. Thus, any propagating transverse electric field distribution can be written as a superposition of the various transverse modes. In perfect waveguides, individual modes do not couple power among each other and coupling of power requires the presence of an appropriate perturbation in the waveguide.

In the case of planar waveguides, the guiding takes place only along the x-direction while along the two other directions, namely, y-direction and z-direction, the propagating wave can diffract just as in free space (see Fig. 4.1a). Since we are primarily interested in the wave guidance due to the refractive index variation along the x-direction, we assume that the wave is of infinite extent along the y-direction and the propagation direction to be along z.

Thus, the fields will have no y-dependence and the propagating mode will be characterized by the following electric field distribution:

$$\vec{E}(x, y, z) = \vec{E}(x) e^{i(\beta z - \omega t)} \tag{4.9}$$

where β is referred to as the propagation constant of the mode. We define an effective index of the mode through the following equation:

$$n_{eff} = \frac{\beta}{k_0} \tag{4.10}$$

For propagation through an infinitely extended medium of refractive index n, the propagation constant will be $\beta = k_0 n$ and, the effective index becomes equal to the refractive index of the medium.

The effective index (which is a function of the frequency of the electromagnetic wave) of the mode determines its phase and group velocity and other characteristics of the mode.

The phase velocity of the mode is given by (see Chap. 2)

$$v_{ph} = \frac{\omega}{\beta} = \frac{c}{n_{eff}} \tag{4.11}$$

with c being the velocity of light in free space.

Similarly, the group velocity of the mode is given by (see Chap. 2)

$$v_{gr} = \frac{d\omega}{d\beta} = \left(\frac{d\beta}{d\omega}\right)^{-1} \tag{4.12}$$

Using Eq. (4.12), it can be shown that (see Chap. 2)

$$v_{gr} = \frac{d\omega}{d\beta} = \frac{c}{N_{eff}} \tag{4.13}$$

where N_{eff} is referred to as the group effective index and is given by

$$N_{eff} = n_{eff} + \omega \frac{dn_{eff}}{d\omega} = n_{eff} - \lambda_0 \frac{dn_{eff}}{d\lambda_0} \tag{4.14}$$

where λ_0 is the free space wavelength.

In bulk media, $n_{eff} = n$, the refractive index; however in optical waveguides, n_{eff} depends both on the media forming the waveguide and on the dimensions (width, depth) of the waveguide. Thus, for a given material medium there is flexibility to modify both the phase and group velocity by appropriately designing the waveguide structure. This technique of dispersion engineering is used in many waveguide devices used both in classical optical systems as well as in integrated quantum photonic devices to achieve desired functionalities.

The corresponding $\vec{H}$ field of the mode will be given by

$$\vec{H}(x, y, z) = \vec{H}(x) e^{i(\beta z - \omega t)} \tag{4.15}$$

Substituting Eqs. (4.9) and (4.15) in Eqs. (4.6) and (4.7), we obtain the following six component equations:

$$-i\beta E_y = i\omega\mu_0 H_x \tag{4.16}$$

$$i\beta E_x - \frac{\partial E_z}{\partial x} = i\omega\mu_0 H_y \tag{4.17}$$

$$\frac{\partial E_y}{\partial x} = i\omega\mu_0 H_z \tag{4.18}$$

$$i\beta H_y = i\omega\epsilon_0 n^2(x) E_x \tag{4.19}$$

$$i\beta H_x - \frac{\partial H_z}{\partial x} = -i\omega\epsilon_0 n^2(x) E_y \tag{4.20}$$

$$\frac{\partial H_y}{\partial x} = -i\omega\epsilon_0 n^2(x) E_z \tag{4.21}$$

Since the field components are independent of y, the y-derivatives have vanished from the equations.

The above set of equations show that the components have split into two independent groups of fields representing two independent set of modes:

- Equations (4.16), (4.18), and (4.20) contain only the field components E_y, H_x, H_z having only a transverse component of the electric field and hence referred to at TE (Transverse Electric) modes. For these modes the other three field components are zero.
- Equations (4.17), (4.19), and (4.21) contain only the field components H_y, E_x, E_z having only a transverse component of the magnetic field and hence referred to at TM (Transverse Magnetic) modes. For these modes the other three field components are zero.

These two sets of modes form an orthogonal, complete set of modes and are independent of each other. Thus, propagation through isotropic planar waveguides can be described in terms of TE and TM modes. The field components must satisfy the usual conditions of continuity of the tangential component of $\vec{E}$ and $\vec{H}$ and normal components of $\vec{D}$ and $\vec{B}$ at all interfaces.

4.3 Step Index Planar Waveguides

We first consider an asymmetric planar optical waveguide with the following refractive index profile referred to as a step refractive index profile (see Fig. 4.2):

$$\begin{aligned} n^2(x) &= n_c^2; \quad & x > 0 \\ &= n_f^2; \quad & -d < x < 0 \\ &= n_s^2; \quad & x < -d \end{aligned} \tag{4.22}$$

and obtain the propagation characteristics of TE and TM modes in such waveguides. The region $x > 0$ is referred to as cover, $-d < x < 0$ as the film and $x < -d$ as the substrate (see Fig. 4.1). We will assume $n_s^2 \geq n_c^2$. If $n_s^2 = n_c^2$, then the waveguide is referred to as a symmetric waveguide.

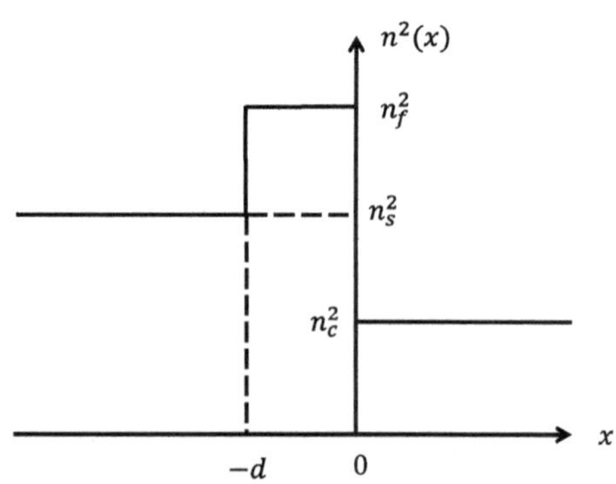

Fig. 4.2 Refractive index profile of an asymmetric step index planar waveguide

4.3.1 TE Modes

In the case of TE modes the only non-zero electric field component is E_y. Combining Eqs. (4.16), (4.18), and (4.20), we obtain the following equation satisfied by E_y:

$$\frac{d^2E_y}{dx^2} + \left(k_0^2 n^2(x) - \beta^2\right)E_y = 0 \tag{4.23}$$

where $k_0 = \frac{\omega}{c}$ is the free space propagation constant.

Substituting for $n^2(x)$ from Eq. (4.22) in Eq. (4.23), we can write the wave equation in each of the regions as follows:

$$\frac{d^2E_y}{dx^2} + \left(k_0^2 n_c^2 - \beta^2\right)E_y = 0; \qquad x > 0 \tag{4.24}$$

$$\frac{d^2E_y}{dx^2} + \left(k_0^2 n_f^2 - \beta^2\right)E_y = 0; \qquad -d < x < 0 \tag{4.25}$$

$$\frac{d^2E_y}{dx^2} + \left(k_0^2 n_s^2 - \beta^2\right)E_y = 0; \qquad x < -d \tag{4.26}$$

For modes to be guided by the waveguide, $E_y \to 0$ for $|x| \to \infty$ with the solutions for E_y outside the film tending to zero as $|x| \to \infty$. For this, the solutions to Eqs. (4.24) and (4.26) must not be oscillatory but decay exponentially. Thus, we obtain the following range of β for guided modes:

$$\beta^2 > k_0^2 n_s^2 \geq k_0^2 n_c^2$$

Also in order to be able to satisfy proper boundary conditions we must also have $\beta^2 < k_0^2 n_f^2$. Thus, we get the following range for β for guided modes:

$$k_0^2 n_f^2 > \beta^2 > k_0^2 n_s^2 \geq k_0^2 n_c^2 \tag{4.27}$$

In terms of effective index of the mode, we must have for guided modes

$$n_f^2 > n_{eff}^2 > n_s^2 \geq n_c^2 \tag{4.28}$$

In order to obtain the solutions for the fields of the mode, we define the following quantities:

$$\kappa_f^2 = k_0^2 n_f^2 - \beta^2 \tag{4.29}$$

$$\gamma_s^2 = \beta^2 - k_0^2 n_s^2 \tag{4.30}$$

$$\gamma_c^2 = \beta^2 - k_0^2 n_c^2 \tag{4.31}$$

where for guided modes κ_f^2, γ_s^2, and γ_c^2 are positive quantities. Thus, the wave equation in the three regions reduce to

$$\frac{d^2 E_y}{dx^2} - \gamma_c^2 E_y = 0; \qquad x > 0 \tag{4.32}$$

$$\frac{d^2 E_y}{dx^2} + \kappa_f^2 E_y = 0; \qquad -d < x < 0 \tag{4.33}$$

$$\frac{d^2 E_y}{dx^2} - \gamma_s^2 E_y = 0; \qquad x < -d \tag{4.34}$$

The solutions of the above equation satisfying the condition that the fields vanish at infinity are given by

$$E_y = Ae^{-\gamma_c x}; \qquad x > 0 \tag{4.35}$$

$$= B\cos\kappa_f x + C\sin\kappa_f x; \qquad -d > x > 0 \tag{4.36}$$

$$= De^{+\gamma_s x}; \qquad x < -d \tag{4.37}$$

where A, B, C, and D are constants of integration. From these solutions, we can obtain the component H_z of the magnetic field in all the regions using Eq. (4.18):

$$H_z = -\frac{i}{\omega\mu_0}(-\gamma_c)Ae^{-\gamma_c x}; \qquad x > 0 \tag{4.38}$$

$$= -\frac{i}{\omega\mu_0}\kappa_f(-B\sin\kappa_f x + C\cos\kappa_f x); \qquad -d > x > 0 \tag{4.39}$$

$$= -\frac{i}{\omega\mu_0}\gamma_s De^{+\gamma_s x}; \qquad x < -d \tag{4.40}$$

We now apply boundary conditions of continuity of the tangential field components E_y and H_z at the two interfaces $x = 0$ and $x = -d$ and obtain the following equations:

$$A = B$$

$$B\cos\kappa_f d - C\sin\kappa_f d = De^{-\gamma_s d}$$

$$-\gamma_c A = \kappa_f C$$

$$\kappa_f \left(B \sin \kappa_f d + C \cos \kappa_f d\right) = \gamma_s D e^{-\gamma_s d}$$

Eliminating the coefficients *A*, *B*, *C*, *D* from the above four equations, we obtain the following eigenvalue equation determining the propagation constant β:

$$\tan \kappa_f d = \frac{\frac{\gamma_s}{\kappa_f} + \frac{\gamma_c}{\kappa_f}}{1 - \left(\frac{\gamma_s}{\kappa_f}\right)\left(\frac{\gamma_c}{\kappa_f}\right)} \tag{4.41}$$

In the above equation for a given waveguide (i.e., given values of the refractive indices of the three regions and the waveguide width) and a given operating wavelength, the only unknown in Eq. (4.41) is the propagation constant β. Thus, solving the above equation will give us the allowed discrete set of values of the propagation constant of guided modes of the planar waveguide.

We can put Eq. (4.41) in a simpler form by defining $\tan \theta_s = \frac{\gamma_s}{\kappa_f}$ and $\tan \theta_c = \frac{\gamma_c}{\kappa_f}$ using which Eq. (4.41) becomes

$$\tan \kappa_f d = \frac{\tan \theta_s + \tan \theta_c}{1 - \tan \theta_s \tan \theta_c} = \tan(\theta_s + \theta_c) \tag{4.42}$$

Thus, $\kappa_f d = \theta_s + \theta_c + m\pi$ or

$$\kappa_f d = \tan^{-1}\left(\frac{\gamma_s}{\kappa_f}\right) + \tan^{-1}\left(\frac{\gamma_c}{\kappa_f}\right) + m\pi; \qquad m = 0, 1, 2, \ldots \tag{4.43}$$

Each value of *m* defines a particular guided TE mode of the waveguide. These are referred to as TE_m modes. For a given wavelength of operation and given set of waveguide parameters, Eq. (4.43) will allow solutions only for a discrete number of *m* values starting from $m = 0$ and thus the given waveguide will support only a discrete and finite number of guided TE modes. The mode corresponding to $m = 0$ is referred to as the fundamental TE mode or TE_0 mode, and has the largest propagation constant. Higher values of *m* lead to higher order TE modes; $m = 1$ corresponds to the first excited mode referred to as TE_1 mode, $m = 2$ to the second excited mode referred to as TE_2 mode, etc.

Using the boundary conditions of continuity of the fields at the two interfaces, we can write all constants in the field in terms of one constant and obtain:

$$E_y = A e^{-\gamma_c x}; \qquad x > 0 \tag{4.44}$$

$$E_y = A\left(\cos \kappa_f x - \frac{\gamma_c}{\kappa_f} \sin \kappa_f x\right); \qquad -d < x < 0 \tag{4.45}$$

$$E_y = Ae^{+\gamma_s(x+d)}\left(\cos\kappa_f d + \frac{\gamma_c}{\kappa_f}\sin\kappa_f d\right); \qquad x < -d \tag{4.46}$$

Note that since we are assuming $n_s \geq n_c$, $\gamma_s \leq \gamma_c$. Thus, the electric field of the mode penetrates deeper into the substrate as compared to that in the cover region.

The power carried by the mode determines the value of the constant A. We can normalize the mode by assuming unit power carried by the mode which will give a definite value of the constant A.

As an example if we consider a planar waveguide with $n_s = 2.18$, $n_f = 2.19$, $n_c = 1.0$, $d = 6$ μm operating at a wavelength of $\lambda_0 = 788$ nm. The effective indices of the TE-guided modes (as obtained by solving Eq. (4.43)) for different values of m are:

$m = 0$:TE_0 mode: $n_{eff} = 2.18920$
$m = 1$:TE_1 mode: $n_{eff} = 2.18684$
$m = 2$:TE_2 mode: $n_{eff} = 2.18304$

Note that all the effective indices of the guided modes lie between n_f and n_s with the TE_0 mode having the largest effective index followed by other TE_m modes. Using the values of n_{eff} we can calculate γ_s, γ_c, and κ_f. Equations (4.44)–(4.46) give the transverse field variations of the different TE modes.

The corresponding un-normalized field distributions of the TE_0, TE_1, and TE_2 modes are shown in Fig. 4.3.

Some points to note:

- TE_0 mode has no node, TE_1 mode has one node, and TE_2 mode has two nodes. In general TE_m mode would have m nodes within the guiding region ($-d < x < 0$) of the waveguide.
- The field penetrates more into the substrate ($x < -6$ μm) than into the cover ($x > 0$) since $n_s > n_c$. In fact since the film index is much larger than that of the cover, the field is almost zero at the film–cover interface.

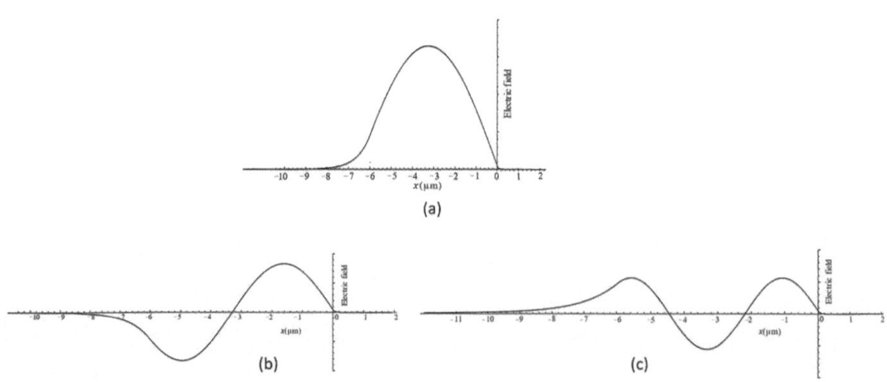

Fig. 4.3 Modal field distributions of (**a**) TE_0, (**b**) TE_1, and (**c**) TE_2 modes of a planar waveguide

- The quantities $\frac{1}{\gamma_s}$ and $\frac{1}{\gamma_c}$ give the penetration depths of the field in the substrate and the cover, respectively, and are the distances from the interface where the field becomes $\frac{1}{e}$ of the field at the corresponding interface (see Eqs. 4.44 and 4.46).
- As the mode approaches cutoff (e.g., when the wavelength of operation increases) γ_s tends to zero and the field penetrates deeper and deeper into the substrate. Thus, the closer a mode is to its cutoff, the deeper is the penetration of the field in the substrate.

Exercise 4.1 A symmetric planar waveguide with $n_f = 1.503$ and $n_s = n_c = 1.5$ supports a TE_0 mode with $n_{eff} = 1.5016$ at a wavelength of 1000 nm. (a) What is the thickness of the waveguide? (b) What is the approximate distance to which the modal field penetrates into the outer regions?

Solution
(a) For this case, $\gamma_c = \gamma_s$, Eq. (4.43) becomes (for $m = 0$)

$$\kappa_f d = 2\tan^{-1}\left(\frac{\gamma_s}{\kappa_f}\right)$$

Substituting the values given, we have

$$\gamma_s = \sqrt{\beta^2 - k_0^2 n_s^2} = k_0\sqrt{n_{eff}^2 - n_s^2} = 0.435\ \mu\text{m}^{-1}$$

and

$$\kappa_f = \sqrt{k_0^2 n_f^2 - \beta^2} = k_0\sqrt{n_f^2 - n_{eff}^2} = 0.4075\ \mu\text{m}^{-1}$$

Hence, we have

$$d = \frac{2}{\kappa_f}\tan^{-1}\left(\frac{\gamma_s}{\kappa_f}\right) = \frac{2}{0.4075}\tan^{-1}\left(\frac{0.435}{0.4075}\right) \approx 4.01\ \mu\text{m}$$

(b) Depth of penetration is given by $\frac{1}{\gamma_s} \approx 2.3\ \mu\text{m}$.

Exercise 4.2 Consider an asymmetric planar waveguide with $n_f = 2.22$, $n_s = 2.2$, $n_c = 1.0$ having a width $d = 4\ \mu\text{m}$ operating at a wavelength of 800 nm. Obtain the effective indices of the TE_0, TE_1, and TE_2 modes. Also obtain the penetration depths of the modes in the substrate and cover regions.

Solution Substituting the values of various parameters in Eq. (4.43), we obtain $n_{eff}(TE_0) \approx 2.2182$, $n_{eff}(TE_1) \approx 2.2129$, and $n_{eff}(TE_2) \approx 2.20454$.

Penetration depth in the substrate and cover are given by $\frac{1}{\gamma_s}$ and $\frac{1}{\gamma_c}$, respectively, and are given by

TE_0 mode: ≈0.45 μm in the substrate and ≈0.06 μm in the cover.

TE_1 mode: ≈0.53 μm in the substrate and ≈0.06 μm in the cover.

TE_2 mode: ≈0.9 μm in the substrate and ≈0.06 μm in the cover.

Note that as the mode number increases, the effective index decreases toward the substrate region and the depth of penetration in the substrate increases. Since the cover refractive index is small compared to the film and substrate refractive indices, the depth of penetration in the cover is small and is almost the same for all the modes.

Exercise 4.3 Consider the waveguide of Exercise 4.2. Show that the TE_0 mode has no node, TE_1 mode has one node, and TE_2 mode has two nodes within the film.

Solution From Eq. (4.45) we note that the field at the film–cover interface is given by A and the field within the film will be zero at values of x satisfying the equation $A\left(\cos\kappa_f x - \frac{\gamma_c}{\kappa_f}\sin\kappa_f x\right) = 0$ or when $\tan\kappa_f|x| = -\frac{\kappa_f}{\gamma_c}$ or for $|x| < d$ satisfying $|x| = \frac{1}{\kappa_f}\left(p\pi - \tan^{-1}\left(\frac{\kappa_f}{\gamma_c}\right)\right); p = 0, 1, 2, \ldots$

For $n_{eff} \approx 2.2182$, $\frac{1}{\kappa_f}\left(p\pi - \tan^{-1}\left(\frac{\kappa_f}{\gamma_c}\right)\right) \approx 4.475p - 0.064$. Thus, zeroes will appear when $|x| \approx 4.475p - 0.064$. Since $|x| < d$, there is no solution and hence there will be no zeroes, and hence this value of n_{eff} corresponds to TE_0 mode.

For $n_{eff} \approx 2.2129$, $\frac{1}{\kappa_f}\left(p\pi - \tan^{-1}\left(\frac{\kappa_f}{\gamma_c}\right)\right) \approx 2.25p - 0.064$. Thus, zeroes will appear when $|x| \approx 2.25p - 0.064$. Thus, there is just one solution corresponding to $p = 1$ giving $|x| \approx 2.19$ μm. Hence, this value of n_{eff} will have a single zero within the core and would correspond to TE_1 mode.

For $n_{eff} \approx 2.20454$, we have $\frac{1}{\kappa_f}\left(p\pi - \tan^{-1}\left(\frac{\kappa_f}{\gamma_c}\right)\right) \approx 1.529p - 0.060$. There are exactly two zeroes one corresponding to $p = 1$ for which $|x| \approx 1.47$ μm and corresponding to $p = 2$ for which $|x| \approx 2.99$ μm. Hence, this mode corresponds to TE_2 mode.

It is a good exercise to verify the location of zeroes by plotting the field distribution within the core for the different effective index values.

Exercise 4.4 Obtain the position of the peak of the electric field of the TE mode in a asymmetric planar waveguide.

Solution The field within the guiding region is given by Eq. (4.45). The peak of the electric field will be at values of x satisfying the following equation:

$$\frac{d}{dx}\left(\cos \kappa_f x - \frac{\gamma_c}{\kappa_f}\sin \kappa_f x\right) = 0$$

which gives us

$$\tan \kappa_f x = -\frac{\gamma_c}{\kappa_f}$$

Considering the example discussed earlier, we have for the TE_0 mode $\kappa_f \approx 0.47\ \mu m^{-1}$, $\gamma_c \approx 15.53\ \mu m^{-1}$. Hence, the peak of the field will be at $x \approx -3.28\ \mu m$ (see Fig. 4.3a).

Exercise 4.5 Consider a symmetric step index planar waveguide with $n_f = 2.19$, $d = 6\ \mu m$ operating at a wavelength of 1 μm. A TE mode is found to have an n_{eff} value of 2.18. (a) Which TE_m mode is this? (b) Obtain the refractive index of the substrate.

Solution

(a) For the given values of parameters since $\kappa_f d = k_0 d\sqrt{n_f^2 - n_{eff}^2} \approx 2.509\pi$, $\kappa_f d$ lies in the range $2\pi < \kappa_f d < 3\pi$, and hence it would correspond to TE_2 mode.
(b) Substituting the values in the eigenvalue equation, $\kappa_f d = 2\tan^{-1}\left(\frac{\gamma_s}{\kappa_f}\right) + 2\pi$, we obtain $n_s \approx 2.17$.

4.3.1.1 Cutoff of Modes

As the wavelength of operation increases, the value of γ_s as obtained by the solution of the eigenvalue equation approaches zero and the penetration depth of the mode in the substrate increases. When γ_s becomes zero, the mode no more decays exponentially in the substrate leading to the cutoff of the mode (we are assuming that $n_s > n_c$). Hence, at cutoff

$$\gamma_s = \sqrt{\beta^2 - k_0^2 n_s^2} = 0 \tag{4.47}$$

i.e.,

$$\beta = k_0 n_s \tag{4.48}$$

Equivalently at cutoff the effective index of the mode becomes equal to the substrate refractive index:

$$n_{eff} = n_s \tag{4.49}$$

For the TE$_0$ mode, which is also referred to as the fundamental TE mode, we have (putting $m = 0$ in Eq. 4.43)

$$\kappa_f d = \tan^{-1}\left(\frac{\gamma_s}{\kappa_f}\right) + \tan^{-1}\left(\frac{\gamma_c}{\kappa_f}\right) \tag{4.50}$$

Thus, at cutoff of the TE$_0$ mode, we have

$$k_0 d\sqrt{\left(n_f^2 - n_s^2\right)} = \tan^{-1}\left(\frac{\sqrt{\left(n_s^2 - n_c^2\right)}}{\sqrt{\left(n_f^2 - n_s^2\right)}}\right)$$

Replacing k_0 by $\frac{2\pi}{\lambda_0}$, we obtain the following expression for the cutoff wavelength λ_c of the TE$_0$ mode:

$$\lambda_c(TE_0) = \frac{2\pi d\sqrt{\left(n_f^2 - n_s^2\right)}}{\tan^{-1}(\sqrt{a})} \tag{4.51}$$

where

$$a = \frac{\left(n_s^2 - n_c^2\right)}{\left(n_f^2 - n_s^2\right)} \tag{4.52}$$

is referred to as the asymmetry parameter. The cutoff wavelength is the value of wavelength above which the given mode will not exist. Each mode has a well-defined cutoff wavelength.

For TE$_m$ modes, the cutoff wavelength will be given by

$$\lambda_c(TE_m) = \frac{2\pi d\sqrt{\left(n_f^2 - n_s^2\right)}}{\tan^{-1}(\sqrt{a}) + m\pi} \tag{4.53}$$

For a given waveguide, as the wavelength of operation increases, the mode approaches cutoff and reaches cutoff when γ_s reaches zero.

For the waveguide parameters used in Fig. 4.3, the effective index values of the TE_2 mode at wavelengths of 900 nm, 1000 nm, and 1010 nm are given by 2.1814,

2.18009, and 2.18002, respectively. As the wavelength increases, the effective index of the mode approaches the refractive index of the substrate (which in this example is 2.18). The cutoff wavelength of this mode is 1017.8 nm.

For symmetric waveguides $n_c = n_s$ and we obtain $a = 0$ and the TE_0 mode is never cut off. Thus, a symmetric waveguide will always support the fundamental symmetric mode. In the case of asymmetric waveguides for which $a \neq 0$, even the TE_0 mode has a finite cutoff.

We define the *V*-number (also referred to as the normalized frequency) of the waveguide by

$$V = k_0 d\sqrt{\left(n_f^2 - n_s^2\right)} = \frac{\omega}{c} d\sqrt{\left(n_f^2 - n_s^2\right)} \tag{4.54}$$

Thus, the value of *V* at cutoff of the TE_0 mode is

$$V_c(TE_0) = \tan^{-1}\left(\sqrt{a}\right) \tag{4.55}$$

In general, the cutoff of the TE_m modes are given by

$$V_c(TE_m) = \tan^{-1}\left(\sqrt{a}\right) + m\pi \tag{4.56}$$

The eigenvalue equation (Eq. 4.43) can be put in a normalized form by defining the normalized propagation constant or normalized effective index by

$$b = \frac{\left(\frac{\beta^2}{k_0^2} - n_s^2\right)}{\left(n_f^2 - n_s^2\right)} = \frac{\left(n_{eff}^2 - n_s^2\right)}{\left(n_f^2 - n_s^2\right)} \tag{4.57}$$

For guided modes $0 < b < 1$. At cutoff $b = 0$ ($n_{eff} = n_s$) and for a well guided mode $b \sim 1$, i.e., ($n_{eff} \approx n_f$). Using Eqs. (4.54) and (4.57) the eigenvalue equation can be put in the following form:

$$V\sqrt{(1-b)} = \tan^{-1}\left(\sqrt{\frac{b}{1-b}}\right) + \tan^{-1}\left(\sqrt{\frac{b+a}{1-b}}\right) + m\pi \tag{4.58}$$

This is a normalized form of the eigenvalue equation. Thus, waveguides having the same value of *a* and *V* will have the same value of the normalized propagation constant *b*.

Figure 4.4 shows the variation of the normalized propagation constant with *V* for TE_0 and TE_1 modes for a waveguide with the following parameters:

$$n_f = 2.1800; n_s = 2.1755; n_c = 1.0; d = 5\ \mu\text{m} \tag{4.59}$$

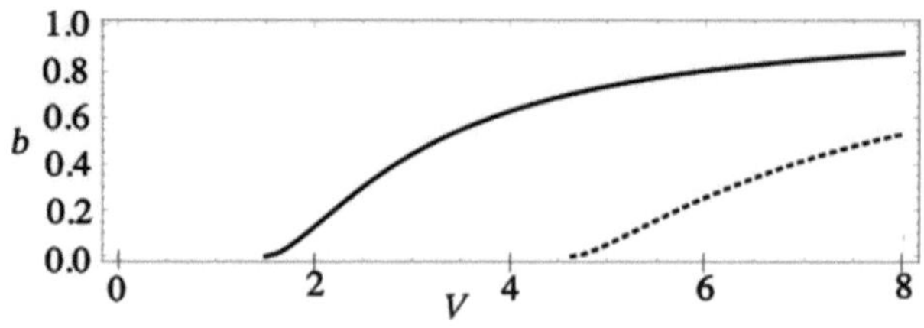

Fig. 4.4 Variation of the normalized propagation constant b with respect to the normalized frequency V for a planar waveguide with the parameter values given by Eq. (4.59). The solid curve corresponds to the TE_0 mode and the dashed curve to TE_1 mode, respectively

The asymmetry parameter of the waveguide has a value $a \approx 190$. The cutoff V-values of the TE_0 and TE_1 modes are ≈ 1.5 and 4.64, respectively (see Fig. 4.4).

If the given waveguide has a V value lying in the following range

$$\tan^{-1}\sqrt{a} < V < \tan^{-1}\sqrt{a} + \pi \tag{4.60}$$

then the waveguide will support only a single TE mode namely the TE_0 mode and such a waveguide is referred to as a single mode waveguide. Most waveguide devices in integrated optics and integrated quantum photonics use single mode waveguides. Note that since V depends on the wavelength of operation, an asymmetric waveguide may not support any mode beyond a certain wavelength. On the other hand symmetric waveguides will always support at least one-guided mode.

Exercise 4.6 Consider a planar waveguide with $n_f = 2.21$; $n_s = 2.20$; $n_c = 1.0$ and with $d = 10$ μm. Obtain the cutoff V value of the TE modes.

Solution For the given waveguide the asymmetry parameter is $a \approx 87.07$. Hence, the cutoff of the TE_m modes will be $V_c^{TE} = \tan^{-1}\sqrt{a} + m\pi \approx 1.464 + m\pi$.

Exercise 4.7 Consider a symmetric step index waveguide with $n_1 = 1.50$ and $n_2 = 1.46$ and $d = 1$ μm. For what range of wavelengths will the waveguide act as a single mode waveguide?

Solution Since the waveguide is a symmetric waveguide, the asymmetry parameter $a = 0$. The fundamental TE_0 mode has no cutoff. The waveguide will support a single TE mode if $V < \pi$ which implies $k_0 d\sqrt{\left(n_f^2 - n_s^2\right)} < \pi$. This gives us the wavelength region where the waveguide will act like a single mode waveguide as $\lambda_0 > 688$ nm.

4.3.1.2 Power Carried by the Mode

The power carried by the mode is obtained by calculating the Poynting vector

$$\vec{S} = \frac{1}{2} Re \left(\vec{E} \times \vec{H}^* \right) \quad (4.61)$$

Using Eqs. (4.16) and (4.18), we will find that $S_x = 0$, $S_y = 0$ and

$$S_z = \frac{\beta}{2\omega\mu_0} |E_y|^2$$

Thus, the total power carried by the mode per unit length along the y-direction is

$$P = \frac{\beta}{2\omega\mu_0} \int_{-\infty}^{+\infty} |E_y|^2 dx \quad (4.62)$$

Substituting for E_y from Eqs. (4.44)–(4.46), we can obtain the relation between the power carried by the mode and the amplitude A of the electric field of the mode. Substituting for the electric field expressions and carrying out the integration we obtain

$$P = \frac{\beta A^2}{4\omega\mu_0} \left(1 + \frac{\gamma_c^2}{\kappa_f^2} \right) \left(d + \frac{1}{\gamma_s} + \frac{1}{\gamma_c} \right) \quad (4.63)$$

The quantity $d_{eff} = \left(d + \frac{1}{\gamma_s} + \frac{1}{\gamma_c} \right)$ is also sometimes referred to as the effective thickness of the waveguide and corresponds to the width over which most of the modal power exists. For a given power P carried by the mode, smaller effective waveguide thickness leads to greater values of the electric field within the guiding region.

Many times it is convenient to define the amplitude A so that the mode is normalized to carry unit power. Thus, in such a case we have

$$P = \frac{\beta}{2\omega\mu_0} \int_{-\infty}^{+\infty} |E_y|^2 dx = 1$$

giving us the value of the constant A:

$$A = \left[\frac{4\omega\mu_0}{\beta} \right]^{1/2} \left(1 + \frac{\gamma_c^2}{\kappa_f^2} \right)^{-1/2} \left(d + \frac{1}{\gamma_s} + \frac{1}{\gamma_c} \right)^{-1/2}$$

As an example if we consider a waveguide with $n_f = 2.239$, $n_s = 2.236$, $n_c = 1.0$, $d = 4$ μm, neglecting the wavelength variation of the refractive indices, for the TE_0

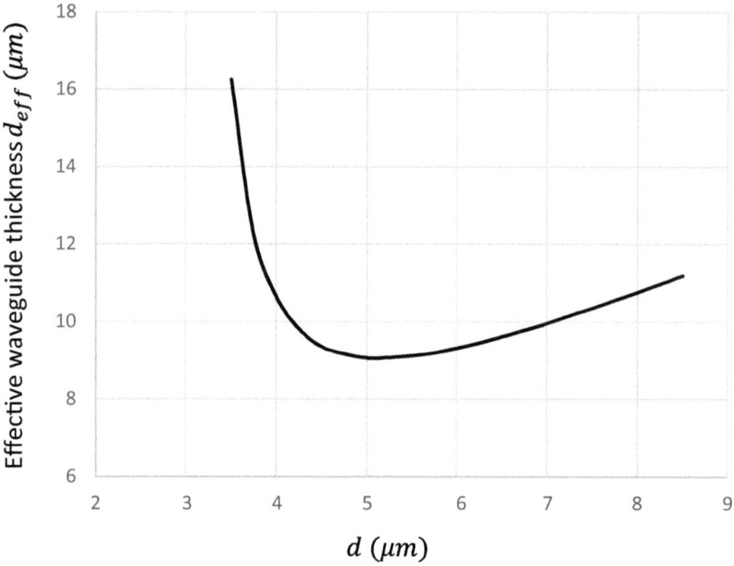

Fig. 4.5 Variation of the effective waveguide thickness $d_{eff} = \left(d + \frac{1}{\gamma_s} + \frac{1}{\gamma_c}\right)$ as a function of the waveguide width d for a waveguide with $n_f = 2.239$, $n_s = 2.236$, $n_c = 1.0$ operating at 1500 nm

mode, we obtain the effective waveguide thickness values of 6.2 μm, 10.6 μm, and 28.9 μm at wavelengths of 1 μm, 1.5 μm, and 1.8 μm, respectively. As the wavelength approaches the cutoff wavelength which in this case is 1.858 μm, the mode size increases rapidly.

For the same waveguide at a wavelength of operation of 1500 nm, the effective waveguide thickness values are 10.8 μm, 9.1 μm, and 10.6 μm at waveguide thicknesses of 8 μm, 5 μm, and 4 μm, respectively. Note that as the waveguide width decreases from a large value, the effective waveguide thickness first decreases and then after reaching a minimum increases again as the mode nears cutoff (see Fig. 4.5).

Since nonlinear effects depend on the electric field of the mode which in turn depends on the cross-sectional area, for a given power launched in a mode, the smaller the mode size, the larger will the electric field be. Small mode size cannot be achieved simply by decreasing the waveguide thickness d since as d decreases, the mode nears cutoff and the value of γ_s tends toward zero, thus increasing the value of effective waveguide thickness. In this context recent work on realizing nanophotonic lithium niobate waveguides on insulator is gaining importance. Such waveguides can exhibit extremely small mode sizes.

In order to show that in such waveguides we can achieve extremely small mode areas of less than 1 μm^2, we assume a symmetric planar waveguide with its core being made of lithium niobate with a refractive index of 2.2 and surrounded by a medium of refractive index 1.5. If the thickness of the film is 500 nm, then at $\lambda_0 = 1500$ nm, the effective index of the fundamental mode is approximately 2.123. The effective waveguide thickness would be

$$\left(d + \frac{2}{k_0\sqrt{n_{eff}^2 - n_s^2}} \approx 0.8\ \mu\text{m} \right)$$

Assuming a similar mode width in the perpendicular direction, the corresponding mode area comes out to be about 0.64 μm^2 which is extremely small.

4.3.2 TM Modes

An exactly similar analysis can be carried out for TM modes which has the field components H_y, E_x, and E_z. We give here thc main results and leave the derivations to the reader (see, e.g., Ghatak and Thyagarajan 1989).

The wave equation satisfied by H_y is given by

$$\frac{d^2H_y}{dx^2} + \left(k_0^2n^2(x) - \beta^2\right)H_y = 0 \tag{4.64}$$

Writing the solution of the equation in each of the regions, we obtain the following field expressions:

$$H_y = Ae^{-\gamma_c x}; \qquad x > 0 \tag{4.65}$$

$$= B\cos\kappa_f x + C\sin\kappa_f x; \qquad -d > x > 0 \tag{4.66}$$

$$= De^{+\gamma_s x}; \qquad x < -d \tag{4.67}$$

From the expressions for H_y, we can use Eq. (4.21) to obtain the expressions for E_z.

Applying appropriate boundary conditions of continuity of H_y and E_z at the two interfaces $x = 0$ and $x = -d$, we obtain the following eigenvalue equation corresponding to TM_m modes:

$$\kappa_f d = \tan^{-1}\left(\frac{n_f^2}{n_s^2}\frac{\gamma_s}{\kappa_f}\right) + \tan^{-1}\left(\frac{n_f^2}{n_c^2}\frac{\gamma_c}{\kappa_f}\right) + m\pi; \qquad m = 0, 1, 2, \ldots \tag{4.68}$$

which in the normalized form is given by

$$V\sqrt{(1-b)} = \tan^{-1}\left(\frac{1}{r_s}\sqrt{\frac{b}{1-b}}\right) + \tan^{-1}\left(\frac{1}{r_c}\sqrt{\frac{b+a}{1-b}}\right) + m\pi \tag{4.69}$$

where

$$r_s = \frac{n_s^2}{n_f^2} \tag{4.70}$$

$$r_c = \frac{n_c^2}{n_f^2} \tag{4.71}$$

The cutoff of the modes, corresponding to $n_{eff} = n_s$ is determined by the following equation:

$$V_c(TM_m) = \tan^{-1}\left(\frac{\sqrt{a}}{r_c}\right) + m\pi \tag{4.72}$$

Note that since $r_c < 1$, $V_c(TM_m) > V_c(TE_m)$. This implies that it is possible to choose waveguide parameters such that the waveguide supports only one TE mode and no TM mode. Such a waveguide is referred to as a single polarization single mode waveguide (see Exercise 4.8).

Exercise 4.8 Consider a planar waveguide with $n_f = 2.21$; $n_s = 2.20$; $n_c = 1.0$ operating at a free space wavelength of 800 nm. For what range of thickness will the waveguide support a single TE mode and no TM mode?

Solution The cutoff wavelength of the TE_0 mode is given by

$$V_c(TE_0) = k_0 d_{c,TE0}\sqrt{\left(n_f^2 - n_s^2\right)} = \tan^{-1}\sqrt{a}$$

and the cutoff wavelength of the TM_0 mode is given by

$$V_c(TM_0) = k_0 d_{c,TM0}\sqrt{\left(n_f^2 - n_s^2\right)} = \tan^{-1}\left(\frac{\sqrt{a}}{r_c}\right)$$

For the given values of the parameters we have $d_{c,TE0} \approx 0.89$ µm and $d_{c,TM0} \approx 0.94$ µm. Hence, to support the TE_0 mode the waveguide thickness must be larger than 0.89 µm and to support the TM_0 mode the waveguide thickness must be larger than 0.94 µm. Thus, to achieve single polarization waveguide that supports only the TE_0 mode the waveguide thickness must satisfy the following: 0.89 µm $< d <$ 0.94 µm.

Exercise 4.9 Consider two symmetric waveguides (a) $n_f = 2.3$, $n_s = 2.2$, $d = 4$ *µm* and (b) $n_f = 2.3$, $n_s = 2.26$, $d = 4$ µm, both operating at a wavelength of 1000 nm. Obtain the effective indices of the TE_0 and TM_0 modes of both the waveguides and check in which case the difference between the effective indices is larger.

Solution Solving the eigenvalue equations for the TE_0 mode and TM_0 modes in both waveguides, we obtain for waveguide (a) $n_{eff}(TE_0) \approx 2.29708$ and $n_{eff}(TM_0) \approx 2.29800$. For waveguide (b), we obtain $n_{eff}(TE_0) \approx 2.29726$ and $n_{eff}(TM_0) \approx 2.29716$. It can be seen that the waveguide with a smaller refractive index difference between the guiding region and the substrate has much smaller difference in effective indices between the TE and TM mode compared to the case when the refractive index difference is much larger.

4.4 Metal-Dielectric Waveguide and the Surface Plasmon Mode

Consider a metal–dielectric interface (silver–air) as shown in Fig. 4.6. It is possible to show that such an interface cannot guide TE modes (required boundary conditions cannot be satisfied) but can guide a TM mode.

We assume the dielectric to have a refractive index n_f and the metal to have a complex dielectric constant given by $K_m = K_{mr} - iK_{mi} = n_m^2$. For metals $Re\left(n_m^2\right) < 0$ and the imaginary part of n_m defines the loss of the metal. For guided TM modes the wave equation is given by

$$\frac{d^2H_y}{dx^2} + \left(k_0^2n^2(x) - \beta^2\right)H_y = 0 \tag{4.73}$$

In order that the interface support a guided mode, β^2 must be greater than $k_0^2n_f^2$. Solving the wave equation in the dielectric and metal regions, we obtain the

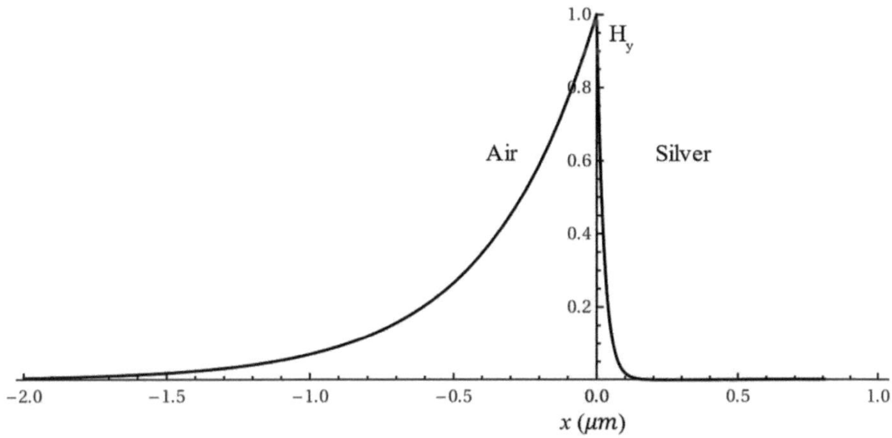

Fig. 4.6 A metal–dielectric interface can act as a waveguide for TM modes. The solid line corresponds to the magnetic field of such a surface plasmon mode at silver–air interface

following solutions with exponential decay in both the dielectric and metal regions so that it will be a mode guided by the interface between the dielectric and the metal:

$$H_y = Ae^{-\gamma_m x}; \qquad x > 0 \tag{4.74}$$

$$H_y = Be^{\gamma_f x}; \qquad x < 0 \tag{4.75}$$

where

$$\gamma_m = \sqrt{(\beta^2 - k_0^2 n_m^2)} \tag{4.76}$$

and

$$\gamma_f = \sqrt{\left(\beta^2 - k_0^2 n_f^2\right)} \tag{4.77}$$

Applying boundary conditions of continuity of H_y and E_z at the interface $x = 0$, we obtain the following eigenvalue equation:

$$\frac{\gamma_m}{\gamma_f} = -\left(\frac{n_m^2}{n_f^2}\right) \tag{4.78}$$

which can be simplified to the following expression for the effective index of the TM mode at the interface:

$$n_{eff} = \frac{\beta}{k_0} = \left(\frac{n_m^2 n_f^2}{n_m^2 + n_f^2}\right)^{1/2} \tag{4.79}$$

Since n_m^2 is negative, for the surface plasmon to exist we must have $n_m^2 + n_f^2 < 0$. Also note that when $n_m^2 + n_f^2 < n_m^2$, Eq. (4.79) implies that the effective index of the mode is $n_{eff} > n_f$, i.e., larger than the refractive index of the dielectric region. In such a case the field will have an exponentially decaying form in the dielectric.

If the imaginary part of the dielectric constant of the metal is very small, then we can make a reasonably good approximation. Thus, writing $n_m^2 = K_{mr} + iK_{mi}$,

$$n_{eff} = \left(\frac{(K_{mr} + iK_{mi})n_f^2}{(K_{mr} + iK_{mi}) + n_f^2}\right)^{1/2} \approx \sqrt{K_{mr}} n_f \left(1 + i\frac{K_{mi}}{2K_{mr}}\right) \frac{1}{\left(K_{mr} + n_f^2\right)^{1/2}}$$
$$\times \left(1 - i\frac{K_{mi}}{2\left(K_{mr} + n_f^2\right)}\right)$$

Keeping only terms of order K_{mi}, we obtain

$$n_{eff} \approx \sqrt{\frac{K_{mr} n_f^2}{\left(K_{mr} + n_f^2\right)} - i \frac{K_{mi} n_f^3}{2\left(K_{mr} + n_f^2\right)\sqrt{K_{mr}\left(K_{mr} + n_f^2\right)}}} \tag{4.80}$$

As an example we consider an interface between silver and air operating at a wavelength of 633 nm. In such a case we have

$$n_m^2 = K_{mr} + iK_{mi} = -15.03 + i1.02$$
$$n_f = 1.00$$

and we obtain

$$n_{eff} = n_{eff,r} + i n_{eff,i} \approx 1.035 + i0.0025$$

The z-dependence of the mode is specified by $e^{i\beta z} = e^{ik_0 n_{eff} z} = e^{ik_0 n_{eff,r} z} e^{-k_0 n_{eff,i} z}$. Hence, the real part of n_{eff} is responsible for phase changes in the field as it propagates and the imaginary part for the losses suffered by the mode. The power loss coefficient is given by

$$2\beta_{im} = 2k_0 n_{eff,i} = 2\frac{2\pi}{\lambda_0} n_{eff,i} \approx 0.0496 \ \mu\text{m}^{-1}$$

The distance over which the power in the mode attenuates by a factor $\frac{1}{e}$ is $\frac{1}{0.0496} \approx 20$ μm. Thus, the propagation length of surface plasmons is quite small as they have a large attenuation coefficient.

Figure 4.6 shows a plot of the corresponding surface plasmon magnetic field for the interface between silver and air at a wavelength of 633 nm. Note the exponential decay of the field in both the regions and the very small penetration depth in metal. Note also that the derivative of the magnetic field component H_y is discontinuous at the interface. The decay length (where the magnetic field decays to 1/e of its value on the interface) within silver and air is about 25 nm and 378 nm, respectively.

Until now we have discussed step index planar waveguides. Planar waveguides can also have graded refractive index profiles. In the next sections, we will briefly discuss graded index planar waveguides; for further details readers are referred to excellent literature on integrated optics.

4.5 Graded Index Planar Waveguides

In contrast to step index planar waveguides which have regions of constant refractive indices, graded index planar waveguides have a transverse coordinate dependent refractive index. Thus, we may characterize a graded index planar waveguide by a refractive index profile given by

$$\begin{aligned} n^2(x) &= n_s^2 + 2n_s \Delta n_s f(x); \quad x<0 \\ &= n_c^2; \quad x>0 \end{aligned} \tag{4.81}$$

where n_s represents the substrate refractive index, Δn_s gives the index change, n_c is the cover refractive index, and $f(x)$ defines the functional dependence of the refractive index on the transverse coordinate. Typical examples of $f(x)$ are

$$f(x) = e^{-x^2/d^2}; \quad x<0 \tag{4.82}$$

$$f(x) = e^{x/d}; \quad x<0 \tag{4.83}$$

$$f(x) = erfc\left(-\frac{x}{d}\right); \quad x<0 \tag{4.84}$$

etc. Equation (4.82) corresponds to a Gaussian refractive index profile, Eq. (4.83) corresponds to an exponential refractive index profile, and Eq. (4.84) corresponds to the complementary error function profile. These profiles are good examples of waveguides fabricated using ion diffusion techniques.

Figure 4.7 shows $f(x)$ corresponding to a Gaussian, exponential, and complementary error function refractive index profiles.

In general, for a graded refractive index profile waveguide, it is not possible to obtain analytical solutions to modal field distributions and their effective indices and various approximate analytical techniques are used for obtaining the propagation characteristics.

As an example we will consider the variational technique to obtain the propagation characteristics of the TE modes of a graded index waveguide. Multiplying Eq. (4.23) by $E_y^*(x)$ and integrating over the entire transverse plane we obtain (since the fields are y-independent, the integration along the y-direction is assumed to be carried over a finite distance and which will get cancelled from the equation)

$$\int_{-\infty}^{+\infty} E_y^*(x) \frac{d^2 E_y}{dx^2} dx + \int_{-\infty}^{+\infty} |E_y|^2 \left(k_0^2 n^2(x) - \beta^2\right) dx = 0$$

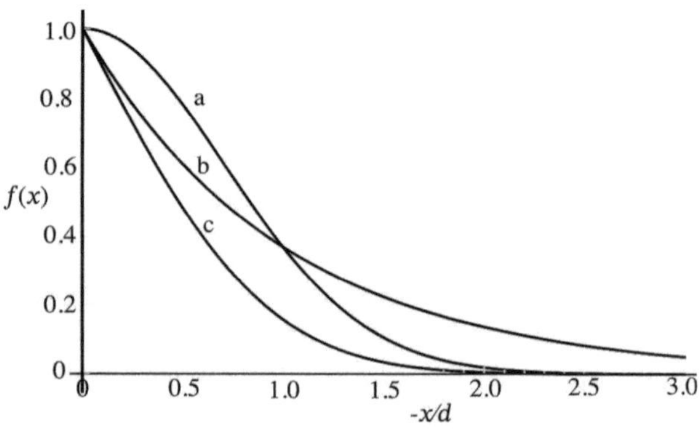

Fig. 4.7 Refractive index profiles corresponding to (**a**) Gaussian, (**b**) exponential, and (**c**) complementary error function as given by Eqs. (4.82)–(4.84)

Integrating the first integral by parts and using the fact that for guided modes both E_y and $\frac{dE_y}{dx}$ vanish at infinity, we obtain the following equation:

$$\beta^2 = \frac{k_0^2 \int n^2(x)|E_y|^2 dx - \int \left|\frac{dE_y}{dx}\right|^2 dx}{\int_{-\infty}^{+\infty} |E_y|^2 dx} \tag{4.85}$$

which is an exact expression relating the propagation constant and the actual transverse field distribution of the mode. We can use Eq. (4.85) to obtain approximate expression for the propagation constant of the mode by using trial modal fields in place of the unknown actual modal fields.

In order to obtain the propagation constant for a given graded index waveguide, we use an approximate method referred to as the variational method in which we assume a trial field $E_{yt}(x)$ with some variational parameters and which closely resembles the actual modal field of the waveguide. If we replace $E_y(x)$ by $E_{yt}(x)$ in Eq. (4.85), we will obtain an approximate expression for the propagation constant given by

$$\beta_t^2 = \frac{k_0^2 \int n^2(x)|E_{yt}|^2 dx - \int \left|\frac{dE_{yt}}{dx}\right|^2 dx}{\int_{-\infty}^{+\infty} |E_{yt}|^2 dx} \tag{4.86}$$

According to the variational principle, if we maximize the value of β_t in Eq. (4.86) by varying the parameters in $E_{yt}(x)$, then the maximum value of β_t will be the closest to the actual propagation constant and the corresponding field distribution will be the closest to the actual field. Of course the accuracy of the method would depend on how close is the approximate trial field is to the actual modal field. Thus, choosing an appropriate trial field is important to obtain accurate propagation constant values.

As an example, we consider a Gaussian refractive index profile as given in Eq. (4.81) with $f(x)$ given by Eq. (4.82). If the refractive index of the cover is much lower than that of the substrate (e.g., air cover for lithium niobate waveguide), then, as seen earlier for step index waveguides, we may assume as a first approximation that the trial field becomes zero at the interface with the cover. In such a case, we may assume a trial field for the fundamental mode to be of the form of a Gaussian and given by

$$\begin{aligned} E_{yt}(x) &= N\frac{\alpha x}{d} e^{-\alpha^2 x^2/d^2}; \qquad x<0 \\ &= 0; \qquad x>0 \end{aligned} \tag{4.87}$$

Here N is a normalization constant and α is a single unknown dimensionless variational parameter. Note that the trial field has been chosen to satisfy the continuity of the field at $x = 0$.

Normalizing the field distribution, we have

$$\frac{N^2\alpha^2}{d^2}\int_{-\infty}^{0} x^2 e^{-2\alpha^2x^2/d^2}dx = 1$$

Since the integrand is an even function of x, we can write

$$\frac{N^2\alpha^2}{2d^2}\int_{-\infty}^{+\infty} x^2 e^{-2\alpha^2x^2/d^2}dx = 1 \tag{4.88}$$

Now

$$\int_{-\infty}^{+\infty} x^2 e^{-2\alpha^2x^2/d^2}dx = \frac{\sqrt{\pi}}{2\left(\frac{2\alpha^2}{d^2}\right)^{3/2}}$$

Substituting in Eq. (4.88), we obtain

$$N = \sqrt{\frac{8\sqrt{2}\alpha}{d\sqrt{\pi}}} \tag{4.89}$$

Substituting for the field and the variational field in Eq. (4.86), we obtain

$$\beta_t^2 = k_0^2\int_{-\infty}^{0}\left[n_s^2 + 2n_s\Delta n_s e^{-x^2/d^2}\right]\left(N\frac{\alpha x}{d}e^{-\alpha^2x^2/d^2}\right)^2 dx$$
$$-\int_{-\infty}^{0}\left[\frac{d}{dx}\left(N\frac{\alpha x}{d}e^{-\alpha^2x^2/d^2}\right)\right]^2 dx$$

where we have used the fact that the trial field is normalized. We open the squares and use the following standard integrals

$$\int_{-\infty}^{+\infty} e^{-px^2}dx = \sqrt{\frac{\pi}{p}} \tag{4.90}$$

$$\int_{-\infty}^{+\infty} x^2 e^{-px^2}dx = \frac{\sqrt{\pi}}{2p^{3/2}} \tag{4.91}$$

$$\int_{-\infty}^{+\infty} x^4 e^{-px^2}dx = \frac{3\sqrt{\pi}}{4p^{5/2}} \tag{4.92}$$

and obtain the following expression for β_t^2:

$$\beta_t^2 = k_0^2 \left[n_s^2 + \frac{4\sqrt{2} n_s \Delta n_s \alpha^3}{(1 + 2\alpha^2)^{3/2}} \right] - \frac{3\alpha^2}{d^2} \tag{4.93}$$

We introduce a normalized propagation constant given by

$$b = \frac{(\beta^2 - k_0^2 n_s^2)}{2k_0^2 n_s \Delta n_s} = \frac{\left(n_{eff}^2 - n_s^2\right)}{2n_s \Delta n_s} \tag{4.94}$$

and obtain the variational expression for the normalized propagation constant for the trial function specified by Eq. (4.87)

$$b_t = \frac{2\sqrt{2}\alpha^3}{(1 + 2\alpha^2)^{3/2}} - \frac{3\alpha^2}{V^2} \tag{4.95}$$

where

$$V = k_0 d \sqrt{2n_s \Delta n_s} \tag{4.96}$$

is the normalized frequency of the waveguide.

Maximizing b_t with respect to the variational parameter α gives us the corresponding value of α and thus the modal field distribution as well as the propagation constant of the fundamental mode of the graded index waveguide.

The optimized values of α for $V = 10$, 8, and 6 come out to be $\alpha_{opt} \approx 2.09$, 1.83, and 1.54, respectively. The corresponding values of the normalized propagation constant are 0.719, 0.655, and 0.553, respectively. If $n_s = 2.1755$, $\Delta n_s = 0.005$, and $\lambda_0 = 800$ nm, the corresponding d values are 8.6 μm, 6.9 μm, and 5.18 μm, respectively. The effective indices can be obtained from Eq. (4.95) and are given by 2.1791, 2.1788, and 2.1783, respectively. Note that all the effective indices lie between $n_s = 2.1755$ and $n_p = 2.1805$ where $n_p = \sqrt{n_s^2 + 2n_s \Delta n_s}$. As the value of V decreases, the effective index reduces toward the substrate refractive index.

Exercise 4.10 Consider a planar waveguide having a Gaussian refractive index profile with $n_s = 2.2$, $\Delta n_s = 0.01$, with $d = 10$ μm. If the operating wavelength is 1318 nm, obtain the effective index of the TE_0 mode and the corresponding field profile.

Solution We have $V = k_0 d \sqrt{2n_s \Delta n_s} = 10$ for which the optimum value of the variational parameter α comes out to be $\alpha_{opt} \approx 2.09$. The corresponding normalized propagation constant is $b = 0.719$ which gives us the following value of effective index:

$$n_{eff} = \sqrt{n_s^2 + 2n_s \Delta n_s b} \approx 2.2072$$

Substituting the values of various parameters the normalized field profile will be given by

$$E_{yt}(x) \approx 0.24xe^{-0.0437x^2}; \quad x < 0$$

with $E_{yt}(x) = 0$; $x > 0$. One can check that the field $E_{yt}(x)$ is normalized.

In the trial field it has been assumed that the field is zero in the cladding region. A better approximation is to use an exponential decaying field in the cover so as to satisfy the boundary conditions of continuity of field and the derivative at the boundary. Thus, the following trial field will be a better approximation:

$$\begin{aligned} E_{yt}(x) &= A\left(1 - \frac{\gamma x}{d}\right)e^{-\alpha^2 x^2/d^2}; \quad & x < 0 \\ &= Ae^{-\gamma x/d}; & x > 0 \end{aligned} \tag{4.97}$$

which now has two parameters α and γ. Here A is the normalization constant. The trial field satisfies the required conditions of continuity of the field as well as its derivative with respect to x at the interface $x = 0$.

Using this trial field it is again possible to obtain analytical expression for the propagation constant. For further details readers may refer to Sharma and Srivastava (2005).

It is possible to get more accurate results by having a trial field with more parameters.

4.6 Modes in Anisotropic Waveguides

Many integrated quantum photonic devices use waveguides fabricated in anisotropic substrates such as lithium niobate ($LiNbO_3$), lithium tantalate ($LiTaO_3$), potassium titanyl phosphate (KTP), etc. The analysis presented earlier has assumed that the media are isotropic and the permittivity of the medium was represented by a scalar quantity.

In anisotropic media the displacement vector and electric field are related through (see Chap. 2)

$$\vec{D} = \overline{\overline{\epsilon}}\vec{E} \tag{4.98}$$

where $\overline{\overline{\epsilon}}$ is the permittivity tensor of the medium. In the principal axis system the permittivity tensor is a diagonal tensor:

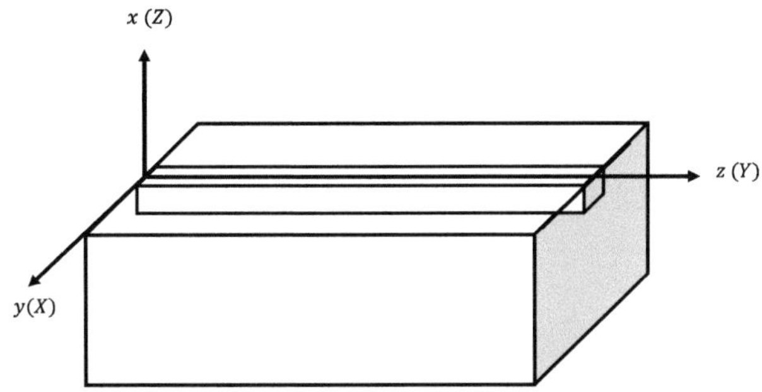

Fig. 4.8 The principal axis system (X, Y, Z) and the laboratory coordinate system (x, y, z) in a Z-cut, Y-propagating lithium niobate waveguide

$$\bar{\bar{\epsilon}} = \begin{pmatrix} \epsilon_X & 0 & 0 \\ 0 & \epsilon_Y & 0 \\ 0 & 0 & \epsilon_Z \end{pmatrix} \tag{4.99}$$

where ϵ_X, ϵ_Y, and ϵ_Z are the principal dielectric permittivities and (X, Y, Z) represent the principal axis system of the anisotropic medium. In isotropic media all the three components are equal to each other, in uniaxial media $\epsilon_X = \epsilon_Y \neq \epsilon_Z$ while in biaxial media $\epsilon_X \neq \epsilon_Y \neq \epsilon_Z$.

We can also write for the dielectric permittivity tensor in anisotropic media

$$\bar{\bar{\epsilon}} = \epsilon_0 \begin{pmatrix} n_X^2 & 0 & 0 \\ 0 & n_Y^2 & 0 \\ 0 & 0 & n_Z^2 \end{pmatrix} \tag{4.100}$$

where n_X, n_Y,and n_Z are the principal refractive indices of the medium.

In most integrated optic devices the directions of the principal axes of the crystal represented by the coordinate system (X, Y, Z) and the laboratory coordinate axes (x, y, z) of the planar or channel waveguide (see Fig. 4.8) which are used to analyze the propagation characteristics, are oriented such that they are parallel (not necessarily identical) to each other. For example in Z-cut, Y-propagating lithium niobate waveguides, the principal axis Z coincides with the coordinate axis x, the principal axis Y axis coincides with the coordinate z axis and the principal axis X coincides with the coordinate y-axis. If the principal axes of the medium point in some arbitrary direction, then the analysis becomes a bit more involved and we will not go into those details here.

In the case where the principal axes and the laboratory coordinate axes are parallel to each other, we will see that we will again obtain two independent sets of modes namely TE and TM modes in planar waveguides. In order to show this we recall Eqs. (4.1) and (4.2):

$$\nabla \times \vec{E} = i\omega\mu_0 \vec{H} \tag{4.101a}$$

$$\nabla \times \vec{H} = -i\omega\vec{D} = -i\omega\bar{\bar{\epsilon}}\vec{E} \tag{4.101b}$$

where we have again assumed time dependence of the form $e^{-i\omega t}$ and have used Eq. (4.98).

We restrict to the case when the principal axes of the anisotropic medium represented by (X, Y, Z) are parallel but not identical to the laboratory system. In the laboratory coordinate system, we have

$$\overline{\overline{\epsilon}}\vec{E} = \begin{pmatrix} \epsilon_x & 0 & 0 \\ 0 & \epsilon_y & 0 \\ 0 & 0 & \epsilon_z \end{pmatrix} \begin{pmatrix} E_x \\ E_y \\ E_z \end{pmatrix} = \begin{pmatrix} \epsilon_x E_x \\ \epsilon_y E_y \\ \epsilon_z E_z \end{pmatrix} \tag{4.102}$$

Here we are writing lower case subscripts for the components of the $\overline{\overline{\epsilon}}$ matrix to take account of the fact that the principal axes of the crystal and the laboratory coordinate system although are parallel to each other but may not coincide. Thus, as we mentioned in the case of Z-cut, Y-propagating lithium niobate waveguides (see Fig. 4.7), $\epsilon_x = \epsilon_Z$, $\epsilon_y = \epsilon_X$, and $\epsilon_z = \epsilon_Y$.

The three components of Eq. (4.101a) remain the same as before for isotropic waveguides, while the three components of Eq. (4.101b) become (assuming, as before $\frac{\partial}{\partial y} = 0$ and a z-dependence given by $e^{i\beta z}$)

$$i\beta H_y = i\omega\epsilon_x E_x \tag{4.103}$$

$$i\beta H_x - \frac{\partial H_z}{\partial x} = -i\omega\epsilon_y E_y \tag{4.104}$$

$$\frac{\partial H_y}{\partial x} = -i\omega\epsilon_z E_z \tag{4.105}$$

We can again see from Eqs. (4.16)–(4.18) and (4.103)–(4.105) that they have split into two independent sets of equations corresponding to TE modes (E_y, H_x, H_z) and TM modes (H_y, E_x, E_z).

Thus, just as in the case of isotropic waveguides, even in the case of anisotropic planar waveguides under the condition that the principal axes of the crystal are parallel to the laboratory coordinate axes, Maxwell's equations separate into two independent and orthogonal groups of modes namely TE and TM modes.

4.6.1 TE Modes

TE modes are specified by E_y, H_x, and H_z. The equations governing TE modes are Eqs. (4.16), (4.18), and (4.104):

$$-i\beta E_y = i\omega\mu_0 H_x \tag{4.106}$$

$$\frac{\partial E_y}{\partial x} = i\omega\mu_0 H_z \tag{4.107}$$

$$i\beta H_x - \frac{\partial H_z}{\partial x} = -i\omega\epsilon_y E_y \tag{4.108}$$

Combining these we obtain the following equation satisfied by the electric field component E_y:

$$\frac{d^2 E_y}{dx^2} + \left(\omega^2 \mu_0 \epsilon_y - \beta^2\right) E_y = 0 \tag{4.109}$$

where $\epsilon_y = \epsilon_0 n_y^2(x)$ and $n_y^2(x)$ represents the x-dependent refractive index profile of the waveguide.

Example As an example, we consider a lithium niobate planar waveguide. Since lithium niobate is a uniaxial medium, we have $\epsilon_X = \epsilon_Y = \epsilon_0 n_o^2$ and $\epsilon_Z = \epsilon_0 n_e^2$. If the planar waveguide is fabricated in a Z–cut, Y–propagating substrate then the laboratory coordinate system and the principal axes are related through $(x \to Z)$, $(y \to X)$, $(z \to Y)$ (see Fig. 4.8). Thus, $n_y^2 = n_X^2 = n_o^2$.

TE Modes

The TE mode will satisfy the following equation (written in the laboratory coordinate system (x, y, z)):

$$\frac{d^2 E_y}{dz^2} + \left(k_0^2 n_o^2(x) - \beta^2\right) E_y = 0 \tag{4.110}$$

where $n_o^2(x)$ is the x-dependent ordinary refractive index profile of the waveguide. The TE modes of such waveguides will be the same as isotropic waveguides with the ordinary refractive index in place of the refractive index. This equation is the same as that for the case of isotropic waveguides except that now the refractive index profile is the ordinary refractive index profile of the uniaxial medium. The eigenvalue equation to determine the propagation constant of the TE modes (see Eq. 4.43) is the same as obtained for isotropic waveguides with $n(x)$ replaced by $n_o(x)$.

4.6.2 TM Modes

Similarly TM modes are governed by the following set of equations (Eqs. 4.17, 4.103, and 4.105):

$$i\beta E_x - \frac{\partial E_z}{\partial x} = i\omega\mu_0 H_y \tag{4.111}$$

$$i\beta H_y = i\omega\epsilon_x E_x = i\omega\epsilon_0 n_x^2 E_x \tag{4.112}$$

$$\frac{\partial H_y}{\partial x} = -i\omega\epsilon_z E_z = -i\omega\epsilon_0 n_z^2 E_z \tag{4.113}$$

where we have used $\epsilon_x = \epsilon_0 n_x^2$ and $\epsilon_z = \epsilon_0 n_z^2$. These equations can be combined to give the following equation:

$$\frac{d^2 H_y}{dx^2} + \frac{n_z^2}{n_x^2}\left(k_0^2 n_x^2 - \beta^2\right) H_y = 0 \tag{4.114}$$

As an example, if we assume that the waveguide is fabricated in a Z-cut, Y-propagating uniaxial substrate, then we have $n_z^2 = n_Y^2 = n_o^2$, $n_y^2 = n_X^2 = n_o^2$ and $n_x^2 = n_Z^2 = n_e^2$ and we have

$$\frac{d^2 H_y}{dx^2} + \frac{n_o^2(x)}{n_e^2(x)}\left(k_0^2 n_e^2(x) - \beta^2\right) H_y = 0 \tag{4.115}$$

It can be seen that this equation is different from the TM mode equation of isotropic waveguides (see Eq. 4.64). This equation has both the ordinary and the extraordinary refractive index profiles.

In order to obtain the TM modes we follow the same procedure as adopted for isotropic waveguides to obtain the propagation characteristics of such anisotropic waveguides. For step index anisotropic waveguides one can again obtain a transcendental equation solving which we can obtain the propagation constant of the TM modes.

The eigenvalue equation determining the propagation constant of the TM modes in a step index planar waveguide in a Z-cut, Y-propagating subtrate with an isotropic cover of refractive index n_c, an anisotropic film with ordinary and extraordinary refractive indices (n_{of}, n_{ef}) and an anisotropic substrate with ordinary and extraordinary refractive indices (n_{os}, n_{es}) and of thickness d, is given by

$$\kappa_f d = \tan^{-1}\left(\frac{n_{of}^2 \gamma_s}{n_{os}^2 \kappa_f}\right) + \tan^{-1}\left(\frac{n_{of}^2 \gamma_c}{n_c^2 \kappa_f}\right) + m\pi; \qquad m = 0, 1, 2\ldots \tag{4.116}$$

Here

$$\kappa_f = \frac{n_{of}}{n_{ef}}\sqrt{k_0^2 n_{ef}^2 - \beta^2} \tag{4.117}$$

$$\gamma_s = \frac{n_{os}}{n_{es}}\sqrt{\beta^2 - k_0^2 n_{es}^2} \tag{4.118}$$

$$\gamma_c = \sqrt{\beta^2 - k_0^2 n_c^2} \tag{4.119}$$

This may be compared with the eigenvalue equation of TM modes in isotropic waveguides (see Eq. 4.68). We can get the corresponding equation for isotropic waveguides by putting $n_{of} = n_{ef} = n_f$ and $n_{os} = n_{es} = n_s$.

Exercise 4.11 Consider a planar waveguide fabricated on a Y-cut X-propagating lithium niobate substrate. Write down the equations satisfied by the fields of the TE and TM modes.

Solution In the given configuration, the laboratory coordinate system (x, y, z) and the principal axis system (X, Y, Z) are related through $(x \rightarrow Y)$, $(y \rightarrow Z)$, $(z \rightarrow X)$ and we have $\epsilon_x = \epsilon_Y = \epsilon_0 n_o^2$, $\epsilon_y = \epsilon_Z = \epsilon_0 n_e^2$, and $\epsilon_z = \epsilon_X = \epsilon_0 n_o^2$.

In this case the equation governing TE modes determined by Eq. (4.109) with ϵ_y replaced by $\epsilon_0 n_e^2$:

$$\frac{d^2 E_y}{dx^2} + \left(k_0^2 n_e^2(x) - \beta^2\right) E_y = 0 \tag{4.120}$$

Thus, the extraordinary refractive index profile determines the propagation characteristic of TE modes in this configuration.

In the case of TM modes, $\epsilon_x = \epsilon_z = \epsilon_0 n_o^2$, the equation satisfied by H_y would be (see Eq. 4.114)

$$\frac{d^2 H_y}{dx^2} + \left(k_0^2 n_o^2(x) - \beta^2\right) H_y = 0 \tag{4.121}$$

which is the same as for isotropic waveguides with the refractive index profile determined by $n_o^2(x)$, the ordinary refractive index profile.

Exercise 4.12 Consider a step index planar waveguide in Y-cut X-propagating lithium niobate operating at a wavelength of 800 nm. The substrate refractive indices are $n_{so} = 2.2553$ and $n_{se} = 2.1755$. The film indices are $n_{fo} = 2.2587$ and $n_{fe} = 2.1785$. The cover refractive index is $n_c = 1.0$. Assuming a waveguide thickness of 5 μm, obtain the effective indices of the TE_0 and TM_0 modes.

Solution Since the substrate is Y-cut X-propagating, the TE modes would correspond to extraordinary polarization and the TM modes to ordinary polarization.

Thus, the eigenvalue equation for TE_0 mode is (see Eqs. 4.120 and 4.43)

$$\kappa_{fe} d = \tan^{-1}\left(\frac{\gamma_{se}}{\kappa_{fe}}\right) + \tan^{-1}\left(\frac{\gamma_{ce}}{\kappa_{fe}}\right) \tag{4.122}$$

where

$$\kappa_{fe} = k_0 \sqrt{\left(n_{fe}^2 - n_{eff,TE}^2\right)};\ \gamma_{se} = k_0 \sqrt{\left(n_{eff,TE}^2 - n_{se}^2\right)};\ \gamma_{ce} = k_0 \sqrt{\left(n_{eff,TE}^2 - n_c^2\right)}$$

Substituting for the various parameters and solving the transcendental equation, we obtain $n_{eff,\ TE} \approx 2.17757$.

The penetration depth of the mode in the substrate is $\frac{1}{\gamma_{se}}$ which is

$$\frac{1}{\gamma_{se}} = \frac{1}{k_0\sqrt{\left(n_{eff,TE}^2 - n_{se}^2\right)}} = \frac{\lambda_0}{2\pi\sqrt{\left(n_{eff,TE}^2 - n_{se}^2\right)}} \approx 1.34\ \mu\text{m}$$

Similarly the penetration depth in the cover is $\frac{1}{\gamma_{ce}}$ which is

$$\frac{1}{\gamma_{ce}} = \frac{\lambda_0}{2\pi\sqrt{\left(n_{eff,TE}^2 - n_c^2\right)}} \approx 0.07\ \mu\text{m}$$

The TM_0 mode will be obtained by solving the following eigenvalue equation (see Eqs. 4.121 and 4.68):

$$\kappa_{fo} d = \tan^{-1}\left(\frac{n_{fo}^2}{n_{so}^2}\frac{\gamma_{so}}{\kappa_{fo}}\right) + \tan^{-1}\left(\frac{n_{fo}^2}{n_c^2}\frac{\gamma_{co}}{\kappa_{fo}}\right) \tag{4.123}$$

where

$$\kappa_{fo} = k_0\sqrt{\left(n_{fo}^2 - n_{eff,TM}^2\right)};\ \gamma_{so} = k_0\sqrt{\left(n_{eff,TM}^2 - n_{so}^2\right)};\ \gamma_{co} = k_0\sqrt{\left(n_{eff,TM}^2 - n_c^2\right)}$$

Substituting for the various parameters and solving the transcendental equation, we obtain $n_{eff,\ TM} \approx 2.25775$.

The penetration depth of the mode in the substrate is $\frac{1}{\gamma_{so}} \approx 1.21\ \mu\text{m}$ and the penetration depth in the cover is $\frac{1}{\gamma_{co}} \approx 0.06\ \mu\text{m}$.

Exercise 4.13 Proton exchange is a standard technique used in the fabrication of waveguides in lithium niobate. This process only increases the extraordinary refractive index in the diffused region while the ordinary refractive index decreases very slightly. Consider planar waveguides formed on a *Z*-cut, *X*-propagating and a *Y*-cut, *X*-propagating substrate. Which polarization modes would these waveguides guide?

Solution Since only the extraordinary refractive index increases, in the case of *Z*-cut, *X*-propagating substrate, this will support only the TM modes while in the case of *Y*-cut, *X*-propagating substrate, it will support only the TE modes. Using this method it is thus possible to realize waveguides that truly support only a single polarization mode and thus find applications in devices where polarization is important.

4.7 Modes in Channel Waveguides

In this section, we will study propagation characteristics of channel waveguides which confine the light in both transverse directions in contrast to planar waveguides which offer confinement only along one transverse dimension. Most integrated optic devices use channel waveguides for realizing various components such as modulators, switches and couplers, nonlinear frequency conversion devices, etc.

Actual modes in such channel waveguides will, in general, have all six components of electric and magnetic fields, to be non-zero. For weakly guiding waveguides, for which the refractive index difference between the guiding region and the substrate is small, we can define TE-like (quasi TE) and TM-like (quasi TM) modes which are characterized by the following major field components (see Fig. 4.9):

TE like: E_y and H_x

TM like: H_y and E_x

In Sect. 4.7.1, we will discuss step index channel waveguides using a simple perturbation method and in Sect. 4.7.2, we will discuss graded index channel waveguides using the variational technique. Both these techniques are also used extensively for solving for bound states in potential wells in quantum mechanics.

4.7.1 Step Index Channel Waveguides

We consider a step index channel waveguide with the following refractive index profile (see Fig. 4.10):

$$\begin{aligned} n(x,y) &= n_f; \quad -d<x<0\&|y|<\frac{w}{2} \\ &= n_s; \quad -d<x<0\&|y|>\frac{w}{2} \text{ and } x<-d \\ &= n_c; \quad x>0 \end{aligned} \tag{4.124}$$

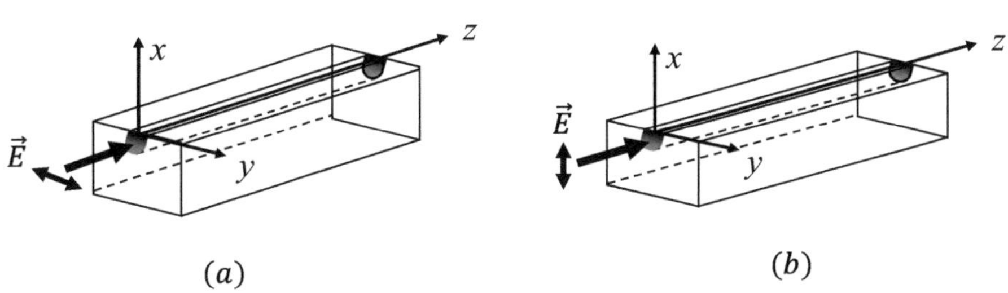

Fig. 4.9 (**a**) TE-like and (**b**) TM-like modes in a channel waveguide

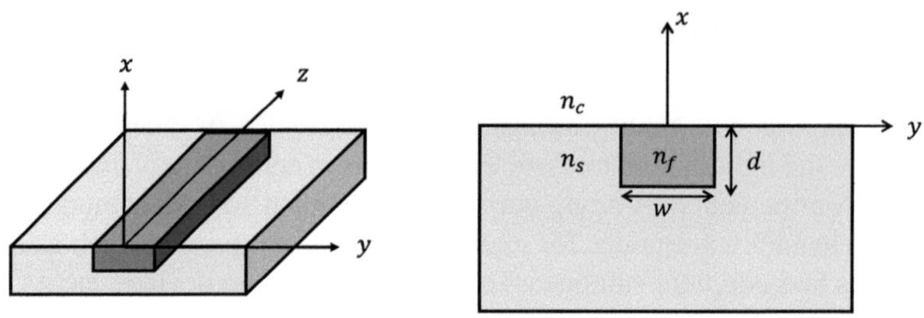

Fig. 4.10 A step index channel waveguide corresponding to the refractive index profile given by Eq. (4.124)

Due to the presence of corner regions it is not possible to get analytical solutions to the waveguide profile and one has to employ approximate techniques to obtain analytical solutions. Also note that the profile is not separable in terms of the two transverse coordinates. Of course it is always possible to obtain the solutions numerically.

Here we employ the perturbation technique in which we consider an equivalent refractive index profile which is separable in *x*- and *y*-coordinates (Kumar et al. 1983, 1984). Thus, we consider the following separable refractive index profile:

$$n_u^2(x,y) = n_x^2(x) + n_y^2(y) - n_f^2 \tag{4.125}$$

where

$$\begin{aligned} n_x^2(x) &= n_f^2; \qquad -d < x < 0 \\ &= n_s^2; \qquad x < -d \\ &= n_c^2; \qquad x > 0 \end{aligned} \tag{4.126}$$

and

$$\begin{aligned} n_y^2(y) &= n_f^2; \qquad |y| < \frac{w}{2} \\ &= n_s^2; \qquad |y| > \frac{w}{2} \end{aligned} \tag{4.127}$$

represent two planar refractive index profiles.

The refractive index profile described by $n_u^2(x,y)$ and the actual profile $n^2(x,y)$ match everywhere except in the corner regions (see Fig. 4.11) and in the corner regions the difference between the actual profile and the approximate profile is $\left(n_f^2 - n_s^2\right)$. Since most waveguides are weakly guiding waveguides $n_s \approx n_f$ and thus this difference is quite small. This small difference can be incorporated in more accurate calculations using the standard perturbation theory of quantum mechanics.

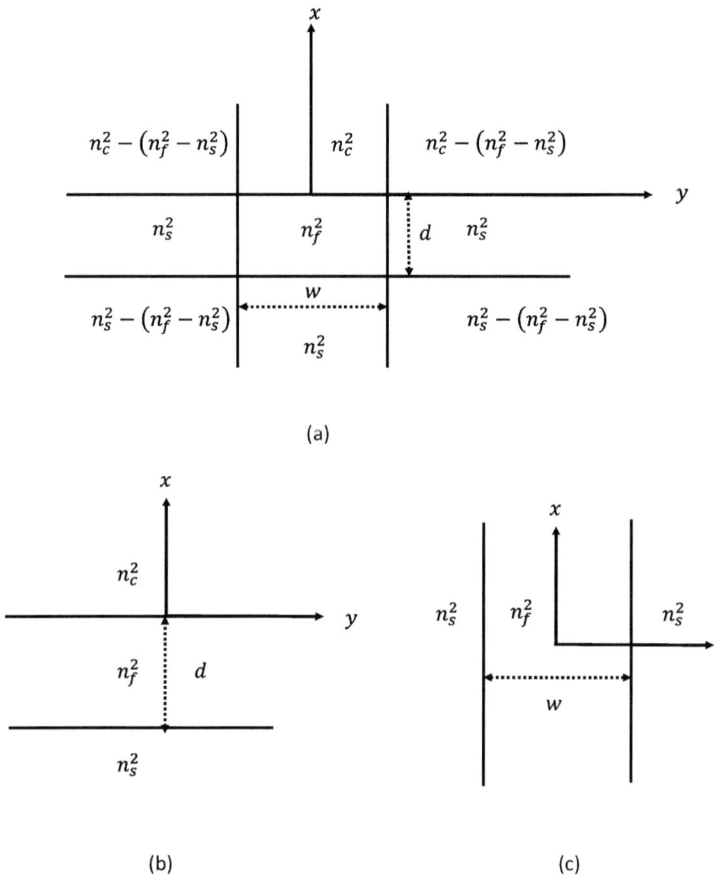

Fig. 4.11 (**a**) Approximate refractive index profile given by Eq. (4.125) representing a rectangular core step index waveguide represented by Eq. (4.124). (**b**) Planar waveguide represented by Eq. (4.126). (**c**) Planar waveguide represented by Eq. (4.127)

Let us first consider TE-like modes (quasi-TE modes) for which the major components of electric and magnetic fields are E_y and H_x. For the refractive index profile given by Eq. (4.125), the field E_y satisfies the following wave equation:

$$\frac{\partial^2 E_y}{\partial x^2} + \frac{\partial^2 E_y}{\partial y^2} + \left(k_0^2 n_u^2(x, y) - \beta^2\right) E_y = 0$$

or

$$\frac{\partial^2 E_y}{\partial x^2} + \frac{\partial^2 E_y}{\partial y^2} + \left(k_0^2 n_x^2(x) + k_0^2 n_y^2(y) - k_0^2 n_f^2 - \beta^2\right) E_y = 0 \tag{4.128}$$

In order to solve the above equation we employ the standard separation of variables technique and write

$$E_y = X(x)Y(y) \tag{4.129}$$

Substituting in Eq. (4.128) and separating the variables, we obtain the following two equations:

$$\frac{d^2X(x)}{dx^2} + \left(k_0^2 n_x^2(x) - \beta_x^2\right)X(x) = 0 \tag{4.130}$$

$$\frac{d^2Y(y)}{dy^2} + \left(k_0^2 n_y^2(y) - \beta_y^2\right)Y(y) = 0 \tag{4.131}$$

and

$$\beta^2 = \beta_x^2 + \beta_y^2 - k_0^2 n_f^2 \tag{4.132}$$

The equations for $X(x)$ and $Y(y)$ represent equations of step index planar waveguides that we had considered in Sect. 4.2. Since we are considering quasi TE modes for which the major field component is E_y, Eq. (4.130) for $X(x)$ would represent the TE mode of the waveguide with profile $n_x^2(x)$ and Eq. (4.131) for $Y(y)$ would represent TM mode of the waveguide with profile $n_y^2(y)$. Note that the refractive index profile is symmetric along the y-direction and asymmetric along the x-direction.

Thus, we have

$$\begin{aligned} X(x) &= A_x e^{-\gamma_{cx} x}; \qquad x > 0 \\ &= B_x \cos \kappa_{fx} x + C_x \sin \kappa_{fx} x; \qquad -d < x < 0 \\ &= D_x e^{\gamma_{sx} x}; \qquad x < -d \end{aligned} \tag{4.133}$$

and

$$\begin{aligned} Y(y) &= A_y e^{-\gamma_{sy} y}; \qquad y > \frac{w}{2} \\ &= B_y \cos \kappa_{fy} y + C_y \sin \kappa_{fy} y; \qquad -\frac{w}{2} < y < \frac{w}{2} \\ &= D_y e^{\gamma_{sy} y}; \qquad y < -\frac{w}{2} \end{aligned} \tag{4.134}$$

where

$$\gamma_{cx}^2 = \beta_x^2 - k_0^2 n_c^2 \tag{4.135}$$

$$\gamma_{sx}^2 = \beta_x^2 - k_0^2 n_s^2 \tag{4.136}$$

$$\kappa_{fx}^2 = k_0^2 n_f^2 - \beta_x^2 \tag{4.137}$$

$$\gamma_{sy}^2 = \beta_y^2 - k_0^2 n_s^2 \tag{4.138}$$

$$\kappa_{fy}^2 = k_0^2 n_f^2 - \beta_y^2 \tag{4.139}$$

and A_x, B_x, etc. are constants.

The corresponding eigenvalue equations to obtain β_x and β_y are given by (see Eqs. 4.43 and 4.68):

$$\kappa_{fx} d = \tan^{-1}\left(\frac{\gamma_{sx}}{\kappa_{fx}}\right) + \tan^{-1}\left(\frac{\gamma_{cx}}{\kappa_{fx}}\right) + (p-1)\pi; \qquad p = 1, 2, \ldots \tag{4.140}$$

$$\kappa_{fy} w = 2\tan^{-1}\left(\frac{n_f^2}{n_s^2}\frac{\gamma_{sy}}{\kappa_{fy}}\right) + (q-1)\pi; \qquad q = 1, 2, \ldots \tag{4.141}$$

Solving Eqs. (4.140) and (4.141) and using Eq. (4.132) give us the effective index of the quasi TE mode of the waveguide. The quasi TE modes of the channel waveguide are referred to as E_{pq}^y modes. The fundamental quasi-TE mode of the channel waveguide corresponds to $p = 1$; $q = 1$ and is referred to as E_{11}^y mode. The corresponding electric field distribution $E_y(x, y)$ can be obtained from Eqs. (4.129), (4.133), and (4.134).

Similarly the quasi-TM modes for which the major electric field component is E_x are referred to as E_{pq}^x modes. Since the major electric field component is E_x in order to solve for these modes, we solve for TE modes for the refractive index profile given by Eq. (4.131) to obtain β_y and for TM modes for the refractive index profile given by Eq. (4.130) to obtain β_x. The corresponding eigenvalue equations specifying the propagation constant would be

$$\kappa_{fx} d = \tan^{-1}\left(\frac{n_f^2}{n_s^2}\frac{\gamma_{sx}}{\kappa_{fx}}\right) + \tan^{-1}\left(\frac{n_f^2}{n_c^2}\frac{\gamma_{cx}}{\kappa_{fx}}\right) + (p-1)\pi; \qquad p = 1, 2, \ldots \tag{4.142}$$

and

$$\kappa_{fy} w = 2\tan^{-1}\left(\frac{\gamma_{sy}}{\kappa_{fy}}\right) + (q-1)\pi; \qquad q = 1, 2, \ldots \tag{4.143}$$

The corresponding field distribution $E_x(x, y)$ can be obtained by using Eq. (4.129), (4.133), and (4.134).

Note that in a planar waveguide the mode numbering begins from 0 while in channel waveguides the mode numbering begins from 1.

In Chap. 3, we discussed the time-independent perturbation theory as applied to quantum mechanical problems to obtain the eigenvalues and eigenfunctions of a potential distribution which is very close to a potential distribution that has exact analytical solutions. The problem of obtaining the modes in the case of optical waveguides is very similar to the problem of obtaining eigensolutions of the Schrödinger equation in quantum mechanics. Hence, the same perturbation method can be applied to obtain the first order correction (due to the difference in refractive index values in the corner regions) to the propagation constant of the actual rectangular core waveguide from the analytically soluble problem using separation of variables technique discussed here (see Kumar et al. 1983). See also Exercise 4.18 where we have discussed application of perturbation theory to optical waveguides.

Figure 4.12 shows density plots showing typical intensity distributions of the fundamental mode E^y_{11} and the first excited mode E^y_{12}, respectively. It may be mentioned that E^y_{pq} modes have $(p - 1)$ zero crossings along the x-direction and $(q - 1)$ zero crossings along the y-direction. Most guided wave devices use the fundamental modes of the waveguides.

The technique presented in this section has also been used to analyze rectangular core anisotropic waveguides with step refractive index profile with the principal axes parallel to the laboratory coordinate axes (Kumar et al. 1984).

Exercise 4.14 Consider a rectangular channel waveguide having a core of refractive index $n_1 = 1.503$ surrounded on all sides by a medium of refractive index $n_2 = 1.500$. The waveguide dimensions are 4 μm × 12 μm (along the x- and y-directions, respectively) and operating at a wavelength of 1 μm (assume that the waveguide is weakly guiding so that TE and TM modes of the same mode number have approximately the same effective indices). Calculate the approximate effective indices of the guided modes E_{11} and E_{12} of the channel waveguide mentioning the mode number.

Solution In order to determine the effective indices of the E_{11} and E_{12} of the channel waveguide, we first need to determine the effective indices of the planar waveguides into which we approximate the channel waveguide.

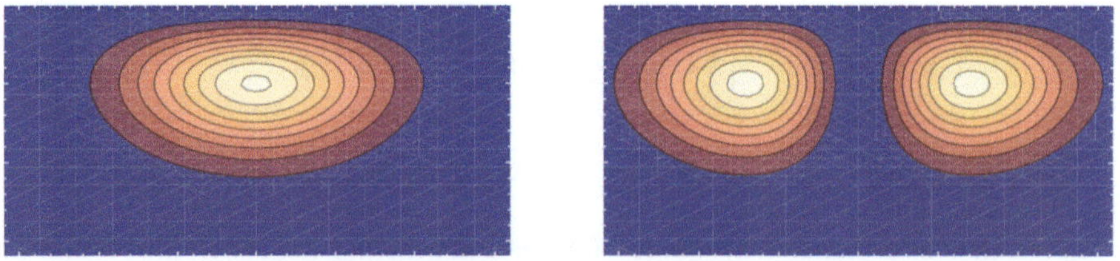

Fig. 4.12 Density plots showing typical intensity patterns of (**a**) the fundamental mode E^y_{11} and (**b**) the first excited mode E^y_{12} in a channel waveguide. The axes are in arbitrary units

First we consider a symmetric planar waveguide of width 4 μm having a guiding region of refractive index $n_f = 1.503$, $n_s = n_c = 1.500$ and operating at a wavelength of 1 μm. In such a case, solving Eq. (4.140) with $p = 1$, we obtain the effective index of the fundamental TE mode of the planar waveguide as $n_{eff,\,x}(p = 1) = 1.50159$.

We then consider a symmetric planar waveguide of width 12 μm having a guiding region of refractive index $n_f = 1.503$, $n_s = n_c = 1.500$ and operating at a wavelength of 1 μm. Solving Eq. (4.143) for $q = 1$ and $q = 2$, we obtain the effective indices of the fundamental and first excited modes of the planar waveguide as $n_{eff,\,y}(q = 1) = 1.50265$ and $n_{eff,\,y}(q = 2) = 1.50222$.

Using the method discussed, the effective index of the E_{11} mode of the channel waveguide would be given by

$$n_{eff}(E_{11}) = \sqrt{\left(n_{eff,x}^2(p=1) + n_{eff,y}^2(q=1) - n_f^2\right)}$$

Substituting the values $n_{eff,\,x}(p = 1) = 1.50159$ and $n_{eff,\,y}(q = 1) = 1.50265$, we obtain $n_{eff}(E_{11}) \approx 1.50124$. Similarly the effective index of the E_{12} mode of the channel waveguide would be given by

$$n_{eff}(E_{12}) = \sqrt{\left(n_{eff,x}^2(p=1) + n_{eff,y}^2(q=2) - n_f^2\right)}$$

Substituting the values $n_{eff,\,x}(p = 1) = 1.50159$ and $n_{eff,\,y}(q = 2) = 1.50222$, we obtain $n_{eff}(E_{12}) \approx 1.50081$. Note that the E_{12} mode of the channel waveguide is quite close to its cutoff.

Exercise 4.15 Consider a channel waveguide fabricated using proton exchange method in lithium niobate. This method produces almost a step index profile. Given $n_f = 2.3$, $n_s = 2.2$, $n_c = 1.0$, $d = 3$ μm and $w = 5$ μm. For operating at a wavelength of 800 nm, obtain the effective index of the E_{11}^x mode of the channel waveguide using the approximate method discussed earlier.

Solution We first note that due to the large refractive index difference between the guiding region and substrate TE and TM modes would have quite different effective indices. Now the E_{11}^x mode has its polarization perpendicular to the substrate and hence is a quasi-TM mode. Thus, to use the method in view of the major field component being along x, we need to obtain the effective index value of the $p = 1$ mode of the x-refractive index profile and the $q = 1$ mode of the y-refractive index profile. These are given below:

For x-refractive index profile: $n_f = 2.3$, $n_s = 2.2$, $n_c = 1.0$, $d = 3\ \mu\text{m}$, at 800 nm: $n_{eff}(p = 1) = 2.29941$

For y-refractive index profile: $n_f = 2.3$, $n_s = 2.2$; $w = 5\ \mu\text{m}$ at 800 nm: $n_{eff}(q = 1) = 2.29819$

Hence, the effective index of the E_{11}^{x} mode would be

$$n_{eff}\left(E_{11}^{x}\right) = \sqrt{(2.29819)^2 + (2.29941)^2 - 2.3^2} \approx 2.2976$$

Note that because of large refractive index difference this waveguide will support multiple modes.

4.7.2 Graded Index Channel Waveguides

In graded index channel waveguides the waveguide has a continuous variation in refractive index in the transverse direction. Just like step index channel waveguides, in general, it is not possible to obtain analytical solutions to the propagation characteristics and one has to employ approximate analytical techniques or use numerical techniques (see, e.g., Suhara et al. 1979). Here we consider the variational method which is one of the methods that is used to solve for modes in graded index channel waveguides. This method was discussed in Sect. 4.4 for analyzing graded index planar waveguides where analytical solutions may not exist and approximate techniques are used to obtain the propagation characteristics.

We assume the waveguide to have a transverse refractive index profile defined by $n^2(x, y)$. Thus, quasi-TE modes will satisfy the following wave equation:

$$\frac{\partial^2 E_y}{\partial x^2} + \frac{\partial^2 E_y}{\partial y^2} + \left(k_0^2 n^2(x, y) - \beta^2\right) E_y = 0 \tag{4.144}$$

Multiplying on both sides of the equation by E_y^* and integrating over the entire transverse cross-section, we obtain

$$\iint E_y^* \nabla_T^2 E_y dxdy + k_0^2 \iint n^2(x, y) \left|E_y\right|^2 dxdy - \beta^2 \iint \left|E_y\right|^2 dxdy = 0 \tag{4.145}$$

where

$$\nabla_T^2 = \frac{\partial^2}{\partial x^2} + \frac{\partial^2}{\partial y^2}$$

is the transverse Laplacian operator. We can simplify Eq. (4.145) to obtain

$$\beta^2 = \frac{\iint E_y^* \nabla_T^2 E_y dxdy + k_0^2 \iint n^2(x,y)|E_y|^2 dxdy}{\iint |E_y|^2 dxdy} \tag{4.146}$$

The above equation is exact in the sense that if we know the solution $E_y(x, y)$, then performing the integrals on the right-hand side will yield us the exact value of propagation constant. Since our objective is actually to determine $E_y(x, y)$ we do not have knowledge of the solution.

As discussed in Sect. 4.4 the variational technique consists of substituting some trial solution for $E_y(x, y)$ with some variational parameters. The trial field is chosen judiciously to match as closely as possible to the actual field profile. The value of β^2 is maximized with respect to the variational parameters which gives us the values of the variational parameters as well as the propagation constant. The value of propagation constant so obtained is always smaller than the actual value of propagation constant. The closer is the match of the trial field with the actual field, the better the approximation for the propagation constant would be.

Equation (4.146) can be transformed using the following

$$\iint E_y^*(x,y) \nabla_T^2 E_y(x,y) dxdy = -\iint |\nabla_T E_y(x,y)|^2 dxdy$$

and we obtain the following expression:

$$\beta^2 = \frac{-\iint |\nabla_T E_y|^2 dxdy + k_0^2 \iint n^2(x,y)|E_y|^2 dxdy}{\iint |E_y|^2 dxdy} \tag{4.147}$$

As an example of a channel waveguide we consider titanium in-diffused waveguide in lithium niobate. The refractive index profile of the waveguide is approximated by Gaussian distributions along both the transverse directions through the following expression:

$$\begin{aligned} n^2(x,y) &= n_s^2 + 2n_s \Delta n_s e^{-x^2/d^2} e^{-y^2/w^2}; \qquad x<0 \\ &= n_c^2; \qquad x>0 \end{aligned} \tag{4.148}$$

where n_s is the bulk refractive index of the substrate, Δn_s is the index increase due to diffusion of titanium ions, d is the depth of the waveguide, and w is the width of the waveguide. The cover refractive index is n_c.

One of the trial fields which can be employed is given by:

$$\begin{aligned} E_{yt}(x,y) &= A\alpha_x \left(\frac{x}{d}\right) e^{-\alpha_x^2 x^2/d^2} e^{-\alpha_y^2 y^2/w^2}; \qquad x<0 \\ &= 0; \qquad x>0 \end{aligned} \tag{4.149}$$

where A is the normalization constant. Here α_x and α_y are the two dimensionless variational parameters and it is assumed that the entire field is within the high index region and the substrate which is a reasonable approximation since the substrate refractive index is much larger than the refractive index of the cover (see also Sect. 4.4 where we had analyzed a graded index planar waveguide).

We normalize the trial field through the following integral:

$$\iint_{-\infty\,-\infty}^{\infty\;\infty} |E_{yt}(x,y)|^2 dx\, dy = 1 \tag{4.150}$$

which becomes

$$\int_0^{\infty} dx \int_{-\infty}^{\infty} dy A^2 \left(\alpha_x\left(\frac{x}{d}\right)e^{-\alpha_x^2x^2/d^2}e^{-\alpha_y^2y^2/w^2}\right)^2 = 1$$

Carrying out the integration, we obtain

$$A = \sqrt{\frac{16\alpha_x\alpha_y}{\pi w d}} \tag{4.151}$$

Hence, the normalized trial field becomes

$$\begin{aligned} E_{yt}(x,y) &= \sqrt{\frac{16\alpha_x\alpha_y}{\pi w d}}\alpha_x\left(\frac{x}{d}\right)e^{-\alpha_x^2x^2/d^2}e^{-\alpha_y^2y^2/w^2}; \quad & x<0 \\ &= 0; & x>0 \end{aligned} \tag{4.152}$$

The trial field given by Eq. (4.152) has a Gaussian distribution along the y-direction (parallel to the substrate–cover interface) along which the waveguide is symmetric and has a Hermite Gaussian distribution along the x-direction (perpendicular to the substrate–cover interface) along which the waveguide is asymmetric.

On substituting Eq. (4.152) in Eq. (4.147) and carrying out the integrations, we obtain the following expression for the effective index of the fundamental mode:

$$n_{eff}^2 = \frac{\beta^2}{k_0^2} = n_s^2 - \frac{\alpha_y^2 d^2 + 3w^2\alpha_x^2}{k_0^2 w^2 d^2} + \frac{8n_s\Delta n_s\alpha_y\alpha_x^3}{\left(2\alpha_x^2+1\right)^{3/2}\sqrt{2\alpha_y^2+1}} \tag{4.153}$$

We can express Eq. (4.153) in terms of the normalized propagation constant defined in Eq. (4.94) and obtain an approximate expression for the normalized propagation constant b_t:

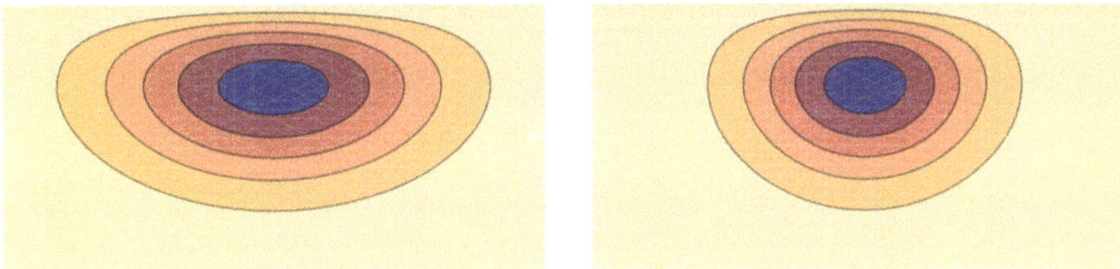

(a) (b)

Fig. 4.13 Field distributions of channel waveguide with $V = 10$, $\lambda = 0.8$ μm, $n_s = 2.2$, $\Delta n_s = 0.05$, $d = 2.7$ μm and (**a**) $w = 5.4$ μm and (**b**) $w = 2.7$ μm obtained by using the variational technique

$$b_t = \frac{4\alpha_y\alpha_x^3}{(2\alpha_x^2+1)^{3/2}\sqrt{2\alpha_y^2+1}} - \frac{\alpha_y^2(d/w)^2+3\alpha_x^2}{V^2} \tag{4.154}$$

where

$$V = k_0 d\sqrt{2n_s\Delta n_s} \tag{4.155}$$

is the normalized frequency. Maximizing b_t with respect to the variational parameters α_x and α_y, we obtain the optimized values of the variational parameters α_x and α_y. Using these optimized values we can obtain the effective index of the fundamental mode of the waveguide as well as the approximate field distribution as given by Eq. (4.152).

As an example, we consider a graded index channel waveguide with $V = 10$. For $w = 2d$, d and 0.5 d, we obtain, respectively, the following optimized values of the variational parameters: $(\alpha_x = 2.07, \alpha_y = 2.97)$, $(\alpha_x = 2.06, \alpha_y = 2.06)$, and $(\alpha_x = 2.02, \alpha_y = 1.39)$. The corresponding normalized propagation constant values are given by $b \approx 0.674$, 0.63, and 0.55, respectively.

Figure 4.13 shows the corresponding field distributions for the following parameter values:

Fig. 4.13 (a): $V = 10$, $\lambda = 0.8$ μm, $n_s = 2.2$, $\Delta n_s = 0.05$, $d = 2.7$ μm, $w = 5.4$ μm

Fig. 4.13 (b): $V = 10$, $\lambda = 0.8$ μm, $n_s = 2.2$, $\Delta n_s = 0.05$, $d = 2.7$ μm, $w = 2.7$ μm,
More details of variational technique applied to optical waveguides can be found in Sharma (2005).

Exercise 4.16 Using the variational field given by Eq. (4.152), obtain the position in the waveguide, where the electric field of the mode attains its highest value?

Solution The modal field is given by Eq. (4.152). Since the y-dependence is given by $e^{-\alpha_y^2 y^2/w^2}$, the maximum field will be at the $y = 0$. The x-dependence is given by $xe^{-\alpha_x^2 x^2/d^2}$ which has its maximum value when $\frac{d}{dx}\left(xe^{-\alpha_x^2 x^2/d^2}\right) = 0$ which gives the value $x_p = -\frac{d}{\sqrt{2}\alpha_x}$. In the example considered above for $V = 10$ and $w = 2d$, $\alpha_x = 2.07$ and thus $x_p = -\frac{d}{\sqrt{2}\alpha_x} = -0.35d$. If the operating wavelength is 800 nm and $n_s = 2.2$ and $\Delta n_s = 0.05$, for $V = 10$, we obtain $d = 2.7$ μm which gives $x_p \approx -0.95$ μm (see Fig. 4.13).

Both the perturbation method discussed in Sect. 4.7.1 and the variational technique can also be applied for devices such as two waveguide or three waveguide directional couplers, etc. and readers are referred to the literature for further details.

4.8 Simple Analogy Between Quantum Mechanics of Potential Wells and TE Modes in Optical Waveguides

The time independent Schrödinger equation describing a particle of mass m in a one-dimensional potential distribution $V(x)$ is given by (see Chap. 3)

$$\frac{d^2\psi}{dx^2} + \frac{2m}{\hbar^2}(E - V(x))\psi = 0 \tag{4.156}$$

where $\psi(x)$ represents the wavefunction and E is the energy eigenvalue.

The one-dimensional wave equation describing the electric field $E_y(x)$ of the TE modes in an optical waveguide having a refractive index profile given by $n(x)$ is (see Eq. 4.23)

$$\frac{d^2E_y}{dx^2} + \left(k_0^2 n^2(x) - \beta^2\right)E_y = 0 \tag{4.157}$$

We now write

$$k_0^2 n^2(x) - \beta^2 = k_0^2 n^2(x) - \beta^2 + k_0^2 n_p^2 - k_0^2 n_p^2 = \left(k_0^2 n_p^2 - \beta^2\right) - k_0^2\left(n_p^2 - n^2(x)\right)$$

Here n_p^2 is a constant and taken to be larger than the maximum value of $n^2(x)$. If we write

$$\left(k_0^2 n_p^2 - \beta^2\right) = \tilde{\beta}^2 \tag{4.158a}$$

$$\left(n_p^2 - n^2(x)\right) = \tilde{n}^2(x) \tag{4.158b}$$

Then Eq. (4.157) becomes

$$\frac{d^2E_y}{dx^2} + \left(\tilde{\beta}^2 - k_0^2\tilde{n}^2(x)\right)E_y = 0 \tag{4.159}$$

Comparing Eqs. (4.156) and (4.159) we find the following equivalence between the Schrödinger equation and the wave equation:

$$\psi(x) < \rightarrow > E_y(x) \tag{4.160}$$

$$\frac{2m}{\hbar^2}E < \rightarrow > \tilde{\beta}^2 = \left(k_0^2 n_p^2 - \beta^2\right) \tag{4.161}$$

$$\frac{2m}{\hbar^2}V(x) < \rightarrow > k_0^2\tilde{n}^2(x) = k_0^2\left(n_p^2 - n^2(x)\right) \tag{4.162}$$

Thus, the energy eigenvalues of a potential well are equivalent to the propagation constant of the TE modes of an equivalent waveguide and the potential distribution is equivalent to the refractive index profile of the waveguide. The wavefunction is similar to the transverse electric field E_y. This analogy is very useful as it allows one to analyze problems in optical waveguide theory or in quantum mechanics by analyzing equivalent problems.

As an example if we take a potential well defined by

$$\begin{aligned} V(x) &= V_c; \quad && x > 0 \\ &= V_f; && -d < x < 0 \\ &= V_s; && x < -d \end{aligned} \tag{4.163}$$

with $V_f < V_s \le V_c$ forming an asymmetric potential well.

To determine the equivalent problem in optical waveguides, we use Eq. (4.162) relating the potential distribution and refractive index profile

$$n^2(x) = n_p^2 - \frac{2m}{\hbar^2 k_0^2}V(x)$$

Hence, the asymmetric potential well is equivalent to the following refractive index profile

$$\begin{aligned} n^2(x) = n_c^2 &= n_p^2 - \frac{2m}{\hbar^2 k_0^2}V_c; \quad && x > 0 \\ = n_f^2 &= n_p^2 - \frac{2m}{\hbar^2 k_0^2}V_f; && -d < x < 0 \\ = n_s^2 &= n_p^2 - \frac{2m}{\hbar^2 k_0^2}V_s; && x < -d \end{aligned} \tag{4.164}$$

Since $V_f < V_s \leq V_c$, we have $n_f^2 > n_s^2 \geq n_c^2$, thus forming a waveguide capable of supporting optical modes.

The eigenvalue equation describing TE modes is given by (see Eq. 4.43). Using the analogy we define the following parameters:

$$\kappa_f^2 = k_0^2 n_f^2 - \beta^2, <\rightarrow> -\frac{2m}{\hbar^2} V_f + \frac{2m}{\hbar^2} E = \frac{2m}{\hbar^2}(E - V_f) \tag{4.165}$$

$$\gamma_s^2 = \beta^2 - k_0^2 n_s^2, <\rightarrow> -\frac{2m}{\hbar^2} E + \frac{2m}{\hbar^2} V_s = \frac{2m}{\hbar^2}(V_s - E) \tag{4.166}$$

$$\gamma_c^2 = \beta^2 - k_0^2 n_c^2, <\rightarrow> -\frac{2m}{\hbar^2} E + \frac{2m}{\hbar^2} V_c = \frac{2m}{\hbar^2}(V_c - E) \tag{4.167}$$

Thus, the eigenvalue equation determining the eigenvalues of the symmetric potential well will be

$$d\sqrt{\frac{2m}{\hbar^2}(E - V_f)} = \tan^{-1}\left(\sqrt{\frac{V_s - E}{E - V_f}}\right) + \tan^{-1}\left(\sqrt{\frac{V_c - E}{E - V_f}}\right) + m\pi; \quad m = 0, 1, 2, \ldots \tag{4.168}$$

The eigenfunctions corresponding to the potential well can similarly be written down from the solutions of the TE modes of the waveguide.

The analogy presented here helps one to apply the various techniques interchangeably between quantum mechanics of potential distributions and modes in optical waveguides. Approximate techniques such as perturbation method, variational method etc. can be used in both the problems in obtaining solutions under different situations.

Exercise 4.17 Consider a harmonic oscillator with a potential distribution given by $V(x) = \frac{1}{2} m\omega^2 x^2$ in the Schrödinger representation. An equivalent optical waveguide is one where the refractive index profile is given by $n^2(x) = n_1^2(1 - \alpha^2 x^2)$, referred to as a parabolic index waveguide. Using the analogy between quantum mechanics and waveguide optics, obtain the propagation constants of the parabolic index waveguide knowing the energy eigenvalues of the harmonic oscillator.

Solution The energy eigenvalues of the harmonic oscillator are given by $E_n = \left(n + \frac{1}{2}\right)\hbar\omega$. Using the analogy the given refractive index profile corresponds to the following potential distribution:

$$\frac{2m}{\hbar^2} V(x) < - > k_0^2\left(n_p^2 - n^2(x)\right) = k_0^2\left(n_p^2 - n_1^2\left(1 - \alpha^2 x^2\right)\right)$$

Thus, we have

$$V(x) = \frac{\hbar^2}{2m} k_0^2 \left(n_p^2 - n_1^2 \right) + \frac{\hbar^2}{2m} k_0^2 n_1^2 \alpha^2 x^2$$

We take $n_p = n_1$,so that we have

$$V(x) = \frac{\hbar^2}{2m} k_0^2 n_1^2 \alpha^2 x^2$$

which represents a harmonic oscillator with frequency ω given by the analogy $\frac{1}{2} m\omega^2 = \frac{\hbar^2}{2m} k_0^2 n_1^2 \alpha^2$ or $\omega = \frac{\hbar}{m} k_0 n_1 \alpha$.

Similarly the corresponding energy is

$$E = \frac{\hbar^2}{2m} \left(k_0^2 n_p^2 - \beta^2 \right)$$

Since for the harmonic oscillator $E_n = \left(n + \frac{1}{2}\right)\hbar\omega$, we have for the parabolic index waveguide

$$\frac{\hbar^2}{2m} \left(k_0^2 n_1^2 - \beta^2 \right) = \left(n + \frac{1}{2} \right) \hbar \times \frac{\hbar}{m} k_0 n_1 \alpha = \left(n + \frac{1}{2} \right) \frac{\hbar^2}{m} k_0 n_1 \alpha$$

Hence,

$$\beta^2 = k_0^2 n_1^2 - (2n + 1) k_0 n_1 \alpha$$

which gives the propagation constants of the various TE modes of the planar parabolic index waveguide (see e.g., Ghatak and Thyagarajan 1998).

Exercise 4.18 Consider an asymmetric planar waveguide with a refractive index profile given by Eq. (4.22). Using time independent perturbation technique discussed in Chap. 3, obtain the change in the effective index of the fundamental mode due to a change in the refractive index of the film from n_f to $n_f + \Delta n_f$ with $\Delta n_f \ll n_f$.

Solution We recall Eq. (3.138) giving the change in the energy eigenvalue of a perturbed Hamiltonian in the Schrödinger representation as

$$\Delta E_n = \int \psi_n^{(0)*} \widehat{W} \psi_n^{(0)} dx$$

In the case of the optical waveguide, the perturbation is given by (see Eq. 4.162)

$$W = \Delta V(x) \approx -\frac{\hbar^2}{2m} k_0^2 2n_f \Delta n_f; \qquad -d < x < 0$$
$$= 0; \qquad x > 0 \; and \; x < -d$$

where we have used $\Delta n_f^2 \approx 2n_f \Delta n_f$ and $\Delta n_p^2 = 0$. The unperturbed wave function corresponding to the region $-d < x < 0$ is

$$\psi_n^{(0)} = A\left(\cos\kappa_f x - \frac{\gamma_c}{\kappa_f}\sin\kappa_f x\right); \qquad -d < x < 0;$$

where

$$\gamma_c^2 = \beta^2 - k_0^2 n_c^2$$
$$\gamma_s^2 = \beta^2 - k_0^2 n_s^2$$
$$\kappa_f^2 = k_0^2 n_f^2 - \beta^2$$

A is the normalization constant satisfying the condition $\int_{-\infty}^{+\infty} |E_y|^2 dx = 1$ (see Eqs. 4.62 and 4.63):

$$A^2 = 2\left(1 + \frac{\gamma_c^2}{\kappa_f^2}\right)^{-1}\left(d + \frac{1}{\gamma_s} + \frac{1}{\gamma_c}\right)^{-1}$$

Thus, we have

$$\Delta E = -\frac{\hbar^2}{2m}\Delta\beta^2 = -\frac{\hbar^2}{2m}k_0^2 2n_f \Delta n_f \int_{-d}^{0} A^2\left(\cos\kappa_f x - \frac{\gamma_c}{\kappa_f}\sin\kappa_f x\right)^2 dx$$

which gives us

$$\Delta\beta^2 = k_0^2 2n_f \Delta n_f P_f$$

where

$$P_f = \int_{-d}^{0} A^2\left(\cos\kappa_f x - \frac{\gamma_c}{\kappa_f}\sin\kappa_f x\right)^2 dx$$

Is the fractional power contained within the film. Hence,

$$\Delta n_{eff}^2 = 2n_f \Delta n_f P_f$$

As an example we consider the parameters $n_f = 2.22$, $n_s = 2.2$, $n_c = 1.0$, $d_f = 4$ μm, and $\lambda_0 = 0.8$ μm. For these values the effective index of the TE_0 mode is $n_{eff} = 2.21822$. We then obtain $\kappa_f = 0.702$ μm^{-1}, $\gamma_s = 2.227$ μm^{-1}, and $\gamma_c = 15.55$ μm^{-1}. Substituting all the values we have $A^2 \approx 9 \times 10^{-4}$ and $P_f \approx 0.996$. Thus, if $\Delta n_f = 0.001$, then $\Delta n_{eff}^2 \approx 4.42 \times 10^{-3}$ giving $\Delta n_{eff} \approx 9.9 \times 10^{-4}$.

We can obtain a more precise value of the effective index by solving for a step index waveguide with a film refractive index $n_f + \Delta n_f$. Solving the transcendental equation assuming $n_f = 2.221$ gives $n_{eff} = 2.21921$. Hence, a more precise numerical evaluation gives $\Delta n_{eff} \approx 9.9 \times 10^{-4}$ which matches very well with first order perturbation calculation.

Hence, a modulation of the refractive index of the film will result in a modulation of the effective index of the mode and a periodic modulation of the film refractive index will lead to a periodic variation of the effective index which can induce coupling of power among two co-propagating modes provided the coupling coefficient between the two modes is non-zero and the required phase matching condition is satisfied (see Chap. 9).

Problems

4.1. Consider a symmetric step index planar waveguide with $n_f = 2.21$, $n_s = n_c = 2.20$, and $d = 8$ μm operating at the free space wavelength of 1500 nm. These refractive index values are typical of waveguides fabricated in lithium niobate which is a very important quantum integrated photonic material. Optical modulators and optical switches in use today in communication systems are usually fabricated on this substrate.

(a) Calculate the number of guided TE and TM modes supported by the waveguide.
(b) By solving the eigenvalue equation, obtain the effective indices of all the guided TE and TM modes.
(c) What thickness range would you choose so that the waveguide supports only a single TE mode?

4.2. Consider two step index symmetric planar waveguides: WG1 $n_f = 1.52$, $n_s = n_c = 1.50$ and $d = 4$ μm. WG2 $n_f = 1.5354$, $n_s = n_c = 1.50$, and $d = 3$ μm. Both the waveguides operate at 1000 nm. Which of these two waveguides will have a larger value of n_{eff} of the TE_0 mode?

4.3. Consider the fundamental TE mode in a symmetric step index planar waveguide with a refractive index profile $n(x)$. Obtain the Poynting vector and show that the power flow in the mode is *only* along the z-direction.

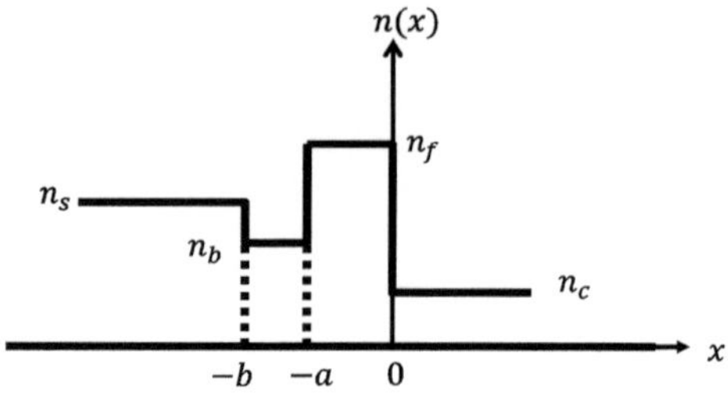

Fig. 4.14 Figure for Problem 4.7

4.4. Show that the TE_0 mode and TE_2 modes of a step index planar waveguide are orthogonal to each other.

4.5. A symmetric planar waveguide with $n_f = 1.503$, $n_s = n_c = 1.50$ supports a TE_0 mode with $n_{eff} = 1.5016$ at a wavelength of 1 μm.

(a) Obtain the thickness of the waveguide.
(b) What is the depth of penetration of the mode within the lower index region?
(c) If the power carried by the TE_0 mode is 10 mW, what is the value of the peak electric field of the mode

4.6. Show that a metal–dielectric interface cannot support TE modes.

4.7. Consider a waveguide having a refractive index profile shown in Fig. 4.14. (a) For $n_s > n_{eff} > n_b$ write down the solutions for $E_y(x)$ in all the regions. (b) Can the waveguide support guided modes with $n_s < n_{eff} < n_b$ (c) For what range of n_{eff} will the modes be guided?

4.8. The TE_1 mode in a planar step index waveguide operating at 1000 nm with $n_f = 2.246$, $n_s = 2.236$, and $n_c = 1.0$ has $n_{eff} = 2.238$. (a) Obtain the thickness of the film. (b) What will be the value of n_{eff} of the mode when it reaches cutoff? (c) What should be the thickness of the film so that it supports only the TE_0 mode?

4.9. Consider a surface plasmon mode at the interface of glass of refractive index 1.5 and three different metals namely aluminum, gold and silver. The dielectric constant of the three metals at a wavelength of 1300 nm is given by, $K\ (aluminum) = -157 + i\ 33.3$; $K(gold) = -40.3 + i\ 2.8$; $K(silver) = -48.9 + i\ 0.56$. Compare the losses in the mode. Calculate the penetration depth of the mode in the dielectric and the metal.

4.10. Consider a symmetric planar waveguide with $n_f = 2.3$, $n_s = n_c = 2.2$, and $d = 2$ μm. Operating at a wavelength of 1000 nm

(a) How many guided TE modes will the waveguide support?
(b) What are the minimum and maximum possible values of n_{eff} of the TE_1 mode?
(c) The effective index of TE_m mode is 2.225. To what value of m does it correspond? Give reasons for your answer.
(d) In what range of wavelengths will the waveguide be single moded?

4.11. Consider a waveguide with $n_f = 1.534$, $n_s = 1.502$ and thickness of 8 μm operating at 1500 nm. The waveguide is covered with gold whose dielectric constant at 1500 nm is given by $K = -25.8 + i1.63$. (a) Obtain the effective index and loss (in dB/cm) of the TE_0 and TM_0 modes of the waveguide. (b) A buffer with a refractive index of 1.446 is now introduced between the metal and the film. Plot the variation of the effective index and the loss of the TE_0 and TM_0 modes as a function of the thickness of the buffer and show that the buffer can significantly reduce the loss of the modes

4.12. Consider a waveguide with a refractive index variation given by

$$n(x) = n_s + \Delta n e^{-x^2/d^2}$$

State the range of the effective index of the guided modes.

4.13. The effective index of the TE_0 mode of a planar waveguide with $n_f = 1.5$, $n_s = 1.48$, and $n_c = 1$ with $d = 3$ μm is given by 1.484. State giving reasons whether the effective index of the E_{11}^y mode of a channel waveguide with the same refractive indices of the film, substrate and cover and depth d but having a width w of 5 μm will be higher or lower than 1.484.

4.14. Consider an anisotropic planar waveguide with a uniaxial film (having ordinary and extraordinary refractive indices n_0 and n_e) and isotropic substrate (n_s) and cover (n_c). Assuming the c-axis of the film to be along the x-direction (perpendicular to the substrate surface), state whether guided TE and TM modes can exist in the following cases:

(a) $n_o = 1.658$, $n_e = 1.485$, $n_s = 1.5$, and $n_c = 1.0$
(b) $n_o = 1.544$, $n_e = 1.553$, $n_s = 1.56$, and $n_c = 1.0$

4.15. Consider a planar waveguide in a uniaxial substrate such as lithium niobate having its optic axis oriented at an angle θ with respect to the substrate–cover interface and lying in the $x-z$ plane. Here the laboratory $x-y-z$ axes are as shown in Fig. 4.1a. Show that even in this case the modes split into two independent sets of modes namely TE and TM modes. Obtain the corresponding wave equation satisfied by the respective modes. Given that the cover is isotropic.

4.16. Consider a waveguide with $n_f = 1.5$, $n_s = 1.48$, $n_c = 1.33$, and $d = 3$ μm operating at $\lambda_0 = 800$ nm.

(a) Obtain the effective index of the TE_0 mode in the waveguide.
(b) Use perturbation theory and obtain the change in effective index if n_c is changed by 0.01. Note the very small change in the effective index which is due to the small fraction of power in the region $x > 0$.

4.17. In Problem 4.16, assume operation at a wavelength of 1000 nm. Consider the TE_1 mode and obtain the change in the effective index of the mode when n_c is changed by 0.001. If this change in effective index is over a length of 2 mm of the waveguide, what is the change in the phase of the output from the waveguide? [Note that in this case the mode under consideration is closer to cutoff and hence has a larger fraction of power in the region $x > 0$ resulting in a larger change in effective index due to change in the refractive index of that region].

References

Boes, A., B. Corcoran, L. Chang, J. Bowers, and A. Mitchell. 2018. Status and potential of lithium niobate on insulator (LNOI) for photonic integrated circuits. *Laser Photonics Rev* 12:1700256.

Ghatak, A., and K. Thyagarajan. 1989. *Optical electronics*. New York: Plenum Publishing Corporation.

———. 1998. *Introduction to fiber optics*. Cambridge: Cambridge University Press.

Kumar, A., K. Thyagarajan, and A. K. Ghatak. 1983. Analysis of rectangular-core dielectric waveguides: an accurate perturbation approach. *Opt Lett* 8:63–65.

Kumar, A., M. R. Shenoy, and K. Thyagarajan. 1984. Modes in anisotropic rectangular waveguides: An accurate and simple perturbation approach. *IEEE Trans Microwave Theory Tech* 32: 1415–1418.

Lin, J., F. Bo, Y. Cheng, and J. Xu. 2020. Advances in on-chip photonic devices based on lithium niobate on insulator. *Opt Express* 8:1910–1936.

Marcuse, D., and P. Liao. 2012. *Theory of dielectric optical waveguides*. USA: Academic Press.

Nishihara, H., M. Haruna, and T. Suhara. 1989. *Optical integrated circuits*. McGraw Hill.

Okamoto, K. 2021. *Fundamentals of optical waveguides*. Academic Press.

Sharma, A. 2005. Guided Wave Optics, Selected Topics, Viva Books, New Delhi

Sharma, E. K., and S. Srivastava. 2005. Chapter 13. Guided wave optics. In *Scalar variational analysis for optical waveguides*, ed. Anurag Sharma. New Delhi: Viva Press.

Suhara, T., and M. Fujimura. 2003. *Waveguide nonlinear-optic devices*. Berlin: Springer.

Suhara, T., Y. Handa, H. Nishihara, and J. Koyama. 1979. Analysis of optical channel waveguides and directional couplers with graded index profile. *J Opt Soc Am* 69:807–815.

Vazimali, M. G., and S. Fathpour. 2022. Applications of thin-film lithium niobate in nonlinear integrated photonics. *Adv Photon* 4:034001.

Zhu, D., L. Shao, M. Yu, R. Cheng, B. Desiatov, C. J. Xin, Y. Hu, J. Holzgrafe, S. Ghosh, A. Shams-Ansari, E. Puma, N. Sinclair, C. Reimer, M. Zhang, and M. Loncar. 2021. Integrated photonics on thin-film lithium niobate waveguide. *Adv Opt Photonics* 13:242–352.

Chapter 5
The Optical Directional Coupler

Abstract This chapter presents a detailed discussion of optical directional couplers, which is one of the important components of integrated quantum photonic circuits. Coupled mode theory is used to analyze two waveguide directional coupler, three waveguide directional coupler, and waveguide arrays. Optical switch using a directional coupler is also presented. The analysis presented in this chapter is used extensively in later chapters.

5.1 Introduction

In Chap. 4, we showed that in an optical waveguide, the electric field of the mode is not completely confined within the high index region of the waveguide but also extends into the surrounding lower refractive index regions of substrate and cover. The evanescent field that exists outside the high index region is very useful in realizing many integrated photonic components. One such very important device is the optical directional coupler. The directional coupler forms one of the building blocks of integrated quantum photonic devices and in optical telecommunication and sensing. These include power dividers, optical switches, optical modulators, wavelength splitters, quantum C-NOT gates, and so on.

In this chapter, we shall discuss the basic working principle of directional coupler and some of its applications.

5.2 Analysis

Consider a pair of waveguides lying close to each other in the coupling region as shown in Fig. 5.1. Since the field of the mode of each of the waveguides extends into the surrounding lower index region, the two waveguides interact with each other through the evanescent field of the neighboring waveguide. This leads to the coupling and exchange of optical power between the two waveguides. Such a device is referred to as an optical directional coupler.

K. Thyagarajan, *Integrated Quantum Photonics*, Graduate Texts in Physics,
https://doi.org/10.1007/978-3-031-85728-7_5

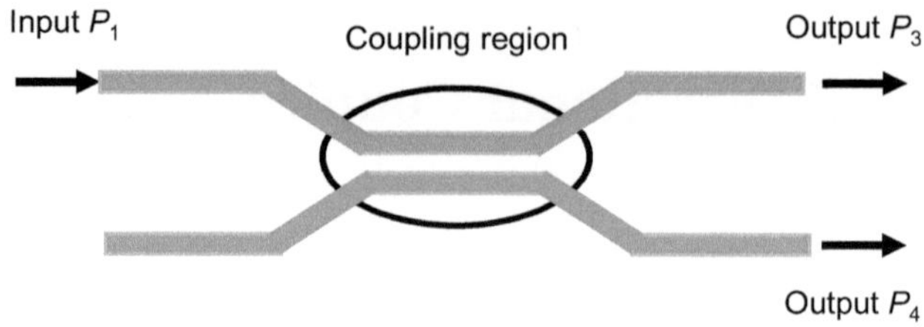

Fig. 5.1 An optical directional coupler consisting of two closely lying waveguides such that in the coupling region their evanescent fields interact, leading to coupling of power among the two waveguides

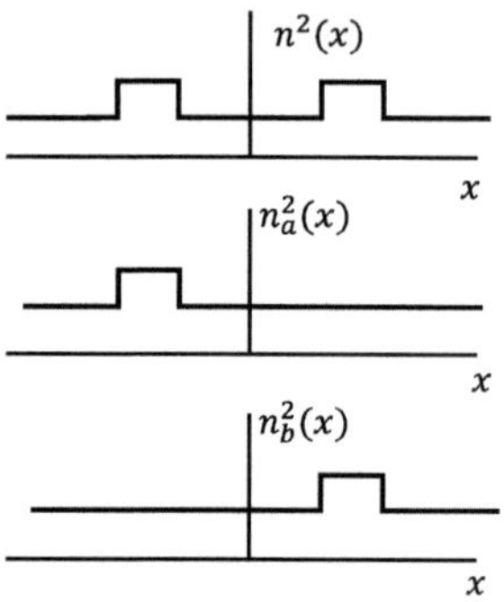

Fig. 5.2 The refractive index profiles $n_a^2(x)$ and $n_b^2(x)$ of two individual waveguides and the refractive index profile $n^2(x)$ of the directional coupler made of the two waveguides. The refractive index profiles are shown in one dimension

Optical directional couplers are very important components in integrated quantum photonics and find applications in modulation, switching, power splitters, entangled photon pair generation, quantum gates for quantum computing, etc.

In order to analyze energy exchange between the two waveguides, we employ the standard coupled-mode theory (see, e.g., Marcuse 1971; Yariv 1973; Hardy and Streifer 1985; Haus et al. 1987). We label the two individual waveguides as a and b and assume that the mode of the waveguide a in the absence of the waveguide b has a propagation constant β_a and a power-normalized transverse electric field profile $E_a(x, y)$. Similarly, we assume that the mode of waveguide b has a propagation constant. β_b and a power-normalized transverse electric field profile given by $E_b(x, y)$. The refractive index profiles of the individual waveguides a and b are represented by $n_a(x, y)$ and $n_b(x, y)$, respectively. We represent the refractive index profile of the composite structure (consisting of the two waveguides) representing the directional coupler by $n(x, y)$. As an example, Fig. 5.2 shows a schematic (in one dimension) of the refractive index profile of individual waveguides and the composite profile consisting of both the waveguides.

Under the coupled mode approximation, we approximate the total transverse electric field in the directional coupler by a linear superposition of modes of waveguides a and b. Thus,

$$E(x,y,z) = A(z)E_a(x,y)e^{i\beta_a z} + B(z)E_b(x,y)e^{i\beta_b z} \tag{5.1}$$

The z-dependent amplitudes $A(z)$ and $B(z)$ account for any energy exchange between the two waveguides. In the absence of any coupling both $A(z)$ and $B(z)$ will be independent of z and the two modes would propagate independently of each other.

The fields $E_a(x,y)$ and $E_b(x,y)$ are assumed to satisfy power normalization conditions (see Chap. 4):

$$\frac{\beta_j}{2\omega\mu_0}\iint \left|E_j(x,y)\right|^2 dxdy = 1; j = a, b \tag{5.2}$$

In view of this $|A(z)|^2$ and $|B(z)|^2$ would represent the powers carried by the modes in waveguides a and b.

Now, since $E_a(x,y)$ and $E_b(x,y)$ are modal fields of waveguides a and b, they satisfy the following equations (see Chap. 4):

$$\nabla_T^2 E_a(x,y) + \left[k_0^2 n_a^2(x,y) - \beta_a^2\right]E_a(x,y) = 0 \tag{5.3}$$

$$\nabla_T^2 E_b(x,y) + \left[k_0^2 n_b^2(x,y) - \beta_b^2\right]E_b(x,y) = 0 \tag{5.4}$$

where

$$\nabla_T^2 = \nabla^2 - \frac{\partial^2}{\partial z^2} = \frac{\partial^2}{\partial x^2} + \frac{\partial^2}{\partial y^2}$$

is the transverse Laplacian.

The total field $E(x,y,z)$ of the directional coupler satisfies the following wave equation:

$$\nabla_T^2 E(x,y,z) + \frac{\partial^2 E}{\partial z^2} + k_0^2 n^2(x,y)E(x,y,z) = 0 \tag{5.5}$$

Substituting for $E(x,y,z)$ from Eq. (5.1) in Eq. (5.5) and using Eqs. (5.3) and (5.4), we obtain

$$\begin{aligned} 2i\beta_a \frac{dA}{dz} E_a e^{i\beta_a z} + 2i\beta_b \frac{dB}{dz} E_b e^{i\beta_b z} + AE_a k_0^2\left[n^2(x,y) - n_a^2(x,y)\right]e^{i\beta_a z} + \\ BE_b k_0^2\left[n^2(x,y) - n_b^2(x,y)\right]e^{i\beta_b z} = 0 \end{aligned} \tag{5.6}$$

In obtaining the above equation, we have assumed that the amplitudes $A(z)$ and $B(z)$ are slowly varying functions of z and hence have neglected terms involving $\frac{\partial^2 A}{\partial z^2}$ and $\frac{\partial^2 B}{\partial z^2}$.

Multiplying Eq. (5.6) by E_a^* and integrating over the entire transverse x-y plane, we obtain

$$2i\beta_a \frac{dA}{dz}\iint |E_a|^2 dxdy e^{i\beta_a z} + k_0^2 A \iint \Delta n_a^2 |E_a|^2 dxdy e^{i\beta_a z} + \\ k_0^2 B \iint \Delta n_b^2 E_a^* E_b dxdy e^{i\beta_b z} = 0 \tag{5.7}$$

where

$$\Delta n_a^2 = n^2(x,y) - n_a^2(x,y) \tag{5.8}$$

and

$$\Delta n_b^2 = n^2(x,y) - n_b^2(x,y) \tag{5.9}$$

and we have ignored the term containing the overlap $\iint E_a^* E_b dxdy$ between the two interacting modal fields. This is a reasonable approximation for weakly interacting waveguides.

We thus obtain the following equation:

$$\frac{dA}{dz} = i\kappa_{aa} A + i\kappa_{ab} B e^{-i(\beta_a - \beta_b)z} \tag{5.10}$$

where

$$\kappa_{aa} = \frac{k_0^2}{2\beta_a} \frac{\iint \Delta n_a^2 |E_a|^2 dxdy}{\iint |E_a|^2 dxdy} = \frac{k_0^2}{4\omega\mu_0} \iint \Delta n_a^2 |E_a|^2 dxdy \tag{5.11}$$

$$\kappa_{ab} = \frac{k_0^2}{2\beta_a} \frac{\iint \Delta n_b^2 E_a^* E_b dxdy}{\iint |E_a|^2 dxdy} = \frac{k_0^2}{4\omega\mu_0} \iint \Delta n_b^2 E_a^* E_b dxdy \tag{5.12}$$

Similarly multiplying Eq. (5.6) by E_b^* and integrating over the entire transverse plane and making similar approximations, we obtain

$$\frac{dB}{dz} = i\kappa_{bb} B + i\kappa_{ba} A e^{i(\beta_a - \beta_b)z} \tag{5.13}$$

where

$$\kappa_{bb} = \frac{k_0^2}{2\beta_b} \frac{\iint \Delta n_b^2 |E_b|^2 dxdy}{\iint |E_b|^2 dxdy} = \frac{k_0^2}{4\omega\mu_0} \iint \Delta n_b^2 |E_b|^2 dxdy \tag{5.14}$$

$$\kappa_{ba} = \frac{k_0^2}{2\beta_b} \frac{\iint \Delta n_a^2 E_b^* E_a dxdy}{\iint |E_b|^2 dxdy} = \frac{k_0^2}{4\omega\mu_0} \iint \Delta n_a^2 E_b^* E_a dxdy \tag{5.15}$$

In the above equations, κ_{aa} and κ_{bb} are referred to as self-coupling coefficients and κ_{ab} and κ_{ba} are referred to as cross-coupling coefficients or just as coupling coefficients. The terms containing the self-coupling coefficients lead to changes in the propagation constants of the two individual waveguides while the term containing the cross-coupling coefficients leads to exchange of energy between the two waveguides.

If we define

$$\tilde{A}(z) = A(z)e^{-i\kappa_{aa}z} \tag{5.16}$$

$$\tilde{B}(z) = B(z)e^{-i\kappa_{bb}z} \tag{5.17}$$

and substitute in Eqs. (5.10) and (5.13), we obtain the following two coupled equations:

$$\frac{d\tilde{A}}{dz} = i\kappa_{ab}\tilde{B}e^{-i(\beta_a+\kappa_{aa}-\beta_b-\kappa_{bb})z} \tag{5.18}$$

$$\frac{d\tilde{B}}{dz} = i\kappa_{ba}\tilde{A}e^{i(\beta_a+\kappa_{aa}-\beta_b-\kappa_{bb})z} \tag{5.19}$$

These equations contain the fact that the presence of waveguide b modifies the propagation constant of waveguide a and vice versa. Thus, the propagation constant of waveguide a gets modified from β_a to $\beta_a + \kappa_{aa}$ and that of waveguide b gets modified from β_b to $\beta_b + \kappa_{bb}$; the additional terms in the coupled mode equations are coming from the self-coupling coefficients.

In the following, we will neglect the effect of the self-coupling terms which can always be taken into account by redefining the propagation constants of the two waveguides. Thus, we obtain the following two coupled equations describing light propagation through the coupled waveguide structure:

$$\frac{dA}{dz} = i\kappa_{ab}Be^{-i\Delta\beta z} \tag{5.20}$$

$$\frac{dB}{dz} = i\kappa_{ba}Ae^{i\Delta\beta z} \tag{5.21}$$

where

$$\Delta\beta = (\beta_a - \beta_b) \tag{5.22}$$

represents the difference in propagation constants of the modes of the two individual waveguides in the absence of any coupling.

If the two interacting waveguides are identical then $\kappa_{ab} = \kappa_{ba}$. In the general case, when the two waveguides are not identical, the cross-coupling coefficients κ_{ab} and κ_{ba} are not equal. If we assume that the waveguides are weakly interacting, energy conservation requires that $(|A|^2 + |B|^2)$ be independent of z. Thus, we have

$$\frac{d(|A|^2 + |B|^2)}{dz} = A^* \frac{dA}{dz} + cc + B^* \frac{dB}{dz} + cc = 0 \tag{5.23}$$

where cc stands for the complex conjugate of the term preceding it. Using Eqs. (5.20) and (5.21) in Eq. (5.23), we obtain

$$\kappa_{ab} = \kappa_{ba} = \kappa \tag{5.24}$$

where we have assumed the coupling coefficients to be real. The above equality is satisfied when the two waveguides are identical or when the interaction between the two waveguides is weak. We shall restrict our discussion to near identical weakly interacting waveguides so that the cross-coupling coefficients satisfy Eq. (5.24) (see, e.g., Haus et al. 1987).

Thus, the coupled mode equations describing light propagation through the coupled waveguide structure are given by

$$\frac{dA}{dz} = i\kappa B e^{-i\Delta\beta z} \tag{5.25}$$

$$\frac{dB}{dz} = i\kappa A e^{i\Delta\beta z} \tag{5.26}$$

5.2.1 Waveguides with Identical Propagation Constants

We first consider the case when the two waveguides have identical propagation constants, that is, $\beta_a = \beta_b$. Thus, the two coupled mode equations become

$$\frac{dA}{dz} = i\kappa B \tag{5.27}$$

$$\frac{dB}{dz} = i\kappa A \tag{5.28}$$

Differentiating Eq. (5.27) with respect to z and using Eq. (5.28), we obtain

$$\frac{d^2A}{dz^2} = i\kappa \frac{dB}{dz} = -\kappa^2 A \tag{5.29}$$

The general solution of Eq. (5.29) is

$$A(z) = A_1 \cos \kappa z + A_2 \sin \kappa z \tag{5.30}$$

Substituting the solution for $A(z)$ in Eq. (5.27), we obtain

$$B(z) = -\frac{i}{\kappa}\frac{dA}{dz} = iA_1 \sin \kappa z - iA_2 \cos \kappa z \tag{5.31}$$

If at the input at $z = 0$, light with unit amplitude is launched in waveguide a with no input in waveguide b, then we have

$$A(z=0) = 1 \tag{5.32}$$

$$B(z=0) = 0 \tag{5.33}$$

Thus, $A_1 = 1$ and $A_2 = 0$. Hence

$$A(z) = \cos \kappa z \tag{5.34}$$

$$B(z) = i \sin \kappa z \tag{5.35}$$

Hence the z-variations of the powers in waveguides a and b are given by

$$P_a(z) = \cos^2 \kappa z \tag{5.36}$$

$$P_b(z) = \sin^2 \kappa z \tag{5.37}$$

The above equations imply that power exchanges periodically between the two waveguides and after a certain propagation distance referred to as coupling length given by the following equation:

$$L_c = \frac{\pi}{2\kappa} \tag{5.38}$$

the entire power launched in waveguide a gets transferred to waveguide b. After another coupling length, the power returns to waveguide a and this exchange of power occurs periodically. The solid curve in Fig. 5.3 shows a plot of the periodic variation of power in the coupled waveguide as a function of z for the case with $\Delta\beta = 0$.

Equations (5.36) and (5.37) show that at a distance $\pi/4\kappa$ which is half the coupling length, the two waveguides contain equal power. Such a device acts as a 3 dB (equal to 50%) power splitter.

It should be noted that at this distance although the power in each waveguide is the same, the fields in the two waveguides have a phase difference of $\pi/2$ (see Eqs. 5.34 and 5.35). This is also required through energy conservation and is similar to the phenomenon observed in the case of a bulk beam splitter wherein there is a

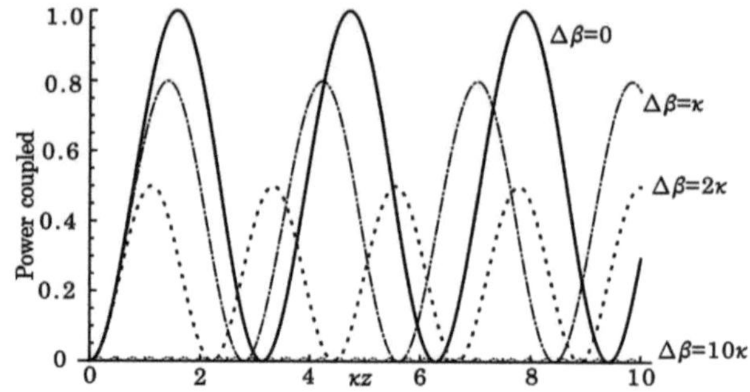

Fig. 5.3 Variation of power in the coupled waveguide with normalized propagation distance z. Solid curve corresponds to $\Delta\beta = 0$, dash-dotted curve to $\Delta\beta = \kappa$, dotted curve to $\Delta\beta = 2\kappa$ and the barely visible curve closest to the x-axis to $\Delta\beta = 10\kappa$

phase difference of $\pi/2$ between the transmitted and reflected beams. We will revisit this aspect in Chap 19 when we discuss the quantum mechanics of beam splitters.

Exercise 5.1 Assuming initial conditions of $A(z = 0) = A(0)$ and $B(z = 0) = B(0)$, show that

$$\begin{pmatrix} A(z) \\ B(z) \end{pmatrix} = \begin{pmatrix} \cos\kappa z & i\sin\kappa z \\ i\sin\kappa z & \cos\kappa z \end{pmatrix} \begin{pmatrix} A(z=0) \\ B(z=0) \end{pmatrix} \tag{5.39}$$

Solution Substituting the initial conditions $A(z = 0) = A(0)$ and $B(z = 0) = B(0)$ in the solutions given by Eqs. (5.30) and (5.31), we obtain $A_1 = A(0)$ and $A_2 = iB(0)$ giving us Eq. (5.39). It is left as an exercise to the reader to show that the 2×2 transfer matrix in Eq. (5.39) is unitary.

5.2.2 *Nonidentical Waveguides*

If the propagation constants of the two waveguides are not equal, then $\Delta\beta \neq 0$. In such a case, we use Eqs. (5.25) and (5.26). Differentiating Eq. (5.25) with respect to z and eliminating $B(z)$ from the equation, we obtain the following differential equation:

$$\frac{d^2A}{dz^2} + i\Delta\beta\frac{dA}{dz} + \kappa^2 A = 0 \tag{5.40}$$

The general solution of the above equation is

$$A(z) = \left(A_1 e^{i\Gamma z} + A_2 e^{-i\Gamma z}\right) e^{-i\delta z} \tag{5.41}$$

where

$$\delta = \frac{\Delta\beta}{2} = \frac{(\beta_a - \beta_b)}{2} \tag{5.42a}$$

$$\Gamma = \sqrt{\left(\kappa^2 + \frac{(\Delta\beta)^2}{4}\right)} = \sqrt{(\kappa^2 + \delta^2)} \tag{5.42b}$$

Using Eqs. (5.41) and (5.25), we obtain

$$B(z) = \frac{1}{\kappa}\left(\{\Gamma - \delta\}\, A_1 e^{i\Gamma z} - \{\Gamma + \delta\} A_2 e^{-i\Gamma z}\right) e^{i\delta z} \tag{5.43}$$

Assuming as before $A(z = 0) = 1$ and $B(z = 0) = 0$; we obtain
$A_1 = \frac{\{\Gamma+\delta\}}{2\Gamma}$; $A_2 = \frac{\{\Gamma-\delta\}}{2\Gamma}$
Thus,

$$A(z) = \left(\cos\Gamma z + i\frac{\delta}{\Gamma}\sin\Gamma z\right)e^{-i\delta z} \tag{5.44}$$

and

$$B(z) = i\frac{\kappa}{\Gamma}\sin\Gamma z e^{i\delta z} \tag{5.45}$$

The power carried by each individual waveguide is now given by

$$P_a(z) = |A(z)|^2 = \cos^2\Gamma z + \frac{\delta^2}{\Gamma^2}\sin^2\Gamma z \tag{5.46}$$

$$P_b(z) = |B(z)|^2 = \frac{\kappa^2}{\Gamma^2}\sin^2\Gamma z \tag{5.47}$$

Note that $P_a(z) + P_b(z) = 1$ showing conservation of power.

For $\delta = 0$, that is, $\Delta\beta = 0$ these equations give the same results as before for waveguides having identical propagation constants.

Figure 5.3 shows the variation of power in the coupled waveguide for $\Delta\beta = 0$, $\Delta\beta = \kappa$, $\Delta\beta = 2\kappa$, and $\Delta\beta = 10\kappa$ showing that as the value of $\frac{\Delta\beta}{\kappa}$ increases the maximum power coupled decreases. It is interesting to notice that if $\Delta\beta \neq 0$ complete power transfer from one waveguide to another is not possible and the maximum power that can be transferred from one waveguide to another is given by

$$P_{b,max} = \frac{\kappa^2}{\Gamma^2} < 1 \tag{5.48}$$

This happens at a distance L_c given by

$$L_c = \frac{\pi}{2\Gamma} = \frac{\pi}{2\sqrt{(\kappa^2 + \delta^2)}} < \frac{\pi}{2\kappa} \tag{5.49}$$

which is less than the coupling length with $\Delta\beta = 0$ (see Fig. 5.3).

Also, the maximum power transferred to the second waveguide will be very small if $\delta \gg \kappa$ or $\Delta\beta \gg 2\kappa$.

Thus, even if two waveguides are lying close to each other which implies a large coupling coefficient κ, if $\Delta\beta$ between the waveguides is sufficiently large and satisfies the condition mentioned above, then not much energy exchange will take place between the two waveguides.

Exercise 5.2 Consider a directional coupler consisting of two nonidentical waveguides having a coupling coefficient of 0.157 mm^{-1}. What is the required $\Delta\beta$ so that the maximum power transferred between the waveguides is 50%? What is the length of the coupling region where this will happen?

Solution The maximum power transfer between the two waveguides is given by (see Eq. 5.47)

$$P_{b,max} = \frac{\kappa^2}{\Gamma^2} = \frac{1}{1 + \frac{(\Delta\beta)^2}{4\kappa^2}}$$

For $P_{b,\ max} = 0.5$, we have $\Delta\beta = 2\kappa$. Given that $\kappa = 0.157$ mm^{-1}, the required $\Delta\beta$ for the maximum power transfer between the two waveguides will be 0.314 mm^{-1}. Note that for typical waveguides $n_{eff} \approx 2.2$ for a wavelength of 800 nm, the propagation constant is $\beta = \frac{2\pi}{\lambda_0} n_{eff} \approx 17 \times 10^3$ mm^{-1}. Hence the effective index difference required is very small compared to the effective index value itself. Note that to achieve $\Delta\beta = 0.314$ mm^{-1} at a wavelength of 800 nm, the required $\Delta n_{eff} \approx 8 \times 10^{-5}$ which is a very small value.

We can put the coupled mode equations in a slightly different form. For this, we first note that the total electric field in the coupler can be written as

$$\begin{aligned} E(x,y,z) = A(z)E_a(x,y)e^{i\beta_a z} + B(z)E_b(x,y)e^{i\beta_b z} = \\ [R(z)E_a(x,y) + S(z)E_b(x,y)]e^{+i\beta_{av} z} \end{aligned} \tag{5.50}$$

where

$$\beta_{av} = \frac{\beta_a + \beta_b}{2} \tag{5.51}$$

and

$$R(z) = A(z)e^{+i\delta z} \tag{5.52}$$

$$S(z) = B(z)e^{-i\delta z} \tag{5.53}$$

Thus,

$$\frac{dA}{dz} = \left(\frac{dR}{dz} - i\delta R\right)e^{-i\delta z} = i\kappa S e^{+i\delta z} e^{-2i\delta z}$$

and we have

$$\frac{dR}{dz} = i\delta R + i\kappa S \tag{5.54}$$

Similarly, we have

$$\frac{dS}{dz} = -i\delta S + i\kappa R \tag{5.55}$$

Equations (5.54) and (5.55) are the same coupled mode equations in a slightly different and useful form. Also, note that if $\beta_a = \beta_b$, then $\delta = 0$ and $R(z) = A(z)$ and $S(z) = B(z)$.

These equations can be solved (see Exercise 5.3) to obtain the following relation connecting R and S at $z = 0$ and at z written in a matrix form:

$$\begin{pmatrix} R(z) \\ S(z) \end{pmatrix} = \begin{pmatrix} P & iQ \\ iQ & P^* \end{pmatrix} \begin{pmatrix} R(z=0) \\ S(z=0) \end{pmatrix} \tag{5.56}$$

where

$$P = \cos\Gamma z + i\frac{(\beta_a - \beta_b)}{2\Gamma}\sin\Gamma z = \cos\Gamma z + i\frac{\delta}{\Gamma}\sin\Gamma z \tag{5.57}$$

$$Q = \frac{\kappa}{\Gamma}\sin\Gamma z \tag{5.58}$$

and $R(z = 0)$ and $S(z = 0)$ are the amplitudes at the input $z = 0$. We can represent the directional coupler by the following transfer matrix relating the output and input of the directional coupler:

$$T = \begin{pmatrix} P & iQ \\ iQ & P^* \end{pmatrix} \tag{5.59}$$

Exercise 5.3 Obtain the solution given in Eqs. (5.57) and (5.58).

Solution Combining Eqs. (5.54) and (5.55), we obtain

$$\frac{d^2R}{dz^2} = -\Gamma^2 R$$

whose general solution is

$$R(z) = a\cos\Gamma z + b\sin\Gamma z$$

Using this solution in Eq. (5.54) and simplifying, we obtain

$$S(z) = \left(i\frac{\Gamma}{\kappa}a - \frac{\delta}{\kappa}b\right)\sin\Gamma z - \left(i\frac{\Gamma}{\kappa}b + \frac{\delta}{\kappa}a\right)\cos\Gamma z$$

Using the initial conditions at $z = 0$, we obtain

$$a = R(z=0); b = i\frac{\kappa}{\Gamma}S(z=0) + i\frac{\delta}{\Gamma}R(z=0)$$

Using these, we obtain Eqs. (5.57) and (5.58).

Exercise 5.4 Show that $T^{\dagger}T = I$ where $T^{\dagger}$ is the adjoint of the matrix T given in Eq. (5.59). What is the physical significance of this condition?

Solution Now

$$T^{\dagger} = \begin{pmatrix} P^* & -iQ^* \\ -iQ^* & P \end{pmatrix}$$

Thus, (since $Q^* = Q$)

$$T^{\dagger}T = \begin{pmatrix} P^* & -iQ \\ -iQ & P \end{pmatrix}\begin{pmatrix} P & iQ \\ iQ & P^* \end{pmatrix} = \begin{pmatrix} |P|^2 + |Q|^2 & 0 \\ 0 & |P|^2 + |Q|^2 \end{pmatrix} = \begin{pmatrix} 1 & 0 \\ 0 & 1 \end{pmatrix} = I$$

Such a matrix is referred to as a unitary matrix and represents energy conservation.

If the directional coupler has $\Delta\beta = 0$, then $\delta = 0$ and $\Gamma = \kappa$, $R(z) = A(z)$ and $S(z) = B(z)$. Thus, Eq. (5.56) for this case becomes

$$\begin{pmatrix} R(z) \\ S(z) \end{pmatrix} = \begin{pmatrix} \cos\kappa z & i\sin\kappa z \\ i\sin\kappa z & \cos\kappa z \end{pmatrix}\begin{pmatrix} R(z=0) \\ S(z=0) \end{pmatrix} \tag{5.60}$$

Exercise 5.5 Equal power is incident on both the input waveguides of a directional coupler with identical waveguides. For what phase difference between the two inputs will the individual powers in the two waveguides remain constant.

Solution Since the powers of the waves incident on the two waveguides are equal, we assume

$$R(z=0)=R_0e^{i\phi_R};S(z=0)=R_0e^{i\phi_S}$$

where ϕ_R and ϕ_S are the phases of the incident waves in the two waveguides. The power in waveguide a is given by

$$P_a=|R(z)|^2=\left|R_0e^{i\phi_R}\cos\kappa z+iR_0e^{i\phi_S}\sin\kappa z\right|^2$$

which gives us

$$P_a=R_0^2(1+\sin 2\kappa z\sin(\phi_R-\phi_S))$$

Hence if the power in the waveguide is to remain independent of z, then we must have

$$\phi_R-\phi_S=m\pi$$

That is, the input amplitudes must be in phase or out of phase by π;the former corresponds to the symmetric normal mode of the coupler and the latter to the antisymmetric normal mode of the coupler (see Sect. 5.2.3).

Exercise 5.6 If the inputs to the directional coupler with $\Delta\beta=0$ are given by $R(z=0)=\tilde{R}_0e^{i\phi_R}$ and $S(z=0)=\tilde{S}_0e^{i\phi_S}$, obtain the power outputs of the directional coupler.

Solution From Eq. (5.56), we have

$$R(z)=\tilde{R}_0e^{i\phi_R}\cos\kappa z+i\tilde{S}_0e^{i\phi_S}\sin\kappa z$$

$$S(z)=i\tilde{R}_0e^{i\phi_R}\sin\kappa z+\tilde{S}_0e^{i\phi_S}\cos\kappa z$$

Thus, the powers carried by the two waveguides will be given by

$$P_a=|R(z)|^2=\tilde{R}_0^2\cos^2\kappa z+\tilde{S}_0^2\sin^2\kappa z-i\tilde{R}_0\tilde{S}_0e^{i(\phi_R-\phi_S)}\cos\kappa z\sin\kappa z$$
$$+i\tilde{R}_0\tilde{S}_0e^{-i(\phi_R-\phi_S)}\cos\kappa z\sin\kappa z$$

which can be simplified to

$$P_a = \tilde{R}_0^2 \cos^2 \kappa z + \tilde{S}_0^2 \sin^2 \kappa z + \tilde{R}_0 \tilde{S}_0 \sin 2\kappa z \sin(\phi_R - \phi_S)$$

Similarly, we have

$$P_b = |S(z)|^2 = \tilde{R}_0^2 \sin^2 \kappa z + \tilde{S}_0^2 \cos^2 \kappa z - \tilde{R}_0 \tilde{S}_0 \sin 2\kappa z \sin(\phi_R - \phi_S)$$

Note that

$$P_a + P_b = \tilde{R}_0^2 + \tilde{S}_0^2$$

independent of z, showing power conservation.

As a special case if we consider $\tilde{R}_0 = \tilde{S}_0$, then

$$P_a = \tilde{R}_0^2 + \tilde{R}_0^2 \sin 2\kappa z \sin(\phi_R - \phi_S) = \tilde{R}_0^2 [1 + \sin 2\kappa z \sin(\phi_R - \phi_S)]$$

Note that even with equal input amplitudes in the two waveguides, the outputs depend on the phase difference between the two inputs.

5.2.3 *Normal Mode Picture*

The above discussion was based on treating the directional coupler problem as an interaction between two adjacent waveguides through evanescent coupling. There is another elegant picture of the coupled waveguide system in terms of normal modes of the system. In this picture, we consider the pair of coupled waveguides as a single waveguide and obtain the normal modes of propagation of the waveguide.

Although obtaining the normal modes of any waveguide requires us to solve Maxwell's equations under appropriate boundary conditions as we did in Chap. 4, in most cases with two or more interacting waveguides, it may not be possible to obtain analytical solutions. In such cases, obtaining normal modes via coupled mode theory may be advantageous as it provides us with an approximate analytical formulation to obtain the normal modes of the coupled structure.

Referring to Eq. (5.1) we had written the following expression for the electric field of a directional coupler:

$$E(x, y, z) = A(z) E_a(x, y) e^{i\beta_a z} + B(z) E_b(x, y) e^{i\beta_b z} \tag{5.61}$$

Earlier we had introduced two new amplitudes $R(z)$ and $S(z)$ (Eqs. 5.52 and 5.53). Equation (5.61) can be rewritten in terms of amplitudes $R(z)$ and $S(z)$ as

$$E(x, y, z) = [R(z) E_a(x, y) + S(z) E_b(x, y)] e^{i\beta_{av} z} \tag{5.62}$$

The coupled equations satisfied by $R(z)$ and $S(z)$ are given by Eqs. (5.54) and (5.55).

In order to obtain the normal modes of the coupled waveguide structure, we postulate solutions of the form

$$R(z) = R_0 e^{i\gamma z} \tag{5.63}$$

and

$$S(z) = S_0 e^{i\gamma z} \tag{5.64}$$

with R_0 and S_0 being independent of z and γ is a constant to be determined. In such a case, the electric field of the coupler evolves according to the following equation:

$$E(x, y, z) = [R_0 E_a(x, y) + S_0 E_b(x, y)] e^{i(\beta_{av}+\gamma)z} \tag{5.65}$$

We now see that assuming solutions of the form given by Eqs. (5.63) and (5.64), the electric field in the directional coupler changes only in phase as it propagates along z; the transverse field distribution represented by $[R_0E_a(x, y) + S_0E_b(x, y)]$ does not change with propagation and hence represents a normal mode of the coupled waveguide structure. The mode has a propagation constant given by $(\beta_{av} + \gamma)$ and is composed of a linear superposition of modes of waveguides a and b with amplitudes R_0 and S_0, respectively.

Substituting the solutions for $R(z)$ and $S(z)$ (Eqs. 5.63 and 5.64) in the coupled amplitude equations (Eqs. 5.54 and 5.55), we obtain

$$i\gamma R_0 = i\delta R_0 + i\kappa S_0 \tag{5.66}$$

$$i\gamma S_0 = -i\delta S_0 + i\kappa R_0 \tag{5.67}$$

From the above two equations, we obtain

$$\frac{R_0}{S_0} = \frac{\kappa}{\gamma - \delta} = \frac{\gamma + \delta}{\kappa} \tag{5.68}$$

which gives us a quadratic equation in γ and gives us two solutions

$$\gamma_\pm = \pm\sqrt{(\kappa^2 + \delta^2)} \tag{5.69}$$

These two values correspond to the two normal modes of the coupled waveguide structure. The propagation constants of the two normal modes are thus (see Eq. 5.65)

$$\beta_+ = \beta_{av} + \gamma_+ = \frac{(\beta_a + \beta_b)}{2} + \left[\kappa^2 + \frac{(\Delta\beta)^2}{4}\right]^{1/2} \tag{5.70}$$

and

$$\beta_- = \beta_{av} + \gamma_- = \frac{(\beta_a + \beta_b)}{2} - \left[\kappa^2 + \frac{(\Delta\beta)^2}{4}\right]^{1/2} \tag{5.71}$$

For these two normal modes, we can evaluate the relative contributions of the modes of individual waveguides a and b as follows (see Eq. 5.68):

(a) For the β_+ normal mode:

$$\frac{S_0}{R_0} = \frac{\gamma_+ - \delta}{\kappa} = \frac{1}{\kappa}\left(\sqrt{\delta^2 + \kappa^2} - \delta\right) \tag{5.72}$$

(b) For the β_- normal mode:

$$\frac{S_0}{R_0} = \frac{\gamma_- - \delta}{\kappa} = -\frac{1}{\kappa}\left(\sqrt{\delta^2 + \kappa^2} + \delta\right) \tag{5.73}$$

We will now discuss some specific cases.

A. *Phase matched waveguides:* $\beta_a = \beta_b = \beta_0$:

In this case $\beta_{av} = \beta_0$ and $\delta = 0$. Thus

$$\beta_+ = \beta_0 + \kappa \tag{5.74}$$

and

$$\beta_- = \beta_0 - \kappa \tag{5.75}$$

The relative amplitudes of the individual modes to the two normal modes are
For the β_+ normal mode: $\frac{S_0}{R_0} = +1$
For the β_- normal mode: $\frac{S_0}{R_0} = -1$

From the above, we may see that for the normal mode with $\beta_+ = \beta_0 + \kappa$, the individual modes of waveguides a and b contribute equally and are in phase. For the normal mode with $\beta_- = \beta_0 - \kappa$, the individual modes of waveguides a and b contribute equally and are out of phase by π.

Figure 5.4 shows the two normal modes of the coupled waveguide structure. As expected, the normal mode having a higher propagation constant has no zero

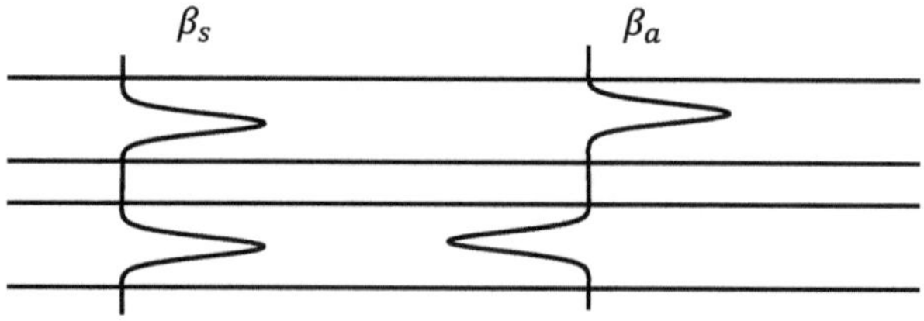

Fig. 5.4 The symmetric and antisymmetric normal modes of a directional coupler consisting of two identical waveguides

crossing and is the fundamental mode of the coupler while the normal mode with the lower propagation constant has a single zero crossing and is the first excited normal mode of the directional coupler.

If the two waveguides are identical, then the β_+ mode would correspond to a symmetric mode and the β_- mode would correspond to an antisymmetric mode.

It may be noted that we can satisfy the condition $\beta_a = \beta_b$ even with two nonidentical waveguides. Of course, this condition will be satisfied only at a particular wavelength. Deviation from this specific phase-matching wavelength will lead to $\beta_a \neq \beta_b$. On the other hand, if the two waveguides are identical, then they will have the same propagation constant irrespective of the operating wavelength. Remember that the coupling coefficient will vary with wavelength and thus, for a given length of the directional coupler, the power splitting ratio between the two waveguides will depend on the wavelength.

5.2.4 Explanation of Coupling of Power Among the Two Waveguides in Terms of the Normal Modes

Let us assume that at $z = 0$ the mode of waveguide a is excited. Thus,

$$E(x, y, z=0) = E_a(x, y) \tag{5.76}$$

We can express the input in terms of the normal modes of the coupler through the following equation:

$$\begin{aligned} E(x, y, z=0) = E_a(x, y) = \frac{1}{2}[E_a(x, y) + E_b(x, y)] + \\ \frac{1}{2}[E_a(x, y) - E_b(x, y)] \end{aligned} \tag{5.77}$$

The first bracketed term on the RHS corresponds to the β_+ normal mode with $S_0 = R_0$ while the second bracketed term on the RHS corresponds to the β_- normal mode with $S_0 = -R_0$. Thus, the input field at $z = 0$ excites both the normal modes with equal amplitudes.

Now each of the two normal modes propagates independently and with propagation constants β_+ and β_-. Hence the transverse field at z will be given by

$$E(x, y, z) = \frac{1}{2}[E_a(x, y) + E_b(x, y)]e^{i(\beta_0+\kappa)z} + \frac{1}{2}[E_a(x, y) - E_b(x, y)]e^{i(\beta_0-\kappa)z}$$

which can be simplified to

$$E(x, y, z) = (E_a \cos \kappa z + iE_b \sin \kappa z)e^{i\beta_0 z} \tag{5.78}$$

This implies that at any value of z, the amplitude of the field in waveguide a will be $\cos\kappa z$ and that in waveguide b will be $i \sin \kappa z$ which is the same as discussed earlier (see Eqs. 5.34 and 5.35).

Note that in the normal mode picture, there is no coupling among the normal modes; it is the interference between the two normal modes that leads to exchange of power among the two waveguides. In contrast, in the coupled mode picture, it is the coupling among the modes of the individual waveguides that leads to exchange of power among the modes. Both these pictures are equivalent.

Note the presence of the factor i in the second term. This represents a $\pi/2$ phase difference between the fields in the two waveguides and is a consequence of energy conservation (see also Chap. 19 on beam splitters).

Now, at $z = \pi/4\kappa$, we have from Eq. (5.78)

$$E(x,y,z) = \frac{1}{\sqrt{2}}(E_a + iE_b)e^{i\beta_0 z} \tag{5.79}$$

which shows equal powers in both waveguides a and b but with a phase difference of $\pi/2$. At this point, the directional coupler behaves as a 3 dB coupler with equal power exiting both the waveguides.

An interesting way of describing the coupler is through a column vector with two rows; row 1 gives the value of R_0 (which is the projection of the normal mode in waveguide a) and row 2 is the value of S_0 (which is the projection of the normal mode in waveguide b). Thus, the total field is represented by the following column vector

$$\begin{pmatrix} R_0 \\ S_0 \end{pmatrix}$$

In such a representation, the two normal modes would be given by

$$\beta_+ \text{ normal mode}: \begin{pmatrix} R_0 \\ S_0 \end{pmatrix} = \frac{1}{\sqrt{2}}\begin{pmatrix} 1 \\ 1 \end{pmatrix} \tag{5.80}$$

$$\beta_- \text{ normal mode}: \begin{pmatrix} R_0 \\ S_0 \end{pmatrix} = \frac{1}{\sqrt{2}}\begin{pmatrix} 1 \\ -1 \end{pmatrix} \tag{5.81}$$

where the $1/\sqrt{2}$ factors ensure that the modes are normalized.

In this notation input with unit amplitude into waveguide a will correspond to the following vector:

$$\begin{pmatrix} 1 \\ 0 \end{pmatrix}$$

which can be represented in terms of the superposition of the two normal modes as

$$\begin{pmatrix}1\\0\end{pmatrix}=\frac{1}{\sqrt{2}}\left\{\frac{1}{\sqrt{2}}\begin{pmatrix}1\\1\end{pmatrix}\right\}+\frac{1}{\sqrt{2}}\left\{\frac{1}{\sqrt{2}}\begin{pmatrix}1\\-1\end{pmatrix}\right\} \tag{5.82}$$

Thus, showing that the input consists of equal amplitudes of the two normal modes which are in phase.

B. *Nonphase-Matched Directional Coupler*

We now consider the case of nonphase-matched directional coupler. In this case $\Delta\beta \neq 0$ and we have

$$\frac{S_0}{R_0}=-\frac{\Delta\beta}{2\kappa}\pm\left(1+\frac{(\Delta\beta)^2}{4\kappa^2}\right)^{1/2} \tag{5.83}$$

Thus, for the normal mode with propagation constant β_+ (Eq. 5.70), we have

$$\begin{pmatrix}R_0\\S_0\end{pmatrix}=N_+\begin{pmatrix}1\\-\frac{\Delta\beta}{2\kappa}+\left(1+\frac{(\Delta\beta)^2}{4\kappa^2}\right)^{1/2}\end{pmatrix} \tag{5.84}$$

with

$$N_+=\left[1+\left\{-\frac{\Delta\beta}{2\kappa}+\left(1+\frac{(\Delta\beta)^2}{4\kappa^2}\right)^{1/2}\right\}^2\right]^{-1/2} \tag{5.85}$$

representing the normalization constant. Note that in this case the amplitudes of modes in waveguides a and b are in phase but have unequal amplitudes. If $\beta_a > \beta_b$ then $\Delta\beta=(\beta_a-\beta_b)>0$ and the amplitude of mode in waveguide b will be less than in waveguide a.

Similarly for the normalized mode with propagation constant β_-, we have

$$\begin{pmatrix}R_0\\S_0\end{pmatrix}=N_-\begin{pmatrix}1\\-\frac{\Delta\beta}{2\kappa}-\left(1+\frac{(\Delta\beta)^2}{4\kappa^2}\right)^{1/2}\end{pmatrix} \tag{5.86}$$

with

$$N_-=\left[1+\left\{-\frac{\Delta\beta}{2\kappa}-\left(1+\frac{(\Delta\beta)^2}{4\kappa^2}\right)^{1/2}\right\}^2\right]^{-1/2} \tag{5.87}$$

Fig. 5.5 Normal modes of a directional coupler consisting of two nonidentical waveguides shown in the figure as two waveguides having different widths

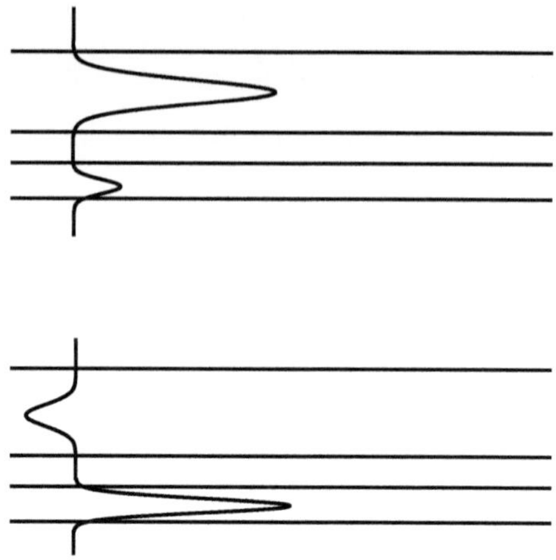

Note that in this case, the amplitudes of modes in waveguides a and b are π out of phase and for $\Delta\beta > 0$, the amplitude of the field in waveguide b will be larger than that in waveguide a.

Figure 5.5 shows a schematic of the fundamental and first excited normal modes of a directional coupler consisting of two nonidentical waveguides shown in the figure with waveguides having different widths.

C. *Highly Nonphase-Matched Waveguides*

Let us consider two highly nonidentical waveguides with $\Delta\beta \gg \kappa$ and assume for specificity that $\beta_a > \beta_b$.

Now for the β_+ normal mode, we have

$$\frac{S_0}{R_0} = -\frac{\Delta\beta}{2\kappa} + \left(1 + \frac{(\Delta\beta)^2}{4\kappa^2}\right)^{1/2} = -\frac{\Delta\beta}{2\kappa} + \frac{\Delta\beta}{2\kappa}\left(1 + \frac{4\kappa^2}{(\Delta\beta)^2}\right)^{1/2} \approx -\frac{\Delta\beta}{2\kappa} + \frac{\Delta\beta}{2\kappa}$$

$$\times \left(1 + \frac{4\kappa^2}{2(\Delta\beta)^2}\right) = \frac{\kappa}{\Delta\beta} \ll 1$$

Thus, for the β_+ normal mode, we have

$$\begin{pmatrix} R_0 \\ S_0 \end{pmatrix} = N_+ \begin{pmatrix} 1 \\ \kappa/\Delta\beta \end{pmatrix} \tag{5.88}$$

with

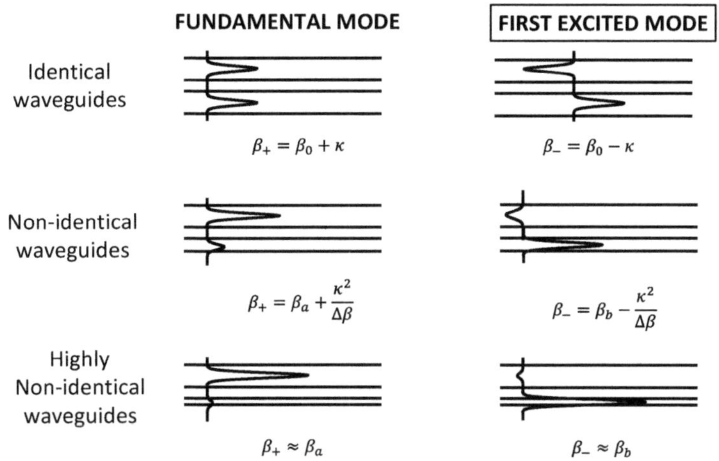

Fig. 5.6 Evolution of the normal modes of a directional coupler as the ratio $\frac{\Delta\beta}{\kappa}$ increases from 0 to $\Delta\beta \gg \kappa$

$$N_{+} = \left(1 + \frac{\kappa^2}{(\Delta\beta)^2}\right)^{-1/2} \tag{5.89}$$

This implies that for this mode, since $S_0 \ll R_0$ the major contribution to the normal mode is from waveguide a. The corresponding field distribution of the normal mode is given in Fig. 5.6.

The corresponding propagation constant is

$$\begin{aligned}\beta_{+} &= \frac{(\beta_a + \beta_b)}{2} + \left[\kappa^2 + \frac{(\Delta\beta)^2}{4}\right]^{1/2} = \frac{(\beta_a + \beta_b)}{2} + \frac{\Delta\beta}{2}\left(1 + \frac{4\kappa^2}{(\Delta\beta)^2}\right)^{1/2} \\ &\approx \beta_a + \frac{\kappa^2}{\Delta\beta}\end{aligned} \tag{5.90}$$

which is almost equal to β_a the propagation constant of waveguide a. Note that waveguide a has the higher propagation constant among the two waveguides.

Similarly for the β_{-} normal mode, we have

$$\frac{S_0}{R_0} = -\frac{\Delta\beta}{2\kappa} - \left(1 + \frac{(\Delta\beta)^2}{4\kappa^2}\right)^{1/2} = -\frac{\Delta\beta}{2\kappa} - \frac{\Delta\beta}{2\kappa}\left(1 + \frac{4\kappa^2}{(\Delta\beta)^2}\right)^{1/2} \approx -\frac{\Delta\beta}{2\kappa} - \frac{\Delta\beta}{2\kappa}$$
$$\times \left(1 + \frac{4\kappa^2}{2(\Delta\beta)^2}\right) = -\frac{\Delta\beta}{\kappa} \gg 1$$

and we may write for the normal mode as

$$\begin{pmatrix} R_0 \\ S_0 \end{pmatrix} = N_- \begin{pmatrix} 1 \\ -\Delta\beta/\kappa \end{pmatrix} = N_+ \begin{pmatrix} \kappa/\Delta\beta \\ -1 \end{pmatrix} \tag{5.91}$$

Thus, in this normal mode, the major contribution comes from waveguide b and the fields in the two waveguides are out of phase by π. The corresponding field distribution is shown in Fig. 5.6.

The corresponding propagation constant is

$$\beta_- = \frac{(\beta_a + \beta_b)}{2} - \left[\kappa^2 + \frac{(\Delta\beta)^2}{4}\right]^{1/2} \approx \beta_b - \frac{\kappa^2}{\Delta\beta} \tag{5.92}$$

which is almost equal to β_b the propagation constant of waveguide b. Note that waveguide b has the lower value of propagation constant among the two waveguides.

Since the value of the coupling coefficient κ depends on the separation between the waveguides, for a given pair of nonphase-matched waveguides ($\Delta\beta \neq 0$), as the separation between the two waveguides increases, κ decreases and for large enough separation the condition $\Delta\beta \gg \kappa$ gets well satisfied and the normal modes evolve into almost individual waveguide modes. Note that if the waveguides are phase matched, the normal modes of the coupler do not change as the separation increases.

5.3 Directional Coupler Switch

One of the very important applications of a directional coupler is in the realization of a directional coupler electro-optic switch (see Chap. 6 for the electro-optic effect). To understand this, we first note that if we have a directional coupler consisting of identical waveguides, then it is possible to couple the entire light incident on one waveguide to the other provided the length of the coupler is equal to the coupling length $L_c = \frac{\pi}{2\kappa}$ (see Eq. 5.38). Thus, if a directional coupler is fabricated with this length, then the light incident in waveguide a exits waveguide b at the output. Such a state is referred to as the cross-state of the optical switch.

For this length of the directional coupler, to achieve the parallel state in which the light exits the same waveguide as at the input, we need to switch the light at the output from waveguide b back to waveguide a keeping the same input. This is achievable if we are able to generate a $\Delta\beta$ between the two waveguides satisfying a condition such that $P_b(z = L_c) = 0$, that is, (see Eq. 5.47)

$$\Gamma L_c = \pi \tag{5.93}$$

Substituting the value of Γ from Eq. (5.42b), this implies that

$$\sqrt{\left(\kappa^2 + \frac{(\Delta\beta)^2}{4}\right)} L_c = \pi \tag{5.94}$$

Using the fact that $L_c = \frac{\pi}{2\kappa}$ this implies we need a $\Delta\beta$ given by

$$\Delta\beta = \pm\frac{\sqrt{3}\pi}{L_c} \tag{5.95}$$

For switching, the waveguides need to be made nonidentical with the value of $\Delta\beta$ given by Eq. (5.95) which will ensure that light incident in waveguide a exits the same waveguide after propagating through a length equal to L_c, resulting in the parallel state.

Using the electro-optic effect (see Chap. 6) the required value of $\Delta\beta$ between the two waveguides can be achieved.

In order to understand this, we start with a directional coupler having a length equal to the coupling length L_c. In such a case power input in waveguide a exits from waveguide b leading to the cross-state. In order to switch the light from the cross-state to the parallel state in which the light exits from waveguide a we need to introduce a $\Delta\beta$ satisfying Eq. (5.95). This is the principle of the electro-optic switch for fast switching of optical signals from one port to another.

As an example, if we consider an optical switch with a coupling length $L_c = 5$ mm then the required Δβ would be about 1.09 mm^{-1}. The induced Δβ depends on the applied voltage through the following relation (see Chap. 6 on electro-optic effect):

$$\Delta\beta = \frac{2\pi}{\lambda_0}\Delta n_{eff} = \frac{2\pi}{\lambda_0} \times \frac{n^3 r}{2}\frac{V}{G}\Gamma \tag{5.96}$$

where V is the externally applied voltage, G is the electrode spacing, Γ is an overlap factor to account for the fact that the applied electric field is not uniform across the modal distribution, n is the effective index of the mode and r is the effective electro-optic coefficient which depends on the polarization state of the mode and the direction of the electric field vis a vis the orientation of the crystal axes of the waveguide. For a discussion on electro-optic effect in waveguides, refer to Chap. 6.

In order to reduce the required drive voltage for switching, the electrode pattern is such that the applied electric field increases the effective index of one waveguide and decreases that of the other waveguide by the same amount (i.e., uses the push-pull effect) which effectively halves the required voltage for switching.

As a typical example, we have

$$n = 2.2, r = 30 \times 10^{-12} \frac{\text{m}}{\text{V}}; \quad G = 10\ \mu\text{m}; \quad \Gamma = 0.5; \quad \lambda_0 = 800\ \text{nm}$$

Thus, the voltage required to achieve $\Delta\beta = 1.09\ \text{mm}^{-1}$ using the push-pull effect is ≈8.7 V which is indeed a very reasonable value.

The directional coupler switch is used extensively in optical communication networks and finds applications also in integrated quantum photonics for switching between entangled states or to achieve other functionalities in quantum optical signal processing in integrated quantum photonics.

5.3.1 *Switched Δβ Coupler*

We must note that although we can adjust $\Delta\beta$ and achieve the parallel state (light exiting the waveguide in which light was launched at the input) of the switch, the cross-state (light exiting the other waveguide) requires us to have the length of the directional coupler to be exactly equal to the coupling length L_c and no adjustment is possible once the switch is fabricated. Thus, any deviation of the length of the coupler from the coupling length during fabrication cannot be adjusted post-fabrication.

This problem can be solved by using the concept of a switched $\Delta\beta$ directional coupler in which both the parallel and cross-states can be achieved by appropriately choosing $\Delta\beta$ using the electro-optic effect (Kogelnik and Schmidt 1976).

To understand the concept of switched $\Delta\beta$ directional coupler switch, we assume the coupler to have a length L and assume it to be divided into two sections with the first half section having a positive value $+\Delta\beta$, i.e, $\beta_a > \beta_b$ and the second half having the same magnitude of $\Delta\beta$ but negative, that is, $\beta_a < \beta_b$. The coupler has switched $\Delta\beta$ between the two halves of the coupler.

We will now present the mathematical analysis to understand the operation of the switched $\Delta\beta$ directional coupler switch.

Now, the amplitudes of the two modes after the first half of the coupler would be (see Eq. 5.56)

$$\begin{pmatrix} R\left(\frac{L}{2}\right) \\ S\left(\frac{L}{2}\right) \end{pmatrix} = \begin{pmatrix} P & iQ \\ iQ & P^* \end{pmatrix} \begin{pmatrix} R(z=0) \\ S(z=0) \end{pmatrix} \tag{5.97}$$

where (see Eqs. 5.57 and 5.58)

$$P = \cos\frac{\Gamma L}{2} + i\frac{(\beta_a - \beta_b)}{2\Gamma}\sin\frac{\Gamma L}{2} = \cos\frac{\Gamma L}{2} + i\frac{\Delta\beta}{2\Gamma}\sin\frac{\Gamma L}{2} \tag{5.98}$$

$$Q = \frac{\kappa}{\Gamma}\sin\frac{\Gamma L}{2} \tag{5.99}$$

and $\beta_a = \beta_b + \Delta\beta$.

The second half of the coupler has a reversed $\Delta\beta$ and hence $\beta_a = \beta_b - \Delta\beta$. Hence for the second half of the coupler, representing the elements of the transfer matrix by $\tilde{P}$ and $\tilde{Q}$, we have

$$\begin{pmatrix} R(L) \\ S(L) \end{pmatrix} = \begin{pmatrix} \tilde{P} & i\tilde{Q} \\ i\tilde{Q} & \tilde{P} \end{pmatrix} \begin{pmatrix} R\left(\frac{\mathrm{L}}{2}\right) \\ S\left(\frac{L}{2}\right) \end{pmatrix}$$

where

$$\tilde{P} = \cos\frac{\Gamma L}{2} + i\frac{(\beta_a - \beta_b)}{2\Gamma}\sin\frac{\Gamma L}{2} = \left(\cos\frac{\Gamma L}{2} - i\frac{\Delta\beta}{2\Gamma}\sin\frac{\Gamma L}{2}\right) = P^*$$

since in the second half of the coupler $\beta_a - \beta_b = -\Delta\beta$ and thus $\tilde{P} = P^*$. Also $\tilde{Q} = Q$. Thus,

$$\begin{pmatrix} R(L) \\ S(L) \end{pmatrix} = \begin{pmatrix} P^* & iQ \\ iQ & P \end{pmatrix} \begin{pmatrix} R\left(\frac{L}{2}\right) \\ S\left(\frac{L}{2}\right) \end{pmatrix} \tag{5.100}$$

Substituting from Eqs. (5.97) in Eq. (5.100), we have

$$\begin{pmatrix} R(L) \\ S(L) \end{pmatrix} = \begin{pmatrix} |P|^2 - Q^2 & 2iP^*Q \\ 2iPQ & |P|^2 - Q^2 \end{pmatrix} \begin{pmatrix} R(z=0) \\ S(z=0) \end{pmatrix} \tag{5.101}$$

Now, with the input in waveguide a we have $R(z = 0) = 1$ and $S(z = 0) = 0$. For achieving the cross-state, we must have $R(z = L) = 0$ and $S(z = L) = 1$.

Using the above equation we have

$$R(L) = \left(|P|^2 - Q^2\right)R(z=0) + 2iP^*QS(z=0) = \left(|P|^2 - Q^2\right)$$

which gives us the condition

$$|P|^2 = Q^2 \tag{5.102}$$

This implies

$$\cos^2\frac{\Gamma L}{2}+\frac{\delta^2}{\Gamma^2}\sin^2\frac{\Gamma L}{2}=\frac{\kappa^2}{\Gamma^2}\sin^2\frac{\Gamma L}{2}$$

or

$$\frac{1}{\sqrt{1+\delta^2/\kappa^2}}\sin\left(\frac{\kappa L}{2}\sqrt{1+\delta^2/\kappa^2}\right)=\frac{1}{\sqrt{2}} \tag{5.103}$$

where $\delta^2=\frac{\Delta\beta^2}{4}=\frac{(\beta_a-\beta_b)^2}{4}$.

Equation (5.103) gives us an equation to compute the value of $\Delta\beta=2\delta$ required to achieve the cross-state for a given length L and coupling coefficient κ.

Equation (5.103) can be put in the following form:

$$\sin\left(\frac{\kappa L}{2}\sqrt{1+\frac{(\delta L)^2}{(\kappa L)^2}}\right)=\frac{1}{\sqrt{2}}\sqrt{1+\frac{(\delta L)^2}{(\kappa L)^2}} \tag{5.104}$$

Note that $\sqrt{1+\frac{(\delta L)^2}{(\kappa L)^2}}\geq 1$.

For a given κL we need to solve Eq. (5.104) to obtain the corresponding value of δL and thus the required $\Delta\beta$.

Figure 5.7 shows a contour plot of κL versus δL satisfying Eq. (5.104). It is apparent that for a range of the coupler length L it is possible to find a value of $\delta=\frac{\Delta\beta}{2}$ which satisfies Eq. (5.104) and thus makes it possible to achieve the cross-state of the switch without the need for fabricating the coupler with a precise length equal to L_c, thus easing fabrication problems.

Let us first consider the case when $\Delta\beta=0$, that is, $\delta=0$. In such a case, Eq. (5.104) gives us

$$\sin\left(\frac{\kappa L}{2}\right)=\frac{1}{\sqrt{2}} \tag{5.105}$$

which implies $\kappa L=\frac{\pi}{2}$ or $L=\frac{\pi}{2\kappa}=L_c$, as discussed earlier (see Fig. 5.7).

As another example, let us assume that the length of the coupler is $L=\sqrt{2}L_c=\frac{\pi}{\sqrt{2}\kappa}$, i.e, $\kappa L=\frac{\pi}{\sqrt{2}}$. Substituting this value in Eq. (5.103) we obtain

$$\frac{1}{\sqrt{1+\delta^2/\kappa^2}}\sin\left(\frac{\pi}{2\sqrt{2}\kappa}\sqrt{1+\delta^2/\kappa^2}\right)=\frac{1}{\sqrt{2}} \tag{5.106}$$

which has the following solution:

$$\delta=\kappa$$

which implies the value of $\Delta\beta$ required is

Fig. 5.7 Plot showing variation of κL with respect to δL. For $\frac{\pi}{2} < \kappa L < \frac{3\pi}{2}$, there is always a value of $\delta = \frac{\Delta\beta}{2}$ for which we can achieve the cross-state of the switch

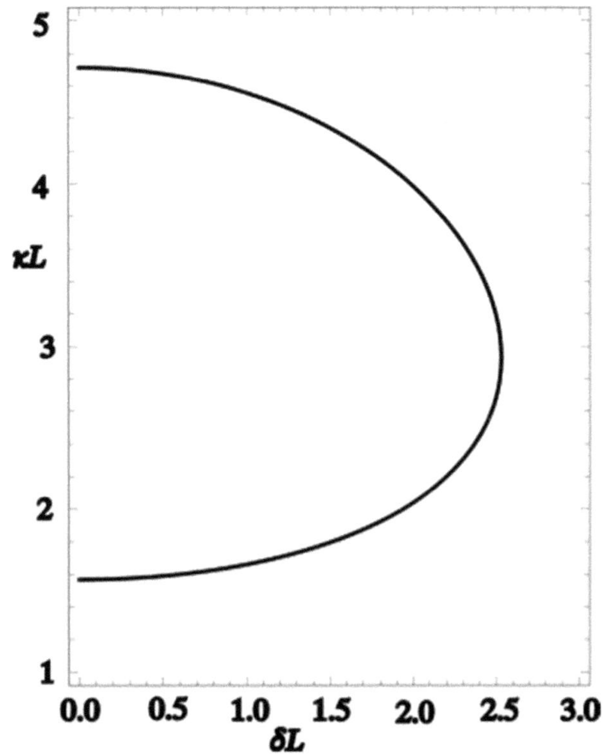

$$\Delta\beta = 2\kappa \tag{5.107}$$

This shows that it is now possible to achieve even the cross-state using a specific value of $\Delta\beta$ which can be induced using the electro optic effect.

If the total length is chosen to be equal to satisfy $L = 2L_c$ implying $\kappa L = \pi$, then the equation to be solved is

$$\frac{1}{\sqrt{1 + \delta^2/\kappa^2}} \sin\left(\frac{\pi}{2}\sqrt{1 + \delta^2/\kappa^2}\right) = \frac{1}{\sqrt{2}} \tag{5.108}$$

whose solution is given by $\delta \approx 0.799\kappa$ or $\Delta\beta \approx 1.597\kappa$.

Exercise 5.7 For the values $L = \frac{\pi}{\sqrt{2}\kappa}$ and $\Delta\beta = 2\kappa$, obtain the amplitudes of the modes in the two waveguides at the midpoint $\frac{L}{2}$ of the coupler and interpret how the coupler is in the cross-state.

Solution Using the values given we have $\Gamma = \sqrt{\delta^2 + \kappa^2} = \sqrt{2}\kappa$ giving $\frac{\Gamma L}{2} = \frac{\sqrt{2}\kappa}{2} \frac{\pi}{\sqrt{2}\kappa} = \frac{\pi}{2}$. Hence

$$P = \cos\frac{\Gamma L}{2} + i\frac{\delta}{\Gamma}\sin\frac{\Gamma L}{2} = \frac{i}{\sqrt{2}}$$

$$Q = \frac{\kappa}{\Gamma}\sin\frac{\Gamma L}{2} = \frac{1}{\sqrt{2}}$$

Since $R(z = 0) = 1$ and $S(z = 0) = 0$, we have using Eq. (5.97)

$$R\left(\frac{L}{2}\right) = \frac{i}{\sqrt{2}};\quad S\left(\frac{L}{2}\right) = \frac{i}{\sqrt{2}}$$

Hence at the midpoint of the switch, the fields in the two waveguides have equal amplitudes. This amplitude distribution has been obtained by having input in the waveguide having higher β ($\Delta\beta > 0$). At this point, the sign of $\Delta\beta$ is changed and the power will now proceed to the waveguide having larger β which happens to be the coupled waveguide. Thus, from this midpoint onwards, the power flows to the waveguide having a higher β which happens to be the coupled waveguide hence leading to the cross-state. It is left as an exercise to verify that if there was no reversal of $\Delta\beta$ in the second half of the coupler, then the coupler will be in the parallel state.

5.3.2 *Wavelength Dependence*

It was shown earlier that complete power transfer from one waveguide to another takes place when $\Delta\beta = 0$. If we consider two nonidentical waveguides, since β_a and β_b depend on wavelength, this condition can be satisfied at a particular center wavelength say $\lambda_0 = \lambda_c$. As we deviate from this wavelength $\Delta\beta$ will no more be equal to zero and thus the power coupled will reduce as given by Eq. (5.47). Choosing a length equal to the coupling length $L_c = \frac{\pi}{2\kappa}$, we have the following expression for the cross-coupled power:

$$P_b(L_c) = \frac{\kappa^2}{\Gamma^2}\sin^2\Gamma L_c = \frac{1}{1 + \frac{(\Delta\beta)^2}{4\kappa^2}}\sin^2\left(\sqrt{1 + \frac{(\Delta\beta)^2}{4\kappa^2}}\frac{\pi}{2}\right) \tag{5.109}$$

At $\Delta\beta = 0$ we have $P_b(L_c) = 1$. As the wavelength deviates from λ_c, $\Delta\beta \neq 0$ and the power coupled will reduce and when $\frac{(\Delta\beta)^2}{4\kappa^2} \approx 0.64$, $P_b(L_c) = 0.5$, i.e., when

$$\Delta\beta \approx \pm\frac{2.5}{L_c} \tag{5.110}$$

Now at any given wavelength λ_0,

$$\Delta\beta(\lambda_0) = \beta_a(\lambda_0) - \beta_b(\lambda_0) \tag{5.111}$$

with $\Delta\beta(\lambda_c) = \beta_a(\lambda_c) - \beta_b(\lambda_c) = 0$. Hence the wavelengths $\lambda_0 = \lambda_c \pm \Delta\lambda$ at which the power transfer will be 0.5 will be given by

$$\Delta\beta(\lambda_c \pm \Delta\lambda) = \beta_a(\lambda_c \pm \Delta\lambda) - \beta_b(\lambda_c \pm \Delta\lambda) \approx \pm\frac{2.5}{L_c}$$

Assuming $\Delta\lambda \ll \lambda_c$ and making Taylor series expansion, we have

$$\left(\beta_a(\lambda_c) \pm \Delta\lambda \left.\frac{d\beta_a}{d\lambda_0}\right|_{\lambda_c}\right) - \left(\beta_b(\lambda_c) \pm \Delta\lambda \left.\frac{d\beta_b}{d\lambda_0}\right|_{\lambda_c}\right) \approx \pm\frac{2.5}{L_c}$$

which on using the fact that $\beta_a(\lambda_c) - \beta_b(\lambda_c) = 0$, gives

$$\Delta\lambda \approx \pm\frac{2.5}{L_c\left(\left.\frac{d\beta_a}{d\lambda_0}\right|_{\lambda_c} - \left.\frac{d\beta_b}{d\lambda_0}\right|_{\lambda_c}\right)} \tag{5.112}$$

and the full width at half maximum will be

$$\Delta\lambda_{FWHM} \approx \frac{5.0}{L_c\left(\left.\frac{d\beta_a}{d\lambda_0}\right|_{\lambda_c} - \left.\frac{d\beta_b}{d\lambda_0}\right|_{\lambda_c}\right)} \tag{5.113}$$

Since

$$\begin{aligned}\left.\frac{d\beta_a}{d\lambda_0}\right|_{\lambda_c} &= \left.\frac{d}{d\lambda_0}\left(\frac{2\pi}{\lambda_0}n_{eff,a}(\lambda_0)\right)\right|_{\lambda_c} = -\frac{2\pi}{\lambda_c^2}\left(n_{eff,a}(\lambda_c) - \lambda_0\left.\frac{dn_{eff,a}}{d\lambda_0}\right|_{\lambda_c}\right)\\ &= -\frac{2\pi}{\lambda_c^2}N_{eff,a}(\lambda_c)\end{aligned} \tag{5.114}$$

Similarly

$$\left.\frac{d\beta_b}{d\lambda_0}\right|_{\lambda_c} = -\frac{2\pi}{\lambda_c^2}N_{eff,b}(\lambda_c) \tag{5.114b}$$

where $N_{eff,a}(\lambda_c)$ and $N_{eff,b}(\lambda_c)$ are the group effective indices of the modes of waveguides a and b, we have

$$\Delta\lambda_{FWHM} \approx \frac{0.8\lambda_c^2}{L_c\left(N_{eff,a}(\lambda_c) - N_{eff,b}(\lambda_c)\right)} \tag{5.115}$$

Exercise 5.8 Consider a directional coupler formed with two nonidentical single-mode waveguides. The effective indices of the two waveguides are described by the following formulae (λ_0 is in μm):

$$n_{ea} = 2.1381 - 0.03(\lambda_0 - 1.53)$$

$$n_{eb} = 2.1379 - 0.04(\lambda_0 - 1.53)$$

If the coupling coefficient is 0.5 cm^{-1}, obtain the center wavelength and bandwidth of the filter by choosing an appropriate length.

Solution The center wavelength will correspond to $n_{ea} = n_{eb}$ which happens when $\lambda_0 = 1.51$ μm.

Since $\kappa = 0.5\ \text{cm}^{-1}$, the length of the coupler required is $L = \frac{\pi}{2\kappa} \approx 3.14$ cm. For the given parameters, at $\lambda_0 = 1.51$ μm we have $N_{eff,a}(\lambda_c) = 2.1834$ and $N_{eff,b}(\lambda_c) = 2.1983$. Thus, the FWHM would be ≈3.9 nm.

5.4 Three Waveguide Directional Coupler

In the previous section, we considered a two-waveguide directional coupler. There are interesting features that are exhibited by couplers with more than two waveguides.

Here we consider a three-waveguide directional coupler with three identical single-mode waveguides but with unequal spacing (see Fig. 5.8). If κ_1 and κ_2 represent the coupling coefficient between waveguides 1 and 2 and between waveguides 2 and 3, respectively, then by using the same procedure as carried out for a two-waveguide coupler, we obtain the following three coupled equations describing the coupling process:

$$\frac{dS_1}{dz} = i\kappa_1 S_2 \tag{5.116}$$

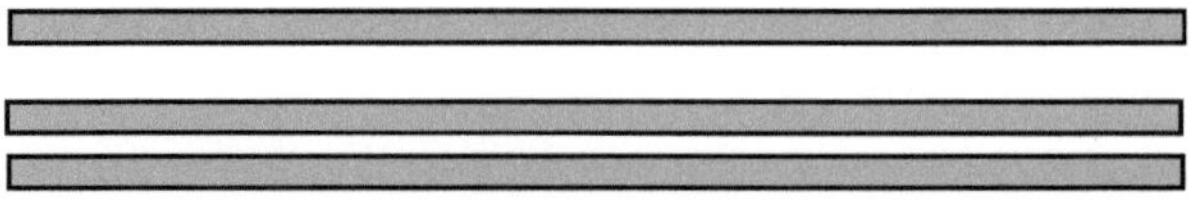

Fig. 5.8 Three-waveguide directional coupler consisting of identical waveguides with unequal spacing between the waveguides

$$\frac{dS_2}{dz} = i\kappa_1 S_1 + i\kappa_2 S_3 \tag{5.117}$$

$$\frac{dS_3}{dz} = i\kappa_2 S_2 \tag{5.118}$$

Here S_1, S_2 and S_3 represent the z-dependent amplitudes of the modes of waveguides 1, 2, and 3, respectively.

In Chap. 25, we will discuss the solution of the above set of coupled equations and relate the quantum mechanical annihilation operators of the output modes with those of the input modes. Such relations are important for analyzing the quantum behavior of such integrated quantum photonic devices.

In order to obtain the normal modes of the three-waveguide coupler following the procedure adopted for a two-waveguide coupler, we postulate the following solution to the three coupled equations:

$$S_j = S_{j0} e^{i\gamma z}; \quad j = 1, 2, 3 \tag{5.119}$$

where the constants S_{j0} represent the amplitude component of the normal mode of the coupler in the three individual waveguides. Substituting in Eqs. (5.116) to (5.118), we obtain the following set of equations:

$$i\gamma S_{10} = i\kappa_1 S_{20} \tag{5.120}$$

$$i\gamma S_{20} = i\kappa_1 S_{10} + i\kappa_2 S_{30} \tag{5.121}$$

$$i\gamma S_{30} = i\kappa_2 S_{20} \tag{5.122}$$

The above three equations can be written in matrix form as

$$\begin{pmatrix} \gamma & -\kappa_1 & 0 \\ \kappa_1 & -\gamma & \kappa_2 \\ 0 & \kappa_2 & -\gamma \end{pmatrix} \begin{pmatrix} S_{10} \\ S_{20} \\ S_{30} \end{pmatrix} = 0 \tag{5.123}$$

which has three solutions:

1. $\gamma = +\sqrt{\kappa_1^2 + \kappa_2^2}$; $S_{20} = \frac{\sqrt{\kappa_1^2+\kappa_2^2}}{\kappa_1} S_{10}$; and $S_{30} = \frac{\kappa_2}{\kappa_1} S_{10}$
2. $\gamma = 0$, $S_{20} = 0$ and $S_{30} = -\frac{\kappa_1}{\kappa_2} S_{10}$
3. $\gamma = -\sqrt{\kappa_1^2 + \kappa_2^2}$; $S_{20} = -\frac{\sqrt{\kappa_1^2+\kappa_2^2}}{\kappa_1} S_{10}$; $S_{30} = \frac{\kappa_2}{\kappa_1} S_{10}$

The corresponding normal mode propagation constants and fields are

1. Mode corresponding to $\gamma = +\sqrt{\kappa_1^2 + \kappa_2^2}$ with propagation constant

$$\beta_{m1} = \beta_0 + \sqrt{\kappa_1^2 + \kappa_2^2} \tag{5.124}$$

this corresponds to the fundamental mode of the three-waveguide coupler with the largest propagation constant among the three normal modes. In vector form with the elements being the component of the normal mode in each of the three waveguides, the normal mode can be written as

$$\begin{pmatrix} S_{10} \\ S_{20} \\ S_{30} \end{pmatrix} = N_{m1} \begin{pmatrix} \kappa_1 \\ \sqrt{\kappa_1^2 + \kappa_2^2} \\ \kappa_2 \end{pmatrix} \tag{5.125}$$

where

$$N_{m1} = \frac{1}{\sqrt{2}} \left(\kappa_1^2 + \kappa_2^2 \right)^{-1/2} \tag{5.126}$$

is the normalization constant. Note that the three amplitudes in the three waveguides are in phase and thus this mode has no zero crossing in its transverse field distribution.

For a three-waveguide coupler with equal spacing among the waveguides, $\kappa_1 = \kappa_2 = \kappa$ and the propagation constant becomes

$$\beta_{m1} = \beta_0 + \sqrt{2}\kappa \tag{5.127}$$

The corresponding normalized field will be represented by the following vector:

$$\begin{pmatrix} S_{10} \\ S_{20} \\ S_{30} \end{pmatrix} = \frac{1}{2} \begin{pmatrix} 1 \\ \sqrt{2} \\ 1 \end{pmatrix} \tag{5.128}$$

Note that the amplitude of the normal mode is higher in the central waveguide and has equal amplitudes in the outer two waveguides.

2. Mode corresponding to $\gamma = 0$ with propagation constant

$$\beta_{m2} = \beta_0 \tag{5.129}$$

This corresponds to the first excited mode of the three-waveguide coupler. This mode has the same propagation constant as an individual noninteracting waveguide. Its normalized field in vector form is

$$\begin{pmatrix} S_{10} \\ S_{20} \\ S_{30} \end{pmatrix} = N_{m2} \begin{pmatrix} \kappa_2 \\ 0 \\ -\kappa_1 \end{pmatrix} \tag{5.130}$$

with

$$N_{m2} = \left(\kappa_1^2 + \kappa_2^2\right)^{-1/2} \tag{5.131}$$

being the normalization constant. Note that this mode has zero contribution from the central waveguide and is referred to as the dark mode. The fields in the two outermost waveguides are out of phase by π. And thus this field would have a single zero crossing corresponding to the first excited mode.

For a three-waveguide coupler with equal spacing among the waveguides, $\kappa_1 = \kappa_2 = \kappa$ and the propagation constant becomes

$$\beta_{m2} = \beta_0$$

The corresponding normalized field will be represented by the following vector:

$$\begin{pmatrix} S_{10} \\ S_{20} \\ S_{30} \end{pmatrix} = \frac{1}{\sqrt{2}} \begin{pmatrix} 1 \\ 0 \\ -1 \end{pmatrix} \tag{5.132}$$

Note that the mode has equal amplitudes in the outermost waveguides and is an antisymmetric mode of the coupler.

3. Mode corresponding to $\gamma = -\sqrt{\kappa_1^2 + \kappa_2^2}$ with propagation constant

$$\beta_{m3} = \beta_0 - \sqrt{\kappa_1^2 + \kappa_2^2} \tag{5.133}$$

this corresponds to the second excited mode of the three-waveguide coupler with the smallest propagation constant among the three normal modes and in vector form the normal mode can be written as

$$\begin{pmatrix} S_{10} \\ S_{20} \\ S_{30} \end{pmatrix} = N_{m3} \begin{pmatrix} \kappa_1 \\ -\sqrt{\kappa_1^2 + \kappa_2^2} \\ \kappa_2 \end{pmatrix} \tag{5.134}$$

where N_{m3} is the normalization constant and is equal to N_{m1}. The field in the central waveguide is π out of phase with respect to the fields in the two outermost waveguides and thus this mode will have two zero crossings in the transverse direction representing the second excited mode of the coupler.

For a three-waveguide coupler with equal spacing among the waveguides, $\kappa_1 = \kappa_2 = \kappa$ and the propagation constant becomes

$$\beta_{m3} = \beta_0 - \sqrt{2}\kappa \tag{5.135}$$

The corresponding normalized field will be represented by the following vector:

$$\begin{pmatrix} S_{10} \\ S_{20} \\ S_{30} \end{pmatrix} = \frac{1}{2}\begin{pmatrix} 1 \\ -\sqrt{2} \\ 1 \end{pmatrix} \tag{5.136}$$

Note that the amplitude of the normal mode is higher in the central waveguide and has equal amplitudes in the outer two waveguides.

Exercise 5.9 Show that the three normal modes of the three waveguide directional coupler are orthogonal to each other.

Solution Corresponding to modes given by Eqs. (5.125) and (5.130), we have

$$N_{m1}\begin{pmatrix} \kappa_1 & \sqrt{\kappa_1^2 + \kappa_2^2} & \kappa_2 \end{pmatrix} N_{m2}\begin{pmatrix} \kappa_2 \\ 0 \\ -\kappa_1 \end{pmatrix} = N_{m1}N_{m2}(\kappa_1\kappa_2 + 0 - \kappa_2\kappa_1) = 0$$

Similarly corresponding to modes given by Eqs. (5.125) and (5.134), we have

$$N_{m1}\begin{pmatrix} \kappa_1 & \sqrt{\kappa_1^2 + \kappa_2^2} & \kappa_2 \end{pmatrix} N_{m3}\begin{pmatrix} \kappa_1 \\ -\sqrt{\kappa_1^2 + \kappa_2^2} \\ \kappa_2 \end{pmatrix} = N_{m1}N_{m3}\left(\kappa_1^2 - \left[\kappa_1^2 + \kappa_2^2\right] + \kappa_2^2\right) = 0$$

And corresponding to modes given by Eqs. (5.130) and (5.134)

$$N_{m2}\begin{pmatrix} \kappa_2 & 0 & -\kappa_1 \end{pmatrix} N_{m3}\begin{pmatrix} \kappa_1 \\ -\sqrt{\kappa_1^2 + \kappa_2^2} \\ \kappa_2 \end{pmatrix} = N_{m2}N_{m3}(\kappa_2\kappa_1 + 0 - \kappa_1\kappa_2) = 0$$

It is interesting to describe the modes by defining the variable θ as

$$\tan\theta = \frac{\kappa_1}{\kappa_2} \tag{5.137}$$

From this, we obtain

$$\gamma = +\sqrt{\kappa_1^2 + \kappa_2^2} = \kappa_2 \sec\theta \tag{5.138}$$

Using these, the three normal modes can be described by:

$$\text{Fundamental mode}: \begin{pmatrix} S_{10} \\ S_{20} \\ S_{30} \end{pmatrix} = \frac{1}{\sqrt{2}} \begin{pmatrix} \sin\theta \\ 1 \\ \cos\theta \end{pmatrix} \tag{5.139}$$

$$\text{First excited mode}: \begin{pmatrix} S_{10} \\ S_{20} \\ S_{30} \end{pmatrix} = \begin{pmatrix} \cos\theta \\ 0 \\ -\sin\theta \end{pmatrix} \tag{5.140}$$

$$\text{Second excited mode}: \begin{pmatrix} S_{10} \\ S_{20} \\ S_{30} \end{pmatrix} = \frac{1}{\sqrt{2}} \begin{pmatrix} \sin\theta \\ -1 \\ \cos\theta \end{pmatrix} \tag{5.141}$$

The value of θ is determined by the ratio of κ_1 to κ_2 which in turn depend on the spacing between waveguides apart from the waveguide characteristics such as transverse dimensions and refractive index difference between the guiding region and surrounding regions.

In Chap. 21, we will discuss how a three-waveguide directional coupler can be used to generate hyperentangled photons. In Chap. 25, we will present a direct solution of the coupled Eqs. (5.116)–(5.118).

Exercise 5.10 Light is launched in the middle waveguide of a three-waveguide directional coupler consisting of identical waveguides with equal spacing. Which modes are excited and with what relative amplitudes?

Solution Since the waveguides are identical and equally spaced, $\kappa_1 = \kappa_2$ and $\theta = \frac{\pi}{4}$ and the three modes are given by Eqs. (5.128), (5.132) and (5.136). We expand the input field in terms of the normal modes and write, with expansion coefficients a, b and c:

$$\begin{pmatrix} 0 \\ 1 \\ 0 \end{pmatrix} = a\frac{1}{2}\begin{pmatrix} 1 \\ \sqrt{2} \\ 1 \end{pmatrix} + b\frac{1}{\sqrt{2}}\begin{pmatrix} 1 \\ 0 \\ -1 \end{pmatrix} + c\frac{1}{2}\begin{pmatrix} 1 \\ -\sqrt{2} \\ 1 \end{pmatrix}$$

which can be solved to obtain $a = \frac{1}{\sqrt{2}}$, $b = 0$, $c = -\frac{1}{\sqrt{2}}$. Thus, the input excites only the fundamental mode and the second excited mode of the three waveguide directional coupler. It is an interesting exercise to evaluate the z-dependence of the power in the three individual waveguides.

5.4.1 Highly Asymmetric Three-Waveguide Coupler

Let us consider a three-waveguide coupler with the spacing between waveguides 2 and 3 to be much larger than the spacing between waveguides 1 and 2 (see Fig. 5.9). In such a case the coupling coefficient among the modes of waveguides 2 and 3 will be much smaller than the coupling coefficient between waveguides 1 and 2 and we will have

$$\kappa_2 \ll \kappa_1 \tag{5.142}$$

Let us now look at the normal modes and their propagation constants in such a case.

1. Fundamental mode (normal mode # 1):
   Propagation constant: $\beta_{m1} = \beta_0 + \sqrt{\kappa_1^2 + \kappa_2^2} \approx \beta_0 + \kappa_1$
   Normalize modal field:

$$\begin{pmatrix} S_{10} \\ S_{20} \\ S_{30} \end{pmatrix} = N_{m1}\kappa_1 \begin{pmatrix} 1 \\ \sqrt{\kappa_1^2 + \kappa_2^2}/\kappa_1 \\ \kappa_2/\kappa_1 \end{pmatrix} \approx \frac{1}{\sqrt{2}} \begin{pmatrix} 1 \\ 1 \\ 0 \end{pmatrix} \tag{5.143}$$

2. First excited mode (normal mode # 2):
   Propagation constant: $\beta_{m2} = \beta_0$
   Normalized modal field:

$$\begin{pmatrix} S_{10} \\ S_{20} \\ S_{30} \end{pmatrix} = -N_{m2}\kappa_1 \begin{pmatrix} -\kappa_2/\kappa_1 \\ 0 \\ 1 \end{pmatrix} \approx \begin{pmatrix} 0 \\ 0 \\ 1 \end{pmatrix} \tag{5.144}$$

3. Second excited mode (normal mode # 3):
   Propagation constant: $\beta_{m3} = \beta_0 - \sqrt{\kappa_1^2 + \kappa_2^2} \approx \beta_0 - \kappa_1$
   Normalized modal field:

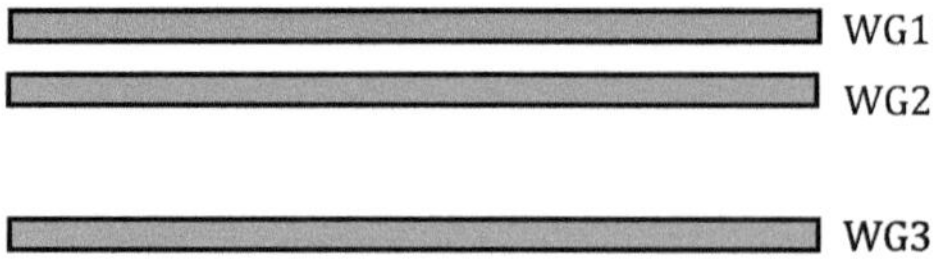

Fig. 5.9 A highly asymmetric three-waveguide directional coupler consisting of three waveguides marked WG1, WG2, and WG3

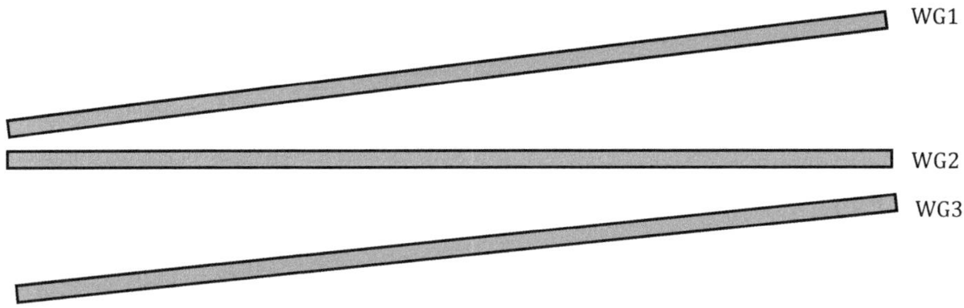

Fig. 5.10 A three-waveguide directional coupler with varying spacing. Waveguides 1 and 2 are very closely lying at the input while at the output waveguides 2 and 3 are closely lying. The angle between the waveguides is very small and highly exaggerated in the figure

$$\begin{pmatrix} S_{10} \\ S_{20} \\ S_{30} \end{pmatrix} = N_{m3}\kappa_1 \begin{pmatrix} 1 \\ -\sqrt{\kappa_1^2+\kappa_2^2}/\kappa_1 \\ \kappa_2/\kappa_1 \end{pmatrix} \approx \frac{1}{\sqrt{2}} \begin{pmatrix} 1 \\ -1 \\ 0 \end{pmatrix} \tag{5.145}$$

It is interesting to note in this case that since waveguide 3 is well separated from the waveguides 1 and 2 which are placed close to each other, the three-waveguide coupler behaves as a two-waveguide coupler of waveguides 1 and 2 and a separate waveguide, waveguide # 3. Mode #1 and mode # 3 correspond to the fundamental and first excited modes of a two-waveguide directional coupler formed by waveguides 1 and 2 and mode # 2 is the mode of the individual waveguide # 3, which is assumed to be well separated from the other two waveguides.

Let us now consider a three-waveguide directional coupler with z-varying separation as shown in Fig. 5.10.

At the input position corresponding to $z = 0$ waveguides 1 and 2 are closely lying while waveguide 3 is separated significantly from the other two waveguides. Hence at $z = 0$, we have $\kappa_2 \ll \kappa_1$.

As z increases waveguide 3 is brought closer gradually to the other two waveguides and the separation between waveguides 1 and 2 becomes equal to the separation between waveguides 2 and 3. Hence in this region $\kappa_2 = \kappa_1$.

After a certain propagation distance, the separation between waveguides 1 and 2 is increased gradually until we have $\kappa_1 \ll \kappa_2$.

We assume that all the transitions are adiabatic transitions so that light launched into any of the normal modes of the coupler evolves and continues to occupy the same normal mode locally as it propagates through the z-dependent coupler.

We now input light into waveguide 3 at $z = 0$. Thus, the input would correspond to the following vector:

$$\begin{pmatrix} S_{10} \\ S_{20} \\ S_{30} \end{pmatrix} = \begin{pmatrix} 0 \\ 0 \\ 1 \end{pmatrix} \tag{5.146}$$

From the above discussions, we can see that this input corresponds to the normal mode # 2 of the three-waveguide coupler at $z = 0$. As this mode propagates since the transition is assumed to be adiabatic, the light will continue to occupy normal mode # 2. Thus, around the center of the structure where the three waveguides are equally spaced, the field will be described by the following vector corresponding to mode # 2 (see Eq. 5.132):

$$\begin{pmatrix} S_{10} \\ S_{20} \\ S_{30} \end{pmatrix} = \frac{1}{\sqrt{2}} \begin{pmatrix} 1 \\ 0 \\ -1 \end{pmatrix} \tag{5.147}$$

Note that the power is distributed equally among the two outermost waveguides with null power in the middle waveguide.

At the end of the coupler since $\kappa_1 \ll \kappa_2$, the field will be described by the following vector corresponding to mode # 2:

$$\begin{pmatrix} S_{10} \\ S_{20} \\ S_{30} \end{pmatrix} = \begin{pmatrix} 1 \\ 0 \\ 0 \end{pmatrix} \tag{5.148}$$

which now corresponds to power in waveguide # 1.

Thus, in this device, light couples from the input waveguide # 3 to waveguide # 1 via waveguide # 2 without exciting waveguide # 2. Such a phenomenon where light couples between two waveguides through an intermediate waveguide that never gets excited is similar to the process referred to as stimulated Raman adiabatic passage (STIRAP) in the field of atom-photon interaction. Note that the spacing between the waveguides is counterintuitive, that is, the input is in a waveguide that is not coupled to the other two waveguides and an adiabatic evolution of the coupling leads to transfer of light to the other outermost waveguide.

Since wavelength does not play any role in the discussion, the coupler will exhibit this characteristic independent of the wavelength of operation.

5.5 $z-$Dependent Directional Couplers

In this section, we will briefly study two waveguide directional couplers in which the coupling coefficient depends on the propagation distance. We first recall the coupled mode equations as given by Eq. (5.54) and (5.55)

$$i\frac{dR}{dz} = -\delta R - \kappa S \tag{5.149a}$$

$$i\frac{dS}{dz} = \delta S - \kappa R \tag{5.149b}$$

We now define a new set of variables $\tilde{R}(z)$ and $\tilde{S}(z)$ through the following equations:

$$\begin{pmatrix} \tilde{R}(z) \\ \tilde{S}(z) \end{pmatrix} = \begin{pmatrix} \cos\theta(z) & -\sin\theta(z) \\ \sin\theta(z) & \cos\theta(z) \end{pmatrix} \begin{pmatrix} R(z) \\ S(z) \end{pmatrix} \tag{5.150}$$

where $\theta(z)$ is currently an unknown quantity and assumed to be z-dependent.

Now, from Eq. (5.150), we obtain

$$i\frac{d\tilde{R}}{dz} = i\frac{d}{dz}[R(z)\cos\theta(z) - S(z)\sin\theta(z)] = i\frac{dR}{dz}\cos\theta(z) - iR(z)\sin\theta(z)\frac{d\theta}{dz}$$
$$-i\frac{dS}{dz}\sin\theta(z) - iS\cos\theta(z)\frac{d\theta}{dz}$$

Substituting for $i\frac{dR}{dz}$ and $i\frac{dS}{dz}$ from Eq. (5.149a) and Eq. (5.149b), we obtain (for compactness we are not mentioning the z-dependence explicitly in the equations)

$$\begin{aligned} i\frac{d\tilde{R}}{dz} &= R(-\delta\cos\theta + \kappa\sin\theta) + S(-\kappa\cos\theta - \delta\sin\theta) - i\tilde{S}(z)\frac{d\theta}{dz} \\ &= T(z) - i\tilde{S}(z)\frac{d\theta}{dz} \end{aligned} \tag{5.151}$$

where

$$T(z) = R(-\delta\cos\theta + \kappa\sin\theta) + S(-\kappa\cos\theta - \delta\sin\theta) \tag{5.152}$$

Inverting Eq. (5.150) we have

$$\begin{pmatrix} R(z) \\ S(z) \end{pmatrix} = \begin{pmatrix} \cos\theta(z) & \sin\theta(z) \\ -\sin\theta(z) & \cos\theta(z) \end{pmatrix} \begin{pmatrix} \tilde{R}(z) \\ \tilde{S}(z) \end{pmatrix} \tag{5.153}$$

Replacing $R(z)$ and $S(z)$ in terms of $\tilde{R}(z)$ and $\tilde{S}(z)$ from Eq. (5.153) in Eq. (5.152) and simplifying, we obtain

$$T(z) = \tilde{R}(-\delta\cos 2\theta + \kappa\sin 2\theta) + \tilde{S}(-\kappa\cos 2\theta - \delta\sin 2\theta) \tag{5.154}$$

We now choose $\theta(z)$ such that

$$\kappa\cos 2\theta = -\delta\sin 2\theta$$

that is,

$$\tan 2\theta = -\frac{\kappa}{\delta} \tag{5.155}$$

Then in such a case, we have

$$T(z) = \tilde{R}\cos 2\theta(-\delta + \kappa \tan 2\theta) = -\Gamma\tilde{R} \tag{5.156}$$

where $\Gamma = \sqrt{\kappa^2 + \delta^2}$ (see Eq. 5.42b). Hence

$$i\frac{d\tilde{R}}{dz} = -\Gamma\tilde{R} - i\frac{d\theta}{dz}\tilde{S}(z) \tag{5.157a}$$

Similarly, we have

$$i\frac{d\tilde{S}}{dz} = \Gamma\tilde{S} + i\frac{d\theta}{dz}\tilde{R}(z) \tag{5.157b}$$

We can now combine Eqs. (5.157a) and (5.157b) and obtain the following matrix equation

$$i\frac{d}{dz}\begin{pmatrix}\tilde{R}\\ \tilde{S}\end{pmatrix} = \begin{pmatrix}-\Gamma & -i\frac{d\theta}{dz}\\ i\frac{d\theta}{dz} & \Gamma\end{pmatrix}\begin{pmatrix}\tilde{R}\\ \tilde{S}\end{pmatrix} \tag{5.158}$$

5.5.1 *z*-Independent Waveguide Coupler

If neither the coupling coefficient κ nor the phase mismatch $\Delta\beta$ vary with z, then θ is independent of z and we will have $\frac{d\theta}{dz} = 0$. Under this condition, the solutions of Eq. (5.158) are

$$\tilde{R}(z) = \tilde{R}(0)e^{i\Gamma z} \tag{5.159}$$

$$\tilde{S}(z) = \tilde{S}(0)e^{-i\Gamma z} \tag{5.160}$$

which represent the two normal modes of the coupled waveguide system propagating independently of each other. The normal modes are linear combination of the modes of individual waveguides

$$\tilde{R}(z) = R\cos\theta - S\sin\theta \tag{5.161}$$

$$\tilde{S}(z) = R\sin\theta + S\cos\theta \tag{5.162}$$

If the waveguides are phase matched then $\delta = 0$ and using Eq. (5.155), we obtain $\theta = -\frac{\pi}{4}$. Thus,

$$\tilde{R} = \frac{1}{\sqrt{2}}(R + S) \tag{5.163}$$

$$\tilde{S} = \frac{1}{\sqrt{2}}(R - S) \tag{5.164}$$

The mode represented by $\tilde{R}$ is the fundamental mode and has the fields in both waveguides to have an equal amplitude which are in phase. Similarly, the mode represented by $\tilde{S}$ corresponds to the first excited mode and has the fields in the two waveguides to have equal amplitude and are out of phase by π. A similar analysis can be carried out to obtain the normal modes for nonphase-matched waveguides for which $\delta \neq 0$.

In particular if $\delta > 0$ (i.e., $\beta_a > \beta_b$) and $\delta \gg \kappa$, that is, the waveguides are highly nonidentical, then we have

$$\tan 2\theta = -\frac{\kappa}{\delta} \ll 1$$

which gives us $\theta \approx -\frac{\kappa}{2\delta}$. Thus, the two normal modes are

$$\tilde{R} = R + \frac{\kappa}{2\delta}S \tag{5.165}$$

and

$$\tilde{S} = -\frac{\kappa}{2\delta}R + S \tag{5.166}$$

Since R and S are proportional to the amplitude of the normal mode in waveguide a and waveguide b, respectively, we find that the normal mode represented by $\tilde{R}$ has a major component in waveguide a (since $\frac{\kappa}{2\delta} \ll 1$) and the amplitudes are in phase in the two waveguides. Similarly, the other normal mode represented by $\tilde{S}$ has its major component in waveguide b and the amplitudes in the two coupled waveguides are out of phase by π. In this case, the fundamental mode of the coupled structure is represented by the mode described by $\tilde{R}$ and the first excited mode by the mode described by $\tilde{S}$. This is consistent with the discussion of a directional coupler consisting of nonidentical waveguides.

Similarly if ($\beta_a < \beta_b$), then $\delta < 0$, then

$$\tan 2\theta = -\frac{\kappa}{\delta} = \frac{\kappa}{|\delta|} \ll 1 \tag{5.167}$$

And we can approximate $\theta \approx \frac{\kappa}{2|\delta|}$. The normal modes in this case are given by

$$\tilde{R} = R - \frac{\kappa}{2|\delta|} S \tag{5.168}$$

and

$$\tilde{S} = \frac{\kappa}{2|\delta|} R + S \tag{5.169}$$

In this case, we find that the normal mode represented by $\tilde{R}$ has a major component in waveguide a (since $\frac{\kappa}{2|\delta|} \ll 1$) and the amplitudes are out of phase in the two waveguides. Similarly, the other normal mode represented by $\tilde{S}$ has its major component in waveguide b and the amplitudes in the two coupled waveguides are in phase. In this case, the fundamental mode of the coupled structure is represented by the mode described by $\tilde{S}$ (with in phase components of fields in both waveguides) and the first excited mode by the mode described by $\tilde{R}$. This is consistent with the discussion of a directional coupler consisting of nonidentical waveguides.

If κ is independent of z while δ is dependent on z, then the modes will be approximately described by $\tilde{R}(z)$ and $\tilde{S}(z)$ if $\frac{d\theta}{dz} \ll \Gamma$. Using Eq. (5.155) for $\tan\theta$ we can show that

$$\frac{d\theta}{dz} = -\frac{\kappa}{2\Gamma^2} \frac{d\delta}{dz} \tag{5.170}$$

Hence if

$$\frac{\kappa}{2\Gamma^2} \frac{d\delta}{dz} \ll \Gamma \tag{5.171}$$

then the modes are approximately described by Eqs. (5.163) and (5.164) and this is referred to as adiabatic transformation. Equation (5.171) can be rewritten as

$$\frac{d\delta}{dz} \ll \frac{2\left(\kappa^2 + \delta^2\right)^{3/2}}{\kappa} \tag{5.172}$$

This gives the condition for adiabaticity of the directional coupler.

Similarly, if δ is independent of z and κ is dependent on z the adiabatic transformation of the mode will take place if

$$\frac{d\kappa}{dz} \ll \frac{2\left(\kappa^2 + \delta^2\right)^{3/2}}{\delta} \tag{5.173}$$

Under adiabatic transformation, power coupled into any particular normal mode of the system will continue to be in that particular normal mode as the system evolves along the propagation direction. Thus, power launched into the fundamental mode of a directional coupler in which the coupling coefficient or the phase mismatch varies with distance, under adiabatic transformation, the power will continue to reside in the local normal mode of the coupler and will not get coupled to other normal modes. This is a very important consideration and many waveguide devices are based on adiabatic transformation for achieving various integrated optic devices.

5.6 Waveguide Arrays

In the previous sections, we discussed waveguide couplers with two and three waveguides. A waveguide array consists of an array of parallel waveguides, which interact through evanescent coupling. Although the array can in general consist of nonidentical waveguides with different spacings, etc., here for simplicity we consider an array of N parallel, identical waveguides which are all equally separated from each other (see Fig. 5.11).

Following a procedure similar to that followed for a two-waveguide or a three-waveguide directional coupler and assuming only nearest neighbor interaction, we will obtain the following set of coupled mode equations (see, e.g., Eqs. (5.116) to (5.118) for a three waveguide coupler):

$$i\frac{dA_m}{dz} = -\kappa(A_{m+1} + A_{m-1}); \qquad m = 1, 2, \ldots N \tag{5.174}$$

where A_m is the amplitude of the mode in the mth waveguide, κ is the coupling coefficient between two adjacent waveguides. In addition, since there are no waveguides before the first waveguide (corresponding to $m = 0$) and after the last waveguide (corresponding to $m = N + 1$), we have

$$A_0 = 0; \quad A_{N+1} = 0 \tag{5.175}$$

In order to determine the normal modes of the array, we substitute a solution of the form implying that the field components in each waveguide of the array are characterized by the same propagation constant

Fig. 5.11 Waveguide array consisting of N identical waveguides with identical spacing

$$A_m(z) = \tilde{A}_m e^{i\delta z} \tag{5.176}$$

where $\tilde{A}_m$ and δ are constants independent of z. Substituting in Eq. (5.174), we obtain

$$\delta \tilde{A}_m e^{i\delta z} = \kappa\left(\tilde{A}_{m+1} e^{i\delta z} + \tilde{A}_{m-1} e^{i\delta z}\right)$$

which gives us the following equation:

$$\frac{\tilde{A}_{m+1} + \tilde{A}_{m-1}}{\tilde{A}_m} = \frac{\delta}{\kappa} \tag{5.177}$$

In the above equation, the right-hand side is independent of m and hence the left-hand side of the above equation should also be independent of m. In addition, we also need to satisfy the conditions given by Eq. (5.175). For this, we try solutions of the form

$$\tilde{A}_m = a_0 \sin m\phi \tag{5.178}$$

where ϕ is an unknown phase. In this case, the left-hand side of Eq. (5.177) gives us

$$\frac{\tilde{A}_{m+1} + \tilde{A}_{m-1}}{\tilde{A}_m} = \frac{a_0 \sin(m+1)\phi + a_0 \sin(m-1)\phi}{a_0 \sin m\phi} = 2\cos\phi \tag{5.179}$$

which is now independent of m. From Eqs. (5.177) and (5.179), we have

$$\delta = 2\kappa \cos\phi \tag{5.180}$$

The solution given by Eq. (5.178) already satisfies the condition $\tilde{A}_0 = 0$. The second condition given by Eq. (5.175) implies

$$\tilde{A}_{N+1} = a_0 \sin(N+1)\phi = 0$$

which gives us the following values of the unknown phase ϕ:

$$\phi = j\frac{\pi}{(N+1)}; \qquad j = 1, 2, 3, \ldots N \tag{5.181}$$

Thus, the complete solution for the jth mode is given by

$$E_j(x, y, z) = \sum_{m=1}^{N} a_0 \sin\left[j\frac{m\pi}{(N+1)}\right] E_m(x, y) e^{i(\beta_0 + \delta_j)z} \tag{5.182}$$

where β_0 is the propagation constant of individual waveguide modes and

$$\delta_j = 2\kappa \cos\left(j\frac{\pi}{(N+1)}\right); \quad j = 1, 2, \ldots N \tag{5.183}$$

Thus, the propagation constant of the jth mode is given by

$$\beta = \beta_0 + \delta_j = \beta_0 + 2\kappa \cos\left(j\frac{\pi}{(N+1)}\right) \tag{5.184}$$

The total field can be normalized and we obtain the jth normal mode solutions as

$$E_j(x, y, z) = \sqrt{\frac{2}{(N+1)}} \sum_{m=1}^{N} E_m(x, y) \sin\left[j\frac{m\pi}{(N+1)}\right] e^{i\left[\beta_0 + 2\kappa \cos\left(j\frac{\pi}{(N+1)}\right)\right]z} \tag{5.185}$$

As an example, we consider a three-waveguide coupler for which $N = 3$. Thus, the propagation constants of the three normal modes are given by

$$\beta_1 = \beta_0 + 2\kappa \cos\left(\frac{\pi}{4}\right) = \beta_0 + \sqrt{2}\kappa \tag{5.186}$$

$$\beta_2 = \beta_0 + 2\kappa \cos\left(\frac{2\pi}{4}\right) = \beta_0 \tag{5.187}$$

$$\beta_3 = \beta_0 + 2\kappa \cos\left(\frac{3\pi}{4}\right) = \beta_0 - \sqrt{2}\kappa \tag{5.188}$$

These are the same as we found in Sect. 5.4. The corresponding normalized normal modes have the following amplitude contributions from individual waveguides

$$\begin{pmatrix} \sin\left[\frac{\pi}{4}\right] \\ \sin\left[\frac{2\pi}{4}\right] \\ \sin\left[\frac{3\pi}{4}\right] \end{pmatrix} = \frac{1}{\sqrt{2}} \begin{pmatrix} \frac{1}{\sqrt{2}} \\ 1 \\ \frac{1}{\sqrt{2}} \end{pmatrix} \tag{5.189}$$

$$\begin{pmatrix} \sin\left[\frac{2\pi}{4}\right] \\ \sin\left[\frac{4\pi}{4}\right] \\ \sin\left[\frac{6\pi}{4}\right] \end{pmatrix} = \frac{1}{\sqrt{2}} \begin{pmatrix} 1 \\ 0 \\ -1 \end{pmatrix} \tag{5.190}$$

$$\begin{pmatrix} \sin\left[\frac{3\pi}{4}\right] \\ \sin\left[\frac{6\pi}{4}\right] \\ \sin\left[\frac{9\pi}{4}\right] \end{pmatrix} = \frac{1}{\sqrt{2}}\begin{pmatrix} \frac{1}{\sqrt{2}} \\ -1 \\ \frac{1}{\sqrt{2}} \end{pmatrix} \tag{5.191}$$

which are the same normal modes as we found in Sect. 5.4.

As another example, we consider a five-waveguide directional coupler with $N = 5$. We can use the formulas derived above and get the following propagation constants of the five normal modes as

$$\beta_1 = \beta_0 + 2\kappa\cos\left(\frac{\pi}{6}\right); \quad \beta_2 = \beta_0 + 2\kappa\cos\left(\frac{2\pi}{6}\right); \quad \beta_3 = \beta_0 + 2\kappa\cos\left(\frac{3\pi}{6}\right);$$
$$\beta_4 = \beta_0 + 2\kappa\cos\left(\frac{4\pi}{6}\right); \quad \beta_5 = \beta_0 + 2\kappa\cos\left(\frac{5\pi}{6}\right)$$

These give us the following values

$$\beta_1 = \beta_0 + \sqrt{3}\kappa; \quad \beta_2 = \beta_0 + \kappa; \quad \beta_3 = \beta_0; \quad \beta_4 = \beta_0 - \kappa; \quad \beta_5 = \beta_0 - \sqrt{3}\kappa \tag{5.192}$$

The normalized modes have the following contributions from individual waveguides.

$$\frac{1}{\sqrt{3}}\begin{pmatrix} 1/2 \\ \sqrt{3}/2 \\ 1 \\ \sqrt{3}/2 \\ 1/2 \end{pmatrix}; \quad \frac{1}{\sqrt{3}}\begin{pmatrix} \sqrt{3}/2 \\ \sqrt{3}/2 \\ 0 \\ -\sqrt{3}/2 \\ -\sqrt{3}/2 \end{pmatrix}; \quad \frac{1}{\sqrt{3}}\begin{pmatrix} 1 \\ 0 \\ -1 \\ 0 \\ 1 \end{pmatrix}; \quad \frac{1}{\sqrt{3}}\begin{pmatrix} \sqrt{3}/2 \\ -\sqrt{3}/2 \\ 0 \\ \sqrt{3}/2 \\ -\sqrt{3}/2 \end{pmatrix};$$
$$\frac{1}{\sqrt{3}}\begin{pmatrix} 1/2 \\ -\sqrt{3}/2 \\ 1 \\ -\sqrt{3}/2 \\ 1/2 \end{pmatrix} \tag{5.193}$$

Figure 5.12 shows the five normal modes of the array with the amplitude contribution from each waveguide of the array to the corresponding normal mode. The structure supports three symmetric and two antisymmetric modes.

The analysis presented in the section can be used for analyzing arrays of waveguides, which are finding greater applications in integrated quantum photonics for

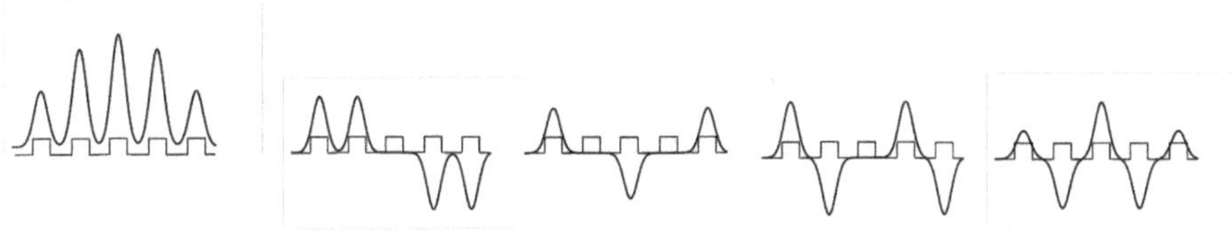

Fig. 5.12 The five normal modes of an array of five ($N = 5$) identical waveguides with equal separation corresponding to (**a**) $j = 1$, (**b**) $j = 2$, (**c**) $j = 3$, (**d**) $j = 4$ and (**e**) $j = 5$

the generation of higher dimensional entanglement (see, e.g., Yang et al. 2024), W-state generation (see, e.g., Perez-Leija et al. (2013), Paul and Thyagarajan (2014)), in quantum random walk (e.g., Perets et al. (2008), Peruzzo et al. (2010), Garanovich et al. (2012)), etc.

Problems

5.1. Choose the correct choice: In a directional coupler, complete power transfer is possible if (a) only if the separation between the waveguides is small (b) only if the waveguides are identical, (c) Even if the waveguides are nonidentical but having the same propagation constant (d) Only for large separation between the waveguides.

5.2. In a directional coupler consisting of identical waveguides, at the input equal powers are launched in both waveguides but with a phase difference of $\pi/2$. What will be the evolution of the powers in the two waveguides as a function of z?

5.3. Consider a directional coupler formed by two single-mode planar waveguides with the following parameters operating at a wavelength of 1000 nm:

$$n_f = 1.503, n_s = 1.50, \quad d = 2\ \mu\text{m}$$

Assume that the center to center spacing between the waveguides is given by D.

(a) Obtain the eigenvalue equations specifying the fundamental and first excited normal modes of the directional coupler.
(b) Plot the variation of the propagation constants of the two normal modes with D.
(c) Use coupled mode theory and determine the variation of the coupling coefficient and propagation constants of the two normal modes with D.
(d) Show that the two methods agree for large values of D.

5.4. Consider a directional coupler with two identical waveguides. The coupler has $L_c = 1$ mm. How much variation in the length of the coupler is permissible such that in the cross-state the power emanating from the same waveguide is less than -30 dB compared to the input power?

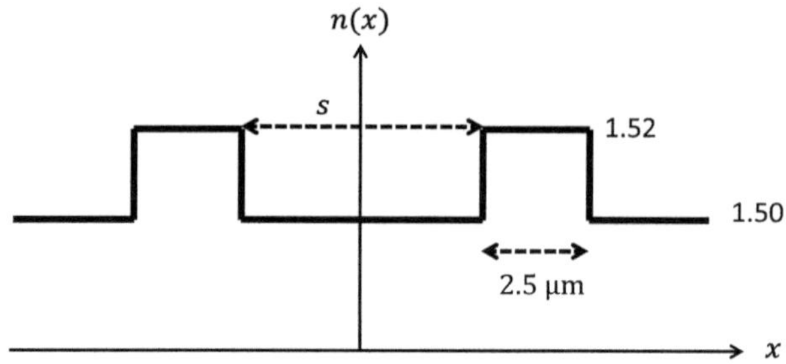

Fig. 5.13 Figure for Problem 5.8

5.5. A directional coupler with a 50% coupling ratio is to be fabricated so that it is least sensitive to fabrication errors in the coupler length. What value of $\frac{\Delta\beta}{\kappa}$ would you choose and why?

5.6. As discussed in Sect. 5.2.3, in the normal mode picture the exchange of power between the two waveguides of the coupler is through interference among the two normal modes. Consider a directional coupler with nonidentical waveguides. Show that the coupling lengths of the coupler obtained through the normal mode picture and through the coupled mode picture are the same.

5.7. Consider a directional coupler made up of two single-mode channel waveguides. Given that the coupling coefficient between the waveguides is $\pi/2\ \text{cm}^{-1}$

(a) What should be the length of the coupler so that the entire power transfer can take place between the waveguides if they are identical?
(b) What is the value of $\Delta\beta$ between the two waveguides so that the maximum transfer of power between the two waveguides is 50%?
(c) What is the corresponding length of the coupler where power launched in waveguide 1 gets divided equally between the two waveguides?

5.8. Consider TE modes of a planar waveguide directional coupler shown in Fig. 5.13 with identical waveguides operating at a wavelength of 1500 nm. Each of the waveguides has a core of refractive index 1.52 with a surrounding medium of refractive index 1.5. The width of each of the waveguide is 2.5 μm.

(a) Write down the field expressions corresponding to the symmetric and antisymmetric normal modes of the structure.
(b) Obtain the eigenvalue equations for the symmetric and antisymmetric normal modes.
(c) Obtain an expression for the coupling coefficient κ of the planar waveguide directional coupler. Using the expression, obtain the propagation constants of the fundamental and first excited normal modes as obtained from the coupled mode theory.
(d) Calculate the effective indices of the fundamental and first excited normal mode using both the normal mode theory and the coupled mode theory and plot the variation of the effective indices of the symmetric and antisymmetric modes as a function of the separation s from s = 0 to 10 μm.

5.9. The coupling length of a directional coupler made of identical waveguides and operating at a wavelength of 1000 nm is 1 mm. What is the difference in effective indices of the symmetric and antisymmetric modes of the coupler?

5.10. At a wavelength of 1500 nm the effective indices of the modes of each individual waveguide is 1.5. If the coupling length of the device is 1 mm, what are the effective indices of the symmetric and antisymmetric normal modes of the coupler?

References

Garanovich, I. L., S. Longhi, A. A. Sukhorukov, and Y. S. Kivshar. 2012. Light propagation and localization in modulated photonic lattices and waveguides. *Phys Rep* 518:1–79.

Hardy, A., and W. Streifer. 1985. Coupled mode theory of parallel waveguides. *J Lightwave Tech* 3: 1135–1146.

Haus, H. A., W. P. Huang, S. Kawakami, and N. A. Whitaker. 1987. Coupled mode theory of optical waveguides. *J Lightwave Tech* 5:16–23.

Kogelnik, H., and R. Schmidt. 1976. Switched directional couplers with alternating $\Delta\beta$. *IEEE J Quant Electron* QE-12:396–401.

Marcuse, D. 1971. The coupling of degenerate modes in two parallel dielectric waveguides. *Bell Syst Tech J* 50:1791–1816.

Paul, S., and K. Thyagarajan. 2014. Generation of N-partite single and two photon W states with enhanced tolerance using waveguide arrays. *J Opt* 16:105503.

Perets, H. B., Y. Lahini, F. Pozzi, M. Sorel, R. Morandotti, and Y. Silberberg. 2008. Realization of quantum walks with negligible decoherence in waveguide lattices. *Phys Rev Lett* 100:170506.

Perez-Leija, A., J. C. Hernandez-Herrejon, and H. Moya-Cessa. 2013. Generating photon-encoded W-states in multiport waveguide-array systems. *Phys Rev A* 87:013842.

Peruzzo, A., M. Lobino, J. C. F. Matthews, N. Matsuda, A. Politi, K. Poulios, X.-Q. Zhou, Y. Lahini, N. Ismail, K. Wörhoff, Y. Bromberg, Y. Silberberg, M. G. Thompson, and J. L. Obrien. 2010. Quantum walks of correlated photons. *Science* 329:1500–1503.

Yang, Y., R. J. Chapman, B. Haylock, F. Lenzini, Y. N. Joglekar, M. Lobino and A. Peruzzo. 2024. Programmable high-dimensional Hamitonian in a photonic waveguide array. *Nat Commun* 15:50. https://doi.org/10.1038/s41467-023-44185-z.

Yariv, A. 1973. Coupled mode theory for guided-wave optics. *IEEE J Quant Electron* 9:919–933.

Chapter 6
Electro-Optic Effect in Optical Waveguides

Abstract Linear electro-optic effect or Pockels effect is one of the most important effects used in integrated quantum photonics to achieve dynamic optical waveguide components such as optical switches and optical modulators. This chapter deals with the Pockels effect in optical waveguides in detail and presents examples of phase modulator, intensity modulator, and polarization converters using the optical waveguide platform

6.1 Introduction

Electro-optic effect refers to the changes in the optical properties such as refractive index, anisotropy, birefringence, and principal axes orientation. of a medium due to an external electric field applied on the medium. If the changes are proportional to the applied electric field, then it is referred to as the linear electro-optic effect or Pockels effect and if the changes are proportional to the square of the applied electric field then it is referred to as the Kerr effect. In this chapter, we restrict our discussion to the Pockels effect as this is the principal electro-optic effect that is used in guided wave optical devices and in integrated quantum photonic devices.

Lithium niobate ($LiNbO_3$) is one of the most important materials in integrated photonics as well as integrated quantum photonics and possesses a very strong electro-optic effect. It is also possible to fabricate low-loss optical waveguides in lithium niobate thus providing a very good platform for the realization of guided wave quantum photonic devices. There are many different techniques for the fabrication of waveguides in lithium niobate; this includes titanium indiffusion, proton exchange technique, and femtosecond laser writing. Also recently thin film lithium niobate waveguides have assumed great importance due to their excellent characteristics. For a recent review of electro-optics in thin film lithium niobate, see Zhang et al. (2021) and Hu et al. (2023).

K. Thyagarajan, *Integrated Quantum Photonics*, Graduate Texts in Physics,
https://doi.org/10.1007/978-3-031-85728-7_6

6.2 Pockels Effect

The equation for the index ellipsoid of an anisotropic medium in the principal axis system in the absence of any external electric field is given by (see Chap. 2)

$$\frac{x^2}{n_x^2}+\frac{y^2}{n_y^2}+\frac{z^2}{n_z^2}=1 \tag{6.1}$$

where n_x, n_y, and n_z represent the principal refractive indices of the medium. In order to make the equations compact, instead of using symbols (x, y, z), we will denote the three Cartesian axes by (x_1, x_2, x_3). In the book as per convenience, we will either use the symbols (x, y, z) or (x_1, x_2, x_3) interchangeably. Thus, the equation of the index ellipsoid can be written as

$$\sum_{i,j=1}^{3}\frac{x_i x_j}{n_{ij}^2}=1 \tag{6.2}$$

with

$$\frac{1}{n_{11}^2}=\frac{1}{n_x^2};\ \frac{1}{n_{22}^2}=\frac{1}{n_y^2};\ \frac{1}{n_{33}^2}=\frac{1}{n_z^2};\frac{1}{n_{ij}^2}=\frac{1}{n_{ji}^2}=0;i\neq j$$

Note that the index ellipsoid equation (Eq. 6.1) does not have cross terms since the equation is written in the principal axis system.

When an external electric field is applied to the crystal then due to the Pockels effect the equation of the index ellipsoid changes and the changes in the coefficients of the index ellipsoid term are described by the following equation:

$$\Delta\left(\frac{1}{n^2}\right)_{ij}=\sum_{k=1}^{3} r_{ijk}F_k;\qquad i,j,k=1,2,3 \tag{6.3}$$

where F_k's represent the components of the applied external electric field. The coefficients r_{ijk} form the elements of a $3 \times 3 \times 3$ electro-optic tensor having 27 elements.

Hence, the index ellipsoid in the presence of the external electric field will be given by

$$\sum_{i,j=1}^{3} x_i x_j\left(\frac{1}{n_{ij}^2}+\Delta\left(\frac{1}{n^2}\right)_{ij}\right)=1$$

Since the indices i, j commute, we can replace the pair of indices i, j by a single index that goes from 1 to 6 as per the standard convention given below:

$$11 \rightarrow 1; 22 \rightarrow 2; 33 \rightarrow 3, 23, 32 \rightarrow 4; 13, 31 \rightarrow 5; 12, 21 \rightarrow 6$$

Using this Eq. (6.3) becomes

$$\Delta\left(\frac{1}{n^2}\right)_i = \sum_{k=1}^{3} r_{ik} F_k; \qquad i = 1, 2, \ldots 6 \tag{6.4}$$

The coefficients r_{ik} form the elements of the contracted linear electro-optic tensor, which can be written in the following matrix:

$$\overline{\overline{r}} = \begin{pmatrix} r_{11} & r_{12} & r_{13} \\ r_{21} & r_{22} & r_{23} \\ r_{31} & r_{32} & r_{33} \\ r_{41} & r_{42} & r_{43} \\ r_{51} & r_{52} & r_{53} \\ r_{61} & r_{62} & r_{63} \end{pmatrix} \tag{6.5}$$

The symmetry of the crystal determines which of the elements of the electro-optic tensor are zero, which of them are equal, which of them are negative of each other, etc. For media possessing inversion symmetry, it can be shown that all elements of the linear electro-optic tensor are zero. Thus, media possessing inversion symmetry do not exhibit Pockels effect. Readers can refer to excellent texts where the form of the tensor is given for crystals belonging to different crystal classes (see, e.g., Saleh and Teich 1991; Yariv 1991a, b).

Equation (6.4) can be written in an expanded form as

$$\overline{\overline{\Delta\left(\frac{1}{n^2}\right)}} = \begin{pmatrix} \Delta\left(\frac{1}{n^2}\right)_1 \\ \Delta\left(\frac{1}{n^2}\right)_2 \\ \Delta\left(\frac{1}{n^2}\right)_3 \\ \Delta\left(\frac{1}{n^2}\right)_4 \\ \Delta\left(\frac{1}{n^2}\right)_5 \\ \Delta\left(\frac{1}{n^2}\right)_6 \end{pmatrix} = \begin{pmatrix} r_{11} & r_{12} & r_{13} \\ r_{21} & r_{22} & r_{23} \\ r_{31} & r_{32} & r_{33} \\ r_{41} & r_{42} & r_{43} \\ r_{51} & r_{52} & r_{53} \\ r_{61} & r_{62} & r_{63} \end{pmatrix} \begin{pmatrix} F_1 \\ F_2 \\ F_3 \end{pmatrix} \tag{6.6}$$

Typical values of the elements of the electro-optic tensor are in the range of 10^{-12} m/V and thus are quite small. Table 6.1 gives typical values of the various electro-optic coefficients of three most commonly used materials in integrated quantum photonics: lithium niobate ($LiNbO_3$), potassium titanyl phosphate (KTP), and lithium tantalate ($LiTaO_3$).

Table 6.1 Electro-optic coefficients of materials used in integrated quantum photonics

Material	Electro-optic coefficients in pm/V
Lithium niobate	$r_{13} = r_{23} = 8.6$; $r_{33} = 30.8$; $r_{22} = -r_{61} = -r_{12} = 3.4$; $r_{51} = r_{42} = 28$
Lithium tantalate	$r_{13} = 8$; $r_{33} = 30$; $r_{22} = 1$; $r_{51} = 20$
KTP	$r_{13} = 8.8$; $r_{23} = 13.8$; $r_{33} = 35$; $r_{51} = 6.9$; $r_{42} = 8.8$

6.2.1 Electro-Optic Effect in Lithium Niobate

We will now discuss the electro-optic effect in lithium niobate. Lithium niobate is a uniaxial crystal and in such a crystal $n_x = n_y = n_o$ and $n_z = n_e$ where n_o and n_e represent the ordinary and extraordinary refractive indices of the crystal.

The electro-optic tensor of lithium niobate is given by

$$\overline{\overline{r}} = \begin{pmatrix} 0 & -r_{22} & r_{13} \\ 0 & r_{22} & r_{13} \\ 0 & 0 & r_{33} \\ 0 & r_{42} & 0 \\ r_{51} & 0 & 0 \\ -r_{22} & 0 & 0 \end{pmatrix} \tag{6.7}$$

Note that due to the different symmetries present in the crystal of lithium niobate, we have $r_{12} = -r_{22}$, $r_{23} = r_{13}$, $r_{61} = -r_{22}$, $r_{42} = r_{51}$ and also the fact that other elements of the matrix have zero values.

As an example, if an external electric field is applied along the z-direction of the crystal (the optic axis direction), then the changes in the index ellipsoid coefficients will be given by (see Eq. (6.6))

$$\Delta\overline{\overline{\left(\frac{1}{n^2}\right)}} = \begin{pmatrix} 0 & -r_{22} & r_{13} \\ 0 & r_{22} & r_{13} \\ 0 & 0 & r_{33} \\ 0 & r_{42} & 0 \\ r_{51} & 0 & 0 \\ -r_{22} & 0 & 0 \end{pmatrix} \begin{pmatrix} 0 \\ 0 \\ F_z \end{pmatrix} = \begin{pmatrix} r_{13}F_z \\ r_{13}F_z \\ r_{33}F_z \\ 0 \\ 0 \\ 0 \end{pmatrix} \tag{6.8}$$

Hence, the index ellipsoid equation in the presence of the applied electric field along the z-direction will be

$$x_1^2\left(\frac{1}{n_o^2} + r_{13}F_z\right) + x_2^2\left(\frac{1}{n_o^2} + r_{13}F_z\right) + x_3^2\left(\frac{1}{n_e^2} + r_{33}F_z\right) = 1 \tag{6.9}$$

In the presence of the external electric field, the equation of the index ellipsoid gets modified. Since the index ellipsoid determines the orientation of the principal axes and the principal refractive indices of the medium, this implies that in the presence of the external electric field, in general, the orientation of the principal axes and their lengths change.

Since there are still no cross terms in Eq. (6.9), for an applied field along the z-direction, this implies that the orientations of the principal axes remain the same; however, the principal refractive indices have changed. The medium still remains uniaxial.

In lithium niobate $r_{13} \approx 8.6 \times 10^{-12}$ m/V and $r_{33} \approx 30.8 \times 10^{-12}$ m/V. For typical applied electric fields (Typical voltage of 10 V across a width of 10 μm), the external electric fields are about $F_z = 10^6$ V/m. Thus, we have $r_{13}F_z = 8.6 \times 10^{-6} \ll 1$ and $r_{33}F_z = 30.8 \times 10^{-6} \ll 1$. Hence, we can make Taylor series approximation and the new principal refractive indices in the presence of the electric field are approximately given by

$$n'_{x_1} = n'_x = \left(\frac{1}{n_o^2} + r_{13}F_z\right)^{-1/2} \approx n_o - \frac{n_o^3 r_{13} F_z}{2} = n'_{x_2} = n'_y \tag{6.10}$$

$$n'_{x_3} = n'_z = \left(\frac{1}{n_e^2} + r_{33}F_z\right)^{-1/2} \approx n_e - \frac{n_e^3 r_{33} F_z}{2} \tag{6.11}$$

Thus, the new principal refractive indices depend linearly on the external electric field. Changing the external electric field changes the principal refractive indices and hence changes the phase changes suffered by the propagating light waves.

If the incident light propagates along the x_1- direction (which is the same as x-direction) then if the incident light is polarized along the x_3-direction (or z- direction), then the phase change suffered by the wave in propagating through a length L would be

$$\phi = \frac{2\pi}{\lambda_0} n'_z L = \frac{2\pi}{\lambda_0}\left(n_e - \frac{n_e^3 r_{33} F_z}{2}\right) L \tag{6.12}$$

which changes with the applied electric field thus leading to phase modulation of the propagating light wave.

The electric field-induced change of phase is given by

$$\Delta\phi = -\frac{2\pi}{\lambda_0} \frac{n_e^3 r_{33} F_z}{2} L \tag{6.13}$$

In the example, we are considering, the application of the external electric field does not lead to any rotation of the principal axes of the crystal. Thus, if the incident light wave is polarized at 45° to the z-direction and propagates along the x-direction, then we first need to break up the polarization state into its components along the principal axes directions, which are the eigenmodes of propagation. Since for propagation along the x-direction, the eigenpolarization states are along the y- and z- directions, we note that the two components of the electric field of the propagating light wave along these two directions have equal magnitude and are in phase at the input. On propagating through a length L of the medium, the phase difference between the two components would be

$$\Delta\phi = \frac{2\pi}{\lambda_0}\left(n'_y - n'_z\right)L = \frac{2\pi}{\lambda_0}(n_o - n_e)L + \frac{\pi}{\lambda_0}\left(n_o^3 r_{13} - n_e^3 r_{33}\right)F_z L$$

Hence, the electric field-induced phase difference between the two eigenmodes is

$$\Delta\phi_E = \frac{\pi}{\lambda_0}\left(n_o^3 r_{13} - n_e^3 r_{33}\right)F_z L$$

Since the phase difference between the two components of the electric field of the light wave depends on the applied electric field F_z, this leads to modulation of the state of polarization of the output light and acts like a polarization modulator.

Polarization modulation can be changed to intensity modulation of the output light wave by passing the output through an appropriately oriented polarizer leading to an electro optic amplitude or intensity modulator.

Exercise 6.1 An electric field is applied along the y-direction in lithium niobate. (a) Write down the equation of the index ellipsoid in the presence of the electric field. (b) Obtain the new principal axes directions and the new principal refractive indices.

Solution For an applied electric field given by $\vec{F} = F_y\hat{y}$, we have

$$\Delta\overline{\overline{\left(\frac{1}{n^2}\right)}} = \begin{pmatrix} 0 & -r_{22} & r_{13} \\ 0 & r_{22} & r_{13} \\ 0 & 0 & r_{33} \\ 0 & r_{42} & 0 \\ r_{51} & 0 & 0 \\ -r_{22} & 0 & 0 \end{pmatrix}\begin{pmatrix} 0 \\ F_y \\ 0 \end{pmatrix} = \begin{pmatrix} -r_{22}F_y \\ r_{22}F_y \\ 0 \\ r_{42}F_y \\ 0 \\ 0 \end{pmatrix} \tag{6.14}$$

Thus, the index ellipsoid equation will be given by

$$\frac{x^2}{n_o^2} + \frac{y^2}{n_o^2} + \frac{z^2}{n_e^2} - r_{22}F_y x^2 + r_{22}F_y y^2 + 2yzr_{42}F_y = 1$$

or

$$x^2\left(\frac{1}{n_o^2} - r_{22}F_y\right) + y^2\left(\frac{1}{n_o^2} + r_{22}F_y\right) + \frac{z^2}{n_e^2} + 2yzr_{42}F_y = 1 \tag{6.15}$$

The presence of the cross term containing yz implies that the new principal axes will correspond to a rotation of the principal axis around the x- axis.

To obtain the new principal axis directions, we consider a rotated coordinate system with (x, y', z') with z' making an angle of θ with the z-axis and obtaining the value of θ which will have no cross terms.

In such a case

$$y = y' \cos\theta - z' \sin\theta \tag{6.16}$$

$$z = y' \sin\theta + z' \cos\theta \tag{6.17}$$

Substituting these in Eq. (6.15), collecting terms containing the product $y'z'$ and putting its coefficient to zero, we obtain the value of $\theta = \theta_0$ with $\tan 2\theta_0 = \frac{2r_{42}F_y}{\left(\frac{1}{n_o^2} - \frac{1}{n_e^2} + r_{22}F_y\right)} \approx \frac{2r_{42}F_y}{\left(\frac{1}{n_o^2} - \frac{1}{n_e^2}\right)}$ neglecting $r_{22}F_y$ in comparison to the other terms. Thus, the new principal axes' direction corresponds to (x, y', z') with $\theta = \theta_0$. Since the right-hand side is ≪1, we can approximate and obtain

$$\theta_0 \approx \frac{r_{42}F_y}{\left(\frac{1}{n_o^2} - \frac{1}{n_e^2}\right)}$$

Using this value of θ, we will obtain the equation for the index ellipsoid in the new principal axis direction and also the new principal refractive indices:

$$x^2\left(\frac{1}{n_o^2} - r_{22}F_y\right) + y'^2\left(\frac{\cos^2\theta_0}{n_y^2} + \frac{\sin^2\theta_0}{n_e^2} + \sin 2\theta_0 r_{42}F_y\right)$$
$$+ z'^2\left(\frac{\sin^2\theta_0}{n_y^2} + \frac{\cos^2\theta_0}{n_e^2} - \sin 2\theta_0 r_{42}F_y\right) = 1 \tag{6.18}$$

where

$$\frac{1}{n_y^2} = \left(\frac{1}{n_o^2} + r_{42}F_y\right) \tag{6.19}$$

The new principal refractive indices can be calculated from the coefficients of (x, y', z') using the specific value of θ_0.

At a wavelength of 1550 nm, we have $n_o = 2.2111$; $n_e = 2.1376$. Using $r_{42} = r_{51} = 28 \times 10^{-12}$ m/V and an applied electric field of 5×10^5 V/m (which corresponds to applying a voltage of 5 V across 10 μm), we obtain $\theta_0 \approx 0.06°$.

Exercise 6.2 The electro-optic tensor of KTP is given by

$$\overline{\overline{r}} = \begin{pmatrix} 0 & 0 & r_{13} \\ 0 & 0 & r_{23} \\ 0 & 0 & r_{33} \\ 0 & r_{42} & 0 \\ r_{51} & 0 & 0 \\ 0 & 0 & 0 \end{pmatrix} \tag{6.20}$$

The typical values of the coefficients are given in Table 6.1. It is also a biaxial crystal. Write down the equation of the index ellipsoid when an electric field along the y- direction is applied.

Solution Since the crystal is biaxial, in the absence of the electric field, the index ellipsoid equation is given by

$$\frac{x^2}{n_x^2} + \frac{y^2}{n_y^2} + \frac{z^2}{n_z^2} = 1 \tag{6.21}$$

The change in the index ellipsoid equation is only due to the term r_{42} and the new index ellipsoid equation will be given by

$$\frac{x^2}{n_x^2} + \frac{y^2}{n_y^2} + \frac{z^2}{n_z^2} + 2yzr_{42}F_y = 1 \tag{6.22}$$

In order to obtain the new principal axes directions and the new principal refractive indices first we note that the cross term in Eq. (6.22) contains the product yz. Hence, the coordinate system in which the index ellipsoid equation will not have cross terms will be rotated with respect to the (x, y, z) axes by some angle θ around the x-axis. Let us choose a coordinate system (x, y', z') with the y'-axis making an angle θ with respect to the y-axis. Equations (6.16) and (6.17) relate the old and new coordinate systems. We substitute in Eq. (6.22) and obtain an equation in terms of (x, y', z'). Putting the coefficient of the $y'z'$ to zero, we obtain the value of θ which will result in an equation without the cross terms. This gives the following equation determining the value of θ:

$$\tan 2\theta = \frac{2r_{42}F_y}{\left(\frac{1}{n_y^2} - \frac{1}{n_z^2}\right)} \tag{6.23}$$

This gives us the following index ellipsoid equation:

$$\frac{x^2}{n_x^2} + y'^2\left(\frac{\cos^2\theta}{n_y^2} + \frac{\sin^2\theta}{n_z^2} + 2\cos\theta\sin\theta r_{42}F_y\right)$$
$$+ z'^2\left(\frac{\sin^2\theta}{n_y^2} + \frac{\cos^2\theta}{n_z^2} - 2\cos\theta\sin\theta r_{42}F_y\right) = 1 \quad (6.24)$$

From the above equation, the new principal refractive indices are $\left(n_x, n_y', n_z'\right)$ with

$$\frac{1}{n_y'^2} = \left(\frac{\cos^2\theta}{n_y^2} + \frac{\sin^2\theta}{n_z^2} + 2\cos\theta\sin\theta r_{42}F_y\right) \quad (6.25)$$

and

$$\frac{1}{n_z'^2} = \left(\frac{\sin^2\theta}{n_y^2} + \frac{\cos^2\theta}{n_z^2} - 2\cos\theta\sin\theta r_{42}F_y\right) \quad (6.26)$$

Taking typical values $n_x = 1.7377$, $n_y = 1.7453$, $n_z = 1.8297$, $r_{42} = 8.8$ pm/V, assuming an electric field of 5×10^5 V/m, (corresponding to applying 5 V across an electrode spacing of 10 μm), we obtain $\theta \approx 8.5 \times 10^{-3\circ}$ which is indeed a very small value.

6.2.2 Changes in the Dielectric Permittivity Tensor

The dielectric permittivity tensor written in the principal axis system and in the absence of the external electric field is given by (see Chap. 2)

$$\overline{\overline{\epsilon}} = \begin{pmatrix} \epsilon_x & 0 & 0 \\ 0 & \epsilon_y & 0 \\ 0 & 0 & \epsilon_z \end{pmatrix} \quad (6.27)$$

The impermeability tensor $\overline{\overline{\eta}}$ is defined through the following equation:

$$\overline{\overline{\epsilon}}\,\overline{\overline{\eta}} = \overline{\overline{\eta}}\,\overline{\overline{\epsilon}} = \epsilon_0 \overline{\overline{I}} \quad (6.28)$$

where $\overline{\overline{I}}$ is the 3 × 3 unit tensor and Eq. (6.4) is nothing but the following equation:

$$\Delta\left(\overline{\overline{\frac{1}{n^2}}}\right) = \Delta\overline{\overline{\eta}} \quad (6.29)$$

Differentiating Eq. (6.28), we have

$$\Delta\overline{\overline{\epsilon}}.\overline{\overline{\eta}} + \overline{\overline{\epsilon}}.\Delta\overline{\overline{\eta}} = 0 \tag{6.30}$$

Multiplying by $\overline{\overline{\epsilon}}$ from the right-hand side and using Eq. (6.28), we obtain the following equation:

$$\Delta\overline{\overline{\epsilon}} = -\frac{1}{\epsilon_0}\overline{\overline{\epsilon}}\,\Delta\left(\overline{\overline{\frac{1}{n^2}}}\right)\overline{\overline{\epsilon}} \tag{6.31}$$

Thus, knowing the value of $\left(\overline{\overline{\frac{1}{n^2}}}\right)$, we can obtain the change in the dielectric permittivity of the medium in the presence of the external electric field.

Exercise 6.3 In Exercise 6.1, we obtained the value of $\left(\overline{\overline{\frac{1}{n^2}}}\right)$ for electric field applied along the *y*-direction in lithium niobate. Obtain the corresponding $\Delta\overline{\overline{\epsilon}}$.

Solution Using Eqs. (6.31) and (6.14), we obtain

$$\Delta\overline{\overline{\epsilon}} =$$

$$-\frac{1}{\epsilon_0}\epsilon_0\begin{pmatrix} n_o^2 & 0 & 0 \\ 0 & n_o^2 & 0 \\ 0 & 0 & n_e^2 \end{pmatrix}\begin{pmatrix} -r_{22}F_y & 0 & 0 \\ 0 & r_{22}F_y & r_{42}F_y \\ 0 & r_{42}F_y & 0 \end{pmatrix}\epsilon_0\begin{pmatrix} n_o^2 & 0 & 0 \\ 0 & n_o^2 & 0 \\ 0 & 0 & n_e^2 \end{pmatrix}$$

Multiplying the matrices, we obtain

$$\Delta\overline{\overline{\epsilon}} = -\epsilon_0\begin{pmatrix} -n_o^4 r_{22}F_y & 0 & 0 \\ 0 & n_o^4 r_{22}F_y & n_o^2 n_e^2 r_{42}F_y \\ 0 & n_o^2 n_e^2 r_{42}F_y & 0 \end{pmatrix} \tag{6.32}$$

The permittivity matrix in the presence of the electric field will be

$$\overline{\overline{\epsilon}} = \epsilon_0\begin{pmatrix} n_o^2 + n_o^4 r_{22}F_y & 0 & 0 \\ 0 & n_o^2 - n_o^4 r_{22}F_y & -n_o^2 n_e^2 r_{42}F_y \\ 0 & -n_o^2 n_e^2 r_{42}F_y & n_e^2 \end{pmatrix}$$

Note the presence of off-diagonal elements in the $\overline{\overline{\epsilon}}$ matrix in the presence of the external electric field.

Thus, in the presence of the external electric field, the permittivity tensor is no more diagonal thus representing the fact that the original principal axes are no more the principal axes in the presence of the electric field. In this case, the applied external electric field not only changes the principal refractive indices but also the orientations of the principal axes.

6.3 Electro-Optic Effect in Waveguides

In this section, we will discuss how the electro-optic effect modifies the waveguide properties and leads to applications in electro-optic devices based on optical waveguides.

As an example, we first consider an x-cut, y-propagating lithium niobate channel waveguide as shown in Fig. 6.1. We will use the crystal principal axes as the coordinate axes for analyzing the characteristics. The waveguide will support a quasi TE mode with its major electric field component oriented along the z-direction, which behaves like an extraordinary wave and a quasi TM mode with its major electric field component oriented along the x-direction, which behaves like an ordinary wave (see Chap. 4). As shown in the figure, there are two planar electrodes across which a voltage is applied leading to an electric field primarily oriented along the z- direction in the waveguide (here we will neglect the component of the applied electric field along the x-direction; more accurate analysis can be performed as detailed in the literature; see, e.g., Marcuse (1975)). Due to the electrode configuration, the electric field F_z will depend on both x and z.

We will now calculate the change in the propagation constant of the mode due to the applied electric field. For this, we start with the following two Maxwell's equations:

$$\nabla \times \vec{E} = -\frac{\partial \vec{B}}{\partial t} = i\omega\mu_0 \vec{H} \tag{6.33}$$

$$\nabla \times \vec{H} = \frac{\partial \vec{D}}{\partial t} = -i\omega \vec{D} \tag{6.34}$$

where we assume time dependence of the form $e^{-i\omega t}$ for all fields.

Taking curl of Eq. (6.33), we have

$$\nabla \times \nabla \times \vec{E} = i\omega\mu_0 \nabla \times \vec{H} = \omega^2 \mu_0 \vec{D}$$

or

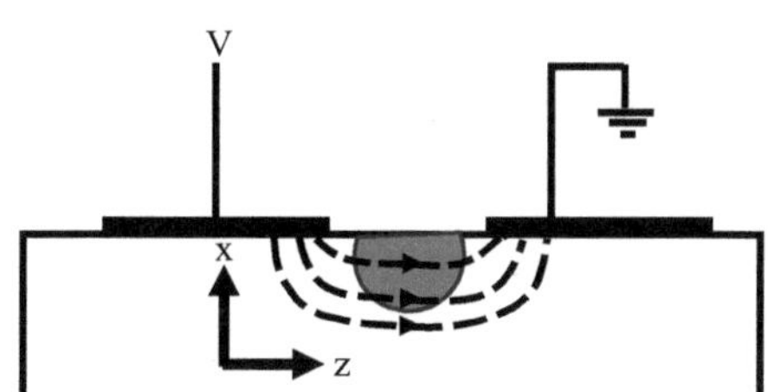

Fig. 6.1 Electric field is applied on a lithium niobate channel waveguide fabricated on an x-cut y-propagating lithium niobate substrate

$$\nabla\left(\nabla.\vec{E}\right) - \nabla^2\vec{E} = \omega^2\mu_0\vec{D} \tag{6.35}$$

In the absence of free charges, we have $\nabla.\vec{D} = \nabla.\left(\epsilon\vec{E}\right) = 0$ which for an almost homogeneous medium with small anisotropy can be approximated by $\nabla.\vec{E} = 0$.

Hence, we have

$$\nabla^2\vec{E} + \omega^2\mu_0\vec{D} = 0$$

Now

$$\vec{D} = \overline{\overline{\epsilon}}\,\vec{E}$$

where $\overline{\overline{\epsilon}}$ is the dielectric permittivity tensor. Hence, the wave equation is given by

$$\nabla^2\vec{E} + \omega^2\mu_0\overline{\overline{\epsilon}}\,\vec{E} = 0 \tag{6.36}$$

In the presence of an external electric field, the permittivity tensor gets modified and we write

$$\overline{\overline{\epsilon}} = \overline{\overline{\epsilon}}_u + \Delta\overline{\overline{\epsilon}} \tag{6.37}$$

where $\overline{\overline{\epsilon}}_u$ is the permittivity tensor in the absence of the external electric field (subscript u stands for unperturbed) and $\Delta\overline{\overline{\epsilon}}$ is the perturbation in the dielectric tensor. Substituting in Eq. (6.36) we have

$$\nabla^2\vec{E} + \omega^2\mu_0\overline{\overline{\epsilon}}_u\vec{E} = -\omega^2\mu_0\Delta\overline{\overline{\epsilon}}\vec{E} \tag{6.38}$$

The right-hand side of Eq. (6.38) denotes the perturbation in the waveguide due to the electro optic effect. In order to obtain the effect of the external electric field on the propagation characteristics of the waveguide, we need to solve Eq. (6.38).

As an example, we note that in the presence of an external electric field along the z-direction in lithium niobate the change in the dielectric permittivity will be given by (see Eqs. 6.8 and 6.31)

$$\Delta\overline{\overline{\epsilon}} = -\frac{1}{\epsilon_0}\epsilon_0\begin{pmatrix} n_o^2 & 0 & 0 \\ 0 & n_o^2 & 0 \\ 0 & 0 & n_e^2 \end{pmatrix}\begin{pmatrix} r_{13}F_z & 0 & 0 \\ 0 & r_{13}F_z & 0 \\ 0 & 0 & r_{33}F_z \end{pmatrix}\epsilon_0\begin{pmatrix} n_o^2 & 0 & 0 \\ 0 & n_o^2 & 0 \\ 0 & 0 & n_e^2 \end{pmatrix}$$

which gives us

$$\Delta\overline{\overline{\epsilon}} = -\epsilon_0 \begin{pmatrix} n_o^4 r_{13} F_z & 0 & 0 \\ 0 & n_o^4 r_{13} F_z & 0 \\ 0 & 0 & n_e^4 r_{33} F_z \end{pmatrix} \tag{6.39}$$

which acts like a dielectric permittivity perturbation on the waveguide and which changes the propagation characteristics of the waveguide. Thus, the permittivity of the medium in the presence of the electric field along the z-direction is given by

$$\overline{\overline{\epsilon}} = \overline{\overline{\epsilon}}_u + \Delta\overline{\overline{\epsilon}} = \epsilon_0 \begin{pmatrix} n_o^2 - n_o^4 r_{13} F_z & 0 & 0 \\ 0 & n_o^2 - n_o^4 r_{13} F_z & 0 \\ 0 & 0 & n_e^2 - n_e^4 r_{33} F_z \end{pmatrix} \tag{6.40}$$

In the configuration, we are considering, the permittivity tensor even in the presence of the external electric field remains diagonal implying that the principal axes have not changed their orientations; only the principal refractive indices have changed. Hence, even in the presence of the external electric field, the quasi TE and quasi TM modes will have extraordinary and ordinary polarization, respectively, and would be the eigenmodes.

Since the electro-optic effect leads to changes in the dielectric permittivity tensor, the effect can be used to modulate the phase, polarization state, or the amplitude of the propagating mode leading to phase modulator, polarization modulator, or intensity modulator. We will now discuss some specific examples of waveguide electro-optic modulator in lithium niobate.

Exercise 6.4 Consider the presence of both the components F_x and F_z in the applied electric field in lithium niobate. Obtain the corresponding $\Delta\overline{\overline{\epsilon}}$.

Solution In the presence of both F_x and F_z, we have

$$\Delta\overline{\overline{\left(\frac{1}{n^2}\right)}} = \begin{pmatrix} 0 & -r_{22} & r_{13} \\ 0 & r_{22} & r_{13} \\ 0 & 0 & r_{33} \\ 0 & r_{42} & 0 \\ r_{51} & 0 & 0 \\ -r_{22} & 0 & 0 \end{pmatrix} \begin{pmatrix} F_x \\ 0 \\ F_z \end{pmatrix} = \begin{pmatrix} r_{13} F_z \\ r_{13} F_z \\ r_{33} F_z \\ 0 \\ r_{51} F_x \\ -r_{22} F_x \end{pmatrix}$$

which implies

$$\Delta\overline{\overline{\left(\frac{1}{n^2}\right)}} = \begin{pmatrix} r_{13} F_z & -r_{22} F_x & r_{51} F_x \\ -r_{22} F_x & r_{13} F_z & 0 \\ r_{51} F_x & 0 & r_{33} F_z \end{pmatrix} \tag{6.41}$$

Hence,

$$\Delta\overline{\overline{\epsilon}} = -\frac{1}{\epsilon_0}\epsilon_0\begin{pmatrix} n_o^2 & 0 & 0 \\ 0 & n_o^2 & 0 \\ 0 & 0 & n_e^2 \end{pmatrix}\begin{pmatrix} r_{13}F_z & -r_{22}F_x & r_{51}F_x \\ -r_{22}F_x & r_{13}F_z & 0 \\ r_{51}F_x & 0 & r_{33}F_z \end{pmatrix}\epsilon_0\begin{pmatrix} n_o^2 & 0 & 0 \\ 0 & n_o^2 & 0 \\ 0 & 0 & n_e^2 \end{pmatrix}$$

which gives

$$\Delta\overline{\overline{\epsilon}} = \begin{pmatrix} n_o^4 r_{13}F_z & -n_o^4 r_{22}F_x & n_o^2 n_e^2 r_{51}F_x \\ -n_o^4 r_{22}F_x & n_o^4 r_{13}F_z & 0 \\ n_o^2 n_e^2 r_{51}F_x & 0 & n_e^4 r_{33}F_z \end{pmatrix} \tag{6.42}$$

The externally applied electric field component F_x leads to off-diagonal components which leads to coupling of power among the eigenmodes of the waveguide.

6.3.1 Electro-Optic Phase Modulator

As an example, we first consider a phase modulator using the same electrode configuration as shown in Fig. 6.1. The quasi TE mode of the channel waveguide corresponds to extraordinary polarization as its main electric field component is along the z-direction or optic axis of the crystal.

In the following analysis, we will use the principal axis system to describe the propagation in the waveguide. Thus, the propagation direction will be along y and the normal to the substrate-cover interface will be along x.

Hence, in the absence of the external electric field, the wave equation satisfied by the electric field component E_z of the quasi TE mode would be (see Chap. 4)

$$\nabla_T^2 E_z + \left(k_0^2 n_e^2 - \beta_u^2\right)E_z = 0 \tag{6.43}$$

where β_u is the unperturbed propagation constant of the mode in the absence of the external electric field and $\nabla_T^2 = \nabla^2 - \frac{\partial^2}{\partial y^2}$ since the propagation direction is chosen to be along the y-direction and we have assumed a y-dependence of the form $e^{i\beta_u y}$.

The external electric field changes the permittivity profile of the waveguide. We assume that the change in the permittivity $\Delta\overline{\overline{\epsilon}}$ to be very small and can be considered to be a perturbation to the waveguide and thus assume that the modal electric field profile does not get modified in the presence of the electric field; only the propagation constant of the waveguide changes due to the applied electric field. This is very similar to the standard perturbation theory in quantum mechanics.

Thus, in the presence of the perturbation due to the external electric field F_z, the electric field $\tilde{E}_z$ of the mode satisfies the equation

$$\nabla_T^2 \tilde{E}_z + \left[k_0^2 \left(n_e^2 - n_e^4 r_{33} F_z\right) - \beta_p^2\right] \tilde{E}_z = 0 \tag{6.44}$$

where $k_0^2 = \omega^2 \epsilon_0 \mu_0$, β_p is the perturbed propagation constant of the mode in the presence of the external electric field and $\tilde{E}_z \approx E_z$. In writing Eq. (6.44), we have used the component ϵ_{zz} from Eq. (6.40).

Multiplying Eq. (6.44) on both sides by E_z and Eq. (6.43) on both sides by $\tilde{E}_z$ and subtracting followed by an integration over the entire $x - z$ plane, we obtain

$$\iint \left(E_z \nabla_T^2 \tilde{E}_z - \tilde{E}_z \nabla_T^2 E_z\right) dxdz = \left(\beta_p^2 - \beta_u^2\right) \iint \tilde{E}_z E_z dxdz + k_0^2 n_e^4 \iint r_{33} F_z \tilde{E}_z E_z dxdz$$

The integral on the left-hand side of the above equation can be shown to be zero using Green's theorem. Hence, we obtain

$$\beta_p^2 = \beta_u^2 - k_0^2 n_e^4 \frac{\iint r_{33} F_z E_z^2 dxdz}{\iint E_z^2 dxdz} \tag{6.45}$$

where we have assumed $\tilde{E}_z \approx E_z$, that is, the modal field distribution remains the same in the presence of the small perturbation and has included the factor r_{33} within the integral to take account of situations when the electro-optic coefficient could depend on the coordinates. In case r_{33} is independent of the coordinates, then we have

$$\beta_p^2 = \beta_u^2 - k_0^2 n_e^4 r_{33} \frac{\iint F_z E_z^2 dxdz}{\iint E_z^2 dxdz} \tag{6.46}$$

Since the second term in the above equation is very small we can approximate and obtain the following expression for the propagation constant in the presence of the external electric field as

$$\beta_p \approx \beta_u - \frac{k_0^2 n_e^4 r_{33}}{2\beta_u} \frac{\iint F_z E_z^2 dxdz}{\iint E_z^2 dxdz} \tag{6.47}$$

Hence, the change in the propagation constant of the mode due to the applied electric field is given by

$$\Delta\beta = \beta_p - \beta_u = -\frac{k_0^2 n_e^4 r_{33}}{2\beta_u} \frac{\iint F_z E_z^2 dxdz}{\iint E_z^2 dxdz} \tag{6.48}$$

Since $\beta_u = k_0 n_{eff}$ where n_{eff} is the effective index of the mode in the absence of the applied electric field, we obtain the following expression for the change in the effective index of the mode due to the electro-optically induced change in the permittivity as

$$\Delta n_{eff} = -\frac{k_0 n_e^4 r_{33}}{2\beta_u} \frac{\iint F_z E_z^2 dxdz}{\iint E_z^2 dxdz} \tag{6.49}$$

Since the waveguide is weakly guiding, on the right-hand side of Eq. (6.49) we may approximate the unperturbed propagation constant of the extraordinarily polarized mode, $\beta_u = k_0 n_e$, where n_e is the extraordinary refractive index of lithium niobate and obtain

$$\Delta n_{eff} \approx -\frac{n_e^3 r_{33}}{2} \frac{\iint F_z E_z^2 dxdz}{\iint E_z^2 dxdz} \tag{6.50}$$

If the potential difference between the two electrode is V and G is the electrode spacing, then we can rewrite Eq. (6.50) as

$$\Delta n_{eff} = -\frac{n_e^3 r_{33}}{2} \frac{V}{G} \Gamma \tag{6.51}$$

where

$$\Gamma = \frac{G}{V} \frac{\iint F_z E_z^2 dxdz}{\iint E_z^2 dxdz} \tag{6.52}$$

represents a dimensionless overlap factor between the applied external electric field distribution $F_z(x, z)$ and the modal field distribution $E_z(x, z)$. Note that Γ is a dimensionless parameter.

Note that if instead of planar electrodes, we had a situation wherein the waveguide was placed between two parallel electrodes with a spacing G, then the electric field $F_z$would be independent of x and z and would simply be equal to $\frac{V}{G}$, the applied voltage divided by the electrode spacing. In this case, we will have $\Gamma = 1$ which is the maximum value of the overlap factor. The value of Γ which is always less than unity indicates the deviation from complete overlap between the applied electric field and the modal field distribution; larger the value of Γ the greater will be the induced phase change due to the applied electric field. This is incorporated in the design of electrodes for optimum performance of electro-optic modulators.

For an actual device, the overlap factor can vary depending on the modal field distributions and the electrode configurations and is typically around 0.7–0.8. Thus, the voltage requirement to achieve a certain phase change due to electro-optic effect will be higher than in the case of bulk.

If the input field at $y = 0$ is described by

$$E_z(y=0) = E_{z0}e^{-i\omega t} \tag{6.53}$$

then the electric field of the mode at the output after propagating through a length L of the waveguide would be

$$E_z(y=L) = E_{z0}e^{i\left(\beta_p L - \omega t\right)} = E_{z0}e^{i\left(k_o n_{eff} L - k_o \Delta n_{eff} L - \omega t\right)} \tag{6.54}$$

Thus, the electro-optically induced phase change suffered by the propagating mode over a length L of the electrode will be given by

$$\Delta\phi = \frac{2\pi}{\lambda_0}\Delta n_{eff} L = -\frac{\pi n_e^3 r_{33}}{\lambda_0}\frac{V}{G}\Gamma L \tag{6.55}$$

The above equation represents the phase change suffered by the extraordinary polarization.

In general, we can write for the phase change as

$$\Delta\phi = -\frac{\pi \overline{n}^3 \overline{r}}{\lambda_0}\frac{V}{G}\Gamma L = -\alpha V \tag{6.56}$$

where

$$\alpha = \frac{\pi \overline{n}^3 \overline{r}}{\lambda_0}\frac{\Gamma L}{G} \tag{6.57}$$

where $\overline{n}$ is the corresponding effective index depending on the polarization state and $\overline{r}$ is the corresponding effective electro-optic coefficient determined by the crystal, the propagation direction, and the polarization state of the propagating mode. Thus,

$$E_z(y=L) = E_{z0}e^{i\left(k_o n_{eff} L - \omega t\right)}e^{-i\alpha V} \tag{6.58}$$

Thus, as the applied voltage V changes, the phase of the output mode changes leading to phase modulation.

The voltage V_0 required to induce a phase change of π at the output is

$$V_0 = \frac{\lambda_0}{\overline{n}^3 \overline{r}}\frac{G}{\Gamma L} \tag{6.59}$$

which is usually written as a voltage-length product:

$$V_0.L = \frac{\lambda_0}{\overline{n}^3 \overline{r}}\frac{G}{\Gamma} \tag{6.60}$$

Note that the required voltage reduces with increased length L of the electrodes, increased overlap factor Γ, higher effective electro-optic coefficient $\overline{r}$. These characteristics are taken into account while designing high-efficiency waveguide modulators. Since $\Gamma < 1$, the voltage required for modulation in the waveguide configuration is larger than that required in bulk. At the same time, propagation through a waveguide overcomes diffraction effects, which limit the length of interaction for a given electrode spacing (see Exercise 6.6).

It may be noted from Eq. (6.60) that for achieving lower voltage-length requirement, lower values of electrode spacing and higher mode overlap integrals are necessary. The electrode spacing G is determined by the mode size and the overlap Γ by the mode distribution and electric field distribution. The technology of lithium niobate on insulator (LNOI) is expected to provide for much improved values of the product as the waveguides are characterized by a large index difference leading to very small mode sizes, resulting in smaller electrode spacing. In this respect waveguides based on thin films of lithium niobate offer low drive voltage and high bandwidth electro optic modulators.

Exercise 6.5 Consider a step-index planar waveguide having a profile given by Eq. (4.22) in which the film and superstrate are made of doped silica and air and the substrate having a finite value of r_{33}. Obtain an expression for Δn_{eff} (as given in Eq. 6.50) assuming the external electric field to be uniform within the substrate.

Solution Since the film and superstate do not possess Pockels effect, we obtain

$$\Delta n_{eff} \approx -\frac{n_e^3 r_{33}}{2} \frac{\iint F_z E_z^2 dxdz}{\iint E_z^2 dxdz} = -\frac{n_e^3 r_{33}}{2} \frac{F_z L}{\iint E_z^2 dxdz} \int_{-\infty}^{-d} E_z^2 dx$$

where L is the width of the electrode in the z-direction which is parallel to the substrate surface. Now

$$\begin{aligned}\int_{-\infty}^{-d} E_z^2 dx &= A^2 \left(\cos \kappa_f d + \frac{\gamma_c}{\kappa_f} \sin \kappa_f d \right)^2 \int_{-\infty}^{-d} e^{+2\gamma_s (x+d)} dx \\ &= \frac{A^2}{2\gamma_s} \left(\cos \kappa_f d + \frac{\gamma_c}{\kappa_f} \sin \kappa_f d \right)^2\end{aligned}$$

where we have used Eq. (4.46). We also have

$$\iint E_z^2 dxdz = \frac{A^2 L}{2} \left(1 + \frac{\gamma_c^2}{\kappa_f^2} \right) \left(d + \frac{1}{\gamma_s} + \frac{1}{\gamma_c} \right)$$

which is obtained in analogy to Eqs. (4.62) and (4.63). Hence, we have

$$\Delta n_{eff} \approx -\frac{n_e^3 r_{33} F_z}{2\gamma_s}\left(1+\frac{\gamma_c^2}{\kappa_f^2}\right)^{-1}\left(d+\frac{1}{\gamma_s}+\frac{1}{\gamma_c}\right)^{-1}\left(\cos\kappa_f d+\frac{\gamma_c}{\kappa_f}\sin\kappa_f d\right)^2 \tag{6.61}$$

Exercise 6.6 Consider a phase modulator of length 20 mm in lithium niobate operating with an input light of wavelength 850 nm and with an input that has extraordinary polarization. Estimate the voltage requirement for inducing a phase shift of π. Assume an electrode spacing of 10 μm and $\Gamma = 0.5$.

Solution Using Eq. (6.59) and putting $\bar{r} = r_{33} = 30 \times 10^{-12}$ m/V, $n = n_e = 2.2$, we have

$$V_0 = \frac{\lambda_0}{\bar{n}^3 \bar{r}} \frac{G}{\Gamma L} = \frac{0.85 \times 10^{-6} \times 10 \times 10^{-6}}{(2.2)^3 \times 30 \times 10^{-12} \times 0.5 \times 20 \times 10^{-3}} \approx 2.7 \text{ V}$$

which is indeed a very small voltage.

It is interesting to note that a Gaussian beam at a wavelength of 850 nm and having a beam waist of 5 μm propagating in a medium of refractive index 2.2 will diffract and increase its transverse dimension, due to diffraction effects, by a factor of $\sqrt{2}$ in a distance of $z = \frac{\pi w_0^2}{\lambda_0} n \approx 0.2$ mm itself (see, e.g., Ghatak and Thyagarajan 1989). Compared to this, since the propagation is in a waveguide, diffraction effects are balanced by the waveguiding effect and the beam width can be maintained over much longer distances showing the clear advantage of using waveguides in place of bulk.

Exercise 6.7 In the electrode configuration shown in Fig. 6.1, if the incident mode has a polarization normal to the substrate-air interface, what will be the phase change suffered by the mode propagating over a length L of the waveguide.

Solution Polarization normal to the substrate-air interface corresponds to a TM mode and in view of the orientation of the substrate, the incident polarization corresponds to an ordinary wave with a refractive index n_o (see Chap. 4). Thus, the change in refractive index will now be determined by the r_{13} coefficient rather than the r_{33} coefficient. Hence, the phase change will be given by (see Eq. 6.55)

$$\Delta\phi = -\frac{\pi n_o^3 r_{13}}{\lambda_0} \frac{V}{G} \Gamma L$$

For lithium niobate at a wavelength of 633 nm $n_o \approx 2.237$, $n_e \approx 2.158$, $r_{13} \approx 8.6 \times 10^{-12}$m/V, $r_{33} \approx 30 \times 10^{-12}$m/V. Thus, $n_o^3 r_{13} \approx 96.3 \times 10^{-12}$ m/V and $n_e^3 r_{33} \approx 301 \times 10^{-12}$ m/V. Assuming the overlap factor in the case with ordinary or extraordinary polarization is almost the same, the voltage requirement of the quasi TE polarization (extraordinary mode) is much smaller than that of the quasi TM ordinary polarization (ordinary mode).

Exercise 6.8 Consider a Y-branch in lithium niobate similar to the one shown in Fig. 7.1. All waveguides are single-mode waveguides. The two arms of the Y-branch are far away and have no interaction between them. In one of the arms of the Y-branch, there is an electro-optic phase modulator. (a) If no voltage is applied to the modulator what will be the powers at the exit of the Y-branch and what will be their phase difference? (b) If the phase modulator introduces a phase difference of π between the two arms, what will be the exiting powers, and what will be their phase difference?

Solution Since the Y-branch is symmetric, there will be equal power in both arms of the Y-branch. (a) when there is no voltage applied to the phase modulator, the two outputs will be in phase. This corresponds to the symmetric state of the two waveguide device (see Chap. 5). (b) When a π difference is generated between the two arms, then the output will correspond to the antisymmetric mode of the two waveguide device with equal powers in both arms but having a phase difference of π between the two arms. What happens if the phase difference generated by the electro-optic modulator corresponds to $\frac{\pi}{2}$?

Figure 6.1 shows a phase modulator configuration realized in x- cut lithium niobate substrate. The two metal electrodes are placed on either side of the waveguide so as to generate an electric field along the z- axis which then allows one to use the maximum electro-optic coefficient r_{33}. It is also possible to design a phase modulator in z- cut y-propagating lithium niobate substrate in which the z- axis (optic axis) is perpendicular to the substrate surface and the propagation direction is along y. In this case, if the electrodes are placed as shown in Fig. 6.1, then the major component of the applied electric field will be along the x- direction and the electro-optic effect will no more be described by Eq. (6.51) and the field will cause coupling among the modes (see Exercise 6.4) rather than a simple phase modulation. In this case in order to again apply electric field predominantly along the z-direction, the electrodes need to be placed on top of the waveguide as shown in Fig. 6.2.

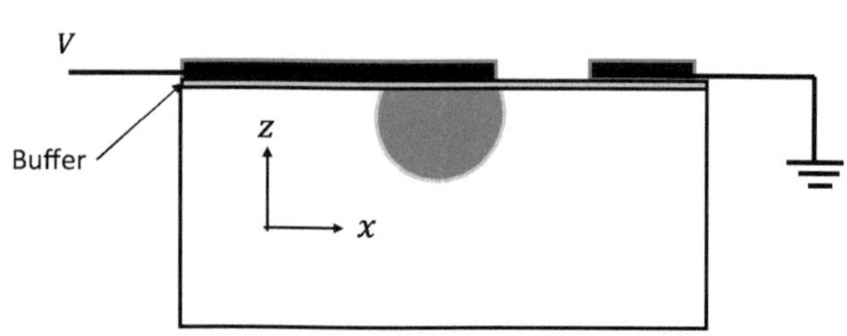

Fig. 6.2 Electrode configuration for phase modulation in a z-cut, y-propagating lithium niobate waveguide

In order to use the maximum electro-optic coefficient namely r_{33}, the mode needs to be polarized along the z- direction, which implies that the mode must be quasi TM mode. We saw in Chap. 4 that having a metal cladding induces a loss in the waveguide due to the finite conductivity of the metal. In order to reduce the effect of metal cladding, usually buffer layer which is a thin film of dielectric is used between the waveguide and the metal electrode which ensures less interaction between the waveguide mode and the metal and thus reduces loss in the waveguide (see Fig. 6.2). Of course using a dielectric buffer layer between the waveguide and the metal electrode will lead to a reduction of the electric field overlapping with the mode which in turn will lead to requirement of slightly higher voltages for phase modulation.

6.3.2 *Electro-Optic Intensity Modulator*

In Sect. 6.3.1, we studied about electro-optic phase modulator using a waveguide. In many applications, we need to modulate the intensity of the light rather than the phase. Thus, we need to convert phase modulation to intensity modulation. This is possible using different arrangements.

In Chap. 7, we will discuss the Mach-Zehnder interferometer modulator which converts phase modulation to intensity modulation by interfering a phase-modulated mode and another mode without any phase modulation. In this case, a change of phase of π in one of the arms of the interferometer leads to full modulation with no output from the device. Also if the voltage is modulated with time, then this leads to amplitude/intensity modulation of the output.

One can also use a directional coupler (see Chap. 5) for modulation or as an optical switch.

6.3.3 *Directional Coupler Optical Switch*

In Chap. 5, we discussed in detail about directional couplers in which closely lying waveguides interact through their evanescent fields and this leads to coupling of optical power from one waveguide to another. We showed that by having a directional coupler of length equal to the coupling length referred to as L_c, power input in, say waveguide # 1, exits from waveguide # 2 which is referred to as the cross state of the coupler. In order to switch the power to waveguide # 1 at the output, referred to as a parallel state of the switch, we need to generate a $\Delta\beta$ given by the following equation between the waveguides (see Eq. 5.95)

$$\Delta\beta = \pm \frac{\sqrt{3}\pi}{L_c} \tag{6.62}$$

For switching operation it must be possible to generate this value of $\Delta\beta$ dynamically. This can be easily accomplished using the electro-optic effect.

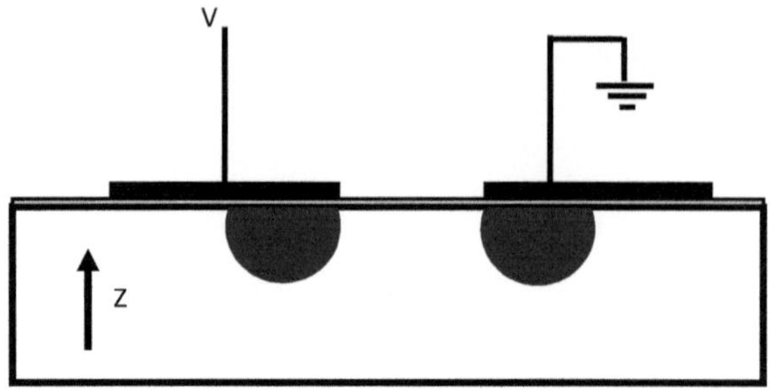

Fig. 6.3 Electro-optic directional coupler switch using the push-pull configuration

Figure 6.3 shows a directional coupler optical switch in *z*- cut, *y*- propagating lithium niobate waveguide. The two waveguides form a directional coupler and the metal electrodes for applying the electric field are placed on top of the two waveguides. In order to avoid loss due to the proximity of the metal to the waveguides, a thin dielectric buffer layer is used.

When a voltage difference is applied between the two electrodes, each of the waveguides will be acted upon by an electric field and due to the design of the electrode pattern, the direction of electric field in the two waveguides will be oppositely oriented. Since the Pockels effect is proportional to the applied electric field, the electric field will result in increase of effective index of the mode of one waveguide and decrease of effective index of the mode of the parallel waveguide. In this case, referred to as push–pull configuration, the required voltage will be halved as compared to applying an electric field on only one of the waveguides (Papuchon et al. 1975).

From Eq. (6.51), we have (assuming that in the two waveguides, the changes are $+\Delta n_{eff}$ and $-\Delta n_{eff}$) the: $\Delta\beta$ generated between the two waveguides due to the electro-optic effect would be

$$\Delta\beta = 2\frac{2\pi}{\lambda_0}\Delta n_{eff} = \frac{2\pi n_e^3 r_{33}}{\lambda_0}\frac{V}{G}\Gamma \tag{6.63}$$

where we have disregarded the sign. The extra factor of 2 is to account for the push-pull effect. Thus, the required voltage for switching would be that voltage which introduces a $\Delta\beta$ given by Eq. (6.63) which equals the $\Delta\beta$ required for switching (see Eq. 6.62). Thus, we obtain the required voltage for switching to be

$$V_s = \frac{\sqrt{3}\lambda_0 G}{2n_e^3 r_{33} L_c \Gamma} \tag{6.64}$$

As an example, we take the following typical values:

$$\lambda_0 = 1550\ \text{nm};\quad n_e = 2.2;\quad r_{33} = 30\frac{\text{pm}}{\text{V}};\quad G = 10\ \mu\text{m};\quad L_c = 30\ \text{mm};\quad \Gamma = 0.5$$

Substituting these values in Eq. (6.64), we obtain the required voltage $V_s \approx 2.8$ V.

In Chap. 5, we have also discussed about switched $\Delta\beta$ directional coupler switch to resolve some practical issues of realizing a directional coupler of an exact length equal to L_c.

6.4 Electro-Optic Polarization Converter

As another example, we consider an electro-optic polarization converter which has many applications in guided wave optics and should find applications in integrated quantum photonics. This is an example of what is referred to as co-directional coupling, which is discussed in detail in Chap. 9.

Figure 6.4 shows the electrode configuration of an electro-optic polarization converter in a z-cut x- propagating lithium niobate substrate.

The electrodes lying on either side of the waveguide produce an electric field primarily oriented along the y-direction in the waveguide. Assuming the electric field to have only the y- component, we first obtain the change $\Delta\overline{\overline{\left(\frac{1}{n^2}\right)}}$ caused by the electric field:

$$\Delta\overline{\overline{\left(\frac{1}{n^2}\right)}} = \begin{pmatrix} 0 & -r_{22} & r_{13} \\ 0 & r_{22} & r_{13} \\ 0 & 0 & r_{33} \\ 0 & r_{42} & 0 \\ r_{51} & 0 & 0 \\ -r_{22} & 0 & 0 \end{pmatrix} \begin{pmatrix} 0 \\ F_y \\ 0 \end{pmatrix} = \begin{pmatrix} -r_{22}F_y \\ r_{22}F_y \\ 0 \\ r_{42}F_y \\ 0 \\ 0 \end{pmatrix} \tag{6.65}$$

which in the de-contracted notation is given by

$$\Delta\overline{\overline{\left(\frac{1}{n^2}\right)}} = \begin{pmatrix} -r_{22}F_y & 0 & 0 \\ 0 & r_{22}F_y & r_{42}F_y \\ 0 & r_{42}F_y & 0 \end{pmatrix} \tag{6.66}$$

Hence, the change in the dielectric permittivity is obtained using Eq. (6.31):

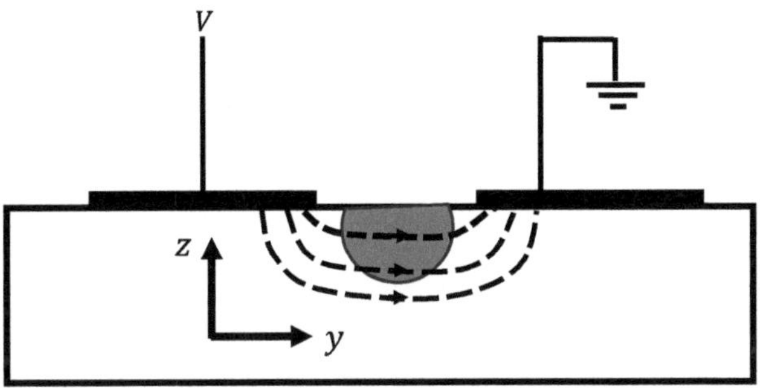

Fig. 6.4 Electro-optic polarization converter configuration in a z-cut x-propagating lithium niobate waveguide

$$\Delta\overline{\overline{\epsilon}} = -\frac{1}{\epsilon_0}\epsilon_0\begin{pmatrix} n_o^2 & 0 & 0 \\ 0 & n_o^2 & 0 \\ 0 & 0 & n_e^2 \end{pmatrix}\begin{pmatrix} -r_{22}F_y & 0 & 0 \\ 0 & r_{22}F_y & r_{42}F_y \\ 0 & r_{42}F_y & 0 \end{pmatrix}\epsilon_0\begin{pmatrix} n_o^2 & 0 & 0 \\ 0 & n_o^2 & 0 \\ 0 & 0 & n_e^2 \end{pmatrix}$$

which gives us

$$\Delta\overline{\overline{\epsilon}} = -\epsilon_0\begin{pmatrix} -n_o^4 r_{22}F_y & 0 & 0 \\ 0 & n_o^4 r_{22}F_y & n_o^2 n_e^2 r_{42}F_y \\ 0 & n_o^2 n_e^2 r_{42}F_y & 0 \end{pmatrix} \tag{6.67}$$

Notice that now the perturbation in the permittivity has off-diagonal elements which will induce coupling among the two eigenpolarizations namely *y*- and *z*-polarizations in this example. The permittivity in the presence of the electric field will be

$$\overline{\overline{\epsilon}} = \epsilon_0\begin{pmatrix} n_o^2 + n_o^4 r_{22}F_y & 0 & 0 \\ 0 & n_o^2 - n_o^4 r_{22}F_y & -n_o^2 n_e^2 r_{42}F_y \\ 0 & -n_o^2 n_e^2 r_{42}F_y & n_e^2 \end{pmatrix} \tag{6.68}$$

The form of the permittivity tensor implies that the original principal axes are no more the principal axes in the presence of the electric field along the *y*-direction. Thus, this component of electric field not only changes the principal refractive indices but also the orientation of the principal axes of the crystal.

The wave equation satisfied by the propagating mode is (see Eq. 6.36)

$$\nabla^2 \vec{E} + \omega^2 \mu_0 \overline{\overline{\epsilon}} \vec{E} = 0 \tag{6.69}$$

In the weakly guiding approximation, we can write for the total electric field as consisting of the electric fields of both the *y*-polarized (quasi TE) and *z*-polarized (quasi TM) propagating modes as

$$\vec{E}(x, y, z) = \hat{y}E_y(x, y, z) + \hat{z}E_z(x, y, z) \tag{6.70}$$

which can be represented in a matrix form as

$$\vec{E} = \begin{pmatrix} 0 \\ E_y \\ E_z \end{pmatrix} \tag{6.71}$$

Substituting from Eqs. (6.68) and (6.71) in Eq. (6.69), we obtain the following equation:

$$\begin{pmatrix} 0 \\ \nabla^2 E_y \\ \nabla^2 E_z \end{pmatrix} + \omega^2 \mu_0 \epsilon_0 \begin{pmatrix} n_o^2 + n_o^4 r_{22} F_y & 0 & 0 \\ 0 & n_o^2 - n_o^4 r_{22} F_y & -n_o^2 n_e^2 r_{42} F_y \\ 0 & -n_o^2 n_e^2 r_{42} F_y & n_e^2 \end{pmatrix} \begin{pmatrix} 0 \\ E_y \\ E_z \end{pmatrix} = 0 \tag{6.72}$$

Hence, we have

$$\nabla^2 E_y + \omega^2 \mu_0 \epsilon_0 \left(\left(n_o^2 - n_o^4 r_{22} F_y \right) E_y - n_o^2 n_e^2 r_{42} F_y E_z \right) = 0$$

or

$$\nabla^2 E_y + \omega^2 \mu_0 \epsilon_0 \left(n_o^2 - n_o^4 r_{22} F_y \right) E_y = \omega^2 \mu_0 \epsilon_0 n_o^2 n_e^2 r_{42} F_y E_z \tag{6.73}$$

Similarly, we obtain

$$\nabla^2 E_z + \omega^2 \mu_0 \epsilon_0 n_e^2 E_z = \omega^2 \mu_0 \epsilon_0 n_o^2 n_e^2 r_{42} F_y E_y \tag{6.74}$$

The terms on the right-hand side of Eqs. (6.73) and (6.74) represent the coupling term and result in coupling between the modes polarized along the y- and z-directions or equivalently quasi TE and quasi TM modes characterized by E_y and E_z, respectively. Note that it is the off-diagonal elements of the perturbation in the permittivity that is responsible for the coupling among the two eigenpolarization states.

Now, in the absence of the applied electric field, we have

$$\nabla^2 E_y + \omega^2 \mu_0 \epsilon_0 n_o^2 E_y = 0 \tag{6.75}$$

and

$$\nabla^2 E_z + \omega^2 \mu_0 \epsilon_0 n_e^2 E_z = 0 \tag{6.76}$$

and the two modes are independent and de-coupled. If β_y and β_z represent the propagation constants of the y-polarized quasi TE mode and z-polarized quasi TM mode, then their x-dependence would be of the form $e^{i\beta_y x}$ and $e^{i\beta_z x}$, respectively, and we would have

$$\frac{\partial^2 E_y}{\partial y^2} + \frac{\partial^2 E_y}{\partial z^2} + \left(k_0^2 n_o^2 - \beta_y^2 \right) E_y = 0 \tag{6.77}$$

and

$$\frac{\partial^2 E_z}{\partial y^2} + \frac{\partial^2 E_z}{\partial z^2} + \left(k_0^2 n_e^2 - \beta_z^2\right) E_z = 0 \tag{6.78}$$

where $k_0^2 = \omega^2 \epsilon_0 \mu_0$. Note that the propagation direction in the waveguide is along x.

The above two equations represent the equations satisfied by the quasi TE mode and quasi TM mode, respectively, in the absence of the perturbation.

We can use the standard coupled mode theory (see Chap. 5) to study the coupling produced by the electro-optic effect. For this, we write the total electric field at any value of x to be given by

$$\vec{E} = \hat{y} E_y(x, y, z) + \hat{z} E_z(x, y, z) = \hat{y} A(x) E_y(y, z) e^{i\beta_y x} + \hat{z} B(x) E_z(y, z) e^{i\beta_z x} \tag{6.79}$$

Here $E_y(y, z)$ and $E_z(y, z)$ represent the transverse modal field distributions of the quasi TE and quasi TM modes, respectively, and $A(x)$ and $B(x)$ represent the x-dependent amplitudes of the two coupled modes which change due to the coupling process. In the absence of coupling, since these are eigenmodes of the waveguide, these two amplitudes will be independent of x; it is the coupling that leads to their $x-$ dependence.

Using the fields E_y and E_z from Eq. (6.79), Eq. (6.73) becomes

$$\begin{aligned} &\nabla^2 \left\{ A(x) E_y(y, z) e^{i\beta_y x} \right\} \\ &\quad + \omega^2 \mu_0 \epsilon_0 \left(n_o^2 A(x) E_y(y, z) e^{i\beta_y x} - n_o^2 n_e^2 r_{42} F_y B(x) E_z(y, z) e^{i\beta_z x} \right) = 0 \end{aligned}$$

where we have neglected the term containing $n_o^4 r_{22} F_y$ in Eq. (6.73). Expanding the Laplacian and assuming $A(x)$ to be a slowly varying function of x which helps us to neglect the term $d^2A(x)/dx^2$, we have

$$\begin{aligned} &A(x) \nabla_T^2 E_y e^{i\beta_y x} + \left[-\beta_y^2 A(x) + 2i\beta_y \frac{dA}{dx} \right] E_y e^{i\beta_y x} \\ &\quad + \omega^2 \mu_0 \epsilon_0 \left(n_o^2 A(x) E_y(y, z) e^{i\beta_y x} - n_o^2 n_e^2 r_{42} F_y B(x) E_z(y, z) e^{i\beta_z x} \right) = 0 \end{aligned}$$

where $\nabla_T^2 = \frac{\partial^2}{\partial y^2} + \frac{\partial^2}{\partial z^2}$. Using the equation satisfied by E_y (Eq. 6.77), we obtain

$$2i\beta_y E_y(y, z) e^{i\beta_y x} \frac{dA}{dx} - k_0^2 n_o^2 n_e^2 r_{42} F_y B(x) E_z(y, z) e^{i\beta_z x} = 0 \tag{6.80}$$

Multiplying by $E_y(y, z)$ and integrating over the entire $y-z$ plane, we obtain

$$2i\beta_y \frac{dA}{dx} e^{i\beta_y x} \iint E_y^2(y, z) dy dz - k_0^2 n_o^2 n_e^2 r_{42} B(x) e^{i\beta_z x} \iint F_y E_y(y, z) E_z(y, z) dy dz = 0 \tag{6.81}$$

where we have neglected the y, z dependence of n_o^2 and n_e^2.

Equation (6.81) can be simplified to give the following equation:

$$\frac{dA}{dx} = -i\frac{k_0^2 n_o^2 n_e^2 r_{42}}{2\beta_y} e^{i\Delta\beta x} \frac{\iint F_y E_y(y,z) E_z(y,z) dydz}{\iint E_y^2(y,z) dydz} B(x)$$

where $\Delta\beta = \beta_z - \beta_y$. We assume that the field is power normalized, that is,

$$\frac{\beta_y}{2\omega\mu_0} \iint E_y^2(y,z) dydz = 1$$

Thus, we obtain the following equation

$$\frac{dA(x)}{dx} = -i\kappa B(x) e^{i\Delta\beta x} \tag{6.82}$$

where

$$\kappa = \frac{k_0^2 n_o^2 n_e^2 r_{42}}{4\omega\mu_0} \iint F_y E_y(y,z) E_z(y,z) dydz \tag{6.83}$$

Thus, the amplitude $A(x)$ of the quasi TE mode depends on the amplitude $B(x)$ of the quasi TM mode, indicating the presence of coupling among the two modes.

Here κ is referred to as the coupling coefficient which depends on the electro-optic coefficient $r_{42} = r_{51}$, the applied electric field distribution $F_y(y,z)$, and the overlap between the applied electric field and the modal field distributions of the two modes.

In a very similar fashion, we obtain the following equation:

$$\frac{dB(x)}{dx} = -i\kappa A(x) e^{-i\Delta\beta x} \tag{6.84}$$

with the normalization condition

$$\frac{\beta_z}{2\omega\mu_0} \iint E_z^2(y,z) dydz = 1 \tag{6.85}$$

Note that the two coupled mode equations are identical to the coupled mode equations of a directional coupler (see Chap. 5) and we will use the conclusions drawn in the case of the directional coupler here to discuss about the efficiency of the coupling process.

For efficient coupling among the two polarizations (quasi TE and quasi TM modes), we must have

$$\Delta\beta = \beta_z - \beta_y = 0 \tag{6.86}$$

Since these correspond to two orthogonally polarized modes which in turn depend on the ordinary and extra ordinary refractive index variations, usually $\Delta\beta \neq 0$ and the coupling of power will be weak or almost negligible.

We resolve this issue using the quasi phase matching technique discussed later in Chap. 11. In this technique, the coupling coefficient κ varies periodically with x. The periodic variation may be brought about either by using a set of periodic electrodes that apply a periodically varying external electric field on the waveguide (see Sec. 6.4.2) or by using the periodic poling technique wherein the orientation of the $z-$axis of the lithium niobate crystal is periodically reversed (see Sec. 6.4.3). The latter technique is discussed in greater detail in Chap. 11 where the same technique is used to have efficient generation of second harmonic frequency 2ω from an input at the fundamental frequency ω using nonlinear interaction.

Assuming

$$\begin{aligned} \kappa(x) &= +\kappa; \quad 0 < x < \frac{\Lambda}{2} \\ &= -\kappa; \quad \frac{\Lambda}{2} < x < \Lambda \end{aligned} \tag{6.87}$$

and

$$\kappa(x + \Lambda) = \kappa(x) \tag{6.88}$$

We can make a Fourier series expansion of $\kappa(x)$ and obtain the following:

$$\kappa(x) = \kappa \sum_{m=-\infty}^{+\infty} G_m e^{-imKx} \tag{6.89}$$

and

$$G_m = \frac{2}{m\pi} \sin\left(\frac{m\pi}{2}\right) e^{\frac{im\pi}{2}} \tag{6.90}$$

Thus, the first two x-dependent terms of the series is given by

$$\kappa(x) = \kappa\left(i\frac{2}{\pi}e^{-iKx} - i\frac{2}{\pi}e^{+iKx} + \ldots\right) \tag{6.91}$$

Here

$$K = \frac{2\pi}{\Lambda} \tag{6.92}$$

is the spatial frequency of the periodically varying coupling coefficient. Using this in Eqs. (6.82) and (6.84), we obtain the following two coupled mode equations:

$$\frac{dA(x)}{dx} = -i\kappa B(x)\left(i\frac{2}{\pi}e^{i(\Delta\beta-K)x} - i\frac{2}{\pi}e^{i(\Delta\beta+K)x} + \ldots\right)$$

$$\frac{dB(x)}{dx} = -i\kappa A(x)\left(i\frac{2}{\pi}e^{-i(\Delta\beta+K)x} - i\frac{2}{\pi}e^{-i(\Delta\beta-K)x} + \ldots\right)$$

If we choose the spatial frequency K to satisfy the following equation:

$$K \approx \Delta\beta \tag{6.93}$$

then the only terms that contribute significantly to the coupling of power between the two modes are the terms containing $e^{i(\Delta\beta-K)x}$ and $e^{-i(\Delta\beta-K)x}$ (see Chap. 11) and neglecting the remaining terms, we obtain the following coupled mode equations:

$$\frac{dA(x)}{dx} = \frac{2\kappa}{\pi}B(x)e^{i(\Delta\beta-K)x} \tag{6.94}$$

$$\frac{dB(x)}{dx} = -\frac{2\kappa}{\pi}A(x)e^{-i(\Delta\beta-K)x} \tag{6.95}$$

If $\Delta\beta < 0$, then we choose the terms with $K \approx -\Delta\beta$ and the subsequent results will remain the same.

For maximum efficiency, we must choose (see Chap. 11)

$$K = \Delta\beta = \frac{2\pi}{\Lambda} \tag{6.96}$$

where Λ is the spatial period of the periodic perturbation and we obtain the following two coupled mode equations:

$$\frac{dA(x)}{dx} = \frac{2\kappa}{\pi}B(x) \tag{6.97}$$

$$\frac{dB(x)}{dx} = -\frac{2\kappa}{\pi}A(x) \tag{6.98}$$

These equations are very similar to the equations we obtained in the case of a directional coupler (see Chap. 5).

If at the input the quasi TE mode is only excited with an amplitude A_0, then we have $A(x = 0) = A_0$ and $B(x = 0) = 0$ and we obtain the following solutions:

$$A(x) = A_0 \cos\left(\frac{2\kappa}{\pi}x\right) \tag{6.99}$$

$$B(x) = -A_0 \sin\left(\frac{2\kappa}{\pi}x\right) \tag{6.100}$$

Thus, the power exchanges periodically between the two modes exactly like a directional coupler (see Eqs. 5.34 and 5.35). Since the powers in the two modes are equal to $|A(x)|^2$ and $|B(x)|^2$, we obtain the power in the two polarizations as

$$P_{TE}(x) = P_{TE}(x=0)\cos^2\left(\frac{2\kappa}{\pi}x\right) \tag{6.101}$$

$$P_{TM}(x) = P_{TE}(x=0)\sin^2\left(\frac{2\kappa}{\pi}x\right) \tag{6.102}$$

The coupling efficiency is given by

$$\eta = \frac{P_{TM}(x)}{P_{TE}(x=0)} = \sin^2\left(\frac{2\kappa}{\pi}x\right) \tag{6.103}$$

Complete power exchange takes place over a length L_c given by

$$L_c = \frac{\pi^2}{4\kappa} \tag{6.104}$$

The required spatial period of maximum efficiency is (see Eq. 6.96)

$$\Lambda = \frac{2\pi}{\Delta\beta} \tag{6.105}$$

Now for weakly guiding waveguides, we may write

$$\beta_y \approx \frac{2\pi}{\lambda_0}n_o \tag{6.106}$$

$$\beta_z \approx \frac{2\pi}{\lambda_0}n_e \tag{6.107}$$

where n_o and n_e are the ordinary and extraordinary refractive indices of lithium niobate at the free space wavelength λ_0. Hence, the required spatial period for maximum coupling is

$$\Lambda \approx \frac{2\pi}{\frac{2\pi}{\lambda_0}(n_o - n_e)} = \frac{\lambda_0}{(n_o - n_e)} \tag{6.108}$$

For lithium niobate $(n_o - n_e) \approx 0.08$ and for $\lambda_0 = 800$ nm, the required spatial period for phase-matched interaction is about 12 μm.

6.4.1 Interaction in Bulk

If the interaction were in bulk crystal with plane wave propagation with $E_y(y, z) = E_y$, then assuming an area of cross-section of the beam to be S, the normalization condition satisfied by E_y implies that

$$\frac{\beta_y}{2\omega\mu_0}\iint E_y^2(y,z)dydz = \frac{\beta_y}{2\omega\mu_0}E_y^2 S = 1$$

which gives us

$$E_y = \sqrt{\frac{2\omega\mu_0}{k_0 n_o S}} \tag{6.109}$$

where we have used $\beta_y = k_0 n_o$ for a plane wave with ordinary polarization. Similarly, the normalization condition satisfied by E_z gives us

$$E_z = \sqrt{\frac{2\omega\mu_0}{k_0 n_e S}} \tag{6.110}$$

with $\beta_z = k_0 n_e$. Hence, in bulk, we obtain

$$\kappa = \frac{k_0^2 n_o^2 n_e^2 r_{42}}{4\omega\mu_0} F_y \sqrt{\frac{2\omega\mu_0}{k_0 n_o S}}\sqrt{\frac{2\omega\mu_0}{k_0 n_e S}} S = \frac{\pi}{\lambda_0}\sqrt{n_o^3 n_e^3}\, r_{42}\frac{V}{G} \tag{6.111}$$

where we have replaced the applied electric field F_y by $\frac{V}{G}$ where V is the applied voltage and G is the electrode spacing.

Hence, in this case the coupling efficiency is given by

$$\eta = \sin^2\left(\frac{2\sqrt{n_o^3 n_e^3}\, r_{42}}{\lambda_0} L \frac{V}{G}\right) \tag{6.112}$$

In the case of the waveguide, we put the coupling efficiency in the form

$$\eta = \sin^2\left(\frac{2\sqrt{n_o^3 n_e^3}\, r_{42}}{\lambda_0} L \frac{V}{G}\Gamma\right) \tag{6.113}$$

where Γ is a dimensionless quantity representing an overlap factor between the applied electric field and the modal field distributions of the interacting modes given by

$$\Gamma = \frac{1}{2}\sqrt{\frac{\epsilon_0}{\mu_0}}\sqrt{n_o n_e}\iint f(y,z)E_y(y,z)E_z(y,z)dydz \tag{6.114}$$

where

$$F_y(y,z) = \frac{G}{V} f(y,z) \tag{6.115}$$

with $f(y, z)$ being a dimensionless function defining the electric field variation with the transverse position coordinates.

The overlap factor lies in the range $0 \le \Gamma \le 1$, with $\Gamma = 0$ representing no overlap and $\Gamma = 1$ representing complete overlap between the applied field and the fields of the interacting waves. Larger the value of Γ stronger the coupling between the two modes. Reader may check that in the case of bulk with complete overlap between the applied field and the propagating wave $\Gamma = 1$.

Let us consider an example where the electric field distributions of the y-polarized and z-polarized modes are power-normalized Gaussian distributions with widths w_y and w_z:

$$E_y(y,z) = \sqrt{\frac{4\omega\mu_0}{\pi\beta_y}} \frac{1}{w_y} e^{-(y^2+z^2)/w_y^2} \tag{6.116}$$

and

$$E_z(y,z) = \sqrt{\frac{4\omega\mu_0}{\pi\beta_z}} \frac{1}{w_z} e^{-(y^2+z^2)/w_z^2} \tag{6.117}$$

If the electric field is independent of position, then $f(y, z) = 1$. In such a case, using $\beta_y \approx \frac{\omega}{c} n_o$, $\beta_z \approx \frac{\omega}{c} n_e$, we have

$$\Gamma = \frac{2 w_y w_z}{\left(w_y^2 + w_z^2\right)} \tag{6.118}$$

If $w_y = w_z$, then $\Gamma = 1$.

As an example, we take the following values:

$$\lambda_0 = 1.5\ \mu\text{m}, n_o = 2.28; n_e = 2.20; r_{42} = 28 \times 10^{-12} \frac{\text{m}}{\text{V}}, G = 10\ \mu\text{m}; \Gamma = 0.9,$$

Substituting these in Eq. (6.112) gives us

$$\eta = \sin^2(37.8\ V.L)$$

The voltage required for complete conversion between the two polarizations is

$$V_0 = \frac{\pi}{2 \times 37.8} \frac{1}{L}$$

Taking a length of 20 mm, for complete conversion the required voltage is 2 V which is a very reasonable value.

Exercise 6.9 Assume that the external applied electric field is given by $F_y = F_{y0} \cos Kx$ where $K = 2\pi/\Lambda$, obtain the variation of power in the two modes.

Solution For the given external electric field variation, we have

$$\kappa = \frac{k_0^2 n_o^2 n_e^2 r_{42}}{4\omega\mu_0} \cos Kx \iint F_{y0} E_y(y,z) E_z(y,z) dydz = \frac{\kappa_0}{2}\left(e^{iKx} + e^{-iKx}\right)$$

where

$$\kappa_0 = \frac{k_0^2 n_o^2 n_e^2 r_{42}}{4\omega\mu_0} \iint F_{y0} E_y(y,z) E_z(y,z) dydz$$

The coupled equations become

$$\frac{dA(x)}{dx} \approx -i\frac{\kappa_0}{2} B(x) \tag{6.119a}$$

$$\frac{dB(x)}{dx} \approx -i\frac{\kappa_0}{2} A(x) \tag{6.119b}$$

Solving them under the boundary conditions of $A(x = 0) = A_0$ and $B(x = 0) = 0$, we obtain the following power variations in the two modes:

$$P_{TE}(x) = P_{TE}(x=0)\cos^2\left(\frac{\kappa_0 x}{2}\right)$$

$$P_{TM}(x) = P_{TE}(x=0)\sin^2\left(\frac{\kappa_0 x}{2}\right)$$

The coupling length for complete power transfer is given by

$$L = \frac{\pi}{\kappa_0}$$

Exercise 6.10 Assume that at the input $x = 0$, both polarization modes are excited with amplitudes $A(0)$ and $B(0)$. Solve the coupled mode equations (Eqs. 6.119a and 6.119b) and show that

$$\begin{pmatrix} A(x) \\ B(x) \end{pmatrix} = \begin{pmatrix} \cos\tilde{\kappa}x & -i\sin\tilde{\kappa}x \\ -i\sin\tilde{\kappa}x & \cos\tilde{\kappa}x \end{pmatrix} \begin{pmatrix} A(0) \\ B(0) \end{pmatrix} \tag{6.120}$$

Also show that the equation satisfies energy conservation. Here $\tilde{\kappa} = \frac{\kappa_0}{2}$.

Solution Differentiating Eq. (6.119a) and substituting from Eq. (6.119b), we obtain

$$\frac{d^2A}{dx^2} = -\tilde{\kappa}^2 A(x)$$

whose general solution is given by

$$A(x) = a\cos\tilde{\kappa}x + b\sin\tilde{\kappa}x$$

where a and b are integration constants. Substituting for $A(x)$ in Eq. (6.119a), we obtain

$$B(x) = -ia\sin\tilde{\kappa}x + ib\cos\tilde{\kappa}x$$

Using the initial conditions at $x = 0$, we obtain

$$A(x) = A(0)\cos\tilde{\kappa}x - iB(0)\sin\tilde{\kappa}x$$

$$B(x) = -iA(0)\sin\tilde{\kappa}x + B(0)\cos\tilde{\kappa}x$$

which is nothing but Eq. (6.120).

Energy conservation implies

$$\frac{d}{dy}\left(|A(x)|^2 + |B(x)|^2\right) = 0$$

Substituting the solutions for $A(x)$ and $B(x)$, we can show that energy conservation is satisfied. It is left as an exercise to solve the coupled mode equations Eqs. (6.94) and (6.95) and obtain similar results.

We have seen that for an efficient coupling, we need to have a periodic interaction between the modes which ensures a periodic coupling coefficient with an appropriate period. There are different techniques to achieve periodic modulation in the coupling coefficient which we shall discuss now.

6.4.2 Periodic Electrode Pattern

We will first discuss a configuration to achieve an applied electric field that is periodic along the x-direction and which will provide us with a periodic coupling

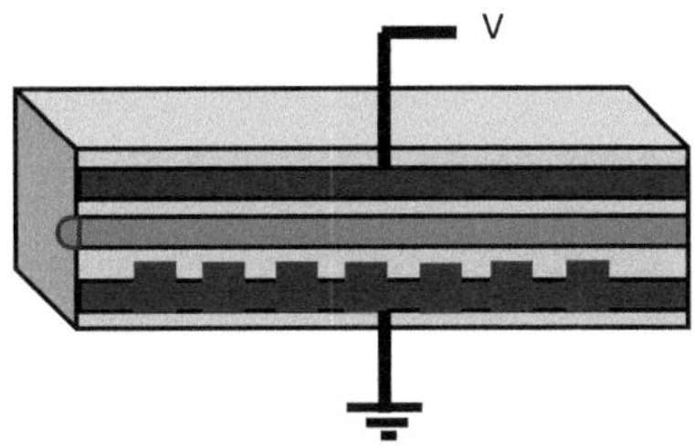

Fig. 6.5 Top view of periodic electrode configuration to generate a periodic variation in the applied electric field that is required for efficient coupling among the quasi TE and quasi TM modes in the channel waveguide in a *z*-cut *x*-propagating lithium niobate waveguide

coefficient. Figure 6.5 shows the top view of a periodic electrode pattern to achieve quasi phase matching among the two modes.

Since the electrode spacing varied periodically along the *x*-direction with a period Λ , we can write for the electric field to be given by

$$\begin{aligned} F_y(x) &= F_1; \qquad |x| < \frac{\Lambda}{4} \\ &= F_2; \qquad \frac{\Lambda}{4} < |x| < \frac{\Lambda}{2} \end{aligned} \tag{6.121}$$

and

$$F_y(x+\Lambda) = F_y(x) \tag{6.122}$$

Here F_1 and F_2 are the *y*-directed applied electric fields (assumed constant) for two different electrode spacings as shown in Fig. 6.5.

If we make a Fourier series expansion of the periodic function $F_y(x)$, we have

$$F_y(x) = \sum_{m=0}^{\infty} C_m e^{-imKx} \tag{6.123}$$

and $K = 2\pi/\Lambda$. We can write Eq. (6.123) in an explicit form as

$$F_y(x) = C_0 + C_1 e^{-iKx} + C_{-1} e^{+iKx} + C_2 e^{-2iKx} + C_{-2} e^{+2iKx} + \ldots.$$

We can easily evaluate all the Fourier coefficients C_m. For example,

$$C_1 = \frac{1}{\Lambda}\int_{-\Lambda/2}^{\Lambda/2} F_y(x) e^{+iKx} dx = \frac{F_1 - F_2}{\pi} \tag{6.124}$$

Substituting the Fourier expansion in the expression for κ, we obtain

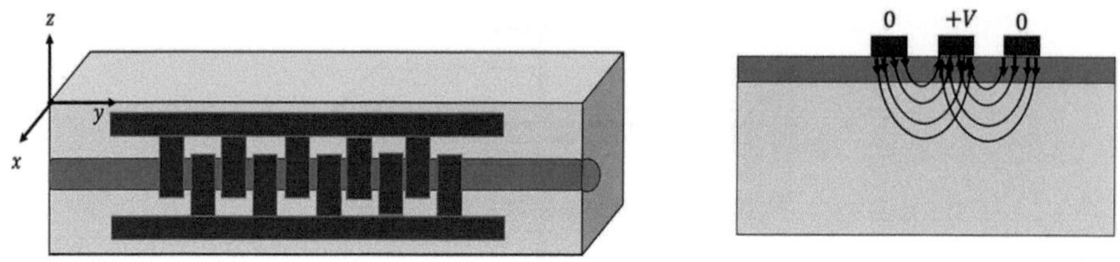

Fig. 6.6 Periodic electrode pattern in an x-cut y-propagating lithium niobate for polarization conversion

$$\kappa = \frac{k_0^2 n_o^2 n_e^2 r_{42}}{4\omega\mu_0} \iint E_y(y,z) E_z(y,z) dy dz \left(C_0 + C_1 e^{-iKx} + C_{-1} e^{+iKx}\right) \tag{6.125}$$

The terms containing C_1 and C_{-1} provide us with the appropriate spatial frequencies for quasi phase matching between the interacting modes.

Figure 6.6 shows another periodic interdigital electrode pattern that can be used in an x-cut y-propagating lithium niobate waveguide. The electric field under the electrode periodically changes from $+x$ direction to $-x$ direction thus compensating the phase mismatch between the TE and TM modes that are being coupled. Compared to the earlier design of periodic electrodes, in this case, we can use a push-pull effect where the electric fields are periodically pointing along $+x$ and $-x$ directions.

We have considered coupling between two orthogonal polarizations using the electro-optic effect. It is also possible to couple among two different spatial modes in a channel waveguide. For example in a z-cut x- propagating waveguide, electro-optic coupling can be used to couple power among the two quasi TM modes $E_{11}^{(z)}$ to $E_{12}^{(z)}$ or between the two quasi TE modes $E_{11}^{(y)}$ to $E_{12}^{(y)}$modes. It should be noted that the spatial modes that are to be coupled are of opposite parity, the $E_{11}^{(z)}$ mode is symmetric along the y-direction and the $E_{12}^{(z)}$ mode is antisymmetric along the y-direction. Similarly the $E_{11}^{(y)}$ mode is symmetric along the y-direction and the $E_{12}^{(y)}$ mode is antisymmetric along the y-direction. Recalling the expression for the coupling coefficient among the interacting modes will be proportional to the overlap integral between the applied electric field and the electric fields of the interacting modes (here y and z are the transverse coordinates)

$$\iint F(y,z) E_{11}^{(z)}(y,z) E_{12}^{(z)}(y,z) dy dz$$

If the applied electric field variation $F(y, z)$ is symmetric along the y-direction, then the coupling coefficient would be zero. Hence, to induce coupling among two opposite parity modes, the applied electric field should not be symmetric. This can be accomplished by appropriately placing the electrodes on the waveguide. In addition, since the modes being coupled have different propagation constants, the applied electric field needs to have a periodic variation of appropriate period along the propagation direction. In Bharadwaj et al. (2015a, b) such a device has been analyzed for mode conversion of photons for integrated quantum photonic applications.

6.4.3 Periodically Polarized Lithium Niobate Waveguide for Quasi Phase Matching

We saw earlier that the presence of the phase term $e^{\pm i\Delta\beta x}$ on the right-hand sides of Eqs. (6.82) and (6.84) leads to poor conversion efficiency. One technique to compensate for this phase mismatch between the two interacting modes is to use periodic electrodes which then lead to a periodic modulation of the coupling coefficient and with an appropriate choice of the spatial frequency of the periodic modulation, it is possible to overcome the effects of phase mismatch.

In Chap. 11, we will study the phenomenon of second harmonic generation where again the issue of phase mismatch between the two interacting frequencies leads to a drastic reduction in efficiency. It is discussed in detail that it is possible to compensate for the phase mismatch by the technique referred to as quasi phase matching. In this technique, the nonlinear coefficient of the material is periodically modulated with an appropriate spatial frequency. It is indeed possible to use the same technique to compensate for the phase mismatch in the application of electro optic devices.

In order to understand this, we note that the electro-optic tensor of a medium is defined in the principal axis system. For example, in the case of lithium niobate, the electro- optic coefficient given by Eq. (6.7) is written in the principal axis system. It is discussed in Chap. 11 that by using special techniques it is possible to periodically reverse the orientation of the z-axis of the crystal resulting in a periodic sign reversal of the electro-optic coefficients. Thus, a periodic domain reversal of lithium niobate leads to a periodic modulation of the electro-optic coefficient r_{51}.

Considering the same configuration discussed in Sect. 6.4.2, a periodic modulation with a period Λ of the electro-optic coefficient is described by the following equation:

$$\begin{aligned} r_{42}(x) &= +r_{42}; \quad 0 < x < \frac{\Lambda}{2} \\ &= -r_{42}; \quad \frac{\Lambda}{2} < x < \Lambda \end{aligned} \tag{6.126}$$

with $r_{42}(x + \Lambda) = r_{42}(x)$. We show in Chap. 11 that a Fourier series of the periodic function gives us

$$r_{42}(x) = r_{42} \sum_{m=-\infty}^{+\infty} G_m e^{-imKx} \tag{6.127}$$

and

$$G_m = \frac{2}{m\pi} \sin\left(\frac{m\pi}{2}\right) e^{\frac{im\pi}{2}} \tag{6.128}$$

Thus, the first two x-dependent terms are

$$r_{42}(x) = r_{42}\left(i\frac{2}{\pi}e^{-iKx} - i\frac{2}{\pi}e^{+iKx} + \ldots\right) \tag{6.129}$$

Hence, we have from Eq. (6.83)

$$\kappa = \frac{k_0^2 n_o^2 n_e^2 r_{42}}{4\omega\mu_0} \iint F_y E_y(y,z) E_z(y,z) dydz \left(i\frac{2}{\pi}e^{-iKx} - i\frac{2}{\pi}e^{+iKx} + \ldots\right) \tag{6.130}$$

giving us a periodic variation in the coupling coefficient.

The use of periodic poling for phase matching the two modes results in high efficiency. At the same time, the effective coupling coefficient has been reduced by a factor $\frac{2}{\pi}$. Note that in this technique the applied electric field is uniform along the x-direction and the periodic modulation is generated in the electro-optic coefficient. This is in contrast to having a uniform electro-optic coefficient and having a periodically varying external electric field that was discussed in Sect. 6.4.2. Periodically poled lithium niobate for polarization conversion has been discussed in many references; see, e.g., Lu et al. (2000), Huo et al. (2010), Inoue and Suhara (2014), Shi et al. (2003), Ding et al. (2019).

6.5 Wavelength Filtering

For efficient power conversion between the two orthogonal modes of the waveguide, we need to satisfy the quasi phase matching condition given by Eq. (6.96). For a given spatial frequency or spatial period, the condition is satisfied only at one specific wavelength. As the wavelength deviates from this value, the conversion efficiency will be reduced.

If $\Delta\beta - K \neq 0$, then we solve Eqs. (6.94) and (6.95) under the condition that $A(x=0) = 1$ and $B(x=0) = 0$. The power carried by the two modes varies as given by

$$P_A(x) = \cos^2\Gamma x + \frac{(\Delta\beta - K)^2}{4\Gamma^2} \sin^2\Gamma x \tag{6.131}$$

$$P_B(x) = \frac{\tilde{\kappa}^2}{\Gamma^2} \sin^2\Gamma x \tag{6.132}$$

where

$$\tilde{\kappa} = \frac{2\kappa}{\pi}; \quad \Gamma^2 = \tilde{\kappa}^2 + \frac{(\Delta\beta - K)^2}{4} \tag{6.133}$$

The coupled mode equations and their solutions are similar to those obtained for a nonphase-matched directional coupler discussed in Sect. 5.2.2 (see Eqs. 5.46 and 5.47).

In order to obtain the bandwidth of the interaction, we assume that at the central wavelength $\lambda_0 = \lambda_c$ we satisfy the quasi phase matching condition:

$$\Delta\beta(\lambda_c) - K = 0 \tag{6.134}$$

At a nearby wavelength $\lambda_0 = \lambda_c \pm \Delta\lambda$, the efficiency will reduce and become

$$\eta = P_B(x = L) = \frac{\tilde{\kappa}^2}{\Gamma^2}\sin^2\Gamma L = \frac{\sin^2\left[\sqrt{1 + \frac{(\Delta\beta - K)^2}{4\tilde{\kappa}^2}}\tilde{\kappa}L\right]}{\left(1 + \frac{(\Delta\beta - K)^2}{4\tilde{\kappa}^2}\right)}$$

We choose $\tilde{\kappa}L = \frac{\pi}{2}$ so that at $\lambda_0 = \lambda_c$, $\Delta\beta - K = 0$ and $\eta = 1$. Thus,

$$\eta = \frac{\sin^2\left[\frac{\pi}{2}\sqrt{1 + \frac{(\Delta\beta - K)^2}{4\tilde{\kappa}^2}}\right]}{\left(1 + \frac{(\Delta\beta - K)^2}{4\tilde{\kappa}^2}\right)} \tag{6.135}$$

The efficiency drops to 50% when $\frac{(\Delta\beta - K)^2}{4\tilde{\kappa}^2} \approx 0.64$ which gives (replacing $\tilde{\kappa}$ by $\frac{\pi}{2L}$)

$$\Delta\beta - K \approx \pm\frac{2.5}{L} \tag{6.136}$$

We will now calculate the wavelength shift from λ_c required to induce this value of $\Delta\beta - K$.

Now at $\lambda_0 = \lambda_c \pm \Delta\lambda$, the phase mismatch will be given by

$$\Delta\beta(\lambda_c \pm \Delta\lambda) - K = \frac{2\pi}{\lambda_c \pm \Delta\lambda}\left[n_o(\lambda_c \pm \Delta\lambda) - n_e(\lambda_c \pm \Delta\lambda)\right] - K \tag{6.137}$$

Assuming $\Delta\lambda \ll \lambda_c$ we can make Taylor series expansion of different terms in Eq. (6.137) and retaining terms of order $\Delta\lambda$, we obtain

$$\Delta\beta(\lambda_c \pm \Delta\lambda) - K \approx \pm 2\pi \frac{\Delta\lambda}{\lambda_c^2} [N_o(\lambda_c) - N_e(\lambda_c)] = \pm 2\pi \frac{\Delta\lambda}{\lambda_c^2} \Delta N(\lambda_c) \qquad (6.138)$$

where

$$N_o(\lambda_c) = n_o(\lambda_c) - \lambda_c \frac{dn_o}{d\lambda}\bigg|_{\lambda_c} \qquad (6.139)$$

and

$$N_e(\lambda_c) = n_e(\lambda_c) - \lambda_c \frac{dn_e}{d\lambda}\bigg|_{\lambda_c} \qquad (6.140)$$

are the group indices of the ordinary and extraordinary modes of the waveguide at $\lambda_0 = \lambda_c$.

From Eqs. (6.136) and (6.138), we obtain

$$2\pi \frac{\Delta\lambda}{\lambda_c^2} \Delta N(\lambda_c) \approx \frac{2.5}{L}$$

giving us

$$\Delta\lambda \approx 0.4 \frac{\lambda_c^2}{L\Delta N(\lambda_c)} \qquad (6.141)$$

and the full width at half maximum (FWHM) is given by

$$\Delta\lambda_{FWHM} \approx 0.8 \frac{\lambda_c^2}{L\Delta N(\lambda_c)} \qquad (6.142)$$

As an example, we take a polarization converter made of lithium niobate for operation at $\lambda_c = 1550$ nm. At this wavelength, we have

$$n_o(\lambda_c) = 2.2111; n_e(\lambda_c) = 2.1376$$
$$N_o(\lambda_c) = 2.2637; N_e(\lambda_c) = 2.1823$$

The period required for perfect quasi phase matching is

$$\Lambda = \frac{\lambda_c}{(n_o - n_e)} \approx 21.1 \ \mu\text{m}$$

If the length of the device is 10 mm, then the FWHM comes out to be

$$\Delta\lambda_{FWHM} \approx 0.8 \frac{\lambda_c^2}{L\Delta N(\lambda_c)} \approx 2.4 \text{ nm}$$

Thus, giving a narrow spectral filter.

Exercise 6.11 Solve Eqs. (6.94) and (6.95) when $\Delta\beta - K \neq 0$ and obtain the efficiency of power transfer.

Solution Equations (6.94) and (6.95) are identical to the problem of directional couplers with phase mismatch and hence can be solved using the same technique. The efficiency of conversion assuming $A(x = 0) = A_0$ and $B(x = 0) = 0$ is given by

$$\eta = \frac{\tilde{\kappa}^2}{\Gamma^2} \sin^2\Gamma x$$

where $\tilde{\kappa} = \frac{2}{\pi}\kappa$ and $\Gamma^2 = \tilde{\kappa}^2 + \frac{(\Delta\beta - K)^2}{4}$.

Exercise 6.12 Consider a waveguide device shown in Fig. 6.7. In the device a two-mode waveguide splits into two single-mode waveguides through a Y-junction and then is made to recombine into a single two-mode waveguide. On one of the arms of the Y, there is an electro-optic phase modulator. If the electro optic phase modulator introduces a phase shift of π, then what will be the output if (a) the fundamental mode is excited at the input and (b) if the first excited mode is excited at the input? Assume the two Y-splitters to be adiabatic (see also Chap. 7).

Solution Since the Y-splitters are adiabatic, in both the cases when the input is in the fundamental mode or in the first excited mode, both arms of the device will have equal powers. If the input is in the fundamental mode, then the two arms will be in phase while if the first excited mode is input, then the two arms will have a phase difference of π. Now if the electro-optic phase modulator introduces a phase difference of π between the two arms, then in case (a) the phase difference of π will lead to the excitation of the first excited mode at the output waveguide and in case (b) the additional phase difference of π will lead

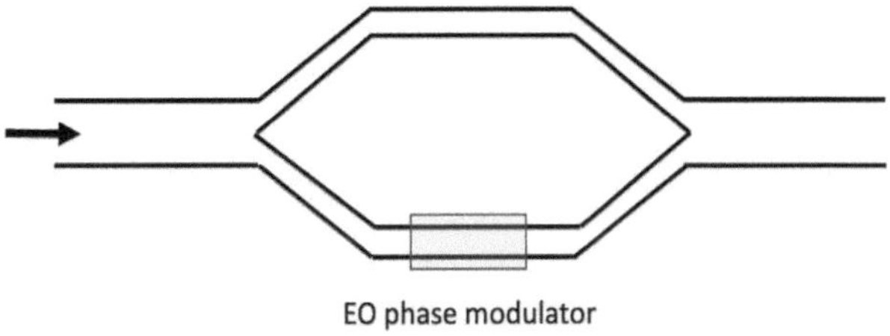

Fig. 6.7 Figure for Exercise 6.12

to the excitation of the fundamental mode at the output waveguide. Thus, the device will act like a spatial mode converter converting the input mode from fundamental to first excited or from first excited mode to the fundamental mode.

Problems

6.1. A uniaxial crystal has the following non-zero electro-optic coefficients: r_{12}, r_{22} and r_{61}. With an external electric field applied along the z-direction, a circularly polarized wave propagates along the z-direction. Will the state of polarization of the wave change? Give reasons for your answer.

6.2. Write down the equation for the index ellipsoid in a uniaxial medium with the only non-zero coefficients of r_{11} , r_{21} , r_{41} , r_{52} and r_{62} with an applied electric field lying in the x-y plane along a direction at 45° to the x – direction.

6.3. The electro-optic tensor of quartz (which is a uniaxial crystal) is given by

$$r = \begin{pmatrix} r_{11} & 0 & 0 \\ -r_{11} & 0 & 0 \\ 0 & 0 & 0 \\ r_{41} & 0 & 0 \\ 0 & -r_{41} & 0 \\ 0 & -r_{11} & 0 \end{pmatrix}$$

Obtain the equation of the index ellipsoid in the presence of an electric field along the z – direction

6.4. Consider a z-cut, x-propagating lithium niobate waveguide. One of the electrode is placed on top of the waveguide and the other electrode is spaced at a distance of G from this electrode. Making appropriate approximations, obtain an expression for the phase change suffered by the quasi TE and quasi TM modes in this waveguide.

6.5. Gallium arsenide is an isotropic medium and has the following non-zero electro-optic coefficients: r_{41}, r_{52} and r_{63}. Obtain the principal axes directions and the principal refractive indices in the presence of an applied electric field along the x – direction.

6.6. Consider an electro-optic waveguide phase modulator and assume that the applied voltage varies as $V = V_1 \sin \Omega t$, with $V_1 \ll V_0$. Show that although the voltage varies sinusoidally with a frequency Ω, the phase modulated output has additional frequency components leading to distortion. Using the following Fourier expansion of $e^{-i\varsigma \sin \Omega t}$,

$$e^{-i\zeta \sin \Omega t} = J_0(\zeta) + 2\sum_{n=1}^{\infty} J_{2n}(\zeta) \cos(2n\Omega t) - 2i\sum_{n=1}^{\infty} J_{2n-1}(\zeta) \sin([2n-1]\Omega t)$$

where $J_m(\zeta)$ represents Bessel function of order m, obtain the ratio of the terms varying with frequency 3Ω and Ω.

6.7. An electro-optic phase modulator using a channel waveguide in y-cut, x-propagating lithium niobate is to be realized.

(a) Draw a schematic diagram representing the waveguide and the electrode geometry to achieve minimum drive voltage.
(b) Obtain an expression for the voltage required to achieve a phase change of π in the mode.

6.8. The transmission through an electro-optic intensity modulator is described by

$$T = \frac{I_{out}}{I_{in}} = \sin^2\left(\frac{\pi V}{2V_\pi}\right)$$

where symbols have their usual meaning. If the applied voltage varies as $V = V_0 \cos \Omega t$ (with $V_0 \ll V_\pi$), then (a) show that the output intensity variation does not have any component at frequency Ω. (b) How would you linearize the modulator, that is, have output intensity variation at the applied frequency Ω?

6.9. Consider an electro-optic intensity modulator having a half wave voltage of 10 V. If the applied voltage varies as $V(t) = 5 + \Delta V \sin \Omega t$ (with $\Delta V \ll 5$ V) what is the maximum permissible value of ΔV so that the ratio of 3Ω component to Ω component in the intensity-modulated light is less than 10^{-4}.

6.10. Consider a lithium niobate phase modulator operating at a wavelength of 800 nm with an applied voltage of $V = V_0 \sin(\pi \times 10^{10} t)$ V. What frequency components would we observe in the output light beam?

6.11. Consider an electro-optic modulator made of lithium niobate consisting of two parts of lengths 3 mm and 1 mm with their optic axes pointing in opposite directions as shown in Fig. 6.8. Obtain an expression for the electro-optically induced phase change suffered by a wave polarized along the z-axis, when a voltage V is applied.

6.12. Consider a polarization converter using bulk periodically poled lithium niobate for operation at a wavelength of 1300 nm with a spectral bandwidth of 2 nm. It is given that at this wavelength $n_o = 2.2204$; $n_e = 2.1455$; $N_o = 2.2744$; $N_e = 2.1917$. (a) What is the required spatial period for quasi phase matching? (b) What should be the length of the interaction region for achieving this bandwidth? (c) What would be the required coupling coefficient for this length? (d) Obtain the corresponding voltage required for achieving complete conversion at the central wavelength of 1300 nm. Assume an electrode spacing of 1 mm.

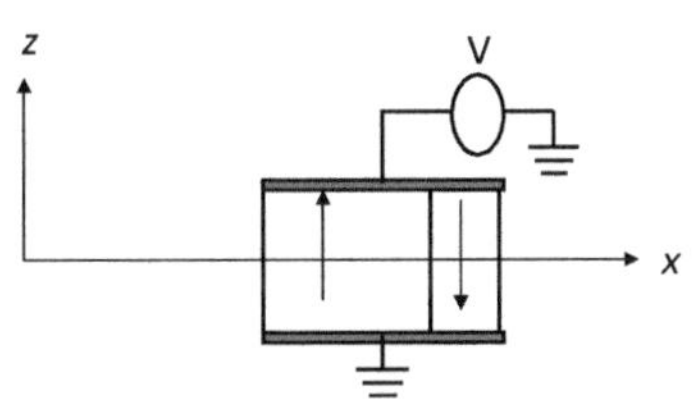

Fig. 6.8 Figure for Problem 6.11

Fig. 6.9 Figure for Problem 6.13

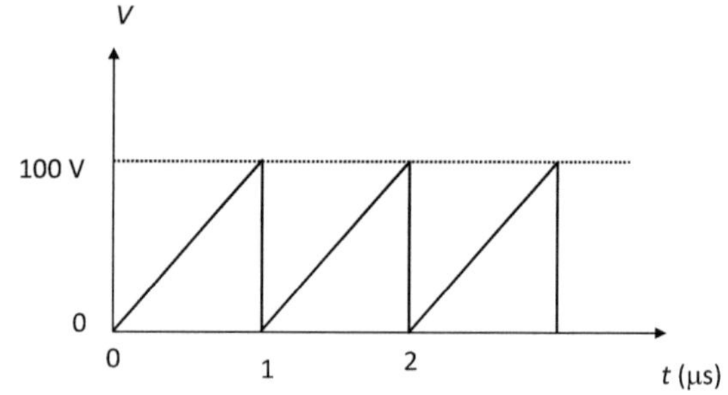

6.13. Consider an electro-optic intensity modulator with a half-wave voltage of 25 V. The output intensity of the modulator in the absence of any applied voltage is zero. If a saw tooth voltage (see Fig. 6.9) with a peak value of 100 V is applied to the modulator, draw a schematic showing the output intensity variation you would get. Mark all important values on the axes.

References

Bharadwaj, D., K. Thyagarajan, M. Jachura, M. Karpiński, and K. Banaszek. 2015a. Scheme for on-chip verification of transverse mode entanglement using the electro-optic effect. *Opt Express* 23:33087–33098.

Bharadwaj, D., K. Thyagarajan, M. Karpinski, and K. Banaszek. 2015b. Generation of higher-dimensional modal entanglement using a three-waveguide directional coupler. *Phys Rev A* 91: 033824.

Ding, T., Y. Zheng, and X. Chen. 2019. Wavelength filtering utilizing periodically poled lithium niobate on insulator ridge waveguide. *J Lightwave Tech* 37:1296–1300.

Ghatak, A., and K. Thyagarajan. 1989. *Optical electronics*. New York: Plenum Publishing Corporation.

Hu, X.-M., Y. Guo, B.-H. Liu, C.-F. Li, and G.-C. Guo. 2023. Progress in quantum teleportation. *Nat Rev Phys* 5:339–353.

Huo, J., K. Liu, and X. Chen. 2010. 1x2 precise electro-optic switch in periodically poled lithium niobate. *Opt Express* 18:15603–15608.

Inoue, T., and T. Suhara. 2014. High-performance electro optic polarization conversion type modulator for short wavelength light using periodically poled MgO:$LiNbO_3$. *Electron Lett* 50:1622–1624. https://doi.org/10.1049/el.2014.2848.

Lu, Y.-Q., Z.-L. Wan, Q. Wang, Y.-X. Xi, and N.-B. Ming. 2000. Electro optic effect of periodically poled optical superlattice $LiNbO_3$ and its applications. *Appl Phys Lett* 77:3719–3721.

Marcuse, D. 1975. Electro optic coupling between TE and TM modes in anisotropic slabs. *IEEE J Quant Electron* 9:759–767.

Papuchon, M., Y. Combemale, X. Mathieu, D. B. Ostrowsky, L. Reiber, A. M. Roy, B. Sejourne, and M. Werner. 1975. Electrically switched optical directional coupler: Cobra. *Appl Phys Lett* 27:289–291.

Saleh, B. E. A., and M. C. Teich. 1991. *Fundamentals of photonics*. Hoboken: Wiley.

Shi, J., X. Chen, Y. Xia, and Y. Chen. 2003. Polarization control by use of the electro optic effect in periodically poled lithium niobate. *Appl Opt* 42:5722–5725.

Yariv, A. 1991a. *Optical electronics*. Oxford: Oxford University Press.

———. 1991b. *Quantum electronics*. Hoboken: Wiley.

Zhang, M., C. Wang, P. Kharel, D. Zhu, and M. Loncar. 2021. Integrated lithium niobate electro-optic modulators: when performance meets scalability. *Optica* 8:652–667.

Chapter 7
Waveguide Y-Splitters and Mach-Zehnder Interferometer

Abstract Waveguide Y-splitters and waveguide-based Mach-Zehnder interferometers are two important components in integrated quantum photonic circuits. This chapter presents detailed analysis of these components and discusses some applications as 3dB couplers, mode splitters, optical modulators, etc.

7.1 Introduction

In Chap. 5, we discussed integrated optic directional couplers which form one of the most important components in integrated optics and integrated quantum photonics. In this chapter, we will discuss two other components namely a Y-splitter/Y-combiner and a Mach-Zehnder interferometer which find applications in many integrated optic circuits and are very useful components of integrated quantum photonic circuits for quantum computation, entangled photon pair generation, etc.

7.2 Y-Splitters and Y-combiners

A Y-splitter as the name implies is a device in which a single waveguide splits into a pair of waveguides forming a Y (see Fig. 7.1). All the waveguides are usually single mode; however, it is possible to have Y-splitters with waveguides supporting mode modes which also find applications as mode splitters, mode combiners, mode converters (see Exercise 6.12), etc.

We first consider a Y-splitter where all the waveguides involved are single-mode waveguides (see Fig. 7.1). In such a device, a single single-mode waveguide splits into a pair of single mode waveguides through a Y-configuration as shown in Fig. 7.1. The pair of single mode waveguides diverge from each other at a small angle and become sufficiently far apart so that there is no evanescent coupling between them. Since the pair of single mode waveguides which are far apart supports two guided modes (see Chap. 5), in this geometry a single-mode waveguide has evolved into a two-mode waveguide consisting of two-single mode waveguides.

K. Thyagarajan, *Integrated Quantum Photonics*, Graduate Texts in Physics,
https://doi.org/10.1007/978-3-031-85728-7_7

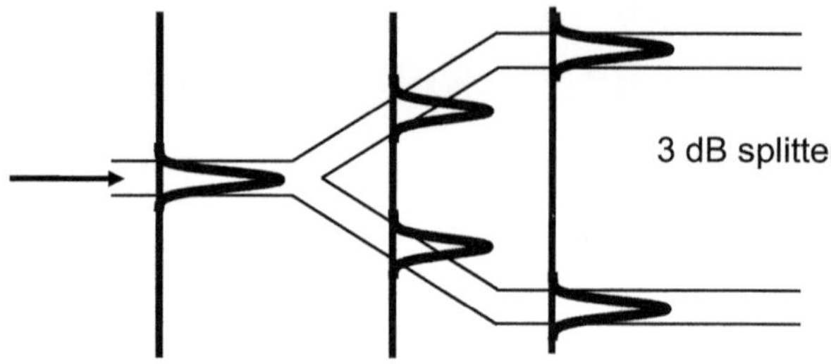

Fig. 7.1 Y-splitter in which a single-mode waveguide splits into two single-mode waveguides. The angle of the Y is shown large for clarity; in practice the angle would be about 1–2°

Recalling our discussions from Chap. 5, we note that if the two waveguides are identical, then the fundamental mode and the first excited mode of the waveguide pair will correspond to symmetric and antisymmetric modes of the device. Since the pair of waveguides can be considered to be a directional coupler with extremely large spacing so that the coupling coefficient κ is almost zero, the two normal modes would then be as shown in Fig. 5.4 and both of them would have equal propagation constants and which are both equal to the propagation constant of the individual waveguides (see Chap. 5).

In fact we can consider the Y-splitter as a z-dependent structure with the separation between the waveguides increasing with distance along the propagation direction. In Chap. 5, we had discussed about adiabatic transitions and if the coupling coefficient or the phase mismatch between the two waveguides changes slowly along the propagation direction, then power launched into a specific normal mode at the input will continue to reside in the same local normal mode as it propagates.

In the Y-splitter, we first have a single waveguide which supports only a single mode of propagation. Thus, light launched into the waveguide will excite the fundamental mode of the waveguide. As the structure evolves along the propagation direction, the single waveguide splits into a pair of waveguides which now support two modes, the fundamental mode and the first excited mode. If the angle made by the Y-junction is very small, then the transition along the Y can be considered to be adiabatic and the launched power would then continue to be in the fundamental normal mode at each value of z (see Chap. 5) and the first excited mode would not get excited.

Figure 7.1 shows such an evolution and as can be seen at the exit of the Y-junction the power will be in the fundamental mode of the pair of waveguides which are far apart to have no evanescent coupling. The fundamental normal mode of a pair of identical waveguides with sufficient spacing which prohibits evanescent coupling would be just a combination of the fundamental modes of each of the waveguides and which are in phase (see Chap. 5). Thus, at the exit end of the Y-splitter, the power in each of the output waveguides will be equal and the fields will be in phase. Thus, an adiabatic transition of the Y-splitter divides the input power in the fundamental mode into two equal parts in the two-output arm of the device. Hence, if the input to the Y-splitter is 1 mW, then the output from each of the output waveguides will be 0.5 mW, assuming no propagation loss and an adiabatic transition. Thus, this structure acts like a 3 dB beam splitter and can be used to divide

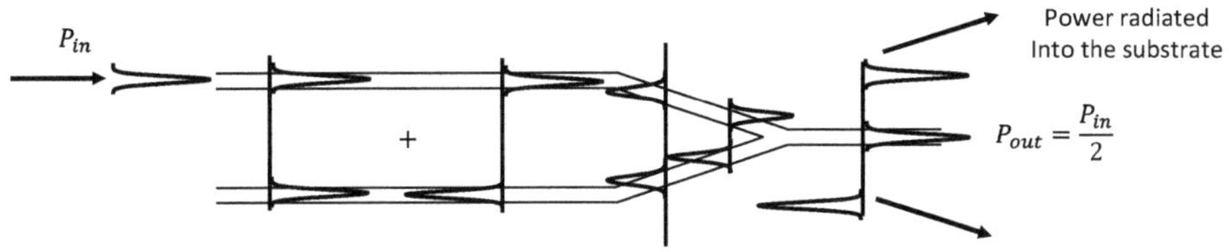

Fig. 7.2 A Y-combiner, the reverse of the Y-splitter shown in Fig. 7.1. Only half of the incident power emerges from the combiner; the other half gets radiated into the substrate

an incident power into two equal parts. Note that the power dividing property will be independent of the wavelength of the input as long as the waveguides are all single mode.

Let us now consider an input to the Y-splitter from the other end of the y-splitter as shown in Fig. 7.2. Again assuming the transition of the waveguide to be adiabatic, input into a particular normal mode will continue to remain in the same normal mode as the structure evolves along the propagation direction. We first note that the normal modes of the input pair of waveguides are combinations of the modes of each individual waveguide; the fundamental mode would be a symmetric combination while the first excited mode would be an antisymmetric combination (see Chap. 5):

$$\text{Fundamental normal mode}: E_s(x,y) = \frac{1}{\sqrt{2}}[E_a(x,y) + E_b(x,y)] \tag{7.1}$$

$$\text{First excited normal mode}: E_{as}(x,y) = \frac{1}{\sqrt{2}}[E_a(x,y) - E_b(x,y)] \tag{7.2}$$

Here, subscripts a and b refer to the two waveguides of the splitter. Since the waveguides are far apart, the coupling coefficient κ would be zero and thus the propagation constants of both the modes would be equal and equal to the propagation constant of each individual waveguide mode. The factors of $\frac{1}{\sqrt{2}}$ in Eqs. (7.1) and (7.2) are to ensure that E_s and E_{as} are power normalized; we have assumed that E_a and E_b are power normalized fields.

Assuming input power in the waveguide a we have

$$E(x,y,z=0) = E_a(x,y) \tag{7.3}$$

which can be written as

$$\begin{aligned} E(x,y,z=0) = \frac{1}{2}[E_a(x,y) + E_b(x,y)] + \\ \frac{1}{2}[E_a(x,y) - E_b(x,y)] = \frac{1}{\sqrt{2}}E_s + \frac{1}{\sqrt{2}}E_{as} \end{aligned} \tag{7.4}$$

Note that the factors $\frac{1}{\sqrt{2}}$ ensure conservation of power; the input field having unit power is equally divided into the fundamental symmetric and the first excited antisymmetric modes, each carrying half the power (see Fig. 7.2).

Assuming the transition to be adiabatic, the power launched in the symmetric normal mode will continue to propagate as the fundamental symmetric normal mode and the power launched in the antisymmetric first excited normal mode will continue to propagate as the antisymmetric first excited normal mode. Since the output waveguide is a single-mode waveguide, it will support only the fundamental mode. Thus, as the field propagates through the device, the power in the fundamental normal mode will exit the waveguide as a guided mode, while the power in the first excited normal mode will get cut off at some point of the Y and will be radiated out of the device into the surrounding. Thus, in this situation the output power will be half of the input power. Note that this device characteristic is independent of wavelength.

It may be noted that the device looks like having two inputs and a single output which is incorrect. One of the arms of the output is the single-mode waveguide, while the other arm is the substrate into which the first excited mode radiates into. The device is still a 2 x 2 device, although the second output arm is not obvious.

Note that if the output is a two-mode waveguide, then the entire power will exit from the output waveguide with equal power being carried by each of the two modes. The device is very similar to a taper in a single waveguide; in this case the taper is such as to split the waveguide into two waveguides.

We now consider equal power input into both waveguides (see Fig. 7.3). If the two inputs are in phase then the input would just correspond to the fundamental mode of the waveguide pair and as the transition is adiabatic, the power will continue to propagate as the fundamental mode and the entire power will exit the waveguide with no loss of power.

On the other hand, if the two inputs are out of phase by π, then the input would correspond to the first excited mode of the waveguide pair. If the output waveguide is a single-mode waveguide, then the power launched at the input will get radiated into the substrate at an appropriate point of the Y-junction and the output power will be zero. On the other hand, if the output waveguide is a two-mode waveguide, then the power will exit the waveguide in the first excited mode (see Fig. 7.3).

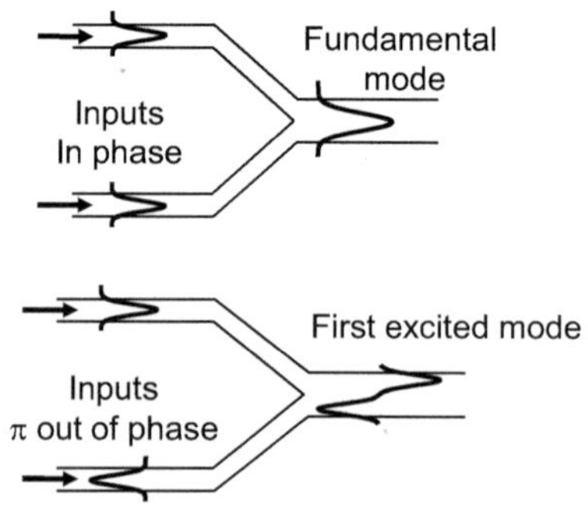

Fig. 7.3 Y-combiner when the output waveguide is a two-mode waveguide

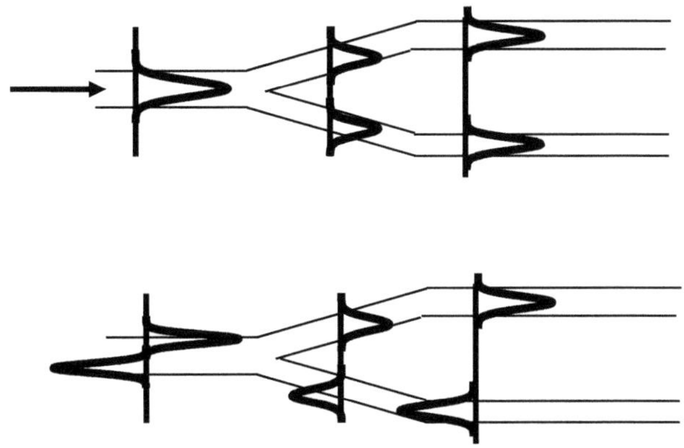

Fig. 7.4 A Y-splitter in which a two-mode waveguide splits into two identical single-mode waveguides

Figure 7.4 shows the case when the input is a two-mode waveguide and when the fundamental mode or the first excited mode is launched at the input. Note that in both cases the output powers are equal and equal to half the input but the fields are in phase in one case and out of phase in the other case.

7.3 Asymmetric Y-Splitter; Spatial Mode Splitter

If the input waveguide supports two modes, then it is possible to design a Y-splitter which will separate the two modes into two different outputs. Such a device will act as a mode splitter and find applications in integrated quantum photonic circuits especially in processing spatial mode entangled photon pairs in waveguides.

An asymmetric Y-splitter is made asymmetric with one of the output waveguides having a higher propagation constant than the other one. Such a structure is shown in Fig. 7.5 where the difference in the output waveguides is shown through two different thicknesses of the waveguides; the waveguides may be made different using different refractive index profiles or any other technique.

Recalling the discussion in Chap. 5, in the asymmetric Y-splitter, we see that the input two-mode waveguide evolves adiabatically to a directional coupler with two non-identical waveguides which then diverge to a large separation implying no evanescent coupling among them. For a non-identical directional coupler with negligible coupling coefficient, the fundamental mode has its field only in the waveguide with a higher propagation constant and the first excited mode has its field only in the waveguide with a lower propagation constant (see Chap. 5). Thus, assuming an adiabatic transition from the input to the output, we see that power in the fundamental mode in the input waveguide will exit from the waveguide having a higher propagation constant, while power in the first excited mode will exit from the waveguide having a lower propagation constant; these correspond to the fundamental and first excited normal modes at the output of the device. Thus, such a device can

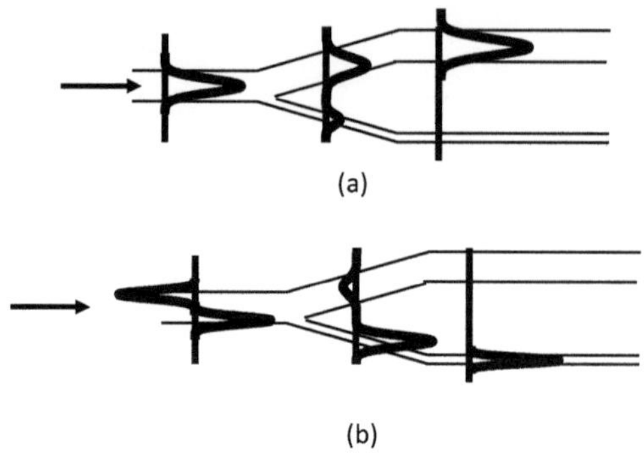

Fig. 7.5 An asymmetric Y-splitter in which a two-mode waveguide splits into two non-identical waveguides which then diverge to a large separation between the output waveguides. (**a**) An input fundamental mode exits from the waveguide having a larger propagation constant. (**b**) An input excited mode exits from the waveguide having a lower propagation constant

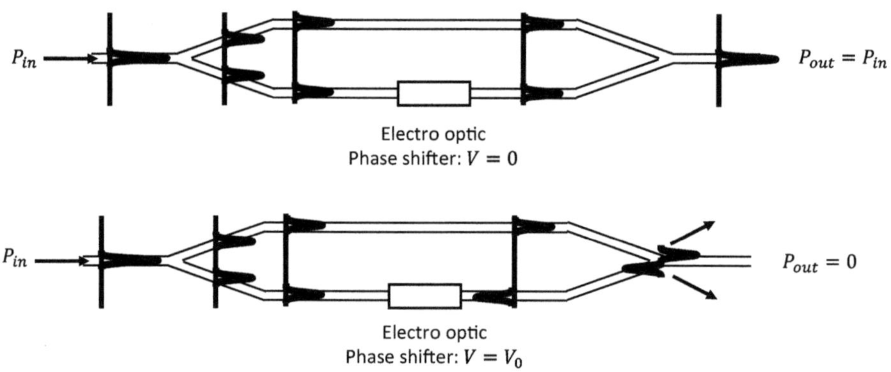

Fig. 7.6 A waveguide Mach-Zehnder interferometer consisting of a Y-splitter and a Y-combiner. All waveguides are single-mode waveguides. Depending on the phase difference between the two waveguides at the combining region, the output varies

separate the two spatial modes into two different outputs and can be very useful in many applications wherein two modes are taking part in the device functioning.

Y-splitters form the building blocks of many important integrated photonic and integrated quantum photonic devices (Sun et al. 2024).

7.4 Mach-Zehnder Interferometer

A guided wave Mach-Zehnder interferometer configuration is shown in Fig. 7.6. It consists of a single-mode waveguide which splits into two single-mode waveguides via a Y-splitter and is sufficiently far apart to have no evanescent wave coupling between them. After a certain propagation length, the two waveguides combine into a single-mode waveguide through a Y-combiner. This is a guided wave form of a bulk Mach-Zehnder interferometer. The Y-splitter and Y-combiners are the two beam splitters used in the bulk Mach-Zehnder interferometer. Note the device has two input and two output ports; the invisible ports are radiation modes into the substrate (see Sect. 7.2).

In order to realize a device capable of dynamic performance, one of the waveguides has an electrode pattern so that an external electric field along an appropriate direction can be applied on the waveguide. In Chap. 6, we studied about the electro optic effect wherein an applied electric field can modify the effective index of the mode of the waveguide and thus modify the phase of the propagating mode. Let the voltage required for changing the phase of the mode by π be represented by V_0.

Let us now analyze light propagation through the interferometer. When light is incident on the device, as discussed earlier, the Y-splitter leads to equal power among the modes of the two arms of the interferometer. If the entire device is symmetric, then the modes will have the same phase as they propagate through the device and when they enter the output Y-combiner, they will be in phase corresponding to the fundamental mode of the waveguide pair. Hence, the entire power will exit the output of the device as the fundamental mode of the output waveguide.

Now, when an electric field is applied on one of the arms of the interferometer, then the phase of the mode propagating in that arm gets changed and the modes propagating in the two arms are no more in phase. Hence, the resultant field distribution does not correspond to the fundamental mode of the waveguide pair. In fact, it will be a linear combination of the fundamental symmetric and the first excited antisymmetric mode of the waveguide pair.

As discussed earlier, since the output waveguide is a single-mode waveguide, as the field enters the Y-combiner, the fraction of power in the fundamental mode will exit the waveguide, while the remaining power in the first excited mode will leak out of the device in to the substrate. Thus, the power at the output will drop.

If the applied voltage is so as to change the phase of the mode in one of the waveguides by π, then the resulting field of the waveguide pair as it enters the Y-combiner will correspond to the first excited antisymmetric mode of the waveguide pair. As discussed earlier, since the output is a single-mode waveguide, the entire power will leak out of the device resulting in zero power exiting in the output waveguide.

For arbitrary phase difference introduced by the electro optic effect, the power will be intermediate between full power and zero power.

Thus, in this device, it is possible to control the output power from the waveguide by application of an electric field on one of the arms of the interferometer. This finds applications in optical modulators, optical switches, etc. and is a very useful device in integrated quantum photonics.

Recall our discussion in Chap. 5 where we had represented the propagating field in a directional coupler in terms of a column matrix. Thus, the fundamental mode of the waveguide pair (which actually corresponds to a directional coupler with zero coupling as they are separated by a large distance to avoid any evanescent coupling) is given by

$$\begin{pmatrix} R_0 \\ S_0 \end{pmatrix} = \frac{1}{\sqrt{2}} \begin{pmatrix} 1 \\ 1 \end{pmatrix} \tag{7.5}$$

and the first excited mode is given by

$$\begin{pmatrix} R_0 \\ S_0 \end{pmatrix} = \frac{1}{\sqrt{2}} \begin{pmatrix} 1 \\ -1 \end{pmatrix} \tag{7.6}$$

In the column matrix, the upper element gives the field in the upper waveguide and the lower element gives the field in the lower waveguide. Since the separation between the two waveguides is large, $\kappa = 0$ and the two guided modes of the waveguide pair have the same propagation constant equal to the propagation constant of individual waveguides.

The input to the device is the fundamental mode of the waveguide which evolves through the Y-splitter into the symmetric mode of the waveguide pair

$$E_{in} = E_0 = \frac{\sqrt{P_{in}}}{\sqrt{2}} \begin{pmatrix} 1 \\ 1 \end{pmatrix} \tag{7.7}$$

where P_{in} is the input power to the device and E_0 is the field (appropriately normalized) before entering the electrode section of the device. The fields are normalized so that the power carried by the field is obtained as:

$$P = \left[\frac{\sqrt{P_{in}}}{\sqrt{2}} \begin{pmatrix} 1 & 1 \end{pmatrix}\right] \left[\frac{\sqrt{P_{in}}}{\sqrt{2}} \begin{pmatrix} 1 \\ 1 \end{pmatrix}\right] = P_{in} \tag{7.8}$$

Now the applied electric field on one of the waveguides (assumed to be the lower waveguide in Fig. 7.6) changes the phase of the light propagating through the waveguide. Thus, after propagating through the electrode section, the field becomes

$$\tilde{E} = \frac{\sqrt{P_{in}}}{\sqrt{2}} \begin{pmatrix} 1 \\ e^{i\phi} \end{pmatrix} \tag{7.9}$$

where $\tilde{E}$ corresponds to the field after the electrode section of the device and ϕ represents the additional phase change induced through the Pockels effect by the applied electric field; this is proportional to the applied electric field (see Chap. 6). If the two arms of the interferometer have different lengths, then ϕ could correspond to the phase difference between the two arms due to this length difference.

We write this field as a superposition of the two normal modes [given by Eqs. (7.5) and (7.6)] of the waveguide pair after the electrode section as follows:

$$\tilde{E} = \frac{\sqrt{P_{in}}}{\sqrt{2}}\begin{pmatrix} 1 \\ e^{i\phi} \end{pmatrix} = \sqrt{P_{in}}e^{i\phi/2}\cos\left(\frac{\phi}{2}\right)\left[\frac{1}{\sqrt{2}}\begin{pmatrix} 1 \\ 1 \end{pmatrix}\right] - i\sqrt{P_{in}}e^{i\phi/2}\sin\left(\frac{\phi}{2}\right)\left[\frac{1}{\sqrt{2}}\begin{pmatrix} 1 \\ -1 \end{pmatrix}\right] \tag{7.10}$$

The first term on the right hand side of Eq. (7.10) corresponds to the symmetric fundamental mode and the second term on the right hand side corresponds to the antisymmetric first excited mode of the waveguide pair. The factors multiplying them give their corresponding amplitudes.

Since the output waveguide is a single-mode waveguide, only the power in the fundamental mode will exit the device; the power in the first excited mode will get radiated out into the substrate.

Thus, the output field will have an amplitude

$$E_{out} = \sqrt{P_{in}}e^{i\phi/2}\cos\left(\frac{\phi}{2}\right)\left[\frac{1}{\sqrt{2}}\begin{pmatrix} 1 \\ 1 \end{pmatrix}\right] \tag{7.11}$$

Assuming no losses, the electric field corresponding to the field that will be coupled into the output single-mode waveguide would be

$$E_{out} = \sqrt{P_{in}}e^{i\phi/2}\cos\left(\frac{\phi}{2}\right)$$

The power exiting will be given by

$$P_{out} = P_{in}\cos^2\left(\frac{\phi}{2}\right) \tag{7.12}$$

Thus, depending on the phase introduced by the electric field, the output power changes. For an applied voltage which generates a value of $\phi = \pi$, the output power becomes zero. Hence, the device behaves as a dynamic modulator. Varying the applied electric field with time varies the phase difference ϕ with time which in turn varies the output power from the device, thus acting as an intensity modulator. Switching ϕ between 0 and π we obtain an electro optic switch in which the output optical power switches from maximum to zero.

In Chap. 6, we showed that the phase change introduced by the electro optic effect in an optical waveguide is given by

$$\phi = \frac{\pi \overline{n}^3 \overline{r}}{\lambda_0}\frac{V}{G}\Gamma L = \alpha V \tag{7.13}$$

where

$$\alpha = \frac{\pi \overline{n}^3 \overline{r}\,\Gamma L}{\lambda_0 \quad G} \tag{7.14}$$

Here, $\overline{r}$ is the effective electro optic coefficient, $\overline{n}$ is the effective index of the mode, Γ is the overlap integral, G is the electrode spacing, and L is the length of the electrode where an external electric field is applied on the waveguide.

Thus, the output power from the interferometer varies with voltage according to the following equation

$$P_{out} = P_{in} \cos^2\left(\frac{\pi V}{2V_0}\right) \tag{7.15}$$

with

$$V_0 = \frac{\pi}{\alpha} = \frac{\lambda_0 G}{\overline{n}^3 \overline{r} \Gamma L} \tag{7.16}$$

which shows the dependence of the output power on the applied voltage. When the applied voltage becomes equal to V_0, the output power becomes zero. Thus, V_0 is the voltage required to introduce a phase change of π in the propagating mode in the waveguide.

Figure 7.7 shows a typical sinusoidal variation in the transmittivity of the device (ratio of transmitted power to incident power) as a function of $\frac{V}{V_0}$. As the applied voltage changes, the output power varies sinusoidally. At $V = V_0$, the phase difference becomes equal to π and the output power goes to zero.

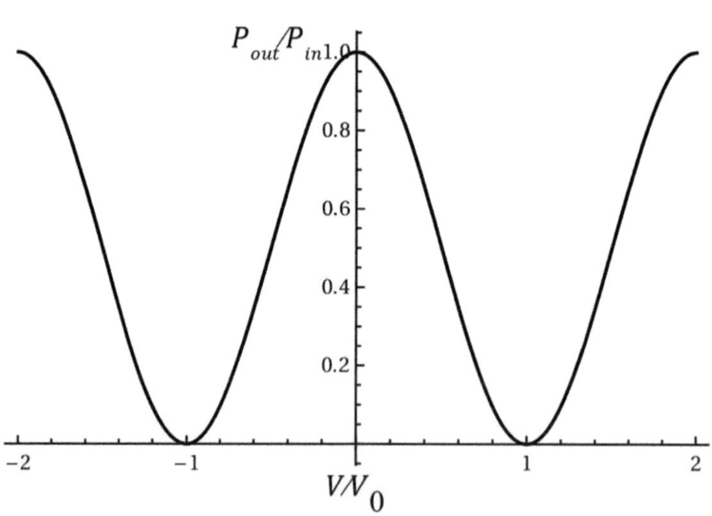

Fig. 7.7 Variation in the transmittance ($\frac{P_{out}}{P_{in}}$) of the Mach-Zehnder interferometer as a function of $\frac{V}{V_0}$ (see Eq. 7.15)

For linear operation, it is desirable to operate around the linear operating point corresponding to a phase difference of $\frac{\pi}{2}$. Thus, applying a constant bias voltage given by $\frac{V_0}{2}$ a phase difference of $\frac{\pi}{2}$ is introduced. The total voltage applied is now given by

$$V = \frac{V_0}{2} + V_m \cos \Omega t \tag{7.17}$$

where V_m is the peak modulation voltage and Ω is the modulation frequency. If $V_m \ll V_0$, then we have approximately

$$P_{out} = P_{in} \cos^2\left(\frac{\pi}{2V_0}\left\{\frac{V_0}{2} + V_m \cos \Omega t\right\}\right) = \frac{P_{in}}{2}\left(1 + \frac{\pi V_m}{V_0} \cos \Omega t\right) \tag{7.18}$$

Hence, the output optical power from the interferometer is modulated at the frequency Ω of the electrical signal. Hence, an electrical signal can be transformed into an optical signal using such a modulator and finds application as optical modulators in optical fiber communication, optical sensing, and integrated quantum photonics.

The device can act as an optical switch; switching the applied voltage from 0 to V_0, the output power switches from full power to zero power.

Note that V_0 depends on the wavelength of the propagating light through direct dependence on wavelength as well as through the wavelength dependence of parameters of the refractive index $\overline{n}$ and the overlap integral Γ. Thus, the bias voltage would be different for different wavelengths.

Such a Mach-Zehnder interferometer device is used for transferring information from an electrical signal to an optical signal for applications in fiber optic communication systems. Cascading Mach-Zehnder interferometers can lead to interesting wavelength filtering characteristics (see, e.g., Horst et al. 2013). It is also used in processing of photons in integrated quantum photonic circuits. By combining with nonlinear effects, such a configuration can be used to generate entangled photon pairs.

Exercise 7.1 Using Eq. (7.15) for the output of a modulator without bias, assuming $V_m \ll V_0$ obtain the output intensity variation.

Solution Substituting $V_m \cos \Omega t$ in Eq. (7.15), we have

$$P_{out} = P_{in} \cos^2\left(\frac{\pi V}{2V_0}\right) = P_{in} \cos^2\left(\frac{\pi V_m}{2V_0} \cos \Omega t\right)$$

which under the condition $V_m \ll V_0$, can be approximated by

$$P_{out} \approx P_{in}\left[1-\left(\frac{\pi V_m}{2V_0}\cos\Omega t\right)^2\right]=P_{in}\left[1-\left(\frac{\pi V_m}{2V_0}\right)^2\cos^2\Omega t\right]$$

We see that for a signal given by $V = V_m \cos \Omega t$, the output power varies as $\cos^2\Omega t$. Hence, under such a condition, the modulator does not behave as a linear device transferring the voltage variation into the output optical power variation.

Exercise 7.2 For linear operation of the Mach-Zehnder interferometer modulator, one needs to operate at the bias point where with no modulating voltage applied, the output from the device is 0.5 (see Eq. 7.18). For this one can use a bias voltage corresponding to $\frac{V_0}{2}$. It is possible to bias the modulator to this point of operation by choosing slightly different lengths of the arms of the interferometer. If n_{eff} is the effective index of the propagating mode what should be the difference in length of the two arms to introduce a phase difference of $\frac{\pi}{2}$ between the two arms? Assume a wavelength of 1500 nm.

Solution For a length difference of Δl between the two arms of the interferometer, the phase difference between the two propagating modes would be $\Delta\phi = \frac{2\pi}{\lambda_0} n_{eff} \Delta l$. Here, λ_0 is the free space wavelength. For this phase difference to be equal to $\frac{\pi}{2}$, we must have

$$\frac{2\pi}{\lambda_0} n_{eff} \Delta l = \frac{\pi}{2}$$

which gives us

$$\Delta l = \frac{\lambda_0}{4n_{eff}}$$

Assuming $n_{eff} = 2.2$, the required length difference at a free space wavelength of 1500 nm would be 170.5 nm which is indeed a very small value.

It must be noted that as the wavelength of the input changes, the phase difference between the two arms will change and thus the operating point will also change. For example if the wavelength changes from 1500 to 1300 nm, assuming that the change in n_{eff} is negligible, the phase difference between the two arms would now be

$$\Delta\phi = \frac{2\pi}{\lambda_0} n_{eff} \Delta l = \frac{2\pi}{850} \times 2.2 \times 170.5 \approx 0.88\pi$$

For this phase difference, the output power with zero modulation voltage would be $\cos^2(\Delta\phi) \approx 0.86$ instead of the required 0.5.

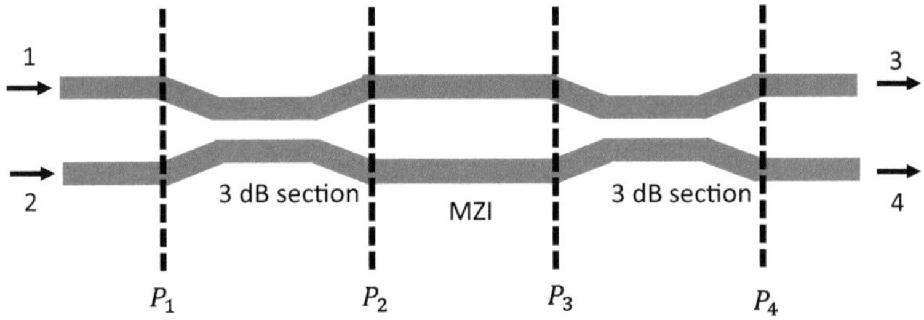

Fig. 7.8 A Mach-Zehnder interferometer design with two inputs and two outputs. Instead of Y-junctions this design uses 3 dB directional couplers to split the power equally among the two waveguides of the interferometer

In a Mach-Zehnder interferometer, there is one usable input and one usable output which are the two arms of the interferometer. It is possible to design a similar interferometer with two input arms and two output arms by using a directional coupler instead of a Y-splitter. Figure 7.8 shows such a configuration wherein the input and output directional couplers are 3 dB couplers so that light input into any of the two input ports will get distributed equally among the two arms of the interferometer and at the output the power distribution between the two output arms will depend on the phase difference introduced in the interferometer.

The device shown in Fig. 7.8 has two inputs and two outputs and the device connecting these can be represented by a transfer matrix between the output and input. For this, we recall our discussion in Chap. 5 where we obtained the relation between the output and input arms of a directional coupler (see Eq. 5.60). With input into one of the waveguides (shown as port 1 in Fig. 7.8), the input state in matrix form represented on Plane P_1 is given by

$$\begin{pmatrix} R(z=0)=1 \\ S(z=0)=0 \end{pmatrix} = \begin{pmatrix} 1 \\ 0 \end{pmatrix} \tag{7.19}$$

After the first 3 dB splitter, the amplitudes on plane P_2 will be given by

$$\begin{pmatrix} R(P_2) \\ S(P_2) \end{pmatrix} = \begin{pmatrix} \frac{1}{\sqrt{2}} & \frac{i}{\sqrt{2}} \\ \frac{i}{\sqrt{2}} & \frac{1}{\sqrt{2}} \end{pmatrix} \begin{pmatrix} 1 \\ 0 \end{pmatrix} = \begin{pmatrix} \frac{1}{\sqrt{2}} \\ \frac{i}{\sqrt{2}} \end{pmatrix} \tag{7.20}$$

If $\Delta\phi$ is the phase difference introduced by the electro optic effect between the two arms of the interferometer, the amplitudes in plane P_3 will be given by

$$\begin{pmatrix} R(P_3) \\ S(P_3) \end{pmatrix} = \begin{pmatrix} e^{i\Delta\phi/2} & 0 \\ 0 & e^{-i\Delta\phi/2} \end{pmatrix} \begin{pmatrix} \frac{1}{\sqrt{2}} \\ \frac{i}{\sqrt{2}} \end{pmatrix} = \begin{pmatrix} \frac{1}{\sqrt{2}} e^{i\Delta\phi/2} \\ \frac{i}{\sqrt{2}} e^{-i\Delta\phi/2} \end{pmatrix} \tag{7.21}$$

where we have assumed that a phase shift of $\Delta\phi/2$ is introduced in each arm. Finally the output on plane P_4 will be given by

$$\begin{pmatrix} R(P_4) \\ S(P_4) \end{pmatrix} = \begin{pmatrix} \frac{1}{\sqrt{2}} & \frac{i}{\sqrt{2}} \\ \frac{i}{\sqrt{2}} & \frac{1}{\sqrt{2}} \end{pmatrix} \begin{pmatrix} \frac{1}{\sqrt{2}} e^{i\Delta\phi/2} \\ \frac{i}{\sqrt{2}} e^{-i\Delta\phi/2} \end{pmatrix}$$
$$= \begin{pmatrix} \frac{1}{2}\left(e^{i\Delta\phi/2} - e^{-i\Delta\phi/2}\right) \\ \frac{i}{2}\left(e^{i\Delta\phi/2} + e^{-i\Delta\phi/2}\right) \end{pmatrix} = i \begin{pmatrix} \sin\frac{\Delta\phi}{2} \\ \cos\frac{\Delta\phi}{2} \end{pmatrix} \tag{7.22}$$

Thus, the powers in the two output ports 3 and 4 will vary as $\sin^2\frac{\Delta\phi}{2}$ and $\cos^2\frac{\Delta\phi}{2}$ respectively showing optical modulation. If the phase difference $\Delta\phi$ is introduced through the electro optic effect, the device becomes an electro optic intensity modulator.

Exercise 7.3 Consider an electro optic modulator using the Mach-Zehnder interferometer configuration using channel waveguides in y-cut x-propagating lithium niobate.

(a) Which quasi mode should onc usc to achieve minimum drive voltage?
(b) Obtain an expression for the voltage required to modulate the output in either of the arms from a maximum to a minimum.

Solution In Chap. 6, we discussed the electro optic effect in lithium niobate and it was mentioned that the largest electro optic coefficient in lithium niobate is r_{33}. In order to use this coefficient, the light wave must be polarized parallel to the crystal z-axis (the optic axis) and the electric field should also be applied along the optic axis. Since the waveguide is fabricated on an y-cut x-propagating substrate, the mode showing minimum drive voltage would the quasi TE mode with its polarization parallel to the $z-$ axis.

The phase shift introduced by the applied electric field is given by (see Eq. 6.55):

$$\Delta\phi = -\frac{\pi n_e^3 r_{33}}{\lambda_0} \frac{V}{G} \Gamma L$$

where all parameters are defined in Chap. 6. The phase change required for complete modulation is $\Delta\phi = \pi$, which gives the required voltage as

$$V = \frac{\lambda_0}{n_e^3 r_{33}} \frac{G}{\Gamma L}$$

Taking typical values of $n_e = 2.2$, $r_{33} = 30$ pm/V, $\Gamma = 0.5$, $L = 10$ mm and $G = 10$ μm, we have for an operating wavelength of 800 nm, the required voltage of 5 V.

Using a similar analysis, it is left as an exercise to obtain the corresponding drive voltage required using the quasi TM mode in the waveguide and compare with the value obtained above.

Let us obtain the transfer matrix of the device shown in Fig. 7.8. The matrix representing the Mach-Zehnder interferometer connecting the output and input ports is a product of the matrices representing the two directional couplers and the phase shifting region and is given by

$$T = \begin{pmatrix} \frac{1}{\sqrt{2}} & \frac{i}{\sqrt{2}} \\ \frac{i}{\sqrt{2}} & \frac{1}{\sqrt{2}} \end{pmatrix} \begin{pmatrix} e^{i\Delta\phi/2} & 0 \\ 0 & e^{-i\Delta\phi/2} \end{pmatrix} \begin{pmatrix} \frac{1}{\sqrt{2}} & \frac{i}{\sqrt{2}} \\ \frac{i}{\sqrt{2}} & \frac{1}{\sqrt{2}} \end{pmatrix}$$
$$= i \begin{pmatrix} \sin\frac{\Delta\phi}{2} & \cos\frac{\Delta\phi}{2} \\ \cos\frac{\Delta\phi}{2} & -\sin\frac{\Delta\phi}{2} \end{pmatrix} \tag{7.23}$$

Hence, we have

$$\begin{pmatrix} R_{out} \\ S_{out} \end{pmatrix} = i \begin{pmatrix} \sin\frac{\Delta\phi}{2} & \cos\frac{\Delta\phi}{2} \\ \cos\frac{\Delta\phi}{2} & -\sin\frac{\Delta\phi}{2} \end{pmatrix} \begin{pmatrix} R_{in} \\ S_{in} \end{pmatrix} \tag{7.24}$$

The 2×2 matrix on the right hand side defines the transfer function of the Mach-Zehnder interferometer using which one can obtain the outputs for any input condition.

Thus, if at the input we have excitation of both the input waveguides with equal amplitude and phase, then we have

$$\begin{pmatrix} R_{in} \\ S_{in} \end{pmatrix} = \frac{1}{\sqrt{2}} \begin{pmatrix} 1 \\ 1 \end{pmatrix} \tag{7.25}$$

and the output would be

$$\begin{pmatrix} R_{out} \\ S_{out} \end{pmatrix} = i \begin{pmatrix} \sin\frac{\Delta\phi}{2} & \cos\frac{\Delta\phi}{2} \\ \cos\frac{\Delta\phi}{2} & -\sin\frac{\Delta\phi}{2} \end{pmatrix} \frac{1}{\sqrt{2}} \begin{pmatrix} 1 \\ 1 \end{pmatrix} = \frac{i}{\sqrt{2}} \begin{pmatrix} \sin\frac{\Delta\phi}{2} + \cos\frac{\Delta\phi}{2} \\ \cos\frac{\Delta\phi}{2} - \sin\frac{\Delta\phi}{2} \end{pmatrix} \tag{7.26}$$

Hence, if $\Delta\phi = +\frac{\pi}{2}$, then

$$\begin{pmatrix} R_{out} \\ S_{out} \end{pmatrix} = i \begin{pmatrix} 0 \\ 1 \end{pmatrix} \tag{7.27}$$

and if $\Delta\phi = -\frac{\pi}{2}$, then

$$\begin{pmatrix} R_{out} \\ S_{out} \end{pmatrix} = i \begin{pmatrix} 1 \\ 0 \end{pmatrix} \tag{7.28}$$

In both cases, at output the power exits from only one of the waveguides.

Note that if there is no phase shift introduced in the intermediate section of the interferometer, in principle the device is equivalent to a directional coupler of length equal to the coupling length. This is so since each of the two directional couplers is a 3 dB coupler which has a length equal to half the coupling length. Thus, in this case the transfer matrix becomes

$$T = \begin{pmatrix} 0 & i \\ i & 0 \end{pmatrix} \tag{7.29}$$

which defines the cross state of the directional coupler (see Chap. 6).

In Chap. 19, we will discuss the behavior of the interferometer with the input being a single photon and show interference of a photon with itself.

Exercise 7.4 Obtain the condition that needs to be satisfied by the matrix T (Eq. 7.23) for energy conservation (see Chap. 6).

Solution From energy conservation we obtain

$$|R_{out}|^2 + |S_{out}|^2 = |R_{in}|^2 + |S_{in}|^2$$

Now

$$|R_{out}|^2 + |S_{out}|^2 = (\, R_{out}^* \quad S_{out}^* \,) \begin{pmatrix} R_{out} \\ S_{out} \end{pmatrix}$$

and

$$|R_{in}|^2 + |S_{in}|^2 = (\, R_{in}^* \quad S_{in}^* \,) \begin{pmatrix} R_{in} \\ S_{in} \end{pmatrix}$$

Thus

$$(\, R_{out}^* \quad S_{out}^* \,) \begin{pmatrix} R_{out} \\ S_{out} \end{pmatrix} = (\, R_{in}^* \quad S_{in}^* \,) T^\dagger T \begin{pmatrix} R_{in} \\ S_{in} \end{pmatrix} = (\, R_{in}^* \quad S_{in}^* \,) \begin{pmatrix} R_{in} \\ S_{in} \end{pmatrix}$$

where $T^\dagger$ is the adjoint (conjugate transpose) of T. Thus, the matrix T must satisfy the following equation:

$$T^\dagger T = I \tag{7.30}$$

Substituting the value of T from Eq. (7.23), Eq. (7.30) can be verified. Such a matrix is a unitary matrix.

Exercise 7.5 Consider the inputs to the MZ interferometer to be

$$R_{in} = \tilde{R}_{in} e^{i\phi_R} \tag{7.31}$$

$$S_{in} = \tilde{S}_{in} e^{i\phi_S} \tag{7.32}$$

where $\tilde{R}_{in}$ and $\tilde{S}_{in}$ are the amplitudes of the two inputs and ϕ_R and ϕ_S their phases. Obtain expression for the output powers from the output arms.

Solution For the input conditions, we have

$$\begin{pmatrix} R_{out} \\ S_{out} \end{pmatrix} = \begin{pmatrix} i\sin\frac{\Delta\phi}{2} & i\cos\frac{\Delta\phi}{2} \\ i\cos\frac{\Delta\phi}{2} & -i\sin\frac{\Delta\phi}{2} \end{pmatrix} \begin{pmatrix} \tilde{R}_{in} e^{i\phi_R} \\ \tilde{S}_{in} e^{i\phi_S} \end{pmatrix} \tag{7.33}$$

which gives us

$$R_{out} = i\tilde{R}_{in} e^{i\phi_R} \sin\frac{\Delta\phi}{2} + i\tilde{S}_{in} e^{i\phi_S} \cos\frac{\Delta\phi}{2} \tag{7.34}$$

Thus, the power output from this arm is given by

$$P_{R,out} = |R_{out}|^2 = \tilde{R}_{in}^2 \sin^2\frac{\Delta\phi}{2} + \tilde{S}_{in}^2 \cos^2\frac{\Delta\phi}{2} + \tilde{R}_{in}\tilde{S}_{in} \sin\Delta\phi \cos(\phi_R - \phi_S) \tag{7.35}$$

Similarly

$$P_{S,out} = |S_{out}|^2 = \tilde{R}_{in}^2 \cos^2\frac{\Delta\phi}{2} + \tilde{S}_{in}^2 \sin^2\frac{\Delta\phi}{2} - \tilde{R}_{in}\tilde{S}_{in} \sin\Delta\phi \cos(\phi_R - \phi_S) \tag{7.36}$$

Note that $P_{R,out} + P_{S,out} = \tilde{R}_{in}^2 + \tilde{S}_{in}^2$ implying power conservation.

(a) If $\tilde{R}_{in} = \tilde{S}_{in}$, the input arms have equal powers then we have

$$P_{R,out} = \tilde{R}_{in}^2[1 + \sin\Delta\phi \cos(\phi_R - \phi_S)] \tag{7.37}$$

$$P_{S,out} = \tilde{R}_{in}^2[1 - \sin\Delta\phi \cos(\phi_R - \phi_S)] \tag{7.38}$$

The output powers in the two arms depend on the phase difference $(\phi_R - \phi_S)$ between the two input fields.

(b) If $\tilde{S}_{in} \gg \tilde{R}_{in}$, and assuming $\Delta\phi = \frac{\pi}{2}$, we can approximate the two output powers as

$$P_{R,out} \approx \frac{1}{2}\tilde{S}_{in}^2 + \frac{1}{\sqrt{2}}\tilde{R}_{in}\tilde{S}_{in} \cos(\phi_R - \phi_S) \tag{7.39}$$

and

$$P_{S,out} \approx \frac{1}{2}\tilde{S}_{in}^2 - \frac{1}{\sqrt{2}}\tilde{R}_{in}\tilde{S}_{in} \cos(\phi_R - \phi_S) \tag{7.40}$$

If the two output powers are detected and converted to electrical currents through photodetectors, then the two output currents will be proportional to the two output powers. Subtracting the output currents, we will obtain

$$\Delta I = \sqrt{2}K\tilde{R}_{in}\tilde{S}_{in} \cos(\phi_R - \phi_S) \tag{7.41}$$

where K is a constant which takes into account the detector response. Thus, the processed output current depends on the phase difference between the two input fields. Such a measurement leads to what is referred to as balanced homodyning (see Chap. 20).

7.5 Tapered Velocity Coupler

The idea of adiabatic mode transformation can be applied to many devices. One such interesting device is the tapered velocity coupler. Figure 7.9 shows the configuration of the device in which a directional coupler is formed between a pair of waveguides referred to as *a* and *b*, one of which has a *z*-dependent structure. If we refer to the waveguide with *z*-dependence as waveguide *a*, then it is so designed that at the input ($z = 0$) $\beta_a < \beta_b$ and at the exit end after a length of interaction *L*, $\beta_a > \beta_b$. Somewhere along the length the two waveguides have the same value of propagation constant. The device is a directional coupler consisting of two non-identical waveguides which evolve in such a fashion that the waveguide having a larger propagation constant switches from one waveguide to the other. In such a case, the fundamental normal mode at the input would correspond to power in waveguide *b* (which has a higher propagation constant) and at the exit end the fundamental normal mode would correspond to power in waveguide *a* (which has a higher propagation constant). At the point, where the two waveguides have the same propagation constant, the fundamental mode would have equal powers in both the waveguides.

The operation of the device can be understood in terms of the evolution of the normal modes of the device. If power is incident on waveguide *b* (see Fig. 7.9), then it would correspond to the fundamental normal mode of the coupler. Now, if the transition is adiabatic, then the incident power will continue to remain in the fundamental mode of the structure at every point along the propagation. Thus, at the exit, the power will still be in the fundamental mode which now happens to have its power in waveguide *a* rather than waveguide *b*. Thus, the power incident in waveguide *b* gets transferred to waveguide *a* at the exit. Note that this operation is independent of the wavelength of operation as long as there is a point somewhere along the interaction region where the propagation constants of the two waveguides become equal and the transition is adiabatic. Thus, this is a broadband coupler coupling light from an input waveguide to another waveguide through adiabatic transfer using directional coupler principle.

Problems

7.1. Consider a Y-combiner where equal power is incident on the two input waveguides but their phase difference is an arbitrary value ϕ. Obtain an expression for the output from the device as a function of ϕ.

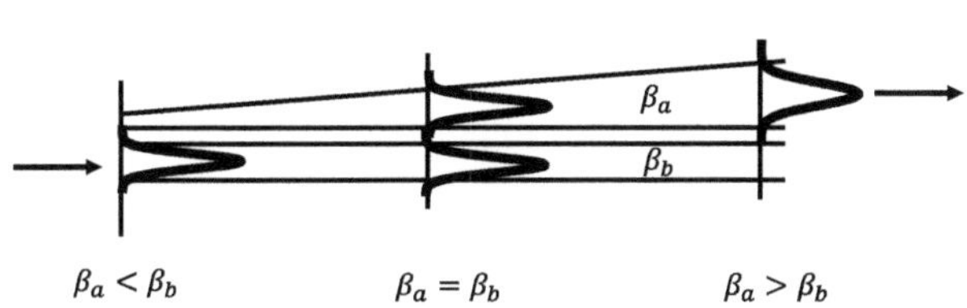

Fig. 7.9 A tapered velocity coupler in which the power continues to propagate in the fundamental mode as it propagates through the device. The coupler is assumed to be adiabatic

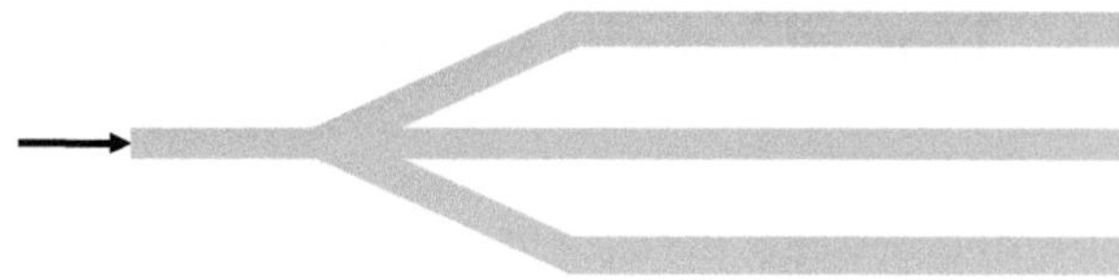

Fig. 7.10 Figure for Problem 7.2

Fig. 7.11 Figure for Problem 7.4

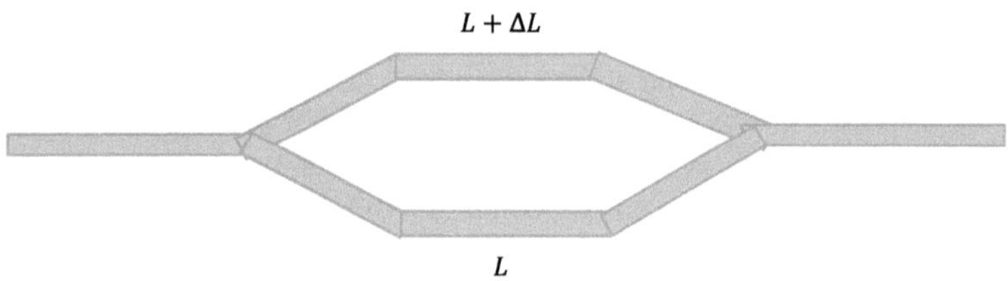

Fig. 7.12 Figure for Problem 7.5

7.2. Consider a splitter in which a single-mode waveguide splits into three identical single-mode waveguides (see Chap. 5 for a discussion on a three waveguide directional coupler) through an adiabatic transition (see Fig. 7.10). (a) If a power P_{in} is incident, then assuming the device to be lossless what will be the powers exiting the three waveguides? (b) If unit power is incident from the other end in the central waveguide or in one of the extreme waveguides in Fig. 7.10, what will be output power in the single waveguide?

7.3. Consider a Y-combiner where unequal power is incident on the two input waveguides and the two fields are in phase. Obtain an expression for the output from the device in terms of the difference in the powers at the input.

7.4. Figure 7.11 shows an integrated optic device with two *Y*–branches with input as shown. All waveguides, including the waveguide between the two Y-branches, are single-mode waveguides. If at the input, the power in one of the input waveguides is P_0 what would be the power exiting each of the two output waveguides?

7.5. Consider a Mach-Zehnder interferometer with unequal arm lengths (see Fig. 7.12) and (a) show that it acts like a wavelength filter. (b) Obtain an expression for the free spectral range (FSR) of the interferometer. (c) For a given filter as the input wavelength is varied, the output power reaches a maximum at two adjacent wavelengths of 1530 and 1590 nm. Assuming an average group index of 1.5, calculate the difference in arm lengths.

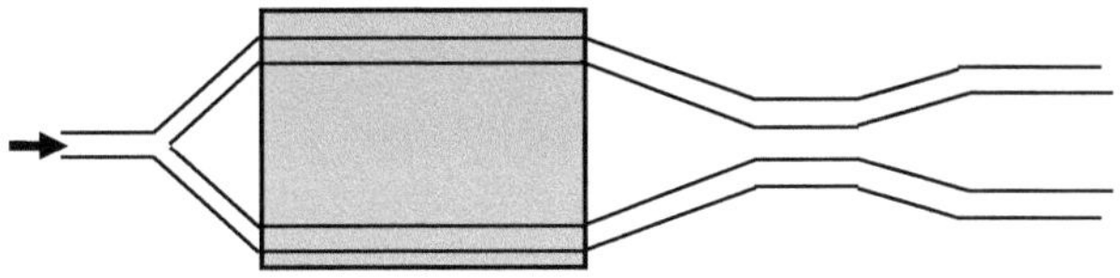

Fig. 7.13 Figure for Problem 7.10

7.6. In continuation of the Problem 7.5, consider three concatenated Mach-Zehnder interferometers with arm length differences of ΔL, $2\Delta L$, and $4\Delta L$. Plot a schematic of the output spectrum that will be observed when a broadband source is used to couple light into the device.

7.7. Suppose the output of the Mach-Zehnder interferometer is a two-mode waveguide instead of a single-mode waveguide. How would the output vary as a function of the applied voltage?

7.8. Referring to Fig. 7.9 if the power is incident on the waveguide with smaller β, then what output do you expect at the output?

7.9. In a Mach-Zehnder interferometer, the minimum voltage required to change from maximum to minimum output is V_0. Plot a schematic of the output power variation with time if the applied voltage has a sawtooth variation given by

$$V(t) = 4V_0 \frac{t}{T}; \quad 0 < t < T$$

$$V(t + T) = V(t)$$

7.10. Consider a device consisting of a Y-splitter and a directional coupler with length $L = \frac{L_c}{2}$ where L_c is the coupling length, as shown in Fig. 7.13. The propagation constants of the modes in the two waveguides in the shaded region of length L are β_1 and β_2. What will be the output from the directional coupler If (a) $(\beta_1 - \beta_2)L = +\frac{\pi}{2}$ and (b) $(\beta_1 - \beta_2)L = -\frac{\pi}{2}$? All waveguides are identical except for the shaded region.

References

Horst, F., W. M. J. Green, S. Assefa, S. M. Shank, Y. A. Vlasov, and B. J. Offrein. 2013. Cascaded Mach-Zenhder wavelength filters in silicon photonics for low loss and flat pass-band WDM (de-)multiplexing. *Opt Express* 21:11652–11658.

Sun, S., Q. Yu, Y. Che, T. Lian, Y. Xie, D. Zhang, and X. Wang. 2024. Mode-insensitive and mode-selective optical switch based on asymmetric Y-junctions and MMI couplers. *Photon Res* 12:423–430.

Chapter 8
Optical Ring Resonators and Arrayed Waveguide Grating

Abstract This chapter discusses the basic operating principles of waveguide ring resonators and arrayed waveguide gratings (AWG) which have important applications as wavelength filters, add-drop filters, wavelength multiplexers, and demultiplexers.

8.1 Introduction

In this chapter, we will discuss ring resonators and arrayed waveguide gratings (AWG) which are very important components in integrated optics and integrated quantum photonics and find wide applications. Both these components act like wavelength filters and are used in many optical communication applications. We will consider simple configurations and discuss the basic principle behind these devices and obtain some of their primary characteristics. Further details of these devices can be found in the excellent literature.

8.2 Optical Ring Resonator

Figure 8.1 shows an example of a ring resonator in which a waveguide in the form of a ring is coupled to a straight waveguide. Coupling between the straight and the ring waveguide happens via evanescent field coupling like a directional coupler (see Chap. 5). Such a device will be shown to exhibit dips in the transmission spectrum and thus it behaves like a spectral filter. Due to this characteristic optical ring resonators find applications in optical communication systems for wavelength division multiplexing and could be used for spectral filtering in integrated quantum photonic circuits. Such ring resonators also find application in sensing through the dependence of the position of the transmission dips on the surrounding medium. They also find very important applications in integrated quantum photonics for the generation of frequency entangled combs.

K. Thyagarajan, *Integrated Quantum Photonics*, Graduate Texts in Physics,
https://doi.org/10.1007/978-3-031-85728-7_8

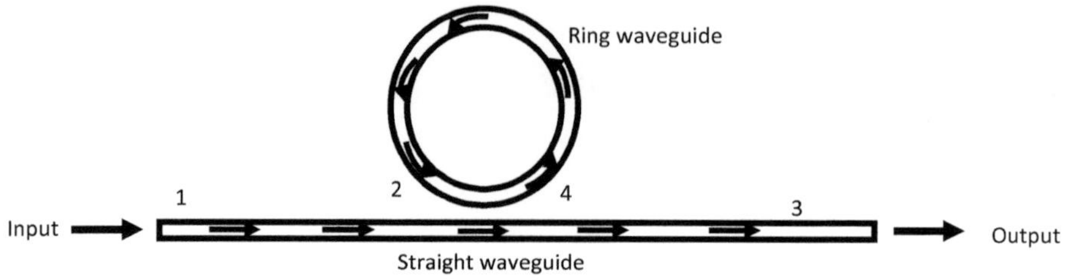

Fig. 8.1 A waveguide ring resonator

8.2.1 *Analysis*

In order to analyze the resonator, we first note that the region where the ring is close to the straight waveguide where evanescent coupling of power can take place between the straight waveguide and the ring is like a directional coupler. If the electric field of the waves in the ports 1 and 2 (see Fig. 8.1) are represented by E_1 and E_2, respectively, and the fields in the ports 3 and 4 are represented by E_3 and E_4, respectively, then we have (see Eq. 5.60)

$$\begin{pmatrix} E_3 \\ E_4 \end{pmatrix} = \begin{pmatrix} \cos \kappa l & i \sin \kappa l \\ i \sin \kappa l & \cos \kappa l \end{pmatrix} \begin{pmatrix} E_1 \\ E_2 \end{pmatrix} \tag{8.1}$$

with κ representing an effective coupling coefficient and l an effective interaction length. We are assuming the straight waveguide and the ring waveguide to be identical. Since the spacing between the two waveguides is varying with distance along the propagation direction, we are referring to effective length and effective coupling coefficient.

Due to the ring resonator geometry, the field coupled at port 4 comes back as input at port 2 after undergoing a phase change and some propagation loss due to propagation through the ring. Thus, we can write for the field entering the coupling region

$$E_2 = E_4 e^{i\gamma(2\pi R)} \tag{8.2}$$

where γ is the propagation constant of the propagating mode of the ring resonator waveguide, and $2\pi R$ represents the circumference of the ring resonator having a radius of R. In writing the above equation, for simplicity, we have neglected the length of the coupling region in comparison to the circumference of the ring.

Now, in general the ring would have losses which are important. To account for the losses, we write

$$\gamma = \gamma_r + i\gamma_i \tag{8.3}$$

where γ_r is the real part of the propagation constant of the mode and γ_i the imaginary part which accounts for the losses in the waveguide. Thus,

$$E_2 = E_4 e^{-\gamma_i L} e^{i\gamma_r L} \tag{8.4}$$

where $L = 2\pi R$. Thus, we have from Eq. (8.1)

$$E_3 = E_1 \cos\kappa l + iE_2 \sin\kappa l \tag{8.5a}$$

$$E_4 = +iE_1 \sin\kappa l + E_2 \cos\kappa l \tag{8.5b}$$

We substitute for E_4 from Eq. (8.4) in Eq. (8.5b) to obtain

$$E_2 e^{\gamma_i L} e^{-i\gamma_r L} = +iE_1 \sin\kappa l + E_2 \cos\kappa l$$

or

$$E_2 = \frac{-i\sin\kappa l}{\cos\kappa l - e^{\gamma_i L} e^{-i\gamma_r L}} E_1$$

Substituting for E_2 in Eq. (8.5a), we have

$$E_3 = E_1\left(\cos\kappa l + \frac{\sin^2\kappa l}{\cos\kappa l - e^{\gamma_i L} e^{-i\gamma_r L}}\right) = E_1 \frac{1 - e^{\gamma_i L} e^{-i\gamma_r L}\cos\kappa l}{\cos\kappa l - e^{\gamma_i L} e^{-i\gamma_r L}}$$

which can be written as

$$E_3 = E_1 \frac{\cos\kappa l - e^{-\gamma_i L} e^{i\gamma_r L}}{1 - \cos\kappa l e^{-\gamma_i L} e^{i\gamma_r L}} \tag{8.6}$$

The power transmittance of the ring resonator is given by

$$T = \left|\frac{E_3}{E_1}\right|^2 = \frac{C^2 + e^{-2\gamma_i L} - 2Ce^{-\gamma_i L}\cos\gamma_r L}{1 + C^2 e^{-2\gamma_i L} - 2Ce^{-\gamma_i L}\cos\gamma_r L} \tag{8.7}$$

where

$$S = \sin\kappa l; \quad C = \cos\kappa l \tag{8.8}$$

Using the fact that $\cos\gamma_r L = 1 - 2\sin^2\frac{\gamma_r L}{2}$, we can rewrite Eq. (8.7) in the form

$$T = \frac{\left(C - e^{-\gamma_i L}\right)^2 + 4Ce^{-\gamma_i L}\sin^2\frac{\gamma_r L}{2}}{\left(1 - Ce^{-\gamma_i L}\right)^2 + 4Ce^{-\gamma_i L}\sin^2\frac{\gamma_r L}{2}} \tag{8.9}$$

Note that the transmission through the ring resonator depends on the incident wavelength primarily through the propagation constant γ_r which is given by

$$\gamma_r = \frac{2\pi}{\lambda_0} n_{eff} \tag{8.10}$$

where λ_0 is the free space wavelength and n_{eff} is the effective index of the propagating mode in the ring waveguide.

The coupling coefficient κ and the loss coefficient γ_i have usually weak wavelength dependence within the range of wavelengths considered.

In order to better appreciate the result, we now consider some special cases.

1. If there is no coupling between the ring and the straight waveguide, then $\kappa = 0$ which implies $C = 1$ and hence the transmittance is $T = 1$ as expected.
2. If there is no loss then $\gamma_i = 0$ and again we obtain $T = 1$. This is because if there are no losses, then the device has a single input port and a single output port and all input energy will exit the waveguide. Energy conservation implies $T = 1$.

Figure 8.2 shows the transmission through a typical ring resonator assuming the following values of the various parameters:

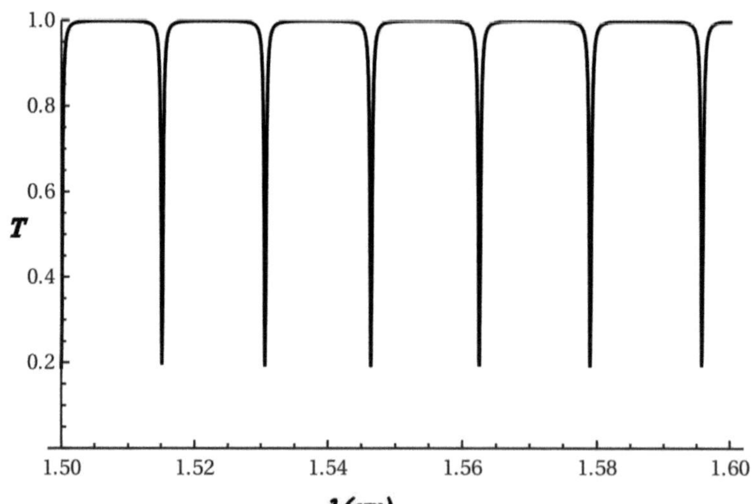

Fig. 8.2 Plot of transmittance of a ring resonator with $C = 0.98; e^{-\gamma_i L} = 0.95; L = 100\ \mu m$; $n_{eff} = 1.5$. Note the sharp resonances

$$C = 0.98;\, e^{-\gamma_i L} = 0.95;\, L = 100\ \mu\text{m};\, n_{eff} = 1.5$$

The phase accumulated by the propagating field over one round trip in the loop is

$$\phi = \gamma_r L = \frac{2\pi}{\lambda_0} n_{eff} L \tag{8.11}$$

where n_{eff} is the effective index at the free space wavelength of λ_0 of the mode of the waveguide in the ring. From Eq. (8.9), we see that whenever

$$\gamma_r L = 2m\pi; \qquad m = 1, 2, \ldots \tag{8.12}$$

then the transmission is minimum. The corresponding frequencies can be obtained as follows

$$\frac{2\pi}{\lambda_0} n_{eff} L = \frac{2\pi\nu}{c} n_{eff} L = 2m\pi$$

giving us

$$\nu = m\frac{c}{n_{eff}(\nu)L}; \qquad m = 1, 2, 3, \ldots \tag{8.13}$$

The frequency spacing between two adjacent minima is referred to as the free spectral range (FSR)—see Sect. 8.2.2.

In terms of wavelength, the transmission minima appear at

$$\lambda_0 = \frac{n_{eff}(\lambda_0)L}{m} \tag{8.14}$$

At these frequencies (or wavelengths), the power transmission through the resonator is given by

$$T_{min} = \frac{(C - e^{-\gamma_i L})^2}{(1 - Ce^{-\gamma_i L})^2} \tag{8.15}$$

Substituting the values of various parameters used in plotting Fig. 8.2, we find $T_{min} \approx 0.19$ (see Fig. 8.2).

The transmission minima are given by Eq. (8.15) and will become zero when

$$C = \cos\kappa l = e^{-\gamma_i L} \tag{8.16}$$

If this condition is satisfied, then the transmission minima will have zero value. This is referred to as critical coupling.

Thus, for example, if the power attenuation in one passage through the ring which is $e^{-2\gamma_i L}$ is 0.2 dB, then we have

$$10\log e^{-2\gamma_i L} = -0.2$$

which implies

$$e^{-2\gamma_i L} = 10^{-0.02} \approx 0.955$$

Hence, for zero transmission we must have

$$\cos\kappa l = \sqrt{0.955} \approx 0.977$$

giving

$$\kappa l \approx 0.215$$

We also have

$$\sin\kappa l \approx 0.213$$

Hence, the power transmission and cross coupling in the directional coupler will be (see Eq. 8.1) $\cos^2\kappa L \approx 0.955$ and $\sin^2\kappa L \approx 0.045$, respectively.

When the optimum coupling condition is satisfied, then the transmission becomes

$$T = \frac{4C^2 \sin^2 \frac{\gamma_r L}{2}}{\left(1 - C^2\right)^2 + 4C^2 \sin^2 \frac{\gamma_r L}{2}} \tag{8.17}$$

Figure 8.3 shows the corresponding transmission spectrum showing that the transmission goes to zero at specific wavelengths.

8.2.2 *Free Spectral Range (FSR)*

In order to obtain a relation for FSR, let us assume that at frequency ν, Eq. (8.12) is satisfied for a particular value of m. Let $\nu + \Delta\nu$ be the frequency corresponding to the value $(m + 1)$. Hence, we have

$$\nu + \Delta\nu = (m+1)\frac{c}{n_{eff}(\nu + \Delta\nu)L} \tag{8.18}$$

Assuming $\Delta\nu \ll \nu$, we have

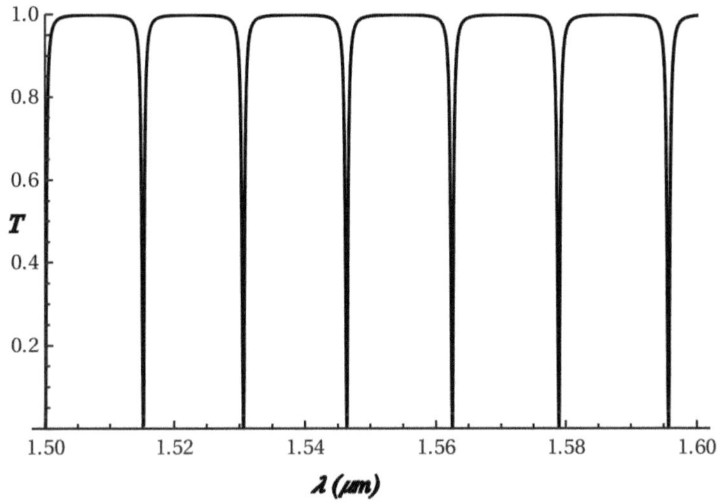

Fig. 8.3 Transmission spectrum of a ring resonator satisfying the critical coupling condition given by Eq. (8.16)

$$(\nu + \Delta\nu)n_{eff}(\nu + \Delta\nu) \approx (\nu + \Delta\nu)\left[n_{eff}(\nu) + \Delta\nu\frac{dn_{eff}(\nu)}{d\nu}\right]$$

which, by ignoring $\Delta\nu^2$ terms, we obtain

$$(\nu + \Delta\nu)n_{eff}(\nu + \Delta\nu) \approx \nu n_{eff}(\nu) + \Delta\nu N_{eff}(\nu) \tag{8.19}$$

where the quantity

$$\left[n_{eff}(\nu) + \nu\frac{dn_{eff}(\nu)}{d\nu}\right] = N_{eff}(\nu) \tag{8.20}$$

is the mode group effective index (see Chap. 4).

Hence, using Eq. (8.19) in Eq. (8.18), we obtain

$$\nu n_{eff}(\nu) + \Delta\nu N_{eff}(\nu) = (m+1)\frac{c}{L}$$

which using Eq. (8.13) gives us the following expression for the free spectral range of the ring resonator:

$$\Delta\nu_{FSR} = \frac{c}{N_{eff}(\nu)L} \tag{8.21}$$

Smaller the value of L, larger is the FSR. The corresponding value in terms of wavelength is given by

$$\Delta\lambda_{FSR} = \frac{\lambda_0^2}{N_{eff}(\lambda_0)L} \tag{8.22}$$

If we assume $L = 1$ cm, $N_{eff} = 1.5$, then we obtain the typical FSR of

$$\Delta\nu_{FSR} = 20 \text{ GHz}$$

At a wavelength of 1500 nm, the corresponding value in wavelength is about 0.15 nm. Smaller values of L will lead to higher values of FSR.

8.2.3 Resonance Linewidth

From Eq. (8.9), we see that when the incident frequency satisfies Eq. (8.13), the transmission will be minimum and given by Eq. (8.15). As the frequency deviates from this value, the transmission will increase leading to a spectral filter characteristic as a band stop filter.

In order to obtain an estimate of the spectral bandwidth of the transmission spectrum, we assume that the coupling is close to the optimum coupling so that the minimum transmission is close to zero, that is, $C \approx e^{-\gamma_i L}$. To obtain the full width at half maximum (FWHM), we look for the frequencies close to the resonance frequency when the transmission becomes 0.5. This will happen when (see Eq. 8.9)

$$\frac{4Ce^{-\gamma_i L}\sin^2\frac{\gamma_r L}{2}}{(1 - Ce^{-\gamma_i L})^2 + 4Ce^{-\gamma_i L}\sin^2\frac{\gamma_r L}{2}} = \frac{1}{2} \tag{8.23}$$

where we have approximately put $C - e^{-\gamma_i L} \approx 0$. Simplifying the above, we obtain

$$\sin\frac{\gamma_r L}{2} = \frac{(1 - Ce^{-\gamma_i L})}{2\sqrt{Ce^{-\gamma_i L}}} \tag{8.24}$$

If we consider a frequency close to a value corresponding to transmission minimum which occurs at $\gamma_r L = 2m\pi$ (see Eq. 8.12), we may replace γ_r in Eq. (8.24) by

$$\gamma_r = \frac{2}{L}(m\pi + \epsilon)$$

where $\epsilon \ll m\pi$. Substituting in Eq. (8.24), we obtain

$$\sin\epsilon = \frac{(1 - Ce^{-\gamma_i L})}{2\sqrt{Ce^{-\gamma_i L}}} \tag{8.25}$$

For sufficiently narrow linewidth $\sin\epsilon \approx \epsilon$ and we have

$$\epsilon = \delta\left(\frac{\gamma_r L}{2}\right) = \frac{(1 - Ce^{-\gamma_i L})}{2\sqrt{Ce^{-\gamma_i L}}} \tag{8.26}$$

Let ν_c be the central frequency for which

$$\gamma_r(\nu_c) = \frac{2m\pi}{L} \tag{8.27}$$

Let $\nu_c + \delta\nu$ be the frequency when $T = 0.5$; this happens when

$$\gamma_r(\nu_c + \delta\nu) = \frac{2}{L}(m\pi + \epsilon) \tag{8.28}$$

Since $\gamma_r = \frac{2\pi\nu}{c} n_{eff}$, we have

$$\gamma_r(\nu_c + \delta\nu) = \frac{2\pi}{c}(\nu_c + \delta\nu) n_{eff}(\nu_c + \delta\nu) \tag{8.29}$$

We make Taylor series expansion and retain terms up to first order in $\delta\nu$ and obtain

$$\gamma_r(\nu_c + \delta\nu) \approx \frac{2\pi}{c}\left[\nu_c n_{eff}(\nu_c) + \delta\nu N_{eff}(\nu_c)\right] \tag{8.30}$$

where

$$N_{eff}(\nu_c) = n_{eff}(\nu_c) + \nu_c \left.\frac{dn_{eff}}{d\nu}\right|_{\nu=\nu_c} \tag{8.31}$$

is the group index of the mode at the central frequency. From Eqs. (8.28) and (8.30), we have

$$\frac{2\pi}{c}\left[\nu_c n_{eff}(\nu_c) + \delta\nu N_{eff}(\nu_c)\right] = \frac{2}{L}(m\pi + \epsilon) \tag{8.32}$$

From Eq. (8.13), we also have

$$\frac{2\pi\nu_c}{c} n_{eff}(\nu_c) = \frac{2m\pi}{L} \tag{8.33}$$

From Eqs. (8.32) and (8.33), we have

$$\epsilon = \frac{\pi N_{eff} L}{c} \delta\nu \tag{8.34}$$

Using Eqs. (8.26) and (8.34), we obtain the following expression for the resonance halfwidth:

$$\delta\nu = \frac{c}{\pi N_{eff} L} \frac{\left(1 - Ce^{-\gamma_i L}\right)}{2\sqrt{Ce^{-\gamma_i L}}} \tag{8.35}$$

Hence, the FWHM is given by

$$\Delta\nu_{FWHM} = 2\delta\nu = \frac{c}{\pi N_{eff} L} \frac{\left(1 - Ce^{-\gamma_i L}\right)}{\sqrt{Ce^{-\gamma_i L}}} \tag{8.36}$$

Expressed in terms of wavelength, we have

$$\Delta\lambda_{FWHM} = \frac{\lambda_c^2}{c} \Delta\nu_{FWHM} = \frac{\lambda_c^2}{\pi N_{eff} L} \frac{\left(1 - Ce^{-\gamma_i L}\right)}{\sqrt{Ce^{-\gamma_i L}}} \tag{8.37}$$

For near zero transmission dips $C = e^{-\gamma_i L}$ (see Eq. 8.16) and we have

$$\Delta\lambda_{FWHM} \approx \frac{\lambda_c^2}{\pi N_{eff} L} \frac{\left(1 - e^{-2\gamma_i L}\right)}{e^{-\gamma_i L}} \tag{8.38}$$

Taking typical values $\gamma_i L = 0.023$, $L = 1$ cm, $N_{eff} = 1.5$, and a wavelength $\lambda_c = 1550$ nm, we obtain

$$\Delta\lambda_{FWHM} \approx 2.4 \text{ pm}$$

which is indeed extremely small.

The Finesse of the resonator is the ratio of FSR to the resonance width

$$F = \frac{\Delta\nu_{FSR}}{\Delta\nu_{FWHM}} = \frac{\pi\sqrt{Ce^{-\gamma_i L}}}{\left(1 - Ce^{-\gamma_i L}\right)} \tag{8.39}$$

Taking typical values $C = e^{-\gamma_i L} = 0.977$ (corresponding to $\gamma_i L = 0.023$), we have $F \approx 68$.

The quality factor Q of the resonator is defined by

$$Q = \frac{\nu}{\Delta\nu_{FWHM}} = \frac{\lambda_c}{\Delta\lambda_{FWHM}} = \frac{\pi N_{eff} L}{\lambda_c} \frac{\sqrt{Ce^{-\gamma_i L}}}{\left(1 - Ce^{-\gamma_i L}\right)} \tag{8.40}$$

The Q of the resonator defines the quality of the resonator, smaller the losses, sharper are the resonances and higher is the Q value.

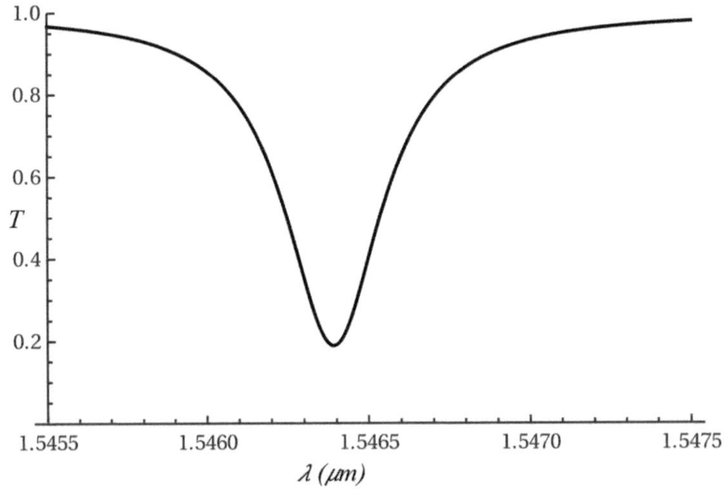

Fig. 8.4 Transmittance versus wavelength of a ring resonator with $C = 0.98; e^{-\gamma_i L} = 0.95; L = 100\ \mu\text{m}; N_{eff} = 1.5$

Taking typical values $\gamma_i L = 0.023$, $L = 1$ cm, $N_{eff} = 1.5$, and a wavelength $\lambda_c = 1550$ nm, we obtain a Q value of 6.5×10^5.

Figure 8.4 shows a typical transmission spectrum of a ring resonator with the following values of various parameters:

$$C = 0.98; e^{-\gamma_i L} = 0.95; L = 100\ \mu\text{m}; N_{eff} = 1.5$$

Note the drop in transmittance at the free space wavelength value of approximately 1546.4 nm.

The ring resonator is very similar to the Fabry Perot interferometer which consists of a pair of highly reflecting mirrors spaced by a certain distance (see, e.g., Ghatak 2024). Light incident normally on the Fabry Perot interferometer suffers multiple reflection between the mirrors and depending on the phase shift in one round trip within the resonator, one obtains transmission peaks. Unlike the Fabry Perot interferometer which has two inputs and two outputs, the ring resonator considered here has a single input and a single output. The equation for transmission dips in the ring resonator is very similar to the equation for transmission maxima in the Fabry Perot interferometer.

Exercise 8.1 The transmission dips correspond to $\gamma_r L = 2m\pi$; $m = 1, 2, \ldots$. Figure 8.4 shows the transmission through a ring resonator with $L = 100\ \mu\text{m}$ and $n_{eff} = 1.5$. Obtain the value of m corresponding to the dip shown in the figure.

Solution Substituting the value $\gamma_r = \frac{2\pi}{\lambda} n_{eff}$ and the values of the parameters, the value of m corresponding to the dip in transmission around 1546 nm is $m = 97$. Note the large order number of the interference.

Exercise 8.2 Consider a ring resonator with the following parameters: $C=0.98; e^{-\gamma_i L}=0.99; L=2$ mm; $N_{eff}=1.5, \lambda \approx 1550$ nm. Calculate the FWHM and the minimum transmission of the ring resonator. Show that there is a dip in transmission between the wavelengths 1562 and 1563 nm and obtain the corresponding value of m.

Solution Substituting the values in Eq. (8.37) we obtain $\Delta\lambda_{FWHM} \approx 7.7$ pm. The minimum transmission obtained using Eq. (8.15) is 0.11. For the given parameters, the wavelengths for minimum transmission are given by $\lambda = \frac{n_{eff} L}{m} = \frac{3000}{m}$ μm. The minimum wavelength of 1562.5 nm appears at $m = 1920$.

Exercise 8.3 The minima of transmission occur at wavelengths given by Eq. (8.14). The position of the minima depends on the effective index of the mode. Thus, any change in the effective index will lead to a change in the position of the minimum. Consider a ring resonator with $C = 0.98$, $e^{-\gamma_i L}=0.95$, $n_{eff} = 1.5$ and $L = 1$ mm. Obtain the change in the effective index which will lead to a change in the wavelength of transmission minimum equal to the FWHM of the resonator. Assume $\lambda_0 = 1500$ nm.

Solution The transmission minima appear at $\frac{\gamma_r L}{2} = m\pi$. Using Eq. (8.26), we have

$$\delta\left(\frac{\gamma_r L}{2}\right) = \frac{L}{2}\delta(\gamma_r) = \frac{\pi L \delta n_{eff}}{\lambda} = \frac{\left(1 - Ce^{-\gamma_i L}\right)}{2\sqrt{Ce^{-\gamma_i L}}}$$

giving us

$$\delta n_{eff} = \frac{\lambda}{\pi L} \frac{\left(1 - Ce^{-\gamma_i L}\right)}{2\sqrt{Ce^{-\gamma_i L}}} \tag{8.41}$$

Substituting the given values, we obtain $\delta n_{eff} \approx 1.7 \times 10^{-5}$. Such a change in effective index could be brought about by changes of temperature or other perturbations and thus such a ring resonator can be used for sensing.

Figure 8.5 shows two transmission curves through a ring resonator. The two plots correspond to two different optical path lengths through the ring resonator with the difference in optical path lengths being $\Delta(n_{eff}L) = 1.5 \times 10^{-8}$ m. This implies that for a length of the ring of $L = 100$ μm and an effective index $n_{eff} = 1.5$, either a change in length of 10^{-8} m or a change in the effective index of $\Delta n_{eff} = 1.5 \times 10^{-4}$ causes significant measurable shift in the transmission spectrum. In Chap. 4, we had discussed the change in effective index of the mode caused due to change in the refractive index of the cover

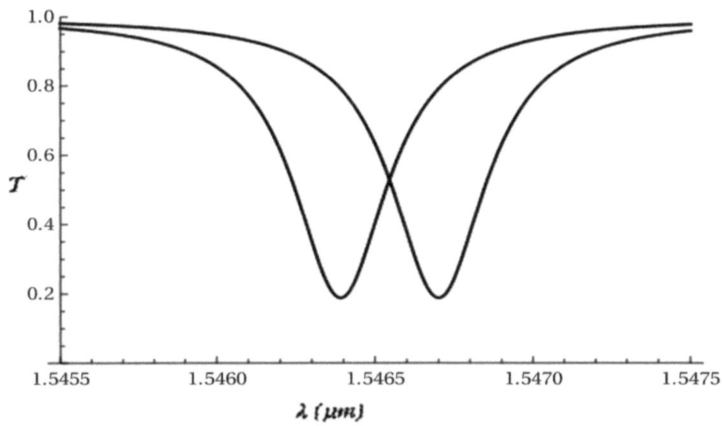

Fig. 8.5 Change in the transmission spectrum of a ring resonator due to a change of optical path length $\Delta(n_{eff}L) = 1.5 \times 10^{-8}$ m

medium. Hence, if the cover medium of the ring resonator changes due to any environmental changes, then it will result in a shift of the resonance spectrum. In fact this can be used to as a very sensitive optical sensor (see, e.g., Kazankiy et al. 2023). It is also possible to dynamically change the phase of the propagating light using the electro optic effect to change the position of the transmission dips (see, e.g., Guarino et al. 2007). See also Problem 8.3.

From Eq. (8.6), it can be seen that the ratio $\frac{E_3}{E_1}$ is complex which implies phase change between the transmitted field and the incident field. Figure 8.6 shows the variation in phase of the transmission coefficient as a function of wavelength around the transmission dip shown in Fig. 8.4. It can be seen that the phase changes rapidly around the transmission dip.

8.2.4 Add-Drop Filter

Figure 8.7 shows a ring resonator configuration that can be used as a wavelength add-drop filter. Input light in Port 1 passes through to output Port 2 and also exits from output Port 3. We will now obtain the output transmission coefficients through output ports 2 and 3 for an input in port 1 assuming both the couplers to be identical with coupling coefficients κ and the ring resonator of length L.

As per our earlier discussions, we obtain the following relations between the different electric field amplitudes of the modes at different points of the device:

At the coupler # 1:

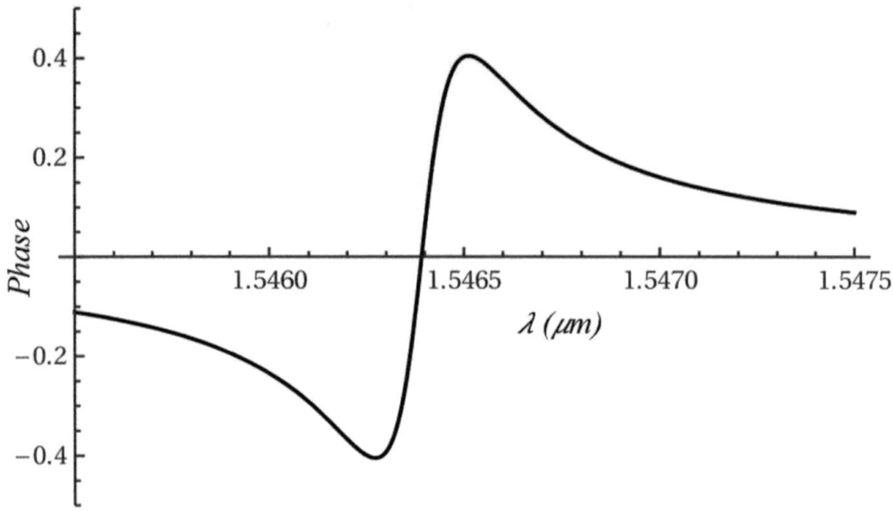

Fig. 8.6 Phase change between the incident and transmission field around a transmission dip

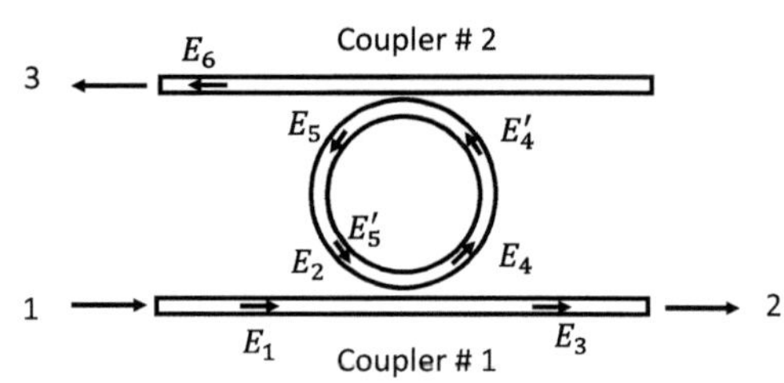

Fig. 8.7 A ring resonator can act like an add-drop filter

$$E_3 = CE_1 + iSE_2$$

$$E_4 = +iSE_1 + CE_2$$

At the coupler # 2:

$$E_5 = CE_4'$$

$$E_6 = +iSE_4'$$

Propagating through the ring:

$$E_4' = E_4 e^{i\zeta}$$

$$E_5' = E_5 e^{i\zeta}$$

We also have

$$E_2 = E_5'$$

Here $\zeta = \gamma\frac{L}{2} = (\gamma_r + i\gamma_i)\frac{L}{2}$, $L = 2\pi R$, R being the radius of the ring and γ_r is the propagation constant of the mode in the ring and γ_i defines the attenuation constant. Also, as before, $S = \sin\kappa l$; $C = \cos\kappa l$, l, being the effective length of the couplers and κ is the effective coupling coefficient.

Combining the equations, we obtain

$$\frac{E_3}{E_1} = \frac{C\left(e^{2i\zeta} - 1\right)}{\left(C^2 e^{2i\zeta} - 1\right)} = \frac{C\left(e^{-\gamma_i L} e^{-i\gamma_r L} - 1\right)}{\left(C^2 e^{-\gamma_i L} e^{-i\gamma_r L} - 1\right)}$$

Hence, the transmission coefficient T_{Th} corresponding to the through port (from Port 1 to Port 3) is given by

$$T_{Th} = \left|\frac{E_3}{E_1}\right|^2 = \frac{C^2\left(1 + e^{-2\gamma_i L} - 2e^{-\gamma_i L}\cos\gamma_r L\right)}{1 + C^4 e^{-2\gamma_i L} - 2C^2 e^{-\gamma_i L}\cos\gamma_r L} \tag{8.42}$$

Similarly, the transmission coefficient from Port 3 (the port for dropping the signal) is given by

$$T_{Dr} = \left|\frac{E_6}{E_1}\right|^2 = \frac{S^4 e^{-\gamma_i L}}{1 + C^4 e^{-2\gamma_i L} - 2C^2 e^{-\gamma_i L}\cos\gamma_r L} \tag{8.43}$$

Note that due to the presence of loss in the system, $T_{Th} + T_{Dr} \neq 1$.
If there is no loss in the system, then $e^{-\gamma_i L} = 1$ and we obtain

$$T_{Th} = \frac{C^2(2 - 2\cos\gamma_r L)}{1 + C^4 - 2C^2\cos\gamma_r L} = \frac{F\sin^2\left(\frac{\gamma_r L}{2}\right)}{1 + F\sin^2\left(\frac{\gamma_r L}{2}\right)} \tag{8.44}$$

and

$$T_{Dr} = \frac{S^4}{1 + C^4 - 2C^2\cos\gamma_r L} = \frac{1}{1 + F\sin^2\left(\frac{\gamma_r L}{2}\right)} \tag{8.45}$$

Here

$$F = \frac{4C^2}{\left(1 - C^2\right)^2} \tag{8.46}$$

is similar to the coefficient of Finesse defined in a Fabry-Perot interferometer.

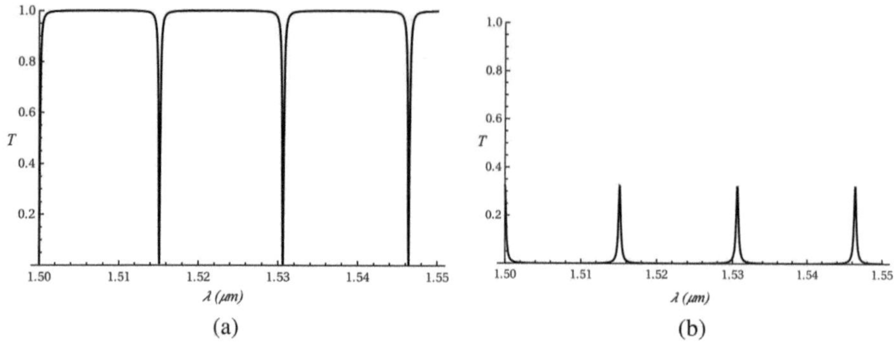

Fig. 8.8 Add-drop filter characteristic of a ring resonator. Variation in transmission through the through channel (Port 2) and the drop channel (Port 3)

Note that in this case, where there is no loss in the ring, we have

$$T_{Th} + T_{Dr} = 1$$

showing power conservation.

Figure 8.8 shows the transmission through the through channel T_{Th} and the drop channel T_{Dr} as a function of wavelength for the following set of parameters:

$$C = 0.98; e^{-\gamma_i L} = 0.97; L = 100\ \mu m; N_{eff} = 1.5$$

Ring resonators find a number of applications in photonics as well as in quantum photonics (see, e.g., Chen and Reano 2012; Vernon and Sipe 2015; Bogaerts et al. 2012; Zhang et al. 2017; Luo et al. 2017; Wang et al. 2021, 2023a, b; He et al. 2022; Ma et al. 2023a, b; Kazakov et al. 2024; Lu et al. 2019).

8.3 Arrayed Waveguide Gratings (AWG)

Arrayed waveguide gratings (AWG) are planar waveguide devices which are very important components for wavelength division multiplexing (WDM) and demultiplexing (Takahashi et al. 1990). In this device, light from a single mode channel waveguide exits into a planar waveguide region (also known as the free propagation region or FPR) and undergoes diffraction as shown in Fig. 8.9. The diffracted wave from the free propagation region then couples into an array of single mode channel waveguides that act like a diffraction grating between the two planar waveguide regions. The evanescent coupling among the channel waveguides is small enough (negligible coupling coefficient) so that the modes propagate through the channel waveguides independently of each other.

The length of each channel waveguide in the waveguide array is increased by a fixed value Δl with respect to the preceding waveguide. Thus, the outputs from the

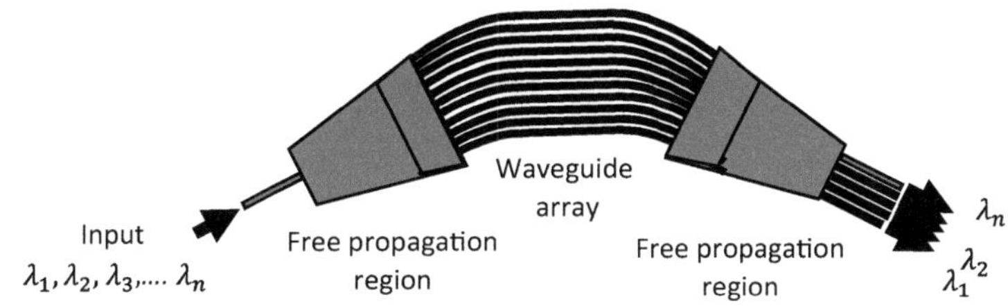

Fig. 8.9 An arrayed waveguide demultiplexer

channel waveguides of the waveguide array differ by constant phase difference of $\phi = \frac{2\pi}{\lambda_0} n_{ch} \Delta l$, where $n_{ch}$is the effective index of the mode of the channel waveguide.

After propagating through different lengths of the channel waveguides, the waves again enter a planar waveguide region where they undergo Fresnel diffraction and get focused on the input ends of the output waveguides. In this process, due to interference effects between the diffracted waves from individual waveguides of the array, different wavelength components in the incident wave focus on different output channel waveguides thus demultiplexing different wavelength components present in the input light. Reverse operation of the device leads to multiplexing of different wavelength channels.

As mentioned above, in the free propagation regions the light wave undergoes Fresnel diffraction along one dimension in the planar waveguide while being confined in the perpendicular direction. The special design with the output ends of the waveguide array and the input ends of the output waveguide array placed along circular arcs with appropriate spacing ensures that the field distribution produced at the output waveguides is the Fourier transform (which indeed corresponds to Fraunhofer diffraction) of the field distribution that exits the waveguide array. A similar effect takes place in the input free propagation region. This is satisfied by an appropriate choice of the length of the free propagation region. The principle behind this is discussed in Sect. 8.4. A similar diffraction takes place in the planar waveguide region corresponding to the input region of the device.

In the next section, we will give a simplified analysis of the operation of the AWG based on the working principle of a diffraction grating. We will obtain equations describing the characteristics of the device such as bandwidth, channel spacing, and number of channels. More detailed discussions on design and analysis of AWG can be found in excellent literature on the subject. See, e.g., Munoz et al. (2002), Smit and Van Dam (1996), Takenouchi and Tsuda (2000), Van Wijk et al. (2023).

8.3.1 *Diffraction from* N *Coherent Point Sources*

In order to understand the basic principle of operation of the AWG device, we consider the operation of a diffraction grating in which light from a number of

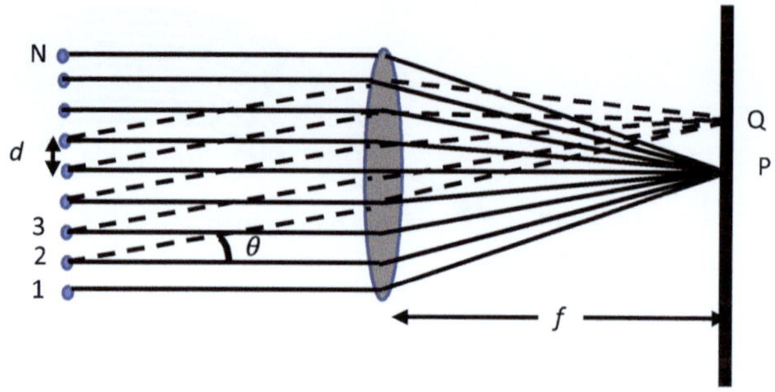

Fig. 8.10 Interference between N point sources at the focal plane of a lens of focal length *f*. P is a point on the axis of the lens and Q is another point where rays traveling at an angle θ with the axis focus at the focal plane

sources diffracts and interferes and in the process leads to the separation of the various wavelengths present in the input light. For this, we consider interference between *N* point sources as shown in Fig. 8.10. We assume that the phase difference between adjacent point sources is ϕ. Light emerging from the point sources is focussed by a lens (to achieve Fraunhofer diffraction) and the intensity pattern is observed on the focal plane of the lens. It may be mentioned that in an actual AWG there is no lens but the device design is such that it leads to Fourier transform of the input field, which is nothing but Fraunhofer diffraction (see Sect. 8.4).

The Fraunhofer diffraction pattern corresponding to *N* point sources with adjacent sources having phase difference of ϕ is given by [see, e.g. Ghatak and Thyagarajan (1978)]

$$I(\theta) = I_0 \frac{\sin^2\left(N\left[\dfrac{\pi d \sin\theta}{\lambda_{pl}} + \dfrac{\phi}{2}\right]\right)}{\sin^2\left(\dfrac{\pi d \sin\theta}{\lambda_{pl}} + \dfrac{\phi}{2}\right)} \tag{8.47}$$

where I_0 is a constant and $\lambda_{pl} = \frac{\lambda_0}{n_{pl}}$ is the wavelength of light in the planar waveguide where n_{pl} is the effective index of the mode in the planar waveguide and λ_0 is the free space wavelength.

The point P shown in Fig. 8.10 corresponds to $\theta = 0$ and the intensity is given by

$$I(\theta) = I_0 \frac{\sin^2\left(\dfrac{N\phi}{2}\right)}{\sin^2\left(\dfrac{\phi}{2}\right)} \tag{8.48}$$

If Δl is so chosen that at a central free space wavelength λ_{0c},

$$\phi = \frac{2\pi}{\lambda_{0c}} n_{ch} \Delta l = 2m\pi; \tag{8.49}$$

for some specific integer value of *m*, then the intensity at P will be a maximum and given by $I(\theta = 0) = N^2 I_0$.

As an example we assume $\lambda_{0c} = 1500$ nm, $n_{ch} = 1.5$, $\Delta l = 50$ μm. This gives us

$$m = \frac{n_{ch}\Delta l}{\lambda_{0c}} = 50$$

Thus, the diffraction order on the axis at P corresponds to the 50th order.

Consider an angle $\delta\theta$ which satisfies the following equation:

$$\frac{\pi d \sin \delta\theta}{\lambda_c} + \frac{\phi}{2} = m\pi + \frac{\pi}{N} \tag{8.50}$$

where $\lambda_c = \frac{\lambda_{0c}}{n_{pl}}$. In such a case, $N\left[\frac{\pi d \sin \delta\theta}{\lambda_c} + \frac{\phi}{2}\right] = Nm\pi + \pi$ and $I(\delta\theta) = 0$. Thus, the intensity which has a maximum value at P (corresponding to $\theta = 0$) falls to zero at $\delta\theta$ given by Eq. (8.50).

Assuming $\delta\theta$ to be very small and using Eq. (8.49) for ϕ, we obtain

$$\delta\theta \approx \frac{\lambda_{0c}}{n_{pl} N d} \tag{8.51}$$

If f represents the focal length of the lens, then the distance between the maximum at P and the first zero in intensity will be

$$\delta x \approx f\delta\theta \approx \frac{\lambda_{0c} f}{n_{pl} N d} \tag{8.52}$$

From Eq. (8.47), we also note that whenever

$$\frac{\pi d \sin \theta}{\lambda_{pl}} + \frac{\phi}{2} = m\pi \tag{8.53}$$

there will be a maximum of intensity. These are the principal maxima of the diffraction pattern. The corresponding angles satisfy the following equation:

$$d \sin \theta = m\lambda_{pl} - \frac{\lambda_{pl}\phi}{2\pi} = \left(m - \frac{\phi}{2\pi}\right)\frac{\lambda_0}{n_{pl}}; \qquad m = \text{integer} \tag{8.54}$$

Thus, if the source emits a single wavelength λ_0, then we will observe a pattern similar to that shown in Fig. 8.11.

In the case of AWG, the phase difference ϕ between adjacent sources is brought about by propagating through channel waveguides of different lengths. If Δl represents the difference in length between adjacent waveguides, then

$$\phi = \frac{2\pi}{\lambda_0} n_{ch} \Delta l = \frac{2\pi\nu}{c} n_{ch}(\nu) \Delta l \tag{8.55}$$

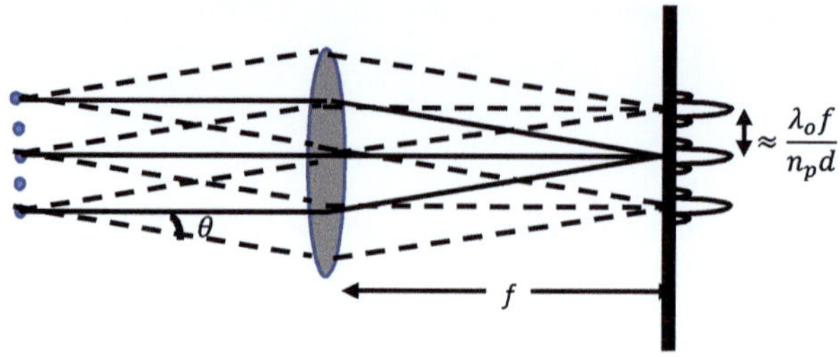

Fig. 8.11 The interference among the N point sources produces a diffraction pattern at the focal plane of the lens here $n_p = n_{pl}$

where $n_{ch}(\nu)$ is the effective index corresponding to the frequency ν of the propagating mode in the channel waveguide. Thus, the frequencies that will have a maximum at P (corresponding to $\theta = 0$) will be

$$\phi = \frac{2\pi\nu}{c} n_{ch}(\nu)\Delta l = 2m\pi$$

or

$$\nu = m\left(\frac{c}{n_{ch}(\nu)\Delta l}\right); \qquad m = 0, 1, 2, \ldots \tag{8.56}$$

where we have written the equation in terms of frequency ν.

We will now discuss some important characteristics of the device.

8.3.2 Free Spectral Range (FSR)

Consider two frequencies ν and $\nu + \Delta\nu$ satisfying the following equations:

$$\frac{2\pi\nu}{c} n_{ch}(\nu)\Delta l = 2m\pi \tag{8.57}$$

and

$$\frac{2\pi(\nu + \Delta\nu)}{c} n_{ch}(\nu + \Delta\nu)\Delta l = 2(m+1)\pi \tag{8.58}$$

Both the above frequencies will have maximum intensity at the point P which will correspond to the mth order for the frequency ν and $(m + 1)$ order for frequency $(\nu + \Delta\nu)$. For proper functioning of the device, we need to ensure that only a specific frequency from the incident light falls on each of the output waveguides. Thus, the value of $\Delta\nu$ satisfying Eq. (8.58) is the largest frequency difference allowed in the device so that only one unique frequency falls on any particular output waveguide.

Assuming $\Delta\nu << \nu$, we expand $n_{ch}(\nu + \Delta\nu)$ and write

$$n_{ch}(\nu + \Delta\nu) \approx n_{ch}(\nu) + \Delta\nu \frac{dn_{ch}}{d\nu}$$

Neglecting terms of order $(\Delta\nu)^2$, we have

$$\begin{aligned}(\nu + \Delta\nu)n_{ch}(\nu + \Delta\nu) &\approx \nu n_{ch}(\nu) + \Delta\nu n_{ch}(\nu) + \nu\Delta\nu \frac{dn_{ch}}{d\nu} \\ &= \nu n_{ch}(\nu) + \Delta\nu\left(n_{ch}(\nu) + \nu\frac{dn_{ch}}{d\nu}\right) = \nu n_{ch}(\nu) + \Delta\nu N_{ch}(\nu)\end{aligned} \tag{8.59}$$

where $N_{ch}(\nu) = n_{ch}(\nu) + \nu\frac{dn_{ch}}{d\nu}$ is the group effective index of the channel waveguide. Hence, using Eq. (8.59) in Eq. (8.58), we have

$$\nu n_{ch}(\nu) + \Delta\nu N_{ch}(\nu) = (m+1)\left(\frac{c}{\Delta l}\right) \tag{8.60}$$

Using Eq. (8.57) in Eq. (8.60), we obtain

$$\Delta\nu_{FSR} = \frac{c}{N_{ch}(\nu)\Delta l} = \frac{n_{ch}(\nu)\nu}{N_{ch}(\nu)m} \tag{8.61}$$

This corresponds to a wavelength range of

$$\Delta\lambda_{FSR} = \frac{\lambda_0^2}{c}\Delta\nu_{FSR} = \frac{\lambda_0^2}{N_{ch}(\nu)\Delta l} \tag{8.62}$$

The quantity $\Delta\nu_{FSR}$ is the frequency free spectral range (FSR) of the device and gives the maximum range of frequencies that can be uniquely demultiplexed or multiplexed by the AWG.

It is interesting to note that larger the group index of the channel waveguide, smaller will be the required length difference Δl for a required $\Delta\lambda_{FSR}$ which implies that the device size would get reduced with waveguides having higher refractive index contrast.

Taking typical values $\Delta l = 50$ μm, $\lambda_0 = 1500$ nm, $N_c \approx n_c \approx 1.5$ and we obtain

$$\Delta\lambda_{FSR} = 30 \text{ nm}$$

Free spectral range is an important characteristic of the device and in order to avoid any cross talk between the different wavelength channels, it is important that the FSR should be larger than the whole frequency range of the incoming channels. Thus, if there are 32 channels spaced by 0.8 nm operating at 1500 nm, then the FSR must be larger than about 26 nm.

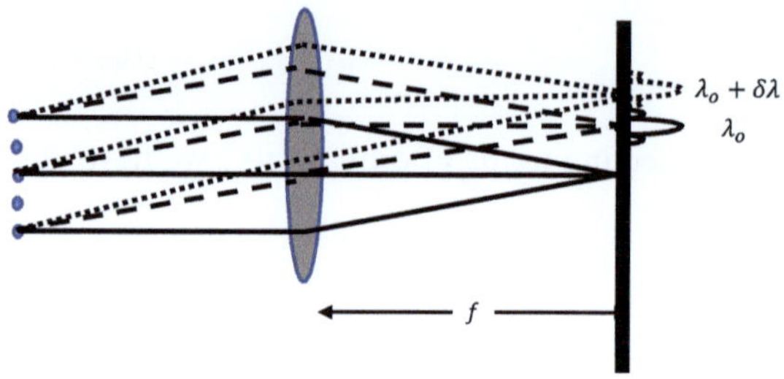

Fig. 8.12 Two adjacent wavelengths focus and form diffraction patterns at their foci. To be able to resolve the two wavelengths, their separation must be larger than their widths as per Rayleigh criterion

8.3.3 *Resolution*

In order to be able to separate adjacent wavelengths, the peak of the interference of an adjacent wavelength $\lambda_0 + \delta\lambda$ must be beyond the first zero of the pattern formed by λ_0 (see Fig. 8.12). Thus, the minimum separation between adjacent wavelengths that can be distinctly separated would be such that the peak of $\lambda_0 + \delta\lambda$ forms at the point where the first zero of λ_0 is formed.

Now, for the wavelength $\lambda_0 + \delta\lambda$ we have

$$\phi = \frac{2\pi}{(\lambda_0 + \delta\lambda)} n_{ch}(\lambda_0 + \delta\lambda)\Delta l \approx \frac{2\pi}{\lambda_0} n_{ch}(\lambda_0)\Delta l - \frac{2\pi\delta\lambda}{\lambda_0^2} N_{ch}(\lambda_0)\Delta l \tag{8.63}$$

where $N_{ch}(\lambda_0) = n_{ch}(\lambda_0) - \lambda_0 \frac{dn_{ch}}{d\lambda_0}$ is the group effective index of the mode in the channel waveguide at the wavelength λ_0.

The central maximum of the wavelength $\lambda_0 + \delta\lambda$ corresponding to the *m*th order will be along an angle $\delta\theta$ satisfying the following equation:

$$\frac{2\pi d n_{pl} \sin\delta\theta}{(\lambda_0 + \delta\lambda)} + \phi \approx \frac{2\pi d n_{pl}\delta\theta}{(\lambda_0 + \delta\lambda)} + \phi \approx \frac{2\pi d n_{pl}\delta\theta}{\lambda_0} + \phi = 2m\pi \tag{8.64}$$

assuming $\delta\theta$ to be small and we have neglected the product $\delta\lambda.\ \delta\theta$.

Substituting for ϕ from Eq. (8.63) in Eq. (8.64), we obtain the angle at which the *m*th order maximum of the wavelength $\lambda_0 + \delta\lambda$ to satisfy the equation:

$$\frac{2\pi d n_{pl}\delta\theta}{\lambda_0} + \frac{2\pi}{\lambda_0} n_{ch}(\lambda_0)\Delta l - \frac{2\pi\delta\lambda}{\lambda_0^2} N_{ch}(\lambda_0)\Delta l = 2m\pi \tag{8.65}$$

Since $\frac{2\pi}{\lambda_0} n_{ch}(\lambda_0)\Delta l = 2m\pi$, we have

$$\frac{2\pi d n_{pl}\delta\theta}{\lambda_0} = \frac{2\pi\delta\lambda}{\lambda_0^2} N_{ch}(\lambda_0)\Delta l$$

which gives us

$$\delta\lambda \approx \frac{\lambda_0 n_{pl} d}{N_{ch}\Delta l}\delta\theta \tag{8.66}$$

The minimum wavelength spacing is obtained by imposing the condition that the central maximum of (λ_0 + $\delta\lambda$) falls at the first minimum of the wavelength λ_0, that is, the $\delta\theta$ giving the angle for the first zero of the principal maximum at P (Eq. 8.51) and the $\delta\theta$ corresponding to the principal maximum of wavelength(λ_0 + $\delta\lambda$) (Eq. 8.66) must be the same.

Substituting for $\delta\theta$ from Eq. (8.51) in Eq. (8.66), we obtain the expression for the minimum separation between adjacent wavelengths that can be resolved to be

$$\delta\lambda = \frac{\lambda_0 n_{pl} d}{N_{ch}\Delta l}\delta\theta = \frac{\lambda_0 n_{pl} d}{N_{ch}\Delta l}\frac{\lambda_0}{n_{pl} N d} = \frac{\lambda_0^2}{N_{ch}\Delta l N} \tag{8.67}$$

which gives the channel spacing between adjacent resolvable channels in the AWG.

The corresponding resolving power of the device is

$$\frac{\lambda_0}{\delta\lambda} = \frac{N_{ch}}{n_{ch}} mN \tag{8.68}$$

where we have used Eq. (8.57).

Equation (8.68) is a very similar expression to that of a diffraction grating with the resolving power depending on the order number m and the number of sources N that are used in forming the diffraction pattern.

If we again assume $\Delta l = 50$ μm, $\lambda_0 = 1500$ nm, $N = 50$ and $N_c \approx n_c \approx 1.5$ we obtain

$$\delta\lambda = 0.6 \text{ nm}$$

We also note that the bandwidth (FSR) of the device is $\Delta\lambda$ and the wavelength spacing is $\delta\lambda$. Thus, the number of channels supported by the AWG will be

$$M = \frac{\Delta\lambda}{\delta\lambda} = \frac{\lambda_0^2}{N_c\Delta l}\frac{N_c\Delta l N}{\lambda_0^2} = N \tag{8.69}$$

Thus, the number of channel waveguides defines the number of wavelength channels over which the AWG will operate.

Figures 8.13 (linear scale) and 8.14 (logarithmic scale) show a typical simulated normalized intensity pattern as a function of wavelength for the following parameter values: $N = 100$; $d = 10$ μm; $n_c = 1.5$; $\Delta l = 50$ μm. Two closely lying peaks of wavelengths are shown. In the AWG, these two wavelength channels will couple to two individual waveguides thus demultiplexing these two wavelengths.

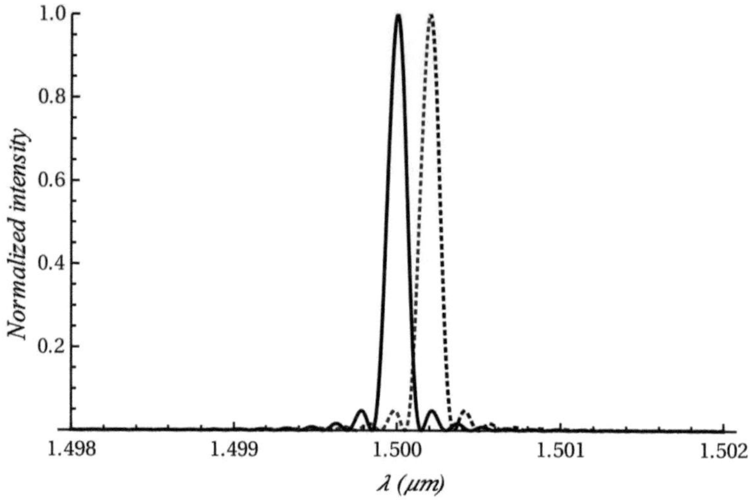

Fig. 8.13 Simulated normalized intensity pattern as a function of wavelength at the input to the output channel waveguides

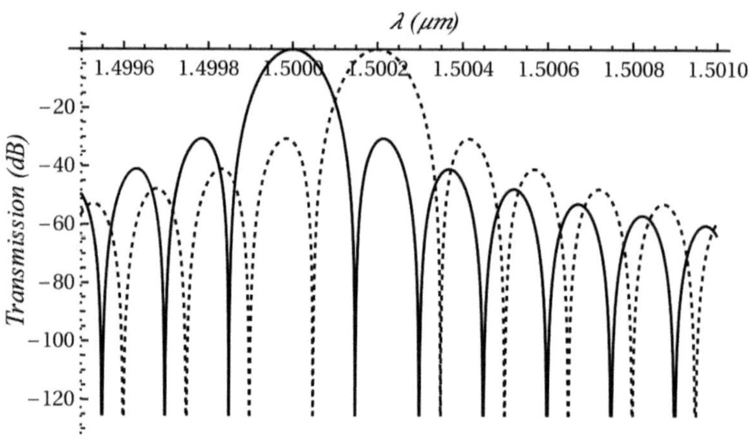

Fig. 8.14 Same as Fig. 8.13 but plotted on a logarithmic scale

For the parameters chosen the wavelength resolution will correspond to 0.3 nm. The central order corresponds to $m = 50$. The FSR of the grating would be 30 nm and the number of channels that the AWG will support will be equal to 100.

It must be mentioned that in an actual device the transmission will be more noisy with greater value of cross talk.

Exercise 8.4 Consider an AWG which needs to operate at a center wavelength of 1550 nm having $\Delta\lambda_{FSR} = 32$ nm with 40 channels in a substrate with a refractive index of 1.5. Estimate the various parameters of the AWG.

Solution From Eq. (8.62), we have

$$\Delta\lambda_{FSR} = \frac{\lambda_0^2}{N_c(\nu)\Delta l} \approx \frac{\lambda_0^2}{n_c(\nu)\Delta l}$$

where we have assumed $N_{ch}(\nu) \approx n_{ch}(\nu)$. Substituting for $\Delta\lambda_{FSR}$, λ_0 and n_{ch}, we obtain $\Delta l \approx 50$ μm. This corresponds to an inter waveguide phase difference of $\Delta\phi = \frac{2\pi}{\lambda_0} n_{ch}\Delta l \approx 97\pi$.

For the central wavelength of 1550 nm, if the diffracted spot is at the center of the output, then the diffraction order for the center wavelength will correspond to

$$m = \frac{\Delta\phi}{2\pi} = 48$$

Thus, the center of the pattern corresponds to the 48th order of diffraction. Here, we have chosen the closest integer for the order number.

Recalling Eq. (8.68), we have

$$\delta\lambda = \frac{n_{ch}}{N_{ch}} \frac{\lambda_0}{mN}$$

Substituting $\delta\lambda = \frac{32}{40} = 0.8$ nm, the required number of waveguides is hence

$$N = \frac{\lambda_0}{m\delta\lambda} = 40$$

where we have chosen an integer closest to the value.

Thus, an AWG with 40 channel waveguides having an effective index of 1.5 and with a difference in length of 50 μm between adjacent waveguides will have a $\Delta\lambda_{FSR} = 32$ nm and support 40 channels with inter-channel separation of 0.8 nm. Figure 8.15 shows the corresponding simulated transmission spectrum for three adjacent channels.

Exercise 8.5 Instead of a one-dimensional periodic array of point sources as discussed earlier, in the AWG, each source is the output from a single mode channel waveguide. Approximating each of the modal fields to have a

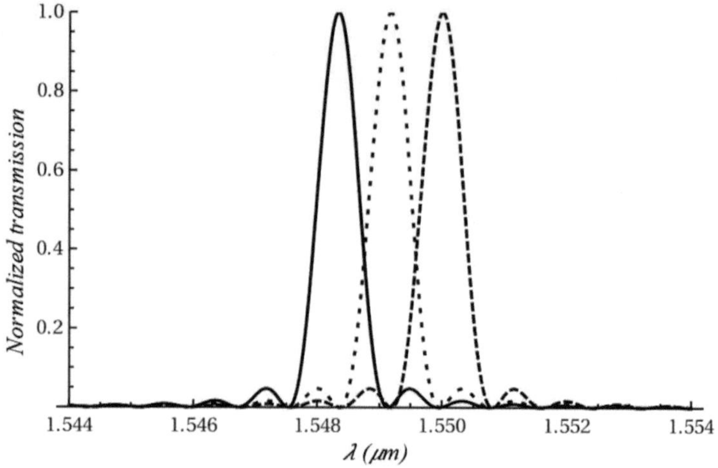

Fig. 8.15 Simulated transmission spectrum of an AWG corresponding to Exercise 8.4

Gaussian distribution given by e^{-x^2/w_0^2}, obtain the Fraunhofer diffraction pattern due to N such equally separated waveguides placed along a line and spaced by a distance d.

Solution If the Fraunhofer diffraction pattern of a single waveguide is given by $I_0(x)$, then the diffraction pattern of a linear array of identical waveguides separated by a distance d and placed along a line is given by (see, e.g., Ghatak and Thyagarajan 1978):

$$I(x) = I_0(x)\left|\sum_{n=1}^{N} e^{iknd}\right|^2 \tag{8.70}$$

with $I_0(x)$ being the diffraction pattern of the output from one waveguide. For a Gaussian field given by e^{-x^2/w_0^2} from the output of each waveguide, we have the intensity distribution at a distance D from the array waveguides,

$$I_0(x) = I(x=0)e^{-2x^2/w^2} \tag{8.71}$$

where

$$w^2 = w_0^2\left(1 + \frac{\lambda^2 D^2}{\pi^2 w_0^4}\right) \approx \frac{\lambda^2 D^2}{\pi^2 w_0^2} \tag{8.72}$$

with $\lambda = \frac{\lambda_0}{n_p}$, λ_0 being the free space wavelength and n_p is the effective index of the propagating mode in the planar waveguide where the wave is diffracting.

In the last step, we have assumed the output channel waveguides to be in the far field of Fraunhofer region of diffraction.

We also have

$$\left| \sum_{n=1}^{N} e^{iknd} \right|^2 = \frac{\sin^2 N\gamma}{\sin^2\gamma} \tag{8.73}$$

where $\gamma = \frac{\pi d \sin\theta}{\lambda}$, θ being the angle of diffraction. If the distance from the output end of the array waveguides to the output waveguides is D, then $\sin\theta \approx \frac{x}{D}$ where x is measured in the plane of the output waveguides. Hence, the pattern falling on the output waveguide will be

$$I(\theta) = I(\theta=0)e^{-2\pi^2 w_0^2 \sin^2\theta/\lambda^2} \frac{\sin^2\left(\frac{N\pi d \sin\theta}{\lambda}\right)}{\sin^2\left(\frac{\pi d \sin\theta}{\lambda}\right)} \tag{8.74}$$

If the phase of each of the Gaussian beams at the exit of the array waveguide increases from the preceding one by ϕ, then we will have

$$I(\theta) = I(\theta=0)e^{-2\pi^2 w_0^2 \sin^2\theta/\lambda^2} \frac{\sin^2\left(\frac{N\pi}{\lambda}\left[d\sin\theta + \frac{\lambda\phi}{2\pi}\right]\right)}{\sin^2\left(\frac{\pi}{\lambda}\left[d\sin\theta + \frac{\lambda\phi}{2\pi}\right]\right)} \tag{8.75}$$

Note that $\lambda = \lambda_0/n_{pl}$. The effect of the Gaussian beam is a reduction in intensity corresponding to larger values of θ. Thus, if we assume $w_0 = 2$ μm, $n_{pl} = 1.5$ and a free space wavelength of $\lambda_0 = 1300$ nm, the intensity decreases by a factor $e^{-105\sin^2\theta}$. Thus, for $\theta = 2°$, the reduction in intensity is $e^{-0.128} \approx 0.88$.

Since the phase difference is introduced by an additional length Δl of the channel waveguide of the array, we have

$$\frac{2\pi}{\lambda}\Delta l = \phi$$

The diffraction order corresponding to $\theta = 0$ will correspond to $m = \frac{\Delta l}{\lambda} = \frac{\Delta l}{\lambda_0} n_{pl}$. Thus, if $\Delta l = 50$ μm, $n_p = 1.5$ and $\lambda_0 = 1.3$ μm, this will correspond to $m = 58$.

Exercise 8.6 Consider an AWG designed for a range of wavelengths from 1530 to 1560 nm. The output of the each waveguide from the waveguide array has a Gaussian spot size $w_0 = 1$ μm. If the peak of the wavelength 1545 nm falls at the center of the output waveguide array with a diffraction order of 80, obtain the relative intensity of the spectra corresponding to wavelengths of 1530 and 1560 nm with respect to the central spot. Given $n_{pl} = 1.5$, $n_{ch} = 1.5$, $d = 8$ μm.

Solution The diffraction pattern of a set of sources emitting a Gaussian field distribution is given by Eq. (8.75). The angle at which the peak of wavelength 1545 nm appears is $\theta = 0$. Hence, we obtain

$$\Delta l = \frac{m\lambda_0}{n_{pl}} \approx 82.4\ \mu\text{m}$$

For a general wavelength λ_0, the maxima of the same order will appear along an angle satisfying the following equation:

$$\frac{\pi}{\lambda_0} n_{pl} d \sin\theta + \frac{\pi}{\lambda_0} n_{ch} \Delta l = 80\pi$$

Substituting for the various parameters, we have the angles corresponding to the wavelengths of 1530 and 1560 nm as $\sin\theta = -0.1$ and $\sin\theta = +0.1$, respectively.

Since the Gaussian beam emanating from individual waveguides has $w_0 = 2$ μm, the relative intensity will be determined by the function $R = e^{-2\pi^2 w_0^2 \sin^2\theta n_{pl}^2/\lambda_0^2}$. Thus, for the wavelengths corresponding to 1530 and 1560 nm, we have 0.83 and 0.83, respectively.

8.3.4 Angular Free Spectral Range

The angular free spectral range θ_{FSR} defines the angular gap between two adjacent orders and represents the angular region over which the wavelengths over the free spectral range will form their spectra. To obtain an expression for this, we first note that for the central wavelength to appear along $\theta = 0$, we have the following equation:

$$\frac{2\pi}{\lambda_c} n_{ch} \Delta l = 2m\pi \tag{8.76}$$

If the angle at which the peak of wavelength $\lambda_c + \Delta\lambda$ appears is $\Delta\theta$, then we have

$$\frac{\pi d \sin\Delta\theta}{\lambda_c + \Delta\lambda} n_{pl} + \frac{\pi}{\lambda_c + \Delta\lambda} n_{ch} \Delta l = m\pi = \frac{\pi}{\lambda_c} n_{ch} \Delta l \tag{8.77}$$

Simplifying we obtain assuming $\Delta\theta$ to be small

$$\frac{d n_{pl} \Delta\theta}{\lambda_c} = n_{ch} \Delta l \frac{\Delta\lambda}{\lambda_c^2} \tag{8.78}$$

where we have neglected the product $\Delta\lambda.\ \Delta\theta$. Hence, the angular difference between two maxima with a wavelength difference of $\Delta\lambda$ will be

$$\Delta\theta = \frac{n_{ch}}{n_{pl}} \frac{\Delta l}{d\lambda_c} \Delta\lambda \tag{8.79}$$

For $\Delta\lambda = \Delta\lambda_{FSR}$, we have

$$\Delta\theta_{FSR} = \frac{n_{ch} \Delta l}{d n_{pl} \lambda_c} \Delta\lambda_{FSR} \tag{8.80}$$

Substituting for $\Delta\lambda_{FSR}$ from Eq. (8.62), we obtain

$$\Delta\theta_{FSR} = \frac{\lambda_c}{d n_{pl}} \tag{8.81}$$

where we have assumed $n_{ch} \approx N_{ch}$. Larger the spacing between the waveguides in the array, smaller will be the angular FSR. Taking $d = 10$ μm, $n_{pl} = 1.5$ and $\lambda_c = 1.545$ μm, we obtain $\Delta\theta_{FSR} \approx 5.9°$.

8.4 Fourier Transform Relationship Between the Waveguide Array and Output Waveguides

In the above discussions, we have considered a lens to be present to generate a Fraunhofer diffraction pattern. In an actual AWG device, the Fraunhofer diffraction is achieved by ensuring that the output of the waveguide array is along a circular arc of appropriate radius and the output channel waveguides to which they couple are also along a circular arc as shown in Fig. 8.16.

In this section, we will show how such a design leads to a Fourier transform relationship (like in Fraunhofer diffraction) through Fresnel diffraction of the wave.

In an AWG the ends of the waveguides constituting the waveguide array are along a circular arc of radius L (marked by the arc P in Fig. 8.17) and the ends of the output waveguides to which the diffracted light couples is also along a circular arc of radius L (marked as arc T in Fig. 8.17). As shown in Fig. 8.17, the centers of the

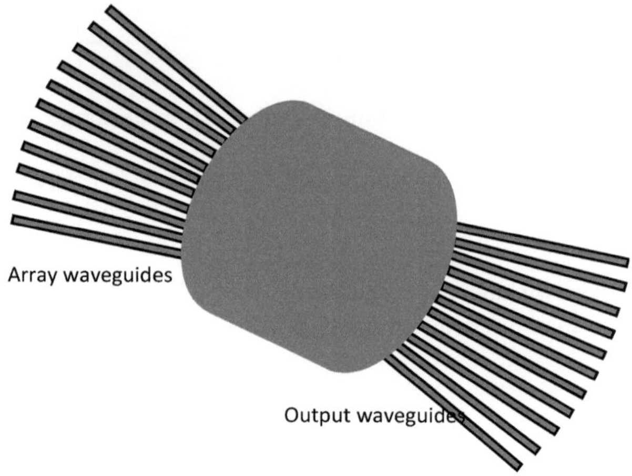

Fig. 8.16 The ends of the array waveguides and output waveguide connected to the planar waveguide are placed along circular arcs of appropriate radii

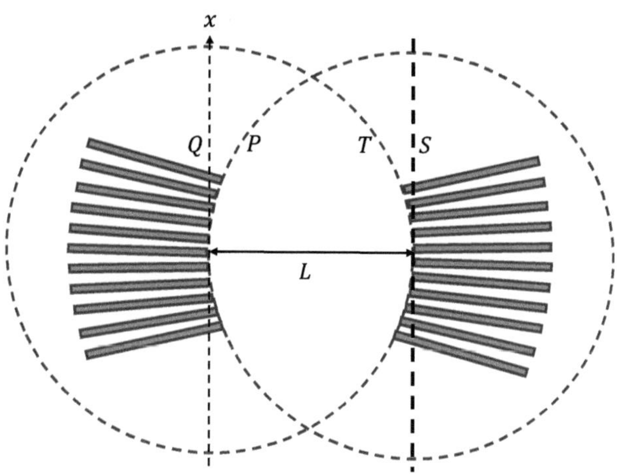

Fig. 8.17 Relation between the output of the waveguide array which is in the form of a circular arc of radius L and the input of the output waveguide array which is also placed along a circular arc of radius L. The separation between the two arrays is also L

waveguide array and the output waveguide array are separated by a distance L. We will now show that the field distribution along the circular arc T is an exact Fourier transform of the field distribution along the circular arc P. The Fourier transformation of the field distribution at the output of the waveguide array results in the spectral decomposition and leads to demultiplexing of the channels.

Let the field distribution on the circular arc P be denoted by $f(x)$. Figure 8.17 shows the curve P (of radius of curvature L) and the straight line Q (which is chosen as the x-axis).

In Fig. 8.18, we have redrawn the circular arc of radius L and the x-axis. Now from Fig. 8.18, it can be seen that

$$AC = ON = OB - NB = L - L\cos\theta$$

For small θ, we have

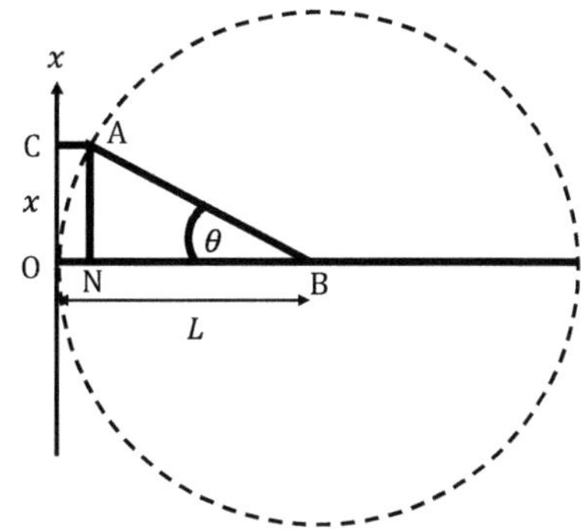

Fig. 8.18 The coordinate system showing relation between the curved line of radius L and the straight line. The field between the curved line and the straight line differs by a phase difference of $e^{-ikx^2/2L}$

$$AC = L\left(1-\sqrt{1-\sin^2\theta}\right) = L\left(1-\sqrt{1-\frac{x^2}{L^2}}\right) \approx \frac{x^2}{2L} \tag{8.82}$$

This length corresponds to an optical phase change of $k\frac{x^2}{2L}$ where k represents the propagation constant of the field in the planar region of the AWG. Hence, if the field distribution on the circular arc shown as P in Fig. 8.17 is $f(x)$, then the field distribution along the line Q can be written as $f(x)e^{-ikx^2/2L}$.

Now this field undergoes *Fresnel diffraction* and propagates towards the line marked S in Fig. 8.17 after propagating through a length L. The Fresnel diffracted field $g(x)$ along the line S is represented by the following integral (see, e.g., Ghatak and Thyagarajan 1978):

$$g(x) \sim \int f(\zeta)e^{-ik\zeta^2/2L}e^{i\frac{k}{2L}(x-\zeta)^2}d\zeta \tag{8.83}$$

which on simplification becomes

$$g(x) \sim e^{ikx^2/2L}\int f(\zeta)e^{-i\frac{kx}{L}\zeta}d\zeta \tag{8.84}$$

Note that the integral on the right hand side of Eq. (8.84) is just the Fourier transform of $f(x)$ with spatial frequency $u = \frac{1}{2\pi}\frac{kx}{L} = \frac{x}{\lambda L}$ where now x is measured on the line S. Since $\frac{x}{L} = \sin\theta$ (see Fig. 8.18), the corresponding spatial frequency is $u = \frac{\sin\theta}{\lambda}$.

Equation (8.84) shows that the Fourier transform is multiplied by an additional x-dependent phase factor $e^{ikx^2/2L}$.

Now, just as we discussed earlier, the separation between the line S and the circular arc (shown as T in Fig. 8.17) of radius of curvature L is again approximately given by $\frac{x^2}{2L}$. This length corresponds to a phase difference of $k\frac{x^2}{2L}$.

Similar to the relation between the field on the line Q and circular arc P, the field $h(x)$ on the circular arc T is related to the field on the straight line S through the following equation:

$$h(x) = g(x)e^{-ikx^2/2L} \tag{8.85}$$

Substituting for $g(x)$ from Eq. (8.85), we obtain the following relation between the fields on the curve P and the curve T:

$$h(x) = g(x)e^{-ikx^2/2L} = \int f(\zeta)e^{i\frac{kx}{L}\zeta}d\zeta \tag{8.86}$$

Thus, we see that the field on the circular arc T of radius L and the field on the circular arc P of radius L which are separated by a distance L are nothing but a Fourier transform pair. The corresponding spatial frequency is $u = \frac{\sin\theta}{\lambda}$.

Thus, in an AWG device, Fourier transforming operation is performed without any lens by choosing the output ends of the waveguide array and the input ends of the output array of waveguides to be along circular arcs of radius of curvature L and having a distance between their centers equal to L.

Since the principal maxima appear when the following condition is satisfied:

$$\frac{\pi d \sin\theta}{\lambda} + \frac{\phi}{2} = m\pi \tag{8.87}$$

we can express the equation in terms of distance measured on the line T by replacing $\sin\theta$ by $\frac{x}{L}$ and obtain

$$x = \frac{\lambda_0 L}{\pi d n_{pl}}\left(m\pi - \frac{\phi}{2}\right) \tag{8.88}$$

Recalling the example considered earlier with the diffraction pattern shown in Fig. 8.15, we can estimate the distance at which the wavelength of 1500.2 nm focuses. The wavelength corresponding to 1500 nm appears at $\theta = 0$, hence at $x = 0$. The diffraction order corresponds to $m = 50$. Hence, we have, assuming $L = 15$ mm:

$$x = \frac{1.5002 \times 10^{-6} \times 15 \times 10^{-3}}{\pi 10 \times 10^{-6} \times 1.5}\left(50\pi - \frac{\pi \times 1.5 \times 50 \times 10^{-6}}{1.5002 \times 10^{-6}}\right) \approx 9.6\ \mu\text{m}$$

Thus, the peak corresponding to the wavelength 1500.2 nm will appear at a distance of about 9.6 μm from the center (which corresponds to the peak position of the wavelength 1500 nm) and couples to an output channel waveguide placed at that position.

Thus, the output end of the waveguide array which is in the form of a circular arc of radius L and the input end of the output waveguide array which is also in the form of a circular arc of radius L and which is at a distance L from the array are Fourier transform pairs. Thus, *Fresnel diffraction* from the end of the waveguide array to the input end of the output waveguide array leads to *Fourier transform* (which is nothing but Fraunhofer diffraction) of the field distribution at the end of the waveguide array. Just like in the case of diffraction grating, light emanating from multiple equally spaced identical waveguides of the waveguide array diffract and interfere on the input of the output waveguide array leading to generation of multiple orders and formation of a spectrum. This leads to demultiplexing of the light incident on the AWG.

Arrayed waveguide gratings find various applications in integrated quantum photonics. They have been used to generate entangled photons (Capmany et al. 2013; Matsuda et al. 2017), used in optical signal processing (Takenouchi and Tsuda 2000), in telecommunications (Smit and Van Dam 1996), etc.

Exercise 8.7 An AWG has to be designed with the following specifications: $\Delta\lambda_{FSR} = 30$ nm, wavelength range: 1530–1560 nm, bandwidth $\delta\lambda = 0.5$ nm, $n_{ch} \approx N_{ch} = 1.5$ and inter-channel spacing of 150 GHz. Obtain the values of Δl and N the number of waveguides in the array. Obtain the diffraction order of the central maximum at 1545 nm. If $L = 10$ mm, obtain the physical separation between adjacent channels separated by 150 GHz assuming $d = 10$ μm.

Solution Using Eq. (5.62) relating the free spectral range and Δl, we have

$$\Delta l = \frac{\lambda_0^2}{N_{ch}\Delta\lambda_{FSR}} = \frac{(1.545 \times 10^{-6})^2}{1.5 \times 30 \times 10^{-9}} = 53.045\ \mu\text{m}$$

This value of Δl corresponds to an order of diffraction at the center of the output waveguide array corresponding to $\theta = 0$ of

$$m = \frac{\phi}{2\pi} = \frac{n_{pl}\Delta l}{\lambda_o} = 51.5$$

In order to have a central maximum at the center of the output waveguide array, we choose the order $m = 52$ at the center which gives a value

$$\Delta l = 52\frac{\lambda_o}{n_{pl}} = 52\frac{1545 \times 10^{-9}}{1.5} = 53.56\ \mu\text{m}$$

The number of waveguides N in the array and the bandwidth are related through Eq. (5.67) and hence we have

$$N = \frac{\lambda_0^2}{N_{ch}\Delta l \delta\lambda} = \frac{(1.545 \times 10^{-6})^2}{1.5 \times 53 \times 10^{-6} \times 0.5 \times 10^{-9}} \approx 60$$

The values of θ for the two extreme wavelengths of 1530 and 1560 nm can be obtained from Eq. (5.78) and are approximately given by 2.98° and 2.98° on either side of the central peak. For $L = 10$ mm, the corresponding physical separations are $\Delta x = 0.93$ mm on either side of the central maximum.

Problems

8.1. Consider a ring resonator shown in Fig. 8.7. Obtain the ratio of power inside the ring to the incident power $|E_4|^2/|E_1|^2$.

8.2. Consider a ring resonator fabricated in z-cut lithium niobate and operating with TM modes at a wavelength of 1500 nm. In order to use the r_{33} electro optic coefficient in lithium niobate, electrodes covering a length $l = \frac{L}{10}$ where L is the length of the ring are placed on the waveguide. Obtain the required electric field to achieve a shift in the resonance wavelength by 10 pm. Approximating the electric field by $E \approx \frac{V}{d}$ where d is the electrode spacing, estimate the corresponding voltage required. Use $n_e = 2.14$, $r_{33} = 30 \times 10^{-12}$ m/V and $d = 10$ μm.

8.3. As discussed, working of arrayed waveguide grating can be understood from the analysis of the diffraction pattern produced by N point sources placed in free space. Consider 11 point sources spaced by $d = 10$ μm and with the phase of the nth point source given by $\phi_n = (n - 1)\frac{130\,\pi}{\lambda_0}$ where λ_0 is measured in μm. The point sources emit wavelength in the range 1530–1565 nm. Light from the point sources is focused by a lens of focal length 1 cm.

   (a) Which wavelength will have a maximum at the center point P and what is the order of the maximum?
   (b) What are the positions where maxima of wavelengths corresponding to 1530 and 1565 nm appear on the focal plane?
   (c) What would be the resolvable wavelength spacing for this arrangement?

8.4. An array of N identical point sources emitting a wavelength of 800 nm are placed on x-axis at $x = 0, 0.5$ μm, 1 μm, 1.5 μm … The phases of the sources are $0, \phi, 2\phi, 3\phi\ldots$ respectively. In the Fraunhofer limit at what angles to the normal to the array will we observe the zero order if ϕ is (1) zero (2) $\pi/6$, (3) $\pi/4$ and (4) $\pi/3$.

8.5. Consider designing an AWG at a center wavelength of 1550 nm, to operate for 16 channels with a channel spacing of 1.6 nm and end waveguide separation of 20 μm. Obtain the required value of Δl and the central order of diffraction. If the device has 150 waveguides in the array, what is the wavelength resolution? Assume $n_{ch} = 1.5$.

References

Bogaerts, W., R. De Heyn, T. V. Vaerenbergh, D. De Vos, S. K. Selvaraja, T. Claes, P. Dumon, P. Biernstman, D. V. Thourhout, and R. Baets. 2012. Silicon microring resonators. *Laser Photonics Rev* 6:47–73.

Capmany, J., J. Mora, C. R. Fernandez-Pousa, and P. Munoz. 2013. Quantum model of light transmission in array waveguide gratings. *Opt Express* 21:14841–14852.

Chen, L., and R. M. Reano. 2012. Compact electric field sensors based on indirect bonding of lithium niobate to silicon microrings. *Opt Express* 20:4032–4038.

Ghatak, A. 2024. *Optics*. 8th ed. India: McGraw Hill.

Ghatak, A. K., and K. Thyagarajan. 1978. *Contemporary optics*. New York: Plenum Publishing Corporation.

Guarino, A., G. Poberaj, D. Rezzonico, R. Degl'Innocenti, and P. Günter. 2007. Electro optically tunable microring resonators in lithium niobate. *Nat Photonics* 1:407–410.

He, X., L. Cortes-Herrera, K. Opong-Mensah, Y. Zhang, M. Song, G. P. Agarwal, and J. Cardenas. 2022. Electrically induced adiabatic frequency conversion in an integrated lithium niobate ring resonator. *Opt Lett* 47:5849–5852.

Kazakov, D., T. P. Letsou, M. Beiser, Y. Zhi, N. Opačak, M. Piccardo, B. Schwarz, and F. Capasso. 2024. Active mid infrared ring resonators. *Nat Commun* 15:607. https://doi.org/10.1038/s41467-023-44628-7.

Kazankiy, N., S. N. Khonina, and M. A. Butt. 2023. A review of photonic sensors based on ring resonator structures: Three widely used platforms and implications of sensing applications. *Micromachines (Basel)* 14:1080.

Lu, J., J. B. Surya, X. Liu, A. W. Bruch, Z. Gong, Y. Xu, and H. X. Tang. 2019. Periodically poled thin film lithium niobate microring resonators with a second harmonic generation efficiency of 250,000%/W. *Optica* 12:1455–1460.

Luo, R., H. Jiang, S. Rogers, H. Liang, Y. He, and Q. Lin. 2017. On chip second harmonic generation and broadband parametric down conversion in a lithium niobate microresonator. *Opt Express* 25:24531–24539.

Ma, Z., J.-Y. Chen, M. Garikapati, Z. Li, C. Tang, Y. M. Sua, and Y.-P. Huang. 2023a. Highly efficient and pure few-photon source on a chip. *Phys Rev Lett* 20:044033.

Ma, J., X. Cheng, N. Zheng, P. Chen, X. Xu, T. Wang, D. Wei, Y. Nie, S. Zhu, M. Xiao, and Y. Zhang. 2023b. Fabrication of 100-nm-period domain structure in lithium niobate on insulator. *Opt Express* 31:37464–37471.

Matsuda, N., H. Nishi, P. Karkus, T. Tsuchizawa, K. Yamada, W. J. Munro, K. Shimuzu, and H. Takesue. 2017. Generation of entangled photons using an arrayed waveguide grating. *J Opt* 19:124005.

Munoz, P., D. Pastor, and J. Capmany. 2002. Modeling and design of arrayed waveguide gratings. *J Lightwave Technol* 20:661–674.

Smit, M. K., and C. Van Dam. 1996. PHASAR based WDM devices: Principles, design and applications. *IEEE J Select Topics Quant Electron* 2:236–250.

Takahashi, H., S. Suzuki, K. Kato, and I. Nishi. 1990. Arrayed waveguide grating for wavelength division multi/demultiplexer with nanometer resolution. *Electron Lett* 26:87–88.

Takenouchi, H., and H. Tsuda. 2000. Analysis of optical signal processing using an arrayed waveguide grating. *Opt Express* 6:124–135.

Van Wijk, A. C., C. R. Doerr, and I. B. Akca. 2023. Custom arrayed waveguide gratings with improved performance. *Adv Photonics Res* 4:2300198.

Vernon, Z., and J. E. Sipe. 2015. Strongly driven nonlinear quantum optics in microring resonators. *Phys Rev A* 92:033840.

Wang, Y., K. D. Jons, and Z. Sun. 2021. Integrated photon pair sources with nonlinear optics. *Appl Phys Rev* 8:011314. https://doi.org/10.1063/5.0030258.

Wang, M., Z. Fang, J. Lin, R. Wu, J. Chen, Z. Liu, H. Zhang, L. Qiao, and Y. Cheng. 2023a. Integrated active lithium niobate photonic devices. *Jpn J Appl Phys* 62:SC0801.

Wang, X., X. Jiao, B. Wang, Y. Liu, S.-P. Xie, M.-Y. Zheng, A. Zhang, and J.-W. Pan. 2023b. Quantum frequency conversion and single-photon detection with lithium niobate nanophotonic chips. *npj Quant Inform* 9:38. https://doi.org/10.1038/s41534-023-00704-w.

Zhang, M., C. Wang, R. Cheng, A. Shams-Ansari, and M. Loncar. 2017. Monolithic ultra high Q lithium niobate microring resonator. *Optica* 4:1536–1537.

Chapter 9
Periodic Interactions in Waveguides

Abstract Periodic interactions in optical waveguides result in exchange of power among different spatial modes and different polarizations in waveguides and are used in realization of many integrated photonic devices. Coupled mode theory analysis of both co-directional as well as contradirectional coupling is presented.

9.1 Introduction

In Chap. 5, we had studied coupling of power among modes propagating in two adjacent waveguides. In this chapter, we will study the process of mode coupling among modes of a single waveguide; coupling of power among modes propagating along the same direction is referred to as co-directional coupling and coupling of power among counter propagating modes is referred to as contradirectional coupling. We will show that efficient transfer of power among modes of the same waveguide needs periodic perturbation in the waveguide characteristics (see Fig. 9.1). Such interactions can also be used to couple power from one polarization to its orthogonal polarization (see Chap. 6). Periodic changes in the waveguide characteristics can be brought about either by periodic modulation in the refractive index of the regions or by periodic modulation of the transverse dimensions of the waveguide. Both these lead to periodic modulation in the effective index of the modes of the waveguide and this may lead to efficient coupling of power among different modes provided additional phase matching conditions are satisfied.

For coupling of power to occur between two modes, there must be a coupling through interaction among the two modes; this is brought about by the periodic modulation in refractive index or the waveguide dimensions. Apart from this, for efficient coupling a resonance condition also referred to as phase matching condition needs to be satisfied. If either of these two conditions are not met, even though there may be modulations in the waveguide characteristics, transfer of power among the modes may not take place.

In Chaps. 11–13, we will discuss about periodic modulation of the nonlinear property of the waveguide which leads to efficient coupling and transfer of power among different frequencies of light. In this chapter, we discuss coupling of power

K. Thyagarajan, *Integrated Quantum Photonics*, Graduate Texts in Physics,
https://doi.org/10.1007/978-3-031-85728-7_9

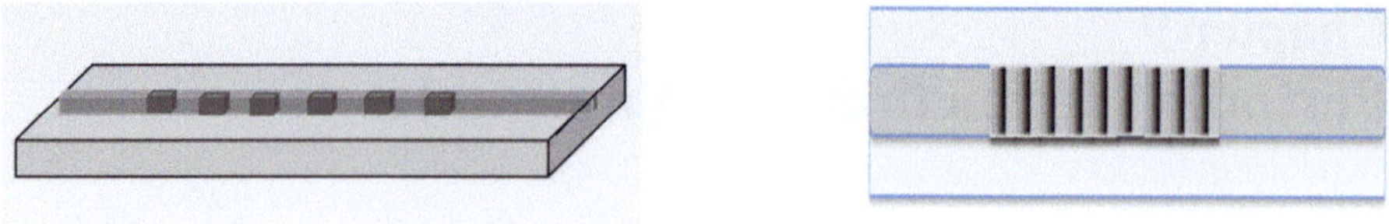

Fig. 9.1 A channel waveguide with periodic modulation of refractive index of the guiding region along the propagation direction: (**a**) periodic waveguide (**b**) top view

among different modes at the same frequency due to modulation of the linear property of the waveguide.

Periodic modulations are very important in integrated photonics and find applications in distributed feedback and distributed Bragg reflector lasers, input-output coupling of power from waveguides, coupling of power among different transverse modes, coupling between modes having different polarizations, fiber Bragg gratings, multilayer coatings for high reflectivity, and so on.

9.2 Coupled Mode Theory

In order to study coupling among modes due to the periodic modulation, we follow a procedure similar to that followed for analyzing the directional coupler in Chap. 5. Let $E_a(x, y)$ and $E_b(x, y)$ represent the transverse modal field distributions of the two modes a and b which are to exchange energy. Assuming that the perturbation to the waveguide caused by the periodic changes in the refractive index or the transverse dimensions is small, we may approximately express the total electric field at any value of z as a superposition of the modes of the unperturbed waveguide:

$$E(x, y, z) = A(z)E_a(x, y)e^{i\beta_a z} + B(z)E_b(x, y)e^{i\beta_b z} \tag{9.1}$$

where β_a and β_b are the propagation constants of the two normal modes of the waveguide which have transverse field distributions described by $E_a(x, y)$ and $E_b(x, y)$. Here we are suppressing the time dependence $e^{-i\omega t}$ of the modes. The amplitudes $A(z)$ and $B(z)$ of the two modes are z dependent due to coupling of power between the two interacting modes.

9.3 Codirectional Coupling

Codirectional coupling corresponds to coupling of power among modes propagating along the same direction in the waveguide. Hence, in this case β_a and β_b have the same sign. In contradirectional coupling which corresponds to coupling of power among modes propagating in opposite directions in the waveguide, the propagation constants β_a and β_b will have opposite signs (see Sect. 9.4).

Note that in the case of directional couplers, the two modes corresponded to the modes of the *two different waveguides* which were exchanging power through evanescent wave coupling. We are now discussing the case when both the modes are normal modes of a *single waveguide* and are by definition orthogonal to each other.

We write the refractive index profile $n_p^2(x, y, z)$ of the perturbed waveguide as

$$n_p^2(x, y, z) = n^2(x, y) + \Delta n^2(x, y, z) \tag{9.2}$$

where $n^2(x, y)$ is the refractive index profile of the unperturbed waveguide without the periodic modulation and $\Delta n^2(x, y, z)$ represents the z-dependent periodic modulation of the refractive index profile.

Since $E_a(x, y)$ and $E_b(x, y)$ are normal modes of the waveguide, they satisfy the following equations (see Chap. 4):

$$\nabla_T^2 E_a(x, y) + \left[k_0^2 n^2(x, y) - \beta_a^2\right] E_a(x, y) = 0 \tag{9.3}$$

and

$$\nabla_T^2 E_b(x, y) + \left[k_0^2 n^2(x, y) - \beta_b^2\right] E_b(x, y) = 0 \tag{9.4}$$

where $\nabla_T^2 = \nabla^2 - \frac{\partial^2}{\partial z^2}$ is the transverse Laplacian.

The total electric field $E(x, y, z)$ of the field propagating through the modulated waveguide satisfies the following wave equation:

$$\nabla^2 E(x, y, z) + k_0^2 n_p^2(x, y, z) E(x, y, z) = 0 \tag{9.5}$$

Substituting for $E(x, y, z)$ from Eq. (9.1) and for $n_p^2(x, y, z)$ from Eq. (9.2) and neglecting $\frac{d^2A}{dz^2}$ and $\frac{d^2B}{dz^2}$ as the amplitudes are assumed to vary slowly with propagation and using Eqs. (9.3) and (9.4), we obtain the following equation (the procedure is very similar to the procedure followed to obtain the coupled mode equations for a directional coupler in Chap. 5):

$$2i\beta_a \frac{dA}{dz} E_a e^{i\beta_a z} + 2i\beta_b \frac{dB}{dz} E_b e^{i\beta_b z} + k_0^2 \Delta n^2(x, y, z)\left(A(z) E_a e^{i\beta_a z} + B(z) E_b e^{i\beta_b z}\right) = 0 \tag{9.6}$$

Multiplying the above equation by E_a^* and integrating over the entire transverse cross-section, we obtain

$$2i\beta_a \frac{dA}{dz} e^{i\beta_a z} \iint |E_a|^2 dxdy + 2i\beta_b \frac{dB}{dz} e^{i\beta_b z} \iint E_a^* E_b dxdy$$
$$+ k_0^2 A(z) e^{i\beta_a z} \iint \Delta n^2(x,y,z) E_a^* E_a dxdy + k_0^2 B(z) e^{i\beta_b z} \iint \Delta n^2(x,y,z) E_a^* E_b dxdy = 0$$

Since the interacting modes, which are normal modes of a single waveguide, are orthogonal to each other, the second integral on the left hand side will be zero. Thus, we obtain

$$\frac{dA}{dz} = i\frac{k_0^2}{2\beta_a} A(z) \frac{\iint \Delta n^2(x,y,z) E_a^* E_a dxdy}{\iint |E_a|^2 dxdy} + i\frac{k_0^2}{2\beta_a} B(z) \frac{\iint \Delta n^2(x,y,z) E_a^* E_b dxdy}{\iint |E_a|^2 dxdy} e^{i(\beta_b - \beta_a)z} \tag{9.7}$$

We define the following coupling coefficients:

$$\kappa_{aa}(z) = \frac{k_0^2}{2\beta_a} \frac{\iint \Delta n^2(x,y,z) |E_a|^2 dxdy}{\iint |E_a|^2 dxdy} \tag{9.8}$$

$$\kappa_{ab}(z) = \frac{k_0^2}{2\beta_a} \frac{\iint \Delta n^2(x,y,z) E_a^* E_b dxdy}{\iint |E_a|^2 dxdy} \tag{9.9}$$

Here κ_{aa} is referred to as the self-coupling coefficient and κ_{ab} is the cross-coupling coefficient. If the field $E_a(x, y)$ is normalized to unit power, then the field satisfies the following equation (see Chap. 4):

$$\frac{\beta_a}{2\omega\mu_0} \iint |E_a|^2 dxdy = 1 \tag{9.10}$$

Hence, we have

$$\kappa_{aa}(z) = \frac{k_0^2}{4\omega\mu_0} \iint \Delta n^2(x,y,z) |E_a|^2 dxdy \tag{9.11}$$

and

$$\kappa_{ab}(z) = \frac{k_0^2}{4\omega\mu_0} \iint \Delta n^2(x,y,z) E_a^* E_b dxdy \tag{9.12}$$

Thus, we obtain

$$\frac{dA}{dz} = i\kappa_{aa}(z) A(z) + i\kappa_{ab}(z) B(z) e^{-i\Delta\beta z} \tag{9.13}$$

and

$$\Delta\beta = (\beta_a - \beta_b) \tag{9.14}$$

Multiplying Eq. (9.6) by E_b^* and following the same procedure as before, we obtain the following equation:

$$\frac{dB}{dz} = i\kappa_{bb}(z)B(z) + i\kappa_{ab}(z)A(z)e^{i\Delta\beta z} \tag{9.15}$$

where

$$\kappa_{bb}(z) = \frac{k_0^2}{4\omega\mu_0}\iint \Delta n^2(x,y,z)|E_b|^2 dxdy \tag{9.16}$$

is again the self-coupling coefficient and we have assumed κ_{ab} to be real.

The field of mode b also satisfies the normalization condition:

$$\frac{\beta_b}{2\omega\mu_0}\iint |E_b|^2 dxdy = 1 \tag{9.17}$$

Thus, the two coupled mode equations describing the evolution of the amplitude of the two interacting modes due to a perturbation in the waveguide are given by Eqs. (9.13) and (9.15) (see also Eqs. (5.10) and (5.13) describing coupling among the modes in a directional coupler). The normalization conditions given by Eqs. (9.10) and (9.17) imply that the powers carried by the two modes at any value of z are equal to $|A(z)|^2$ and $|B(z)|^2$.

For a constant z-independent perturbation, the self-coupling coefficients κ_{aa} and κ_{bb} lead to just a change of the propagation constant of the modes. Also these two terms do not lead to significant effects when coupling of power among two different modes due to a z-dependent perturbation is considered (see Exercise 9.1).

The coupled equations when the effects of self-coupling coefficients can be neglected are given by

$$\frac{dA}{dz} = i\kappa_{ab}(z)B(z)e^{-i\Delta\beta z} \tag{9.18a}$$

$$\frac{dB}{dz} = i\kappa_{ab}(z)A(z)e^{i\Delta\beta z} \tag{9.18b}$$

These equations are similar to the coupled mode equations of an optical directional coupler discussed in Chap. 5 except the coupling coefficient is in this case taken to be $z-$dependent and the two modes that are interacting are the modes of the same waveguide instead of being modes of two different waveguides.

9.3.1 Constant Coupling Coefficient

If Δn^2 is independent of z, then the perturbation is constant along the length of the waveguide. All the coupling coefficients are then independent of z. If the two modes have different propagation constants, then $\Delta\beta \neq 0$. The two resulting equations are the same as that of a directional coupler with non-phase matched waveguides (see Chap. 5). In this case, the self-coupling coefficients κ_{aa} and κ_{bb} which will be independent of z, just lead to a change in the propagation constants of the two interacting modes (see Exercise 9.1). Here we neglect the effect of self-coupling terms which can be taken into account by redefining the propagation constants of the modes.

The solutions to these coupled mode equations will also be the same as that in the case of directional couplers. Thus, If we assume the initial conditions $A(z = 0) = 1$ and $B(z = 0) = 0$, that is, unit power is assumed to be incident in mode a with no input in mode b, the amplitudes of the modes vary as follows:

$$A(z) = \left(\cos \Omega z + i\frac{\Delta\beta}{2\Omega} \sin \Omega z\right) e^{-i\Delta\beta z/2} \tag{9.19a}$$

$$B(z) = i\frac{\kappa_{ab}}{\Omega} \sin \Omega z e^{i\Delta\beta z/2} \tag{9.19b}$$

where

$$\Omega = \sqrt{\left(\kappa_{ab}^2 + \frac{(\Delta\beta)^2}{4}\right)} \tag{9.20}$$

The power in the modes b and a will vary with z as follows:

$$P_b(z) = \frac{\kappa_{ab}^2}{\Omega^2} \sin^2 \Omega z \tag{9.21}$$

and

$$P_a(z) = \cos^2 \Omega z + \frac{(\Delta\beta)^2}{4\Omega^2} \sin^2 \Omega z = 1 - P_b(z) \tag{9.22}$$

Since the interacting modes are two different modes of the same waveguide, in general $\Delta\beta \neq 0$. In addition if $\Delta\beta \gg \kappa_{ab}$, $\Omega \gg \kappa_{ab}$ and the maximum power coupled in mode b would be much less than unity. Thus, in this case, the exchange of power between the two modes of the *unperturbed waveguide* will be very weak.

As in the case of the directional coupler, this poor exchange of power is due to the presence of the terms $e^{-i\Delta\beta z}$ and its complex conjugate in Eqs. (9.18a) and (9.18b). In order to achieve reasonable exchange of power between the two modes with different propagation constants, we need to get rid of this term and this is possible through periodic coupling.

It should be pointed out that in the presence of a z-independent perturbation, we can consider the perturbed refractive index profile given by $n_p^2(x,y) = n^2(x,y) + \Delta n^2(x,y)$ as the profile of another waveguide with z-independent refractive index profile and obtain the normal modes of such a waveguide. The coupling we are discussing is the coupling among modes of the unperturbed waveguide due to the perturbation. There is actually no coupling of power among the normal modes of the perturbed waveguide. The formulation presented here with z-independent refractive index profiles is useful in obtaining the propagation constants of waveguides with refractive index profiles that are very close to waveguides with refractive index profiles that have analytical solutions. Such a technique is referred to as perturbation technique and is used extensively in quantum mechanics (see Chap. 3) and in integrated photonics (see Exercise 9.2).

It is interesting to note that even if the interacting modes have identical propagation constants, there may be no coupling of power between the two modes due to the coupling coefficient κ_{ab} among the modes being zero. Thus, in weakly guiding circular core waveguides (optical fibers) or in symmetric square cross-section waveguides, multiple modes may have the same propagation constant but do not couple power among themselves due to the vanishing of the coupling coefficient. Thus, efficient exchange of power among modes needs both a finite coupling coefficient as well as phase matching, here represented by satisfying the condition $\Delta\beta = 0$.

Exercise 9.1 Show that the self-coupling coefficients κ_{aa} and κ_{bb} lead to only changes in the propagation constant of the interacting modes.

Solution We define two new variables:

$$\tilde{A}(z) = A(z)e^{-i\kappa_{aa}z}$$

and

$$\tilde{B}(z) = B(z)e^{-i\kappa_{bb}z}$$

Now

$$\frac{dA}{dz} = \frac{d\tilde{A}}{dz}e^{i\kappa_{aa}z} + i\kappa_{aa}\tilde{A}e^{i\kappa_{aa}z}$$

Substituting in Eq. (9.13), we obtain

$$\frac{d\tilde{A}}{dz}e^{i\kappa_{aa}z} + i\kappa_{aa}\tilde{A}e^{i\kappa_{aa}z} = i\kappa_{aa}(z)\tilde{A}e^{i\kappa_{aa}z} + i\kappa_{ab}(z)\tilde{B}(z)e^{i\kappa_{bb}z}e^{-i\Delta\beta z}$$

which simplifies to

$$\frac{d\tilde{A}}{dz} = i\kappa_{ab}(z)\tilde{B}(z)e^{-i\Delta\tilde{\beta}z}$$

where

$$\Delta\tilde{\beta} = \Delta\beta + \kappa_{aa} - \kappa_{bb} = (\beta_a + \kappa_{aa}) - (\beta_b + \kappa_{bb}) = \tilde{\beta}_a - \tilde{\beta}_b$$

and

$$\tilde{\beta}_a = (\beta_a + \kappa_{aa}); \tilde{\beta}_b = (\beta_b + \kappa_{bb})$$

which are the modified propagation constants due to self-perturbation. Similarly one can show that

$$\frac{d\tilde{B}}{dz} = i\kappa_{ab}(z)\tilde{A}(z)e^{i\Delta\tilde{\beta}z}$$

Thus, the effect of the self-coupling terms is just to modify the individual propagation constants of the modes and can be easily taken care of by redefining the propagation constants of the individual modes.

Also note that if Δn^2 is independent of z, then we now have a perturbed waveguide having a z-independent refractive index profile given by $n^2(x, y) = n^2(x, y) + \Delta n^2(x, y)$. It is then possible to find the normal modes of this perturbed waveguide with propagation constant of modes approximately given by $\tilde{\beta}_a$ and $\tilde{\beta}_b$ (see Exercise 9.2).

Exercise 9.2 Consider a step index planar waveguide with a refractive index profile given by

$$\begin{aligned} n^2(x) &= n_1^2; |x| < a \\ &= n_2^2; |x| > a \end{aligned}$$

We consider a position independent perturbation given by Δn^2. Obtain the modified propagation constant of the perturbed waveguide.

Solution From Eq. (9.11), we have

$$\kappa_{aa} = \frac{k_0^2}{4\omega\mu_0}\iint \Delta n^2(x, y, z)|E_a|^2 dxdy = \frac{k_0^2}{4\omega\mu_0}\Delta n^2 \iint |E_a|^2 dxdy = \frac{k_0^2}{2\beta_a}\Delta n^2$$

where we have used Eq. (9.10). Thus, we have

$$\tilde{\beta}_a = \beta_a + \kappa_{aa} = \beta_a + \frac{k_0^2}{2\beta_a}\Delta n^2 \tag{9.23}$$

The right hand side of the above equation is an approximation for

$$\tilde{\beta}_a = \left(\beta_a^2 + k_0^2 \Delta n^2\right)^{1/2} \tag{9.24}$$

In fact the perturbation considered in this problem leads to a step index waveguide with a refractive index profile given by

$$\begin{aligned} n_p^2(x) &= n_1^2 + \Delta n^2; |x| < a \\ &= n_2^2 + \Delta n^2; |x| > a \end{aligned}$$

It is easy to show that the propagation constant of this waveguide and the original waveguide described by $n^2(x)$ are related through Eqs. (9.24) and (9.23) is an approximation to this for $\Delta n^2 \ll n_1^2$.

In fact when the perturbation is independent of z, then the propagation constant β_p of the perturbed waveguide can be approximated by the following expression:

$$\beta_p \approx \beta_u + \frac{k_0^2}{4\omega\mu_0} \iint \Delta n^2(x, y)|E_a|^2 dxdy$$

where β_u is the propagation constant of the unperturbed waveguide.

9.3.2 *Periodically Varying Coupling Coefficient*

We saw in Sect. 9.3.1 that for a constant z-independent perturbation, the coupling between two modes of the waveguide is weak due to the difference in propagation constants of the two modes. We can achieve efficient exchange of power among the two non-phase matched modes by having the cross-coupling coefficient κ_{ab} to be z dependent. In order to show this, let us assume a sinusoidal variation with z of the refractive index modulation $\Delta n^2(x, y, z)$:

$$\Delta n^2(x, y, z) = \Delta n_0^2(x, y) \sin Kz \tag{9.25}$$

where $K = \frac{2\pi}{\Lambda}$ is the spatial frequency of the modulation along the z-direction and Λ is the period of modulation and $\Delta n_0^2(x, y)$ is assumed independent of z (see Fig. 9.2). Thus, we have

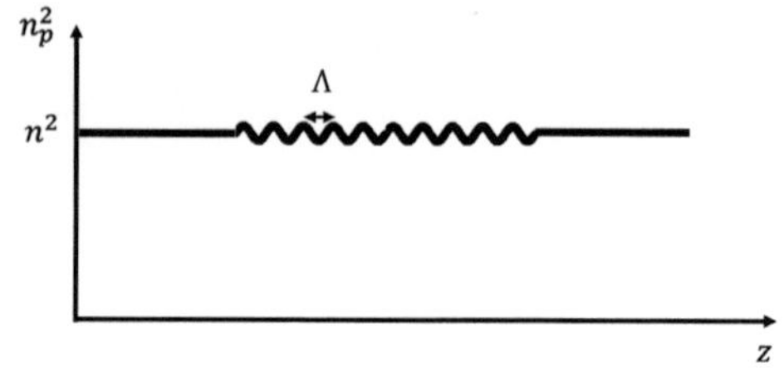

Fig. 9.2 Representation of a sinusoidal refractive index perturbation of a waveguide

$$\kappa_{ab}(z) = \left(\frac{k_0^2}{4\omega\mu_0}\iint \Delta n_0^2(x,y)E_a^*E_b dxdy\right)\sin Kz = \kappa_{ab}\sin Kz \tag{9.26}$$

where

$$\kappa_{ab} = \frac{k_0^2}{4\omega\mu_0}\iint \Delta n_0^2(x,y)E_a^*E_b dxdy \tag{9.27}$$

is the z-independent coupling coefficient. Similarly we will have

$$\kappa_{aa}(z) = \kappa_{aa}\sin Kz \tag{9.28a}$$

$$\kappa_{bb}(z) = \kappa_{bb}\sin Kz \tag{9.28b}$$

where κ_{aa} and κ_{bb} will be z-independent coupling self-coupling coefficients.

A modulation in the coupling coefficients can be achieved through either a periodic modulation of the index of the different regions of the waveguide or by having a periodic variation in the transverse dimensions of the waveguide.

In such a case, the coupled equations (Eqs. 9.18a and 9.18b) become

$$\frac{dA}{dz} = i\kappa_{ab}\sin KzB(z)e^{-i\Delta\beta z} = +\frac{\kappa_{ab}}{2}B(z)\left(e^{-i(\Delta\beta-K)z} - e^{-i(\Delta\beta+K)z}\right) \tag{9.29}$$

$$\frac{dB}{dz} = i\kappa_{ab}\sin KzA(z)e^{i\Delta\beta z} = +\frac{\kappa_{ab}}{2}A(z)\left(e^{i(\Delta\beta+K)z} - e^{i(\Delta\beta-K)z}\right) \tag{9.30}$$

If we choose the spatial frequency of the periodic variation in the coupling coefficient to satisfy the following equation

$$K \approx \Delta\beta = (\beta_a - \beta_b) \tag{9.31}$$

then as is shown in Exercise 9.3, the terms that will lead to significant power transfer among the two modes are those with the exponential factor $e^{\pm i(\Delta\beta - K)z} \approx 1$. Retaining only those terms, the coupled equations become

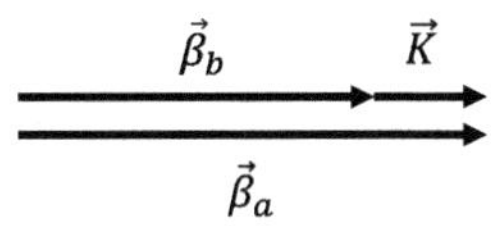

Fig. 9.3 Vector diagram representing phase matched interaction showing co-directional coupling among two co propagating modes of the waveguide

$$\frac{dA}{dz} = \kappa B(z) e^{-i(\Delta\beta - K)z} \tag{9.32}$$

$$\frac{dB}{dz} = -\kappa A(z) e^{i(\Delta\beta - K)z} \tag{9.33}$$

where $\kappa = \frac{\kappa_{ab}}{2}$.

If the spatial frequency of the periodic modulation is chosen such that $K = \Delta\beta$, then the two coupled equations become:

$$\frac{dA}{dz} = \kappa B(z) \tag{9.34}$$

$$\frac{dB}{dz} = -\kappa A(z) \tag{9.35}$$

Figure 9.3 shows a diagrammatic representation referred to as the vector diagram of the quasi phase matching condition

$$\Delta\beta - K = \beta_a - \beta_b - K = 0 \tag{9.36}$$

Here, the propagation constant vectors of the two modes are represented by vectors along the z-direction and represented by $\vec{\beta}_a$ and $\vec{\beta}_b$ and the spatial frequency of the periodic grating with variations along the z-direction represented by a vector $\vec{K}$ oriented along the z-direction. Such vector diagrams are very interesting diagrammatic representation of the quasi phase matching condition.

We first consider the solutions under phase matching. Differentiating Eq. (9.35) with respect to z, we obtain

$$\frac{d^2B}{dz^2} = -\kappa \frac{dA}{dz} = -\kappa^2 B$$

whose general solution is

$$B(z) = a \cos \kappa z + b \sin \kappa z \tag{9.37}$$

Substituting in Eq. (9.35) gives us

$$A(z) = a \sin \kappa z - b \cos \kappa z \tag{9.38}$$

At the input if power is incident in mode a with no input in mode b, we have $A(z = 0) = A_0$ and $B(z = 0) = 0$. Using these initial conditions, we obtain the following solutions:

$$A(z) = A_0 \cos(\kappa z) \tag{9.39}$$

$$B(z) = -A_0 \sin(\kappa z) \tag{9.40}$$

The power propagating through each mode is then given by

$$P_a(z) = |A(z)|^2 = P_a(0) \cos^2(\kappa z) \tag{9.41}$$

$$P_b(z) = |B(z)|^2 = P_a(0) \sin^2(\kappa z) \tag{9.42}$$

showing periodic and complete exchange of energy between the two modes just like the case of a directional coupler.

In this case, complete power transfer from the input mode to the coupled mode is possible and the length (coupling length) required is given by

$$L_c = \frac{\pi}{2\kappa} \tag{9.43}$$

The required spatial period to achieve complete transfer of power from one mode to the other is (see Eq. 9.36)

$$\Lambda = \frac{2\pi}{K} = \frac{2\pi}{\Delta\beta} = \frac{2\pi}{(\beta_a - \beta_b)} \tag{9.44}$$

Note that larger the difference $(\beta_a - \beta_b)$, smaller will be the spatial period required for efficient coupling of power. The period Λ also depends on the wavelength; hence as we will see later, as the wavelength changes, the power coupling reduces (see Sect. 9.3.2).

If $\Delta\beta - K \neq 0$, then instead of Eqs. (9.34) and (9.35) we need to solve Eqs. (9.32) and (9.33) with the initial conditions $A(z = 0) = A_0 = 1$ and $B(z = 0) = 0$. These equations are similar to the coupled mode equations that we had obtained for a directional coupler (see Eqs. 5.20 and 5.21). Following the same procedure as for the directional coupler (see Sect. 5.2), we get the following expression for the powers in modes a and b (see Exercise 9.4):

$$P_a(z) = \cos^2\Omega z + \frac{\Gamma^2}{4\Omega^2} \sin^2\Omega z \tag{9.45}$$

$$P_b(z) = \frac{\kappa^2}{\Omega^2} \sin^2 \Omega z \tag{9.46}$$

where

$$\Omega = \sqrt{\left(\kappa^2 + \frac{\Gamma^2}{4}\right)} = \sqrt{\left(\kappa^2 + \frac{(\Delta\beta - K)^2}{4}\right)} \tag{9.47}$$

Note that

$$P_a(z) + P_b(z) = \cos^2 \Omega z + \frac{1}{\Omega^2}\left(\kappa^2 + \frac{\Gamma^2}{4}\right) \sin^2 \Omega z = 1 \tag{9.48}$$

as required by conservation of energy.

The above analysis shows that when $\Delta\beta \neq K$, complete power transfer does not take place. The maximum power coupled is (for unit power incident)

$$P_{b,max}\left(z = \frac{\pi}{2\Omega}\right) = \frac{\kappa^2}{\Omega^2} = \frac{1}{\sqrt{1 + \frac{(\Delta\beta - K)^2}{4\kappa^2}}} \tag{9.49}$$

Thus, if $(\Delta\beta - K) \gg 2\kappa$, then $P_{b,\,max} \ll 1$ and the coupling does not lead to much power transfer.

As in the case of any coupling process, two conditions need to be satisfied for efficient power transfer:

(a) The coupling coefficient between the two modes must be non-zero, that is, $\kappa \neq 0$ and
(b) The propagation constant of the two modes must satisfy appropriate phase matching condition $\Delta\beta - K = 0$.

For example, if the waveguide has a symmetric refractive index profile and the index modulation is also symmetric, then even if the phase matching condition is satisfied, there would be no coupling of power between the fundamental symmetric mode and the first excited antisymmetric mode since the integral contained in the coupling coefficient κ_{ab} given by Eq. (9.12) would be zero. At the same time if the coupling coefficient is non-zero but the propagation constant difference between the modes is such that $(\Delta\beta - K) \gg 2\kappa$, then again there will be negligible power coupling among the two modes.

Periodic coupling to achieve power transfer between modes propagating along the same direction (co directional coupling) is used in the realization of many integrated optic devices. In Chap. 6, we discussed an example of polarization converter using the electro optic effect which is used to transfer power between two orthogonal polarization modes in a waveguide.

Exercise 9.3 Use Eqs. (9.26) and (9.28a) in Eq. (9.13), integrate Eq. (9.29) from $z-\frac{\Delta z}{2}$ to $z+\frac{\Delta z}{2}$ and show that the different terms on the right hand side of Eq. (9.29) leads, in general, to negligible contribution.

Solution We substitute and integrate Eq. (9.13) from $\left(z-\frac{\Delta z}{2}\right)$ to $\left(z+\frac{\Delta z}{2}\right)$:

$$\int_{z-\frac{\Delta z}{2}}^{z+\frac{\Delta z}{2}} \frac{dA}{dz}dz = i\kappa_{aa}\int_{z-\frac{\Delta z}{2}}^{z+\frac{\Delta z}{2}} \sin Kz A(z)dz + i\kappa_{ab}\int_{z-\frac{\Delta z}{2}}^{z+\frac{\Delta z}{2}} \sin Kz B(z)e^{-i\Delta\beta z}dz$$

We assume that $A(z)$ and $B(z)$ are slowly varying compared to the sine terms and integrate by taking them out of the integral and obtain

$$\begin{aligned} A\left(z+\frac{\Delta z}{2}\right) - A\left(z-\frac{\Delta z}{2}\right) &= 2i\kappa_{aa}A(z)\sin Kz\frac{\sin\left(\frac{K\Delta z}{2}\right)}{K} \\ &+ \kappa_{ab}B(z)e^{-i(\Delta\beta-K)z}\frac{\sin(\Delta\beta-K)\Delta z/2}{(\Delta\beta-K)} \\ &- \kappa_{ab}B(z)e^{-i(\Delta\beta-K)z}\frac{\sin(\Delta\beta+K)\Delta z/2}{(\Delta\beta+K)} \end{aligned}$$

Now, $\Delta\beta = \frac{2\pi}{\lambda_0}\Delta n_{eff}$ where Δn_{eff} is the difference in propagation constants between the two modes. The maximum value of Δn_{eff} is the index difference between the guiding region and surrounding which is typically approximately 0.005. Thus, at a wavelength of 800 nm we have $\Delta\beta = \frac{2\pi}{0.8\times 10^{-6}}0.005 \approx 3.9\times 10^4\ \text{m}^{-1}$. Assuming $K \approx \Delta\beta$, we note that the maximum value of the following terms:

$$\frac{\sin\left(\frac{K\Delta z}{2}\right)}{K} \le \frac{1}{K} \approx 2.5\times 10^{-5}\ \text{m}$$

$$\frac{\sin(\Delta\beta+K)\Delta z/2}{(\Delta\beta+K)} \le \frac{1}{(\Delta\beta+K)} \approx 1.3\times 10^{-5}\ \text{m}$$

The term on the right hand side of Eq. (9.29) with the exponential factor $e^{-i(\Delta\beta-K)z}$ will contribute significantly and increase linearly with Δz if $\Delta\beta - K = 0$ since

$$\frac{\sin(\Delta\beta-K)\Delta z/2}{(\Delta\beta-K)} \approx \frac{\Delta z}{2}$$

Thus, the contributions from all the terms except the term with the factor $e^{-i(\Delta\beta-K)z}$ will be negligible and this term can contribute significantly if $\Delta\beta \approx K$.

Exercise 9.4 Solve Eqs. (9.32) and (9.33) with the initial conditions $A(z = 0) = 1$ and $B(z = 0) = 0$ and obtain Eqs. (9.45) and (9.46).

Solution Differentiating Eq. (9.33) with respect to z and using Eqs. (9.32) and (9.33), we obtain the following differential equation satisfied by $B(z)$:

$$\frac{d^2B}{dz^2} - i\Gamma\frac{dB}{dz} + \kappa^2 B = 0$$

where $\Gamma = \Delta\beta - K$. The general solution of the equation is

$$B(z) = e^{i\Gamma z/2}\left(ae^{i\Omega z} + be^{-i\Omega z}\right)$$

where a and b are constants of integration and

$$\Omega = \sqrt{\left(\kappa^2 + \frac{\Gamma^2}{4}\right)} = \sqrt{\left(\kappa^2 + \frac{(\Delta\beta - K)^2}{4}\right)}$$

Substituting the solution for $B(z)$ in Eq. (9.33), we obtain

$$A(z) = -\frac{i}{\kappa}e^{-i\Gamma z/2}\left[a\left(\Omega + \frac{\Gamma}{2}\right)e^{i\Omega z} - b\left(\Omega - \frac{\Gamma}{2}\right)e^{-i\Omega z}\right]$$

Using the initial conditions, we obtain the following solutions:

$$A(z) = e^{-\frac{i\Gamma z}{2}}\left(\cos\Omega z + i\frac{\Gamma}{2\Omega}\sin\Omega z\right) \tag{9.50}$$

$$B(z) = -\frac{\kappa}{\Omega}e^{\frac{i\Gamma z}{2}}\sin\Omega z \tag{9.51}$$

Hence, the powers in the two modes are $P_a(z) = |A(z)|^2$ and $P_b(z) = |B(z)|^2$ and are given by Eqs. (9.45) and (9.46)

Exercise 9.5 Assume that the fundamental and first excited modes of a graded index channel waveguide can be described by the following equations:

$$E_a(x, y) = Ae^{-(x^2+y^2)/2w_0^2} \tag{9.52}$$

and

$$E_b(x, y) = Bxe^{-(x^2+y^2)/2w_0^2} \tag{9.53}$$

where A and B are normalization constants. The perturbation is assumed to be uniform in the (x, y) direction and given by

$$\Delta n^2(x, y, z) = \Delta n_0^2 \sin Kz \tag{9.54}$$

where Δn_0^2 is a constant. Obtain the coupling coefficient between the two modes of the waveguide.

Solution The coupling coefficient is given by Eq. (9.12). Substituting from Eqs. (9.52)–(9.54) in Eq. (9.12), we obtain

$$\kappa_{ab}(z) = \frac{k_0^2}{4\omega\mu_0} AB\Delta n_0^2 \sin Kz \iint xe^{-(x^2+y^2)/w_0^2} dxdy = 0$$

since the x-integrand is odd. Thus, although there is a periodic perturbation, the coupling coefficient is zero and there will be no transfer of power among the two modes.

Exercise 9.6 Consider a planar waveguide with a film index n_f for $-d < x < 0$ a substrate refractive index n_s for $x < -d$ and a cover refractive index n_c for $x > 0$. The waveguide supports two TE modes, TE_0 and TE_1 with effective indices n_{e0} and n_{e1} respectively. Consider a perturbation $\Delta n^2(x, y, z) = \Delta n_0^2 \sin Kz; x > 0$ and $\Delta n^2(x, y, z) = 0; x < 0$. Obtain an expression for the coupling coefficient.

Solution Since the perturbation is only in the region $x > 0$, we have (see Chap. 4)

$$\kappa_{ab}(z) = \frac{k_0^2}{4\omega\mu_0} \Delta n_0^2 \sin KzA_0A_1l_y \int_0^\infty e^{-\gamma_{c0}x} e^{-\gamma_{c1}x} dx$$

where A_0 and A_1 are the normalization constants of the TE_0 and TE_1 modes respectively and $\gamma_{c0} = k_0\sqrt{n_{e0}^2 - n_c^2}$ and $\gamma_{c1} = k_0\sqrt{n_{e1}^2 - n_c^2}$ and $l_y = \int_0^{l_y} dy$ where we have integrated over a length l_y in the $y-$direction. Hence

$$\kappa_{ab}(z) = \frac{k_0^2}{4\omega\mu_0} \Delta n_0^2 \sin KzA_0A_1l_y \frac{1}{(\gamma_{c0} + \gamma_{c1})} \tag{9.55}$$

Exercise 9.7 The effective indices of the TE_m modes of an asymmetric planar step index waveguide operating at a wavelength of 800 nm are given by $n_{eff}(TE_0) \approx 2.2182$, $n_{eff}(TE_1) \approx 2.2129$ and $n_{eff}(TE_2) \approx 2.2045$. A periodic perturbation is applied on the waveguide only within the core of the waveguide. (a) Which pair of modes among these can be coupled? (b) What will be the required period of periodic modulation for coupling power among the TE_0 and TE_1 and TE_0 and TE_2 modes?

Solution

(a) Since the waveguide is an asymmetric planar waveguide and the perturbation is only within the core of the waveguide, the coupling coefficients between any pair of these three modes will be non-zero and it is possible to achieve periodic coupling of power among any pair of modes.

(b) The spatial period required for coupling from TE_0 and TE_1 mode will be

$$\Lambda = \frac{2\pi}{(\beta_a - \beta_b)} = \frac{\lambda_0}{(n_{eff}(TE_0) - n_{eff}(TE_1))} \approx 150.9\ \mu\text{m}$$

The spatial period required for coupling from TE_0 and TE_2 mode will be

$$\Lambda = \frac{2\pi}{(\beta_a - \beta_b)} = \frac{\lambda_0}{(n_{eff}(TE_0) - n_{eff}(TE_2))} \approx 58.4\ \mu\text{m}$$

9.3.3 Bandwidth of the Coupling Process

Maximum efficiency of coupling of power between the two interacting modes takes place when $\Delta\beta = K$. Since $\Delta\beta$ depends on the wavelength of the propagating mode, for a given spatial frequency of the periodic modulation, that is, for a given value of K, one particular wavelength (central wavelength) will exactly satisfy this equation. As one deviates from this wavelength, we will no more satisfy the condition $\Delta\beta = K$ and the efficiency of power coupling will decrease as per Eq. (9.46). Hence, the interaction process would have a certain bandwidth over which the coupling will have a value greater than a chosen value. This will define the spectral bandwidth of the interaction process. We will now obtain an expression for the spectral bandwidth.

The difference in the propagation constants of the two modes is given by

$$\Delta\beta = \beta_a - \beta_b = \frac{2\pi}{\lambda_0}\left(n_{eff,a}(\lambda_0) - n_{eff,b}(\lambda_0)\right)$$

where λ_0 is the free space wavelength and $n_{eff,\ a}(\lambda_0)$ and $n_{eff,\ b}(\lambda_0)$ are the effective indices of the two interacting modes at the free space wavelength λ_0. Note that $\Delta\beta$ is a function of wavelength due to its explicit dependence on wavelength and also due to the wavelength dependence of the effective indices of the two modes. Thus, the condition $\Delta\beta = K$ can be satisfied at one chosen wavelength and as the wavelength changes, this is no more satisfied and the coupling of power reduces.

We will now evaluate the bandwidth of the interaction process. For this we assume that the periodic perturbation is chosen with such a spatial frequency K that at the central wavelength $\lambda_0 = \lambda_c$, the condition $\Delta\beta - K = 0$ is satisfied. Thus, we have

$$\Delta\beta - K = \frac{2\pi}{\lambda_c}\left(n_{eff,a}(\lambda_c) - n_{eff,b}(\lambda_c)\right) - K = 0 \tag{9.56}$$

which using $K = \frac{2\pi}{\Lambda}$ gives us

$$\Lambda = \frac{\lambda_c}{\Delta n_{eff}(\lambda_c)} \tag{9.57}$$

where

$$\Delta n_{eff}(\lambda_c) = \left(n_{eff,a}(\lambda_c) - n_{eff,b}(\lambda_c)\right) \tag{9.58}$$

At the central wavelength λ_c since $\Delta\beta(\lambda_c) = K$ complete power transfer among the modes can take place if the length of interaction is chosen be equal to the coupling length given by Eq. (9.43)

$$L_c = \frac{\pi}{2\kappa} \tag{9.59}$$

Thus, if the device length is chosen to be equal to L_c then for any wavelength λ_0, the fraction of power transferred to mode b from mode a is given by (see Eq. 9.46)

$$P_b(L_c) = \frac{\kappa^2}{\Omega^2}\sin^2\Omega L_c \tag{9.60}$$

For this length, the power transferred to mode b will become zero if $\Omega L_c = \pi$ or

$$\kappa^2 + \frac{(\Delta\beta(\lambda_0) - K)^2}{4} = \frac{\pi^2}{L_c^2}$$

which using Eq. (9.59) gives us

$$\Delta\beta(\lambda_0) - K = \pm\sqrt{3}\frac{\pi}{L_c} \tag{9.61}$$

Now as the input wavelength changes, the value of $\Delta\beta$ changes and the power transferred also reduces.

At wavelengths $\lambda_0 = \lambda_c \pm \Delta\lambda_0$, we have

$$\Delta\beta(\lambda_0) - K = \frac{2\pi}{(\lambda_c \pm \Delta\lambda_0)}\left(n_{eff,a}(\lambda_c \pm \Delta\lambda_0) - n_{eff,b}(\lambda_c \pm \Delta\lambda_0)\right) - K$$

Assuming $\Delta\lambda_0 \ll \lambda_c$, we can make Taylor series expansion and write

$$n_{eff,a}(\lambda_c \pm \Delta\lambda_0) = n_{eff,a}(\lambda_c) \pm \Delta\lambda_0 \frac{dn_{eff,a}(\lambda_o)}{d\lambda_0}\bigg|_{\lambda_o=\lambda_c} \tag{9.62}$$

Similarly for $n_{eff,\,b}(\lambda_c \pm \Delta\lambda_0)$. Thus, we have

$$\begin{aligned}\Delta\beta(\lambda_0) - K \approx{}& \frac{2\pi}{\lambda_c}\left(1 \mp \frac{\Delta\lambda_0}{\lambda_c}\right)\\ &\times\left[n_{eff,a}(\lambda_c) - n_{eff,b}(\lambda_c) \pm \Delta\lambda_0\left(\frac{dn_{eff,a}(\lambda_o)}{d\lambda_0}\bigg|_{\lambda_o=\lambda_c} - \frac{dn_{eff,b}(\lambda_o)}{d\lambda_0}\bigg|_{\lambda_o=\lambda_c}\right)\right] - K\end{aligned}$$

Keeping terms of order $\Delta\lambda_0$ and using Eq. (9.56), we obtain

$$\Delta\beta(\lambda_0) - K \approx \pm\frac{2\pi}{\lambda_c^2}\left(\Delta n_{eff} - \lambda_c \frac{d\Delta n_{eff}}{d\lambda_0}\bigg|_{\lambda_o=\lambda_c}\right)\Delta\lambda_0 \tag{9.63}$$

Since the power transferred becomes zero when Eq. (9.61) is satisfied, the wavelength deviation from the central wavelength λ_c for the transferred power to become zero is given by

$$\Delta\lambda_0 = \pm\frac{\sqrt{3}}{2}\frac{\lambda_c^2}{L_c\left(\Delta n_{eff} - \lambda_c \frac{d\Delta n_{eff}}{d\lambda_0}\Big|_{\lambda_o=\lambda_c}\right)} = \pm\frac{\sqrt{3}}{2}\frac{\lambda_c^2}{L_c \Delta N_{eff}(\lambda_c)} \tag{9.64}$$

where

$$\Delta N_{eff}(\lambda_c) = \left(\Delta n_{eff} - \lambda_c \frac{d\Delta n_{eff}}{d\lambda_0}\bigg|_{\lambda_o=\lambda_c}\right) \tag{9.65}$$

is the difference in group indices of the two modes at the central wavelength λ_c. The fractional bandwidth is defined as

$$\frac{\Delta\lambda_0}{\lambda_c} = \pm\frac{\sqrt{3}}{2}\frac{\lambda_c}{L_c \Delta N_{eff}(\lambda_c)} \tag{9.66}$$

If we assume $\Delta N_{eff} \approx \Delta n_{eff}$, then

$$\frac{\Delta\lambda_0}{\lambda_c} \approx \pm\frac{\sqrt{3}}{2}\frac{\lambda_c}{L_c \Delta n_{eff}} = \pm\frac{\sqrt{3}}{2}\frac{\Lambda}{L_c} \tag{9.67}$$

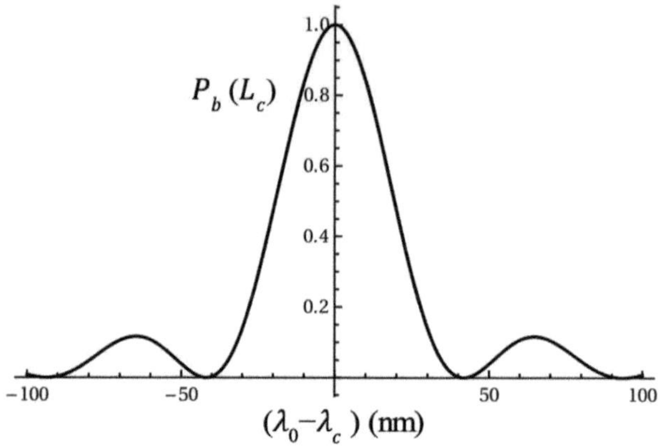

Fig. 9.4 Plot of variation of the coupled power with the deviation of wavelength from the central wavelength for $L_c = 10$ mm, $\lambda_c = 1550$ nm, $\Delta N_{eff}(\lambda_c) = 0.005$

where we have used Eq. (9.57). Thus, the spectral width between the two zeroes in coupling will be $\sqrt{3}\frac{\Lambda}{L_c}$. Equation (9.67) gives the approximate bandwidth of the coupling process.

Figure 9.4 gives a plot of $P_b(L_c)$ as a function of the deviation from the central wavelength for the following values of parameters:

$$L_c = 10 \text{ mm}, \lambda_c = 1550 \text{ nm}, \Delta N_{eff}(\lambda_c) = 0.005$$

It can be seen that at $\lambda_0 = \lambda_c$, the complete power gets coupled and as the wavelength deviates from this wavelength, the coupled power reduces and reaches zero at particular wavelengths on either side of the peak. The bandwidth $\Delta\lambda_0$ comes out to be about 42 nm.

Since the length of interaction is L_c and the period of the spatial modulation is Λ, we note that the number of spatial periods of modulation within the length of interaction is $N = \frac{L_c}{\Lambda}$. Thus, the fractional bandwidth given by Eq. (9.67) is inversely proportional to the number of periods of the modulation within the length of interaction of the device. This is very similar to that in a diffraction grating where the resolving power of the grating is proportional to the number of exposed slits in the grating (see, e.g., Ghatak 2024).

Exercise 9.8 If the difference in effective indices between the two interacting modes at a wavelength of 1000 nm is 0.002, what is the required spatial period for maximum coupling? If the coupling length is 10 mm, what is the corresponding coupling coefficient? What would be the bandwidth of the coupling process?

Solution The spatial period required for maximum coupling is

$$\Lambda = \frac{\lambda_c}{\Delta n_{eff}} = \frac{10^{-6}}{0.002} m = 0.5 \text{ mm}$$

For a coupling length of 10 mm, the required coupling coefficient is $\kappa = \frac{\pi}{2L_c} = 157\text{ m}^{-1}$. The corresponding bandwidth (wavelength spacing between the two zeroes) is (assuming $\Delta N_{eff} \approx \Delta n_{eff}$) approximately 86 nm. The power coupling will be maximum at 1000 nm and will drop to zero around the wavelengths 957 nm and 1043 nm. Thus, this coupling mechanism is not extremely wavelength selective. In Sect. 9.4, we will show that compared to codirectional coupling, contradirectional coupling is highly wavelength selective.

Exercise 9.9 For efficient coupling between two co-propagating guided modes with effective indices 1.5086 and 1.5046 at $\lambda_0 = 0.6$ μm, what should be the periodicity of the periodic modulation? If one desires to have a bandwidth (wavelength difference between the two zeroes on either side of the maximum) of 20 nm, what should be the coupling coefficient and the corresponding interaction length?

Solution From Eq. (9.57), we find the required periodicity as

$$\Lambda = \frac{2\pi}{(\beta_a - \beta_b)} = \frac{\lambda_0}{\Delta n_{eff}} = \frac{0.6}{0.004} = 0.15 \text{ mm}$$

The bandwidth of the coupling process is given by

$$\frac{\Delta\lambda_0}{\lambda_0} \approx \frac{\sqrt{3}}{2} \frac{\Lambda}{L_c}$$

Substituting the values, we obtain $L_c \approx 7.79$ mm.
The coupling coefficient required is $\kappa = \frac{\pi}{2L_c} \approx 0.2\text{ mm}^{-1}$.

It should be pointed out that the coupling coefficient is proportional to the overlap integral between the perturbation $\Delta n^2(x, y, z)$ and the modal fields $E_a(x, y)$ and $E_b(x, y)$ of the interacting modes (see Eq. 9.12). Thus, depending on the symmetry in the waveguide modes and the perturbation, it is possible that the coupling coefficient becomes zero. In this case, there will be no power coupling even if the phase matching condition is satisfied.

Thus, for example, in a symmetric planar waveguide, with a symmetric perturbation coupling between TE_0 and TE_2 mode can take place; however, there would be no coupling between TE_0 and TE_1modes even if one satisfies the condition $\Delta\beta = K$.

Also if the two interacting modes are orthogonally polarized, then no coupling can take place if the perturbation is a scalar perturbation, that is, the $\Delta\epsilon$ caused by the perturbation is diagonal. In order to achieve coupling among two orthogonal polarizations, the perturbation must induce off diagonal elements in the permittivity tensor of the medium. This is discussed in greater detail in Chap. 6.

In a typical channel waveguide, the waveguide is symmetric along the direction parallel to the surface of the substrate (see Chap. 4). Thus, to couple among two different spatial modes one must appropriately choose the perturbation as well as the period of the perturbation.

The overlap integral also points out the fact that any perturbation in the superstate or the substrate of the waveguide leads to smaller coupling coefficient compared to the case if the perturbation is within the core of the waveguide since the magnitude of the electric fields of the interacting modes would have smaller values in the outside region. At the same time it is possible to induce coupling by introducing modulations in the medium surrounding the waveguide and the perturbation needs not be within the guiding region of the waveguide.

In Chap. 6, we discussed a polarization converter using the electro optic effect. This device is precisely based on co-directional coupling among the two orthogonal modes of the waveguide and used periodic electrodes to achieve phase matching.

Exercise 9.10 In continuation of Exercise 9.5, we consider coupling of power from the TE_0 mode to the TE_2 mode in a graded index channel waveguide. The transverse field distribution of the TE_0 mode is given by Eq. (9.52) and it is given that the field distribution of the TE_2 mode is given by

$$E_b(x,y) = B\left(2\frac{x^2}{w_0^2} - 1\right)e^{-(x^2+y^2)/2w_0^2} \tag{9.68}$$

(a) Show that the two modes are orthogonal to each other.

(b) If the index perturbation is described by the following expression

$$\Delta n_0^2(x,y) = \Delta n_0^2 e^{-x^2/d^2} \tag{9.69}$$

obtain an expression for the coupling coefficient between the TE_0 mode and the TE_2 mode.

(c) What would be the coupling coefficient between the TE_0 and TE_1 mode given in Exercise 9.5?

Solution In order to show that the modes are orthogonal to each other, we evaluate the following integral

$$\int_{-\infty}^{+\infty} E_a(x,y)E_b(x,y)dxdy = AB\int_{-\infty}^{+\infty} dx\left(2\frac{x^2}{w_0^2} - 1\right)e^{-x^2/w_0^2}\int_{-\infty}^{+\infty} dye^{-y^2/w_0^2}$$

Now

$$\int_{-\infty}^{+\infty} dx\left(2\frac{x^2}{w_0^2}-1\right)e^{-x^2/w_0^2}=\frac{2}{w_0^2}\int_{-\infty}^{+\infty} dx x^2 e^{-x^2/w_0^2}-\int_{-\infty}^{+\infty} dx e^{-x^2/w_0^2}$$
$$=\sqrt{\pi}w_0-\sqrt{\pi}w_0=0$$

Hence, the modes are orthogonal to each other.

(a) The coupling coefficient is given by

$$\kappa=\frac{\kappa_{ab}}{2}=\frac{k_0^2}{8\omega\mu_0}\iint\Delta n_0^2(x,y)E_a^*E_b dxdy$$
$$=\frac{k_0^2}{8\omega\mu_0}AB\Delta n_0^2\iint e^{-x^2/d^2}\left(2\frac{x^2}{w_0^2}-1\right)e^{-(x^2+y^2)/w_0^2}dxdy$$

On performing the integration we obtain

$$\kappa=-\frac{k_0^2}{8\omega\mu_0}AB\Delta n_0^2\frac{\pi w_0 d}{\left(1+\frac{d^2}{w_0^2}\right)^{3/2}} \tag{9.70}$$

(b) The coupling coefficient between the TE_0 and TE_1 modes will be

$$\kappa=\frac{k_0^2}{8\omega\mu_0}AB\Delta n_0^2\iint xe^{-x^2/d^2}e^{-(x^2+y^2)/w_0^2}dxdy=0$$

implying that, in this case, there would be no coupling between the two modes TE_0 and TE_1.

Exercise 9.11 A co directional coupling is to be designed in lithium niobate waveguide having ordinary and extraordinary effective indices of $n_{eff,o}=2.23$ and $n_{eff,e}=2.15$. (a) For operation at a wavelength of 1500 nm, what period of the periodic index variation would be required? (a) If the filter is to be used to separate wavelengths of 1500 nm and 1501 nm with negligible cross-talk, what should be the minimum length of the device?

Solution

(a) The required spatial period is given by Eq. (9.57):

$$\Lambda=\frac{\lambda_c}{\Delta n_{eff}}=\frac{1.5}{0.08}=18.75\ \mu\text{m}$$

(b) For resolving wavelengths of 1500 and 1501 nm, the required bandwidth $\Delta\lambda_0$ given by Eq. (9.67) must have a maximum value of 1 nm. Using Eq. (9.67), we obtain the required minimum length of

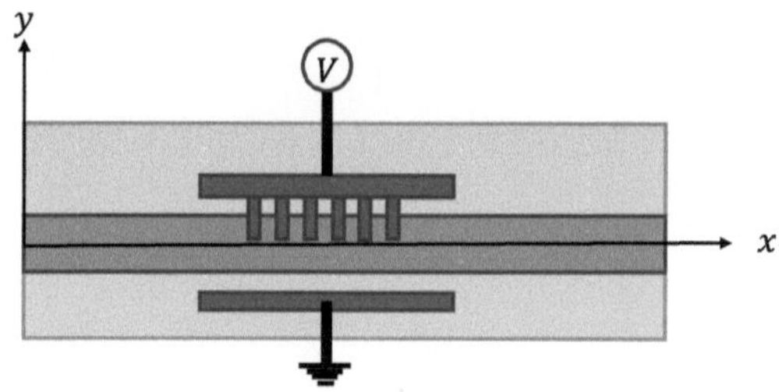

Fig. 9.5 Electrode pattern to induce asymmetric periodic electric field on the waveguide

$$L_c = \frac{\sqrt{3}}{2}\frac{\Lambda\lambda_0}{\Delta\lambda_0} = \frac{\sqrt{3}}{2}\frac{18.75\times10^{-6}\times1.5\times10^{-6}}{10^{-9}} \approx 24\ \text{mm}$$

Exercise 9.12 Consider a channel waveguide in z-cut x-propagating KTP which can support two spatial modes $E_{11}^{(z)}(y,z)$ and $E_{21}^{(z)}(y,z)$ modes. It is desired to use the electro optic effect (see Chap. 6) to induce dynamic coupling among the two modes (a) Show that it is not possible to induce power exchange between the modes by using a symmetric electrode pattern which induces a perturbation which is independent of y. (b) Design an asymmetric electrode pattern that can be used to induce efficient coupling among the two modes (Bharadwaj et al. 2015a, b).

Solution

(a) Since the two modes $E_{11}^{(z)}(y,z)$ and $E_{21}^{(z)}(y,z)$ which need to be coupled have opposite parity, with $E_{11}^{(z)}(y,z)$ being symmetric along y and $E_{21}^{(z)}(y,z)$ being antisymmetric along y, the integral $\iint \Delta n^2(x,y,z)E_{11}^{(z)*}(y,z)E_{21}^{(z)}(y,z)dydz$ vanishes if the induced perturbation is symmetric and hence the coupling coefficient $\kappa_{ab} = 0$, thus no power exchange can take place between the modes.

(b) Since the fields have major electric field components along the z-direction, it is possible to use the r_{33} coefficient of KTP to induce periodic changes in the refractive index of the waveguide for coupling the two modes. A digital electrode pattern as shown in Fig. 9.5 can be used to induce an asymmetric periodic index changes in the region $y > 0$ in the waveguide.

Assuming an electric field of the form $\vec{F} = \hat{x}F_x + \hat{y}F_y + \hat{z}F_z$, the equation for the index ellipsoid in the presence of the electric field will be given by (see Chap. 6)

$$\frac{x^2}{(n_x+\Delta n_x)^2} + \frac{y^2}{(n_y+\Delta n_y)^2} + \frac{z^2}{(n_z+\Delta n_z)^2} + 2yzr_{42}F_y + 2xzr_{51}F_x = 1$$

where $\Delta n_x \approx -n_x^3 r_{13} F_z/2$, $\Delta n_y \approx -n_y^3 r_{23} F_z/2$ and $\Delta n_z \approx -n_z^3 r_{33} F_z/2$. The two cross-terms will try to induce coupling among the two orthogonal polarizations but since the period required for coupling among the modes is different, these will not be able to induce any coupling.

The z-polarized modes in the waveguide which have their major electric field component along the z-direction, the electro optic coefficient $r_{33}(\approx 36$ pm/V) will induce changes through the F_z component of the applied electric field. Since the changes in the refractive index are not uniform in the entire cross-section of the waveguide, the overlap integral $\iint \Delta n^2(x,y,z) E_{11}^{(z)*}(y,z) E_{21}^{(z)}(y,z) dydz$ will be non-zero and thus will induce coupling among the two modes.

Also since the electrodes have a periodic pattern, the changes in the effective index will also be periodic. By an appropriate choice of the period, it is possible satisfy the phase matching condition and thus to induce efficient power coupling among the $E_{11}^{(z)}(y,z)$ and $E_{21}^{(z)}(y,z)$ modes. The required spatial period will be given by

$$\Lambda = \frac{\lambda_0}{n_{eff,11} - n_{eff,21}}$$

where $n_{eff,\,11}$ and $n_{eff,\,21}$ are the effective indices of the $E_{11}^{(z)}(y,z)$ and $E_{21}^{(z)}(y,z)$ modes, respectively. For a waveguide having a width of 3 μm and a depth of 2 μm with $\Delta n = 0.02$, the required period is $\Lambda \approx 105\ \mu m$.

In Bharadwaj et al. (2015a, b), such a device was proposed for testing entanglement among generated photon pairs using spontaneous parametric down conversion (see Chap. 23).

9.4 Contradirectional Coupling

In the case of contradirectional coupling, that is, coupling between modes propagating in opposite directions along the waveguide, β_a and β_b have opposite signs. Let us assume that the input mode a is propagating along the $+z$ direction and the coupled mode b is propagating along the $-z$ direction. Note that both modes may be the same transverse mode propagating along opposite directions. If the waveguide is a single-mode waveguide both propagation constants will have the same magnitude being the same mode. We will consider this case here. The analysis presented can be easily generalized for coupling among different spatial modes propagating along opposite directions.

The total electric field will be given by

$$E(x,y,z) = A(z) E_a(x,y) e^{i\beta_a z} + B(z) E_b(x,y) e^{-i\beta_b z} \tag{9.71}$$

where $A(z)$ and $B(z)$ are the amplitudes of the modes propagating along the $+z$ and $-z$ directions, $E_a(x, y)$ and $E_b(x, y)$ are their field distributions and β_a and β_b their propagation constants. Note that the exponential z-dependent terms have opposite signs in the two terms representing modes propagating along opposite directions (the time dependence which is not specified explicitly in the equations is $e^{-i\omega t}$). If the interacting modes correspond to the same transverse spatial mode but propagating in opposite directions, then $\beta_b = \beta_a$. At the same time, the two modes which are propagating along opposite directions are orthogonal to each other even if they correspond to the same transverse field profiles.

Here we assume that the interacting modes are the same spatial modes propagating in opposite directions. In order to obtain the coupled mode equations describing the variation of the amplitudes of the two interacting modes, we employ the same procedure as outlined to obtain Eqs. (9.18a) and (9.18b). Noting that in this case $\Delta\beta = \beta_a - (-\beta_a) = 2\beta_a$, we obtain the following two coupled mode equations describing the evolution of the amplitudes of the two modes:

$$\frac{dA}{dz} = i\kappa(z)B(z)e^{-2i\beta_a z} \tag{9.72}$$

$$\frac{dB}{dz} = -i\kappa(z)A(z)e^{2i\beta_a z} \tag{9.73}$$

with $\kappa(z)$ given by (see Eq. 9.12)

$$\kappa(z) = \kappa_{ab}(z) = \frac{k_0^2}{4\omega\mu_0} \iint \Delta n^2(x, y, z) E_a^* E_a dxdy \tag{9.74}$$

where as compared to Eq. (9.12), $E_b(x, y)$ is replaced by $E_a(x, y)$ since both interacting modes are assumed to be the same spatial mode but propagating along opposite directions. Note the negative sign in Eq. (9.73) as compared to the coupled mode equation in co-directional interaction which has come about due to the factor $e^{-i\beta_b z}$ in view of the contradirectional nature of coupling.

Assuming a sinusoidal perturbation with spatial frequency K, that is, $\Delta n^2(x, y, z) = \Delta n_0^2(x, y) \sin Kz$, then we will have

$$\kappa(z) = \kappa_0 \sin Kz \tag{9.75}$$

where

$$\kappa_0 = \frac{k_0^2}{4\omega\mu_0} \iint \Delta n_0^2(x, y) E_a^* E_a dxdy$$

We substitute in Eq. (9.72) and Eq. (9.73) to get

$$\frac{dA}{dz} = \frac{\kappa_0}{2} B(z)\left(e^{-i(2\beta_a - K)z} - e^{-i(2\beta_a + K)z}\right) \tag{9.76}$$

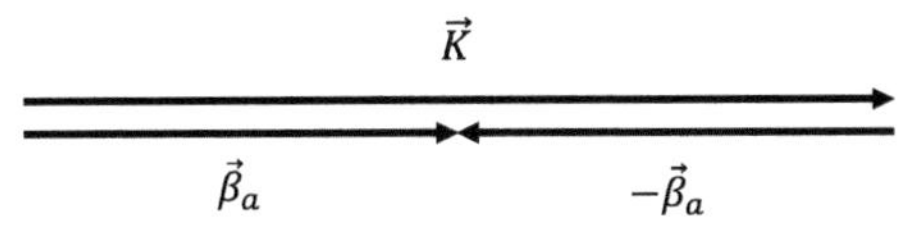

Fig. 9.6 Vector diagram showing phase matching in contradirectional coupling among two counter propagating modes

$$\frac{dB}{dz} = -\frac{\kappa_0}{2}A(z)\left(e^{i(2\beta_a+K)z} - e^{i((2\beta_a-K))z}\right) \tag{9.77}$$

As discussed earlier, if we choose $K \approx 2\beta_a$, then we can neglect the nonphase matched terms (the second bracketed term in Eq. (9.76) and the first bracketed term in Eq. (9.77)), then we have

$$\frac{dA}{dz} = \tilde{\kappa}B(z)e^{-i(2\beta_a-K)z} \tag{9.78}$$

$$\frac{dB}{dz} = \tilde{\kappa}A(z)e^{i((2\beta_a-K))z} \tag{9.79}$$

where

$$\tilde{\kappa} = \frac{\kappa_0}{2} \tag{9.80}$$

Note that compared to the case of codirectional coupling, the two coupled equations in the case of contradirectional coupling have the same sign and this leads to hyperbolic solutions to the coupling rather than oscillatory solutions.

We first assume the following condition

$$K = 2\beta_a \tag{9.81}$$

which will now correspond to the phase matching condition for the interaction. Under phase matching the coupled mode equations are given by:

$$\frac{dA}{dz} = \tilde{\kappa}B(z) \tag{9.82}$$

$$\frac{dB}{dz} = \tilde{\kappa}A(z) \tag{9.83}$$

Recall that the phase matching condition for efficient codirectional coupling was that the spatial frequency of the modulation must equal the difference in propagation constants of the interacting modes (see Eq. 9.44). Since in this case the interacting modes have propagation constants β_a and $-\beta_a$, the same condition implies Eq. (9.81).

Figure 9.6 shows the vector diagram corresponding to contradirectional coupling.

As boundary conditions we assume that mode with amplitude A propagating in the $+z$ direction (incident mode) is incident at $z = 0$ with unit amplitude. We also assume that there is no power incident at the end of the interaction region corresponding to $z = L$ in the mode B propagating in the $-z$ direction (coupled mode), that is,

$$A(z=0)=1 \tag{9.84}$$

$$B(z=L)=0 \tag{9.85}$$

In order to obtain the solutions, we differentiate Eq. (9.83) and use Eq. (9.82) to obtain

$$\frac{d^2B}{dz^2}=\tilde{\kappa}\frac{dA(z)}{dz}=\tilde{\kappa}^2B \tag{9.86}$$

whose general solution is

$$B(z)=a\cosh\tilde{\kappa}z+b\sinh\tilde{\kappa}z \tag{9.87}$$

Using this solution in Eq. (9.83), we obtain

$$A(z)=a\sinh\tilde{\kappa}z+b\cosh\tilde{\kappa}z \tag{9.88}$$

Using the boundary conditions, we have

$$A(z)=\cosh(\tilde{\kappa}z)-\tanh(\tilde{\kappa}L)\sinh(\tilde{\kappa}z) \tag{9.89}$$

$$B(z)=\sinh(\tilde{\kappa}z)-\tanh(\tilde{\kappa}L)\cosh(\tilde{\kappa}z) \tag{9.90}$$

The amplitude of the counter propagating mode (propagating along $-z$ direction) at $z = 0$ (the input of the waveguide) is

$$B(0)=-\tanh(\tilde{\kappa}L) \tag{9.91}$$

and the amplitude of the incident mode exiting the waveguide at $z = L$ is given by

$$A(L)=\text{sech}(\tilde{\kappa}L) \tag{9.92}$$

The reflectivity of the periodic perturbed region or the periodic structure is defined by the ratio of the power in the counter propagating mode exiting the waveguide at $z = 0$ to the power of the incident mode at $z = 0$:

$$R=\left|\frac{B(0)}{A(0)}\right|^2=\tanh^2(\tilde{\kappa}L) \tag{9.93}$$

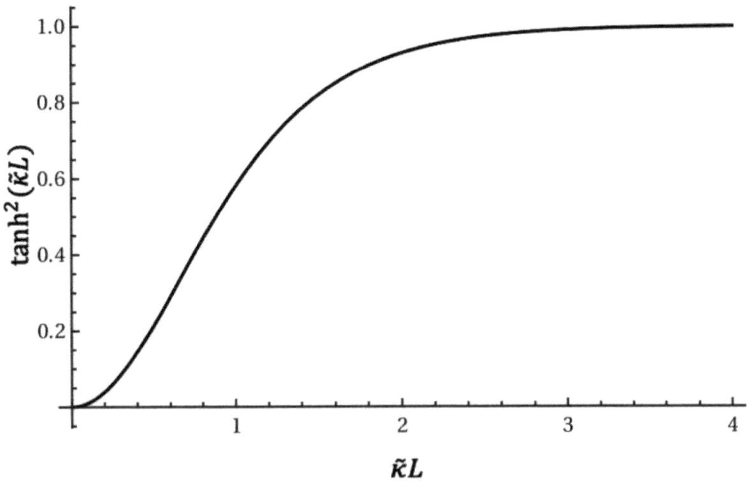

Fig. 9.7 Variation in reflectivity as given by Eq. (9.93) with $\tilde{\kappa}L$. Note that as the value of $\tilde{\kappa}L$ increases the reflectivity asymptotically approaches unity

Similarly the transmittivity is defined by the ratio of the power in the incident propagating mode exiting the waveguide at $z = L$ to the power of the incident mode at $z = 0$:

$$T = \left| \frac{A(L)}{A(0)} \right|^2 = \mathrm{sech}^2(\tilde{\kappa}L) \tag{9.94}$$

Note that $R + T = 1$ as required by energy conservation.

Figure 9.7 shows the variation of reflectivity with $\tilde{\kappa}L$.

It is interesting to note that the power transfer in the contradirectional interaction from the incident mode to the reflected mode (contrapropagating mode) can never be 100%; the longer the length of interaction, the greater the reflectivity. This is contrary to the codirectional case we discussed earlier where it is possible to transfer full power from the incident mode to the coupled mode and the power exchange is periodic unlike the contradirectional case where the power transfer is monotonic.

If the effective index of the mode is n_{eff}, then the required period for efficient contradirectional coupling satisfying the phase matching condition $K = 2\beta_a$ is given by (see Eq. 9.81)

$$\Lambda = \frac{2\pi}{K} = \frac{2\pi}{2\beta_a} = \frac{\lambda_0}{2n_{eff}} \tag{9.95}$$

where λ_0 is the free space wavelength. Note that the period required for contradirectional coupling is half the effective wavelength of the propagating mode within the waveguide.

As an example if $n_{eff} = 2$ and $\lambda_0 = 1500$ nm, the required period is 375 nm. This is much smaller than the period required for codirectional coupling.

If $(K - 2\beta_a) \neq 0$, then we can solve the coupled equations with the same boundary conditions and obtain the following expression for the reflectivity (see Exercise 9.14):

$$R = \frac{\tilde{\kappa}^2 \sinh^2 \Omega L}{\Omega^2 \cosh^2 \Omega L + \frac{\Gamma^2}{4} \sinh^2 \Omega L} \tag{9.96}$$

where

$$\Gamma = 2\beta_a - K; \tag{9.97}$$

$$\Omega^2 = \tilde{\kappa}^2 - \frac{\Gamma^2}{4} \tag{9.98}$$

It is instructive to obtain an approximate expression for the coupling coefficient $\tilde{\kappa}$. For this, we represent the index perturbation $\Delta n_0^2(x, y)$ by the following expression:

$$\Delta n_0^2(x, y) = \Delta n_0^2 f(x, y) \tag{9.99}$$

where Δn_0^2 is a constant independent of coordinates and $f(x, y)$ gives the functional dependence of the refractive index modulation in the waveguide. Thus, we have

$$\tilde{\kappa} = \frac{\kappa_0}{2} = \frac{k_0^2 \Delta n_0^2}{8\omega\mu_0} \iint f(x, y) E_a^* E_a dxdy \tag{9.100}$$

Since the field E_a satisfies the normalization condition given by Eq. (9.10), we can write Eq. (9.100) as

$$\tilde{\kappa} = \frac{k_0^2 \Delta n_0^2}{8\omega\mu_0} \frac{\iint f(x, y) E_a^* E_a dxdy}{\frac{\beta_a}{2\omega\mu_0} \iint |E_a|^2 dxdy} = \frac{k_0^2 \Delta n_0^2}{4\beta_a} \frac{\iint f(x, y) E_a^* E_a dxdy}{\iint |E_a|^2 dxdy} \tag{9.101}$$

The quantity

$$I = \frac{\iint f(x, y) E_a^* E_a dxdy}{\iint |E_a|^2 dxdy} \tag{9.102}$$

Is a dimensionless quantity and specifies the overlap between the refractive index modulation represented by $f(x, y)$ and the modal field distribution represented by $E_a(x, y)$. Note that if the refractive index modulation is uniform throughout the waveguide cross-section, then $f(x, y) = 1$ and $I = 1$. On the other hand, if the refractive index modulation has no overlap with the modal field, then $I = 0$.

Hence, we have

$$\tilde{\kappa} = \frac{k_0^2 \Delta n_0^2}{4\beta_a} I \tag{9.103}$$

We can further approximate by writing $\beta_a = k_0 n_{eff} \approx k_0 n_0$, where n_0 is the refractive index of the guiding region. We can also write $\Delta n_0^2 = 2n_0 \Delta n_0$. Substituting these in Eq. (9.103), we obtain an approximate and useful expression for the coupling coefficient:

$$\tilde{\kappa} \approx \frac{\pi \Delta n_0 I}{\lambda_0} \tag{9.104}$$

Exercise 9.13 Consider contradirectional coupling in a waveguide with an effective index variation given by the following equation

$$n_{eff}(z) = 1.5 + 0.01 \cos 5\pi z$$

where z is measured in micrometers. At what free space wavelength would the efficiency of coupling be maximum?

Solution For the given grating $K = 5\pi$ μm^{-1}. Hence, the period of the index modulation is $\Lambda = \frac{2\pi}{K} = 0.4$ μm. Using Eq. (9.95), we have the required wavelength as $\lambda_0 = 2n_{eff}\Lambda = 2 \times 1.5 \times 0.4 = 1.2$ μm.

Exercise 9.14 For the case when $(K - 2\beta_a) \neq 0$, solve the coupled mode equations with same boundary conditions of the incidence of the mode at $z = 0$ with unit amplitude and the contradirectional mode having zero amplitude at $z = L$, and obtain the expression given by Eq. (9.96) for the reflectivity.

Solution Combining Eqs. (9.78) and (9.79) and eliminating $A(z)$ from the equations, we obtain

$$\frac{d^2B}{dz^2} - i\Gamma \frac{dB}{dz} - \tilde{\kappa}^2 B = 0 \tag{9.105}$$

where Γ is defined in Eq. (9.97). It is instructive to compare Eq. (9.105) with the corresponding equation for B(z) obtained in Exercise 9.4 for the co-directional interaction case. The general solution of Eq. (9.105) is

$$B(z) = e^{\frac{i\Gamma z}{2}}\left(ae^{\Omega z} + be^{-\Omega z}\right) \tag{9.106}$$

where Ω is defined in Eq. (9.98). Using the boundary condition $B(z = L) = 0$, we obtain $b = -ae^{2\Omega L}$. Hence,

$$B(z) = 2ae^{\Omega L}e^{\frac{i\Gamma z}{2}} \sinh \Omega(z - L) \tag{9.107}$$

Substituting in Eq. (9.79), we obtain

$$A(z) = \frac{2a}{\kappa}e^{\Omega L}e^{-\frac{i\Gamma z}{2}}\left[\Omega\cosh\Omega(\mathrm{z} - \mathrm{L}) + i\frac{\Gamma}{2}\sinh\Omega(\mathrm{z} - \mathrm{L})\right] \tag{9.108}$$

Using the boundary condition $A(z = 0) = 1$, we have

$$a = \frac{\tilde{\kappa}}{2}e^{-\Omega L}\frac{1}{\Omega\cosh\Omega\mathrm{L} - i\frac{\Gamma}{2}\sinh\Omega\mathrm{L}}$$

Hence,

$$B(z) = \frac{\tilde{\kappa}}{\Omega\cosh\Omega\mathrm{L} - i\frac{\Gamma}{2}\sinh\Omega\mathrm{L}}e^{\frac{i\Gamma z}{2}}\sinh\Omega(z - L) \tag{9.109}$$

Since $B(z)$ is the amplitude of the reflected mode, the reflection coefficient would be

$$R = |B(z = 0)|^2 = \frac{\tilde{\kappa}^2\sinh^2\Omega\mathrm{L}}{\Omega^2\cosh^2\Omega\mathrm{L} + \frac{\Gamma^2}{4}\sinh^2\Omega\mathrm{L}} \tag{9.110}$$

Under phase matched operation, $\Gamma = 0$ and $\Omega = \tilde{\kappa}$ and Eq. (9.110) reduces to Eq. (9.93).

9.4.1 Bandwidth

We now wish to calculate the bandwidth of the contradirectional interaction. For the wavelength satisfying the condition given by Eq. (9.95), the reflectivity will be maximum. Deviation from this wavelength leads to $(K - 2\beta_a) \neq 0$ and the reflectivity drops.

The maximum reflectivity corresponds to $\Gamma = 0$ and is given by $\tanh^2(\tilde{\kappa}L)$ and occurs for a wavelength $\lambda_0 = \lambda_c$ given by (see Eq. (9.95)):

$$\lambda_c = 2n_{eff}(\lambda_c)\Lambda \tag{9.111}$$

where $n_{eff}(\lambda_c)$ is the effective index of the mode at the wavelength λ_c.

For a given grating with a period Λ defined by Eq. (9.111), as the input wavelength changes from λ_c the phase matching condition is no more satisfied and the reflectivity drops. As the wavelength changes, $\Gamma = 2\beta_a - K = \frac{4\pi n_{eff}(\lambda_0)}{\lambda_0} - \frac{2\pi}{\Lambda}$ keeps increasing in magnitude and when $\Gamma^2 > 4\tilde{\kappa}^2$, then Ω^2 becomes negative (see Eq. (9.98)) and Ω becomes pure imaginary. In this case writing $\Theta^2 = -\Omega^2 = \frac{\Gamma^2}{4} - \tilde{\kappa}^2$, Eq. (9.96) becomes

$$R = \frac{\tilde{\kappa}^2 \sin^2\Theta L}{\Theta^2 \cos^2\Theta L + \frac{\Gamma^2}{4}\sin^2\Theta L} \tag{9.112}$$

Thus, when $\Theta L = \pm\pi$, the reflectivity drops to zero. This implies that

$$\Gamma = 2\beta_a(\lambda_0) - K = \pm\frac{2}{L}\left(\pi^2 + \tilde{\kappa}^2 L^2\right)^{1/2} \tag{9.113}$$

Assuming $\lambda_0 = \lambda_c \pm \Delta\lambda$, with $\Delta\lambda \ll \lambda_c$, we have

$$\beta_a(\lambda_0) = \frac{2\pi}{\lambda_0} n_{eff}(\lambda_0) = \frac{2\pi}{\lambda_c \pm \Delta\lambda} n_{eff}(\lambda_c \pm \Delta\lambda)$$

Making a Taylor series expansion and neglecting higher order terms in $\Delta\lambda$, we have

$$\beta_a(\lambda_0) \approx \frac{2\pi}{\lambda_c}\left[n_{eff}(\lambda_c) \mp \frac{\Delta\lambda}{\lambda_c} N_{eff}(\lambda_c)\right] = \beta_a(\lambda_c) \mp \frac{2\pi}{\lambda_c}\frac{\Delta\lambda}{\lambda_c} N_{eff}(\lambda_c)$$

where $N_{eff}(\lambda_c) = n_{eff}(\lambda_c) - \lambda_c \frac{dn_{eff}}{d\lambda}$ is the group index of the mode. Thus,

$$\Gamma = 2\beta_a(\lambda_0) - K = 2\beta_a(\lambda_c) \mp \frac{4\pi}{\lambda_c}\frac{\Delta\lambda}{\lambda_c} N_{eff}(\lambda_c) - K = \mp\frac{4\pi}{\lambda_c}\frac{\Delta\lambda}{\lambda_c} N_{eff}(\lambda_c) \tag{9.114}$$

where we have used the fact that $2\beta_a(\lambda_c) - K = 0$.

Substituting in Eq. (9.113), we obtain the wavelength deviation from the central wavelength where the reflectivity reduces to zero as

$$\frac{\Delta\lambda}{\lambda_c} = \frac{\lambda_c}{2LN_{eff}}\left(1 + \frac{\tilde{\kappa}^2 L^2}{\pi^2}\right)^{1/2} \tag{9.115}$$

If we assume $N_{eff} \approx n_{eff}$, then we obtain an approximate expression for $\Delta\lambda$:

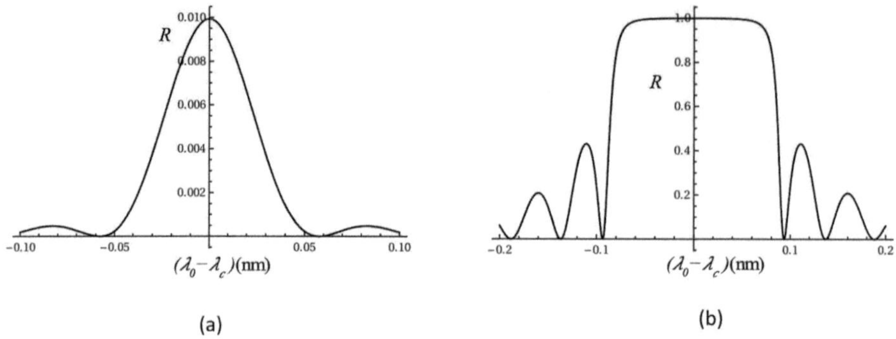

Fig. 9.8 Reflectivity spectra of two periodic modulations having the same length with (**a**) $\tilde{\kappa}L = 0.1$ and (**b**) $\tilde{\kappa}L = 4$

$$\frac{\Delta\lambda}{\lambda_c} \approx \frac{\Lambda}{L}\left(1 + \frac{\tilde{\kappa}^2 L^2}{\pi^2}\right)^{1/2} \tag{9.116}$$

The wavelength spacing between the zeros of reflectivity on either side of the reflectivity peak in reflectivity is $2\Delta\lambda$. The quantity $\Delta\lambda$ gives an approximate value for the bandwidth corresponding to full width at half maximum (FWHM) of the reflection spectrum.

If $\tilde{\kappa}L \ll \pi$, that is, either a weak coupling or a small length or both, then the bandwidth given by Eq. (9.116) can be approximated by

$$\frac{\Delta\lambda}{\lambda_c} = \frac{\Lambda}{L} \tag{9.117}$$

In this case, the fractional bandwidth is again equal to the ratio of the spatial grating period to the length of periodic interaction region. The bandwidth decreases as the length of the interaction region increases.

On the other hand if $\tilde{\kappa}L \gg \pi$, that is, strong coupling or large length or both, the bandwidth is approximately given by

$$\frac{\Delta\lambda}{\lambda_c} = \frac{\tilde{\kappa}\,\Lambda}{\pi} \tag{9.118}$$

In this case, the bandwidth is independent of the length of the device for a given periodic perturbation which defines the value of $\tilde{\kappa}$.

Figure 9.8a, b gives a comparison of the reflectivity spectrum from two contradirectional coupling cases with $L = 10$ mm, $N_{eff} = 2.18$, $\lambda_c = 1550$ nm corresponding to $\tilde{\kappa}L = 0.1$ and $\tilde{\kappa}L = 4.0$. Note the difference in the shape of the reflectivity spectra apart from the peak value of reflectivity and the bandwidth.

Exercise 9.15 I wish to design a contradirectional device having a length of 10 mm and with a peak reflectivity of 93% at a center wavelength of 1550 nm. What should be the coupling coefficient and what will be the corresponding bandwidth of the device?

Solution If the peak reflectivity is 93%, then we have

$$\tanh^2(\tilde{\kappa}L) = 0.93$$

which can be solved to give

$$\tilde{\kappa}L = \frac{1}{2}\ln\left(\frac{1+\sqrt{0.93}}{1-\sqrt{0.93}}\right) \approx 2.0$$

For the given device length of 10 mm, the required coupling coefficient is $\tilde{\kappa} = 0.2\ \text{mm}^{-1}$.

Figure 9.9 shows a plot of reflectivity R versus $\Delta\lambda = (\lambda_0 - \lambda_c)$ for $\tilde{\kappa}L = 2.0$. The peak reflectivity will be $R = \tanh^2(\tilde{\kappa}L) \approx 0.93$. Assuming a central wavelength of 1550 nm and $N_{eff} = 2.18$, we have using Eq. (9.115), $\Delta\lambda \approx 0.065$ nm which is indeed very small.

Hence, contradirectional coupling using periodic interaction is highly wavelength selective and can be used to separate very closely lying wavelengths. This is also the basic principle of fiber Bragg gratings which find wide applications in optical fiber

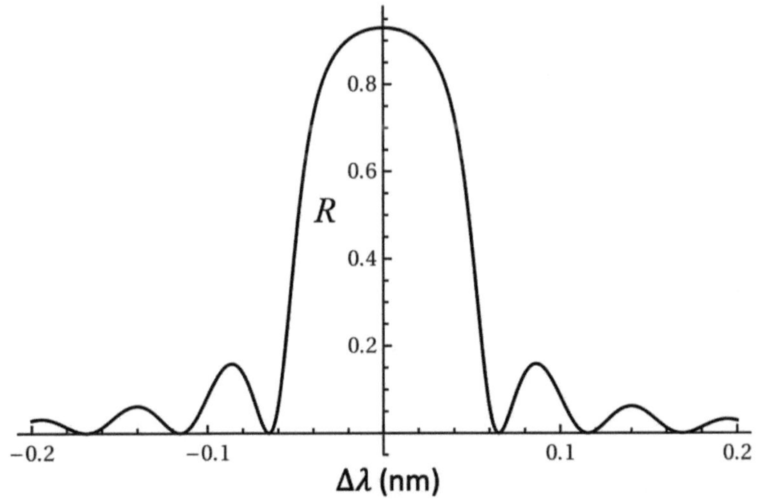

Fig. 9.9 Variation in reflectivity R with $\Delta\lambda = (\lambda_0 - \lambda_c)$ for $\tilde{\kappa}L = 2.0$ as given by Eq. (9.96)

communication systems and optical fiber sensing and distributed feedback and distributed Bragg grating lasers.

Exercise 9.16 I wish to design a contradirectional coupler operating at a center wavelength of 1550 nm and having a bandwidth of 1 nm and a peak reflectivity of 95%. If the effective index of the mode is 1.47 obtain the required grating period, length and coupling coefficient

Solution For obtaining a peak reflectivity of 95%, we have

$$\tanh(\tilde{\kappa}L) = \sqrt{0.95}$$

which gives $\tilde{\kappa}L \approx 2.178$.

We have the expression for the required bandwidth using the value of $\tilde{\kappa}L$ obtained above

$$\Delta\lambda = \frac{\Lambda}{L}\lambda_c\left(1 + \frac{\tilde{\kappa}^2 L^2}{\pi^2}\right)^{1/2} = 1.217\frac{\Lambda}{L}\lambda_c$$

The required grating period would be given by

$$\Lambda = \frac{\lambda_0}{2n_{eff}} = \frac{1.55}{2 \times 1.47} \approx 0.527\ \mu\text{m}$$

From the above equations, we obtain for the required length of interaction to be

$$L = 1.217\frac{\Lambda}{\Delta\lambda}\lambda_c \approx 0.994\ \text{mm}$$

This gives us the required coupling coefficient as

$$\tilde{\kappa} \approx \frac{2.178}{L} \approx 2.19\ \text{mm}^{-1}$$

Problems

9.1. Consider a planar waveguide with the following refractive indices: $n_f = 1.52$, $n_s = 1.50$, $n_c = 1$ and $d = 6$ μm and operating at a free space wavelength of 800 nm. (a) How many TE modes does this waveguide support? (b) Obtain their corresponding effective indices.(c) The thickness of the waveguide is varied periodically with a period Λ and with $\Delta d = 0.1$ μm. (d) Obtain the value of Λ for efficient coupling from the fundamental mode to the other modes supported by the waveguide, assuming codirectional interaction. (e) What are the corresponding coupling coefficients and what is the required interaction length for complete conversion among the modes?

9.2. Consider a directional coupler made of two symmetric single mode planar waveguides with $n_f = 2.0$, $n_s = 1.95$ and $d = 1$ μm. The center to center spacing between the two waveguides is 2 μm. (a) Obtain the effective indices of the two normal modes of the directional coupler. (b) Suggest a geometry which can be used to induce coupling among the two normal modes. (c) What will be the required period of the coupling process for efficient conversion?

9.3. Contradirectional coupling is described by Eqs. (9.78) and (9.79). Show that these equations satisfy power conservation. Hint: remember that the two interacting modes are propagating in opposite directions.

9.4. A waveguide having the following periodic variation in effective index has maximum reflectivity of 0.81 at 0.6 μm.

$$n_{eff}(z) = 1.5 + 0.001 \sin\left(\frac{2\pi z}{\Lambda}\right)$$

(a) Obtain the value of Λ.
(b) What is the length of interaction?

Assume $I = 0.8$.

9.5. It is proposed to design two contradirectional coupling interactions having peak reflectivity of 90% at 1500 nm and with bandwidths of 0.1 nm and 10 nm. The medium has a refractive index of 1.5. Obtain the coupling coefficient and length of interaction for the two cases and compare them.

9.6. Consider two contradirectional coupling interactions in two identical waveguides having the same reflection coefficient and same peak wavelength of reflectivity. Reflector R_1 has an interaction length of 1 mm while reflector R_2 has an interaction length of 10 mm. Which one of them has a larger bandwidth?

9.7. Consider a step index channel waveguide similar to that shown in Fig. 9.1 fabricated in an isotropic substrate. The waveguide supports the fundamental mode $E^x_{11}(x, y)$ and two excited modes $E^x_{12}(x, y)$ and $E^x_{21}(x, y)$. The core refractive index varies periodically with the propagation distance z according to $n_{core}(x, y, z) = n_f + \Delta n \sin Kz$ where n_f and Δn are constants. Is it possible to couple power input in the E^x_{11} either of these modes and if so under what conditions?

9.8. I wish to design a contradirectional reflector with a periodic refractive index variation across the entire cross-section of a channel waveguide. The effective index of the fundamental mode of the waveguide is 2.21. The reflector is to reflect 64% of the light at the center wavelength of 1500 nm with a bandwidth (spacing between the two reflection minima on either side of the peak) of 1 nm. Calculate

(a) The required length of the periodic region
(b) The required index modulation Δn and
(c) The required length of the periodic modulation.

9.9. Consider a channel waveguide of length L in a z-cut x-propagating lithium niobate waveguide. The effective indices of the quasi TE and quasi TM modes are given by

$$n_{eff,o} = 2.26 + 0.001 \sin Kx$$

$$n_{eff,e} = 2.20 + 0.001 \sin Kx$$

where x is measured in μm and $\kappa L \gg 1$. When both the quasi modes are excited at the input, maximum reflectivity is observed at 440 nm and at a nearby longer wavelength. Assume $I = 0.8$.

(a) Obtain the period of refractive index modulation.
(b) What is the wavelength of the other reflection peak?
(c) Show that the peaks of reflection will be non-overlapping and very well resolved.

9.10. The reflectivity from a waveguide having periodic index modulation has a peak reflectivity of 50% at $\lambda_0 = 1550$ nm and the reflectivity becomes zero at $\lambda_0 = 1550.1$ nm and $\lambda_0 = 1549.9$ nm. If the effective index of the waveguide mode is $n_{eff} = 1.5$

(a) Calculate the period, the length and the peak refractive index modulation of the device.
(b) What modification would one do to decrease the bandwidth of the reflector without changing the peak reflectivity?

Assume $I = 1.0$.

References

Bharadwaj, D., K. Thyagarajan, M. Jachura, M. Karpiński, and K. Banaszek. 2015a. Scheme for on-chip verification of transverse mode entanglement using the electro-optic effect. *Opt Express* 23:33087–33098.

Bharadwaj, D., K. Thyagarajan, M. Karpinski, and K. Banaszek. 2015b. Generation of higher-dimensional modal entanglement using a three-waveguide directional coupler. *Phys Rev A* 91: 033824.

Ghatak, A. 2024. *Optics*. 8th ed. India: McGraw Hill.

Chapter 10
Nonlinear Optical Polarization

Abstract Nonlinear optical effects play a very important role in integrated quantum photonics and are used in the generation of entangled photon pairs, squeezed states, etc. This chapter presents discussion on the generation of second-order nonlinear polarization in a medium that results in the generation of second harmonic and sum and difference frequency. Calculation of the generated nonlinear polarization dependent on the nonlinear coefficient of the medium the state of polarization of the propagating modes is presented. The results presented here are used in later chapters.

10.1 Introduction

At low light intensity levels, media behave in a linear fashion, that is, the electric polarization created in the medium by the propagating light wave is proportional to the electric field of the light wave. This leads to all linear optical phenomena such as refraction, reflection, and dispersion, which form the basis of linear optics. The linear dependence is only an approximation which is valid as long as the electric field inside the medium is very small compared to the electric fields holding the atoms together in the medium. At higher light intensities, the electric field of the light wave becomes comparable to the electric field inside the medium and the linear relationship between the electric polarization and the electric field is no more valid. These additional terms lead to nonlinear optical effects due to the nonlinear response of the medium to the electric field. The second-order nonlinear optical effects lead to second harmonic generation, sum, and difference frequency generation, parametric amplification, etc. Third-order nonlinear effects lead to intensity dependent refractive index, self-phase modulation, cross phase modulation, four wave mixing, etc. which play an important role in optical fiber communication systems. Both second-order and third-order nonlinear optical effects are widely used in integrated quantum photonics for the generation of nonclassical states of light including entangled photon pairs. In later chapters, we will discuss the application of second-order nonlinear effects in integrated quantum photonics.

Nonlinear optical effects play a very important role in integrated quantum photonics. Here we will discuss only second-order nonlinear effects which are

K. Thyagarajan, *Integrated Quantum Photonics*, Graduate Texts in Physics,
https://doi.org/10.1007/978-3-031-85728-7_10

important effects in realizing nonclassical states of light and in frequency conversion of signals. In this chapter, we will discuss about the second-order nonlinear polarization generated in media such as lithium niobate and KTP which are two very important substrates used in integrated quantum photonics. The results of this chapter will be used in later chapters while studying second harmonic generation, parametric amplification, spontaneous parametric down conversion, etc. in waveguides.

10.2 Second-Order Nonlinear Polarization

In a linear medium, the electric polarization $\vec{P}$ is assumed to be a linear function of the electric field $\vec{E}$:

$$\vec{P} = \epsilon_0 \chi \vec{E} \tag{10.1}$$

where ϵ_0 is the permittivity of free space and χ is the linear electric susceptibility of the medium. This equation implies that the time dependence of the polarization and the electric field is the same. Thus, if the incident wave has a frequency ω, then the polarization will also have the same frequency ω and no new frequencies are generated in the polarization. Since polarization is nothing but dipole moment per unit volume, this implies that the dipoles are oscillating at the frequency of the incident wave. Oscillating dipoles radiate electromagnetic waves and hence the waves generated by the oscillating dipoles are at the same frequency as the incident wave and no new frequencies are generated by the medium.

At high optical intensities (which corresponds to high electric fields of the light waves), all media behave in a nonlinear fashion. Thus, Eq. (10.1) gets modified to

$$\vec{P} = \epsilon_0 \chi \vec{E} + \epsilon_0 \chi^{(2)} \vec{E}\,\vec{E} + \epsilon_0 \chi^{(3)} \vec{E}\,\vec{E}\,\vec{E} + \ldots \tag{10.2}$$

The above equation in component form reads as follows:

$$P_j = \epsilon_0 \sum_{k=1}^{3} \chi_{jk} E_k + \epsilon_0 \sum_{k=1}^{3} \sum_{l=1}^{3} \chi_{jkl}^{(2)} E_k E_l + \epsilon_0 \sum_{k=1}^{3} \sum_{l=1}^{3} \sum_{m=1}^{3} \chi_{jklm}^{(3)} E_k E_l E_m + \ldots; \quad j = 1, 2, 3 \tag{10.3}$$

where E_k is the kth component of the total electric field within the medium. Here χ_{jk} are the components (9 components) of the linear dielectric susceptibility, $\chi_{jkl}^{(2)}$ are the components (27 components) of the second-order nonlinear susceptibility, $\chi_{jklm}^{(3)}$ are the components (81 components) of the third-order nonlinear susceptibility, and so on.

If we use the Einstein convention of repeated indices being summed over, then we can write the above equation in a more compact form

$$P_j = \epsilon_0 \chi_{jk} E_k + \epsilon_0 \chi_{jkl}^{(2)} E_k E_l + \epsilon_0 \chi_{jklm}^{(3)} E_k E_l E_m + \ldots \quad (10.4)$$

In Chap. 2, we had used an equation similar to Eq. (10.4) with only the first term leading to linear optics. In Eq. (10.4), the second term which corresponds to second-order nonlinearity leads to second harmonic generation, sum and difference frequency generation, parametric amplification, etc., while the third term which corresponds to third-order nonlinear effects is responsible for third harmonic generation, self-phase modulation, cross phase modulation, four wave mixing, etc. In media possessing a center of inversion symmetry, all elements of the second-order susceptibility tensor $\chi_{jkl}^{(2)}$ are zero and the first nonzero nonlinear term is the third-order nonlinearity. Optical fibers do not exhibit second-order nonlinear effects; the third-order effects play a significant role in high data rate optical fiber communication systems.

We shall focus here on second-order nonlinear optical effects that are present in media not possessing a center of inversion symmetry. The second-order nonlinear polarization is defined through the following equation:

$$P_j^{(NL)} = 2\epsilon_0 \sum_{k,l=1}^{3} d_{jkl} E_k E_l; \qquad j = 1, 2, 3 \quad (10.5)$$

where we have replaced $\chi_{jkl}^{(2)}$ by $2d_{jkl}$ consistent with the literature. Using the standard procedure of repeated indices being summed over, we rewrite Eq. (10.5) as:

$$P_j^{(NL)} = 2\epsilon_0 d_{jkl} E_k E_l; \qquad j, k, l = 1, 2, 3 \quad (10.6)$$

Since the indices take values from 1 to 3, the tensor d_{jkl} has 27 elements and is a $3 \times 3 \times 3$ matrix.

Since the electric field components E_k and E_l commute, the nonlinear tensor d_{jkl} satisfies the following equation:

$$d_{jkl} = d_{jlk} \quad (10.7)$$

Thus, we can use the following contraction of indices:

$$11 \to 1; \quad 22 \to 2; \quad 33 \to 3; \quad 23, 32 \to 4; \quad 13, 31 \to 5, 12, 21 \to 6 \quad (10.8)$$

and write the nonlinear tensor d_{jkl} as a 6×3 matrix as follows:

$$\bar{\bar{d}} = \begin{pmatrix} d_{11} & d_{12} & d_{13} & d_{14} & d_{15} & d_{16} \\ d_{21} & d_{22} & d_{23} & d_{24} & d_{25} & d_{26} \\ d_{31} & d_{32} & d_{33} & d_{34} & d_{35} & d_{36} \end{pmatrix} \tag{10.9}$$

Thus, the contracted nonlinear tensor has 18 components. Just like the electro optic tensor introduced in Chap. 6, the nonlinear optical tensor elements need to satisfy the symmetry requirements of the crystal. Depending on the symmetries, some elements of the tensor are zero, some are equal to each other, while others are equal to each other but of opposite sign.

10.3 Second Harmonic Generation

We first consider second harmonic generation (i.e., generation of the frequency 2ω from an incident wave at frequency ω). We assume that the fields are monochromatic and correspond to homogeneous plane waves propagating along the z-direction. In such a case, the kth component of the electric fields at frequencies ω and 2ω is given by

$$E_k(\omega, z) = \frac{1}{2}\left(E_k^{(\omega)} e^{i(k_1 z - \omega t)} + c.c\right) \tag{10.10a}$$

$$E_k(2\omega, z) = \frac{1}{2}\left(E_k^{(2\omega)} e^{i(k_2 z - 2\omega t)} + c.c\right) \tag{10.10b}$$

where k_1 and k_2 represent the propagation constants of the waves corresponding to frequencies ω and 2ω, respectively, and $E_k^{(\omega)}$ and $E_k^{(2\omega)}$ correspond to the kth components of the complex electric field amplitudes at frequencies ω and 2ω, respectively.

Hence, the kth and l^{th} components of the **total** electric field within the medium assuming only waves at frequencies ω and 2ω to be present in the medium will be

$$E_k = \frac{1}{2}\left(E_k^{(\omega)} e^{i(k_1 z - \omega t)} + c.c\right) + \frac{1}{2}\left(E_k^{(2\omega)} e^{i(k_2 z - 2\omega t)} + c.c\right) \tag{10.11a}$$

$$E_l = \frac{1}{2}\left(E_l^{(\omega)} e^{i(k_1 z - \omega t)} + c.c\right) + \frac{1}{2}\left(E_l^{(2\omega)} e^{i(k_2 z - 2\omega t)} + c.c\right) \tag{10.11b}$$

In writing the above equations, we have assumed that the kth and l^{th} components of the electric fields at the individual frequencies to have the same propagation constant which may not always be true. For example, one component of the field at ω could correspond to the ordinary polarization, while the other could correspond to extra ordinary polarization. In such cases, we would need to use different propagation constants for the two components. Here to keep the analysis simple, we will assume that both of the components have the same propagation constant; it is straightforward to use different values of propagation constant and calculate the nonlinear polarization (see Exercise 10.1).

We substitute for E_k and E_l in Eq. (10.6) and collect terms on the right hand side that correspond to frequency 2ω and obtain the following expression for the nonlinear polarization generated at the second harmonic (frequency 2ω):

$$P_j^{(2\omega)} = \frac{1}{2}\left(\epsilon_0 d_{jkl} E_k^{(\omega)} E_l^{(\omega)} e^{2i(k_1 z - \omega t)} + c.c.\right) = \frac{1}{2}\left(\tilde{P}_j^{(2\omega)} e^{2i(k_1 z - \omega t)} + c.c.\right) \tag{10.12}$$

where

$$\tilde{P}_j^{(2\omega)} = \epsilon_0 d_{jkl} E_k^{(\omega)} E_l^{(\omega)} \tag{10.13}$$

Equation (10.12) gives the three Cartesian components of the nonlinear polarization generated at frequency 2ω and $\tilde{P}_j^{(2\omega)}$ represents the amplitude of the complex nonlinear polarization at frequency 2ω.

Thus, for example, the x-component of the nonlinear polarization will be given by (where we have contracted the last two indices in the elements of the d tensor)

$$\begin{aligned}\frac{1}{\epsilon_0}\tilde{P}_x^{(2\omega)} = \frac{1}{\epsilon_0}\tilde{P}_1^{(2\omega)} &= d_{11} E_x^{(\omega)} E_x^{(\omega)} + d_{12} E_y^{(\omega)} E_y^{(\omega)} + d_{13} E_z^{(\omega)} E_z^{(\omega)} + 2d_{14} E_y^{(\omega)} E_z^{(\omega)} \\ &+ 2d_{15} E_x^{(\omega)} E_z^{(\omega)} + 2d_{16} E_x^{(\omega)} E_y^{(\omega)}\end{aligned} \tag{10.14}$$

Note that the factors of 2 in the last three terms are due to commutability of the last two indices k and l of the nonlinear tensor d_{jkl}.

Similarly we can obtain expressions for the y- and z-components of the amplitude of the generated complex nonlinear polarization. The three components of the nonlinear polarization can be put in the form of a matrix relation as given below:

$$\begin{pmatrix} \tilde{P}_x^{(2\omega)} \\ \tilde{P}_y^{(2\omega)} \\ \tilde{P}_z^{(2\omega)} \end{pmatrix} = \epsilon_0 \begin{pmatrix} d_{11} & d_{12} & d_{13} & d_{14} & d_{15} & d_{16} \\ d_{21} & d_{22} & d_{23} & d_{24} & d_{25} & d_{26} \\ d_{31} & d_{32} & d_{33} & d_{34} & d_{35} & d_{36} \end{pmatrix} \begin{pmatrix} E_x^{(\omega)} E_x^{(\omega)} \\ E_y^{(\omega)} E_y^{(\omega)} \\ E_z^{(\omega)} E_z^{(\omega)} \\ 2E_y^{(\omega)} E_z^{(\omega)} \\ 2E_x^{(\omega)} E_z^{(\omega)} \\ 2E_x^{(\omega)} E_y^{(\omega)} \end{pmatrix} \tag{10.15}$$

Thus, the electric field components of the incident light wave at frequency ω and the nonlinear tensor elements of the medium determine the components of the nonlinear polarization generated at the frequency 2ω. Knowing the components of the electric field of the wave at frequency ω, it is possible to evaluate the components of the nonlinear polarization that is generated by the incident wave.

As an example, the nonlinear tensor for lithium niobate is given by

$$\overline{\overline{d}} = \begin{pmatrix} 0 & 0 & 0 & 0 & d_{15} & -d_{22} \\ -d_{22} & d_{22} & 0 & d_{15} & 0 & 0 \\ d_{31} & d_{31} & d_{33} & 0 & 0 & 0 \end{pmatrix} \tag{10.16}$$

where, as can be seen some elements are zero while others are equal and some are equal in magnitude but of opposite sign ($d_{21} = -d_{22}; d_{24} = d_{15}; d_{32} = d_{31}; d_{16} = -d_{22}$). These relationships can indeed be obtained by using the symmetry properties of lithium niobate.

Similarly the nonlinear tensor for KTP (potassium titanyl phosphate) is

$$\overline{\overline{d}} = \begin{pmatrix} 0 & 0 & 0 & 0 & d_{15} & 0 \\ 0 & 0 & 0 & d_{24} & 0 & 0 \\ d_{31} & d_{32} & d_{33} & 0 & 0 & 0 \end{pmatrix} \tag{10.17}$$

The nonlinear tensor is written in the principal axis system of the crystal. For calculating the elements of the tensor in any other coordinate system, one needs to use the transformation properties of the tensor.

The values of the various tensor elements of lithium niobate, KTP, and lithium tantalate are given in Table 10.1. The magnitudes of the tensor elements are of the order of $10^{-12} \frac{\text{m}}{\text{V}}$.

Note that for lithium niobate the largest nonlinear coefficient is $d_{33} = d_{333} = d_{zzz}$. Using this component in the nonlinear interaction leads to higher efficiencies of the process and would require lower powers for operation.

Example 10.1 As an example, let us consider the incidence of a wave polarized along the z-direction and propagating along the x-direction at frequency ω in lithium niobate (see Fig. 10.1). This will correspond to an extraordinary polarization. Here (x, y, z) corresponds to the principal axis system of the crystal. Thus, we have

$$E_x^{(\omega)} = 0; \quad E_y^{(\omega)} = 0; \quad E_z^{(\omega)} \neq 0 \tag{10.18}$$

Thus, the nonlinear polarization generated will have components given by

Table 10.1 Values of elements of d tensor of important nonlinear optical materials in integrated quantum photonics

Medium	Nonlinear tensor elements in pm/V
Lithium niobate ($LiNbO_3$)	$d_{33} \approx 33; d_{31} = d_{15} \approx 6; d_{22} \approx 3$
Potassium titanyl phosphate ($KTiOPO_4$): KTP	$d_{31} \approx 6.5; d_{32} \approx 5; d_{33} \approx 13.7; d_{24} \approx 7.6; d_{15} \approx 6.1$
Lithium tantalate ($LiTaO_3$)	$d_{33} \approx 21; d_{22} \approx 2; d_{31} \approx 1$

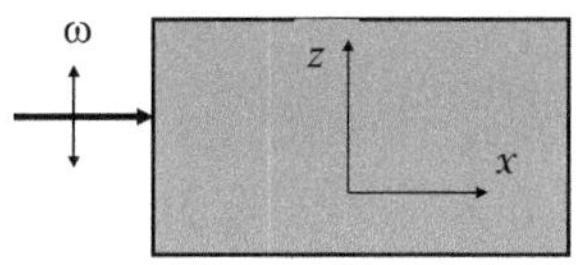

Fig. 10.1 A fundamental wave at frequency ω having extraordinary polarization propagating along the x-direction in lithium niobate

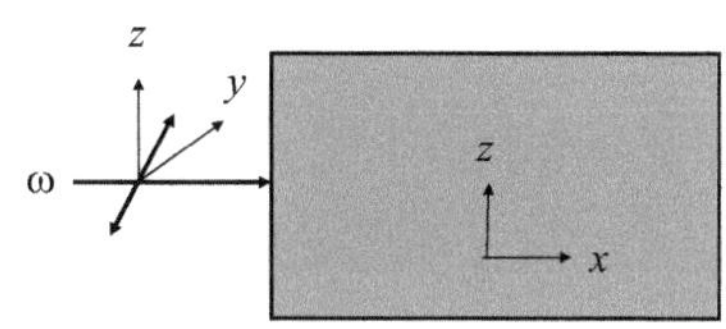

Fig. 10.2 Fundamental wave propagating along the x-direction with an input polarization at 45° to the $z-$ axis

$$\begin{pmatrix} \widetilde{P}_x^{(2\omega)} \\ \widetilde{P}_y^{(2\omega)} \\ \widetilde{P}_z^{(2\omega)} \end{pmatrix} = \epsilon_0 \begin{pmatrix} 0 & 0 & 0 & 0 & d_{15} & -d_{22} \\ -d_{22} & d_{22} & 0 & d_{15} & 0 & 0 \\ d_{31} & d_{31} & d_{33} & 0 & 0 & 0 \end{pmatrix} \begin{pmatrix} 0 \\ 0 \\ E_z^{(\omega)} E_z^{(\omega)} \\ 0 \\ 0 \\ 0 \end{pmatrix}$$

$$= \begin{pmatrix} 0 \\ 0 \\ d_{33} E_z^{(\omega)} E_z^{(\omega)} \end{pmatrix} \tag{10.19}$$

Hence, the nonlinear polarization in this case has only the z-component. Since polarization is the source of electromagnetic field, this component of the polarization at frequency 2ω can only generate the z-component of the second harmonic field. The efficient generation of the second harmonic requires an additional condition referred to as the phase matching condition which will be discussed later in Chap. 11.

Example 10.2 As another example, we again consider propagation along the x-direction but the input polarization is at 45° to the z-axis (see Fig. 10.2). Thus, the input electric field has both y- and $z-$ components:

$$E_x^{(\omega)} = 0; \quad E_y^{(\omega)} \neq 0; \quad E_z^{(\omega)} \neq 0 \tag{10.20}$$

and we have

$$\begin{pmatrix} \tilde{P}_x^{(2\omega)} \\ \tilde{P}_y^{(2\omega)} \\ \tilde{P}_z^{(2\omega)} \end{pmatrix} = \epsilon_0 \begin{pmatrix} 0 & 0 & 0 & 0 & d_{15} & -d_{22} \\ -d_{22} & d_{22} & 0 & d_{15} & 0 & 0 \\ d_{31} & d_{31} & d_{33} & 0 & 0 & 0 \end{pmatrix} \begin{pmatrix} 0 \\ E_y^{(\omega)} E_y^{(\omega)} \\ E_z^{(\omega)} E_z^{(\omega)} \\ 2E_y^{(\omega)} E_z^{(\omega)} \\ 0 \\ 0 \end{pmatrix}$$
$$= \begin{pmatrix} 0 \\ d_{22} E_y^{(\omega)} E_y^{(\omega)} + 2d_{15} E_y^{(\omega)} E_z^{(\omega)} \\ d_{31} E_y^{(\omega)} E_y^{(\omega)} + d_{33} E_z^{(\omega)} E_z^{(\omega)} \end{pmatrix} \tag{10.21}$$

which now shows that the nonlinear polarization has both y and z-components. The y-component of the nonlinear polarization can generate the y-component of the second harmonic wave which corresponds to an ordinary wave at 2ω and the z-component of the nonlinear polarization can generate the z-component of the second harmonic which now corresponds to an extraordinary wave at 2ω.

Since the waves are propagating along the x-direction, the y-component of electric field will be an ordinary wave with a refractive index $n_o(\omega)$, while the z-component will be an extraordinary wave with a refractive index $n_e(\omega)$. Thus, the propagation vector k_1 corresponding to the electric field component E_y will be $k_o^{(\omega)} = \frac{\omega}{c} n_o(\omega)$ and the propagation vector k_1 corresponding to the electric field component E_z will be $k_e^{(\omega)} = \frac{\omega}{c} n_e(\omega)$. This is precisely what was mentioned in the paragraph after Eqs. (10.11a) and (10.11b); see Exercise 10.1.

At the same time, notice that in the term $d_{33} E_z^{(\omega)} E_z^{(\omega)}$ in the z-component of the nonlinear polarization at 2ω, both components of the fundamental wave at frequency ω have extraordinary polarizations and characterized by the same propagation constant $k_e^{(\omega)} = \frac{\omega}{c} n_e(\omega)$. Similarly in the term $d_{31} E_y^{(\omega)} E_y^{(\omega)}$ both components of the fundamental wave at frequency ω have ordinary polarizations and characterized by the same propagation constant $k_o^{(\omega)} = \frac{\omega}{c} n_o(\omega)$.

Exercise 10.1 Consider second harmonic generation in the configuration shown in Fig. 10.2. What is the velocity of propagation of the y-component of the nonlinear polarization which mixes the y- and z- components of the electric field at ω and which can generate a wave at the second harmonic that is polarized along the y-direction? What is velocity of the electromagnetic wave at frequency 2ω?

Solution From Eq. (10.21) we see that the term $2d_{15} E_y^{(\omega)} E_z^{(\omega)}$ mixes the two components of the wave at frequency ω to generate y-component of the nonlinear polarization at 2ω. Since the propagation constants of the components $E_y^{(\omega)}$ (which corresponds to an ordinary wave) and $E_z^{(\omega)}$ (which corresponds to an extraordinary wave) are $k_o^{(\omega)} = \frac{\omega}{c} n_o(\omega)$ and $k_e^{(\omega)} = \frac{\omega}{c} n_e(\omega)$, respectively, the nonlinear polarization term will have the factor $e^{i\left(\left[k_o^{(\omega)}+k_e^{(\omega)}\right]x-2\omega t\right)}$. Thus, the velocity of propagation of the nonlinear polarization at 2ω would be $v_{nl} = \frac{2\omega}{\left(k_o^{(\omega)}+k_e^{(\omega)}\right)} = \frac{2c}{n_o(\omega)+n_e(\omega)}$. The velocity of the electromagnetic wave at 2ω with its polarization along the y-direction (corresponding to an ordinary wave) that the nonlinear polarization is trying to generate will be $v_{em} = \frac{2\omega}{k_o^{(2\omega)}} = \frac{c}{n_o(2\omega)}$ which, in general, is different from the velocity of the nonlinear polarization at the frequency 2ω.

As an example if the fundamental wave corresponds to a wavelength of 1560 nm and the second harmonic to 780 nm, then considering lithium niobate, the y-component will correspond to ordinary polarization and the z-component to extraordinary polarization. Thus, we have $n_o(\omega) = 2.2108$, $n_e(\omega) = 2.1373$ and $n_o(2\omega) = 2.2580$. Thus, the velocity of the nonlinear polarization is $v_{nl} = \frac{c}{2.1741}$ and the velocity of the electromagnetic wave that it is generating will be $v_{em} = \frac{c}{2.2580}$. Note that these two velocities are not equal. We will show in Chap. 11 that this difference in the two velocities leads to a drastic reduction in the efficiency of second harmonic generation.

Exercise 10.2 Consider SHG in lithium niobate with the waves propagating along the optic axis. If the incident wave at the fundamental frequency is polarized along the x-direction, obtain the nonlinear polarization that is generated at the second harmonic.

Solution Since the wave at ω is polarized along the x-direction, we have

$$E_x^{(\omega)} \neq 0; \quad E_y^{(\omega)} = 0; \quad E_z^{(\omega)} = 0$$

Hence

$$\begin{pmatrix} \widetilde{P}_x^{(2\omega)} \\ \widetilde{P}_y^{(2\omega)} \\ \widetilde{P}_z^{(2\omega)} \end{pmatrix} = \epsilon_0 \begin{pmatrix} 0 & 0 & 0 & 0 & d_{15} & -d_{22} \\ -d_{22} & d_{22} & 0 & d_{15} & 0 & 0 \\ d_{31} & d_{31} & d_{33} & 0 & 0 & 0 \end{pmatrix} \begin{pmatrix} E_x^{(\omega)} E_x^{(\omega)} \\ 0 \\ 0 \\ 0 \\ 0 \\ 0 \end{pmatrix}$$

$$= \begin{pmatrix} 0 \\ -d_{22} E_x^{(\omega)} E_x^{(\omega)} \\ d_{31} E_x^{(\omega)} E_x^{(\omega)} \end{pmatrix}$$

Thus, the nonlinear polarization at 2ω has components along the y-direction and $z-$ direction.

Exercise 10.3 Consider second harmonic generation in KDP (potassium dihydrogen phosphate) which is a uniaxial crystal. The d tensor of KDP is given by

$$\bar{\bar{d}} = \begin{pmatrix} 0 & 0 & 0 & d_{14} & 0 & 0 \\ 0 & 0 & 0 & 0 & d_{25} & 0 \\ 0 & 0 & 0 & 0 & 0 & d_{36} \end{pmatrix} \tag{10.22}$$

Assume that the fundamental is an ordinary wave polarized along the y-direction and propagating along the x-direction. Obtain the nonlinear polarization generated in the medium.

Solution The electric field components of an ordinary wave polarized along the y-direction and propagating along the x-direction are given by

$$E_x^{(\omega)} = 0; \quad E_y^{(\omega)} \neq 0; \quad E_z^{(\omega)} = 0 \tag{10.23}$$

The corresponding nonlinear polarization generated is

$$\begin{pmatrix} \widetilde{P}_x^{(2\omega)} \\ \widetilde{P}_y^{(2\omega)} \\ \widetilde{P}_z^{(2\omega)} \end{pmatrix} = \epsilon_0 \begin{pmatrix} 0 & 0 & 0 & d_{14} & 0 & 0 \\ 0 & 0 & 0 & 0 & d_{25} & 0 \\ 0 & 0 & 0 & 0 & 0 & d_{36} \end{pmatrix} \begin{pmatrix} 0 \\ E_y^{(\omega)} E_y^{(\omega)} \\ 0 \\ 0 \\ 0 \\ 0 \end{pmatrix} = \begin{pmatrix} 0 \\ 0 \\ 0 \end{pmatrix} \tag{10.24}$$

which implies that in spite of the crystal being nonlinear, there is no nonlinear polarization generated in the medium due to the particular chosen propagation direction and polarization. Thus, there will be no generation of the second harmonic in such a situation.

We saw in Exercise 10.3 that for the particular medium, propagation direction and polarization state of the incident wave, there was no nonlinear polarization generated

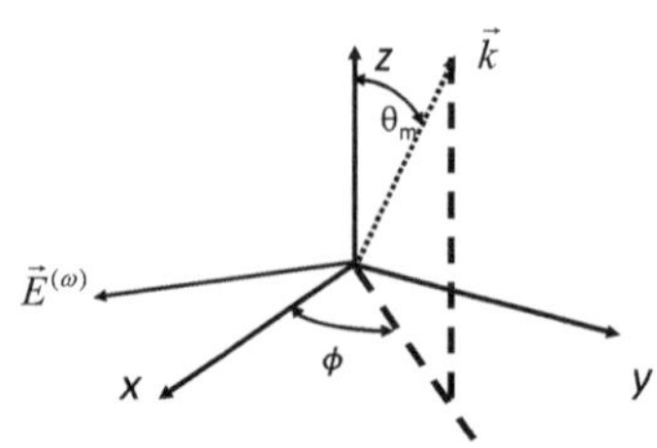

Fig. 10.3 Second harmonic generation in KDP with the incident wave at frequency ω propagating along the direction shown in the figure with its electric field lying in the $x - y$ plane

in the medium. In this example, in order to generate a nonlinear polarization we would need at least two nonzero Cartesian components of the electric field due to the form of the d tensor of KDP. Thus, if we still assume that the fundamental wave is an ordinary wave but now propagating along a direction with polar angles (θ, ϕ) as shown in Fig. 10.3, then we have

$$E_x^{(\omega)} \neq 0; \quad E_y^{(\omega)} \neq 0; \quad E_z^{(\omega)} = 0 \tag{10.25}$$

and the nonlinear polarization would be given by

$$\begin{pmatrix} \widetilde{P}_x^{(2\omega)} \\ \widetilde{P}_y^{(2\omega)} \\ \widetilde{P}_z^{(2\omega)} \end{pmatrix} = \epsilon_0 \begin{pmatrix} 0 & 0 & 0 & d_{14} & 0 & 0 \\ 0 & 0 & 0 & 0 & d_{25} & 0 \\ 0 & 0 & 0 & 0 & 0 & d_{36} \end{pmatrix} \begin{pmatrix} E_x^{(\omega)} E_x^{(\omega)} \\ E_y^{(\omega)} E_y^{(\omega)} \\ 0 \\ 0 \\ 0 \\ 2E_x^{(\omega)} E_y^{(\omega)} \end{pmatrix}$$

$$= \begin{pmatrix} 0 \\ 0 \\ 2\epsilon_0 d_{36} E_x^{(\omega)} E_y^{(\omega)} \end{pmatrix} \tag{10.26}$$

Thus,

$$\vec{\widetilde{P}}^{(2\omega)} = \hat{z} 2\epsilon_0 d_{36} E_x^{(\omega)} E_y^{(\omega)}$$

showing that there is now nonlinear polarization generated with a z-component. This nonlinear polarization can generate an electromagnetic wave at the second harmonic which will have extraordinary polarization provided phase matching condition is satisfied (see Chap. 11).

Exercise 10.4 Consider SHG in KDP whose d-tensor is given in Eq. (10.22). The fundamental wavelength is 800 nm at which $n_o = 1.5015$ and $n_e = 1.4633$. The wave at ω propagates along a direction making an angle of 45° with the y-axis and with its $\vec{k}$ lying in the $y-z$ plane and polarized in the $y - z$ plane. (a) Obtain the nonlinear polarization generated at 2ω. (b) Which corresponding effective nonlinear coefficient will be responsible for the SHG?

Solution The polarization of the wave lies in the plane containing the propagation vector and the optic axis of the crystal and hence represents an extraordinary wave. The $\vec{D}$ of the wave will be perpendicular to the $\vec{k}$ of the wave and hence will make an angle of 45° with the optic axis. If n_o and n_e are the ordinary and extraordinary refractive indices of KDP at the given frequency, then we have (see Chap. 2)

$$\frac{D_y}{D_z} = \frac{n_o^2 E_y}{n_e^2 E_z} = -1$$

which gives $\frac{E_y}{E_z} \approx -0.95$. Hence, the $\vec{E}$ of the wave makes an angle of ≈43.52° with the optic axis. Hence, the wave at ω will have the following electric field components:

$$E_x^{(\omega)} = 0; E_y^{(\omega)} \approx -0.725E^{(\omega)}; E_z^{(\omega)} \approx 0.689E^{(\omega)}$$

Hence, we have for the nonlinear polarization

$$\begin{pmatrix} \widetilde{P}_x^{(2\omega)} \\ \widetilde{P}_y^{(2\omega)} \\ \widetilde{P}_z^{(2\omega)} \end{pmatrix} = \epsilon_0 \begin{pmatrix} 0 & 0 & 0 & d_{14} & 0 & 0 \\ 0 & 0 & 0 & 0 & d_{25} & 0 \\ 0 & 0 & 0 & 0 & 0 & d_{36} \end{pmatrix} \begin{pmatrix} 0 \\ E_y^{(\omega)2} \\ E_z^{(\omega)2} \\ 2E_y^{(\omega)}E_z^{(\omega)} \\ 0 \\ 0 \end{pmatrix}$$

$$= \begin{pmatrix} 2d_{14}E_y^{(\omega)}E_z^{(\omega)} \\ 0 \\ 0 \end{pmatrix}$$

Thus, the generated nonlinear polarization has only an $x-$ component. The nonlinear coefficient d_{14} is responsible for the second harmonic generation.

Exercise 10.5 Assume second harmonic generation by an ordinary plane wave at frequency ω propagating along the x-direction in lithium niobate. (a) What will be the nonlinear polarization component that can generate an extraordinary wave at the second harmonic? (b) What will be the velocities of the nonlinear polarization and the electromagnetic wave at the second harmonic? (c) What is the required condition under which these two velocities will become equal to each other?

Solution

(a) An ordinary wave propagating along the x-direction will have its polarization along the y-direction. Hence

$$E_x^{(\omega)} = 0; E_y^{(\omega)} \neq 0; E_z^{(\omega)} = 0$$

The component of the nonlinear polarization that will be responsible for the generation of the second harmonic wave having extraordinary polarization will be the z-component. Thus, we have from Eq. (10.13)

$$\tilde{P}_z^{(2\omega)} = \epsilon_0 d_{zyy} E_y^{(\omega)} E_y^{(\omega)} = \epsilon_0 d_{32} E_y^{(\omega)} E_y^{(\omega)}$$

(b) Since the incident wave at ω is an ordinary wave, its propagation constant will be given by $\frac{\omega}{c} n_o(\omega)$. Thus, the velocity of the nonlinear polarization at the second harmonic will be (see Eq. 10.12)

$$v_{nl} = \frac{c}{n_o(\omega)}$$

The velocity of the extraordinary wave at 2ω will be

$$v_{em} = \frac{2\omega}{k_e(2\omega)} = \frac{2\omega}{\left(\frac{2\omega}{c} n_e(2\omega)\right)} = \frac{c}{n_e(2\omega)}$$

(c) For the velocities to be equal, we must have $n_e(2\omega) = n_o(\omega)$.

10.4 Nonlinear Polarization at ω

In the above discussion, we evaluated the nonlinear polarization generated at the second harmonic frequency 2ω from an incident wave at frequency ω. The generated second harmonic wave can mix nonlinearly with the wave at frequency ω and generate a nonlinear polarization at frequency ω through the process of difference frequency generation. Note that this is in addition to the linear polarization at frequency ω that is generated within the medium due to the incident wave. Thus, at frequency ω we have two contributions to the polarization; the linear part due to linear susceptibility and the nonlinear part due to the nonlinear susceptibility. We will now evaluate the nonlinear polarization that is generated at frequency ω.

For this we substitute Eqs. (10.11a) and (10.11b) in Eq. (10.6) and obtain the terms which vary at a frequency ω to obtain the following expression for the nonlinear polarization at frequency ω:

$$P_j^{(\omega)} = \frac{1}{2}\left(2\epsilon_0 d_{jkl} E_k^{(2\omega)} E_l^{(\omega)*} e^{i([k_2 - k_1]z - \omega t)} + c.c.\right) = \frac{1}{2}\left(\tilde{P}_j^{(\omega)} e^{i([k_2 - k_1]z - \omega t)} + c.c.\right) \tag{10.27}$$

where

$$\tilde{P}_j^{(\omega)} = 2\epsilon_0 d_{jkl} E_k^{(2\omega)} E_l^{(\omega)*} \tag{10.28}$$

is the jth component of the complex nonlinear polarization at frequency ω.

Thus, mixing of the generated electromagnetic wave at frequency 2ω and the incident wave at frequency ω gives rise to the nonlinear polarization at frequency ω. In fact, the generated nonlinear polarization at frequency ω is actually the difference frequency between waves at frequency 2ω and at ω (see Chap. 12). Thus, the wave at frequency ω feeds the wave at frequency 2ω through the nonlinear polarization at frequency 2ω and similarly the wave at frequency 2ω mixes with the wave at frequency ω and generates nonlinear polarization at frequency ω. This process could lead to either increase in the power or decrease in the power of the electromagnetic wave depending on the relative phase between the nonlinear polarization and the electromagnetic wave. We will use these discussions in Chap. 11 on second harmonic generation as well as in Chaps. 12 and 13 on sum and difference frequency generation and parametric amplification.

Example 10.3 We revert back to Example 10.1 and obtain the nonlinear polarization at the frequency ω. In the example, an extraordinary wave at frequency ω generates nonlinear polarization at 2ω and the generated nonlinear polarization has only the extraordinary component. Hence, this nonlinear polarization will generate an electromagnetic wave at frequency 2ω which will have an extraordinary polarization, that is, the electromagnetic wave at frequency 2ω will have only its z-component. Thus, in this example both the wave at frequency ω and the wave at frequency 2ω have only the z-component; the x- and y- components of the electric fields are zero. Thus, in Eq. (10.28) the only nonzero terms will have j and k equal to 3 (i.e., z-component). Hence, we have

$$\tilde{P}_1^{(\omega)} = \tilde{P}_x^{(\omega)} = 2\epsilon_0 d_{133} E_3^{(2\omega)} E_3^{(\omega)*} = 2\epsilon_0 d_{13} E_3^{(2\omega)} E_3^{(\omega)*} = 0 \tag{10.29}$$

since in lithium niobate $d_{13} = 0$. Similarly since $d_{23} = 0$ we will have $\tilde{P}_y^{(\omega)} = 0$. The only nonzero component of the nonlinear polarization at frequency ω will be the $z-$ component:

$$\tilde{P}_3^{(\omega)} = \tilde{P}_z^{(\omega)} = 2\epsilon_0 d_{333} E_3^{(2\omega)} E_3^{(\omega)*} = 2\epsilon_0 d_{33} E_3^{(2\omega)} E_3^{(\omega)*} \tag{10.30}$$

Hence, the nonlinear polarization at frequency ω has only the extraordinary component. This polarization term will add to the linear polarization at frequency ω to drive the electromagnetic wave at frequency ω.

10.5 Nonlinear Polarization in Sum and Difference Frequency Generation

In Chap. 12, we will study sum and difference frequency generation. In this case, we will have three interacting waves with different frequencies. The frequencies are referred to as pump with frequency ω_p, signal with frequency ω_s, and idler with frequency ω_i. The three frequencies satisfy the following condition:

$$\omega_p = \omega_s + \omega_i \tag{10.31}$$

Sum frequency generation corresponds to the generation of nonlinear polarization at frequency ω_p from waves incident at frequencies ω_s and ω_i. Difference frequency generation corresponds to the case when the wave incident at frequency ω_p mixes with waves at frequency ω_s or ω_i to generate nonlinear polarization at the difference frequency ω_i or ω_s.

To analyze this process, we will consider the presence of waves at all three frequencies through the medium. The corresponding k^{th} component of the electric fields is described by

$$E_k(\omega_p, z) = \frac{1}{2}\left(E_k^{(\omega_p)} e^{i(k_p z - \omega_p t)} + c.c\right) \tag{10.32}$$

$$E_k(\omega_s, z) = \frac{1}{2}\left(E_k^{(\omega_s)} e^{i(k_s z - \omega_s t)} + c.c\right) \tag{10.33}$$

$$E_k(\omega_i, z) = \frac{1}{2}\left(E_k^{(\omega_i)} e^{i(k_i z - \omega_i t)} + c.c\right) \tag{10.34}$$

where k_p, k_s, and k_i are the propagation constants at frequencies ω_p, ω_s, and ω_i, respectively. The kth and l^{th} components of the total electric field within the medium will be given by

$$E_k = \frac{1}{2}\left(E_k^{(\omega_p)} e^{i(k_p z - \omega_p t)} + c.c\right) + \frac{1}{2}\left(E_k^{(\omega_s)} e^{i(k_s z - \omega_s t)} + c.c\right) + \frac{1}{2}\left(E_k^{(\omega_i)} e^{i(k_i z - \omega_i t)} + c.c\right) \tag{10.35}$$

$$E_l = \frac{1}{2}\left(E_l^{(\omega_p)} e^{i(k_p z - \omega_p t)} + c.c\right) + \frac{1}{2}\left(E_l^{(\omega_s)} e^{i(k_s z - \omega_s t)} + c.c\right) + \frac{1}{2}\left(E_l^{(\omega_i)} e^{i(k_i z - \omega_i t)} + c.c\right) \tag{10.36}$$

where again for simplicity we are assuming that the kth and l^{th} components of the electric fields have the same propagation constant.

For compactness, we define

$$\theta_p = (k_p z - \omega_p t) \tag{10.37a}$$

$$\theta_s = (k_s z - \omega_s t) \tag{10.37b}$$

$$\theta_i = (k_i z - \omega_i t) \tag{10.37c}$$

and note that

$$\theta_s + \theta_i = (k_s + k_i)z - (\omega_s + \omega_i)t = (k_s + k_i)z - \omega_p t \tag{10.38a}$$

Similarly

$$\theta_p - \theta_i = (k_p - k_i)z - \omega_s t \tag{10.38b}$$

and

$$\theta_p - \theta_s = (k_p - k_s)z - \omega_i t \tag{10.38c}$$

Substituting in Eq. (10.6), we can obtain the nonlinear polarization generated in the medium at the pump, signal, and idler frequencies. Thus,

$$\begin{aligned} P_j^{(NL)} &= 2\epsilon_0 d_{jkl} E_k E_l \\ &= 2\epsilon_0 d_{jkl}\left[\frac{1}{2}\left(E_k^{(\omega_p)} e^{i\theta_p} + c.c\right) + \frac{1}{2}\left(E_k^{(\omega_s)} e^{i\theta_s} + c.c\right) + \frac{1}{2}\left(E_k^{(\omega_i)} e^{i\theta_i} + c.c\right)\right] \\ &\times \left[\frac{1}{2}\left(E_l^{(\omega_p)} e^{i\theta_p} + c.c\right) + \frac{1}{2}\left(E_l^{(\omega_s)} e^{i\theta_s} + c.c\right) + \frac{1}{2}\left(E_l^{(\omega_i)} e^{i\theta_i} + c.c\right)\right] \end{aligned} \tag{10.39}$$

To obtain the nonlinear polarization generated at the pump, signal, and idler frequencies, we need to collect terms having frequencies ω_p, ω_s, and ω_i, respectively, from the right-hand side of Eq. (10.39) for which we can use Eqs. (10.37a)–(10.37c). These can be obtained easily and are given by

$$P_j^{(NL)}(\omega_p) = \frac{1}{2}\epsilon_0 d_{jkl}\left[\left(E_k^{(\omega_s)} E_l^{(\omega_i)} + E_k^{(\omega_i)} E_l^{(\omega_s)}\right) e^{i[(k_s+k_i)z - \omega_p t]} + c.c\right] \tag{10.40}$$

$$P_j^{(NL)}(\omega_s) = \frac{1}{2}\epsilon_0 d_{jkl}\left[\left(E_k^{(\omega_p)} E_l^{(\omega_i)*} + E_k^{(\omega_i)*} E_l^{(\omega_p)}\right) e^{[(k_p - k_i)z - \omega_s t]} + c.c\right] \tag{10.41}$$

$$P_j^{(NL)}(\omega_i) = \frac{1}{2}\epsilon_0 d_{jkl}\left[\left(E_k^{(\omega_p)} E_l^{(\omega_s)*} + E_k^{(\omega_s)*} E_l^{(\omega_p)}\right) e^{[(k_p - k_s)z - \omega_i t]} + c.c\right] \tag{10.42}$$

Using the fact that $d_{jkl} = d_{jlk}$, we define:

$$\tilde{P}_j^{(\omega_p)} = 2\epsilon_0 d_{jkl} E_k^{(\omega_s)} E_l^{(\omega_i)} \tag{10.43}$$

$$\tilde{P}_j^{(\omega_s)} = 2\epsilon_0 d_{jkl} E_k^{(\omega_p)} E_l^{(\omega_i)*} \tag{10.44}$$

$$\tilde{P}_j^{(\omega_i)} = 2\epsilon_0 d_{jkl} E_k^{(\omega_p)} E_l^{(\omega_s)*} \tag{10.45}$$

and obtain

$$P_j^{(NL)}(\omega_p) = \frac{1}{2}\left[\tilde{P}_j^{(\omega_p)} e^{i\left[(k_s+k_i)z-\omega_p t\right]} + c.c\right] \tag{10.46}$$

$$P_j^{(NL)}(\omega_s) = \frac{1}{2}\left[\tilde{P}_j^{(\omega_s)} e^{\left[(k_p-k_i)z-\omega_s t\right]} + c.c\right] \tag{10.47}$$

$$P_j^{(NL)}(\omega_i) = \frac{1}{2}\left[\tilde{P}_j^{(\omega_i)} e^{\left[(k_p-k_s)z-\omega_i t\right]} + c.c\right] \tag{10.48}$$

These equations for nonlinear polarization are very useful in obtaining the effective nonlinear coefficient involved in the nonlinear interaction process which depends on the crystal through its nonlinear tensor, the propagation direction of the waves, and the polarization state of the interacting waves.

Exercise 10.6 In a lithium niobate crystal consider an input consisting of waves at two frequencies: one of frequency ω_s having extraordinary polarization and the other of frequency ω_i having ordinary polarization. The waves are propagating along the y-axis of the crystal. Obtain the nonlinear polarization that will lead to the generation of an electromagnetic wave having ordinary polarization at the sum frequency $\omega_s + \omega_i = \omega_p$. What will be the velocity of the corresponding nonlinear polarization?

Solution The input consists of the following fields: $E_z^{(\omega_s)}$ (extraordinary polarization) and $E_x^{(\omega_i)}$ (ordinary polarization). The nonlinear polarization that will lead to the generation of the ordinary wave at the sum frequency will be $\tilde{P}_x^{(\omega_p)}$, the x-component of the nonlinear polarization, which will be (see Eq. 10.43)

$$\tilde{P}_x^{(\omega_p)} = 2\epsilon_0 d_{xkl} E_k^{(\omega_s)} E_l^{(\omega_i)}$$

Now, the only nonzero elements in the first row of the d tensor of lithium niobate are $d_{15} = d_{xxz} = d_{xzx}$ and $d_{16} = d_{xxy} = d_{xyx}$. Thus, we will have

$$\tilde{P}_x^{(\omega_p)} = 2\epsilon_0 d_{xxz} E_x^{(\omega_s)} E_z^{(\omega_i)} + 2\epsilon_0 d_{xzx} E_z^{(\omega_s)} E_x^{(\omega_i)} + 2\epsilon_0 d_{xxy} E_x^{(\omega_s)} E_y^{(\omega_i)}$$
$$+ 2\epsilon_0 d_{xyx} E_y^{(\omega_s)} E_x^{(\omega_i)}$$

Since the only nonzero components of the incident fields are $E_z^{(\omega_s)}$ and $E_x^{(\omega_i)}$, we have

$$\tilde{P}_x^{(\omega_p)} = 2\epsilon_0 d_{xxz} E_z^{(\omega_s)} E_x^{(\omega_i)} = 2\epsilon_0 d_{15} E_z^{(\omega_s)} E_x^{(\omega_i)}$$

The propagation constant of the field corresponding to $E_z^{(\omega_s)}$ will be $k_e(\omega_s)$ and that corresponding to $E_x^{(\omega_i)}$ will be $k_o(\omega_i)$. Hence, the velocity of the nonlinear polarization will be

$$v_{pol} = \frac{\omega_p}{k_e(\omega_s) + k_o(\omega_i)}$$

Exercise 10.7 Consider the propagation of waves along the x-direction with extraordinary polarization at frequencies ω_p and ω_s in lithium niobate. Which components of the nonlinear polarization at the frequency $\omega_i = \omega_p - \omega_s$ will be nonzero? What will be the velocity of the corresponding nonlinear polarization?

Solution Since the waves propagating along the x-direction at ω_p and ω_s have extraordinary polarizations, the nonzero electric field components at these two frequencies are $E_z^{(\omega_p)}$ and $E_z^{(\omega_s)}$. From Eq. (10.45), we have

$$\tilde{P}_x^{(\omega_i)} = 2\epsilon_0 d_{13} E_z^{(\omega_p)} E_z^{(\omega_s)*} = 0$$
$$\tilde{P}_y^{(\omega_i)} = 2\epsilon_0 d_{23} E_z^{(\omega_p)} E_z^{(\omega_s)*} = 0$$
$$\tilde{P}_z^{(\omega_i)} = 2\epsilon_0 d_{33} E_z^{(\omega_p)} E_z^{(\omega_s)*}$$

since $d_{13} = 0$ and $d_{23} = 0$. Hence, the nonlinear polarization at ω_i has only the z-component which will generate electromagnetic wave at the frequency ω_i having extraordinary polarization.

The velocity of the nonlinear polarization would be

$$v_{nl} = \frac{\omega_i}{k_e(\omega_p) - k_e(\omega_s)}$$

Exercise 10.8 Consider the propagation along the y-direction in KTP of an x-polarized wave at frequency ω_p and a z-polarized wave at frequency ω_s. What will be the nonlinear polarization generated at the difference frequency $\omega_i = \omega_p - \omega_s$?

Solution From Eq. (10.45), we have using the d tensor of KTP (see Eq. 10.17):

$$\tilde{P}_x^{(\omega_i)} = 2\epsilon_0 d_{xxz} E_x^{(\omega_p)} E_z^{(\omega_s)*} + 2\epsilon_0 d_{xzx} E_z^{(\omega_p)} E_x^{(\omega_s)*} = 2\epsilon_0 d_{xxz} E_x^{(\omega_p)} E_z^{(\omega_s)*}$$

$$\tilde{P}_y^{(\omega_i)} = 2\epsilon_0 d_{yyz} E_y^{(\omega_p)} E_z^{(\omega_s)*} + 2\epsilon_0 d_{yzy} E_z^{(\omega_p)} E_y^{(\omega_s)*} = 0$$

$$\tilde{P}_z^{(\omega_i)} = 2\epsilon_0 d_{zxx} E_x^{(\omega_p)} E_x^{(\omega_s)*} + 2\epsilon_0 d_{zyy} E_y^{(\omega_p)} E_y^{(\omega_s)*} + 2\epsilon_0 d_{zzz} E_z^{(\omega_p)} E_z^{(\omega_s)*} = 0$$

Hence, the nonlinear polarization at ω_i has only the x-component and hence can produce an idler polarized along the x-direction provided the phase matching condition is satisfied (see Chap. 12).

Problems

10.1. What are all the possible wavelengths that can be generated using inputs at wavelengths of 800 and 1200 nm in a nonlinear medium?

10.2. A constant electric field E_0 (zero frequency) is applied along the z-axis of $LiNbO_3$. Obtain the z-component of the generated nonlinear polarization at ω when an extraordinary wave at frequency ω and propagation constant k_e is propagating through the medium along the y-direction. Interpret the result. Thus, the refractive index of a medium can be changed by applying an external electric field; this is the linear electro optic effect or Pockels effect (see Chap. 6).

10.3. If the second harmonic generation takes place with electric field components at frequency ω having different polarization states, then we need to use different propagation constant values in Eqs. (10.11a) and (10.11b). In such a case, these equations become

$$E_k = \frac{1}{2}\left(E_k^{(\omega)} e^{i(k_{1k}z-\omega t)} + c.c\right) + \frac{1}{2}\left(E_k^{(2\omega)} e^{i(k_{2k}z-2\omega t)} + c.c\right)$$

$$E_l = \frac{1}{2}\left(E_l^{(\omega)} e^{i(k_{1l}z-\omega t)} + c.c\right) + \frac{1}{2}\left(E_l^{(2\omega)} e^{i(k_{2l}z-2\omega t)} + c.c\right)$$

with k_{1k} and k_{1l} representing the propagation constants of the field at frequency ω corresponding to the kth and l^{th} components of the electric field and similarly for the frequency 2ω. Obtain the corresponding nonlinear polarization generated at frequency 2ω and at frequency ω.

10.4. From Eq. (10.39) obtain expressions for the nonlinear polarization given by Eqs. (10.40)–(10.42).

10.5. Consider a waveguide fabricated in KTP which is described by the nonlinear tensor given in Eq. (10.17). The propagation direction in the waveguide is along x-axis. Assume that at the input we have the fundamental frequency ω with its electric field making an angle θ with the z- axis and lying in the $y-z$ plane. Obtain the nonlinear polarization generated at 2ω. For what value of θ will the component of nonlinear polarization along the y-direction be maximum?

10.6. Consider second harmonic generation in KTP with the incident light at ω polarized along y-direction and propagating along the x-direction. The nonlinear tensor of KTP is given by Eq. (10.17). Starting from the equation for the nonlinear polarization

(a) Obtain the components of the nonlinear polarization generated at 2ω.
(b) What will be the state of polarization of the generated second harmonic wave?

10.7. An optical fiber made of silica exhibits third-order nonlinearity in which the nonlinear polarization is given by

$$P^{(NL)} = \epsilon_0 \chi^{(3)} E^3$$

where E is the total electric field. Assuming the incidence of waves at frequencies ω_1 and ω_2 (with propagation constants β_1 and β_2) obtain an expression for the nonlinear polarization generated at frequency ω_1.

10.8. The extraordinary refractive index of lithium niobate at wavelengths of 800 nm and 400 nm are given by 2.1755 and 2.3321, respectively. Consider second harmonic generation with both frequencies having extraordinary polarization. What are the velocities of the nonlinear polarization generated at the second harmonic frequency and the electromagnetic wave at frequency 2ω?

10.9. The d matrix of potassium niobate (a biaxial crystal) is given by

$$d = \begin{pmatrix} 0 & 0 & 0 & 0 & d_{15} & 0 \\ 0 & 0 & 0 & d_{24} & 0 & 0 \\ d_{31} & d_{32} & d_{33} & 0 & 0 & 0 \end{pmatrix} \quad n_x = 2.12, \quad n_y = 2.22, \quad n_z = 2.26$$

Consider propagation along the y-direction in the medium with the wave at ω having both x and z components. Obtain the nonlinear polarization that will be generated 2ω

10.10. Consider lithium niobate with two adjacent regions with their principal axes as shown in Fig. 10.4. Express the d tensor of the two regions in the

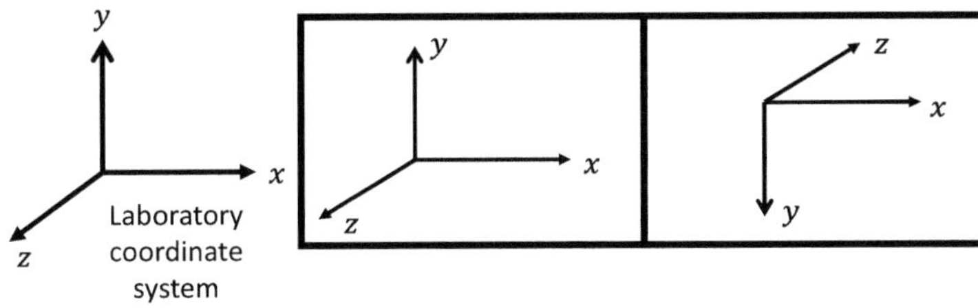

Fig. 10.4 Figure for Problem 10.10

laboratory coordinate system shown. Determine which of the elements of the d tensor change sign between the adjacent regions and which ones do not.

10.11. Consider second harmonic generation in lithium niobate for propagation along the y-directon with the wave at ω being an ordinary wave and that at 2ω being an extraordinary wave. Obtain the component of the nonlinear polarization that will be responsible for the generation of the second harmonic.

10.12. Consider difference frequency generation in lithium niobate with the pump propagating along the y- axis of the crystal as an ordinary wave. The signal and idler are assumed to be orthogonally polarized and propagating along the same direction. Obtain expressions for the nonlinear polarization $P_j^{(NL)}$ at the pump, signal, and idler frequencies when (a) signal has ordinary polarization and idler has extraordinary polarization and (b) signal has extraordinary polarization and idler has ordinary polarization. (c) In both cases, obtain the velocities of the nonlinear polarization at the pump, signal, and idler frequencies.

10.13. Consider second harmonic generation in lithium niobate. If the wave at ω is an ordinary wave with its $\vec{k}$ lying in the $x-y$ plane and making an angle of ϕ with the x-axis, obtain the components of the nonlinear polarization generated at 2ω.

10.14. A pump wave having extraordinary polarization and frequency ω_p propagates along the y-axis of lithium niobate. Which element of the d- tensor would be responsible for generating a difference frequency when a signal having ordinary polarization is propagating along the same direction?

10.15. Consider sum frequency generation with a signal at frequency ω_s propagating along the $+x$ direction and an idler wave propagating along the $-x$ direction in lithium niobate. All waves are assumed to have extraordinary polarization. What is the velocity of the nonlinear polarization produced at the sum frequency propagating along the $+x$ direction?

Chapter 11
Second Harmonic Generation

Abstract This chapter discusses the phenomenon of second harmonic generation (SHG) in optical waveguides and the basic concepts of phase matching and quasi phase matching. Coupled mode equations are derived leading to expression for efficiency of SHG. Also discussed are bandwidths of second harmonic generation and the effective nonlinear coefficient that is responsible for the second harmonic generation.

11.1 Introduction

Ever since the first demonstration of SHG was made by Franken and his co-workers in 1961 by focusing a 3 kW ruby laser pulse ($\lambda_0 = 6943$ Å) on a quartz crystal, the field of nonlinear optics has grown enormously and today finds applications in many branches of science and technology. In Chap. 10, we discussed and obtained expressions for nonlinear polarization generated by the second-order nonlinearity in the medium. This nonlinear polarization is responsible for second harmonic generation, sum and difference frequency generation, parametric amplification, etc.

In this chapter, we will discuss in greater detail the generation of second harmonic electromagnetic wave at frequency 2ω from an incident electromagnetic wave at frequency ω. For simplicity, we will present a scalar analysis of the interaction. In Sect. 11.7, we will briefly discuss how to incorporate the vector nature of the fields and the nonlinear polarization in the analysis.

Let us begin by a simple argument as to how the incident light at frequency ω can lead to the generation of light at 2ω.

Let us consider an incident wave at frequency ω propagating along the z -direction in a nonlinear medium the electric field of which is given by

$$E = E_0 \cos(kz - \omega t + \phi) \tag{11.1}$$

where k is the propagation constant of the wave and ω is the frequency of the wave. In such a case for a linear medium, we have the following expression for the electric polarization generated in the medium

K. Thyagarajan, *Integrated Quantum Photonics*, Graduate Texts in Physics,
https://doi.org/10.1007/978-3-031-85728-7_11

$$P = \epsilon_0 \chi E_0 \cos(kz - \omega t + \phi) \tag{11.2}$$

where χ is the linear susceptibility. The polarization is an oscillatory function of time having only one component of frequency which is the same as that of the input electromagnetic wave. Since polarization is nothing but dipole moment per unit volume, this implies that the dipoles of the medium are oscillating at the frequency corresponding to the incident wave. These oscillating dipoles representing the polarization generate electromagnetic waves at the same frequency as the propagating wave which adds to the wave that is incident from outside the medium.

Now in the presence of second-order nonlinearity, we can write for the polarization as (see Chap. 10)

$$P = \epsilon_0 \chi E + \epsilon_0 \chi^{(2)} E^2 \tag{11.3}$$

where the second term on the right hand side is the nonlinear term and again for simplicity we have written a scalar equation and $\chi^{(2)}$ represents the effective second-order nonlinear susceptibility of the medium (see Chap. 10). Now, if we assume an incident wave as given by Eq. (11.1), then the total polarization will be given by

$$P = \epsilon_0 \chi E_0 \cos(kz - \omega t + \phi) + \epsilon_0 \chi^{(2)} E_0^2 \cos^2(kz - \omega t + \phi)$$

which can be written as

$$P = \epsilon_0 \chi E_0 \cos(kz - \omega t + \phi) + \frac{\epsilon_0}{2} \chi^{(2)} E_0^2 + \frac{\epsilon_0}{2} \chi^{(2)} E_0^2 \cos\ 2(kz - \omega t + \phi) \tag{11.4}$$

Now we see that the polarization generated in the medium contains three terms, one varying at frequency ω, one varying at frequency 2ω, and a third term which is time independent. The time independent term leads to what is referred to as optical rectification and the term at frequency 2ω will lead to the generation of a wave at frequency 2ω which is the second harmonic of the incident wave. The term at frequency ω corresponds to the linear susceptibility term.

Similarly if two electromagnetic waves at frequencies ω_1 with propagation constant k_1 and ω_2 with propagation constant k_2 are incident in the medium, then the total incident electromagnetic field can be written as

$$E = E_1 \cos(k_1 z - \omega_1 t + \phi_1) + E_2 \cos(k_2 z - \omega_2 t + \phi_2) \tag{11.5}$$

Substituting in Eq. (11.3), we will find that the generated polarization has additional frequency terms such as $\omega_1 + \omega_2$ and $\omega_1 - \omega_2$; the former leads to what is referred to as sum frequency generation and the latter to difference frequency generation. These will be discussed in Chap. 12.

The various new frequency components of the polarization mean that the dipoles are oscillating not only at the frequency of the incident waves but also at new frequencies. The oscillating dipoles will then radiate at these new frequencies. Thus, in the presence of nonlinearity the medium tends to generate waves at new frequencies which are related to the frequencies of the incident waves.

Nonlinear optical effects play a very important role in today's science, engineering, and technology by providing us with coherent sources at various frequencies, parametric amplifiers, and a source of quantum states of light, etc. In Chap. 23, we will discuss the quantum mechanical treatment of the nonlinear interaction process and show how this leads to the generation of various nonclassical states of light such as entangled states, squeezed states, and so on.

Here we will discuss only the second-order nonlinear effects that are present in media that do not possess a center of inversion symmetry.

11.2 Nonlinear Polarization

In Chap. 10, we discussed about the nonlinear polarization at the second harmonic that is generated in the crystal. The components and the magnitudes of the nonlinear polarization depend on the crystal, the polarization state of the incident fundamental wave, and the propagation direction. We also obtained the different components of the nonlinear polarization using the vector nature of the electromagnetic wave as well as the polarization. In order to understand the basic principles of second harmonic generation, we will consider a scalar formulation with the understanding that the nonlinear coefficient is an effective nonlinear coefficient determined by the crystal, the propagation direction, and the polarization state of the incident wave. The vector aspects of the interaction process will be discussed in Sect. 11.7.

We consider the effects of second-order nonlinearity and write for the nonlinear polarization generated in a medium by the following scalar equation:

$$P^{(NL)} = 2\epsilon_0 d E^2 \tag{11.6}$$

where as in Chap. 10, we have replaced $\chi^{(2)}$ by $2d$, with d representing an effective nonlinear coefficient taking part in the interaction. The value of the effective nonlinear coefficient will depend on the medium, the direction of propagation, and the polarization states of the various interacting modes.

We saw in Chap. 10 that the nonlinearity in the medium is described by the nonlinearity tensor $\overline{\overline{d}}$ which relates the nonlinear polarization vector to the electric field vectors. In order to keep the fundamental analysis simpler, we are using a scalar relation here. Later in Sect. 11.7, we will discuss specific examples of how the effective nonlinear coefficient (which is related to the elements of the $\overline{\overline{d}}$ tensor) can be obtained for a given crystal, polarization states of the interacting fields, and the direction of propagation in the medium.

11.3 Wave Equation

In this section, we will derive the wave equation describing the propagation of waves in the presence of nonlinearity. We start with Maxwell's equations in a medium with no free charges and no free currents:

$$\nabla \times \vec{E} = -\frac{\partial \vec{B}}{\partial t} = -\mu_0 \frac{\partial \vec{H}}{\partial t} \tag{11.7}$$

$$\nabla \times \vec{H} = \frac{\partial \vec{D}}{\partial t} \tag{11.8}$$

$$\nabla . \vec{D} = 0 \tag{11.9}$$

$$\nabla . \vec{B} = 0 \tag{11.10}$$

Using

$$\vec{D} = \epsilon_0 \vec{E} + \vec{P} = \epsilon_0 \vec{E} + \epsilon_0 \chi \vec{E} + \vec{P}_{NL} = \epsilon \vec{E} + \vec{P}_{NL} \tag{11.11}$$

where $\vec{P}_{NL}$ is the nonlinear polarization and $\epsilon = \epsilon_0(1 + \chi)$ represents the permittivity of the medium. Taking curl of Eq. (11.7) and using Eq. (11.8), we have

$$\nabla \times \nabla \times \vec{E} = -\mu_0 \frac{\partial \nabla \times \vec{H}}{\partial t} = -\mu_0 \frac{\partial^2 \vec{D}}{\partial t^2}$$

Expanding the left hand side, we have

$$\nabla \left(\nabla \cdot \vec{E} \right) - \nabla^2 \vec{E} = -\mu_0 \frac{\partial^2 \vec{D}}{\partial t^2}$$

Assuming $\nabla \cdot \vec{E} \approx 0$, and substituting from Eq. (11.11), we obtain

$$\nabla^2 \vec{E} = \epsilon \mu_0 \frac{\partial^2 \vec{E}}{\partial t^2} + \mu_0 \frac{\partial^2 \vec{P}_{NL}}{\partial t^2} \tag{11.12}$$

The above equation represents the wave equation satisfied by the electromagnetic waves propagating in the medium. The additional term $\mu_0 \frac{\partial^2 \vec{P}_{NL}}{\partial t^2}$ on the right hand side of Eq. (11.12) is the nonlinear polarization term which acts as a source for the generation of electromagnetic waves at new frequencies.

Now, each frequency component propagating within the medium needs to satisfy the wave equation. Thus, if we consider second harmonic generation, then the wave equations needs to be satisfied for the frequencies ω and 2ω. The wave equation describing the wave at frequency ω is given by

$$\nabla^2 \vec{E}^{(\omega)} = \epsilon(\omega)\mu_0 \frac{\partial^2 \vec{E}^{(\omega)}}{\partial t^2} + \mu_0 \frac{\partial^2 \vec{P}_{NL}^{(\omega)}}{\partial t^2} \tag{11.13}$$

where $\vec{E}^{(\omega)}$ is the electric field of the fundamental frequency ω, $\vec{P}_{NL}^{(\omega)}$ is the nonlinear polarization at the frequency ω, and $\epsilon(\omega)$ is the permittivity of the medium at frequency ω. Similarly the second harmonic at frequency 2ω needs to satisfy the following equation:

$$\nabla^2 \vec{E}^{(2\omega)} = \epsilon(2\omega)\mu_0 \frac{\partial^2 \vec{E}^{(2\omega)}}{\partial t^2} + \mu_0 \frac{\partial^2 \vec{P}_{NL}^{(2\omega)}}{\partial t^2} \tag{11.14}$$

where $\vec{E}^{(2\omega)}$ is the electric field of the waves at frequency 2ω, $\epsilon(2\omega)$ represent the dielectric permittivity of the medium at frequency 2ω, and $\vec{P}_{NL}^{(2\omega)}$ is the nonlinear polarization at the frequency 2ω.

11.4 Second Harmonic Generation in Waveguides

In this section, we will discuss second harmonic generation in optical waveguides made in a medium possessing second-order nonlinearity. Examples include waveguides fabricated in lithium niobate, lithium tantalate, KTP, etc., in which high-quality optical waveguides with low propagation losses can be fabricated using different techniques. Waveguides fabricated in these crystals are used extensively in integrated quantum photonics for the generation and processing of nonclassical states of light.

We consider a channel optical waveguide (see Chap. 4) in which we launch light in a spatial mode having a frequency ω. This wave will generate nonlinear polarization at frequency 2ω which in turn will generate electromagnetic waves at frequency 2ω. Thus, we assume that at any point within the waveguide, light waves at the fundamental frequency ω and at the second harmonic frequency 2ω are propagating in guided modes along the z-direction.

The electric field of the propagating mode at frequency ω is written as

$$E^{(\omega)}(x,y,z,t) = \frac{1}{2}\left[A_1(z)\psi_1(x,y)e^{i(\beta_1 z - \omega t)} + c.c.\right] \tag{11.15}$$

where β_1 is the propagation constant of the mode at frequency ω and $\psi_1(x,y)$ is the corresponding transverse modal electric field distribution, and $A(z)$ is the z-

dependent complex amplitude of the mode due to the conversion of energy from the fundamental to the second harmonic. In the absence of nonlinearity, A_1 will be independent of z.

Similarly the electric field of the propagating mode at frequency 2ω is written as

$$E^{(2\omega)}(x,y,z,t) = \frac{1}{2}\left[A_2(z)\psi_2(x,y)e^{i(\beta_2 z - 2\omega t)} + c.c.\right] \tag{11.16}$$

where β_2 is the propagation constant of the mode at frequency 2ω and $\psi_2(x, y)$ is the transverse modal electric field distribution at the frequency 2ω, and $A_2(z)$ is the z-dependent complex amplitude of the mode due to coupling of power between the two frequencies.

Thus, the total electric field at any value of z is given by

$$\begin{aligned} E = E^{(\omega)}(x,y,z,t) + E^{(2\omega)}(x,y,z,t) = \frac{1}{2}\left[A_1(z)\psi_1(x,y)e^{i(\beta_1 z - \omega t)} + c.c.\right] \\ + \frac{1}{2}\left[A_2(z)\psi_2(x,y)e^{i(\beta_2 z - 2\omega t)} + c.c.\right] \end{aligned} \tag{11.17}$$

We will assume that the amplitudes $A_1(z)$ and $A_2(z)$ are slowly varying with distance so that we will be able to neglect their second derivative with respect to z.

When we discussed nonlinear polarization in Chap. 10, we were considering plane wave propagation in nonlinear media while here we are dealing with waveguides and thus the fields are expressed as modes of propagation with a transverse field distribution with a specific propagation constant which depends on the waveguide parameters. The modal field distributions will depend on the particular spatial modes being used for the interaction. This flexibility is sometimes used to achieve efficient conversion. Also when waves propagate in the bulk, they suffer from diffraction effects, while in the case of waveguides, due to confinement, the waves can be confined to small cross-section over much longer interaction lengths. This is one of the advantages of using waveguides for nonlinear optical effects.

Substituting from Eq. (11.17) in Eq. (11.6), we obtain

$$P^{(NL)} = \frac{1}{2}\epsilon_0 d\left[\left(A_1(z)\psi_1(x,y)e^{i(\beta_1 z - \omega t)} + c.c.\right) + \left(A_2(z)\psi_2(x,y)e^{i(\beta_2 z - 2\omega t)} + c.c.\right)\right]^2 \tag{11.18}$$

On opening the brackets, we will get various terms at different frequencies. From these terms, we find the terms at frequency 2ω and obtain the following expression for the nonlinear polarization at 2ω:

$$P_{NL}^{(2\omega)} = \frac{1}{2}\epsilon_0\left(dA_1^2(z)\psi_1^2(x,y)e^{2i(\beta_1 z - \omega t)} + c.c.\right) \tag{11.19}$$

It is interesting to note that the nonlinear polarization is also propagating in the form of a wave. The phase velocity of this wave is $\frac{2\omega}{2\beta_1} = \frac{\omega}{\beta_1}$ which is also the phase velocity of the electromagnetic wave at frequency ω. At the same time note that the phase velocity of the propagating electromagnetic mode at frequency 2ω is $\frac{2\omega}{\beta_2}$ (see Eq. 11.16) which, in general, is different from the phase velocity of the nonlinear polarization that is responsible for its creation. This difference leads to drop in efficiency of the interaction process.

Similarly the nonlinear polarization at frequency ω would be

$$P_{NL}^{(\omega)} = \frac{1}{2}\epsilon_0 d\left(2A_2(z)A_1^*(z)\psi_1(x,y)\psi_2(x,y)e^{i((\beta_2-\beta_1)z-\omega t)} + c.c.\right) \tag{11.20}$$

Substituting the expressions for the electromagnetic field (Eq. 11.16) and the nonlinear polarization in the wave equation (Eq. 11.14) satisfied by the second harmonic, we have

$$\frac{1}{2}\nabla^2\left(A_2\psi_2(x,y)e^{i(\beta_2 z-2\omega t)} + c.c.\right) = \frac{1}{2}\epsilon(2\omega)\mu_0\frac{\partial^2}{\partial t^2}\left(A_2\psi_2(x,y)e^{i(\beta_2 z-2\omega t)} + c.c.\right)$$
$$+\frac{1}{2}\mu_0\epsilon_0 d\frac{\partial^2}{\partial t^2}\left(A_1^2\psi_1^2(x,y)e^{2i(\beta_1 z-\omega t)} + c.c.\right)$$

We neglect $\frac{d^2A_2}{dz^2}$ and use the fact that the mode corresponding to the second harmonic at frequency 2ω satisfies the following equation (see Eq. 4.144):

$$\left(\nabla^2 - \frac{\partial^2}{\partial z^2}\right)\psi_2(x,y) + \left(\frac{4\omega^2}{c^2}n^2(2\omega,x,y) - \beta_2^2\right)\psi_2(x,y) = 0 \tag{11.21}$$

where $n^2(2\omega,x,y)$ represents the refractive index profile of the waveguide at the frequency 2ω and $\epsilon(2\omega) = \epsilon_0 n^2(2\omega,x,y)$. Equating the coefficient of the term $e^{-2i\omega t}$ on both sides, we arrive at the following equation (neglecting the term $\frac{d^2A_2}{dz^2}$):

$$2i\beta_2\frac{dA_2}{dz}\psi_2(x,y)e^{i\beta_2 z} = \mu_0\epsilon_0(-4\omega^2)d(x,y)A_1^2\psi_1^2(x,y)e^{2i\beta_1 z} \tag{11.22}$$

Multiplying both sides by $\psi_2(x,y)$ and integrating over the $x-y$ plane, we obtain (we assume that the modal field distribution are real functions of the transverse coordinates)

$$\frac{dA_2}{dz} = i\frac{\omega}{cn_{eff}(2\omega)}A_1^2\frac{\iint d(x,y)\psi_1^2(x,y)\psi_2(x,y)dxdy}{\iint\psi_2(x,y)\psi_2(x,y)dxdy}e^{-i(\beta_2-2\beta_1)z} \tag{11.23}$$

where we have assumed that the nonlinear coefficient $d(x,y)$ could also have dependence on the transverse coordinates. Here $n_{eff}(2\omega)$ is the effective index of the mode corresponding to frequency 2ω and is given by

$$\beta_2 = \frac{2\omega}{c} n_{eff}(2\omega) \tag{11.24}$$

Defining

$$\Delta\beta = \beta_2 - 2\beta_1 \tag{11.25}$$

we have

$$\frac{dA_2}{dz} = i\frac{\omega}{cn_2} A_1^2 \frac{\iint d(x,y)\psi_1^2(x,y)\psi_2(x,y)dxdy}{\iint \psi_2(x,y)\psi_2(x,y)dxdy} e^{-i\Delta\beta z} \tag{11.26}$$

where we have written $n_2 = n_{eff}(2\omega)$. If we assume that the fields $\psi_1(x,y)$ and $\psi_2(x,y)$ are normalized as defined below

$$\iint \psi_1(x,y)\psi_1(x,y)dxdy = 1 \tag{11.27}$$

and

$$\iint \psi_2(x,y)\psi_2(x,y)dxdy = 1 \tag{11.28}$$

then we have

$$\frac{dA_2}{dz} = i\frac{\omega}{cn_2} A_1^2 \iint d(x,y)\psi_1^2(x,y)\psi_2(x,y)dxdy e^{-i\Delta\beta z} \tag{11.29}$$

We define

$$\overline{d} = \iint d(x,y)\psi_1^2(x,y)\psi_2(x,y)dxdy \tag{11.30}$$

Since in general the nonlinear coefficient could depend on the transverse spatial coordinates, we write

$$d(x,y) = d_{eff}\tilde{d}(x,y) \tag{11.31}$$

where d_{eff} is the material nonlinear coefficient determined by the polarization states of the interacting modes and the direction of propagation while the dimensionless function $\tilde{d}(x,y)$ describes the transverse spatial dependence of the nonlinear coefficient. Hence, we have

$$\overline{d} = d_{eff} \iint \tilde{d}(x,y)\psi_1^2(x,y)\psi_2(x,y)dxdy = d_{eff}\Gamma \tag{11.32}$$

where

$$\Gamma = \iint \tilde{d}(x,y)\psi_1^2(x,y)\psi_2(x,y)dxdy \tag{11.33}$$

In view of Eqs. (11.27) and (11.28), the quantity Γ^2 will have units of inverse of area. Thus, the quantity $\frac{1}{\Gamma^2}$ represents an effective area determined by the overlap of the modal field distributions of the interacting modes and the transverse variation in the nonlinear coefficient. Larger the overlap between the fields and the nonlinear coefficient variation, smaller will be the effective area and higher will be the nonlinear effect.

Note that the dimension of $\overline{d}$ is different from that of the nonlinear coefficient d. From Eq. (11.29), we have

$$\frac{dA_2}{dz} = i\frac{\omega d_{eff}\Gamma}{cn_2}A_1^2 e^{-i\Delta\beta z} \tag{11.34}$$

Similarly the equation satisfied by the fundamental frequency ω will be (using Eq. 11.20)

$$\frac{1}{2}\nabla^2\left(A_1\psi_1(x,y)e^{i(\beta_1 z-\omega t)} + c.c.\right) = \frac{1}{2}\epsilon(\omega)\mu_0\frac{\partial^2}{\partial t^2}\left(A_1\psi_1(x,y)e^{i(\beta_1 z-\omega t)} + c.c.\right)$$
$$+\frac{1}{2}\mu_0\epsilon_0 d\frac{\partial^2}{\partial t^2}\left(2A_2A_1^*\psi_2(x,y)\psi_1(x,y)e^{i((\beta_2-\beta_1)z-\omega t)} + c.c.\right)$$

We neglect $\frac{d^2A_1}{dz^2}$ and use the fact that the mode corresponding to the fundamental satisfies the following equation

$$\left(\nabla^2 - \frac{\partial^2}{\partial z^2}\right)\psi_1(x,y) + \left(\frac{\omega^2}{c^2}n^2(\omega,x,y) - \beta_1^2\right)\psi_1(x,y) = 0$$

where $n^2(\omega,x,y)$ represents the refractive index profile of the waveguide at the frequency ω and $\epsilon(\omega) = \epsilon_0 n^2(\omega,x,y)$. Equating the coefficients of the terms with $e^{-i\omega t}$ time dependence, we arrive at the following equation:

$$2i\beta_1\frac{dA_1}{dz}\psi_1(x,y)e^{i\beta_1 z} = \mu_0\epsilon_0(-\omega^2)d(x,y)2A_2A_1^*\psi_2(x,y)\psi_1(x,y)e^{i(\beta_2-\beta_1)z} \tag{11.35}$$

We multiply both sides by $\psi_1(x, y)$ and integrate over the entire $x - y$ plane and obtain the following equation:

$$\frac{dA_1}{dz} = i\frac{\omega d_{eff}\Gamma}{cn_1}A_2A_1^* e^{i\Delta\beta z} \tag{11.36}$$

where we have used

$$\beta_1 = \frac{\omega}{c}n_{eff}(\omega) = \frac{\omega}{c}n_1$$

and we have used Eq. (11.32) defining $\overline{d}$ and the normalization defined by Eq. (11.27).

Equations (11.34) and (11.36) are the coupled mode equations describing the process of second harmonic generation in an optical waveguide. In the coupled mode equations, the quantity Γ accounts for the overlap between the modal field distribution of the interacting modes with the nonlinearity coefficient of the medium. If Γ happens to be equal to zero, then there will be no generation of second harmonic even though the medium is nonlinear.

Since the conversion efficiencies are usually small, we will solve the equation under the no-pump depletion approximation. Thus, we will assume that the amplitude E_1 of the input mode corresponding to frequency ω does not change much and will assume that it almost remains constant. For example if the input power at the fundamental frequency corresponds to 1 W and out of this 10 mW is converted to the second harmonic then at the output the power in the fundamental will still be 0.99 W. The electric fields associated with 1 and 0.99 W are almost the same and this difference will be neglected under the no pump depletion approximation. At the same time, the power of the second harmonic increases from 0 to 10 mW.

We assume that at the input to the waveguide at $z = 0$, only the mode at the fundamental frequency ω is excited with an amplitude A_{10}; then we have

$$A_1(z=0) = A_{10}; \quad A_2(z=0) = 0 \tag{11.37}$$

In such a case, we need to solve only Eq. (11.34) neglecting the variation in A_1 with z. Integrating Eq. (11.34) from $z = 0$ to z, we have

$$A_2(z) = i\frac{\omega d_{eff}\Gamma}{cn_2}A_1^2\int_0^z e^{-i\Delta\beta z'}dz' = i\frac{\omega d_{eff}\Gamma}{cn_2}A_1^2\frac{e^{-i\Delta\beta z} - 1}{-i\Delta\beta}$$
$$= i\frac{\omega d_{eff}\Gamma}{cn_2}A_1^2 e^{-i\Delta\beta z/2}\frac{-2i\sin\left(\frac{\Delta\beta z}{2}\right)}{-i\Delta\beta}$$

or

$$A_2(z) = i\frac{\omega d_{eff}\Gamma}{cn_2}A_1^2 z e^{-i\Delta\beta z/2}\left[\frac{\sin\left(\frac{\Delta\beta z}{2}\right)}{\frac{\Delta\beta z}{2}}\right] \tag{11.38}$$

which gives the amplitude of the mode at the frequency 2ω at any value z.

Now the power carried by the second harmonic mode is given by (see Chap. 4)

$$P(2\omega) = \frac{\beta_2}{2(2\omega)\mu_0}|A_2(z)|^2 \iint \psi_2(x,y)\psi_2(x,y)dxdy \tag{11.39}$$

Replacing β_2 by Eq. (11.24) and using the normalization condition given by Eq. (11.28), we have

$$P(2\omega) = \frac{n_2}{2c\mu_0}|A_2(z)|^2 \tag{11.40}$$

Similarly the power carried by the fundamental frequency is

$$P(\omega) = \frac{n_1}{2c\mu_0}|A_1(z)|^2 \tag{11.41}$$

Thus, we have

$$P(2\omega) = \frac{2\mu_0\omega^2}{cn_1^2 n_2}d_{eff}^2\Gamma^2 z^2 P^2(\omega)\left(\frac{\sin\left(\frac{\Delta\beta z}{2}\right)}{\frac{\Delta\beta z}{2}}\right)^2 \tag{11.42}$$

We define the efficiency of second harmonic generation

$$\eta = \frac{P(2\omega)}{P(\omega)} = \frac{2\mu_0\omega^2}{cn_1^2 n_2}d_{eff}^2\Gamma^2 z^2 P(\omega)\left(\frac{\sin\left(\frac{\Delta\beta z}{2}\right)}{\frac{\Delta\beta z}{2}}\right)^2$$

or

$$\eta = \eta_0 z^2 P(\omega)\left(\frac{\sin\left(\frac{\Delta\beta z}{2}\right)}{\frac{\Delta\beta z}{2}}\right)^2 \tag{11.43}$$

where

$$\eta_0 = \frac{2\mu_0\omega^2}{cn_1^2 n_2}d_{eff}^2\Gamma^2 \tag{11.44}$$

Note that the efficiency of conversion to the second harmonic varies periodically as a function of propagation distance z. For a given $\Delta\beta$, the maximum conversion from the fundamental to the second harmonic happens at a length $z = L_c$ when $\frac{\Delta\beta L_c}{2} = \frac{\pi}{2}$ which gives us the length

$$L_c = \frac{\pi}{\Delta\beta} = \frac{\pi}{\beta_2 - 2\beta_1} \tag{11.45}$$

This length is referred to as the *coherence length* for second harmonic generation and is the minimum distance at which the conversion from the fundamental to the second harmonic is maximum. This is a very important quantity. Under phase matching $\Delta\beta = 0$ and the coherence length tends to infinity.

From Eq. (11.43), we note the following. The conversion efficiency from fundamental to the second harmonic is

- Proportional to the square of the effective nonlinear coefficient d_{eff} which depends on the material, the polarization states of the interacting modes, and the direction of propagation.
- Proportional to the overlap factor Γ which depends on the transverse dependence $\tilde{d}(x, y)$ of the nonlinear coefficient and also the transverse modal field distributions $\psi_1(x, y)$ and $\psi_2(x, y)$ of the interacting modes. Thus, for example, if $\tilde{d}(x, y)$ is a constant over the entire cross-section, then the efficiency will depend on $\iint \psi_1^2(x, y)\psi_2(x, y)dxdy$. Thus, if $\psi_1(x, y)$ the mode at the fundamental frequency is a symmetric transverse spatial mode and $\psi_2(x, y)$ the mode at the second harmonic is an antisymmetric transverse spatial mode, the value of the integral will vanish and there will be no conversion from fundamental to the second harmonic. Thus, when using waveguides for nonlinear interactions, it is very important to estimate the overlap integrals which will effectively determine the nonlinearity responsible for the conversion.
- Proportional to the power carried by the fundamental wave. This is a consequence of the nonlinear interaction.
- Proportional to the function $\left(\frac{\sin\left(\frac{\Delta\beta z}{2}\right)}{\frac{\Delta\beta z}{2}}\right)^2$. The maximum value of this function is unity and that happens when $\Delta\beta = 0$. Thus, for maximum efficiency we must have

$$\Delta\beta = \beta_2 - 2\beta_1 = 0$$

or

$$\beta_2 = 2\beta_1 \tag{11.46}$$

which implies

$$n_2 = n_1$$

that is, the effective indices of the interacting modes at ω and 2ω must be equal. In such a case, the amplitude of the second harmonic grows linearly with z and is given by (see Eq. 11.38)

$$A_2(z) = i\frac{\omega d_{eff}\Gamma}{cn_2}A_1^2 z \tag{11.47}$$

The condition given by Eq. (11.46) is referred to as *phase matching condition* and represents the condition for maximum efficiency of second harmonic generation. Note that Eq. (11.47) which shows a linear increase in the second harmonic field with propagation is valid only for small efficiencies. For higher efficiencies, the two coupled equations for $A_1(z)$ and $A_2(z)$ must be solved simultaneously without assuming no pump depletion approximation.

Phase matching condition is a very important condition in nonlinear interactions and special techniques need to be used to satisfy this condition for maximizing the efficiency.

In the picture of photons, second harmonic generation is the fusion, in the presence of a medium, of two photons of frequency ω to form a single photon of frequency 2ω. In this picture, the phase matching condition is nothing but momentum conservation condition since the momenta of the photons corresponding to the fundamental and the second harmonic frequencies are simply $\hbar\beta_1$ and $\hbar\beta_2$, respectively. Figure 11.1 shows the vector diagram demonstrating momentum conservation.

- We can write Eq. (11.43) in the following form (assuming $\Delta\beta = 0$) which corresponds to the maximum possible efficiency in the given structure

$$\frac{\eta}{P(\omega)z^2} = \frac{2\mu_0\omega^2}{cn_1^2 n_2} d_{eff}^2 \Gamma^2 \tag{11.48}$$

It can be noted that the right hand side of the above equation depends on the waveguide and material properties through the effective indices of the interacting modes, the effective area which depends on the transverse modal field distributions of the interacting modes as well as the nonlinear coefficient which will be determined by the medium and the polarization states of the fundamental and second harmonic modes. Thus, for a comparison among different second

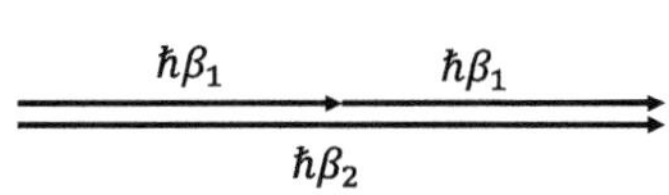

Fig. 11.1 Vector diagram showing momentum conservation in the second harmonic generation process

harmonic generation configurations the efficiency is quoted in terms of the unit %/W/cm^2.

- Under phase-matched operation, for low efficiencies, the efficiency increases quadratically with propagation distance.

11.4.1 Interaction in Bulk

Instead of waveguides if we analyze second harmonic generation for plane waves propagating in bulk media and if we assume the cross-section area of the beams to be S, then in view of the normalization conditions (Eqs. 11.27 and 11.28), we have (neglecting diffraction effects) and assuming the nonlinear coefficient to be independent of the transverse coordinates,

$$\psi_1(x,y) = \frac{1}{\sqrt{S}}; \quad \psi_2(x,y) = \frac{1}{\sqrt{S}} \tag{11.49}$$

and

$$\Gamma = \frac{1}{\sqrt{S}} \tag{11.50}$$

Hence, the efficiency for second harmonic generation is given by

$$\eta = \frac{P(2\omega)}{P(\omega)} = \frac{2\mu_0\omega^2}{cn_1^2 n_2} d_{eff}^2 z^2 \frac{P(\omega)}{S} \left(\frac{\sin\left(\frac{\Delta kz}{2}\right)}{\frac{\Delta kz}{2}} \right)^2 \tag{11.51}$$

where we have replaced β by k which is the propagation constant in the bulk medium. Equation (11.51) is the standard expression for efficiency of second harmonic generation using plane waves in bulk (see, for example, Yariv 1991; Ghatak and Thyagarajan 1998). Note that in the case of bulk, the efficiency depends on power per unit area $\frac{P(\omega)}{S}$ of the fundamental frequency. Thus, focusing a beam will result in increased efficiency; of course diffraction effects will lead to reduced effective lengths of interaction and there is an optimum focusing. This is one of the advantages of using optical waveguides which overcome diffraction effects through wave guidance.

Exercise 11.1 Show that Eqs. (11.34) and (11.36) satisfy energy conservation.

Solution The power in the fundamental mode and the second harmonic modes are given by Eqs. (11.40) and (11.41). Thus, conservation of power requires

$$\frac{d}{dz}(P(\omega)+P(2\omega))=0$$

Substituting from Eqs. (11.40) and (11.41), we obtain

$$\begin{aligned}\frac{d}{dz}(P(\omega)+P(2\omega)) &= \frac{d}{dz}\left(\frac{n_1}{2c\mu_0}|A_1(z)|^2+\frac{n_2}{2c\mu_0}|A_2(z)|^2\right)\\ &= \frac{1}{2c\mu_0}\left(n_1\frac{d|A_1(z)|^2}{dz}+n_2\frac{d|A_2(z)|^2}{dz}\right)\end{aligned}$$

Now

$$n_1\frac{d|A_1(z)|^2}{dz}+n_2\frac{d|A_2(z)|^2}{dz}=n_1\left(A_1^*\frac{dA_1}{dz}+c.c.\right)+n_2\left(A_2^*\frac{dA_2}{dz}+c.c.\right)$$

where $c.\ c.$ represents complex conjugate. Substituting from Eqs. (11.34) and (11.36), we have

$$\begin{aligned}&n_1\left(i\frac{\omega d_{eff}\Gamma}{cn_1}A_2A_1^{*2}e^{i\Delta\beta z}-i\frac{\omega d_{eff}\Gamma}{cn_1}A_2^*A_1^2e^{-i\Delta\beta z}\right)\\ &+n_2\left(i\frac{\omega d_{eff}\Gamma}{cn_2}A_2^*A_1^2e^{-i\Delta\beta z}-i\frac{\omega d_{eff}\Gamma}{cn_2}A_2A_1^{*2}e^{i\Delta\beta z}\right)=0\end{aligned}$$

which shows power conservation.

Exercise 11.2 The phase-matched SHG efficiency of a 2 cm long crystal is 0.5% at an input power of 100 mW. What will be the output power at 2ω when the input power at ω is 200 mW?

Solution From Eq. (11.43), it follows that the efficiency of second harmonic generation is proportional to the power of the incident wave. Thus, when the input power at the fundamental frequency is doubled from 100 to 200 mW, the efficiency also gets doubled from 0.5% to 1%. Thus, the output power at the second harmonic for an input power of 200 mW at the fundamental frequency will be 2 mW. Note that the power at the second harmonic for 100 mW input power was 0.5 mW and when the input power is doubled, the output power at the second harmonic increases by a factor of 4.

Exercise 11.3 The phase-matched SHG efficiency of a 1 cm long waveguide is 1%. For what value of $\Delta\beta$ will the SHG efficiency drop to zero?

Solution From Eq. (11.43), we see that, for an interaction length L, the efficiency of SHG will become zero when $\frac{\Delta\beta L}{2}=\pi$ or $\Delta\beta=\frac{2\pi}{L}$. Thus for $L=1$ cm, the efficiency drops to zero for $\Delta\beta\approx 6.28\times 10^2\ \mathrm{m}^{-1}$. Note that the value of $\Delta\beta$ is very small compared to the value of β itself, which assuming a typical effective index of 2 and a wavelength of $\lambda_f = 800$ nm is $\beta = \frac{2\pi}{\lambda_f} n_1 \approx 1.57 \times 10^7\ \text{m}^{-1}$.

Exercise 11.4 The effective indices of the modes of a channel waveguide at wavelengths of 1 and 0.5 μm are given by 2.17 and 2.27, respectively. For a mode excited at 1 μm, what is the velocity of the nonlinear polarization generated at the frequency corresponding to the wavelength of 1 μm?

Solution The nonlinear polarization generated at the fundamental frequency is given by Eq. (11.20). The velocity of the nonlinear polarization is given by

$$v_{nl}^{(\omega)} = \frac{\omega}{\beta_2 - \beta_1} = \frac{\omega}{\left(\frac{2\omega}{c} n_2 - \frac{\omega}{c} n_1\right)} = \frac{c}{(2n_2 - n_1)} \approx 1.266 \times 10^8\ \text{m/s}$$

The electromagnetic wave at the fundamental frequency is propagating with a velocity $\frac{c}{n_1} \approx 1.38 \times 10^8$ m/s, which is not the same as the velocity of the nonlinear polarization that is trying to generate the wave, implying that phase matching condition is not satisfied resulting in lowering of efficiency of conversion.

Exercise 11.5 Consider SHG in a channel waveguide in z-cut, x-propagating lithium niobate with a refractive index profile varying as given by (see Chap. 4)

$$\begin{aligned} n^2(z, y) &= n_b^2 + 2n_b \Delta n e^{\left(-z^2/h^2\right)} e^{\left(-y^2/w^2\right)}; \quad && z < 0 \\ &= n_c^2; && z > 0 \end{aligned}$$

where n_b, Δn, n_c, h, and w are constants. Note that in Chap. 4 the coordinate system we used was different; here we are using a coordinate system matching the principal axes of the material of the waveguide. Is it possible to achieve second harmonic generation from the E_{11} mode at ω to the E_{12} mode at 2ω? Here the first subscript corresponds to the mode number along the z-direction (perpendicular to the substrate-cover interface) and the second subscript to that along the y-direction (parallel to the substrate-cover interface). Assume that the nonlinear coefficient $\tilde{d}$ is independent of y and z.

Solution For conversion from ω to 2ω in the waveguide, we need to ensure that the modal overlap Γ is nonzero (see Eq. 11.33). Since the waveguide is symmetric along the y-direction, the mode E_{11} will be symmetric along the y-direction, while the mode E_{12} will be antisymmetric along the y-direction. The overlap integral would be given by $\Gamma = \iint \tilde{d}(z, y) \psi_{11}^2(z, y, \omega) \psi_{12}(z, y, 2\omega) dz dy = \tilde{d} \iint \psi_{11}^2(z, y, \omega) \psi_{12}(z, y, 2\omega) dz dy$ which in this case would be zero and hence there would no conversion even if phase matching condition is satisfied. What would happen if the modes of ω and 2ω are interchanged?

As an example, let us consider second harmonic generation using bulk lithium niobate with the fundamental and second harmonic being extraordinary waves. Thus, we have $\Delta\beta = \Delta k = \frac{2\omega}{c}(n_e(2\omega) - n_e(\omega))$. Also for this case the value of the nonlinear coefficient is $d = d_{33} = 30 \times 10^{-12}$ m/V. Let the wavelength of the incident wave be 1000 nm and let the input power be 1 W and the area of cross-section be 1 mm^2 (unfocussed beam). For this wavelength, we take $n_e(\omega) \approx 2.15$, $n_e(2\omega) \approx 2.25$.

Substituting the values in Eq. (11.51), we have

$$\eta = 2.58\, z^2 \left(\frac{\sin\left(\frac{\Delta kz}{2}\right)}{\frac{\Delta kz}{2}} \right)^2$$

with z being measured in meters. Also, it must be noted that the above equation is valid as long as $\eta \ll 1$. When the phase matching condition is not being satisfied, we find that the maximum efficiency happens when the length of interaction satisfies the following equation: $\frac{\Delta kz}{2} = \frac{\pi}{2}$.

For the given values $\Delta k = k_2 - 2k_1 = \frac{4\pi}{\lambda_f}(n_e(2\omega) - n_e(\omega)) = 4\pi \times 10^5\ \mathrm{m}^{-1}$ where λ_f is the free space wavelength of the fundamental wave. Thus, the maximum efficiency is achieved at a minimum length of $z = \frac{\pi}{\Delta k} = 2.5$ μm! At this length, the efficiency comes out to be $\eta = 2.58 \times \frac{4}{(\Delta k)^2} \approx 6.5 \times 10^{-12}$ implying a power in the second harmonic of 6.5 picowatts. Beyond this length, the power in the second harmonic will start to flow towards the fundamental and thus decrease and at 5 μm, the power in the second harmonic will go to zero. After this, the power in the second harmonic will again rise until a propagation length of 7.5 μm and so on. The maximum power in the second harmonic would be only 6.5 pW.

For the same set of values, if phase matching could be achieved, then we would have $\left(\frac{\sin\left(\frac{\Delta\beta z}{2}\right)}{\frac{\Delta\beta z}{2}} \right)^2 = 1$ giving $\eta \approx 2.58z^2$. Hence, taking a length as $z = 1$ cm, we have $\eta \approx 2.58 \times 10^{-4}$ which implies an output power of 0.26 mW! This shows how important it is to ensure that the phase matching condition is satisfied.

Figure 11.2 shows the variation in efficiency with length of interaction for phase-matched and nonphase-matched operations for an input power $P(\omega) = 100$ mW. Phase-matched efficiency increases as the square of the length of interaction, while if phase matching condition is not satisfied, then the efficiency oscillates between a maximum value and zero. It is interesting to note that since energy needs to be conserved, when the power in the second harmonic is decreasing, power flows from the second harmonic back to the fundamental.

Figure 11.3 shows a plot of efficiency $\frac{\eta}{\eta_0}$ with $\frac{\Delta\beta z}{2}$ and shows the drop in efficiency as the value of $\Delta\beta$ becomes nonzero. Note that when $\Delta\beta z = 2\pi$, the efficiency drops to zero (see Eq. 11.43). The width of the curve in Fig. 11.3 gives the bandwidth of the interaction process. Thus, as the length of interaction increases, the value of $\Delta\beta$ required to reduce the efficiency to zero decreases thus making the interaction more critical.

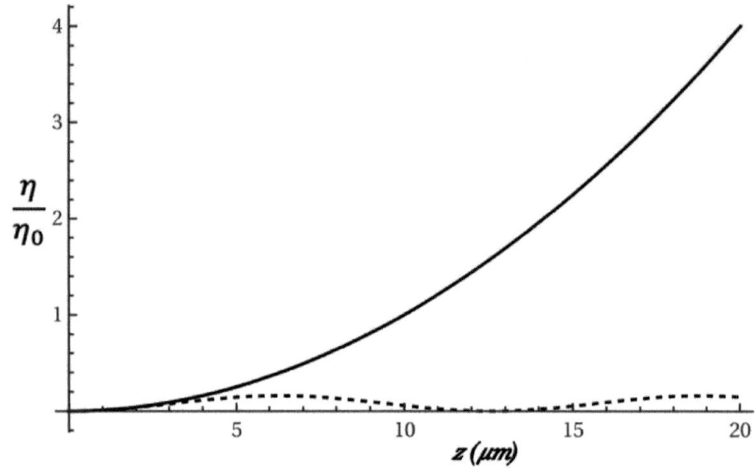

Fig. 11.2 Shows the variation in normalized efficiency for the phase-matched (solid line) and non-phase-matched (dashed line) interaction for $\Delta\beta = 0.5\ \mu\text{m}^{-1}$. As it can be seen, deviation from phase matching results in oscillatory variation and a drastic reduction in efficiency of conversion

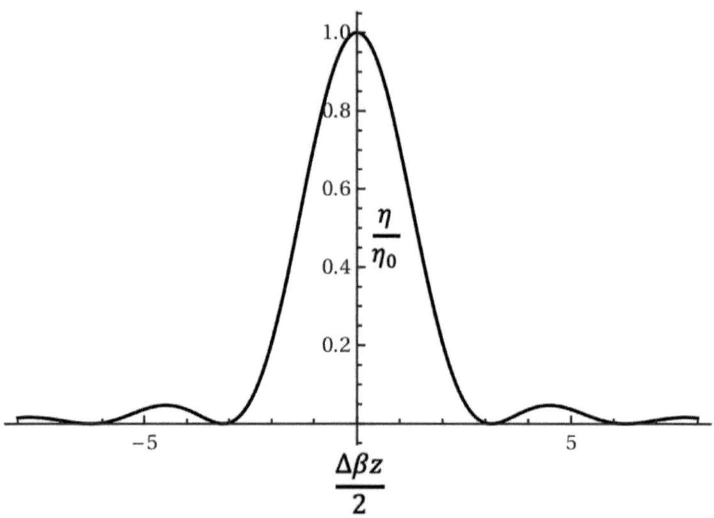

Fig. 11.3 Variation in normalized efficiency $\frac{\eta}{\eta_0}$ with $\frac{\Delta\beta z}{2}$

Figure 11.4 shows a schematic of the variation in the nonlinear polarization at 2ω and the electromagnetic wave at frequency 2ω under phase mismatch. It can be seen that under nonphase-matched interaction, since the speed of propagation of the nonlinear polarization and that of the electromagnetic wave that it is trying to generate are not equal, the two get out of phase as they propagate. When the phase difference becomes π, then the energy flows from the second harmonic to the

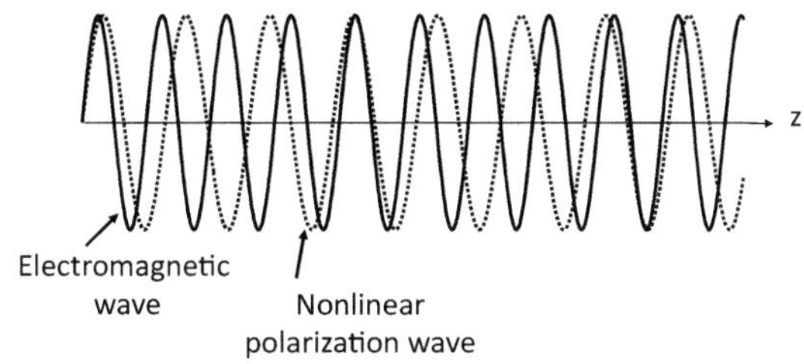

Fig. 11.4 Schematic of the z-variation in nonlinear polarization (dashed curve) at the second harmonic and the electromagnetic wave (solid curve) at the second harmonic when phase matching condition is not satisfied

fundamental, instead of from fundamental to the second harmonic, and thus the power in the second harmonic decreases until it becomes zero. And then the process begins again.

The phase shift suffered by the nonlinear polarization wave in propagating through a length L is

$$\phi_{NL} = 2\beta_1 L = 2\frac{\omega}{c} n_1 L \tag{11.52}$$

The phase shift suffered by the electromagnetic wave at 2ω in propagating through the same length L is

$$\phi_{EM} = \beta_2 L = \frac{2\omega}{c} n_2 L \tag{11.53}$$

Thus, the phase difference between the electromagnetic wave at 2ω and the nonlinear polarization (which is the source of the electromagnetic wave at 2ω) over a length L is

$$\Delta\phi = \phi_{EM} - \phi_{NL} = \frac{2\omega}{c}(n_2 - n_1)L = \Delta\beta L \tag{11.54}$$

The length required for accumulating a phase difference of π is

$$L = \frac{\pi}{\Delta\beta} \tag{11.55}$$

which is nothing but the coherence length. Thus, coherence length is nothing but the distance over which the accumulated phase difference between the nonlinear polarization and the electromagnetic wave that it is trying to generate becomes π.

This shows that in the nonphase-matched case the efficiency periodically varies with the length of interaction. Maximum efficiency is achieved when the length of interaction satisfies $\Delta\beta L_c = \pi$, that is, for a length

$$L_c = \frac{\pi}{\Delta\beta} = \frac{\pi}{\beta_2 - 2\beta_1} = \frac{\pi c}{2\omega(n_2 - n_1)} \tag{11.56}$$

If the free space wavelength of the fundamental frequency is λ_f, then $\omega = \frac{2\pi c}{\lambda_f}$ and the coherence length is given by

$$L_c = \frac{\lambda_f}{4(n_2 - n_1)} \tag{11.57}$$

One of the important techniques for achieving phase matching involves birefringence phase matching. In this technique, the anisotropic property of the crystal is used and by choosing orthogonal polarization states of the fundamental and second harmonic frequency, it is possible to achieve phase matching under certain conditions (see, e.g., Ghatak and Thyagarajan 1998; Yariv 1991).

A more interesting technique is referred to as quasi phase matching (QPM) and involves periodically reversing the sign of the nonlinear coefficient with an appropriate period to achieve increased efficiency. This will be discussed in Sect. 11.6.

Exercise 11.6 Consider SHG for a fundamental wavelength of 1 μm over a crystal of length 1 cm. Estimate the maximum allowed value of $\Delta\beta$ so that the reduction in peak efficiency due to nonphase-matched operation is not less than 81% of the phase-matched case.

Solution From Eq. (11.42), we find that when $\left(\frac{\sin\left(\frac{\Delta\beta z}{2}\right)}{\frac{\Delta\beta z}{2}}\right)^2 = 0.81$, the efficiency will drop to 81% of the value when phase matching is satisfied. We can solve this equation and find that this happens when $\frac{\Delta\beta z}{2} \approx 0.785$. For the interaction length given, the allowed maximum value of $\Delta\beta$ is $1.57 \times 10^2\ \mathrm{m}^{-1}$. Taking typical values of $\lambda_f = 1000$ nm, $n_{eff} \approx 2.2$, we have $\beta = \frac{2\pi}{\lambda_f} n_{eff} \approx 13.8 \times 10^6\ \mathrm{m}^{-1}$. Thus, the allowed $\Delta\beta$ for keeping the efficiency close to the maximum is much smaller than the value of β itself.

11.5 Effective Nonlinear Coefficient

As an example to appreciate the role of modal field distributions of the interacting modes, we approximate the modal fields at the fundamental frequency and the second harmonic through Gaussian approximation (see Chap. 4). We assume that both at the fundamental frequency ω and the second harmonic frequency 2ω the interacting modes correspond to the fundamental spatial mode.

We approximate the fundamental spatial mode at ω by

$$\psi_1(x, y) = \frac{\sqrt{2}}{\left(\pi w_{fx} w_{fy}\right)^{1/2}} \exp\left(-\frac{x^2}{w_{fx}^2} - \frac{y^2}{w_{fy}^2}\right) \tag{11.58}$$

where w_{fx} and w_{fy} are the widths of the modal distributions along the x- and y-directions, respectively, and the factor multiplying the Gaussian function is the normalization constant in order to satisfy Eq. (11.27).

Similarly, we can write for the spatial modal field of the second harmonic which is assumed to have different Gaussian widths since its wavelength is half that of the fundamental

$$\psi_2(x,y)=\frac{\sqrt{2}}{\left(\pi w_{sx}w_{sy}\right)^{1/2}}\exp\left(-\frac{x^2}{w_{sx}^2}-\frac{y^2}{w_{sy}^2}\right) \tag{11.59}$$

where w_{sx} and w_{sy} are the widths of the modal distributions along the x- and y-directions, respectively, and the factor multiplying the Gaussian function is the normalization constant in order to satisfy Eq. (11.28).

On carrying out the integration and assuming that the nonlinear coefficient d is independent of coordinates, we have

$$\tilde{d}=\iint d(x,y)\psi_1^2(x,y)\psi_2(x,y)dxdy=2d\sqrt{\frac{2}{\pi}}\frac{\sqrt{w_{sx}w_{sy}}}{\left(w_{fx}^2+2w_{sx}^2\right)^{1/2}\left(w_{fy}^2+2w_{sy}^2\right)^{1/2}}$$

Thus,

$$\Gamma=2\sqrt{\frac{2}{\pi}}\frac{\sqrt{w_{sx}w_{sy}}}{\left(w_{fx}^2+2w_{sx}^2\right)^{1/2}\left(w_{fy}^2+2w_{sy}^2\right)^{1/2}} \tag{11.60}$$

This shows the dependence of the overlap factor Γ and hence the effective nonlinear coefficient on the widths of the modal field distributions at the fundamental frequency and the second harmonic.

As an example let us assume $w_{fx}=w_{fy}=5$ μm and $w_{sx}=w_{sy}=3$ μm. In such a case, we have $\Gamma=0.11\ \mu\text{m}^{-1}$ and the effective area $\frac{1}{\Gamma^2}$ becomes about 81 μm^2.

Exercise 11.7 Show that $\psi_1(x,y)$ given by Eq. (11.58) satisfies the normalization condition.

Solution We have

$$\iint\psi_1^2(x,y)dxdy=\frac{2}{\pi w_{fx}w_{fy}}\int_{-\infty}^{+\infty}\exp\left(-\frac{2x^2}{w_{fx}^2}\right)dx\int_{-\infty}^{+\infty}\exp\left(-\frac{2y^2}{w_{fy}^2}\right)dy$$

$$=\frac{2}{\pi w_{fx}w_{fy}}\sqrt{\frac{\pi}{2}}w_{fx}\sqrt{\frac{\pi}{2}}w_{fy}=1$$

where we have used the standard integral

$$\int_{-\infty}^{+\infty} e^{-px^2+qx}dx = \sqrt{\frac{\pi}{p}}e^{q^2/4p}$$

11.6 Quasi Phase Matching

As discussed earlier, phase matching is extremely important for any efficient nonlinear interaction. Recently, the technique of *quasi phase matching* (QPM) which was originally proposed by Armstrong et al. (1961) has become a convenient way to achieve high efficiency nonlinear conversion at any desired wavelength in any nonlinear material possessing second-order nonlinearity. QPM relies on the fact that the phase mismatch in the two interacting beams can be compensated by periodically readjusting the phase of interaction through periodically modulating the nonlinear characteristics of the medium at a spatial frequency equal to the wavevector mismatch of the two interacting waves (see Fejer et al. 1992). Thus, in SHG when the nonlinear polarization at 2ω and the electromagnetic wave at 2ω have an accumulated phase difference of π (see Fig. 11.4), then the sign of the nonlinear coefficient is reversed so that the energy flowing from the nonlinear polarization at frequency 2ω to the electromagnetic wave at frequency 2ω can add constructively with the existing energy. Thus, by properly choosing the period of the spatial modulation of the nonlinear coefficient, one could achieve increased conversion efficiency. This scheme is called quasi phase matching (QPM) and is being very widely studied for application to nonlinear interactions in bulk and in waveguides.

In a ferroelectric material such as lithium niobate, the signs of the nonlinear coefficients are linked to the direction of the spontaneous polarization. Thus, a periodic reversal of the ferroelectric domains of the crystal can be used for achieving QPM. This is the currently used technique to obtain high efficiency SHG and other nonlinear interactions in $LiNbO_3$, $LiTaO_3$, and KTP. The most popular technique today to achieve periodic domain reversal in $LiNbO_3$ is the technique of electric field poling (Yamada et al. 1993; Myers and Bosenberg 1997). In this method, a high electric field pulse is applied to properly oriented lithium niobate crystal using lithographically defined electrode patterns to produce a permanent periodic domain reversed pattern. Such a periodically domain reversed $LiNbO_3$ crystal with the periodically reversed domains going through the entire depth of the crystal is also referred to as PPLN (pronounced *piplin*) and is now commercially available.

To clarify the concept of quasi phase matching, let us first assume that the nonlinear coefficient d and hence $d_{eff}(z)$ varies sinusoidally with a period Λ. In such a case, we have

$$d_{eff}(z) = d_{eff0} \sin Kz \tag{11.61}$$

where d_{eff0} is the amplitude of modulation of the nonlinear coefficient and $K = \frac{2\pi}{\Lambda}$ represents the spatial frequency of the periodic modulation. For easier understanding, we are assuming the modulation to be sinusoidal; in general, the modulation will be periodic but not sinusoidal and this will be discussed later. Any periodic modulation can be written as a superposition of sinusoidal and co-sinusoidal variations. Thus, our discussion is valid for one of the Fourier component of the variation.

With a variation given by Eq. (11.61), Eq. (11.34) for the variation in the amplitude of the second harmonic becomes

$$\frac{dA_2}{dz} = i\frac{\omega d_{eff}(z)\Gamma}{cn_2}A_1^2 e^{-i\Delta\beta z} = i\frac{\omega d_{eff0}\Gamma}{cn_2}\sin KzA_1^2 e^{-i\Delta\beta z}$$

which can be written as

$$\frac{dA_2}{dz} = \frac{\omega d_{eff0}\Gamma}{2cn_2}A_1^2\left(e^{i(K-\Delta\beta)z} - e^{-i(K+\Delta\beta)z}\right) \tag{11.62}$$

If we choose the spatial frequency K such that $K = \Delta\beta$, then we have

$$\frac{dA_2}{dz} = \frac{\omega d_{eff0}\Gamma}{2cn_2}A_1^2\left(1 - e^{-i2\Delta\beta z}\right)$$

Integrating the above equation under no pump depletion approximation and assuming $E_2(z = 0) = 0$ give us

$$A_2(z) = \frac{\omega d_{eff0}\Gamma}{2cn_2}A_1^2\left(z - \frac{e^{-i2\Delta\beta z} - 1}{-i2\Delta\beta}\right) \tag{11.63}$$

The second term in the brackets will contribute very little to the efficiency (see Exercise 11.8) as discussed before and thus we can approximate the second harmonic by

$$A_2(z) = \frac{\omega d_{eff0}\Gamma}{2cn_2}A_1^2 z \tag{11.64}$$

which now shows that the amplitude of the second harmonic grows linearly with distance z (similar to the case of perfect phase matching) and hence the power will grow quadratically with the distance. Although the efficiency of second harmonic grows quadratically with z, it must be noted that due to the sinusoidal variation in the nonlinear coefficient, the effective nonlinear coefficient has been reduced by a factor of 2 from d_{eff0} to $\frac{d_{eff0}}{2}$ (see Eq. 11.47) which implies a reduction in efficiency by a factor of 4 compared to a perfect phase-matched interaction.

Exercise 11.8 Compare the magnitudes of the two terms within the bracket in Eq. (11.63).

Solution The bracketed term in Eq. (11.63) can be written as

$$\left(z-\frac{e^{-i2\Delta\beta z}-1}{-i2\Delta\beta}\right)=z-e^{-i\Delta\beta z}\frac{\sin\Delta\beta z}{\Delta\beta}$$

If we consider lithium niobate and assume second harmonic generation from a wavelength of 1000–500 nm using extraordinary polarization, we have $n_1 = 2.1591$ and $n_2 = 2.2486$. Hence, we have

$$\Delta\beta=\frac{4\pi}{\lambda_f}(n_2-n_1)\approx 1.1\times 10^6\ \mathrm{m}^{-1}$$

For $\Delta\beta \neq 0$, the magnitude of the maximum value of the second term is $\frac{1}{\Delta\beta}\approx 9\times 10^{-7}$ m, which is negligible compared to typical value of the first term z of 10 mm. Thus, the second term in Eq. (11.63) does not contribute significantly to the efficiency.

The condition $K = \Delta\beta$ gives us the period of the poling to be

$$\Lambda=\frac{2\pi}{K}=\frac{2\pi}{\Delta\beta} \tag{11.65}$$

In terms of the fundamental wavelength λ_f, we have

$$\Lambda=\frac{\lambda_f}{2(n_2-n_1)} \tag{11.66}$$

11.6.1 Periodic Poling

In a practically realized periodically poled crystal, the nonlinear coefficient d varies periodically (not sinusoidally) along the propagation direction. As a specific example, we consider the following variation:

$$\begin{aligned} d(z) &= +d; \quad 0<z<\frac{\Lambda}{2} \\ d(z) &= -d; \quad \frac{\Lambda}{2}<z<\Lambda \end{aligned} \tag{11.67}$$

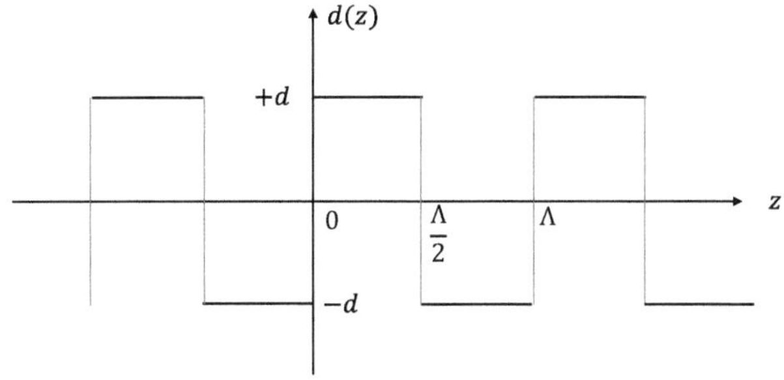

Fig. 11.5 Periodic variation with a period Λ of the nonlinear coefficient in a periodically poled crystal. The variation shown has a duty cycle of 50%

and

$$d(z+\Lambda)=d(z) \tag{11.68}$$

where d is the appropriate nonlinear coefficient determined by the polarization states of the interacting waves, the nonlinear tensor of the crystal, and the direction of propagation. Figure 11.5 shows such a variation for a duty cycle of 0.5. Since d varies periodically, the effective nonlinear coefficient will also vary periodically with the same period (see Eq. 11.31).

We can make a Fourier series expansion of $d(z)$ and we have

$$d(z)=d\sum_{m=-\infty}^{+\infty}G_m e^{-imKz}=d\left(G_0+G_1e^{-iKz}+G_{-1}e^{iKz}+G_2e^{-2iKz}+G_{-2}e^{2iKz}+\cdots\right) \tag{11.69}$$

where G_m is the mth Fourier coefficient. We have written down the first few terms only. It can be shown that the magnitude of the Fourier coefficient G_1 is (see Exercise 11.9)

$$G_1=i\frac{2}{\pi};\quad G_{-1}=-i\frac{2}{\pi} \tag{11.70}$$

Considering first-order QPM with the grating period close to satisfying $K=\Delta\beta$, we need to solve the following equation (see Eq. 11.34):

$$\frac{dA_2}{dz}=\frac{\omega\Gamma}{cn_2}\frac{2d_{eff}}{\pi}A_1^2e^{-i(\Delta\beta-K)z} \tag{11.71}$$

where we have kept only the term that with the phase factor $e^{-i(\Delta\beta-K)z}$ which will become unity when QPM condition is exactly satisfied. Integrating Eq. (11.71) from 0 to L and assuming $E_2(z=0)=0$, we have

$$A_2(L)=\frac{\omega\Gamma}{cn_2}\frac{2d_{eff}}{\pi}A_1^2e^{-\frac{i(\Delta\beta-K)L}{2}}L\frac{\sin\left[\frac{(\Delta\beta-K)L}{2}\right]}{\frac{(\Delta\beta-K)L}{2}} \tag{11.72}$$

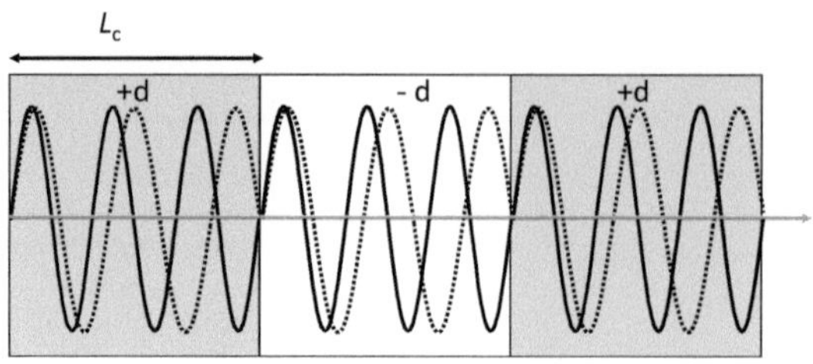

Fig. 11.6 Schematic of how in QPM, when the electromagnetic field at 2ω (solid curve) and the nonlinear polarization at 2ω (dashed curve) get out of phase by π, the nonlinear coefficient reverses in sign to bring them back in phase

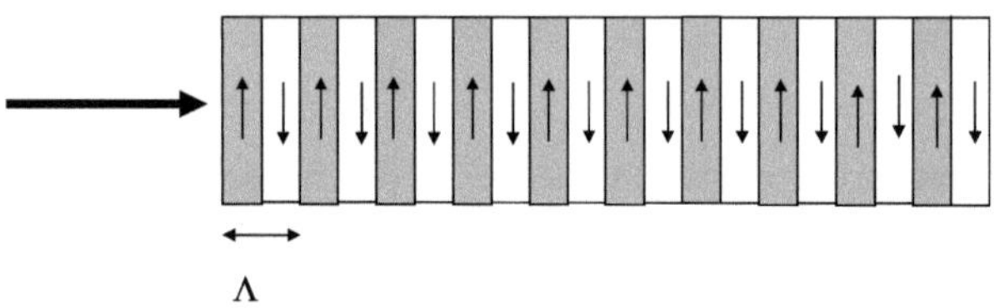

Fig. 11.7 Quasi phase matching is achieved by periodically modulating the nonlinear coefficient of the medium. In lithium niobate this corresponds to periodically reversing the direction of the z-axis of the crystal. The arrows show the direction of the optic axis of the crystal

Comparing with Eq. (11.38) we can see that due to QPM, we now have an effective nonlinear coefficient which is smaller by a factor $2/\pi$ compared to the case when the nonlinear coefficient was independent of z and $\Delta\beta$ is now replaced by $(\Delta\beta - K)$. Maximum efficiency occurs when $(\Delta\beta - K) = 0$ and the efficiency reduces as the value of $(\Delta\beta - K)$ deviates from zero.

The amplitude of the second harmonic when $K = \Delta\beta$ will be given by

$$A_2(L) = \frac{\omega\Gamma}{cn_2}\frac{2d_{eff}}{\pi}A_1^2 L \tag{11.73}$$

Thus, as compared to the perfectly phase-matched situation with $\Delta\beta = 0$ (Eq. 11.47), in this case the effective nonlinear coefficient has reduced by a factor of $\frac{2}{\pi}$. Since the efficiency is proportional to the square of the nonlinear coefficient, this implies that the efficiency reduces by a factor of $\left(\frac{2}{\pi}\right)^2 \approx 0.4$. Thus, compared to perfectly phase-matched case, the efficiency reduces by a factor of about 0.4 due to quasi phase matching.

Figure 11.6 shows a schematic representation of QPM process in which when the phase difference between the nonlinear polarization and the electromagnetic field that it is generating becomes π, the sign of the nonlinear coefficient is reversed which results in bringing the two fields back in phase.

Figure 11.7 shows a schematic of a domain reversed crystal where the arrows point towards the principal z-axis direction. Figure 11.8 shows the corresponding vector diagram when the first-order QPM condition is exactly satisfied.

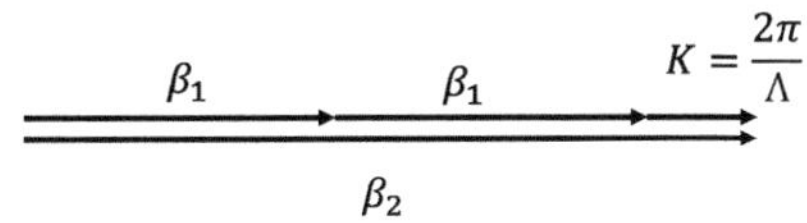

Fig. 11.8 Vector diagram corresponding to first-order QPM

For lithium niobate for a fundamental wavelength of 1060 nm, assuming bulk crystal, we have $n_e(\omega) = 2.1561$; $n_e(2\omega) = 2.2355$ giving us a domain reversal period of approximately 6.7 μm.

The duty cycle of the periodic modulation is defined by the ratio of the width of the region having $+d$ to that of the period of the modulation. Thus, in the discussions given above, the duty cycle is assumed to be 50%. Different Fourier components are referred to as different orders of quasi phase matching and have different spatial frequency components. Thus, if the coefficient corresponding to $m = 1$ is used for quasi phase matching, then it corresponds to first-order QPM when the component corresponding to the factor e^{iKz} is used in QPM. Similarly the Fourier coefficient of the term e^{iKmz} which has a spatial frequency mK is used, then it is referred to as the mth order QPM. From Exercise 11.9, it follows that for the chosen duty cycle of 50%, the even order Fourier coefficients are absent. Third-order QPM corresponds to the spatial frequency $3K$ which is thrice that corresponding to the first-order QPM. However, notice that the magnitude of the corresponding Fourier coefficient is $d_3 = d\frac{2}{3\pi}$ and if hence is this coefficient is used in the nonlinear interaction, this would lead to an effective nonlinear coefficient which is $\frac{2}{3\pi} \approx 0.21$ times the actual nonlinear coefficient resulting in an efficiency of only 4.5% of the efficiency that would be possible using the entire d coefficient.

As an example consider an interaction requiring a QPM period of 3 μm for the first-order QPM process. This corresponds to a spatial frequency of $= \frac{2\pi}{3} \approx 2.09$ μm^{-1}. If fabrication of a period of 3 μm is difficult, it is possible to use the third-order QPM for the interaction. If the period of the fabricated grating is Λ_3 with the corresponding spatial frequency K_3, then we must have $3K_3 = K$ which implies

$$3\frac{2\pi}{\Lambda_3} = \frac{2\pi}{3}$$

Thus, the required QPM period is

$$\Lambda_3 = 9 \ \mu\text{m}$$

which is three times the period required for first-order QPM and which is much more easier to fabricate. However, as mentioned before using this would result in a drop of efficiency as compared to first-order QPM of

$$\left(\frac{d_3}{d_1}\right)^2 = \left(\frac{2/3\pi}{2/\pi}\right)^2 \approx \frac{1}{9}$$

a significant reduction indeed.

Here we have considered a QPM grating with a single constant period. We have seen that it is a Fourier component of the periodic modulation of the nonlinear coefficient that is responsible for enhanced efficiency of nonlinear interaction. It is possible to choose a variation in the nonlinear coefficient that has multiple spatial periods (see Chap. 23) or more complex grating profiles. This is referred to as domain engineering and provides us with possible designs to achieve different nonlinear interactions (see, e.g., Norton and de Sterke 2004; Asobe et al. 2005; Thomas et al. 2016; Tambasco et al. 2016; Babai-Hemati et al. 2024).

Exercise 11.9 Consider a periodic domain reversed crystal with a period Λ with the nonlinear coefficient $+d$ over a length l and a nonlinear coefficient $-d$ over the length $(l - \Lambda)$ corresponding to a duty cycle of l/Λ. Make a Fourier series expansion as given by Eq. (11.69) and obtain the Fourier coefficients.

Solution We write for the nonlinear coefficient variation

$$\begin{aligned} d(z) &= +d; \quad 0<z<l \\ &= -d; \quad l<z<\Lambda \end{aligned} \tag{11.74}$$

Writing the Fourier series

$$d(z) = d \sum_{n=-\infty}^{+\infty} G_n e^{-inKz} \tag{11.75}$$

Multiplying both sides by e^{+imKz} and integrating from 0 to Λ, we have

$$\int_0^\Lambda d(z) e^{+imKz} dz = d \sum_{m=-\infty}^{+\infty} G_n \int_0^\Lambda e^{+imKz} e^{-inKz} dz$$

The integral on the right hand side will equal Λ for $n = m$ and zero otherwise. Hence

$$G_m = \frac{1}{\Lambda}\left(\int_0^l e^{+imKz} dz - \int_l^\Lambda e^{+imKz} dz\right)$$

Carrying out the integration and using the fact that $K\Lambda = 2\pi$, we obtain

$$G_m = \frac{2}{m\pi} \sin\left(\frac{m\pi l}{\Lambda}\right) e^{im\pi l/\Lambda} \tag{11.76}$$

A duty cycle of 0.5 corresponds to $\frac{l}{\Lambda} = 0.5$. Thus, in this case we obtain $G_m = \frac{2}{m\pi} \sin\left(\frac{m\pi}{2}\right) e^{im\pi/2}$ giving $G_1 = i\frac{2}{\pi}$, $G_{-1} = -i\frac{2}{\pi}$; $G_3 = -i\frac{2}{3\pi}$; $G_{-3} = -i\frac{2}{3\pi}$, etc.

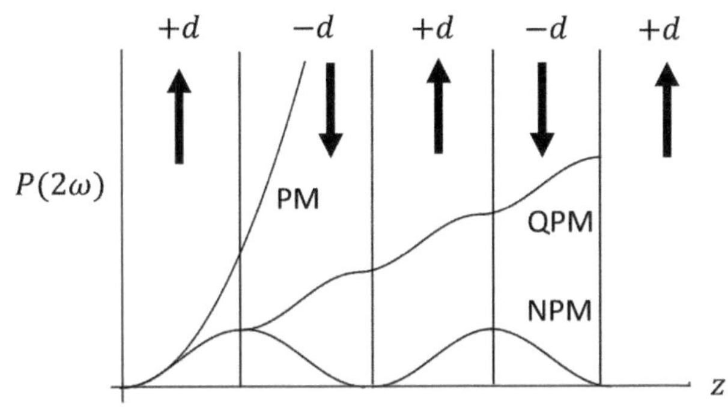

Fig. 11.9 Comparison of SHG power generation using perfect phase-matched operation (PM), nonphased operation (NPM) and quasi phase-matched operation (QPM)

Exercise 11.10 If I wish to use second-order QPM what duty cycle of domain reversal should I choose to maximize the efficiency?

Solution For second-order QPM, we will use the Fourier coefficient G_2. For maximizing this we must have $\sin\left(\frac{2\pi l}{\Lambda}\right) = 1$ which implies $\frac{l}{\Lambda} = 0.25$ or 0.75. Hence, for second-order QPM, the duty cycle of the domain reversal should be either 0.25 or 0.75.

Figure 11.9 shows a schematic of the comparison of the variation in power in the second harmonic between perfect phase-matched operation, nonphase-matched operation, and quasi phase-matched operations. As can be seen, although QPM is not as effective as perfect phase matching, it provides us with a technique to achieve high efficiencies as the power in the second harmonic monotonically increases.

Exercise 11.11 The nonlinear tensor of KTP (which is an important crystal in integrated quantum photonics) is given by (see Chap. 10)

$$\bar{\bar{d}} = \begin{pmatrix} 0 & 0 & 0 & 0 & d_{15} & 0 \\ 0 & 0 & 0 & d_{24} & 0 & 0 \\ d_{31} & d_{32} & d_{33} & 0 & 0 & 0 \end{pmatrix}$$

Consider SHG in a waveguide fabricated in such a waveguide. If the propagation direction is along x, obtain the nonlinear polarization for different input polarizations and the corresponding quasi phase matching conditions that need to be satisfied for efficient SHG.

Solution Since propagation direction is along the x-direction, the incident mode can be predominantly y-polarized or predominantly z-polarized or could have excitation of both components. Thus, we assume the incident mode at the fundamental frequency to have both the components. If we represent the components by $E_y^{(\omega)}$ and $E_z^{(\omega)}$, then the nonlinear polarization generated will be given by

$$\begin{pmatrix} \tilde{P}_x \\ \tilde{P}_y \\ \tilde{P}_z \end{pmatrix} = \epsilon_0 \begin{pmatrix} 0 & 0 & 0 & 0 & d_{15} & 0 \\ 0 & 0 & 0 & d_{24} & 0 & 0 \\ d_{31} & d_{32} & d_{33} & 0 & 0 & 0 \end{pmatrix} \begin{pmatrix} 0 \\ E_y^{(\omega)} E_y^{(\omega)} \\ E_z^{(\omega)} E_z^{(\omega)} \\ 2E_y^{(\omega)} E_z^{(\omega)} \\ 0 \\ 0 \end{pmatrix}$$

$$= \epsilon_0 \begin{pmatrix} 0 \\ 2d_{24} E_y^{(\omega)} E_z^{(\omega)} \\ d_{32} E_y^{(\omega)2} + d_{33} E_z^{(\omega)2} \end{pmatrix}$$

If the incident mode is y-polarized, then $E_z^{(\omega)} = 0$ and the nonlinear polarization has only the z-component and given by

$$\tilde{P}_z = \epsilon_0 \, d_{32} E_y^{(\omega)2}$$

This polarization will generate a second harmonic mode polarized along the z-direction having an effective index $n_{eff,z}(\omega)$. Since the fundamental mode is y-polarized, its effective index would be $n_{eff,y}(\omega)$. Thus, the quasi phase matching condition would be

$$\frac{2\omega}{c} n_{eff,z}(2\omega) - 2\frac{\omega}{c} n_{eff,y}(\omega) - K_1 = 0$$

where K_1 is the corresponding spatial frequency of the QPM grating. This interaction will lead to a second harmonic polarized along the z-direction and the interaction will be determined by the nonlinear coefficient d_{32}. This corresponds to Type I interaction.

Similarly if the incident mode is z-polarized, then $E_y^{(\omega)} = 0$ and the nonlinear polarization has only the z-component and given by

$$\tilde{P}_z = \epsilon_0 d_{33} E_z^{(\omega)2}$$

The corresponding QPM condition would be given by

$$\frac{2\omega}{c} n_{eff,z}(2\omega) - 2\frac{\omega}{c} n_{eff,z}(\omega) - K_2 = 0$$

where K_2 is the corresponding spatial frequency of the QPM grating. This interaction will lead to a second harmonic polarized along the z-direction and the interaction will be determined by the nonlinear coefficient d_{33}. This corresponds to Type 0 interaction.

If the incident mode at the fundamental frequency has both y- and z-components, then the nonlinear polarization has both y- and z-components given by

$$\tilde{P}_y = 2\epsilon_0 d_{24} E_y^{(\omega)} E_z^{(\omega)}$$

$$\tilde{P}_z = \epsilon_0 d_{32} E_y^{(\omega)^2} + \epsilon_0 d_{33} E_z^{(\omega)^2}$$

The y-component $\tilde{P}_y$ will generate a y-polarized second harmonic which will be determined by the nonlinear coefficient d_{24}. In this interaction, the input has both orthogonal polarization components leading to a second harmonic along y-direction. This is referred to as Type II interaction.

The corresponding QPM condition is given by

$$\frac{2\omega}{c} n_{eff,y}(2\omega) - \frac{\omega}{c} n_{eff,z}(\omega) - \frac{\omega}{c} n_{eff,y}(\omega) - K_3 = 0$$

The z-component $\tilde{P}_z$ will generate a z-polarized second harmonic which will be determined by the nonlinear coefficient d_{32} and d_{33}.

Exercise 11.12 I wish to achieve SHG using QPM in a z-cut, x-propagating waveguide in lithium niobate. The incident mode at ω is a quasi TE mode, while the generated mode at 2ω is a quasi TM mode. What is the spatial period for first-order QPM? Given the values of the effective indices: at 1000 nm: $n_o = 2.236$, $n_e = 2.159$; at 500 nm: $n_o = 2.341$ and $n_e = 2.249$.

Solution For the given crystal orientation, the quasi TE mode will be an ordinary polarization and the quasi TM mode will be having extraordinary polarization. Thus, the QPM condition is given by

$$K = \frac{2\omega}{c} n_{eff,e}(2\omega) - \frac{2\omega}{c} n_{eff,o}(\omega) = \frac{2\omega}{c} \left(n_{eff,e}(2\omega) - n_{eff,o}(\omega) \right)$$

Thus, the effective index of the interacting mode at ω is $n_{eff,\,o}(\omega) = 2.236$ and the effective index of the generated mode at 2ω will be $n_{eff,\,e}(2\omega) = 2.249$. Hence, the required first-order QPM period would be $\Lambda = \frac{1}{2(2.249 - 2.236)} \approx 38.5\ \mu\text{m}$.

11.7 Bandwidth of the Interaction

As we have seen, for a given fundamental frequency ω and its second harmonic 2ω, the effective indices of the waveguide at these two frequencies determine the spatial frequency or the spatial period of quasi phase matching. If a device is fabricated with this grating period, then it will possess high efficiency for SHG corresponding to the design frequency (or center frequency). As the input frequency deviates from this design frequency, for the fixed spatial frequency of the QPM grating, we would no more satisfy the QPM condition exactly and the efficiency would then drop. When the input frequency changes to a value such that $(\Delta\beta - K)\frac{L}{2} = \pi$, the efficiency drops to zero (see Eq. 11.72). Thus, the nonlinear interaction is characterized by a bandwidth which defines the frequency range of the input fundamental frequency over which the second harmonic generation is significant.

Let ω_f represent the fundamental frequency and $\omega_s = 2\omega_f$ represent the second harmonic frequency. In order to estimate the bandwidth, we first note that at the design fundamental frequency ω_{f0}, and the corresponding second harmonic frequency ω_{s0} the following QPM condition is exactly satisfied:

$$\Delta\beta(\omega_{f0}) - K = \beta(\omega_{s0}) - 2\beta(\omega_{f0}) - K = 0 \tag{11.77}$$

As we have discussed earlier, when the fundamental frequency changes to a value such that $\Delta\beta - K = \pm\frac{2\pi}{L}$, then the second harmonic efficiency becomes zero. Let the corresponding fundamental frequency be $\omega_f = \omega_{f0} \pm \Delta\omega_f$ with a corresponding second harmonic frequency of $\omega_s = \omega_{s0} \pm \Delta\omega_s$ where $\Delta\omega_s = 2\Delta\omega_f$. Thus, the efficiency becomes zero when

$$\Delta\beta(\omega_{f0} + \Delta\omega_f) - K = \beta(\omega_{s0} + \Delta\omega_s) - 2\beta(\omega_{f0} + \Delta\omega_f) - K = \pm\frac{2\pi}{L}$$

Assuming $\Delta\omega_f \ll \omega_{f0}$ and $\Delta\omega_s \ll \omega_{s0}$, we can make Taylor series expansion and write

$$\left[\beta(\omega_{s0}) + \Delta\omega_s \frac{d\beta}{d\omega}\bigg|_{\omega=\omega_{s0}}\right] - 2\left[\beta(\omega_{f0}) + \Delta\omega_f \frac{d\beta}{d\omega}\bigg|_{\omega=\omega_{f0}}\right] - K = \pm\frac{2\pi}{L}$$

which using Eq. (11.77) becomes

$$\Delta\omega_s \frac{d\beta}{d\omega}\bigg|_{\omega=\omega_{s0}} - 2\Delta\omega_f \frac{d\beta}{d\omega}\bigg|_{\omega=\omega_{f0}} = \pm\frac{2\pi}{L}$$

or

$$2\Delta\omega_f \left(\frac{d\beta}{d\omega}\bigg|_{\omega=\omega_{s0}} - \frac{d\beta}{d\omega}\bigg|_{\omega=\omega_{f0}}\right) = \pm\frac{2\pi}{L} \tag{11.78}$$

Now, $\frac{d\beta}{d\omega}$ is related to the group index $N_{eff}(\omega)$ through the following equation (see Eq. 4.13):

$$\frac{d\beta}{d\omega} = \frac{N_{eff}(\omega)}{c}$$

where

$$N_{eff}(\omega) = \left(n_{eff} + \omega \frac{dn_{eff}}{d\omega}\right)$$

Thus, Eq. (11.78) becomes

$$\frac{2\Delta\omega_f}{c}\left(N_{eff}(\omega_{s0}) - N_{eff}(\omega_{f0})\right) = \pm\frac{2\pi}{L}$$

or

$$\Delta\omega_f = \pm\frac{\pi c}{L\left(N_{eff}(\omega_{s0}) - N_{eff}(\omega_{f0})\right)} \tag{11.79}$$

which gives us an estimate of the bandwidth of the interaction process. The SHG peaks at $\omega_f = \omega_{f0}$ and the efficiency drops to zero at a fundamental frequency $\omega_f = \omega_{f0} \pm \Delta\omega_f$ with $\Delta\omega_f$ given by Eq. (11.79).

We can write Eq. (11.79) in terms of wavelength by using the fact that

$$\omega_f = \frac{2\pi c}{\lambda_f}$$

where λ_f is the free space wavelength corresponding to the fundamental frequency. Hence,

$$\Delta\omega_f = -\frac{2\pi c}{\lambda_f^2}\Delta\lambda_f$$

and Eq. (11.79) becomes

$$\Delta\lambda_f = \pm\frac{\lambda_{f0}^2}{2L\left(N_{eff}(\lambda_{f0}) - N_{eff}\left(\frac{\lambda_{f0}}{2}\right)\right)} \tag{11.80}$$

Thus, the second harmonic has maximum efficiency at the incident wavelength λ_{f0} and the efficiency drops to zero when the incident wavelength changes to $\lambda_{f0} \pm \Delta\lambda_f$. Hence, $\Delta\lambda_f$ gives an estimate of the bandwidth of the interaction process.

We also note that

$$\left\{\frac{\sin\left[\frac{(\Delta\beta - K)L}{2}\right]}{\frac{(\Delta\beta - K)L}{2}}\right\}^2 = 0.5 \tag{11.81}$$

when $\frac{(\Delta\beta - K)L}{2} \approx 1.39$ or when $(\Delta\beta - K)L \approx 2.78$. Using a similar procedure as above, we can thus obtain the full width at half maximum (FWHM) of the interaction process to be

$$\Delta\lambda_{f,FWHM} = \frac{0.44\lambda_f^2}{L\left(N_{eff}(\lambda_{f0}) - N_{eff}\left(\frac{\lambda_{f0}}{2}\right)\right)} \tag{11.82}$$

This equation can also be written as

$$\Delta\lambda_{f,FWHM} = \frac{0.44\lambda_{f0}}{L\left|\frac{n_{eff}(\lambda_{s0}) - n_{eff}(\lambda_{f0})}{\lambda_{f0}} + \left.\frac{dn_{eff}}{d\lambda}\right|_{\lambda=\lambda_{f0}} - \frac{1}{2}\left.\frac{dn_{eff}}{d\lambda}\right|_{\lambda=\lambda_{s0}}\right|} \tag{11.83}$$

We immediately note that longer interaction lengths are characterized by smaller interaction bandwidths. Thus, increasing the length of interaction to increase efficiency also decreases the bandwidth and any slight deviation of the input frequency from the design frequency would lead to drop in efficiency.

The bandwidth also depends on the group indices (which in turn depend on the dispersion, $\frac{dn_{eff}}{d\lambda}$, of the effective indices) of the waveguide modes at the fundamental frequency and the second harmonic frequency. The dispersion of the effective indices depends both on the refractive index of the medium at the two frequencies but also on the waveguide geometry and index increase. In principle, it is possible to modify the bandwidth for a given material using waveguide engineering.

We may also note that if instead of quasi phase matching, we had conventional phase matching, then this would imply $n_{eff}(\lambda_{s0}) = n_{eff}(\lambda_{f0})$ and the bandwidth expression would be

$$\Delta\lambda_f = \frac{0.44\lambda_{f0}}{L\left|\left.\frac{dn_{eff}}{d\lambda}\right|_{\lambda=\lambda_{f0}} - \frac{1}{2}\left.\frac{dn_{eff}}{d\lambda}\right|_{\lambda=\lambda_{s0}}\right|} \tag{11.84}$$

Exercise 11.13 Show that the group index can also be written in terms of free space wavelength λ_0 as

$$N_{eff}(\lambda) = \left(n_{eff} - \lambda_0\frac{dn_{eff}}{d\lambda_0}\right) \tag{11.85}$$

Solution Using $\omega = \frac{2\pi c}{\lambda_0}$, we have

$$\omega \frac{dn_{eff}}{d\omega} = \frac{2\pi c}{\lambda_0}\left(-\frac{\lambda_0^2}{2\pi c}\frac{dn_{eff}}{d\lambda_0}\right) = -\lambda_0 \frac{dn_{eff}}{d\lambda_0}$$

Substituting in the equation for N_{eff}, we obtain Eq. (11.85).

11.8 Calculation of the Effective Nonlinear Coefficient

We have mentioned earlier that the nonlinear coefficient that is responsible for the second harmonic generation depends on the medium, the states of polarization of the fundamental and the second harmonic and also the propagation direction. In this section, we will discuss about the calculation of the effective nonlinear coefficient.

As an example, we will discuss second harmonic generation in lithium niobate. In Chap. 10, we obtained expressions for the nonlinear polarization created in lithium niobate. Here we will discuss specific examples and show how the polarization state of the interacting waves and the propagation direction define the specific nonlinear coefficient responsible for the second harmonic.

We first consider the case of a channel waveguide fabricated in z-cut (i.e., the optic axis of the crystal is perpendicular to the surface of the substrate) lithium niobate with the propagation direction being along the x-direction. We also assume the incident wave at ω to be an extraordinary wave, that, it is polarized along the z-direction with its electric field perpendicular to the surface of the substrate. This case was discussed in Example 10.1 for plane wave propagation in bulk.

Since the interaction is taking place in a waveguide and the wave at the fundamental frequency is polarized along the z-direction, being an extraordinary polarization, we have for the electric field of the fundamental mode

$$\vec{E}^{(\omega)}(x,y,z,t) = \frac{1}{2}\left[\hat{z}A_1(x)\psi_{1e}(y,z)e^{i\left(\beta_e^{(\omega)}x-\omega t\right)} + c.c.\right] \tag{11.86}$$

where the subscript e corresponds to the fact that the mode at frequency ω is an extraordinary mode and the propagation constant of the mode is given by $\beta_e^{(\omega)}$ and $\psi_{1e}(y,z)$ represents the corresponding modal field of the extraordinary mode in the waveguide. Note that we are using the principal axes directions as the coordinate axes and the mode is propagating along the x-direction and the transverse coordinates are y and z.

Since the electric field of the mode at the fundamental frequency has only the z-component, the nonlinear polarization at the second harmonic will have only the z-component and will be given by (see Example 10.1)

$$\vec{P}^{(2\omega)} = \hat{z}P_z^{(2\omega)} = \frac{1}{2}\left(\hat{z}\epsilon_0 d_{33}A_1^2\psi_{1e}^2(y,z)e^{2i\left(\beta_e^{(\omega)}x-\omega t\right)} + c.c.\right) \tag{11.87}$$

Comparing this with Eq. (11.19) shows that in this case the nonlinear coefficient responsible for the second harmonic generation is d_{33} and $\beta_1 = \beta_e^{(\omega)}$. Since the z-

component of the nonlinear polarization generates an extraordinary polarized mode at the second harmonic frequency with a propagation constant given by $\beta_e^{(2\omega)}$, the phase matching condition will be given by

$$\beta_e^{(2\omega)} = 2\beta_e^{(\omega)} \tag{11.88}$$

and the spatial period for efficient quasi phase matching will be given by

$$\Lambda = \frac{2\pi}{\beta_e^{(2\omega)} - 2\beta_e^{(\omega)}} \tag{11.89}$$

In terms of effective indices of the modes defined by

$$\beta_e^{(\omega)} = \frac{\omega}{c} n_{eff,e}^{(\omega)} \tag{11.90}$$

and

$$\beta_e^{(2\omega)} = \frac{2\omega}{c} n_{eff,e}^{(2\omega)} \tag{11.91}$$

Equation (11.89) can be written as

$$\Lambda = \frac{\pi c}{\omega \left(n_{eff,e}^{(2\omega)} - n_{eff,e}^{(\omega)} \right)}$$

or in terms of the wavelength λ_f of the fundamental wave,

$$\Lambda = \frac{\lambda_f}{2\left(n_{eff,e}^{\left(\frac{\lambda_f}{2}\right)} - n_{eff,e}^{(\lambda_f)} \right)} \tag{11.92}$$

The effective indices at the fundamental frequency and the second harmonic frequency depend on the waveguide characteristics. Thus, apart from the medium, the waveguide profile can also be used to achieve quasi phase matching in waveguides.

We now consider another example corresponding to Example 10.2 in which the incident polarization of the fundamental wave excites both the ordinary polarized mode (polarization along the y-direction parallel to the substrate surface) and the extraordinary polarized mode (polarization along the z-direction) perpendicular to the substrate surface.

Now the electric field of the incident fundamental mode will have both the ordinary and the extraordinary components and will be given by

$$\vec{E}^{(\omega)}(x,y,z,t)=\frac{1}{2}\left[\widehat{z}A_{1e}(x)\psi_{1e}(y,z)e^{i\left(\beta_e^{(\omega)}x-\omega t\right)}+c.c.\right]$$
$$+\frac{1}{2}\left[\widehat{y}A_{1o}(x)\psi_{1o}(y,z)e^{i\left(\beta_o^{(\omega)}x-\omega t\right)}+c.c.\right] \tag{11.93}$$

where $\psi_{1o}(y,z)$ represents the transverse spatial modal field of the ordinary polarization and $\beta_o^{(\omega)}$ is the propagation constant of the mode. The factors $A_{1e}(x)$ and $A_{1o}(x)$ represent the corresponding amplitudes of the extraordinary and ordinary components of the incident wave, respectively. The subscript o represents ordinary polarization.

As discussed earlier, in this case the nonlinear polarization will have both y and z components. The y- and z-components of the nonlinear polarization will now be given by

$$P_y^{(2\omega)}=\frac{1}{2}\left(\epsilon_0 d_{22}A_{1o}^2\psi_{1o}^2(y,z)e^{2i\left(\beta_o^{(\omega)}x-\omega t\right)}+c.c.\right)$$
$$+\frac{1}{2}\left(2\epsilon_0 d_{15}A_{1o}A_{1e}\psi_{1o}(y,z)\psi_{1e}(y,z)e^{i\left(\left[\beta_o^{(\omega)}+\beta_e^{(\omega)}\right]x-2\omega t\right)}+c.c.\right) \tag{11.94}$$

$$P_z^{(2\omega)}=\frac{1}{2}\left(\epsilon_0 d_{31}A_{1o}^2\psi_{1o}^2(y,z)e^{2i\left(\beta_o^{(\omega)}x-\omega t\right)}+c.c.\right)$$
$$+\frac{1}{2}\left(\epsilon_0 d_{33}A_{1e}^2\psi_{1e}^2(y,z)e^{2i\left(\beta_e^{(\omega)}x-\omega t\right)}+c.c.\right) \tag{11.95}$$

The y-component of the nonlinear polarization will be able to efficiently generate the y-component (corresponding to the ordinary polarization) of the second harmonic field with a propagation constant $\beta_o^{(2\omega)}$ if the following phase matching condition is satisfied

$$\beta_o^{(2\omega)}=2\beta_o^{(\omega)} \tag{11.96}$$

or the following quasi phase matching condition with a periodic domain reversal with a spatial frequency K_1

$$\beta_o^{(2\omega)}=2\beta_o^{(\omega)}+K_1 \tag{11.97}$$

and the effective nonlinear coefficient responsible for this would be d_{22}. In this case, the extraordinary component of the incident field would play no role in the process.

The y-component of the nonlinear polarization can also generate the y -component of the second harmonic field provided the following phase matching condition is satisfied:

$$\beta_o^{(2\omega)}=\beta_o^{(\omega)}+\beta_e^{(\omega)} \tag{11.98}$$

or by using quasi phase matching which would need a spatial frequency K_2 satisfying the following equation:

$$\beta_o^{(2\omega)} = \beta_o^{(\omega)} + \beta_e^{(\omega)} + K_2 \tag{11.99}$$

In this case, the effective nonlinear coefficient responsible for the second harmonic generation would be d_{15} and the process would involve the amplitudes and field profiles of both the ordinary and extraordinary input modes corresponding to the fundamental frequency.

In a similar fashion, the z-component of the nonlinear polarization can lead to the generation of the extraordinary wave at the second harmonic frequency requiring either of the following quasi phase matching conditions:

$$\beta_e^{(2\omega)} = 2\beta_o^{(\omega)} + K_3 \tag{11.100}$$

with a quasi phase-matched grating with spatial frequency K_3 and which uses the nonlinear coefficient d_{31} or the following quasi phase matching condition

$$\beta_e^{(2\omega)} = 2\beta_e^{(\omega)} + K_4 \tag{11.101}$$

with a quasi phase-matched grating with spatial frequency K_4 which uses the nonlinear coefficient d_{33}.

Note that in both the cases involving $P_z^{(2\omega)}$, it is either the ordinary or the extraordinary polarization components at the fundamental frequency that creates the nonlinear polarization and there is no mixing of the two modes. Thus, these processes are no different from just exciting either the ordinary mode or the extraordinary mode at the fundamental frequency.

The same process that leads to second harmonic generation which is an up conversion process (i.e., increase in frequency of the light from ω to 2ω) also leads to down conversion process (where the higher frequency light at 2ω gets converted to lower frequency light at ω.) The various phase matching conditions that we have discussed above are also the same phase matching conditions for the down conversion process. We will discuss the down conversion process in greater detail in Chap. 12.

Exercise 11.14 Consider a y-cut, x-propagating lithium niobate waveguide with modes at 800 having effective indices of 2.17757 and 2.25775 corresponding to z-polarized mode and y-polarized modes, respectively. At 400 nm, the effective indices are 2.3355 and 2.4428 corresponding to z-polarized mode and y-polarized modes, respectively. (a) For an input TE_0 mode at 800 nm, what is the QPM period required for Type 0 SHG? (b) Obtain the corresponding spatial period required for Type I interaction with input light corresponding to TM_0 mode. (c) What is the corresponding spatial period required for Type II interaction with TE_0 and TM_0 mode inputs at 800 nm?

Solution

(a) Type 0 interaction (*zzz*) (with both the fundamental and the second harmonic modes corresponding to extraordinary polarization) which in this case corresponds to all being in the TE_0 modes. Hence, the QPM condition is

$$K_1 = \beta_2(TE_0) - 2\beta_1(TE_0)$$

giving the corresponding spatial period as

$$\Lambda_1 = \frac{\lambda_f}{2\left[n_{eff}^{(2\omega)}(TE_0) - n_{eff}^{(\omega)}(TE_0)\right]} = \frac{0.8}{2(2.3355 - 2.17757)} \approx 2.53\ \mu\text{m}$$

This interaction is determined by the $d_{33} = d_{zzz}$ nonlinear coefficient of lithium niobate.

(b) In the case of Type I (*zyy*) interaction (with the fundamental having ordinary polarization and the second harmonic having extraordinary polarization), the QPM condition will be

$$K_2 = \beta_2(TE_0) - 2\beta_1(TM_0)$$

which gives the corresponding spatial period as

$$\Lambda_2 = \frac{\lambda_f}{2\left[n_{eff}^{(2\omega)}(TE_0) - n_{eff}^{(\omega)}(TM_0)\right]} = \frac{0.8}{2(2.3355 - 2.25775)} \approx 5.14\ \mu\text{m}$$

This interaction is determined by the nonlinear coefficient $d_{32} = d_{zyy}$.

(c) In the case of Type II (*yyz*) interaction (with the fundamental having both ordinary and extraordinary polarization and the second harmonic having ordinary polarization), the QPM condition will be

$$K_3 = \beta_2(TM_0) - \beta_1(TE_0) - \beta_1(TM_0)$$

which gives the corresponding spatial period as

$$\Lambda_3 = \frac{\lambda_f}{\left[2n_{eff}^{(2\omega)}(TM_0) - n_{eff}^{(\omega)}(TE_0) - n_{eff}^{(\omega)}(TM_0)\right]}$$

$$= \frac{0.8}{(2 \times 2.4428 - 2.17757 - 2.25775)} \approx 1.8\ \mu\text{m}$$

This interaction is determined by the nonlinear coefficient $d_{24} = d_{yyz}$.

Since the required grating periods are small, it is feasible to use third-order QPM to achieve efficient SHG. For example for first-order Type 0 QPM, the required period calculated is 2.53 μm. The required period for third-order QPM would be 7.59 μm, which is easier to realize.

Exercise 11.15 Consider SHG for a fundamental wavelength of 1600 nm in a y-cut, x-propagating lithium niobate waveguide. The effective indices of the modes at 800 are 2.17757 and 2.25775 corresponding to z-polarized mode and y-polarized modes, respectively. At 1600 nm, the effective indices are 2.1367 and 2.21051 corresponding to z-polarized mode and y-polarized modes, respectively. (a) For an input TE_0 mode at 1600 nm, what is the QPM period required for Type 0 SHG? (b) Obtain the corresponding spatial period required for Type I interaction with input light corresponding to TM_0 mode. (c) What is the corresponding spatial period required for Type II interaction with TE_0 and TM_0 modes?

Solution For Type 0 interaction, the required QPM period is

$$\Lambda_1 = \frac{\lambda_f}{2\left[n_{eff}^{(2\omega)}(TE_0) - n_{eff}^{(\omega)}(TE_0)\right]} = \frac{1.6}{2(2.17757 - 2.1367)} \approx 19.6\ \mu\text{m}$$

For Type I interaction, the required QPM period is

$$\Lambda_2 = \frac{\lambda_f}{2\left[n_{eff}^{(2\omega)}(TE_0) - n_{eff}^{(\omega)}(TM_0)\right]} = \frac{1.6}{2|2.17757 - 2.21051|} \approx 24.3\ \mu\text{m}$$

For Type II interaction, the required QPM period is

$$\Lambda_3 = \frac{\lambda_f}{\left[2n_{eff}^{(2\omega)}(TM_0) - n_{eff}^{(\omega)}(TE_0) - n_{eff}^{(\omega)}(TM_0)\right]}$$
$$= \frac{1.6}{(2 \times 2.25775 - 2.1367 - 2.21051)} \approx 9.5\ \mu\text{m}$$

These QPM periods are the same for the reverse process of generation of 1600 nm from 800 nm using parametric down conversion that will be discussed in Chaps. 12 and 13.

Problems

11.1. The effective index of the modes of a waveguide at wavelengths of 800 and 400 nm is given by 2.176 and 2.334, respectively. What are the velocities of the nonlinear polarization generated at 2ω and the electromagnetic wave at 2ω?

11.2. In an SHG experiment, the second harmonic conversion efficiency is 1% when the input power is 1 W. If the input wavelength is 800 nm, how many photons at the second harmonic frequency are exiting per unit time from the medium?

11.3. Consider SHG generation with two nonlinear crystals having lengths of L and $2L$. In which of these two cases will the efficiency decrease faster from its corresponding maximum value as the temperature is varied?

11.4. Show that the nonlinear polarization generated at the fundamental frequency ω is given by Eq. (11.20).

11.5. I wish to generate second harmonic in $LiNbO_3$ waveguide in a z-cut y-propagating substrate The incident mode at a wavelength of 1.5 μm has an ordinary polarization and that at 0.75 μm has an extraordinary polarization. The effective indices of the modes of the waveguide at 1.5 and 0.75 μm are given below:

$$n_o(1.5\ \mu\text{m}) = 2.214; \quad n_e(1.5\ \mu\text{m}) = 2.141$$
$$n_o(0.75\ \mu\text{m}) = 2.264; \quad n_e(0.75\ \mu\text{m}) = 2.183$$

If I wish to use QPM, what spatial periods would be required to achieve first-order and third-order QPM? (a) If both the modes at the fundamental and second harmonic are extraordinary polarizations, what will be the corresponding QPM period? In both the cases, which nonlinear coefficient will be involved in the interaction?

11.6. In an experiment on SHG, first-order quasi phase matching is achieved by having $d(z) = d$ for $0 < z < \frac{\Lambda}{2}$ and $d(z) = 0$ for $\frac{\Lambda}{2} < z < \Lambda$ and $d(z + \Lambda) = d(z)$. Make a Fourier series expansion and obtain the value of effective nonlinear coefficient for first-order QPM. How does it compare with the d variation considered in Sec. 11.6?

11.7. Consider second harmonic generation (SHG) for a fundamental wavelength of 1 μm over a crystal of length L. The efficiency of SHG is given by Eq. (11.43).

(a) Plot the variation in $\frac{\eta}{\eta_0}$ with L in the range 0–5 cm for $\Delta\beta = 0$, $\Delta\beta = \frac{\pi}{5}$ cm^{-1} and $\Delta\beta = \frac{\pi}{20}$ cm^{-1}.

(b) Plot the variation in $\frac{\eta}{\eta_0}$ with $\Delta\beta$ for $L = 5$ mm, $L = 10$ mm and $L = 50$ mm.

(c) What can one conclude from these plots?

11.8. The d matrix of potassium niobate (a biaxial crystal) is given

$$d = \begin{pmatrix} 0 & 0 & 0 & 0 & d_{15} & 0 \\ 0 & 0 & 0 & d_{24} & 0 & 0 \\ d_{31} & d_{32} & d_{33} & 0 & 0 & 0 \end{pmatrix}$$

At a wavelength of 926 nm and 463 nm, the principal refractive indices of the crystal are $n_z(\omega) = 2.270$ and $n_z(2\omega) = 2.447$.

(a) Which nonlinear coefficient will be involved in the Type 0 interaction?
(b) What is the corresponding spatial period required for efficient second harmonic generation of 463 nm?

11.9. Consider SHG in a periodically poled *z*-cut *y*-propagating lithium niobate crystal; here *xyz* represents the principal axes of the crystal.

(a) To use the periodic variation in d_{31}, what should be the polarization state of the ω and 2ω waves?
(b) Obtain an expression for the corresponding period Λ of domain inversion required for third-order QPM.

11.10. Consider SHG for a fundamental wavelength of 1000 nm in a waveguide of length 20 mm. What is the maximum allowed value of $(n_{eff}(2\omega) - n_{eff}(\omega))$ for which the maximum efficiency reduces due to nonphase-matched operation to 50%.

11.11. KTP is a biaxial crystal and the dispersion of the refractive indices $n_y(\lambda)$ and $n_z(\lambda)$ are given by the following equations (Fan et al. 1987)

$$n_y(\lambda) = \sqrt{2.19229 + \frac{0.83547}{(1 - 0.04970\lambda^{-2})} - 0.01621\lambda^2}$$

$$n_z(\lambda) = \sqrt{2.25411 + \frac{1.06543}{(1 - 0.05486\lambda^{-2})} - 0.02140\lambda^2}$$

with λ is the free space wavelength in micrometers.

(a) For an input fundamental wavelength of 1000 nm, obtain the required QPM period for efficient SHG for Type 0 interaction.
(b) What will be the corresponding QPM required for Type II phase matching?

11.12. At high efficiencies, we cannot make the no pump depletion approximation and we need to solve the two coupled equations simultaneously. Solve Eqs. (11.34) and (11.36) under phase-matched operation and obtain the variation in efficiency of SHG with the length of the medium. Show that the solution matches with the no pump depletion approximation for low efficiencies.

11.13. In a waveguide, the variation in $\Delta\beta$ (in units of μm^{-1}) with wavelength λ (measured in μm) is approximately given by the following formula in the region from 990 to 1010 nm (λ_0 is free space wavelength measured in μm):

$$\Delta\beta = \beta_2 - 2\beta_1 = 4.7921 - 3.666\lambda_0$$

(a) What should be the period of first-order QPM for efficient SHG at a fundamental wavelength of 1000 nm?
(b) For a device length of 1 mm estimate the wavelength at which the SHG efficiency will fall to zero.

References

Armstrong, J. A., N. Bloembergen, J. Ducuing, and P. S. Pershan. 1961. Interactions between light waves in nonlinear dielectric. *Phys Rev* 127:1918–1939.

Asobe, M., O. Tadanaga, H. Miyazawa, Y. Nishida, and H. Suzuki. 2005. Multiple quasi-phase matched device using continuous phase modulation of $\chi^{(2)}$ grating and its application to variable wavelength conversion. *IEEE J Quant Electron* 41:1540–1547.

Babai-Hemati, J., F. V. Brugh, H. Hermann, and C. Silberhorn. 2024. Tailored second harmonic generation in Ti-diffused PPLN waveguides using micro-heaters. *Opt Express* 32:6876–6886.

Fan, T. Y., C. E. Huang, B. Q. Hu, R. C. Eckardt, Y. X. Fan, R. L. Byer. and R. S. Feigelson. 1987. Second harmonic generation and accurate index of refraction measurements in flux-grown $KTiOPO_4$, *Appl Opt* 26:2390–2394.

Fejer, M. M., G. A. Magel, D. H. Jundt, and R. L. Byer. 1992. Quasi phase matched second harmonic generation: tuning and tolerances. *IEEE J Quant Electron* 28:2631–2654.

Ghatak, A., and K. Thyagarajan. 1998. *Introduction to fiber optics*. Cambridge: Cambridge University Press.

Myers, L. E., and W. R. Bosenberg. 1997. Periodically poled lithium niobate and quasi phase matched optical parametric oscillators. *IEEE J Quant Electron* 33:1663–1672.

Norton, A. H., and C. M. de Sterke. 2004. Aperiodic 1-dimensional structures for quasi-phase matching. *Opt Express* 12:841–846.

Tambasco, J.-L., A. Boes, L. G. Helt, M. J. Steel, and A. Mitchell. 2016. Domain engineering algorithm for practical and effective photon sources. *Opt Express* 24:19616–19626.

Thomas, A., M. V. Camp, O. Minaeva, D. Simon, and A. V. Sergienko. 2016. Spectrally engineered broadband photon source for two-photon quantum interferometry. *Opt Express* 24:24947–24958.

Yamada, M., N. Nada, M. Saitoh, and K. Watanabe. 1993. First order quasi-phase matched $LiNbO_3$ waveguide periodically poled by applying an external field for efficient blue second harmonic generation. *Appl Phys Lett* 62:435–436.

Yariv, A. 1991. *Optical electronics*. Oxford: Oxford University Press.

Chapter 12
Sum and Difference Frequency Generation

Abstract Sum frequency generation (SFG) and difference frequency generation (DFG) in optical waveguides are discussed in detail in this chapter. Coupled mode equations for three wave interaction are derived and solved under phase matched and non-phase matched interaction and quasi phase matched interaction. Difference frequency generation is shown to lead to optical amplification of the signal through nonlinear interaction. Expressions for the bandwidth of SFG and DFG are also derived.

12.1 Introduction

In Chap. 11 we discussed second harmonic generation in waveguides in which an incident wave at a frequency ω due to second order nonlinearity of the medium generates power at the second harmonic at a frequency 2ω. In this chapter we will consider the incidence of waves at two different frequencies in the waveguide and show that the same second order nonlinearity can lead to the generation of sum frequency (SFG) or difference frequency (DFG). Second harmonic generation is the degenerate case of sum frequency generation when both the input frequencies happen to be equal.

To be specific we represent the three interacting waves, the two incident waves and the newly generated wave with the following nomenclature: we call the waves as *pump* having a frequency ω_p, the *signal* having a frequency ω_s and an *idler* having a frequency ω_i. The three frequencies are assumed to satisfy the following equation:

$$\omega_p = \omega_s + \omega_i \tag{12.1}$$

Thus the pump has the highest frequency. By convention $\omega_s > \omega_i$. The second order nonlinearity in the medium induces coupling among these three frequencies. The process in which input waves at ω_s and ω_i add to generate wave at $\omega_p = \omega_s + \omega_i$ is referred to as sum frequeny generation (SFG). Similarly input waves at frequency ω_p and ω_s generate waves at frequency $\omega_i = \omega_p - \omega_s$ or waves at frequency ω_p and ω_i generate waves at frequency $\omega_s = \omega_p - \omega_i$ are referred to as difference frequency

K. Thyagarajan, *Integrated Quantum Photonics*, Graduate Texts in Physics,
https://doi.org/10.1007/978-3-031-85728-7_12

generation (DFG). We must note that it is the same set of coupled equations that is responsible for describing all these processes. Depending on the input, we will have either sum or difference frequency generation. It is worth noting that if $\omega_s = \omega_i$ then sum frequency generation is nothing but second harmonic generation.

Sum frequency generation can play a major part in quantum information science and technology by providing a technique to convert photons from the infrared region to the near visible region. Quantum communication uses photons in the telecommunication wavelength band of 1550 nm where optical fibers exhibit lowest loss while most of the other quantum processes such as quantum computation, quantum memory or metrology tend to use the wavelength range of 600–900 nm. For this sum frequency generation is a very important technique that can help in preserving the quantum nature of the photons while converting them from the infrared to the near visible range (see, e.g., Kaiser et al. 2019; Tournier et al. 2009).

Difference frequency generation finds applications in converting photons at lower wavelengths to photons at higher wavelengths and in parametric processes for the generation of non-classical states of light such as entangled states and parametric amplification.

We will first obtain the coupled mode equations describing the process on nonlinear mixing among the three waves, pump, signal and idler and then consider sum frequency generation and difference frequency generation processes.

12.2 Coupled Mode Equations

We start with the nonlinear wave equation (see Eq. 11.12):

$$\nabla^2 \vec{E} = \epsilon\mu_0 \frac{\partial^2 \vec{E}}{\partial t^2} + \mu_0 \frac{\partial^2 \vec{P}_{NL}}{\partial t^2} \tag{12.2}$$

which is satisfied by all the individual frequencies propagating through the medium.

As in Chap. 11 we will use a scalar formulation for understanding the fundamental physics behind the interaction process. The nonlinear polarization is given by (see Eq. 11.6):

$$P^{(NL)} = 2\epsilon_0 d E^2 \tag{12.3}$$

where E is the total electric field within the waveguide. Since at any point within the waveguide, we will have waves at pump, signal and idler frequencies, we write the total electric field as

$$E = E_p + E_s + E_i \tag{12.4}$$

where

$$E_p = \frac{1}{2}\left[A_p(z)\psi_p(x,y)e^{i\left(\beta_p z - \omega_p t\right)} + c.c.\right] \tag{12.5}$$

$$E_s = \frac{1}{2}\left[A_s(z)\psi_s(x,y)e^{i(\beta_s z - \omega_s t)} + c.c.\right] \tag{12.6}$$

$$E_i = \frac{1}{2}\left[A_i(z)\psi_i(x,y)e^{i(\beta_i z - \omega_i t)} + c.c.\right] \tag{12.7}$$

where $\psi_p(x, y)$, $\psi_s(x, y)$ and $\psi_i(x, y)$ represent the normalized transverse modal field distributions of the interacting modes corresponding to pump, signal and idler respectively, β_p, β_s and β_i represent the corresponding propagation constants and $A_p(z)$, $A_s(z)$ and $A_i(z)$ represent the z-dependent amplitudes of these waves. We are assuming that the modes are propagating along the z-direction.

The modal fields satisfy the following normalization conditions:

$$\iint \psi_p(x,y)\psi_p(x,y)dxdy = 1 \tag{12.8a}$$

$$\iint \psi_s(x,y)\psi_s(x,y)dxdy = 1 \tag{12.8b}$$

$$\iint \psi_i(x,y)\psi_i(x,y)dxdy = 1 \tag{12.8c}$$

where all the field profiles are assumed to be real.

Substituting for the total electric field in Eq. (12.3) and collecting terms at the frequencies corresponding to pump, signal and idler we have the following expressions for the corresponding nonlinear polarizations:

$$P_{NL}^{(\omega_p)} = \frac{1}{2}\epsilon_0 d\left(2A_s A_i \psi_s(x,y)\psi_i(x,y)e^{i\left((\beta_s+\beta_i)z - \omega_p t\right)} + c.c.\right) \tag{12.9a}$$

$$P_{NL}^{(\omega_s)} = \frac{1}{2}\epsilon_0 d\left(2A_p A_i^* \psi_p(x,y)\psi_i(x,y)e^{i\left((\beta_p-\beta_i)z - \omega_s t\right)} + c.c.\right) \tag{12.9b}$$

$$P_{NL}^{(\omega_i)} = \frac{1}{2}\epsilon_0 d\left(2A_p A_s^* \psi_p(x,y)\psi_s(x,y)e^{i\left((\beta_p-\beta_s)z - \omega_i t\right)} + c.c.\right) \tag{12.9c}$$

Following a procedure similar to our discussion in the case of second harmonic generation in Chap. 11 we can obtain the following three coupled mode equations describing coupling among the three frequencies (see Exercise 12.2):

$$\frac{dA_p}{dz} = i\frac{\omega_p d_{eff}\Gamma}{cn_p}A_s A_i e^{-i\Delta\beta z} \tag{12.10a}$$

$$\frac{dA_s}{dz} = i\frac{\omega_s d_{eff}\Gamma}{cn_s}A_p A_i^* e^{i\Delta\beta z} \tag{12.10b}$$

$$\frac{dA_i}{dz} = i\frac{\omega_i d_{eff}\Gamma}{cn_i}A_p A_s^* e^{i\Delta\beta z} \tag{12.10c}$$

where

$$\Delta\beta = \beta_p - \beta_s - \beta_i \tag{12.11}$$

and the effective interaction area is given by (see Eq. 11.33)

$$\Gamma = \iint \tilde{d}(x,y)\psi_p(x,y)\psi_s(x,y)\psi_i(x,y)dxdy \tag{12.12}$$

with (see Eq. 11.31)

$$d(x,y) = d_{eff}\tilde{d}(x,y) \tag{12.13}$$

where d_{eff} is the effective material nonlinear coefficient determined by the polarization states of the interacting modes and the direction of propagation (see Chap. 10) while the dimensionless function $\tilde{d}(x,y)$ describes the transverse spatial dependence of the nonlinearity.

We also have

$$\beta_p = \frac{\omega_p}{c} n_p \tag{12.14a}$$

$$\beta_s = \frac{\omega_s}{c} n_s \tag{12.14b}$$

$$\beta_i = \frac{\omega_i}{c} n_i \tag{12.14c}$$

with n_p, n_s and n_i being the *effective indices* of the modes corresponding to pump, signal and idler respectively.

Equations (12.10a)–(12.10c) describe the three coupled equations describing the three wave interaction that can be used to analyze sum frequency generation, difference frequency generation and parametric amplification. The coupling among the modes at the three frequencies is brought about by the nonlinearity in the medium.

Exercise 12.1 Obtain Eq. (12.9b) for the nonlinear polarization generated at the signal frequency ω_s.

Solution Substituting from Eq. (12.4) in Eq. (12.3) we have

$$P^{(NL)} = 2\epsilon_0 dE^2 = 2\epsilon_0 d\left(E_p + E_s + E_i\right)^2$$

Using Eqs. (12.5)–(12.7) in the above equation we obtain

$$P^{(NL)} = \frac{1}{2}\epsilon_0 d\Big[\Big(A_p(z)\psi_p(x,y)e^{i(\beta_p z-\omega_p t)} + c.c.\Big) + \Big(A_s(z)\psi_s(x,y)e^{i(\beta_s z-\omega_s t)} + c.c.\Big) + \Big(A_i(z)\psi_i(x,y)e^{i(\beta_i z-\omega_i t)} + c.c.\Big)\Big]^2$$

Opening the square we will obtain many terms. We collect terms which have a time dependence at a frequency ω_s. Since the three frequencies satisfy Eq. (12.1), terms containing products of $A_p(z)$ and $A_i^*(z)$ or its complex conjugate will have this time dependence. Collecting these terms we obtain Eq. (12.9b). Similarly collecting other appropriate terms we will obtain Eqs. (12.9a) and (12.9c).

Exercise 12.2 Obtain Eq. (12.10c).

Solution The wave equation satisfied by the electric field at the idler frequency ω_i is given by:

$$\nabla^2\vec{E}_i = \epsilon(\omega_i)\mu_0\frac{\partial^2\vec{E}_i}{\partial t^2} + \mu_0\frac{\partial^2\vec{P}_{NL}^{(\omega_i)}}{\partial t^2}$$

where $\epsilon(\omega_i)$ is the dielectric permittivity at the frequency ω_i. Substituting for the nonlinear polarization generated at the idler frequency and using Eqs. (12.7) and (12.9c) and by equating terms from both sides having a time dependence of the form $e^{-i\omega_i t}$, we obtain the following equation:

$$\nabla^2\Big(A_i(z)\psi_i(x,y)e^{i(\beta_i z-\omega_i t)}\Big) = \epsilon(\omega_i)\mu_0\frac{\partial^2}{\partial t^2}\Big(A_i(z)\psi_i(x,y)e^{i(\beta_i z-\omega_i t)}\Big) + \mu_0\frac{\partial^2}{\partial t^2} \times\Big(\epsilon_0 d\Big(2A_pA_s^*\psi_p(x,y)\psi_s(x,y)e^{i((\beta_p-\beta_s)z-\omega_i t)}\Big)\Big)$$

We simplify the above equation and neglect the term $\frac{d^2A_i}{dz^2}$ and use the following wave equation satisfied by $\psi_i(x,y)$ (see Eq. 11.21):

$$\left(\nabla^2 - \frac{\partial}{\partial z^2}\right)\psi_i(x,y) + \left[\frac{\omega_i^2}{c^2}n^2(\omega_i,x,y) - \beta_i^2\right]\psi_i(x,y) = 0$$

where $n^2(\omega_i,x,y)$ represents the refractive index profile at the idler frequency ω_i. Thus we obtain

$$2i\beta_i\psi_i(x,y)\frac{dA_i}{dz} = -\frac{\omega_i^2}{c^2}d2A_pA_s^*\psi_p(x,y)\psi_s(x,y)e^{i((\beta_p-\beta_s-\beta_i)z)}$$

Multiplying both sides by $\psi_i(x,y)$ and integrating over the entire transverse plane and using the normalization relation (Eq. 12.8c) and Eq. (12.14c) we obtain Eq. (12.10c). Similar procedure will give us Eqs. (12.10a) and (12.10b).

It may be mentioned here that if quasi phase matching is used in the three wave mixing process and if the periodic domain reversal has 50% duty cycle and first order QPM is used, then in Eqs. (12.10a)–(12.10c), we need to replace d_{eff} by $\frac{2}{\pi} d_{eff}$ and $\Delta\beta$ by $\Delta\beta - K$, where K is the spatial frequency of the QPM grating (see Chap. 11).

In order to obtain an important relation between the amplitudes of the three interacting waves we calculate the following quantity:

$$\frac{n_s}{\omega_s}\frac{d|A_s|^2}{dz} = \frac{n_s}{\omega_s}\left(A_s^*\frac{dA_s}{dz} + A_s\frac{dA_s^*}{dz}\right) = \frac{n_s}{\omega_s}\left(A_s^*\frac{dA_s}{dz} + c.c.\right)$$

where $c.\ c.$ represents complex conjugate of the earlier term. Substituting from Eq. (12.10b) we have

$$\frac{n_s}{\omega_s}\frac{d|A_s|^2}{dz} = \frac{n_s}{\omega_s}\left(A_s^* i\frac{\omega_s d_{eff}\Gamma}{cn_s}A_pA_i^*e^{i\Delta\beta z} + cc\right) = i\frac{d_{eff}\Gamma}{c}\left(A_pA_s^*A_i^*e^{i\Delta\beta z} - cc\right)$$

Similarly

$$\frac{n_i}{\omega_i}\frac{d|A_i|^2}{dz} = i\frac{d_{eff}\Gamma}{c}\left(A_pA_s^*A_i^*e^{i\Delta\beta z} - cc\right)$$

and

$$\frac{n_p}{\omega_p}\frac{d|A_p|^2}{dz} = -i\frac{d_{eff}\Gamma}{c}\left(A_pA_s^*A_i^*e^{i\Delta\beta z} - cc\right)$$

Thus we obtain the following very important relation:

$$-\frac{n_p}{\omega_p}\frac{d|A_p|^2}{dz} = \frac{n_s}{\omega_s}\frac{d|A_s|^2}{dz} = \frac{n_i}{\omega_i}\frac{d|A_i|^2}{dz} \tag{12.15}$$

Equation (12.15) is referred to as Manley Rowe relation and is a very important result in nonlinear optics.

In order to understand Eq. (12.15), we recall our discussion leading to Eq. (11.41) for the power carried by the mode. Following a similar procedure, we obtain the power carried by the pump mode as $\frac{n_p}{2c\mu_0}|A_p|^2$. Since the energy of each pump photon is $\hbar\omega_p$, the number of pump photons N_p crossing the waveguide per unit time at any value of z is $\frac{n_p}{2c\mu_0}\frac{|A_p|^2}{\hbar\omega_p}$ and the rate at which the number of pump photons are *lost* is given by

$$\frac{dN_p}{dz} = -\frac{n_p}{2c\mu_0}\frac{1}{\hbar\omega_p}\frac{d|A_p|^2}{dz} \tag{12.16}$$

The negative sign corresponds to loss of pump photons. Similarly the rate of *generation* of signal and idler photons are given by

$$\frac{dN_s}{dz} = \frac{n_s}{2c\mu_0}\frac{1}{\hbar\omega_s}\frac{d|A_s|^2}{dz} \tag{12.17}$$

and

$$\frac{dN_i}{dz} = \frac{n_i}{2c\mu_0}\frac{1}{\hbar\omega_i}\frac{d|A_i|^2}{dz} \tag{12.18}$$

Thus Eq. (12.15) implies that for every pump photon lost, one signal photon and one idler photon gets generated or for every pump photon generated one signal photon and one idler photon get lost. Thus the nonlinear interaction process between the pump, signal and idler can be visualized as splitting of a pump photon to a signal and idler pair of photons or the merging of a signal and idler pair of photons to form a pump photon.

Recall our discussion of second harmonic generation where we found that for maximum efficiency of conversion, we must satisfy a phase matching condition. In an identical manner we will find that for efficient sum or difference frequency generation we need to satisfy the following phase matching condition:

$$\Delta\beta = \beta_p - \beta_s - \beta_i = 0 \tag{12.19a}$$

or the following quasi phase matching condition:

$$\Delta\beta - K = \beta_p - \beta_s - \beta_i - K = 0 \tag{12.19b}$$

where K is the spatial frequency of the quasi phase matched grating.

Figure 12.1 shows the corresponding vector diagram for perfect phase matching. The vector diagram corresponding to perfect quasi phase matching is given in Fig. 12.4.

We will now solve the coupled mode equations under phase matching condition $\Delta\beta = 0$ (or the quasi phase matching condition $\Delta\beta - K = 0$) which become

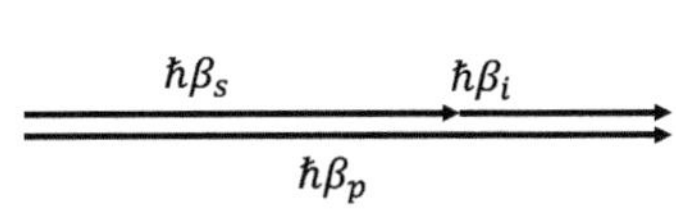

Fig. 12.1 Vector diagram for three wave interaction between a pump, signal, and idler

$$\frac{dA_p}{dz} = i\frac{\omega_p d_{eff}\Gamma}{cn_p} A_s A_i \tag{12.20a}$$

$$\frac{dA_s}{dz} = i\frac{\omega_s d_{eff}\Gamma}{cn_s} A_p A_i^* \tag{12.20b}$$

$$\frac{dA_i}{dz} = i\frac{\omega_i d_{eff}\Gamma}{cn_i} A_p A_s^* \tag{12.20c}$$

The three coupled equations describe all three wave interaction processes. In the case of QPM, d_{eff} needs to be replaced by the corresponding Fourier coefficient (see Sect. 11.6); e.g., for first order QPM d_{eff} gets replaced by $\frac{2}{\pi}d_{eff}$. Solving these equations would give us the evolution of the powers at the three frequencies.

12.3 Sum Frequency Generation

In order to consider sum frequency generation, at the input, we assume the excitation of modes at ω_s and ω_i which through the nonlinear interaction generate the mode at the frequency $\omega_p = \omega_s + \omega_i$. We also assume that the mode at ω_s has high power and assume that the conversion efficiency is small so that we may neglect the variation in the amplitude A_s of the mode at ω_s. This is referred to as the no depletion approximation. Thus the two equations that need to be solved are the equations Eqs. (12.20a) and (12.20c).

We also assume the initial conditions $A_i(z = 0) = A_{i0}$ and $A_p(z = 0) = 0$, that is, at the input we have a strong power (assumed to be almost constant and independent of z) in the mode at the signal frequency ω_s and a weak input in the mode at the idler frequency ω_i.

Differentiating Eq. (12.20a) and substituting from Eq. (12.20c) we obtain

$$\frac{d^2A_p}{dz^2} = -\frac{\omega_p\omega_i d_{eff}^2\Gamma^2}{c^2 n_p n_i}|A_s|^2 A_p \tag{12.21}$$

If we define

$$\gamma^2 = \frac{\omega_p\omega_i d_{eff}^2\Gamma^2}{c^2 n_p n_i}|A_s|^2 \tag{12.22}$$

then Eq. (12.21) becomes

$$\frac{d^2A_p}{dz^2} = -\gamma^2 A_p \tag{12.23}$$

whose general solution is given by

$$A_p(z) = C_1 \sin \gamma z + C_2 \cos \gamma z \tag{12.24}$$

where C_1 and C_2 are constants of integration. Since $A_p(z = 0) = 0$ we have $C_2 = 0$. Thus

$$A_p(z) = C_1 \sin \gamma z \tag{12.25}$$

Substituting in Eq. (12.20a) and simplifying we have

$$A_i(z) = -i\frac{cn_p}{\omega_p d_{eff} A_s \Gamma}\frac{dA_p}{dz} = -i\frac{cn_p \gamma}{\omega_p d_{eff} A_s \Gamma} C_1 \cos \gamma z$$

If we assume A_s to be real, we can substitute for γ from Eq. (12.22) and obtain

$$A_i(z) = -i\sqrt{\frac{n_p \omega_i}{n_i \omega_p}} C_1 \cos \gamma z \tag{12.26}$$

Using $A_i(z = 0) = A_{i0}$ gives us the value of the constant C_1:

$$C_1 = i\sqrt{\frac{n_i \omega_p}{n_p \omega_i}} A_{i0} \tag{12.27}$$

Hence we have the following solutions:

$$A_i(z) = A_{i0} \cos \gamma z \tag{12.28}$$

$$A_p(z) = i\sqrt{\frac{n_i \omega_p}{n_p \omega_i}} A_{i0} \sin \gamma z \tag{12.29}$$

The power carried by the pump and idler modes are given by (see Chap. 11)

$$P(\omega_p) = \frac{n_p}{2c\mu_0} |A_p(z)|^2 \tag{12.30}$$

$$P(\omega_i) = \frac{n_i}{2c\mu_0} |A_i(z)|^2 \tag{12.31}$$

Substituting the expressions from Eqs. (12.28) and (12.29) we obtain

$$P(\omega_p) = \frac{\omega_p}{\omega_i}\frac{n_i}{2c\mu_0} |A_{i0}|^2 \sin^2 \gamma z = \frac{\omega_p}{\omega_i} P_i(0) \sin^2 \gamma z \tag{12.32}$$

$$P(\omega_i) = \frac{n_i}{2c\mu_0} |A_{i0}|^2 \cos^2 \gamma z = P_i(0) \cos^2 \gamma z \tag{12.33}$$

where $P_i(0) = \frac{n_i}{2c\mu_0} |A_{i0}|^2$ is the incident power at the frequency ω_i.

Equations (12.32) and (12.33) show the generation of power at the sum frequency ω_p and the oscillatory exchange of power between the modes at frequencies ω_i and ω_p. It can also be seen that complete power exchange from the frequency ω_i to ω_p is possible. Indeed after a distance

$$L = \frac{\pi}{2\gamma} \tag{12.34}$$

the entire power in the mode at frequency ω_i gets converted to the mode at frequency ω_p. Beyond this, the power flows back again from the mode at frequency ω_p to the mode at frequency ω_i.

Figure 12.2 shows the variation of power at the frequencies ω_p and ω_i for a signal wavelength of 800 nm and an idler of 1500 nm (corresponding to $\omega_s \approx 2.36 \times 10^{15}\ \text{s}^{-1}$ and $\omega_i \approx 1.26 \times 10^{15}\ \text{s}^{-1}$) as a function of z. The corresponding sum frequency would be $\omega_p \approx 3.62 \times 10^{15}\ \text{s}^{-1}$ corresponding to a free space wavelength of about 521.7 nm). The figure shows the oscillatory exchange of power between the two modes at frequency ω_p and at frequency ω_i. In the figure, it is assumed that $P_i(z = 0) = 1$. The maximum power in the sum frequency corresponding to the pump is higher by a factor $\frac{\omega_p}{\omega_i}$ of the power input at the idler frequency which in Fig. 12.1 corresponds to about 2.87. This comes about since one photon at the idler (with energy $\hbar\omega_i$) and one photon of the signal (with energy $\hbar\omega_s$) combine to form one photon (with energy $\hbar\omega_p$) at the pump frequency. Also since the energy per photon at the pump is $\hbar\omega_p$ and that of an idler photon is $\hbar\omega_i$ for the same number of photons the energy of the pump will be higher by the factor $\frac{\omega_p}{\omega_i}$ as shown by the solutions. Note that although in the analysis we have assumed no depletion of power at the signal frequency ω_s, for every photon generated at the pump frequency ω_p one photon at signal frequency is lost. Hence the power at the signal would decrease; but since the signal power is assumed to be large and the

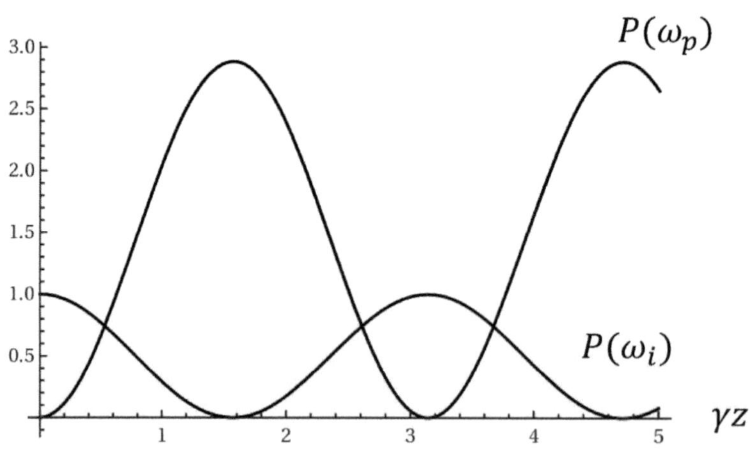

Fig. 12.2 Variation of power at the input idler frequency and the generated pump frequency as a function of γz. Note the periodic coupling of power among the two modes

conversion is assumed to be weak, the change in signal power with propagation is being neglected.

Although usually sum frequency generation takes place between the fundamental modes of channel waveguides, depending on the phase matching and modal overlap integrals it is possible to have sum frequency generation among different spatial modes also (see, e.g., Karpinsky et al. 2009).

Exercise 12.3 Consider sum frequency generation with input strong power at ω_s and a small power at ω_i. Obtain the variation of power at $\omega_p = \omega_s + \omega_i$ and at ω_i when $\Delta\beta \neq 0$, that is, under non phase matched interaction. Assume no-pump depletion approximation.

Solution For this we need to consider Eqs. (12.10a) and (12.10c). Differentiating Eq. (12.10a) we obtain

$$\frac{d^2A_p}{dz^2} = i\sigma_p A_s\left(\frac{dA_i}{dz} - i\Delta\beta A_i\right)e^{-i\Delta\beta z}$$

where

$$\sigma_p = \frac{\omega_p d_{eff}\Gamma}{cn_p}$$

Substituting for $\frac{dA_i}{dz}$ from Eq. (12.10c) and for A_i from Eq. (12.10a) we get

$$\frac{d^2A_p}{dz^2} + i\Delta\beta\frac{dA_p}{dz} + \sigma_p\sigma_i|A_s|^2A_p = 0 \tag{12.35}$$

where

$$\sigma_i = \frac{\omega_i d_{eff}\Gamma}{cn_i} \tag{12.36}$$

The general solution to Eq. (12.35) is

$$A_p(z) = e^{-i\Delta\beta z/2}\left(ae^{i\Omega z} + be^{-i\Omega z}\right)$$

where a and b are constants to be determined from the initial conditions and

$$\Omega = \sqrt{\gamma^2 + \frac{\Delta\beta^2}{4}} \tag{12.37}$$

$$\gamma^2 = \sigma_p \sigma_i |A_s|^2 = \frac{\omega_p \omega_i d_{eff}^2 \Gamma^2}{c^2 n_p n_i} |A_s|^2 \tag{12.38}$$

(see Eq. 12.22). Since $A_p(z = 0) = 0$, we have $b = -a$. Hence

$$A_p(z) = 2iae^{-i\Delta\beta z/2} \sin \Omega z$$

Substituting for $A_p(z)$ in Eq. (12.10a) we obtain

$$A_i(z) = \frac{2a}{\sigma_p A_s} e^{+i\Delta\beta z/2} \left(-i\frac{\Delta\beta}{2} \sin \Omega z + \Omega \cos \Omega z \right)$$

If $A_i(z = 0) = A_{i0}$, then we get

$$a = \frac{\sigma_p A_s A_{i0}}{2\Omega}$$

$$A_i(z) = A_{i0} e^{+i\Delta\beta z/2} \left(\cos \Omega z - i\frac{\Delta\beta}{2\Omega} \sin \Omega z \right)$$

and (using the value of γ), we have

$$A_p(z) = i\sqrt{\frac{n_i \omega_p}{n_p \omega_i}} \frac{A_{i0}}{\sqrt{1 + \frac{\Delta\beta^2}{4\gamma^2}}} e^{-i\Delta\beta z/2} \sin(\Omega z)$$

Hence the powers in the idler and the pump after an interaction length L are given by

$$P_i(L) = \frac{n_i}{2c\mu_0} |A_i(L)|^2 = P_i(z=0) \left(\cos^2 \Omega L + \frac{\Delta\beta^2}{4\Omega^2} \sin^2 \Omega L \right) \tag{12.39}$$

$$\begin{aligned} P_p(L) &= \frac{n_p}{2c\mu_0} |A_p(L)|^2 = \left(\frac{\omega_p}{\omega_i}\right) P_i(z=0) \frac{1}{\left(1 + \frac{\Delta\beta^2}{4\gamma^2}\right)} \sin^2 \Omega L \\ &= \left(\frac{\omega_p}{\omega_i}\right) P_i(z=0) \gamma^2 L^2 \frac{\sin^2 \Omega L}{(\Omega L)^2} \end{aligned} \tag{12.40}$$

The maximum power transfer takes place at a distance corresponding to $\Omega L = \frac{\pi}{2}$. Under phase matching or quasi phase matching we have $\Omega = \gamma$ and complete conversion will take place when $\gamma L = \frac{\pi}{2}$.

It is interesting to note that these solutions are very similar to the case of the directional coupler wherein exchange of energy takes place between two parallel waveguides due to evanescent coupling which is a linear process (see Chap. 5). In fact the coupled equations are the same and their solutions are also the same; of course the terms correspond physically to different parameters.

Exercise 12.4 Show from the expressions for $P_p(L)$ and $P_i(L)$ obtained in Exercise 12.3 that the number of pump photons generated is equal to the number of idler photons annihilated.

Solution Since no pump photons are incident $P_p(z=0)=0$, the number of pump photons generated is given by

$$\Delta N_p(L)=\frac{P_p(L)}{\hbar\omega_p}=\frac{P_i(z=0)}{\hbar\omega_i}\frac{1}{\left(1+\frac{\Delta\beta^2}{4\gamma^2}\right)}\sin^2\Omega L$$

The idler power decreases by $[P_i(z=0)-P_i(L)]$ and hence the number of idler photons annihilated will be given by

$$\Delta N_i(L)=\frac{P_i(z=0)-P_i(L)}{\hbar\omega_i}=\frac{P_i(z=0)}{\hbar\omega_i}\left[1-\left(\cos^2\Omega L+\frac{\Delta\beta^2}{4\Omega^2}\sin^2\Omega L\right)\right]$$

which on simplification gives

$$\Delta N_i(L)=\frac{P_i(z=0)}{\hbar\omega_i}\frac{1}{\left(1+\frac{\Delta\beta^2}{4\gamma^2}\right)}\sin^2\Omega L$$

where we have used Eq. (12.37) for Ω^2. Thus we find that $\Delta N_p(L)=\Delta N_i(L)$.

Exercise 12.5 Complete conversion from ω_i to ω_p takes place in a phase matched sum frequency generation process with input signal power of 1 W (at a wavelength of 800 nm) and an idler power of 1 mW (at a wavelength of 1500 nm). Obtain the power generated at the sum frequency and also the drop in power of the signal frequency.

Solution Since complete conversion takes place, the number of idler photons converted is given by

$$N_i=\frac{P_i(z=0)}{\hbar\omega_i}\approx 7.6\times 10^{15}$$

The number of pump photons generated and the number of signal photons annihilated must both be equal to N_i. Thus the power generated at the pump frequency will be

$$P_p = N_i \hbar \omega_p$$

and the decrease in power at the signal frequency would be

$$P_s = N_i \hbar \omega_s$$

The pump wavelength corresponds to

$$\frac{1}{\lambda_p} = \frac{1}{\lambda_s} + \frac{1}{\lambda_i}$$

which gives us $\lambda_p = 521.7$ nm. Thus the generated pump power will be approximately 2.89 mW and input signal power decreases by 1.89 mW. Note that the signal power has dropped from 1000 to 998.11 mW, a very small reduction in power of signal. At the same time the power generated at the pump frequency is equal to the sum of the loss in signal power and idler power, which is just a confirmation of energy conservation. It is the change in amplitude of the signal mode due to the small reduction in signal power which is neglected in the no-pump depletion approximation.

12.3.1 Bandwidth in Sum Frequency Generation

We recall the power generated at the sum frequency (see Eq. 12.40) and write it in the following form for a length of interaction given by L:

$$P_p(L) = \frac{n_p}{2c\mu_0} |A_p(L)|^2 = \left(\frac{\omega_p}{\omega_i}\right) P_i(z=0) \gamma^2 L^2 \frac{\sin^2 \Omega L}{(\Omega L)^2} \tag{12.41}$$

where Ω is given in Eq. (12.37).

To achieve efficient sum frequency generation corresponding to a given value of $\omega_s = \omega_{s0}$ and $\omega_i = \omega_{i0}$, which will generate the sum frequency $\omega_p = \omega_{p0} = \omega_{s0} + \omega_{i0}$, one needs to satisfy either the phase matching condition $\Delta\beta = 0$ or the quasi phase matching condition $\Delta\beta - K = 0$ (both of which correspond to $\Omega = \gamma$) assuming first order QPM for the given set of frequencies. In such a setup the question that arises is for a given high power at a well-defined frequency ω_{s0}, what is the range of frequencies around the frequency ω_{i0} where we can achieve efficient sum frequency generation. In other words we wish to find out the bandwidth of the sum frequency generation process.

At the input frequencies ω_{s0} and ω_{i0} the phase matching or the quasi phase matching condition is exactly satisfied and we have

$$\Delta\beta - K = \beta(\omega_{p0}) - \beta(\omega_{s0}) - \beta(\omega_{i0}) - K = 0 \tag{12.42}$$

with the required spatial frequency K obtained from the above equation and $\Delta\beta$ is given by Eq. (12.19a). If we required complete conversion from ω_{i0} to ω_{p0}, then we can see from Eq. (12.32) that the required length of interaction is (see Eq. 12.34)

$$L = \frac{\pi}{2\gamma} \tag{12.43}$$

Now, keeping the signal frequency constant, if the input frequency ω_i deviates from ω_{i0}, then we would no more be satisfying the quasi phase matching condition and the efficiency would drop as discussed in Exercise 12.4. When the frequency changes from ω_{i0} to $\omega_i = \omega_{i0} + \Delta\omega_i$ with the corresponding sum frequency changing from ω_{p0} to $\omega_p = \omega_{p0} + \Delta\omega_p$ such that $\Omega L = \pi$, then the efficiency of sum frequency generation becomes zero (see Eq. 12.40 where we have replaced $\Delta\beta$ by $(\Delta\beta - K)$ corresponding to quasi phase matching) which gives us

$$\sqrt{\gamma^2 + \frac{(\Delta\beta - K)^2}{4}} = \frac{\pi}{L}$$

Simplifying we have

$$\Delta\beta - K = \pm\sqrt{\frac{4\pi^2}{L^2} - 4\gamma^2} = \pm\sqrt{3}\frac{\pi}{L} \tag{12.44}$$

Corresponding to an input idler frequency ω_i, we have

$$\Delta\beta - K = \beta(\omega_p) - \beta(\omega_{s0}) - \beta(\omega_i) - K \tag{12.45}$$

Since the incident frequency ω_{s0} which has high power is assumed to be constant and we need to satisfy $\omega_p = \omega_s + \omega_i$, we also have $\Delta\omega_p = \Delta\omega_i$.

Assuming that $\Delta\omega_i \ll \omega_{i0}$, we can may Taylor series expansion in Eq. (12.45) and obtain

$$\Delta\beta - K = \beta(\omega_{p0}) + \Delta\omega_p \frac{d\beta}{d\omega}\bigg|_{\omega=\omega_{p0}} - \beta(\omega_{s0}) - \beta(\omega_{i0}) - \Delta\omega_i \frac{d\beta}{d\omega}\bigg|_{\omega=\omega_{i0}} - K$$

which using Eq. (12.42) becomes

$$\Delta\beta - K = \left(\frac{d\beta}{d\omega}\bigg|_{\omega=\omega_{p0}} - \frac{d\beta}{d\omega}\bigg|_{\omega=\omega_{i0}}\right)\Delta\omega_i$$

Since $\frac{d\beta}{d\omega} = \frac{N_{eff}}{c}$ (see Eq. 4.13), we have

$$\Delta\beta - K = \frac{1}{c}\left(N_{eff}(\omega_{p0}) - N_{eff}(\omega_{i0})\right)\Delta\omega_i \tag{12.46}$$

where $N_{eff}(\omega_{p0}) = c\frac{d\beta}{d\omega}\big|_{\omega=\omega_{p0}}$ and $N_{eff}(\omega_{i0}) = c\frac{d\beta}{d\omega}\big|_{\omega=\omega_{i0}}$ are the group indices of the modes at the pump and idler frequencies respectively.

Using Eqs. (12.44) and (12.46) we obtain for the deviation of the idler frequency from ω_{i0} when the efficiency of the sum frequency becomes zero as

$$\Delta\omega_i = \pm\sqrt{3}\frac{\pi c}{L}\frac{1}{\left(N_{eff}(\omega_{p0}) - N_{eff}(\omega_{i0})\right)} \tag{12.47}$$

Hence the frequency width between the two zeroes (around the maximum efficiency) of efficiency will be given by

$$\Delta\omega_{izz} = 2\sqrt{3}\frac{\pi c}{L}\frac{1}{\left(N_{eff}(\omega_{p0}) - N_{eff}(\omega_{i0})\right)} \tag{12.48}$$

In terms of wavelength the bandwidth becomes

$$\Delta\lambda_{izz} = \frac{\sqrt{3}}{L}\frac{\lambda_{i0}^2}{\left(N_{eff}(\lambda_{p0}) - N_{eff}(\lambda_{i0})\right)} \tag{12.49}$$

Equation (12.49) gives us an estimate of the bandwidth of the sum frequency generation process. Note that as usual, longer lengths imply smaller bandwidths and the difference in the group velocities (or group effective indices) at the input frequency ω_{i0} and the generated sum frequency ω_{p0} also decides the bandwidth. The group effective index depends on the characteristics of the material used through material dispersion but also depends on the waveguide characteristics through waveguide dispersion. Thus it is possible to have some control on the bandwidth through control of the waveguide characteristics.

Sum frequency generation leading to up conversion of frequency of photons is used in integrated quantum photonic applications as single photon detectors, interface between quantum memories and other quantum systems for conversion of telecom band photons to the near visible band where much more efficient detection systems are available (for a review of frequency up-conversion refer to Ma et al. 2012). Efficient quantum frequency upconverters using periodically poled lithium niobate waveguides and waveguides based on lithium niobate on insulators (LNOI) have been studied (see, e.g., Tournier et al. 2009; Wang et al. 2023; Murakami et al. 2023; Kaiser et al. 2019; Vazimali and Fathpour 2022)

Exercise 12.6 Obtain the value of $\Delta\beta$ when the power generated at the output at the sum frequency drops to half the value with $\Delta\beta - K = 0$.

Solution From Eq. (12.41) we find that the power generated at the pump frequency when $\Delta\beta = 0$ is given by

$$P_p(L) = \left(\frac{\omega_p}{\omega_i}\right) P_i(z=0) \sin^2\gamma L$$

which has a maximum value given by $\left(\frac{\omega_p}{\omega_i}\right) P_i(z=0)$ at $\gamma L = \frac{\pi}{2}$.

For this length if we require the power generated at the pump frequency to be half of this value then we obtain

$$\gamma^2 L^2 \frac{\sin^2 \Omega L}{(\Omega L)^2} = \frac{1}{2}$$

or

$$\frac{\sin^2 \Omega L}{(\Omega L)^2} = \frac{2}{\pi^2} \approx 0.2026$$

which happens when $\Omega L \approx 2.01$ which implies

$$\gamma^2 + \frac{(\Delta\beta - K)^2}{4} \approx \frac{4.04}{L^2}$$

which using $\gamma = \frac{\pi}{2L}$ gives $\Delta\beta - K = \pm \frac{2.508}{L}$. Hence the total 3 dB bandwidth is given by twice this value $\Delta\beta - K = \frac{5.016}{L}$. Using Eq. (12.46) this implies a total 3 dB bandwidth of

$$\Delta\omega_{3\text{dB}} = \frac{5.016\, c}{L} \left| \left(N_{eff}(\omega_{p0}) - N_{eff}(\omega_{i0})\right| \right.^{-1}$$

which is smaller than $\Delta\omega_{izz}$ given in Eq. (12.48).

Exercise 12.7 Consider sum frequency generation in bulk lithium niobate of length 10 mm for the generation of power at a wavelength of 700 nm with a high power signal input at a wavelength of 1310 nm. All the interacting waves are assumed to have extraordinary polarization. The extraordinary refractive indices and group refractive indices at the pump, signal, and idler wavelengths are given below:

λ (nm)	$n(\lambda)$	$N(\lambda)$
700	2.1891	2.3056
1310	2.1451	2.1912
1503.3	2.1389	2.1835

(a) Show that the idler wavelength comes out to be 1503.3 nm.
(b) What is the first order QPM period required for efficient sum frequency generation?

(c) Obtain the bandwidth corresponding to the idler wavelength for efficient sum frequency generation. What will be the corresponding bandwidth of the generated sum frequency? Assume that the input signal wavelength remains constant.
(d) Which nonlinear coefficient will be involved in the sum frequency generation?

Solution

(a) The idler wavelength is obtained from the following equation:

$$\frac{1}{\lambda_i} = \frac{1}{\lambda_p} - \frac{1}{\lambda_s}$$

and comes out to be 1503.3 nm.

(b) The QPM period is given by

$$\frac{1}{\Lambda} = \frac{n(\lambda_p)}{\lambda_p} - \frac{n(\lambda_s)}{\lambda_s} - \frac{n(\lambda_i)}{\lambda_i}$$

which on substitution of the values gives us $\Lambda \approx 14.9$ μm.

(c) The bandwidth can be obtained using Eq. (12.49) and comes out to be 3.2 nm. Since ω_s remains constant, from the equation $\omega_p = \omega_s + \omega_i$, we obtain $\Delta\omega_p = \Delta\omega_i$. Using the equation $\omega = \frac{2\pi c}{\lambda}$, we obtain $\Delta\lambda_p = \frac{\lambda_p^2}{\lambda_i^2}\Delta\lambda_i$ giving us $\Delta\lambda_p \approx 0.7$ nm.

(d) Since all the waves have extraordinary polarization, the coefficient that is responsible for the sum frequency generation is $d_{zzz} = d_{33}$.

12.4 Difference Frequency Generation and Amplification

Difference frequency generation corresponds to generation of one of the lower frequencies by nonlinear mixing of the modes at the pump frequency ω_p with the other frequency. For example mode at pump frequency ω_p can interact with the signal at frequency ω_s and generate the frequency $\omega_i = \omega_p - \omega_s$ as the difference frequency. We will show that in this process the signal mode gets amplified and this acts like an optical amplifier and is referred to as optical parametric amplifier. Optical parametric amplifiers will be discussed in greater detail in Chap. 13. Difference frequency generation which produces photons of lower frequency from photons of higher frequency (frequency down conversion) (e.g., Takesue 2010) has applications in converting photons with frequency in the visible range (wavelength of 780 nm) to photons in the telecommunication band (wavelength of 1530–1565 nm) for applications in quantum communications (see, e.g., Chou et al. 1998; Murakami et al. 2023), for signal amplification over a broad wavelength band (Kuznetsov et al. 2024) etc.

For difference frequency generation we assume a strong pump mode at frequency ω_p and an input mode at frequency ω_s. We will again assume no pump depletion approximation, that is, the amplitude of the pump mode will be assumed to be independent of z and hence under phase matching or quasi phase matching, we need to consider only the equations corresponding to the signal and idler (Eqs. 12.20b and 12.20c).

Differentiating Eq. (12.20c) and substituting from Eq. (12.20b) we obtain

$$\frac{d^2A_i}{dz^2} = i\frac{\omega_i d_{eff}\Gamma}{cn_i}A_p\frac{dA_s^*}{dz} = \frac{\omega_i\omega_s d_{eff}^2\Gamma^2}{c^2n_in_s}\left|A_{p0}\right|^2A_i \tag{12.50}$$

Defining

$$g^2 = \frac{\omega_i\omega_s d_{eff}^2\Gamma^2}{c^2n_in_s}\left|A_{p0}\right|^2 \tag{12.51}$$

Equation (12.50) becomes

$$\frac{d^2A_i}{dz^2} = g^2A_i \tag{12.52}$$

The general solution of the above equation is

$$A_i(z) = C_1\sinh gz + C_2\cosh gz \tag{12.53}$$

where C_1 and C_2 are integration constants. Note that unlike the case of sum frequency generation in this case the solutions are not oscillatory but hyperbolic functions.

Since at the input there is no idler i.e., $A_i(z = 0) = 0$ we get $C_2 = 0$. Hence

$$A_i(z) = C_1\sinh gz \tag{12.54}$$

Substituting in Eq. (12.20c) and assuming that the pump amplitude is real we have

$$A_s^*(z) = -i\frac{cn_i}{\omega_i d_{eff}\Gamma A_P}\frac{dA_i}{dz} = -i\frac{cn_ig}{\omega_i d_{eff}\Gamma A_p}C_1\cosh gz \tag{12.55}$$

Let the amplitude of the input signal mode be A_{s0}. Then we have

$$C_1 = i\sqrt{\frac{\omega_in_s}{\omega_sn_i}}A_{s0}^* \tag{12.56}$$

Thus we obtain the following solutions for the amplitudes of the signal and idler modes

$$A_s^*(z) = A_{s0}^*\cosh gz \tag{12.57}$$

$$A_i(z) = i\sqrt{\frac{\omega_i n_s}{\omega_s n_i}} A_{s0}^* \sinh gz \tag{12.58}$$

The powers at the signal and idler frequencies are

$$P_s(z) = \frac{n_s}{2c\mu_0}|A_s(z)|^2 = \frac{n_s}{2c\mu_0}|A_{s0}|^2 \cosh^2 gz = P_s(0)\cosh^2 gz \tag{12.59}$$

and

$$P_i(z) = \frac{n_i}{2c\mu_0}|A_i(z)|^2 = \frac{n_i}{2c\mu_0}\left(\frac{\omega_i n_s}{\omega_s n_i}\right)|A_{s0}|^2 \sinh^2 gz = \left(\frac{\omega_i}{\omega_s}\right)P_s(0)\sinh^2 gz \tag{12.60}$$

where $P_s(0)$ is the power at the input signal frequency. Since both $\cosh^2 gz$ and $\sinh^2 gz$ are increasing functions of z, it is apparent that as the difference frequency ω_i (idler) gets generated, the input signal at frequency ω_s gets amplified $[P_s(z) > P_s(0)]$.

The factor $\left(\frac{\omega_i}{\omega_s}\right)$ is due to the fact that the number of signal photons generated must be equal to the number of idler photons generated. The number of signal photons incident per unit time is $\frac{P_s(0)}{\hbar\omega_s}$ and thus the number of photons generated in a distance z by the nonlinear process will be $\frac{P_s(z)-P_s(0)}{\hbar\omega_s} = \frac{P_s(0)}{\hbar\omega_s}\sinh^2 gz$. The number of idler photons generated over a length z is $\frac{P_i(z)}{\hbar\omega_i} = \frac{P_s(0)}{\hbar\omega_s}\sinh^2 gz$ which is equal to the number of signal photons generated in the interaction.

Figure 12.3 shows the variation in the power of the idler and signal frequencies with gz for the same set of frequencies considered for Fig. 12.2. In the figure, it is assumed that $P_s(0) = 1$.

We see that the generation of the difference frequency is accompanied by the amplification of the signal wave. It is interesting to note that the input signal gets amplified irrespective of the phase of the input wave. Such an amplifier is called *phase insensitive parametric amplifier*.

What happens if the input consisted of all the three interacting waves, namely the pump, signal and idler. This will be discussed in detail in Chap. 13 where we show that depending on the relative phases of the three inputs, the signal may get amplified or attenuated. Such an amplifier is referred to as a *phase sensitive parametric amplifier* which is a very important element in integrated quantum photonics as it has very attractive noise characteristics (see Chap. 22).

Chapter 13 discusses parametric amplifiers in greater detail where we will also address the question as to what happens if in the difference frequency generation process the signal frequency is exactly half of the pump frequency, that is, $\omega_s = \frac{\omega_p}{2}$ which also implies $\omega_i = \omega_p - \omega_s = \frac{\omega_p}{2}$. Thus both the signal and idler have a frequency which is exactly half of the pump frequency leading to what is referred to as a *degenerate parametric amplifier*. This will correspond to reverse of the second harmonic generation process and is very interesting from the point of view of parametric amplification.

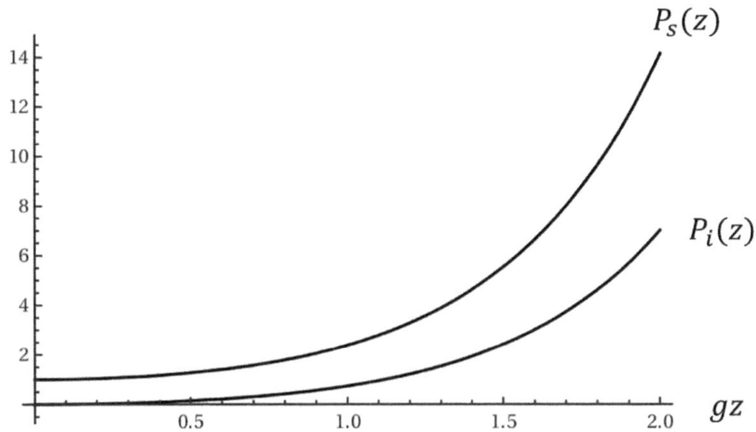

Fig. 12.3 The z-variation of the powers in the input signal and generated idler frequencies. Note that as the idler gets generated, the signal gets amplified

12.4.1 Nonphase-Matched Difference Frequency Generation

We now consider difference frequency generation with $\Delta\beta \neq 0$. For this we need to solve Eqs. (12.10b) and (12.10c). We also assume the no-pump depletion approximation with input pump at ω_p and a weak signal at ω_s.

Differentiating Eq. (12.10c) with respect to z and using Eqs. (12.10b) and (12.10c) we obtain the following equation:

$$\frac{d^2A_i}{dz^2} - i\Delta\beta\frac{dA_i}{dz} - \sigma_s\sigma_i|A_p|^2A_i = 0 \tag{12.61}$$

where

$$\sigma_i = \frac{\omega_i d_{eff}\Gamma}{cn_i}; \quad \sigma_s = \frac{\omega_s d_{eff}\Gamma}{cn_s} \tag{12.62}$$

The general solution of Eq. (12.61) is

$$A_i(z) = e^{+i\Delta\beta z/2}(a\sinh\Omega z + b\cosh\Omega z) \tag{12.63}$$

where a and b are constants of integration and

$$\Omega^2 = g^2 - \frac{\Delta\beta^2}{4}; \quad g^2 = \sigma_s\sigma_i|A_p|^2 \tag{12.64}$$

Since $A_i(z = 0) = 0$ we have $b = 0$ and we obtain the solution

$$A_i(z) = ae^{+i\Delta\beta z/2} \sinh \Omega z \tag{12.65}$$

Substituting the solution for $A_i(z)$ in Eq. (12.10c) and using the fact that $A_s(z = 0) = A_{s0}$ which is the input signal amplitude, we obtain the following solution for $A_s^*(z)$:

$$A_s^*(z) = A_{s0}^* e^{-i\Delta\beta z/2} \left(\cosh \Omega z + i \frac{\Delta\beta}{2\Omega} \sinh \Omega z \right) \tag{12.66}$$

and

$$A_i(z) = i \frac{\sigma_i}{\Omega} A_p A_{s0}^* e^{+i\Delta\beta z/2} \sinh \Omega z \tag{12.67}$$

Using the expressions for the powers at the two frequencies we obtain the following expressions for the power in the signal and the idler at a length L as

$$P_s(L) = P_{s0} \left(\cosh^2 \Omega L + \frac{\Delta\beta^2}{4\Omega^2} \sinh^2 \Omega L \right) \tag{12.68}$$

$$P_i(L) = \frac{\omega_i}{\omega_s} \frac{1}{\left(1 - \frac{\Delta\beta^2}{4g^2} \right)} P_{s0} \sinh^2 \Omega L = \frac{\omega_i}{\omega_s} \frac{g^2}{\Omega^2} P_{s0} \sinh^2 \Omega L \tag{12.69}$$

Note that if $g^2 > \frac{\Delta\beta^2}{4}$, then Ω is real and the power in the idler increases with propagation and simultaneously the signal also gets amplified. Thus even if the phase matching condition is not exactly satisfied (i.e., even if $\Delta\beta \neq 0$), it is still possible to generate the idler and achieve signal amplification provided the input pump power (note that g^2 depends on the pump power) is sufficiently large. If the pump power is such that $g^2 < \frac{\Delta\beta^2}{4}$, then Ω becomes pure imaginary and the hyperbolic solutions reduce to oscillatory solutions which imply that there is no signal amplification and the idler power has a sinusoidal variation with propagation.

It is to be noted that the number of signal photons generated in the process is

$$N_s(L) = \frac{P_s(L) - P_s(0)}{\hbar\omega_s} = \frac{P_s(0)}{\hbar\omega_s} \frac{g^2}{\Omega^2} \sinh^2 \Omega L \tag{12.70}$$

The number of idler photons generated is equal to

$$N_i(L) = \frac{P_i(L) - P_i(0)}{\hbar\omega_i} = \frac{P_i(L)}{\hbar\omega_i} = \frac{P_s(0)}{\hbar\omega_s} \frac{g^2}{\Omega^2} \sinh^2 \Omega L \tag{12.71}$$

which is exactly equal to the number of signal photons generated in the waveguide. For every idler photon generated, one signal photon is also created. These solutions also reduce to the solution obtained under phase matching when $\Delta\beta = 0$.

The bandwidth of difference frequency generation process will be discussed in Chap. 13.

12.5 Quasi Phase Matching

In Chap. 11 we discussed the concept of quasi phase matching wherein it was shown that the phase mismatch between the interacting waves can be compensated by a periodic reversal of the nonlinear coefficient. Even for sum frequency and difference frequency generation processes, it is possible to use QPM to achieve high efficiency nonlinear interaction.

In a uniform medium the phase matching condition for efficient sum frequency or difference frequency generation is given by Eq. (12.19a). Depending on the medium and the wavelengths involved in the process, it may not be always possible to satisfy the phase matching condition. Also one may not have a choice to employ the largest nonlinear coefficient of the material due to phase matching requirements. In such cases it is possible to employ quasi phase matching technique to achieve high efficiency.

The spatial frequency required for QPM for the three wave interaction between the pump having frequency ω_p and propagation constant β_p, a signal with a frequency ω_s and having a propagation constant β_s and an idler with a frequency ω_i with a propagation constant β_i is given by

$$K = \Delta\beta = \beta_p - \beta_s - \beta_i \tag{12.72}$$

or

$$\frac{2\pi}{\Lambda} = \frac{\omega_p}{c} n_{eff}(\omega_p) - \frac{\omega_s}{c} n_{eff}(\omega_s) - \frac{\omega_i}{c} n_{eff}(\omega_i) \tag{12.73}$$

where $n_{eff}(\omega_p)$, $n_{eff}(\omega_s)$ and $n_{eff}(\omega_i)$ are the effective indices of the modes at the pump, signal and idler frequencies respectively.

We can write Eq. (12.73) in terms of the free space wavelengths of the pump (λ_p), signal (λ_s) and idler (λ_i):

$$\frac{1}{\Lambda} = \frac{n_{eff}(\lambda_p)}{\lambda_p} - \frac{n_{eff}(\lambda_s)}{\lambda_s} - \frac{n_{eff}(\lambda_i)}{\lambda_i} \tag{12.74}$$

The three wavelengths are related through the following equation:

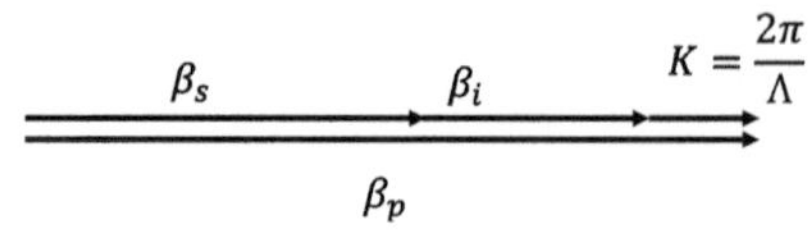

Fig. 12.4 Vector diagram corresponding to QPM for interaction among pump, signal and idler frequencies

$$\frac{1}{\lambda_p} = \frac{1}{\lambda_s} + \frac{1}{\lambda_i} \tag{12.75}$$

For a given waveguide and the interacting frequencies and their corresponding polarizations, it is possible to calculate the required spatial period for quasi phase matching (see Exercise 12.8). Figure 12.4 shows the corresponding vector diagram for a quasi-phase matched interaction process involving pump, signal and idler modes.

Under QPM, the equations derived earlier remain the same except that if m^{th} order QPM is employed, then we need to replace the effective nonlinear coefficient with the mth Fourier component of the expansion of the nonlinear coefficient and instead of $\Delta\beta$ we would have $\Delta\beta - mK$, K being the spatial frequency of the QPM grating. Thus using the first order QPM would result in an effective nonlinear coefficient which is less by a factor $\frac{2}{\pi}$ compared to the case when no nonlinear grating is used for the interaction (see Chap. 11) and $\Delta\beta$ gets replaced by $\Delta\beta - K$.

Exercise 12.8 The equation describing the wavelength variation of the ordinary and extraordinary refractive index of bulk lithium niobate are given by (λ is measured in μm):

$$n_o^2(\lambda) = 4.905 + \frac{0.11768}{\lambda^2 - 0.0475} - 0.027169\lambda^2 \tag{12.76}$$

$$n_e^2(\lambda) = 4.582 + \frac{0.099169}{\lambda^2 - 0.044432} - 0.02195\lambda^2 \tag{12.77}$$

We consider three wave interaction in bulk lithium niobate and assume that the pump and signal wavelengths to be 800 and 1550 nm propagating along the x-direction.

(a) Write down the phase matching condition corresponding to three wave interaction using quasi phase matching assuming the waves at pump, signal and idler to be extraordinary waves. Using the above equations, calculate the required QPM period. Which nonlinear coefficient would be involved in this process?

(b) What will be the phase matching condition if the pump is an ordinary wave, signal is an extraordinary wave and idler is an ordinary wave. Obtain the required QPM period in this case. Which nonlinear coefficient would be involved in this process?
(c) What will be the phase matching condition if the pump is an ordinary wave, signal is an ordinary wave and idler is an extraordinary wave. Obtain the required QPM period in this case. Which nonlinear coefficient would be involved in this process?

Solution For the given pump and signal wavelengths, the wavelength of the idler is given by $\lambda_i = \frac{\lambda_p \lambda_s}{\lambda_s - \lambda_p} \approx 1653.3$ nm.

From Eqs. (12.76) and (12.77) we obtain the ordinary and extraordinary refractive indices at the three wavelengths which are given in the following Table 12.1.

(a) The quasi phase matching condition when all the waves are extraordinary is given by

$$\frac{1}{\Lambda} = \frac{n_{ep}}{\lambda_p} - \frac{n_{es}}{\lambda_s} - \frac{n_{ei}}{\lambda_i}$$

where the additional subscripts differentiate between pump (p), signal (s) and idler (i). Substituting the values from Table 12.1 we obtain the required period to be 20.4 μm.

Since all the interacting waves are extraordinarily polarized along the z-direction, the nonlinear coefficient $d_{zzz} = d_{33}$ will be responsible for this process.

(b) In this case the QPM condition would be given by

$$\frac{1}{\Lambda} = \frac{n_{op}}{\lambda_p} - \frac{n_{es}}{\lambda_s} - \frac{n_{oi}}{\lambda_i}$$

Substituting the values from the Table 12.1 gives us the required period to be 9.59 μm.

In this case the pump and idler have ordinary polarization (along y) and the signal has extraordinary polarization (along z). Hence the nonlinear coefficient $d_{yyz} = d_{24}$ will be responsible for this process.

Table 12.1 Ordinary and extraordinary refractive indices of $LiNbO_3$ at the pump, signal, and idler wavelengths

λ (nm)	n_o	n_e
800	2.25527	2.17588
1550	2.21127	2.13806
1653.3	2.20784	2.13515

(c) For this case the QPM condition would be

$$\frac{1}{\Lambda} = \frac{n_{op}}{\lambda_p} - \frac{n_{os}}{\lambda_s} - \frac{n_{ei}}{\lambda_i}$$

Substituting the values from the Table 12.1 gives us the required period to be 9.9 μm.

In this case the pump and signal have ordinary polarization (along y) and the idler has extraordinary polarization (along z). Hence the nonlinear coefficient $d_{yyz} = d_{24}$ will be responsible for this process.

Thus we see that the required QPM period depends on the polarization state of the three interacting waves.

Problems

12.1. Give all possible wavelengths that can be generated using inputs at 800 and 1200 nm in a medium with second order nonlinearity. State the corresponding phase matching and quasi phase matching conditions.

12.2. Show that the solutions given by Eqs. (12.30) and (12.31) satisfy the condition on photon number conservation.

12.3. Show that the coupled mode equations given by Eqs. (12.10a)–(12.10c) satisfy power conservation.

12.4. Using the classical coupled mode equations Eqs. (12.20a)–(12.20c) show that with an input pump $A_p(z = 0) = A_{p0}$ if at the input $A_s(z = 0) = 0$ and $A_i(z = 0) = 0$, then no signal or idler can be generated by the medium. The generation of signal and idler with only pump input represents the process of spontaneous parametric down conversion and is discussed in Chap. 22.

12.5. We use the Gaussian approximation for the modal fields of a channel waveguide and represent the modes at the pump (p), signal (s) and idler (i) by the following fields:

$$\psi_j(x, y) = \frac{\sqrt{2}}{\left(\pi w_{jx} w_{jy}\right)^{1/2}} \exp\left(-\frac{x^2}{w_{jx}^2} - \frac{y^2}{w_{jy}^2}\right); \qquad j = p, s, i$$

where w_{jx} and w_{jy} correspond to the modal widths along the x- and y-directions respectively. (a) Show that the field distribution satisfies the normalization condition. (b) Using the given field distribution obtain an expression for the effective area given by $\frac{1}{\Gamma^2}$ where Γ is given by Eq. (12.12). Check that $\frac{1}{\Gamma^2}$ has indeed dimensions of area. Assume that the nonlinear coefficient d is independent of position coordinates, that is, $\tilde{d}(x, y) = 1$.

12.6. Consider sum frequency generation in which light waves at wavelengths of 800 and 1300 nm are incident on the waveguide. The power of 800 nm is 1 W while that of 1300 nm is 1 mW. If during the sum frequency generation process the entire power of the 1300 nm light is converted to the sum frequency, what will be the power of the sum frequency? Also what will be the power output of the signal at 800 nm?

12.7. In a difference frequency generation using *z*-cut, *x*-propagating lithium niobate the pump, signal and idler are all extraordinary modes of the waveguide. Which element of the $\overline{\overline{d}}$ tensor would be involved in the interaction process?

12.8. Consider sum frequency generation process with 2 W of power at 980 nm and 20 mW of power at 1500 nm incident in a nonlinear waveguide. If at the output the power exiting at 1500 nm is 15 mW, what is the power exiting at the sum frequency?

12.9. Two extraordinarily polarized waves at wavelengths of 1000 nm (with power of 1 W) and 1500 nm (with a power of 1 mW) are coupled into a waveguide in *z*-cut, *y*-propagating lithium niobate waveguide. The waveguide is periodically poled with a period 10.1 μm. Which new wavelength will be generated most efficiently at the output of the waveguide and why? The effective index values of the fundamental mode of the waveguide is given in Table 12.2 given below:

12.10. A pump wave having extraordinary polarization and frequency ω propagates along the *y*-direction in an optical waveguide fabricated in *z*-cut lithium niobate. Which element of the nonlinearity tensor would be responsible for difference frequency generation of a signal which has ordinary polarization propagating along the waveguide?

12.11. Consider quasi phase matched sum frequency generation in a lithium niobate waveguide fabricated in a *z*-cut, *y*-propagation crystal. The input signal wave having frequency ω_S has an ordinary polarization while the idler at frequency ω_I has an extraordinary polarization.

(a) Obtain the *x*- and *z*-components of the nonlinear polarization at the sum frequency ω_P.
(b) What will be the state of polarization of the generated sum frequency?
(c) Express the required spatial period of the quasi phase matched grating for efficient sum frequency generation in terms of the appropriate effective indices.

Table 12.2 Effective indices of a waveguide at different wavelengths for Problem 12.9

λ (nm)	n_{eff}
500	2.2486
600	2.2105
750	2.1816
1000	2.1591
1500	2.1390
2000	2.1253
3000	2.0957

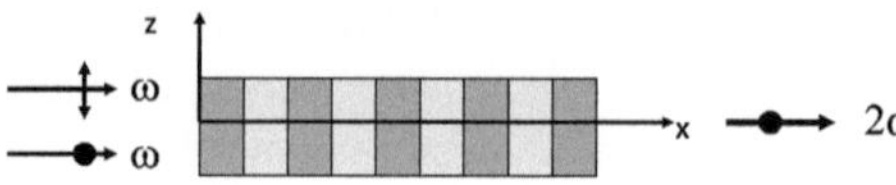

Fig. 12.5 For Problem 12.15

12.12. Consider sum frequency generation in lithium niobate using input modes at 1000 nm and 1500 nm with powers of 1 W and 1 mW, respectively. (a) What is the wavelength of the output sum frequency? (b) If it is required to use the d_{33} nonlinear coefficient what would be the period of the first order quasi phase matched grating that would maximize the efficiency? (c) For this obtain the length where maximum power at the sum frequency will be obtained. Assume $\frac{1}{\Gamma^2} = 7 \times 10^{-10}\ \text{m}^2$. (d) What is the maximum possible power that is generated at the sum frequency? Use effective index values given in Table 12.2.

12.13. A pump having extraordinary polarization and a frequency ω_P propagates along the y-axis of a lithium niobate crystal. Which element of the d tensor would be responsible for amplification of a signal at a frequency $\omega_P/2$, having ordinary polarization and propagating along y-direction? Obtain expression for the corresponding nonlinear polarization.

12.14. Using $\chi^{(2)}$ nonlinearity, I wish to generate a wavelength of 3 μm from a laser beam of wavelength 1 μm. What process can lead to this and what other input wave you would need.

12.15. Consider sum frequency generation in z-cut, x-propagating lithium niobate waveguide in which the input consists of a strong mode having power P_e with extraordinary polarization and a weak mode having power P_o with ordinary polarization (see Fig. 12.5). Both of them have the same frequency ω. (a) which nonlinear coefficient will be involved in the process? (b) What will be the quasi phase matching condition for the process?

References

Chou, M. H., J. Hauden, M. A. Arbore, and M. M. Fejer. 1998. 1.5 μm band wavelength conversion based on difference frequency generation in $LiNbO_3$ waveguides with integrated coupling structures. *Opt Lett* 23:1004–1006.

Kaiser, F., P. Vergyris, A. Martin, D. Aktas, M. P. De Micheli, O. Alibart, and S. Tanzilli. 2019. Quantum optical frequency up-conversion for polarization entangled qubits: towards interconnected quantum information devices. *arXiv*:1901.09826v2 [quant-ph] (accessed March 18, 2024).

Karpinsky, M., C. Tadzewicz, and K. Banaszek. 2009. Experimental characterization of three wave mixing in a multimode nonlinear KTiPO4 waveguide. *Appl Phys Lett* 94:181105.

Kuznetsov, N., A. Nardi, A. Davydova, M. Churaev, J. Riemensberger, P. Seidler, and T. J. Kippenberg. 2024. An ultra-broadband photonic-chip based travelling wave parametric amplifier. *arXiv*: 2404.08609v1 [physics.optics].

Ma, L., O. Slattery, and X. Tang. 2012. Single photon frequency up-conversion and its applications. *Phys Rep* 521:69–94.

Murakami, S., R. Fujimoto, T. Kobayashi, R. Ikuta, A. Inoue, T. Umeki, S. Miki, F. China, H. Terai, R. Kasahara, T. Mukai, N. Imoto, and T. Yamamoto. 2023. Quantum frequency conversion using 4-port fiber pigtailed PPLN module. *Opt Express* 31:29271–29279.

Takesue, H. 2010. Single-photon frequency down-conversion experiment. *Phys Rev A* 82:013833.

Tournier, M., O. Alibart, F. Doutre, S. Tascu, M. P. de Micheli, D. B. Ostrowsky, K. Thyagarajan, and S. Tanzilli. 2009. Up-conversion detectors at 1550 nm for quantum communications: review and recent advances. *EAS Publ Ser* 37:311–339. https://doi.org/10.1051/eas/0937039.

Vazimali, M. G., and S. Fathpour. 2022. Applications of thin-film lithium niobate in nonlinear integrated photonics. *Adv Photon* 4:034001.

Wang, M., Z. Fang, J. Lin, R. Wu, J. Chen, Z. Liu, H. Zhang, L. Qiao, and Y. Cheng. 2023. Integrated active lithium niobate photonic devices. *Jpn J Appl Phys* 62:SC0801.

Chapter 13
Optical Parametric Amplification

Abstract This chapter discusses the phenomenon of optical parametric amplification using second-order nonlinear effects in optical waveguides. Degenerate and nondegenerate parametric amplifiers are discussed and expressions for gain coefficient in the two cases are derived. Both phase-sensitive and phase-insensitive amplifiers and the bandwidth of amplification are discussed.

13.1 Introduction

Parametric amplifiers have a long history and find applications in many areas such as mechanics, electronics, and optics. A classic example of a parametric amplifier is a child on a swing. If the swing has a natural frequency of oscillation of Ω, then a child on the swing can change the amplitude of oscillation of the swing by pumping the swing (by sitting and standing periodically) at a frequency of 2Ω. For the child to amplify the swing, the child must sit and stand at the right phase of oscillation of the swing. On the other hand, if the child sits and stands at exactly the opposite phase, then the swing will get attenuated. Thus, this is an example of an amplifier which is phase sensitive.

In this chapter, we will consider an optical parametric amplifier. Parametric amplification is achievable using second-order nonlinear interaction as a difference frequency generation process as we discussed in Chap. 12. The amplification can be either phase insensitive or phase sensitive depending on the input conditions. In the case of phase-insensitive amplification, the input signal gets amplified through the nonlinear process independent of the phase of the signal. This was discussed in Chap. 12. Optical amplifiers such as erbium doped fiber amplifier or Raman fiber amplifiers are other examples of phase-insensitive amplifiers.

On the other hand in the case of phase-sensitive amplifier, the input signal can either get amplified or attenuated depending on the relative phase of the signal. Phase-sensitive amplifiers are very interesting since unlike phase-insensitive amplifiers where the amplifiers always add noise to the signal during the amplification process, phase-sensitive amplifiers can amplify the signal without adding any noise (see Chap. 22). Thus, phase-sensitive amplifiers are very interesting from the point

K. Thyagarajan, *Integrated Quantum Photonics*, Graduate Texts in Physics,
https://doi.org/10.1007/978-3-031-85728-7_13

of view of optical fiber communication and other applications where noise is very important (see, e.g., Kazama et al. 2021; Ledezma et al. 2022; Yan et al. 2022).

In Chap. 12, we discussed about sum frequency generation and difference frequency generation. As we will see parametric amplification takes place through a difference frequency generation process. If the signal and idler are identical, then this results in what is referred to as degenerate parametric amplifier. On the other hand, if the signal and idler are different, then we have what is referred to as nondegenerate parametric amplifier. The degenerate parametric amplifier is a phase-sensitive amplifier, while the nondegenerate amplifier can be either phase insensitive or phase sensitive.

In Chap. 22, we will discuss spontaneous parametric down conversion (SPDC) wherein the pump photon can spontaneously generate lower frequency photon pairs. This is a purely quantum mechanical phenomena. In this chapter, in the degenerate case we are assuming inputs both at the pump and at the signal or in the nondegenerate case inputs in all the three modes and will show amplification of the signal.

13.2 Degenerate Parametric Amplifier

We will start with a discussion of what is referred to as degenerate parametric amplifier. In this, the signal frequency is exactly half of the pump frequency (both signal and idler are identical) and is exactly the reverse of second harmonic generation that we discussed in Chap. 11.

We begin with the following two coupled mode equations that describe the nonlinear interaction in the presence of a pump at frequency 2ω and a signal at frequency ω. (Note that in Chap. 11 on second harmonic generation, the input at frequency ω was referred to as the pump). Since in the interaction there are only two frequencies ω and 2ω, we will use the coupled mode equations describing second harmonic generation to study this interaction (see Chap. 11). The only difference would be in second harmonic generation the incident pump frequency is ω, while in the parametric amplifier process the inputs are at both ω and 2ω. Thus, the difference frequency would also be at the frequency ω, and hence this process is defined as a degenerate process. The case when the input is only at frequency 2ω generates the frequency ω (referred to as spontaneous parametric down conversion) will be discussed in Chap. 22 since this process requires a quantum mechanical analysis.

The coupled mode equations are given by (see Eqs. 11.34 and 11.36)

$$\frac{dA_p}{dz} = i\frac{\omega d_{eff}\Gamma}{cn_p}A_s^2 e^{-i\Delta\beta z} \tag{13.1}$$

$$\frac{dA_s}{dz} = i\frac{\omega d_{eff}\Gamma}{cn_s}A_pA_s^*e^{i\Delta\beta z} \tag{13.2}$$

where A_s (signal at frequency ω) and A_p (pump at frequency 2ω) are the complex amplitudes of the modes having frequencies ω (here referred to as the signal) and 2ω (here referred to as the pump), respectively; n_p and n_s are the effective indices of the interacting waveguide modes at the pump and signal frequencies respectively. Here we have replaced subscripts 1 and 2 by s and p signifying that these are the signal and pump, respectively.

The effective nonlinear coefficient is defined as before by Eqs. (11.31) and (11.32) and the overlap integral is defined by Eq. (11.33) with the effective area of interaction being $\frac{1}{\Gamma^2}$ with the subscripts 1 and 2 replaced by s and p, respectively.

Also, the phase mismatch is given by (see Eq. 11.25)

$$\Delta\beta = \beta_p - 2\beta_s \tag{13.3}$$

We assume perfect phase matched or quasi phase matched operation so that we can put $\Delta\beta = 0$ or $\Delta\beta - K = 0$, and hence the coupled mode equations become

$$\frac{dA_p}{dz} = i\frac{\omega d_{eff}\Gamma}{cn_p}A_s^2 \tag{13.4}$$

$$\frac{dA_s}{dz} = i\frac{\omega d_{eff}\Gamma}{cn_s}A_pA_s^* \tag{13.5}$$

In order to analyze the degenerate parametric amplifier, we assume that at the input of the waveguide, we have a strong pump at frequency 2ω and a weak signal at frequency ω. We will also neglect the depletion of the pump and thus assume A_p to be a constant.

We first differentiate Eq. (13.5) with respect to z and obtain

$$\frac{d^2A_s}{dz^2} = i\frac{\omega d_{eff}\Gamma}{cn_s}A_p\frac{dA_s^*}{dz}$$

which using the complex conjugate of Eq. (13.5) becomes

$$\frac{d^2A_s}{dz^2} = i\frac{\omega d_{eff}\Gamma}{cn_s}A_p\left(-i\frac{\omega d_{eff}\Gamma}{cn_s}A_p^*A_s\right) = \left(\frac{\omega d_{eff}\Gamma}{cn_s}\right)^2|A_p|^2A_s = \gamma^2A_s \tag{13.6}$$

where

$$\gamma^2 = \kappa^2|A_p|^2 = \left(\frac{\omega d_{eff}\Gamma}{cn_s}\right)^2|A_p|^2 \tag{13.7}$$

with

$$\kappa = \frac{\omega d_{eff} \Gamma}{c n_s} \tag{13.8}$$

The general solution of Eq. (13.6) is

$$A_s(z) = a \cosh \gamma z + b \sinh \gamma z \tag{13.9}$$

where a and b are integration constants. We assume the following initial conditions of the signal and pump

$$A_s(z=0) = A_{s0} e^{-i\phi_{s0}} \tag{13.10}$$

$$A_p(z=0) = A_{p0} e^{-i\phi_{p0}} \tag{13.11}$$

where A_{s0} and A_{p0} represent the input amplitudes of the signal and pump, respectively, and ϕ_{s0} and ϕ_{p0} their initial phases. Using the initial condition given by Eq. (13.10), we have

$$a = A_{s0} e^{-i\phi_{s0}}$$

From Eq. (13.5), we note that

$$\left.\frac{dA_s}{dz}\right|_{z=0} = i \frac{\omega d_{eff} \Gamma}{c n_s} A_p(z=0) A_s^*(z=0)$$

Substituting from Eq. (13.10) and (13.11), we obtain

$$b\gamma = i \frac{\omega d_{eff} \Gamma}{c n_s} A_{p0} A_{s0} e^{-i\phi_{p0}} e^{i\phi_{s0}}$$

which gives us

$$b = i\, A_{s0} e^{i\phi_{s0}} e^{-i\phi_{p0}}$$

Hence, we obtain

$$A_s(z) = A_{s0} e^{-i\phi_{s0}} \left[\cosh\, \gamma z + i e^{-i\Delta\phi} \sinh\, \gamma z\right] \tag{13.12}$$

where

$$\Delta\phi = \phi_{p0} - 2\phi_{s0} \tag{13.13}$$

Equation (13.12) gives the evolution of the signal as a function of z. It is interesting to note that the evolution depends on the phase difference. We will first consider two special values of ϕ.

Signal Amplification

First let us consider the following value of $\Delta\phi$:

$$\Delta\phi = +\frac{\pi}{2} \tag{13.14}$$

then we have

$$A_s(z) = A_{s0}e^{-i\phi_{s0}}[\cosh \gamma z + \sinh \gamma z] = A_{s0}e^{-i\phi_{s0}}e^{\gamma z} \tag{13.15}$$

which shows exponential increase in signal amplitude leading to amplification. Since the power in the signal is proportional to $|A_s(z)|^2$, we note that in this case the signal power increases exponentially with z as $e^{2\gamma z}$. Thus, the system behaves as an amplifier for the signal at frequency ω with an amplification factor of 2γ.

Signal Attenuation

We next consider the following value of $\Delta\phi$:

$$\Delta\phi = -\frac{\pi}{2} \tag{13.16}$$

then we have

$$A_s(z) = A_{s0}e^{-i\phi_{s0}}[\cosh \gamma z - \sinh \gamma z] = A_{s0}e^{-i\phi_{s0}}e^{-\gamma z} \tag{13.17}$$

which shows attenuation of the signal. In this case, the signal power decreases as $e^{-2\gamma z}$ and the system behaves as an attenuator rather than as an amplifier!

Thus, when both the pump and signal (which has half the frequency of the pump leading to a degenerate case) are incident in the waveguide, the device behaves as a phase-sensitive amplifier. For specific values of the phase of the signal, it either gets amplified or attenuated.

Exercise 13.1 Show using the coupled mode equations (assuming no pump depletion approximation) that even if phase matching for the second harmonic generation is satisfied, if the input consists only of the frequency 2ω, there can be no generation of the frequency ω.

Solution Equation (13.9) gives the solution of the coupled mode equation. If no mode at signal frequency is incident, then $A_s(z = 0) = 0$. This also implies from Eq. (13.5) that $\frac{dA_s}{dz}\big|_{z=0} = 0$. Using these initial conditions in Eq. (13.9), we see that $a = 0$ and $b = 0$. Hence, $A_s(z) = 0$, thus implying that there is no generation of the mode at frequency ω. Thus, according to the coupled equation, without an input at the signal frequency, there is no generation of the mode at the signal frequency in the case of a degenerate parametric amplifier. It is also to be noted that in the absence of an input at frequency ω, there is no nonlinear polarization generated at the frequency ω and hence no generation of the frequency ω. The generation of signal in the degenerate case is a purely quantum phenomenon and is referred to as spontaneous parametric down conversion (SPDC) which will be discussed in Chap. 22.

Note also from Eq. (13.17) that if $A_{s0} = 0$, then $A_s(z) = 0$, that is, there is no generation of the wave at frequency ω.

13.2.1 Input Signal with Arbitrary Phase

In the above section, we discussed the cases when the signal phase had two specific values with respect to the pump phase. Let us now consider a more general case when the input signal has an arbitrary phase and assume that the phase of the pump is zero, that is, $\phi_{p0} = 0$. In this case, we have $\Delta\phi = -2\phi_{s0}$.

Let us consider an input signal field given by

$$E_s(z=0) = A_{s0}\psi_s(x,y)\cos(\omega_s t + \chi) \tag{13.18}$$

where χ represents the phase of the signal. By writing $\cos(\omega_s t + \chi) = \cos\left(\omega_s t - \frac{\pi}{4} + \chi + \frac{\pi}{4}\right)$, we expand the cosine function and write the Eq. (13.18) on as follows:

$$E_s(z=0) = A_{s0}\psi_s(x,y)\cos\left(\chi + \frac{\pi}{4}\right)\cos\left(\omega_s t - \frac{\pi}{4}\right) + A_{s0}\psi_s(x,y)\sin\left(\chi + \frac{\pi}{4}\right)\cos\left(\omega_s t + \frac{\pi}{4}\right) \tag{13.19}$$

Here, we have replaced $\sin\left(\omega_s t - \frac{\pi}{4}\right)$ by $-\cos\left(\omega_s t + \frac{\pi}{4}\right)$.

The signal field has been split into two parts; the first part corresponds to $\cos\left(\omega_s t - \frac{\pi}{4}\right)$ which has $\phi_{s0} = -\frac{\pi}{4}$ and the second part corresponds to $\cos\left(\omega_s t + \frac{\pi}{4}\right)$ which has $\phi_{s0} = \frac{\pi}{4}$. Hence, the phase difference between the two components of the signal wave is $\frac{\pi}{2}$ and thus they correspond to the two quadrature components of the input signal field.

Thus, for the quadrature component $\cos\left(\omega_s t - \frac{\pi}{4}\right)$ of the signal, we have

$$\Delta\phi = \phi_{p0} - 2\phi_{s0} = 0 + \frac{\pi}{2} = +\frac{\pi}{2}$$

As discussed earlier when $\Delta\phi = +\frac{\pi}{2}$, this particular quadrature of the signal gets amplified.

Similarly for the quadrature component $\cos\left(\omega_s t + \frac{\pi}{4}\right)$, we have

$$\Delta\phi = \phi_{p0} - 2\phi_{s0} = 0 - \frac{\pi}{2} = -\frac{\pi}{2}$$

Again as discussed earlier when $\Delta\phi = -\frac{\pi}{2}$, this particular quadrature of the signal gets attenuated.

Thus, we see that for any given signal one quadrature component gets amplified, while the other quadrature component gets attenuated. This phase-sensitive nature of the amplifier is responsible for very interesting noise properties of phase-sensitive amplifiers (see Chap. 22).

The above discussions correspond to the case when there is a single signal mode of a given polarization and spatial mode at a frequency which is half of the pump frequency. As an example if the nonlinear interaction is taking place in a z-cut lithium niobate and both the pump and signal are extraordinarily polarized, then the amplifier will behave as a phase-sensitive amplifier.

In Sect. 13.3, we will discuss a case when the signal has a frequency half of the pump but is input in two orthogonal polarization states thus leading to a nondegenerate amplifier.

Exercise 13.2 Consider a degenerate phase-sensitive parametric amplifier. What is the maximum value of Δβ so that the signal undergoes amplification?

Solution Consider a quasi phase-matched degenerate phase-sensitive amplifier with $\Delta\beta \neq 0$. In this case, assuming no pump depletion approximation, we need to solve the following equation for the generated signal at ω (see Eq. 13.2)

$$\frac{dA_s}{dz} = i\frac{\omega d_{eff}\Gamma}{cn_s} A_p A_s^* e^{i\Delta\beta z} = i\kappa A_p A_s^* e^{i\Delta\beta z}$$

where $\kappa = \frac{\omega d_{eff}\Gamma}{cn_s}$. Under no pump depletion approximation, we assume A_p to be a constant. Differentiating the above equation with respect to z, we obtain

$$\frac{d^2A_s}{dz^2} = i\kappa A_p \frac{dA_s^*}{dz} e^{i\Delta\beta z} - \kappa A_p A_s^* \Delta\beta e^{i\Delta\beta z} = i\kappa A_p\left(-i\kappa A_p^* A_s e^{-i\Delta\beta z}\right)e^{i\Delta\beta z} + i\Delta\beta\frac{dA_s}{dz}$$

which simplifies to

$$\frac{d^2A_s}{dz^2} - i\Delta\beta\frac{dA_s}{dz} - \kappa^2|A_p|^2 A_s = 0$$

The general solution of the above equation is

$$A_s(z) = e^{i\frac{\Delta\beta z}{2}}\left(ae^{\Omega z} + be^{-\Omega z}\right)$$

where $\Omega = \sqrt{\gamma^2 - \frac{(\Delta\beta)^2}{4}}$ where $\gamma^2 = \kappa^2|A_p|^2$.

The for $\gamma^2 > \frac{(\Delta\beta)^2}{4}$, the solution for $A_s(z)$ is in terms of hyperbolic functions implying amplification. On the other hand when $\frac{(\Delta\beta)^2}{4} > \gamma^2$, then Ω^2 becomes negative and the solution for $A_s(z)$ becomes sinusoidal functions. Thus, even when $\Delta\beta \neq 0$, it is possible to achieve amplification provided that the pump power is sufficient to achieve $\gamma^2 > \frac{(\Delta\beta)^2}{4}$.

13.3 Nondegenerate Phase-Sensitive Parametric Amplifier

In Chap. 12, we discussed difference frequency generation with an input pump and an input signal wave. In this case, we showed that the difference frequency process leads to amplification of the signal irrespective of its phase and simultaneously an idler wave is also generated. Such an amplifier behaves as a *phase-insensitive parametric amplifier*. Here we discuss the case when along with the input pump mode, modes at signal and idler are both excited in the waveguide. We will show that depending on the phase relationship between the phases of the pump, signal, and idler, the signal may get amplified or attenuated.

At any point in the medium, we assume the presence of three waves namely, the pump (ω_p), the signal (ω_s), and the idler (ω_i). These three frequencies are related to the following equation:

$$\omega_p = \omega_s + \omega_i \tag{13.20}$$

We will also assume that the interaction is quasi phase-matched satisfying the following equation:

$$\beta_p = \beta_s + \beta_i + K \tag{13.21}$$

where β_p, β_s, and β_i represent the propagation constants of the waveguide mode corresponding to the pump, signal, and idler frequencies, respectively.

We will again assume no-pump depletion and hence the two coupled mode equations that will determine the evolution of the signal and idler are given by (see Eqs. 12.20b and 12.20c)

$$\frac{dA_s}{dz} = i\frac{\omega_s d_{eff}\Gamma}{cn_s}A_pA_i^* \tag{13.22}$$

$$\frac{dA_i}{dz} = i\frac{\omega_i d_{eff}\Gamma}{cn_i}A_pA_s^* \tag{13.23}$$

As input conditions, we assume

$$A_p(z=0)=A_{p0}=|A_{p0}|e^{-i\phi_{p0}} \tag{13.24}$$

$$A_s(z=0)=A_{s0}=|A_{s0}|e^{-i\phi_{s0}} \tag{13.25}$$

$$A_i(z=0)=A_{i0}=|A_{i0}|e^{-i\phi_{i0}} \tag{13.26}$$

where $|A_{p0}|$, $|A_{s0}|$, and $|A_{i0}|$ are the amplitudes of the pump, signal, and idler modes and ϕ_{p0}, ϕ_{s0}, and ϕ_{i0} are the phases of the corresponding modes at the input.

Assuming no pump depletion, we will solve Eqs. (13.22) and (13.23). Differentiating Eq. (13.22) with respect to z, we obtain

$$\frac{d^2A_s}{dz^2}=i\frac{\omega_s d_{eff}\Gamma}{cn_s}A_p\frac{dA_i^*}{dz}=g^2A_s \tag{13.27}$$

where

$$g^2=\frac{\omega_i\omega_s d_{eff}^2\Gamma^2}{n_in_sc^2}|A_{p0}|^2 \tag{13.28}$$

Note that in the degenerate case $\omega_i=\omega_s$, $n_i=n_s$, and g^2 given by Eq. (13.28) becomes equal to γ^2 given in Eq. (13.7).

The general solution of Eq. (13.27) is

$$A_s(z)=a\cosh gz+b\sinh gz \tag{13.29}$$

Substituting this solution in Eq. (13.22), we obtain

$$A_i^*(z)=-i\sqrt{\frac{\omega_in_s}{\omega_sn_i}}e^{i\phi_{p0}}(a\sinh gz+b\cosh gz) \tag{13.30}$$

Using the initial conditions given by Eqs. (13.25) and (13.26), we obtain the following solutions:

$$A_s(z)=A_{s0}\cosh gz+i\sqrt{\frac{\omega_sn_i}{\omega_in_s}}A_{i0}^*e^{-i\phi_{p0}}\sinh gz \tag{13.31}$$

$$A_i^*(z)=A_{i0}^*\cosh gz-i\sqrt{\frac{\omega_in_s}{\omega_sn_i}}A_{s0}e^{i\phi_{p0}}\sinh gz \tag{13.32}$$

where A_{s0} and A_{i0} are *complex amplitudes* of the signal and idler at the input of the waveguide (see Eqs. 13.25 and 13.26). These are the general solution to the coupled mode equations and describe the evolution of the signal and idler as a function of z.

As an example let us assume the following input condition relating the amplitudes of the signal and idler:

$$|A_{i0}| = \sqrt{\frac{\omega_i n_s}{\omega_s n_i}} |A_{s0}| \tag{13.33}$$

In order to understand what this relation implies, we note that the input photon flux at the signal and idler is given by

$$N_s(0) = \frac{P_s(0)}{\hbar\omega_s} = \frac{1}{2c\mu_0} \frac{n_s}{\hbar\omega_s} |A_{s0}|^2 \tag{13.34}$$

and

$$N_i(0) = \frac{P_i(0)}{\hbar\omega_i} = \frac{1}{2c\mu_0} \frac{n_i}{\hbar\omega_i} |A_{i0}|^2 \tag{13.35}$$

Condition given by Eq. (13.33) simply implies that we are considering a case when the flux of photons incident at the signal and idler frequencies is equal to each other.

For this case, we have

$$A_s(z) = |A_{s0}| e^{-i\phi_{s0}} \cosh gz + i\sqrt{\frac{\omega_s n_i}{\omega_i n_s}} |A_{i0}| e^{i\phi_{i0}} e^{-i\phi_{p0}} \sinh gz$$

or

$$A_s(z) = |A_{s0}| e^{-i\phi_{s0}} \left[\cosh gz + e^{-i(\Delta\phi - \pi/2)} \sinh gz\right] \tag{13.36}$$

where

$$\Delta\phi = \phi_{p0} - \phi_{s0} - \phi_{i0} \tag{13.37}$$

Now, if the phases of the pump, signal and idler satisfy the following condition

$$\Delta\phi = \pi/2 \tag{13.38}$$

then

$$A_s(z) = |A_{s0}| e^{-i\phi_{s0}} e^{gz} \tag{13.39}$$

which shows amplification of the signal wave. The signal power variation with z is given by

$$P_s(z) = \frac{n_s}{2c\mu_0} |A_s(z)|^2 = \frac{n_s}{2c\mu_0} |A_{s0}|^2 e^{2gz} = P_s(0) e^{2gz} \tag{13.40}$$

where $P_s(0)$ is the incident signal power. In this case, the signal power grows exponentially as e^{2gz}.

On the other hand if the phases of the pump, signal, and idler satisfy the following condition

$$\Delta\phi = -\pi/2 \tag{13.41}$$

then

$$A_s(z) = |A_{s0}|e^{-i\phi_{s0}}e^{-gz} \tag{13.42}$$

which shows attenuation. The signal power variation with z is given by

$$P_s(z) = \frac{n_s}{2c\mu_0}|A_s(z)|^2 = P_s(0)e^{-2gz} \tag{13.43}$$

In this case, the signal power decays exponentially as e^{-2gz}.

Hence, in the presence of all three waves (pump, signal, and idler) at the input, the signal may get amplified or attenuated depending on the relative phase of the input waves. Hence, such an amplifier is termed phase-sensitive parametric amplifier and has very interesting quantum characteristics as will be discussed in Chap. 22.

Figure 13.1 shows parametric amplification in an optical waveguide fabricated in a domain engineered lithium niobate crystal which demonstrates the phase-sensitive nature of the amplification process. The input consists of three waves having wavelengths 1546 nm, $\lambda_s = 1541.4$ nm and $\lambda_i = 1550.6$ nm which are coupled into the waveguide with a fixed phase relationship. The wave at the wavelength of 1546 nm first undergoes second harmonic generation to generate a wave at $\lambda_p = 773$ nm which then acts as the pump wave and interacts with the signal and amplifies it. It also shows results of the same amplifier operating as a phase-insensitive amplifier.

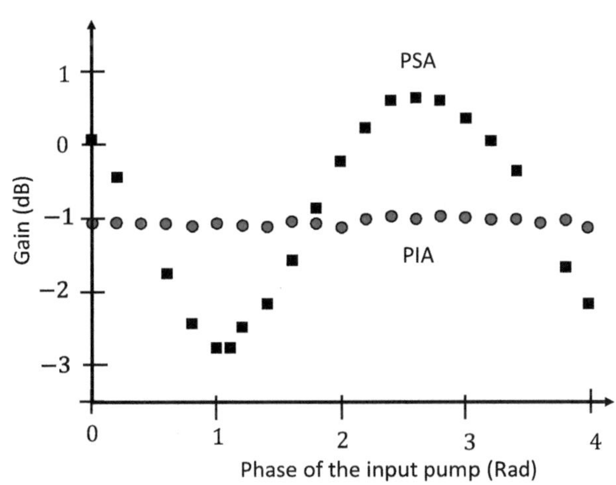

Fig. 13.1 Comparison of gain measurements in a phase-sensitive amplifier (PSA) and a phase-insensitive amplifier (PIA) as a function of the input phase of the pump in a periodically poled lithium niobate. (Adapted from Lee et al. 2009)

For later reference, we rewrite Eqs. (13.31) and (13.32) in a slightly convenient modified form by defining the following new variables:

$$\tilde{A}_s(z) = \sqrt{\frac{n_s}{\omega_s}} A_s(z) \tag{13.44}$$

$$\tilde{A}_i(z) = \sqrt{\frac{n_i}{\omega_i}} A_i(z) \tag{13.45}$$

Note that in terms of the new variables, the signal power is given by

$$P_s(z) = \frac{n_s}{2c\mu_0} |A_s(z)|^2 = \omega_s \frac{1}{2c\mu_0} \left|\tilde{A}_s(z)\right|^2 \tag{13.46}$$

and

$$P_i(z) = \frac{n_i}{2c\mu_0} |A_i(z)|^2 = \omega_i \frac{1}{2c\mu_0} \left|\tilde{A}_i(z)\right|^2 \tag{13.47}$$

Now the flux of signal photons at any value of z is given by $\frac{P_s(z)}{\hbar\omega_s}$ which is now proportional to $\left|\tilde{A}_s(z)\right|^2$. Similarly the flux of idler photons at any value of z will be given by $\frac{P_i(z)}{\hbar\omega_i}$ which is proportional to $\left|\tilde{A}_i(z)\right|^2$. Thus, the new variables $\tilde{A}_s(z)$ and $\tilde{A}_i(z)$ are proportional to photon fluxes rather than power in the electromagnetic wave. We will use these equations for quantum description of this nonlinear process.

With these definitions, Eqs. (13.31) and (13.32) become

$$\tilde{A}_s(z) = \tilde{A}_{s0} \cosh gz + i\tilde{A}_{i0}^* e^{-i\phi_{p0}} \sinh gz \tag{13.48}$$

$$\tilde{A}_i^*(z) = -i\tilde{A}_{s0} e^{i\phi_{p0}} \sinh gz + \tilde{A}_{i0}^* \cosh gz \tag{13.49}$$

which treat both the signal and idler in a nice symmetric fashion. We can put these equations in a matrix form as below:

$$\begin{pmatrix} \tilde{A}_s(z) \\ \tilde{A}_i^*(z) \end{pmatrix} = \begin{pmatrix} \cosh gz & ie^{-i\phi_{p0}} \sinh gz \\ -ie^{i\phi_{p0}} \sinh gz & \cosh gz \end{pmatrix} \begin{pmatrix} \tilde{A}_{s0} \\ \tilde{A}_{i0}^* \end{pmatrix} \tag{13.50}$$

The 2 × 2 matrix in Eq. (13.50) defines the nonlinear element connecting the amplitudes of the signal and idler at the input with that at the output.

Exercise 13.3 Consider a parametric amplifier with pump at ω_p, signal ω_s, and idler ω_i. At the input, all three waves are incident with

$$A_p(0) = u_{p0};\quad A_s(0) = u_{s0};\quad A_i(0) = iu_{i0} = iu_{s0}$$

where u_{p0}, u_{s0}, and u_{i0} are real quantities. Determine whether the waves at ω_s and ω_i get amplified or attenuated during the interaction process. Neglect pump depletion.

Solution For the given input conditions, $\phi_{p0} = 0$, $\phi_{s0} = 0$, and $\phi_{i0} = -\frac{\pi}{2}$. Hence, $\phi_{p0} - \phi_{s0} - \phi_{i0} = \pi/2$. Hence, as per Eq. (13.36) the signal will be amplified.

Exercise 13.4 Using Eq. (13.50), show that $|\tilde{A}_s(z)|^2 - |\tilde{A}_i(z)|^2 = |\tilde{A}_s(0)|^2 - |\tilde{A}_i(0)|^2$.

Solution We have

$$|\tilde{A}_s(z)|^2 = \left|\tilde{A}_{s0}\cosh gz + i\tilde{A}_{i0}^* e^{-i\phi_{p0}}\sinh gz\right|^2 = |\tilde{A}_{s0}|^2\cosh^2 gz + |\tilde{A}_{i0}|^2\sinh^2 gz$$

Similarly

$$|\tilde{A}_i(z)|^2 = \left| -i\tilde{A}_{s0}e^{i\phi_{p0}}\sinh gz + \tilde{A}_{i0}^*\cosh gz\right|^2 = |\tilde{A}_{s0}|^2\sinh^2 gz + |\tilde{A}_{i0}|^2\cosh^2 gz$$

Hence,

$$|\tilde{A}_s(z)|^2 - |\tilde{A}_i(z)|^2 = |\tilde{A}_{s0}|^2(\cosh^2 gz - \sinh^2 gz) + |\tilde{A}_{i0}|^2(\sinh^2 gz - \cosh^2 gz) = |\tilde{A}_{s0}|^2 - |\tilde{A}_{i0}|^2$$

13.4 Nondegenerate Phase-Insensitive Parametric Amplifier

In Sect. 13.3, we considered all three waves to be input into the waveguide. This leads to a phase-sensitive amplifier. A parametric amplifier operates in a phase-insensitive manner when at the input we have only two of the waves, for example, the pump and signal waves, and there is no input corresponding to the idler. This was discussed in Chap. 12 and we obtained the following equations representing the power of the signal as a function of z:

$$P_s(z) = \frac{n_s}{2c\mu_0}|A_s(z)|^2 = \frac{n_s}{2c\mu_0}|A_{s0}|^2\cosh^2 gz = P_s(0)\cosh^2 gz \tag{13.51}$$

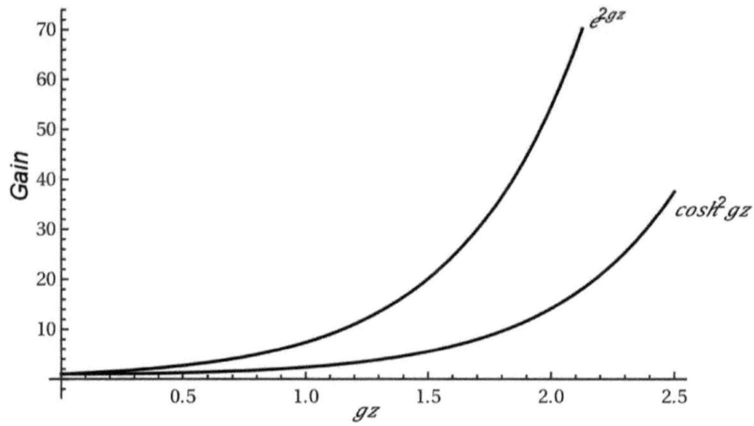

Fig. 13.2 Signal gain in a phase-sensitive and a phase-insensitive parametric amplifier

where $P_s(0)$ is the power input at the signal frequency into the waveguide. This can also be obtained from Eq. (13.50) in which we assume the incidence only of the signal with the input idler amplitude being zero, that is, $\tilde{A}_{i0}^* = 0$.

Since cosh$gz > 1$, we see that the amplitude of the signal always grows with propagation. In fact since $\cosh gz = \frac{e^{gz}+e^{-gz}}{2}$, for large z we find that the signal amplitude will grows exponentially. In this case, the phase of the pump or that of the signal has no role to play and the amplifier operates as a phase-insensitive amplifier.

Figure 13.2 shows a comparison of the signal gain in a phase-sensitive amplifier (see Eq. 13.40) and a phase-insensitive amplifier (see Eq. 13.51). As can be seen, the signal power increases much more rapidly in a phase-sensitive amplifier as compared to a phase-insensitive amplifier.

The gain coefficient in the amplification process is given by g:

$$g^2 = \frac{\omega_i \omega_s d_{eff}^2 \Gamma^2}{c^2 n_i n_s} |A_{p0}|^2 \tag{13.52}$$

Thus, the gain coefficient depends on the effective nonlinear coefficient d_{eff}, the overlap factor Γ and most importantly the pump power. Expressing $|A_{p0}|^2$ in terms of pump power P_p given by

$$P_p = \frac{n_p}{2c\mu_0} |A_{p0}|^2 \tag{13.53}$$

we have

$$g^2 = \frac{2\mu_0 \omega_i \omega_s d_{eff}^2 \Gamma^2}{c n_i n_s n_p} P_p \tag{13.54}$$

Such an amplifier corresponds to a phase-insensitive amplifier since irrespective of the phase of the input signal, the signal always undergoes amplification.

In Chap. 12, we also discussed nondegenerate phase-insensitive parametric amplifier (with the signal and idler photons have different frequencies) with inputs at the pump frequency and the signal frequency.

Normally, phase-insensitive amplification is achieved when the signal and idler frequencies are different $\omega_s \neq \omega_i$ and the input consists of only two of the waves, for example, the pump and the signal waves.

Phase-insensitive operation is also possible if the signal and idler have the same frequency but they correspond to two different modes such as two orthogonal polarization states. Thus, at the input we could have a pump mode at frequency ω_p and two signal modes of identical frequency ($=\omega_p/2$) but having orthogonal polarization states.

For example, let us assume that the waveguide is fabricated in KTP and the propagation direction corresponds to the principal x-axis direction. Using the nonlinear tensor of KTP, we can see that the nonlinear coefficient responsible for interaction between a pump wave at frequency ω_p polarized along the y-direction, a signal wave at frequency $\omega_s = \omega_p/2$ polarized along the z-direction, and another signal at frequency $\omega_s = \omega_p/2$ polarized along the y-direction is d_{24} which is non zero (see also Problem 12.16) and thus it is possible for this nonlinear interaction to take place provided the corresponding phase matching or quasi phase matching condition is satisfied. The quasi phase matching condition is given by

$$\beta_p^{(y)} = \beta_s^{(z)} + \beta_s^{(y)} + K \tag{13.55}$$

where $\beta_p^{(y)}$ is the propagation constant of the y-polarized mode at the pump frequency, $\beta_s^{(y)}$ is the propagation constant of the y-polarized mode at signal frequency (which is half the pump frequency), and $\beta_s^{(z)}$ is the propagation constant of the z-polarized mode at the same signal frequency. Note that now the signal and idler are not identical and are different, although both of them have the same frequency.

Figure 13.3 shows the interaction which can be quasi phase matched.

If the input to the crystal is a pump mode polarized along y-direction and a signal at frequency $\omega_s = \frac{\omega_p}{2}$, which is polarized along z-direction (see Fig. 13.3), then the parametric amplifier will behave as a phase-insensitive amplifier since the input has

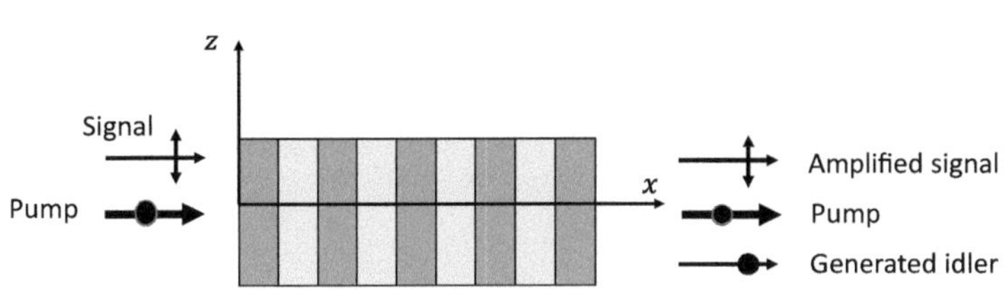

Fig. 13.3 A phase-insensitive amplifier can be realized even with the signal having half the pump frequency provided the signal interacts as two different polarizations but having the same frequency

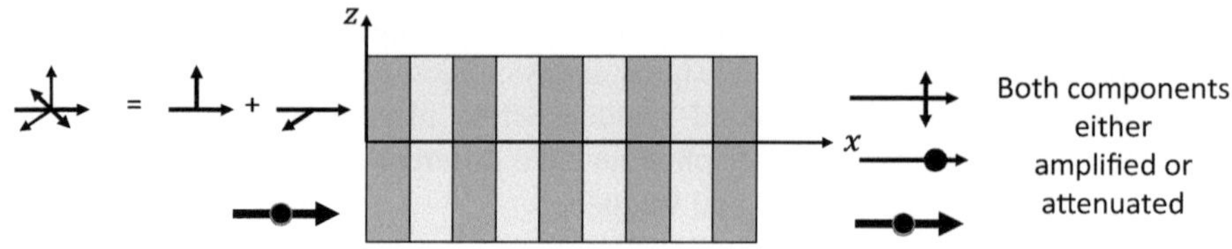

Fig. 13.4 Phase-sensitive operation of the amplifier with input at both signal polarization states (solid arrows) along with the pump wave (dashed arrow)

only one of the interacting modes at frequency ω_s apart from the pump. Note that although the frequencies of the two down converted modes (signal and idler) are the same, they correspond to different modes of propagation.

This same amplifier configuration will act as a phase-sensitive amplifier if both the polarization states of signal and idler which are two different modes (having two different polarization states) of the waveguide are incident along with the pump as shown in Fig. 13.4. This launch of modes is possible when the input polarization state of signal is at 45° to the z-axis. This is similar to a phase-sensitive operation with both signal and idler input in the waveguide along with the pump with both signal and idler having the same frequency but being distinguished by their different polarization states. In this case, the amplification process will depend on the relative phase of the input waves with respect to the pump wave.

Replacing phase-insensitive optical amplifiers (such as erbium doped fiber amplifiers) by phase-sensitive optical parametric amplifiers has been shown to drastically improve the performance of fiber optic communication system. The impressive results show that it is possible to extend the reach of optical fiber communication from 800 to 4500 km by using phase-sensitive amplifiers (Olsson et al. 2018).

Exercise 13.5 A phase-insensitive parametric amplifier operates with a pump wavelength of 1 μm and a signal wavelength of 1.55 μm. If the input pump power is 1 W and an input signal power of 1 mW is converted to 2 mW at the output, obtain the output power at the idler and pump frequencies.

Solution For the given pump and signal wavelengths, the idler wavelength comes out to be 2.818 μm. The increase in power of the signal wavelength is 1 mW which corresponds to $N_s = \frac{\Delta P_s}{\hbar\omega_s} \approx 7.8 \times 10^{15}$ signal photons. Since photon numbers are conserved, the number of idler photons generated is also equal to N_s and the number of pump photons lost is also N_s. Thus, the output power at idler is given by $P_i = N_s\hbar\omega_i = \frac{\lambda_s}{\lambda_i}\Delta P_s \approx 0.55$ mW. The decrease in pump power is given by $P_p = N_s\hbar\omega_p = \frac{\lambda_s}{\lambda_p}\Delta P_s = 1.55$ mW. Note that the power lost in the pump is equal to the gain in signal and idler power showing energy conservation.

13.4.1 Amplification Bandwidth of a Nondegenerate Phase-Insensitive Parametric Amplifier

In Exercise 13.7, it is shown that for a nonphase-matched nondegenerate parametric amplifier, for a given interaction length L when the phase mismatch $\Delta\beta$ satisfies the following equation

$$\Delta\beta L = \pm 2\sqrt{\left(g^2L^2 + \pi^2\right)} \tag{13.56}$$

the idler is no more generated implying that there are no idler photons which in turn imply that there are no additional signal photons generated. Assuming quasi phase matching, this happens when

$$\Delta\beta - K = \pm \frac{2}{L}\sqrt{\left(g^2L^2 + \pi^2\right)} \tag{13.57}$$

Now under quasi phase matching

$$\Delta\beta - K = \beta_p - \beta_s - \beta_i - K \tag{13.58}$$

At the input signal frequency ω_{s0} (with a fixed pump at frequency ω_{p0}), the quasi phase matching condition is exactly satisfied and we have

$$\Delta\beta - K = \beta\left(\omega_{p0}\right) - \beta(\omega_{s0}) - \beta(\omega_{i0}) - K = 0 \tag{13.59}$$

with the required spatial frequency K obtained from the above equation.

Now if the input signal frequency ω_s deviates from ω_{s0}, then we would no more be satisfying the quasi phase matching condition and as per our earlier discussion, the signal amplification would reduce. When the frequency changes from ω_{s0} to $\omega_s = \omega_{s0} + \Delta\omega_s$ with the corresponding idler frequency changing from ω_{i0} to $\omega_i = \omega_{i0} + \Delta\omega_i$ such that $\Delta\beta$ satisfies Eq. (13.57), then there would be no idler generation and no signal amplification.

Using Eq. (13.58), we have for a signal wavelength ω_s and an idler wavelength ω_i,

$$\Delta\beta - K = \beta\left(\omega_{p0}\right) - \beta(\omega_s) - \beta(\omega_i) - K \tag{13.60}$$

Since the pump frequency ω_{p0} is assumed to be constant and we need to satisfy

$$\omega_{p0} = \omega_s + \omega_i = \omega_{s0} + \Delta\omega_s + \omega_{i0} + \Delta\omega_i = \omega_{p0} + \Delta\omega_s + \Delta\omega_i$$

Hence, we have

$$\Delta\omega_s = -\Delta\omega_i \tag{13.61}$$

Assuming that $\Delta\omega_s \ll \omega_{s0}$, and $\Delta\omega_i \ll \omega_{i0}$ we can have Taylor series expansion in Eq. (13.60) and obtain

$$\Delta\beta - K = \beta(\omega_{p0}) - \beta(\omega_{s0}) - \Delta\omega_s \frac{d\beta}{d\omega}\bigg|_{\omega=\omega_{s0}} - \beta(\omega_{i0}) - \Delta\omega_i \frac{d\beta}{d\omega}\bigg|_{\omega=\omega_{i0}} - K$$

Recalling from Chap. 4 that the group index is given by

$$N_{eff} = c\frac{d\beta}{d\omega} \tag{13.62}$$

using Eq. (13.60), we have

$$\Delta\beta - K = -\left(\frac{d\beta}{d\omega}\bigg|_{\omega=\omega_{s0}} - \frac{d\beta}{d\omega}\bigg|_{\omega=\omega_{i0}}\right)\Delta\omega_s = -\frac{1}{c}\left(N_{eff}(\omega_{s0}) - N_{eff}(\omega_{i0})\right)\Delta\omega_s \tag{13.63}$$

Substituting in Eq. (13.57), we obtain

$$\Delta\omega_s = \Delta\omega_i = \pm\frac{2}{L}\sqrt{(g^2L^2 + \pi^2)}\frac{c}{\left(N_{eff}(\omega_{s0}) - N_{eff}(\omega_{i0})\right)} \tag{13.64}$$

If we represent the frequency bandwidth between the two zeroes of amplification around the peak gain as $\Delta\omega_{s,zz}$, we have

$$\Delta\omega_{s,zz} = \frac{4c}{L}\sqrt{(g^2L^2 + \pi^2)}\frac{1}{\left(N_{eff}(\omega_{s0}) - N_{eff}(\omega_{i0})\right)} \tag{13.65}$$

We can express the bandwidth in terms of wavelength by noting that since $\omega_s = \frac{2\pi c}{\lambda_s}$, we have

$$\Delta\omega_s = -\frac{2\pi c}{\lambda_s^2}\Delta\lambda_s \tag{13.66}$$

Hence, we have

$$\Delta\lambda_{s,zz} = \frac{2\lambda_{s0}^2}{\pi L}\sqrt{(g^2L^2 + \pi^2)}\frac{1}{\left(N_{eff}(\lambda_{s0}) - N_{eff}(\lambda_{i0})\right)} \tag{13.67}$$

Equation (13.67) gives us an estimate of the bandwidth of the parametric amplification process and the difference frequency generation process.

If the gain of the parametric amplifier is large at the phase matching wavelength, then we can approximate $\sqrt{(g^2L^2 + \pi^2)} \approx gL$ and the gain bandwidth for such a case is given approximately by

$$\Delta\lambda_{s,zz} = \frac{2g\lambda_{s0}^2}{\pi} \frac{1}{\left(N_{eff}(\lambda_{s0}) - N_{eff}(\lambda_{i0})\right)} \tag{13.68}$$

or in terms of frequency

$$\Delta\omega_{s,zz} = 4gc \frac{1}{\left(N_{eff}(\omega_{s0}) - N_{eff}(\omega_{i0})\right)} \tag{13.69}$$

Note that in this approximation, the gain bandwidth becomes independent of length and depends on the strength of the coupling between the interacting waves through γ and the dispersive properties of the waveguide through the group refractive index at the signal and idler wavelengths.

On the other hand, if the amplifier gain is small, then we can approximate $\sqrt{(g^2L^2 + \pi^2)}$ by π and obtain

$$\Delta\omega_{s,BW} = \frac{4\pi c}{L} \frac{1}{\left(N_{eff}(\omega_{s0}) - N_{eff}(\omega_{i0})\right)} \tag{13.70}$$

which depends inversely on the length of the device.

> ***Exercise 13.6*** A phase-matched degenerate optical parametric amplifier has a gain coefficient of 5 m^{-1}. What will be the gain coefficient if $\Delta\beta = 1$ m^{-1}?
>
> ***Solution*** The gain coefficient when $\Delta\beta \neq 0$ and $g > \frac{\Delta\beta}{2}$ is given by $\Omega = \sqrt{g^2 - \frac{\Delta\beta^2}{4}}$. Hence, in this case we obtain the gain coefficient as 4.97 m^{-1}.

We write Eq. (12.69) as

$$P_i(L) = \frac{\omega_i}{\omega_s} g^2 L^2 P_{s0} \frac{\sinh^2 \Omega L}{(\Omega L)^2} \tag{13.71}$$

Figure 13.5 shows the variation in $\frac{\sinh^2 \Omega L}{(\Omega L)^2}$ with $\Delta\beta L$ for $g^2L^2 = 10$. For $\Delta\beta L = 0$, $\frac{\sinh^2 \Omega L}{(\Omega L)^2} = \frac{\sinh^2 gL}{(gL)^2} \approx 13.9$. Note that when $\Delta\beta = 0$, that is, at phase matching, $\Omega = g$ and the signal gain is given by $\sinh^2 gL$ (see Eq. 12.68) which in the example we are considering is 139 which is 21 dB.

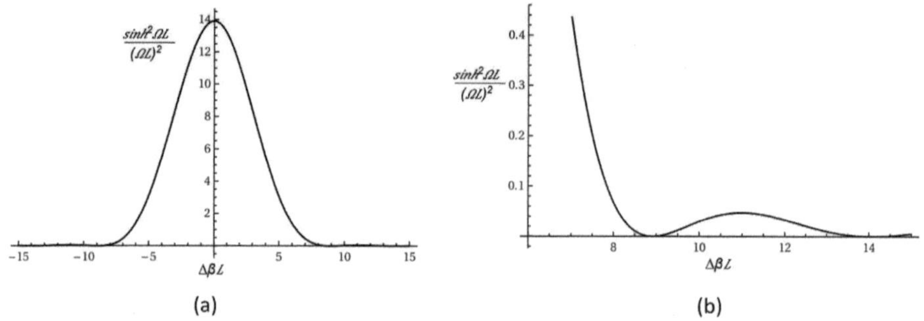

Fig. 13.5 (**a**) Variation in $\frac{\sinh^2\Omega L}{(\Omega\mathrm{L})^2}$ with $\Delta\beta L$ for $g^2L^2 = 10$. (**b**) Expanded figure showing oscillatory behavior for $\Omega^2 = g^2 - \frac{\Delta\beta^2}{4} < 0$

Exercise 13.7 As the phase mismatch $\Delta\beta$ increases, when $\frac{\Delta\beta}{2} = g$, $\Omega = 0$ and $\frac{\sinh^2\Omega L}{(\Omega\mathrm{L})^2} = 1.0$. Beyond this value of $\Delta\beta$, $\Omega^2 < 0$ and Ω become imaginary. For imaginary value of Ω, $\sinh\Omega\mathrm{L} = i \sin(|\Omega|L)$ and the variation become oscillatory as can be seen from Fig. 13.5b. What is the minimum value of $\Delta\beta L$ when $\frac{\sinh^2\Omega L}{(\Omega\mathrm{L})^2} = 0$?

Solution For $\Omega^2 < 0$, we define $\tilde{\Omega}^2 = -\Omega^2$. Thus,

$$\frac{\sinh^2\Omega L}{(\Omega\mathrm{L})^2} = \frac{\sinh^2\left(\mathrm{i}\tilde{\Omega}L\right)}{\left(i\tilde{\Omega}\mathrm{L}\right)^2} = \frac{\sin^2\left(\tilde{\Omega}L\right)}{\left(\tilde{\Omega}\mathrm{L}\right)^2}$$

which becomes zero when

$$\sin\left(\tilde{\Omega}L\right) = 0$$

which gives us $\tilde{\Omega}L = m\pi; m = 1, 2, \ldots$. Hence, the first zero appears when $\tilde{\Omega}L = \pi$ which gives

$$\tilde{\Omega}^2L^2 = \pi^2$$

Substituting $\tilde{\Omega}^2 = \frac{\Delta\beta^2}{4} - g^2$, we have

$$\frac{(\Delta\beta L)^2}{4} - g^2L^2 = \pi^2$$

Hence,

$$\Delta\beta L = \pm 2\sqrt{(g^2L^2 + \pi^2)}$$

Using the value of $g^2L^2 = 10$, we obtain $\Delta\beta L \approx 8.92$ (see Fig. 13.5). For this value of $\Delta\beta$, no idler is being generated which also implies that there are no additional signal photons being generated and the gain of the amplifier becomes unity.

Exercise 13.8 For a parametric amplifier at 1550 nm, a pump at 800 nm is used. Assuming the refractive indices at pump, signal, and idler wavelengths to be as given in Table 12.1. obtain the required pump power to achieve a value of $g^2L^2 = 10$ for an interaction length of 5 cm. Assume all the modes to have extraordinary polarizations and $d_{eff} = d_{33}$. Take the mode overlap area as 50 μm^2.

Solution For the given length, the required value is $g^2 = 4 \times 10^3$ m^{-2}. Using Eq. (13.54) for g^2, the required pump power P_p comes out to be about 190 mW. If first order QPM is used, then the nonlinear coefficient reduces by a factor of $\frac{2}{\pi}$ and the required power increses by a factor of $\frac{\pi^2}{4}$ and becomes 470 mW.

Problems

13.1. A phase-matched phase-insensitive parametric amplifier has a gain coefficient of 5 m^{-1}. What will be the gain coefficient if $\Delta\beta = 1$ m^{-1}? What is the maximum value of $\Delta\beta$ for which the amplifier will provide gain?

13.2. It is desired to design a phase-insensitive optical parametric amplifier for a signal wavelength of 1500 nm based on a periodically poled lithium niobate waveguide with the pump wavelength of 800 nm.

(a) To use d_{33} coefficient, what should be the states of polarization of the pump, signal and idler?
(b) What is the required period for first order QPM?
(c) Obtain the signal gain over 1 cm of the waveguide if pumped by 1 W of pump power.
(d) For an input signal power of 10 mW, plot the output signal power variation with signal wavelength and thus estimate the gain bandwidth of the amplifier.

For the calculation, approximate the effective indices of the modes using the following equations for the ordinary and extraordinary refractive indices of lithium niobate:

$$n_o^2(\lambda) = 4.9048 + \frac{0.11768}{(\lambda_0^2 - 0.0475)} - 0.027169\lambda^2$$

$$n_e^2(\lambda) = 4.582 + \frac{0.099169}{(\lambda_0^2 - 0.044432)} - 0.02195\lambda^2$$

Here λ_0 is the free space wavelength in μm. Assume $\Gamma^2 = 10^9\ \text{m}^{-2}$.

13.3. Consider a degenerate parametric amplifier in a KTP waveguide with the incident light at 2ω propagating along the x-direction and polarized along the y-direction. The propagation constants of the modes in the waveguide corresponding to frequency ω and 2ω and polarized along the y- and z-directions are given by β_{1y}, β_{1z}, β_{2y} and β_{2z}, respectively.

(a) If in addition to the input mode at 2ω, the mode at ω polarized along y is also incident, will this behave as a phase-sensitive or a phase-insensitive amplifier?

(b) Obtain the nonlinear polarization that will lead to the amplification of the signal at ω polarized along the y-direction with the idler polarized along the z-direction. What is the corresponding quasi phase matching condition?

13.4. I wish to achieve parametric amplification using lithium niobate crystal with a signal at 1.2 μm and a pump at 0.8 μm travelling along the same direction and an idler travelling in the reverse direction with all waves being extraordinary waves. If the effective indices of the waveguide at the pump, signal and idler wavelengths are 2.1755, 2.1493 and 2.1143, respectively, what is the quasi phase matching period required?

13.5. Consider a degenerate parametric amplifier with a pump field given by

$$E_p(x, y, z, t) = A_{20}\psi_p(x, y)\cos\left(\beta_2 z - 2\omega t + \frac{\pi}{4}\right)$$

For what input phase of the signal at frequency ω will the signal get amplified?

13.6. A pump having extraordinary polarization and frequency 2ω propagates along the y-direction in a z-cut lithium niobate waveguide. Which element of the d tensor would be responsible for amplification of a signal at frequency ω having ordinary polarization?

13.7. Consider a parametric amplifier in periodically poled lithium niobate with the pump at frequency ω_p as an ordinary mode and propagating along the y-direction in a z-cut lithium niobate. The QPM period of the periodic poling is so as to satisfy the phase matching condition with the signal and idler having the same frequency ($\omega_p/2$) but having orthogonal polarizations. Show that

(a) If the input wave at $\omega_p/2$ is an ordinary wave, the amplifier acts like a phase-insensitive amplifier and.

Table 13.1 Refractive indices and group refractive indices at the pump, signal and idler wavelengths

	$n(\lambda_0)$	$N(\lambda_0)$
Pump	2.2388	2.4752
Signal	2.1755	2.2642
Idler	2.1395	2.1840

(b) If it is linearly polarized at 45° to the optic axis, the amplifier acts as a phase-sensitive amplifier.

13.8. Consider a parametric amplifier in periodically poled lithium niobate waveguide with the pump as an ordinary wave and propagating along the crystal y-axis. The signal and idler are assumed to be orthogonally polarized and propagating also along the y-axis. Starting from the expression for nonlinear polarization, obtain expressions for the nonlinear polarization at the signal and idler frequencies for both situations (signal with ordinary polarization and idler with extraordinary polarization and vice versa). From these expressions, obtain the corresponding first order QPM periods required for both the processes.

13.9. In a parametric amplifier with a pump at 980 nm and signal at 1500 nm, what is the wavelength of the idler? If the generated idler power is 10 mW, what will be the difference in the signal power between the output and input?

13.10. Consider a lithium niobate based phase-insensitive parametric amplifier having a gain coefficient of 5 m^{-1} and a length of 5 cm and operating with the nonlinear coefficient d_{33}. The pump wavelength is 520 nm and the signal wavelength is 800 nm. Calculate the gain bandwidth $\Delta\lambda_{s,zz}$ of the amplifier. The refractive indices and the group refractive indices at the pump, signal, and idler wavelengths are given in Table 13.1.

References

Kazama, T., T. Umeki, S. Shimizu, T. Kashiwazaki, K. Enbutsu, R. Kasahara, Y. Miyamoto, and K. Watanabe. 2021. Over 30 dB gain and 1 dB noise figure phase-sensitive amplification using a pump-combiner-integrated fiber I/O PPLN module. *Opt Express* 29:28824–28834.

Ledezma, L., R. Sekine, Q. Guo, R. Nehra, S. Jahani, and A. Marandi. 2022. Intense optical parametric amplification in dispersion-engineered nanophotonic lithium niobate waveguides. *Optica* 9:303–308.

Lee, K. J., F. Parmigiani, S. Liu, J. Kakande, P. Petropoulos, K. Gallo, and D. Richardson. 2009. Phase sensitive amplification based on quadratic cascading in a periodically poled lithium niobate waveguide. *Opt Express* 17:20393–20400.

Olsson, S. L. I., H. Eliasson, E. Astra, M. Karlsson, and P. A. Andrekson. 2018. Long haul optical transmission link using low-noise phase sensitive amplifiers. *Nat Commun* 9 (1): 2513. https://doi.org/10.1038/s41467-018-04956-5.

Yan, Z., H. He, H. Liu, M. Iu, O. Ahmed, E. Chen, P. Blakey, Y. Akasaka, T. Ikeuchi, and A. S. Helmy. 2022. $\chi^{(2)}$ based AlGaAs phase sensitive amplifier with record gain, noise and sensitivity. *Optica* 9:56–60.

Chapter 14
Quantization of the Electromagnetic Field and the Concept of Photon

Abstract This chapter deals with the quantization of the electromagnetic field and introduces the concept of photon as a quantum of excitation of the electromagnetic field by showing that each mode of the electromagnetic field behaves like a linear harmonic oscillator. Annihilation and creation operators of individual modes of the electromagnetic field are introduced and an expression for the electric field operator is obtained.

14.1 Introduction

In this chapter starting from Maxwell's equations in free space, we will quantize the electromagnetic field and show that the electromagnetic field can be considered to be a linear superposition of an infinite number of modes of the field (with each mode corresponding to a specific frequency, propagation direction and a particular state of polarization) with each mode *behaving* like a harmonic oscillator. Taking analogy from the quantum picture of harmonic oscillators, we see that the energy of each electromagnetic field mode can only increase or decrease in quanta of energy $\hbar\omega$ where ω is the frequency of the mode. Such a quantum of energy is referred to as the photon. Thus, photons are nothing but quanta of excitation of the electromagnetic field. We will then discuss some important quantum states of the electromagnetic field. In Chaps. 15–18, we will discuss different states of the electromagnetic field and look at some of their important properties. Quantization of electromagnetic field is discussed in many books which the reader may refer to for further details; for example, Yariv (1991), Mandel and Wolf (1995), Gerry and Knight (2005), Garrison and Chiao (2008), Grynberg et al. (2010), Thyagarajan and Ghatak (2011), Cohen-Tannoudji et al. (2019).

K. Thyagarajan, *Integrated Quantum Photonics*, Graduate Texts in Physics,
https://doi.org/10.1007/978-3-031-85728-7_14

14.2 Maxwell's Equations

We start with the Maxwell's equations in free space with no free charges ($\rho_f = 0$) and no free currents $\left(\vec{J}_f = 0\right)$:

$$\nabla \times \vec{E} = -\frac{\partial \vec{B}}{\partial t} = -\mu_0 \frac{\partial \vec{H}}{\partial t} \tag{14.1}$$

$$\nabla \times \vec{H} = \epsilon_0 \frac{\partial \vec{E}}{\partial t} \tag{14.2}$$

$$\nabla \cdot \vec{E} = 0 \tag{14.3}$$

$$\nabla \cdot \vec{B} = 0 \tag{14.4}$$

Since $\nabla \cdot \vec{B} = 0$, we can write

$$\vec{B} = \mu_0 \vec{H} = \nabla \times \vec{A} \tag{14.5}$$

where $\vec{A}$ is the vector potential. Substituting from Eq. (14.5) in Eq. (14.1), we obtain

$$\nabla \times \vec{E} = -\frac{\partial\left(\nabla \times \vec{A}\right)}{\partial t} = -\nabla \times \frac{\partial \vec{A}}{\partial t}$$

Thus

$$\nabla \times \left(\vec{E} + \frac{\partial \vec{A}}{\partial t}\right) = 0 \tag{14.6}$$

implying that

$$\vec{E} + \frac{\partial \vec{A}}{\partial t} = -\nabla \phi \tag{14.7}$$

where ϕ is the scalar potential. Applying the operator $\vec{\nabla} \cdot$ in Eq. (14.7) we obtain, using Eq. (14.3)

$$\nabla^2 \phi + \frac{\partial\left(\vec{\nabla} \cdot \vec{A}\right)}{\partial t} = 0 \tag{14.8}$$

We choose the Coulomb gauge and assume

$$\phi = 0 \tag{14.9}$$

and since $\vec{B}$ remains unchanged if $\vec{A}$ is replaced by $\vec{A} + \nabla\chi$, where χ is an arbitrary scalar function, we choose χ such that

$$\vec{\nabla}\cdot\vec{A} = 0 \tag{14.10}$$

Hence, we obtain

$$\vec{E} = -\frac{\partial\vec{A}}{\partial t} \tag{14.11}$$

$$\vec{B} = \nabla\times\vec{A} \tag{14.12}$$

Now

$$\nabla\times\nabla\times\vec{A} = \nabla\times\vec{B} = \mu_0\nabla\times\vec{H} = \epsilon_0\mu_0\nabla\times\frac{\partial\vec{E}}{\partial t} = -\epsilon_0\mu_0\frac{\partial^2\vec{A}}{\partial t^2}$$

Since

$$\nabla\times\nabla\times\vec{A} = \nabla\left(\nabla\cdot\vec{A}\right) - \nabla^2\vec{A} = -\nabla^2\vec{A}$$

where we have used Eq. (14.10), we have

$$\nabla^2\vec{A} = \epsilon_0\mu_0\frac{\partial^2\vec{A}}{\partial t^2} \tag{14.13}$$

which is the equation satisfied by the vector potential $\vec{A}$. Now using separation of variables technique in Cartesian coordinates, we obtain the solutions of Eq. (14.13) in the form of plane waves (see Exercise 14.1)

$$\vec{A} \sim \widehat{e}\, e^{\pm i\vec{k}.\vec{r}}\, e^{\pm i\omega t} \tag{14.14}$$

which represent plane wave solutions corresponding to frequency ω, with a propagation direction defined by $\vec{k}$ and polarization state defined by $\widehat{e}$. Since $\vec{\nabla}\cdot\vec{A} = 0$, we have

$$\vec{k}\cdot\widehat{e} = 0 \tag{14.15}$$

showing that the waves are transverse. For every $\vec{k}$, there are two orthogonal orientations of the polarization state $\widehat{e}$ corresponding to two independent modes of propagation.

Exercise 14.1 Use separation of variables technique and show that the solution of Eq. (14.13) in Cartesian coordinates is given by Eq. (14.14).

Solution We use separation of variables technique and write

$$\vec{A}(x,y,z,t) = \hat{e}X(x)Y(y)Z(z)T(t)$$

Substituting in Eq. (14.13) and dividing by $X(x)Y(y)Z(z)T(t)$, we have

$$\frac{1}{X}\frac{d^2X}{dx^2} + \frac{1}{Y}\frac{d^2Y}{dy^2} + \frac{1}{Z}\frac{d^2Z}{dz^2} = \epsilon_0\mu_0\frac{1}{T}\frac{d^2T}{dt^2}$$

The variables have separated out and we write

$$\frac{1}{X}\frac{d^2X}{dx^2} = -k_x^2; \quad \frac{1}{Y}\frac{d^2Y}{dy^2} = -k_y^2; \quad \frac{1}{Z}\frac{d^2Z}{dz^2} = -k_z^2; \quad \frac{1}{c^2}\frac{1}{T}\frac{d^2T}{dt^2} = -k^2$$

where $k^2 = k_x^2 + k_y^2 + k_z^2$ and $c^2 = \frac{1}{\epsilon_0\mu_0}$. Solving the equations we obtain the solution given in Eq. (14.14) with $\omega = ck$ and $\vec{k} = k_x\hat{x} + k_y\hat{y} + k_z\hat{z}$.

In free space, all values of ω are allowed and all propagation directions are allowed. For each propagation direction, $\hat{e}$ can take two independent directions perpendicular to the direction of propagation. The total electromagnetic field is a superposition of all frequencies, all propagation directions, and the two orthogonal polarization states. A given frequency, direction of propagation, and a state of polarization define a mode of the electromagnetic field.

In order to simplify the analysis and discretize the problem, we consider a box with a finite volume. Imposing boundary conditions on the walls of the box, we will find a discrete set of allowed frequencies instead of a continuous range. This leads to the total solution to be a discrete sum of frequencies rather than an integral. If the dimensions of the box are allowed to go to infinity, the sum will transform into an integral over all allowed frequencies.

Two different types of boundary conditions can be considered: one in which we require the fields to vanish at the boundaries (assuming conducting walls) leading to standing wave solutions and the other in which we require the fields to be identical on opposite walls of the box leading to propagating solutions. The latter one is also referred to as periodic boundary conditions. Here we will employ the periodic boundary conditions to obtain propagating wave solutions.

We consider a cube of side L oriented along the Cartesian axes (see Fig. 14.1). Periodic boundary conditions imply

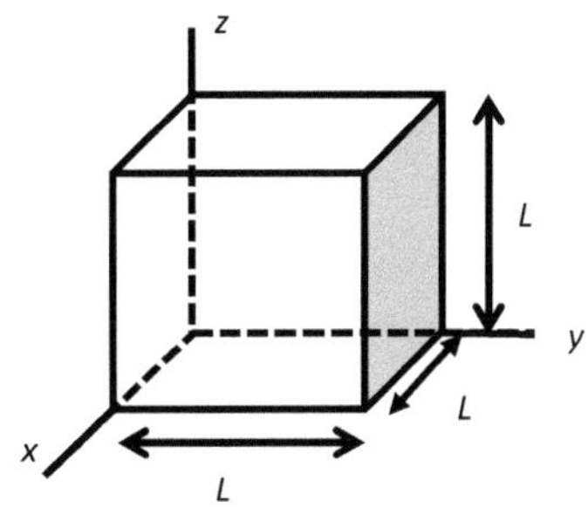

Fig. 14.1 Cube of side L used for applying periodic boundary conditions

$$\vec{A}(x=L, y=0, z=0) = \vec{A}(x=0, y=0, z=0) \tag{14.16a}$$

$$\vec{A}(x=0, y=L, z=0) = \vec{A}(x=0, y=0, z=0) \tag{14.16b}$$

$$\vec{A}(x=0, y=0, z=L) = \vec{A}(x=0, y=0, z=0) \tag{14.16c}$$

Applying these conditions on the solutions, we obtain

$$k_x = \frac{2\pi n_x}{L}; \quad n_x = 0, \quad \pm 1, \quad \pm 2, \ldots. \tag{14.17a}$$

$$k_y = \frac{2\pi n_y}{L}; \quad n_y = 0, \quad \pm 1, \quad \pm 2, \ldots. \tag{14.17b}$$

$$k_z = \frac{2\pi n_z}{L}; \quad n_z = 0, \quad \pm 1, \quad \pm 2, \ldots. \tag{14.17c}$$

where, the propagation vector is given by

$$\vec{k} = \hat{x}k_x + \hat{y}k_y + \hat{z}k_z \tag{14.18}$$

thus forming a discrete set of possible solutions.

The corresponding frequency of the electromagnetic wave would be

$$\omega = ck = c\left(k_x^2 + k_y^2 + k_z^2\right)^{1/2} = \frac{2\pi c}{L}\left(n_x^2 + n_y^2 + n_z^2\right)^{1/2} \tag{14.19}$$

These represent a discrete set of frequencies determined by the integers n_x, n_y, and n_z.

Hence, the general solution of Eq. (14.13) is

$$\vec{A}(\vec{r}, t) = \sum_{m=-\infty}^{+\infty} \left[q_m(t)\vec{A}_m(\vec{r}) + q_m^*(t)\,\vec{A}_m^*(\vec{r})\right] \tag{14.20}$$

where we have used a single index m to represent the combination of (n_x, n_y, n_z) and a particular polarization $\widehat{e}$. The second term in Eq. (14.20) is the complex conjugate of the first term to ensure that the vector potential is a real quantity.

Here

$$\vec{A}_m(\vec{r}) = \widehat{e}_m e^{i\vec{k}_m\cdot\vec{r}} \tag{14.21}$$

$$q_m(t) = |q_m| e^{-i\omega_m t} \tag{14.22}$$

We also assume that

$$k_{-m} = -k_m \tag{14.23a}$$

$$\omega_{-m} = \omega_m \tag{14.23b}$$

Thus, for example if we consider a mode specified by $m = 1$, then the corresponding terms in the sum given by Eq. (14.20) would be

$$q_1(t)\vec{A}_1(\vec{r}) + q_1^*(t)\,\vec{A}_1^*(\vec{r}) = e^{-i\omega_1 t}e^{i\vec{k}_1\cdot\vec{r}} + e^{i\omega_1 t}e^{-i\vec{k}_1\cdot\vec{r}} = e^{i\left(\vec{k}_1\cdot\vec{r} - \omega_1 t\right)} + e^{-i\left(\vec{k}_1\cdot\vec{r} - \omega_1 t\right)}$$

with both terms representing waves propagating along the $+\vec{k}_1$ direction.

On the other hand, using Eqs. (14.23a) and (14.23b) the term corresponding to $m = -1$ gives us

$$q_{-1}(t)\vec{A}_{-1}(\vec{r}) + q_{-1}^*(t)\,\vec{A}_{-1}^*(\vec{r}) = e^{-i\omega_{-1}t}e^{i\vec{k}_{-1}\cdot\vec{r}} + e^{i\omega_{-1}t}e^{-i\vec{k}_{-1}\cdot\vec{r}}$$
$$= e^{-i\omega_1 t}e^{-i\vec{k}_1\cdot\vec{r}} + e^{i\omega_1 t}e^{i\vec{k}_1\cdot\vec{r}} = e^{i\left(\vec{k}_1\cdot\vec{r} + \omega_1 t\right)} + e^{-i\left(\vec{k}_1\cdot\vec{r} + \omega_1 t\right)}$$

which now correspond to the same frequency as the case for $m = 1$ but now represent waves propagating in the $-\vec{k}_1$ direction. Thus, the terms corresponding to $+m$ and $-m$ represent two different modes propagating in opposite directions.

Hence, the solution given by Eq. (14.20) represents the general solution with each term representing a mode of propagation of the electromagnetic field satisfying the periodic boundary conditions that we have imposed. Each value of m corresponds to a particular mode of propagation specified by

- A specific frequency
- Specific polarization
- Specific direction of propagation

Orthogonality of the modes implies

$$\iiint \vec{A}_m(\vec{r})\cdot\vec{A}_p^{\,*}(\vec{r})d\tau = V\delta_{mp} \tag{14.24}$$

where V represents the volume of integration.

Using Eqs. (14.11) and (14.12), we obtain

$$\vec{E} = -\frac{\partial\vec{A}}{\partial t} = \sum \vec{E}_m \tag{14.25}$$

and

$$\vec{H} = \frac{1}{\mu_0}\nabla\times\vec{A} = \sum \vec{H}_m \tag{14.26}$$

where

$$\vec{E}_m = i\omega_m\left[q_m(t)\vec{A}_m(\vec{r}) - q_m^*(t)\ \vec{A}_m^{\,*}(\vec{r})\right] \tag{14.27}$$

and

$$\vec{H}_m = \frac{i}{\mu_0}\vec{k}_m\times\left[q_m(t)\vec{A}_m(\vec{r}) - q_m^*(t)\ \vec{A}_m^{\,*}(\vec{r})\right] \tag{14.28}$$

Now the total energy of the electromagnetic wave is given by

$$H = \frac{1}{2}\iiint\left(\epsilon_0\vec{E}\cdot\vec{E} + \mu_0\vec{H}\cdot\vec{H}\right)d\tau \tag{14.29}$$

After some mathematical steps, we obtain (see Exercise 14.2)

$$\frac{1}{2}\iiint\epsilon_0\vec{E}\cdot\vec{E}\,d\tau = -\frac{1}{2}\epsilon_0 V\sum\omega_m^2\left(q_m q_{-m} + q_m^* q_{-m}^* - 2q_m q_m^*\right) \tag{14.30a}$$

and

$$\frac{1}{2}\iiint\mu_0\vec{H}\cdot\vec{H}\,d\tau = \frac{1}{2}\epsilon_0 V\sum\omega_m^2\left(q_m q_{-m} + q_m^* q_{-m}^* + 2q_m q_m^*\right) \tag{14.30b}$$

Hence, the total energy of the electromagnetic waves will be

$$H = 2\epsilon_0 V\sum\omega_m^2 q_m q_m^* = \sum H_m \tag{14.31}$$

where

$$H_m = 2\epsilon_0 V \omega_m^2 q_m q_m^* \tag{14.32}$$

represents the energy of the *m*th mode.

Exercise 14.2 Obtain Eq. (14.30a).

Solution We have

$$\frac{1}{2}\iiint \epsilon_0 \vec{E}\cdot\vec{E}\,d\tau = -\frac{\epsilon_0}{2}\iiint d\tau \sum_m \sum_n \omega_m \omega_n \left[q_m(t)\vec{A}_m(\vec{r}) - q_m^*(t)\,\vec{A}_m^*(\vec{r})\right]$$
$$\cdot\left[q_n(t)\vec{A}_n(\vec{r}) - q_n^*(t)\,\vec{A}_n^*(\vec{r})\right]$$

whose right hand side is rewritten as

$$-\frac{\epsilon_0}{2}\sum_m\sum_n \omega_m\omega_n \Bigg[q_m(t)q_n(t)\iiint d\tau \vec{A}_m(\vec{r})\cdot\vec{A}_n(\vec{r}) - q_m(t)q_n^*(t)\iiint d\tau\vec{A}_m(\vec{r})\cdot\vec{A}_n^*(\vec{r})$$
$$- q_m^*(t)q_n(t)\iiint d\tau\vec{A}_m^*(\vec{r})\cdot\vec{A}_n(\vec{r}) + q_m^*(t)q_n^*(t)\iiint d\tau\vec{A}_m^*(\vec{r})\cdot\vec{A}_n^*(\vec{r})\Bigg]$$

We now simplify the equation using the normalization given by Eq. (14.24). We also note that $\vec{A}_n(\vec{r}) = \vec{A}_{-n}^*(\vec{r})$ and $\vec{A}_m(\vec{r}) = \vec{A}_{-m}^*(\vec{r})$. Thus, we have

$$\frac{1}{2}\iiint \epsilon_0 \vec{E}\cdot\vec{E}\,d\tau = -\frac{\epsilon_0 V}{2}\sum_m\sum_n \omega_m\omega_n \Big[q_m(t)q_n(t)\delta_{n,-m} - q_m(t)q_n^*(t)\delta_{n,m}$$
$$- q_m^*(t)q_n(t)\delta_{n,m} + q_m^*(t)q_n^*(t)\delta_{n,-m}\Big]$$

Thus,

$$\frac{1}{2}\iiint \epsilon_0 \vec{E}\cdot\vec{E}\,d\tau = -\frac{1}{2}\epsilon_0 V \sum_m \omega_m^2\left(q_m(t)q_{-m}(t) - 2q_m(t)q_m^*(t) + q_m^*(t)q_{-m}^*(t)\right)$$

In a similar fashion Eq. (14.30b) can also be derived.

We now introduce two dimensionless variables

$$Q_m(t) = \sqrt{\epsilon_0 V}\left[q_m(t) + q_m^*(t)\right] \tag{14.33}$$

and

$$P_m(t) = \frac{1}{i}\sqrt{\epsilon_0 V \omega_m^2}\left[q_m(t) - q_m^*(t)\right] \tag{14.34}$$

We can invert the above equation to obtain

$$q_m(t) = \frac{1}{\sqrt{4\epsilon_0 V \omega_m^2}}\left[\omega_m Q_m(t) + iP_m(t)\right] \tag{14.35a}$$

$$q_m^*(t) = \frac{1}{\sqrt{4\epsilon_0 V \omega_m^2}}\left[\omega_m Q_m(t) - iP_m(t)\right] \tag{14.35b}$$

We now write the total energy of the electromagnetic field in terms of the variables $Q_m(t)$ and $P_m(t)$:

$$H = \sum H_m \tag{14.36}$$

with

$$H_m = \frac{P_m^2}{2} + \frac{\omega_m^2 Q_m^2}{2} \tag{14.37}$$

which resembles the energy of a linear harmonic oscillator with unit mass and with $Q_m(t)$ representing the position and $P_m(t)$ representing the momentum (see Sect. 3.4).

Thus, the electromagnetic field can be considered to be a collection of an infinite number of linear harmonic oscillators, with each mode being represented by an independent harmonic oscillator. It must be emphasized here that the results obtained are completely classical and through mathematical procedure the energy of the electromagnetic field has been put in a suggestive form resembling the energy of a harmonic oscillator.

14.3 Quantizing the Electromagnetic Field

In Chap. 3, we had discussed the quantization of the linear harmonic oscillator. Since the electromagnetic field is equivalent to an infinite number of independent harmonic oscillators, we can quantize the electromagnetic field by quantizing the independent harmonic oscillator modes that represent the electromagnetic field.

In order to quantize the electromagnetic field, we follow the standard approach of quantization of a linear harmonic oscillator (see Chap. 3). For this, we replace the quantities $Q_m(t)$ and $P_m(t)$ by operators $\widehat{Q}_m(t)$ and $\widehat{P}_m(t)$ which satisfy the following commutation relation:

$$\left[\widehat{Q}_m(t), \widehat{P}_m(t)\right] = i\hbar \tag{14.38a}$$

$$\left[\widehat{Q}_m(t), \widehat{P}_{m'}(t)\right] = 0 \quad \text{if } m \neq m' \tag{14.38b}$$

$$\left[\widehat{Q}_m(t), \widehat{Q}_{m'}(t)\right] = 0 \tag{14.38c}$$

$$\left[\widehat{P}_m(t), \widehat{P}_{m'}(t)\right] = 0 \tag{14.38d}$$

It may be noted that these operators are time dependent and are in the Heisenberg picture (see Chap. 21).

We also introduce the following operators referred to as the annihilation and creation operators (see Eqs. 3.52a and 3.52b):

$$\widehat{a}_m(t) = \frac{1}{\sqrt{2\hbar\omega_m}}\left[\omega_m\widehat{Q}_m(t) + i\widehat{P}_m(t)\right] \tag{14.39a}$$

$$\widehat{a}_m^\dagger(t) = \frac{1}{\sqrt{2\hbar\omega_m}}\left[\omega_m\widehat{Q}_m(t) - i\widehat{P}_m(t)\right] \tag{14.39b}$$

The operators $\widehat{Q}_m$ and $\widehat{P}_m$ are Hermitian operators.

Since $\widehat{Q}_m(t)$ and $\widehat{P}_m(t)$ satisfy the commutation relations given in Eqs. (14.38a)–(14.38d), we have

$$\left[\widehat{a}_m(t), \widehat{a}_m^\dagger(t)\right] = \frac{1}{2\hbar\omega_m}\left[\left\{\omega_m\widehat{Q}_m(t) + i\widehat{P}_m(t)\right\}, \left\{\omega_m\widehat{Q}_m(t) - i\widehat{P}_m(t)\right\}\right]$$

which after simplification gives us (see Eq. 3.54)

$$\left[\widehat{a}_m(t), \widehat{a}_m^\dagger(t)\right] = 1 \tag{14.40}$$

Comparing Eqs. (14.35a), (14.35b) and (14.39a), (14.39b), we see that

$$q_m(t) = \frac{1}{\sqrt{4\epsilon_0 V\omega_m^2}}\left[\omega_m Q_m(t) + iP_m(t)\right] = \sqrt{\frac{\hbar}{2\epsilon_0 V\omega_m}}\widehat{a}_m$$

and

$$q_m^*(t) = \frac{1}{\sqrt{4\epsilon_0 V\omega_m^2}}\left[\omega_m Q_m(t) - iP_m(t)\right] = \sqrt{\frac{\hbar}{2\epsilon_0 V\omega_m}}\widehat{a}_m^\dagger$$

Since $q_m(t)$ has a time dependence of the form $e^{-i\omega_m t}$, we have for the time dependence of the operator $\widehat{a}_m$:

$$\widehat{a}_m(t) = \widehat{a}_m e^{-i\omega_m t} \tag{14.41}$$

Similarly

$$\widehat{a}_m^\dagger(t) = \widehat{a}_m^\dagger e^{+i\omega_m t} \tag{14.42}$$

where $\widehat{a}_m = \widehat{a}_m(0)$ and $\widehat{a}_m^\dagger = \widehat{a}_m^\dagger(0)$ and are time independent operators. We also have

$$\left[\widehat{a}_m, \widehat{a}_m^\dagger\right] = 1 \tag{14.43}$$

Inverting Eqs. (14.39a) and (14.39b), we obtain (see Eqs. 3.53a and 3.53b)

$$Q_m(t) = \sqrt{\frac{\hbar}{2\omega_m}}\left(\widehat{a}_m(t) + \widehat{a}_m^\dagger(t)\right)$$

and

$$P_m(t) = \sqrt{\frac{\hbar\omega_m}{2}}\frac{\left(\widehat{a}_m(t) - \widehat{a}_m^\dagger(t)\right)}{i}$$

Using Eqs. (14.39a) and (14.39b), we can express the total energy of the electromagnetic field and obtain for the Hamiltonian operator

$$\widehat{H} = \sum \widehat{H}_m = \sum \frac{\widehat{P}_m^2}{2} + \frac{\omega_m^2 \widehat{Q}_m^2}{2} = \sum \frac{1}{2}\hbar\omega_m\left\{\widehat{a}_m^\dagger(t)\widehat{a}_m(t) + \widehat{a}_m(t)\widehat{a}_m^\dagger(t)\right\}$$

which using the commutation relation satisfied by $\widehat{a}_m(t)$ and $\widehat{a}_m^\dagger(t)$ can be written as

$$\widehat{H} = \sum \hbar\omega_m\left\{\widehat{a}_m^\dagger(t)\widehat{a}_m(t) + \frac{1}{2}\right\} = \sum \hbar\omega_m\left\{\widehat{a}_m^\dagger\widehat{a}_m + \frac{1}{2}\right\} \tag{14.44}$$

which is the quantum mechanical Hamiltonian operator for the total electromagnetic field.

We define a number operator of the mth mode as follows (see Eq. 3.71):

$$\widehat{N}_m = \widehat{a}_m^\dagger\widehat{a}_m \tag{14.45}$$

Thus, the total number operator for the total electromagnetic field is given by

$$\widehat{N} = \sum \widehat{a}_m^\dagger\widehat{a}_m \tag{14.46}$$

Using a procedure which is exactly the same as carried out for a linear harmonic oscillator (see Chap. 3), we obtain the following eigenvalues of energy corresponding to the mth electromagnetic mode as

$$E_m = \left(n_m + \frac{1}{2}\right)\hbar\omega_m; \quad n_m = 0, 1, 2, \ldots. \tag{14.47}$$

which represents the state of excitation of the mth mode of the electromagnetic field. Eq. (14.47) implies that the energy of the mth mode of the electromagnetic field having a frequency ω_m can only increase or decrease in quanta of excitation of energy $\hbar\omega_m$. *Such a quantum of energy is referred to as the photon.*

If there is no excitation of the mth mode, that is, $n_m = 0$, the energy of the mode is $\frac{1}{2}\hbar\omega_m$ which is referred to as the zero point energy of the mode.

The total energy of the electromagnetic field would be

$$E = \sum \hbar\omega_m \left\{ n_m + \frac{1}{2} \right\} \tag{14.48}$$

Corresponding to the mth mode, we have the following:

- Annihilation operator $\widehat{a}_m$
- Creation operator: $\widehat{a}_m^\dagger$
- Number operator: $\widehat{a}_m^\dagger \widehat{a}_m$
- Energy operator: $\hbar\omega_m \left\{ \widehat{a}_m^\dagger \widehat{a}_m + \frac{1}{2} \right\}$

The annihilation operator $\widehat{a}_m$ of the mth mode decreases the quantum of excitation of the mode by $\hbar\omega_m$ or one photon, while the creation operator $\widehat{a}_m^\dagger$ increases the quantum of excitation of the mode by $\hbar\omega_m$ or by one photon (see Chap. 3).

Considering just one mode of the electromagnetic field, we will represent the quantum state corresponding to the mth mode having n_m photons as

$$|\psi\rangle = |n_m\rangle \tag{14.49}$$

From the equations of the linear harmonic oscillator (see Chap. 3), we also have

$$\widehat{a}_m |n_m\rangle = \sqrt{n_m} |n_m - 1\rangle \tag{14.50a}$$

$$\widehat{a}_m^\dagger |n_m\rangle = \sqrt{n_m + 1} |n_m + 1\rangle \tag{14.50b}$$

$$\widehat{a}_m^\dagger \widehat{a}_m |n_m\rangle = n_m |n_m\rangle \tag{14.50c}$$

$$|n_m\rangle = \frac{1}{\sqrt{n_m!}} \left(\widehat{a}_m^\dagger \right)^{n_m} |n_m = 0\rangle \tag{14.50d}$$

The different number states of a given mode are orthonormal:

$$\langle n_m | p_m \rangle = \delta_{np}$$

A general electromagnetic field state also referred to as a multimode state having n_1 photons in mode 1, n_2 photons in mode 2, n_3 photons in mode 3, etc. will be represented by (see Chap. 16)

$$|\psi\rangle = |n_1\rangle|n_2\rangle|n_3\rangle\ldots.|n_m\rangle\ldots. \tag{14.51}$$

or simply as

$$|\psi\rangle = |n_1, n_2, n_3\ldots..n_m, \ldots.\rangle \tag{14.52}$$

We will assume that these states are orthonormal states. Hence,

$$\langle n_1, n_2, n_3\ldots..n_m, \ldots.|n_1, n_2, n_3\ldots..n_m, \ldots.\rangle = 1 \tag{14.53}$$

and

$$\langle n_1', n_2', n_3'\ldots..n_m', \ldots.|n_1, n_2, n_3\ldots..n_m, \ldots.\rangle = \delta_{n_1',n_1}\delta_{n_2',n_2}\delta_{n_3',n_3}\ldots\delta_{n_m',n_m}\ldots \tag{14.54}$$

We also have

$$\widehat{H}|n_1, n_2, n_3\ldots..n_m, \ldots.\rangle = \sum \hbar\omega_m\left\{n_m + \frac{1}{2}\right\}|n_1, n_2, n_3\ldots..n_m, \ldots.\rangle \tag{14.55}$$

A vacuum state is one in which there is no excitation of any mode or there are no photons in any mode; thus, $n_m = 0$ for all values of m. A multimode vacuum state (also referred to as radiation vacuum) is represented by

$$|\psi\rangle = |n_1 = 0, n_2 = 0, n_3 = 0\ldots..n_m = 0, \ldots.\rangle \tag{14.56}$$

We must note that

$$\widehat{a}_m|n_1, n_2, n_3\ldots..n_m, \ldots.\rangle = \sqrt{n_m}|n_1, n_2, n_3\ldots..(n_m - 1), \ldots.\rangle \tag{14.57}$$

and

$$\widehat{a}_m^\dagger|n_1, n_2, n_3\ldots..n_m, \ldots.\rangle = \sqrt{n_m + 1}|n_1, n_2, n_3\ldots..(n_m + 1), \ldots.\rangle \tag{14.58}$$

Note that

$$|n_1, n_2, n_3\ldots..n_m, \ldots.\rangle = \frac{1}{\sqrt{n_1!n_2!n_3!\ldots n_m!, \ldots}}\left(\widehat{a}_1^\dagger\right)^{n_1}\left(\widehat{a}_2^\dagger\right)^{n_2}\left(\widehat{a}_3^\dagger\right)^{n_3}\ldots.\left(\widehat{a}_m^\dagger\right)^{n_m}\ldots$$
$$|n_1 = 0, n_2 = 0, n_3 = 0\ldots..n_m = 0, \ldots.\rangle \tag{14.59}$$

The number states form a complete set of states and hence the general state of the electromagnetic field can be represented by a linear combination of the photon number states

$$|\psi\rangle = \sum C_{n_1,n_2,n_3.....n_m,....}|n_1, n_2, n_3.....n_m, ...\rangle \tag{14.60}$$

where $C_{n_1,n_2,n_3.....n_m,....}$ represent the expansion coefficient and are arbitrary complex numbers. The quantity $|C_{n_1,n_2,n_3.....n_m,....}|^2$ represents the probability of finding n_1 photons in mode 1, n_2 photons in mode 2, n_3 photons in mode 3, ...,n_m photons in mode m, etc.

Exercise 14.3 Consider two modes of the electromagnetic field one of which is horizontally polarized and the other vertically polarized and both of them having the same frequency and same propagation direction. What will be the total number operator? What will be the corresponding quantum state with one horizontally polarized photon and one vertically polarized photon?

Solution Referring the horizontal and vertical polarization modes by subscripts H and V, we have for the total number operator

$$\widehat{N} = \widehat{a}_H^\dagger \widehat{a}_H + \widehat{a}_V^\dagger \widehat{a}_V$$

The quantum state with one horizontally polarized photon and one vertically polarized photon will be represented by the following state:

$$|\psi\rangle = |1_H, 1_V\rangle$$

If we call mode with horizontal polarization by the subscript 1 and that with the vertical polarization by subscript 2 comparing the above state with Eq. (14.60), the state would be $c_{1_1 1_2}|1_1, 1_2\rangle$.

Exercise 14.4 Differentiate between states $|\psi_1\rangle = |1_H, 1_V\rangle$ and $|\psi_2\rangle = \frac{1}{\sqrt{2}}(|1_H\rangle + |1_V\rangle)$. Obtain the number of photons in the two states.

Solution The state $|1_H, 1_V\rangle$ is a short form for $|1_H, 1_V, 0, 0, 0....\rangle$ and is a state in which two modes, one corresponding to horizontal polarization and one to vertical polarization are occupied by one photon each and all other modes are empty. On the other hand, the state $\frac{1}{\sqrt{2}}(|1_H\rangle + |1_V\rangle)$ is a short form of $\frac{1}{\sqrt{2}} \times (|1_H, 0_V, 0, 0, 0...\rangle + |0_H, 1_V, 0, 0, 0, ...\rangle)$ and is a superposition state of the two following states: (1) one photon in the horizontal polarization and all other states are empty and (2) one photon in the vertical polarization and all other states are empty. If we mention only the states that are being considered with all other modes being empty, we have the following short form: $\frac{1}{\sqrt{2}} \times (|1_H, 0_V\rangle + |0_H, 1_V\rangle)$

To obtain the number of photons in the modes, we operate on the two states by the total photon number operator (we are writing the number operators only for the states that are being considered; other modes are all empty):

$$\hat{N}|\psi_1\rangle = \left(\hat{a}_H^\dagger \hat{a}_H + \hat{a}_V^\dagger \hat{a}_V\right)|1_H, 1_V\rangle = \hat{a}_H^\dagger \hat{a}_H|1_H, 1_V\rangle + \hat{a}_V^\dagger \hat{a}_V|1_H, 1_V\rangle = |1_H, 1_V\rangle + |1_H, 1_V\rangle = 2|1_H, 1_V\rangle = 2|\psi_1\rangle$$

Thus, the state is an eigenstate of the total photon number operator and is occupied by two photons.

We also have

$$\hat{N}|\psi_2\rangle = \left(\hat{a}_H^\dagger \hat{a}_H + \hat{a}_V^\dagger \hat{a}_V\right)\frac{1}{\sqrt{2}}(|1_H, 0_V\rangle + |0_H, 1_V\rangle)$$
$$= \frac{1}{\sqrt{2}}\hat{a}_H^\dagger \hat{a}_H(|1_H, 0_V\rangle + |0_H, 1_V\rangle) + \frac{1}{\sqrt{2}}\hat{a}_V^\dagger \hat{a}_V(|1_H, 0_V\rangle + |0_H, 1_V\rangle)$$

which simplifies to

$$\hat{N}|\psi_2\rangle = \frac{1}{\sqrt{2}}(|1_H, 0_V\rangle + 0) + \frac{1}{\sqrt{2}}(0 + |0_H, 1_V\rangle) = \frac{1}{\sqrt{2}} \times (|1_H, 0_V\rangle + |0_H, 1_V\rangle) = |\psi_2\rangle$$

Hence, this state is also an eigenstate of the total photon number operator but is occupied by just one photon. The two states are quite different.

Exercise 14.5 Consider a state with four occupied modes 1 to 4: $|\psi\rangle = |2_1, 4_2, 3_3, 1_4\rangle$. The subscripts represent mode numbers. What is the value of $\hat{a}_2^{\dagger 2}|\psi\rangle$?

Solution Since $\hat{a}_2^{\dagger 2}$ operates only on Mode # 2, we have

$$\hat{a}_2^{\dagger 2}|\psi\rangle = |2_1\rangle\left(\hat{a}_2^{\dagger 2}|4_2\rangle\right)|3_3\rangle|1_4\rangle = |2_1\rangle\left(\hat{a}_2^\dagger\sqrt{5}|5_2\rangle\right)|3_3\rangle|1_4\rangle = |2_1\rangle\left(\sqrt{30}|6_2\rangle\right) \times |3_3\rangle|1_4\rangle = \sqrt{30}|2_1, 6_2, 3_3, 1_4\rangle$$

Exercise 14.6 Write down an expression for a state consisting of two modes in which the probability of finding one photon in each mode is $\frac{1}{4}$ and the probability of finding two photons in each mode is $\frac{3}{4}$.

Solution The state will be given by

$$|\psi\rangle = \frac{1}{\sqrt{2}}|1_1, 1_2\rangle + \frac{\sqrt{3}}{2}|2_1, 2_2\rangle$$

Check that the state is normalized.

Since the electric field of the mth mode is given by Eq. (14.27), we use Eqs. (14.35a), (14.35b) and (14.39a), (14.39b) and obtain the electric field operator as

$$\widehat{E}_m(t) = i\sqrt{\frac{\hbar\omega_m}{2\epsilon_0 V}}\left(\widehat{a}_m(t)e^{i\vec{k}_m\cdot\vec{r}} - \widehat{a}_m^\dagger(t)e^{-i\vec{k}_m\cdot\vec{r}}\right) \tag{14.61}$$

Since the operators are time dependent, this expression for the electric field is in the Heisenberg picture.

Substituting the time dependence of the annihilation and creation operators (see Eqs. 14.41 and 14.42), we have

$$\widehat{E}_m(t) = i\sqrt{\frac{\hbar\omega_m}{2\epsilon_0 V}}\left(\widehat{a}_m e^{-i\left(\omega_m t - \vec{k}_m\cdot\vec{r}\right)} - \widehat{a}_m^\dagger e^{i\left(\omega_m t - \vec{k}_m\cdot\vec{r}\right)}\right) \tag{14.62}$$

Expanding the exponentials in terms of sine and cosine functions and simplifying, we obtain the following expression for the electric field operator

$$\widehat{E}_m(t) = \sqrt{\frac{2\hbar\omega_m}{\epsilon_0 V}}\left(\widehat{X}_{1m}\sin\left(\omega_m t - \vec{k}_m\cdot\vec{r}\right) - \widehat{X}_{2m}\cos\left(\omega_m t - \vec{k}_m\cdot\vec{r}\right)\right) \tag{14.63}$$

where

$$\widehat{X}_{1m} = \frac{\widehat{a}_m + \widehat{a}_m^\dagger}{2} \tag{14.64}$$

and

$$\widehat{X}_{2m} = \frac{\widehat{a}_m - \widehat{a}_m^\dagger}{2i} \tag{14.65}$$

are referred to as quadrature operators as they are coefficients of sine and cosine dependences as given in Eq. (14.63). Note that $\widehat{a}_m$ is not a Hermitian operator, while $\widehat{X}_{1m}$ and $\widehat{X}_{2m}$ are both Hermitian operators and thus are observables. The quadrature operators are important in quantum picture of light and will be used in later chapters.

Exercise 14.7 Consider a state consisting of two modes one with frequency ω_1 and propagation constant k_1 and another with frequency ω_2 and propagation constant k_2. Write down an expression for the total electric field operator.

Solution The total electric field operator will be

$$\widehat{E}(t) = i\sqrt{\frac{\hbar\omega_1}{2\epsilon_0 V}}\left(\widehat{a}_1 e^{-i\left(\omega_1 t - \vec{k}_1\cdot\vec{r}\right)} - \widehat{a}_1^\dagger e^{i\left(\omega_1 t - \vec{k}_1\cdot\vec{r}\right)}\right) + i\sqrt{\frac{\hbar\omega_2}{2\epsilon_0 V}} \times \left(\widehat{a}_2 e^{-i\left(\omega_2 t - \vec{k}_2\cdot\vec{r}\right)} - \widehat{a}_2^\dagger e^{i\left(\omega_2 t - \vec{k}_2\cdot\vec{r}\right)}\right)$$

where $\widehat{a}_1$ and $\widehat{a}_2$ represent the annihilation operators of the two modes and $\widehat{a}_1^\dagger$ and $\widehat{a}_2^\dagger$ represent the creation operators corresponding to the two modes.

Exercise 14.8 Show that $\widehat{X}_{1m}$ and $\widehat{X}_{2m}$ are Hermitian operators.

Solution Using Eqs. (14.64) and (14.65), we have

$$\widehat{X}_{1m}^\dagger = \frac{\widehat{a}_m^\dagger + \widehat{a}_m}{2} = \widehat{X}_{1m}$$

and

$$\widehat{X}_{1m}^\dagger = \frac{\widehat{a}_m^\dagger - \widehat{a}_m}{2(-i)} = \widehat{X}_{2m}$$

Thus, $\widehat{X}_{1m}$ and $\widehat{X}_{2m}$ are Hermitian operators and hence represent observables.

In the Schrödinger picture, the corresponding electric field operator would be

$$\widehat{E}_m = i\sqrt{\frac{\hbar\omega_m}{2\epsilon_0 V}}\left(\widehat{a}_m e^{i\vec{k}_m\cdot\vec{r}} - \widehat{a}_m^\dagger e^{-i\vec{k}_m\cdot\vec{r}}\right) \tag{14.66}$$

which will be time independent.

We will use the formulation developed in this chapter in Chaps. 16–18 to discuss different quantum states of the electromagnetic field.

Exercise 14.9 Obtain the commutation relation between the quadrature operators $\widehat{X}_1$ and $\widehat{X}_2$.

Solution

$$\left[\widehat{X}_1, \widehat{X}_2\right] = \left[\frac{\widehat{a}+\widehat{a}^\dagger}{2}, \frac{\widehat{a}-\widehat{a}^\dagger}{2i}\right] = \frac{1}{4i}\left[\left(\widehat{a}+\widehat{a}^\dagger\right), \left(\widehat{a}-\widehat{a}^\dagger\right)\right]$$
$$= \frac{1}{4i}\left\{-\left[\widehat{a}, \widehat{a}^\dagger\right] + \left[\widehat{a}^\dagger, \widehat{a}\right]\right\} = \frac{i}{2} \tag{14.67}$$

Exercise 14.10 Obtain an expression for the Hamiltonian operator $\widehat{H}_m$ in terms of $\widehat{X}_{1m}$ and $\widehat{X}_{2m}$.

Solution From Eqs. (14.64) and (14.65), we have

$$\widehat{a}_m = \widehat{X}_{1m} + i\widehat{X}_{2m} \tag{14.68}$$

$$\widehat{a}_m^\dagger = \widehat{X}_{1m} - i\widehat{X}_{2m} \tag{14.69}$$

From Eqs. (14.68) and (14.69), we obtain

$$\widehat{a}_m^\dagger \widehat{a}_m = \left(\widehat{X}_{1m} - i\widehat{X}_{2m}\right)\left(\widehat{X}_{1m} + i\widehat{X}_{2m}\right) = \widehat{X}_{1m}^2 + \widehat{X}_{2m}^2 - \frac{1}{2}$$

where we have used Eq. (14.67). Hence,

$$\widehat{H}_m = \hbar\omega_m\left\{\widehat{a}_m^\dagger \widehat{a}_m + \frac{1}{2}\right\} = \hbar\omega_m\left(\widehat{X}_{1m}^2 + \widehat{X}_{2m}^2\right) \tag{14.70}$$

14.4 Momentum of the Electromagnetic Field and Its Quantization

In classical electromagnetic theory, the total linear momentum of the electromagnetic field in a volume is given by

$$\vec{P} = \epsilon_0\mu_0 \int \vec{E} \times \vec{H} d\tau \tag{14.71}$$

We will first express this in terms of the modes we have obtained earlier. In order to do so, we note that corresponding to the mth mode, the electric and magnetic fields are given by (see Eqs. 14.27 and 14.28)

$$\vec{E}_m = i\omega_m\left[q_m(t)\vec{A}_m(\vec{r}) - q_m^*(t)\ \vec{A}_m^*(\vec{r})\right] \tag{14.72}$$

and

$$\vec{H}_m = \frac{i}{\mu_0}\vec{k}_m \times \left[q_m(t)\vec{A}_m(\vec{r}) - q_m^*(t)\,\vec{A}_m^*(\vec{r})\right] \tag{14.73}$$

Writing $\vec{A}_m(\vec{r}) = \vec{e}_m A_m(\vec{r})$, where $\vec{e}_m$ is the unit vector along the direction of the electric field, we have

$$\vec{E}_m = i\omega_m \vec{e}_m\left[q_m(t)A_m(\vec{r}) - q_m^*(t)\,A_m^*(\vec{r})\right] \tag{14.74}$$

and

$$\vec{H}_m = \frac{i}{\mu_0}\vec{e}_m^{\,\prime}\left[q_m(t)A_m(\vec{r}) - q_m^*(t)\,A_m^*(\vec{r})\right] \tag{14.75}$$

where

$$\vec{e}_m^{\,\prime} = \vec{k}_m \times \vec{e}_m \tag{14.76}$$

Hence,

$$\int \vec{E}\times\vec{H}\,d\tau = -\frac{1}{\mu_0}\int\sum\sum\omega_m\left(\vec{e}_m\times\vec{e}_n^{\,\prime}\right)\left[q_m(t)A_m(\vec{r}) - q_m^*(t)\,A_m^*(\vec{r})\right]$$
$$\times\left[q_n(t)A_n(\vec{r}) - q_n^*(t)\,A_n^*(\vec{r})\right]d\tau$$

Using the following orthogonality conditions

$$\iiint A_m(\vec{r})A_n^*(\vec{r})d\tau = V\delta_{m,n} \tag{14.77}$$

and

$$\iiint A_m(\vec{r})A_n(\vec{r})d\tau = V\delta_{m,-n} \tag{14.78}$$

we have

$$\int \vec{E}\times\vec{H}\,d\tau = -\frac{V}{\mu_0}$$
$$\times\sum\omega_m\left[\left(\vec{e}_m\times\vec{e}_{-m}^{\,\prime}\right)q_m q_{-m} - 2\left(\vec{e}_m\times\vec{e}_m^{\,\prime}\right)q_m q_m^* + \left(\vec{e}_m\times\vec{e}_{-m}^{\,\prime}\right)q_m^* q_{-m}^*\right] \tag{14.79}$$

Now

$$\left(\vec{e}_m \times \vec{e}'_m\right) = \vec{k}_m \tag{14.80}$$

and

$$\left(\vec{e}_m \times \vec{e}'_{-m}\right) = \vec{k}_{-m} = -\vec{k}_m \tag{14.81}$$

Hence, we have

$$\vec{P} = \epsilon_0 V \sum \omega_m \vec{k}_m \left(q_m q_{-m} + q^*_m q^*_{-m} + 2 q_m q^*_m\right) \tag{14.82}$$

Now, in the sum the term corresponding to m will be

$$\omega_m \vec{k}_m \left(q_m q_{-m} + q^*_m q^*_{-m} + 2 q_m q^*_m\right)$$

and the term corresponding to $-m$ would be

$$\begin{aligned} &\omega_{-m} \vec{k}_{-m} \left(q_{-m} q_m + q^*_{-m} q^*_m + 2 q_{-m} q^*_{-m}\right) \\ &= -\omega_m \vec{k}_m \left(q_m q_{-m} + q^*_m q^*_{-m} + 2 q_{-m} q^*_{-m}\right) \end{aligned}$$

where we have used the fact that $\omega_{-m} = \omega_m$ and $\vec{k}_{-m} = -\vec{k}_m$.

Hence, in the sum in Eq. (14.82) when we take into account the terms with m and $-m$ the term that will survive is $2q_m q^*_m$ corresponding to m and $-m$. Thus, we obtain

$$\vec{P} = \epsilon_0 V \sum \omega_m \vec{k}_m 2 q_m q^*_m \tag{14.83}$$

Using the same quantization procedure used for obtaining an expression for the Hamiltonian, we obtain the following expression for the momentum operator of the electromagnetic field as

$$\widehat{\vec{P}} = \sum \frac{\hbar \vec{k}_m}{2} \left(\widehat{a}^\dagger_m \widehat{a}_m + \widehat{a}_m \widehat{a}^\dagger_m\right) = \sum \hbar \vec{k}_m \widehat{a}^\dagger_m \widehat{a}_m \tag{14.84}$$

where in the last step we have used the fact that $\vec{k}_{-m} = -\vec{k}_m$.

Equation (14.84) gives the total momentum operator of the quantized electromagnetic field. Note that like the Hamiltonian operator, the momentum operator is expressed in terms of the number operator $\widehat{N}_m = \widehat{a}^\dagger_m \widehat{a}_m$. Hence, the Hamiltonian operator and the momentum operator commute with each other.

For a single mode in a state $|n\rangle$, the momentum is given by

$$\hbar \vec{k} \hat{a}^\dagger \hat{a} |n\rangle = n\hbar \vec{k} |n\rangle \tag{14.85}$$

showing that the state is an eigenstate of the momentum operator with an eigenvalue $n\hbar \vec{k}$. Thus, we can ascribe a momentum $\hbar \vec{k}$ to each photon.

Problems

14.1. Verify that the quadrature operators $\hat{X}_1$ and $\hat{X}_2$ are dimensionless.

14.2. Consider a mode of the electromagnetic field and show that $\left[\hat{a}^\dagger, \hat{H}\right] = -\hbar\omega\, \hat{a}^\dagger$, where we have dropped the subscript of mode number for compactness.

14.3. Show that the Hamiltonian operator $\hat{H}$ and the momentum operator $\vec{\hat{P}}$ commute with each other.

14.4. A single mode electromagnetic wave is in a state given by $|\psi\rangle = \frac{1}{2}|2\rangle + \frac{\sqrt{3}}{2}|3\rangle$. Show that the state is normalized. Obtain the expectation value of the energy in this state.

14.5. Obtain the number of photons in the two mode state given by $|\psi\rangle = |2_i, 2_j\rangle$ where subscripts i and j correspond to the two occupied modes. Also obtain the number of photons in the two mode state described by $|\psi\rangle = \frac{\sqrt{3}}{2}|2_i\rangle + \frac{1}{2}|2_j\rangle$.

References

Cohen-Tannoudji, C., B. Diu, and F. Laloë. 2019. *Quantum mechanics*. Vol. I, II and III. Hoboken: Wiley.

Garrison, J. C., and R. Y. Chiao. 2008. *Quantum optics*. Oxford: Oxford University Press.

Gerry, C., and P. Knight. 2005. *Introductory quantum optics*. Cambridge: Cambridge University Press.

Grynberg, G., A. Aspect, and C. Fabre. 2010. *Introduction to quantum optics: From the semi-classical approach to quantized light*. Cambridge: Cambridge University Press.

Mandel, L., and E. Wolf. 1995. *Optical coherence and quantum optics*. Cambridge: Cambridge University Press.

Thyagarajan, K., and A. Ghatak. 2011. *Lasers: Fundamentals and applications*. Springer.

Yariv, A. 1991. *Quantum electronics*. Hoboken: Wiley.

Chapter 15
Single-Mode Photon States

Abstract This chapter discusses single-mode photon states, number states, vacuum state, coherent states, and superposition states. Expectation value and variance of electric field in these states is discussed. Coherent state is expressed in terms of number states and the probability of detecting a certain number of photons in the coherent state is presented. Also presented are diagrammatic representation of different quantum states.

15.1 Introduction

In Chap. 14, we had quantized the electromagnetic field and had shown that each mode of the electromagnetic field behaves like a linear harmonic oscillator. The energy of each mode having a frequency ω can increase or decrease only in quanta of energy $\hbar\omega$. This quantum of excitation is referred to as the photon. The energy of such a mode can have the values $\left(n + \frac{1}{2}\right)\hbar\omega$ and in such a case we refer to the mode being occupied by n photons. We had obtained expressions describing the electric field operator, the Hamiltonian operator, the number operator, etc.

In this chapter, we will focus on a single mode of the radiation field and discuss various single-mode quantum states of light. In Chap. 14, we had seen that a mode of the electromagnetic field is defined by a specific frequency, direction of propagation, and a specific polarization. Such a single mode can be in different quantum states one of which, namely the coherent state, we will see, is closest to a classical state of radiation. There are other states such as number states, squeezed states, etc. which have no classical analogues and have very interesting properties that find applications in the field of quantum information science and technology. Squeezed states will be discussed in detail in Chap. 18.

We will begin with the simplest of single-mode states, namely the number state.

K. Thyagarajan, *Integrated Quantum Photonics*, Graduate Texts in Physics,
https://doi.org/10.1007/978-3-031-85728-7_15

15.2 Number State

Let us consider a single mode of the electromagnetic field having a frequency ω and propagating along the *z*-direction. If the field has precisely *n* photons in the mode, then the energy of the field is given by

$$E=\left(n+\frac{1}{2}\right)\hbar\omega;\quad n=0,1,2,\ldots. \tag{15.1}$$

Since we are considering a single-mode state, we are not writing any subscript on the various quantities. Since the state is characterized by a definite number of photons, it is referred to as a number state and is written as

$$|\psi\rangle=|n\rangle \tag{15.2}$$

We are here assuming that all other modes are unoccupied meaning that there are no photons in any of the other modes and only this mode under consideration is occupied by *n* photons. Equation (15.2) is a short form of writing (assuming a discrete representation of the modes)

$$|\psi\rangle=|0,0,0,\ldots n,0,0,\ldots.\rangle \tag{15.3}$$

meaning that all modes are empty with no excitation except the mode under consideration which is assumed to be occupied by *n* photons.

Since the state has a well-defined number of photons, it is an eigenstate of the energy operator and is characterized by a well-defined energy:

$$\widehat{H}|\psi\rangle=\hbar\omega\left(\widehat{a}^{\dagger}\widehat{a}+\frac{1}{2}\right)|n\rangle=\hbar\omega\left(n+\frac{1}{2}\right)|n\rangle \tag{15.4}$$

Since $\widehat{a}^{\dagger}\widehat{a}|n\rangle=n|n\rangle$ (see Eq. 14.50c), the state is also an eigenstate of the number operator:

$$\widehat{N}|\psi\rangle=\widehat{a}^{\dagger}\widehat{a}|n\rangle=n|n\rangle \tag{15.5}$$

Since this state is an eigenstate of the two operators, it implies that a measurement of the energy in the mode will give a precise value and a measurement of the number of photons in the mode will also give a precise value.

15.2.1 Expectation Value of Electric Field

We will now calculate the expectation value of the electric field in the number state. The electric field operator for the mode is given by (see Eq. 14.62)

$$\widehat{E} = i\sqrt{\frac{\hbar\omega}{2\varepsilon_0 V}}\left[\widehat{a}e^{i(kz-\omega t)} - \widehat{a}^{\dagger}e^{-i(kz-\omega t)}\right] \tag{15.6}$$

Note that since the mode we are considering is propagating along the z-direction, we have $\vec{k}\cdot\vec{r} = kz$. If we operate on the state with the electric field operator, we obtain

$$\begin{aligned}\widehat{E}|n\rangle &= i\sqrt{\frac{\hbar\omega}{2\varepsilon_0 V}}\left[\widehat{a}e^{i(kz-\omega t)} - \widehat{a}^{\dagger}e^{-i(kz-\omega t)}\right]|n\rangle \\ &= i\sqrt{\frac{\hbar\omega}{2\varepsilon_0 V}}\left[\sqrt{n}|n-1\rangle e^{i(kz-\omega t)} - \sqrt{n+1}|n+1\rangle e^{-i(kz-\omega t)}\right]\end{aligned} \tag{15.7}$$

where we have used the properties of the annihilation and creation operators on the number state namely (see Eqs. 14.50a and 14.50b)

$$\widehat{a}|n\rangle = \sqrt{n}|n-1\rangle \tag{15.8}$$

$$\widehat{a}^{\dagger}|n\rangle = \sqrt{n+1}|n+1\rangle \tag{15.9}$$

From Eq. (15.7), we see that the number state is **not** an eigenstate of the electric field operator.

We now calculate the expectation value of the electric field operator:

$$\begin{aligned}\left\langle\widehat{E}\right\rangle = \langle n|\widehat{E}|n\rangle &= i\sqrt{\frac{\hbar\omega}{2\varepsilon_0 V}}\left[\langle n|\widehat{a}|n\rangle e^{i(kz-\omega t)} - \langle n|\widehat{a}^{\dagger}|n\rangle e^{-i(kz-\omega t)}\right] \\ &= i\sqrt{\frac{\hbar\omega}{2\varepsilon_0 V}}\left[\sqrt{n}\langle n|n-1\rangle e^{i(kz-\omega t)} - \sqrt{n+1}\langle n|n+1\rangle e^{-i(kz-\omega t)}\right] = 0\end{aligned} \tag{15.10}$$

since $|n\rangle$ form orthonormal states (see Eq. 14.54).

Thus, although the state contains n photons, the expectation value of the electric field in this state is zero. Thus, measurement of the electric field in an ensemble of identically prepared number states will give different values and the average of all the measurements will be zero.

In order to better understand the state, we now calculate the expectation value of the square of the electric field operator:

$$\left\langle \widehat{E}^2 \right\rangle = \langle n|\widehat{E}^2|n\rangle = -\frac{\hbar\omega}{2\varepsilon_0 V}\langle n|\left[\widehat{a}e^{i(kz-\omega t)} - \widehat{a}^\dagger e^{-i(kz-\omega t)}\right]^2|n\rangle \tag{15.11}$$

When we expand the brackets, we will get the following terms: $\langle n|\widehat{a}^2|n\rangle$, $\langle n|\widehat{a}^{\dagger 2}|n\rangle$, $\langle n|\widehat{a}\widehat{a}^\dagger|n\rangle$ and $\langle n|\widehat{a}^\dagger\widehat{a}|n\rangle$.

Now,

$$\langle n|\widehat{a}^2|n\rangle = \langle n|\widehat{a}\sqrt{n}|n-1\rangle = \sqrt{n}\sqrt{n-1}\langle n|n-2\rangle = 0 \tag{15.12}$$

Similarly we have

$$\langle n|\widehat{a}^{\dagger 2}|n\rangle = 0 \tag{15.13}$$

and

$$\langle n|\widehat{a}^\dagger\widehat{a}|n\rangle = n \tag{15.14}$$

$$\langle n|\widehat{a}\widehat{a}^\dagger|n\rangle = \langle n|\left(\widehat{a}^\dagger\widehat{a} + 1\right)|n\rangle = (n+1) \tag{15.15}$$

Substituting these values in Eq. (15.11), we have

$$\left\langle \widehat{E}^2 \right\rangle = \frac{\hbar\omega}{\varepsilon_0 V}\left(n + \frac{1}{2}\right) \tag{15.16}$$

which is nonzero. Thus, the number state has zero expectation value of the electric field, while the expectation valuc of the square of the electric field is nonzero.

We define the variance in the electric field by

$$(\Delta E)^2 = \left\langle \widehat{E}^2 \right\rangle - \left\langle \widehat{E} \right\rangle^2 = \frac{\hbar\omega}{\varepsilon_0 V}\left(n + \frac{1}{2}\right) \tag{15.17}$$

which is nonzero.

The number state cannot thus represent a classical electromagnetic field. Since the number state has a precise value of number of photons or energy, the phase of the electromagnetic field that it represents is completely random. Thus, although we have an ensemble of identical states, when a measurement of the electric field is carried out, the phase of the electric field is completely random and can vary over the entire Argand plane thus leading to a null value of expectation value. At the same time, since the electric field is not zero, we obtain a finite variance of the electric field.

Figure 15.1 is a pictorial representation of the electric field corresponding to the number state. In the ensemble, all phase values from 0 to 2π are possible with equal probabilities and thus leading to zero expectation value.

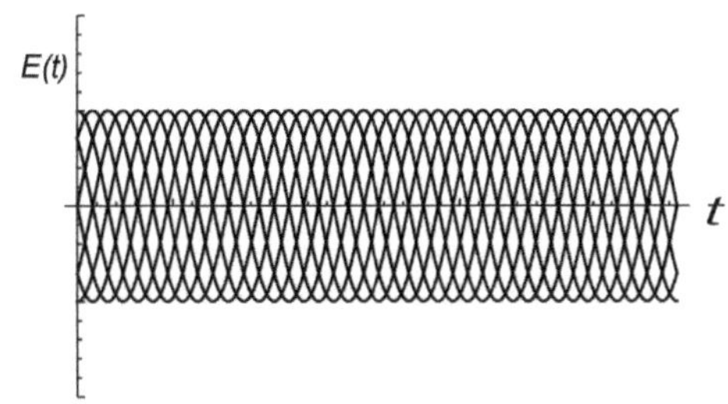

Fig. 15.1 Pictorial representation of the electric field corresponding to a number state

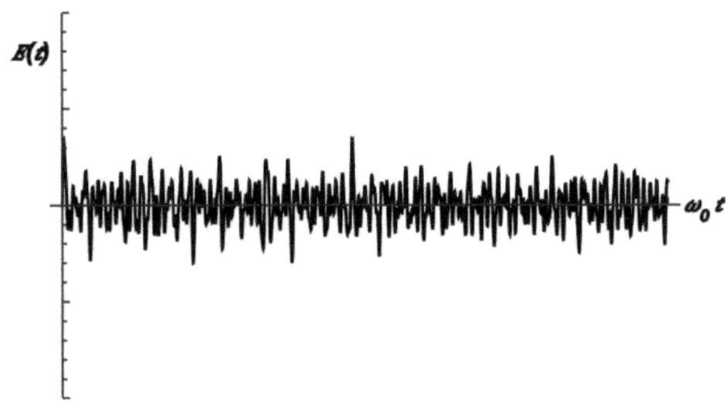

Fig. 15.2 A pictorial representation of the vacuum state having random fluctuations with an average value of zero

15.3 Vacuum State

A radiation vacuum state is a state with no photon in any of the modes of the electromagnetic field. It would correspond to the single-mode number state discussed above with $n = 0$. We will represent this with

$$|\psi\rangle = |0, 0, 0, \ldots n = 0, 0, 0, \ldots.\rangle = |0\rangle \tag{15.18}$$

Putting $n = 0$ in the evaluation carried out in Sect. 15.2 we see that expectation value of the electric field is zero and the variance is still nonzero and given by

$$(\Delta E)^2 = \frac{\hbar\omega}{2\varepsilon_0 V} \tag{15.19}$$

Thus, although there are no photons in any mode, there are still fluctuations in the electric field; this is referred to as zero point fluctuation. Thus, radiation vacuum is not free from electric field. It is these zero point fluctuations that are responsible for effects such as spontaneous emission from atoms and spontaneous parametric down conversion.

Figure 15.2 gives a pictorial representation of a vacuum state which has random fluctuations in the electric field with an average value of zero and a finite variance.

15.4 Superposition States

As an example, we consider the following superposition state of a single mode of the radiation field:

$$|\psi\rangle = \frac{1}{\sqrt{2}}|0\rangle + \frac{1}{\sqrt{2}}|1\rangle \tag{15.20}$$

The state is a coherent superposition of the vacuum state $|0\rangle$ and the one photon state $|1\rangle$.

The first question that we ask is "Is this an eigenstate of the number operator?," that is, does the state have a precise number of photons? In order to check, this we operate the state with the number operator:

$$\widehat{N}|\psi\rangle = \widehat{a}^{\dagger}\widehat{a}|\psi\rangle = \widehat{a}^{\dagger}\widehat{a}\left(\frac{1}{\sqrt{2}}|0\rangle + \frac{1}{\sqrt{2}}|1\rangle\right) = \frac{1}{\sqrt{2}}|1\rangle \tag{15.21}$$

Where we have used the fact that $\widehat{a}^{\dagger}\widehat{a}|0\rangle = 0$. Thus, we see that the given state is not an eigenstate of the number operator. That means that the state does not have a precise number of photons.

We next calculate the probability of obtaining 0 photons when a measurement of the number of photons is carried out. This is given by

$$P(0) = |\langle 0|\psi\rangle|^2 = \frac{1}{2}$$

Similarly the probability of obtaining one photon is

$$P(1) = |\langle 1|\psi\rangle|^2 = \frac{1}{2}$$

The probability of obtaining two photons is

$$P(2) = |\langle 2|\psi\rangle|^2 = 0$$

And similarly the probability of obtaining any number of photons other than zero and one is zero. Thus, the given state is a superposition state of two states, one state with zero photons and one state with one photon. In view of the multiplicative factors, the probability of finding zero or one photon in the state is 0.5 and the probability of finding any other number of photons is zero. Thus, when measurements are carried out on the number of photons in an ensemble of identically prepared states, the measured value will be either 0 or 1 with equal probabilities.

The above example is a superposition state of only two states of the single mode. One can define superposition states with any number of photon number states with the total probability being unity. Thus, we can write a general superposition state in a single radiation mode as

$$|\psi\rangle = \sum_{n=0}^{\infty} c_n |n\rangle \tag{15.22}$$

where c_n is in general a complex amplitude and the probability of finding n photons in this mode is $|c_n|^2$ which need to satisfy the equation

$$\sum_{n=0}^{\infty} |c_n|^2 = 1 \tag{15.23}$$

Exercise 15.1 Consider a single-mode superposition state descried by

$$|\psi\rangle = \frac{1}{2}|n\rangle + \frac{\sqrt{3}}{2}|n+1\rangle$$

(a) Is this an eigenstate of the number operator?
(b) Obtain the expectation value of the energy in this state.

Solution
(a) To check whether this is an eigenstate of the number operator, we calculate

$$\widehat{N}|\psi\rangle = \widehat{a}^{\dagger}\widehat{a}|\psi\rangle = \widehat{a}^{\dagger}\widehat{a}\left(\frac{1}{2}|n\rangle + \frac{\sqrt{3}}{2}|n+1\rangle\right) = \frac{n}{2}|n\rangle + \frac{\sqrt{3}}{2}(n+1)|n+1\rangle$$

Thus $|\psi\rangle$ is not an eigenstate of the number operator.
(b) Expectation value of energy is given by

$$\hbar\omega\langle\psi|\left(\widehat{a}^{\dagger}\widehat{a} + \frac{1}{2}\right)|\psi\rangle = \hbar\omega\left(n + \frac{3}{4}\right)$$

15.5 Single-Mode Coherent State

We now consider a special single-mode superposition state referred to as the coherent state.

A coherent state is a state which is an eigenstate of the annihilation operator. Thus, if we represent the coherent state by $|\alpha\rangle$, then we have

$$\widehat{a}|\alpha\rangle = \alpha|\alpha\rangle \tag{15.24}$$

with α being the eigenvalue of the state. In general, α can be complex. In view of the above equation, we also have

$$\langle\alpha|\widehat{a}^{\dagger} = \alpha^{*}\langle\alpha| \tag{15.25}$$

We assume that the state is a normalized state and hence,

$$\langle\alpha|\alpha\rangle = 1 \tag{15.26}$$

Note that the coherent state is not an eigenstate of the number operator since

$$\widehat{N}|\alpha\rangle = \widehat{a}^{\dagger}\widehat{a}|\alpha\rangle = \alpha\widehat{a}^{\dagger}|\alpha\rangle \neq \cdots|\alpha\rangle \tag{15.27}$$

Thus, the state does not contain a well-defined number of photons and can be expressed as a superposition of the number states. We can evaluate the expectation value of the number of photons $\overline{n}$ in this state as

$$\langle\alpha|\widehat{N}|\alpha\rangle = \overline{n} = \langle\alpha|\widehat{a}^{\dagger}\widehat{a}|\alpha\rangle = |\alpha|^{2} \tag{15.28}$$

This is the average number of photons in the coherent state.

In order to evaluate the variance in the photon number, we first calculate

$$\langle\alpha|\widehat{N}^{2}|\alpha\rangle = \langle\alpha|\widehat{a}^{\dagger}\widehat{a}\widehat{a}^{\dagger}\widehat{a}|\alpha\rangle = \langle\alpha|\widehat{a}^{\dagger}\left(1 + \widehat{a}^{\dagger}\widehat{a}\right)\widehat{a}|\alpha\rangle = |\alpha|^{2} + |\alpha|^{4} \tag{15.29}$$

where we have used the fact that $\left[\widehat{a}, \widehat{a}^{\dagger}\right] = 1$. Thus, the variance is given by

$$(\Delta N)^{2} = \left\langle\widehat{N}^{2}\right\rangle - \left\langle\widehat{N}\right\rangle^{2} = |\alpha|^{2} = \overline{n} \tag{15.30}$$

and the standard deviation becomes

$$\Delta N = |\alpha| = \sqrt{\overline{n}} \tag{15.31}$$

Hence, the ratio of the standard deviation to the average becomes

$$\frac{\Delta N}{\langle N\rangle} = \frac{1}{|\alpha|} = \frac{1}{\sqrt{\overline{n}}} \tag{15.32}$$

Thus, the fractional fluctuations in the number of photons in the coherent state reduces as the average number of photons increases.

Let us consider a beam of coherent source having a frequency ν and a power P, then the average number of photons crossing per unit time would be $\overline{n}=\frac{P}{h\nu}$. Let us consider a perfect photodetector which converts every photon to an electron. In such a case, the average current measured by the photodetector would be $\overline{\imath}=q\overline{n}=\frac{qP}{h\nu}$ where q is the charge of the electron. If the photodetector measurement time is T seconds, then the average number of photons detected in this time interval would be $N=\overline{n}T=\frac{PT}{h\nu}$. Since the noise in the photon number fluctuations over a time interval T is $\sqrt{N}=\sqrt{\frac{PT}{h\nu}}$, the fluctuations in the measured current would be $\Delta i=\frac{q\sqrt{N}}{T}=q\sqrt{\frac{P}{h\nu T}}=\sqrt{\frac{q\overline{\imath}}{T}}$. If we define the bandwidth of detection by B, then $B=\frac{1}{2T}$ and we have $\Delta i=\sqrt{2q\overline{\imath}B}$. This represents what is referred to as shot noise in the detection process and is proportional to the square root of the average current or the average optical power and the square root of the bandwidth of the detector.

As an example, let us consider a laser operating at a wavelength of $\lambda_0=1500$ nm and emitting a power $P=1$ mW. This corresponds to an average of $\overline{N}_{ph}=\frac{P}{\left(hc/\lambda_0\right)}=7.5\times10^{15}$ photons crossing per second. A photodetector with a measurement time of 1 μs will receive on an average 7.5×10^9 photons with a noise of $\sqrt{7.5\times10^9}\approx8.7\times10^4$ photons. The noise power measured by the photodetector would then correspond to $P_n=8.7\times10^4\frac{hc}{\lambda_0}\times10^6\approx12$ nW. A perfect photodetector would give an average current output of $7.5\times10^{15}\times1.6\times10^{-19}=1.21$ mA with a current noise of $8.7\times10^4/10^{-6}\times1.6\times10^{-19}=14$ nA.

We can also evaluate the expectation value of the electric field in this state. We thus have

$$\left\langle\widehat{E}\right\rangle=\langle\alpha|\widehat{E}|\alpha\rangle=i\sqrt{\frac{\hbar\omega}{2\varepsilon_0V}}\left[\langle\alpha|\widehat{a}|\alpha\rangle e^{i(kz-\omega t)}-\langle\alpha|\widehat{a}^{\dagger}|\alpha\rangle e^{-i(kz-\omega t)}\right]$$

which, using Eqs. (15.24) and (15.25), becomes

$$\left\langle\widehat{E}\right\rangle=\langle\alpha|\widehat{E}|\alpha\rangle=i\sqrt{\frac{\hbar\omega}{2\varepsilon_0V}}\left[\alpha e^{i(kz-\omega t)}-\alpha^*e^{-i(kz-\omega t)}\right]\tag{15.33}$$

If we write

$$\alpha=|\alpha|e^{i\phi}\tag{15.34}$$

then we have

$$\left\langle\widehat{E}\right\rangle=-\sqrt{\frac{2\hbar\omega}{\varepsilon_0V}}|\alpha|\sin(kz-\omega t+\phi)\tag{15.35}$$

which is very similar to the electric field of a classical electromagnetic wave having a certain amplitude depending on $|\alpha|$ and a particular phase. *Thus, the coherent state is the quantum mechanical analog of a classical electromagnetic wave.*

We now wish to calculate the variance in the electric field. For this, we first evaluate $\left\langle \widehat{E}^2 \right\rangle$:

$$\left\langle \widehat{E}^2 \right\rangle = \langle\alpha|\widehat{E}^2|\alpha\rangle = -\frac{\hbar\omega}{2\varepsilon_0 V}\langle\alpha|\left[\widehat{a}e^{i(kz-\omega t)} - \widehat{a}^\dagger e^{-i(kz-\omega t)}\right]^2|\alpha\rangle \tag{15.36}$$

We expand the brackets and use the following expressions:

$$\langle\alpha|\widehat{a}^2|\alpha\rangle = \alpha^2 = |\alpha|^2 e^{2i\phi} \tag{15.37}$$

$$\langle\alpha|\widehat{a}^{\dagger 2}|\alpha\rangle = \alpha^{*2} = |\alpha|^2 e^{-2i\phi} \tag{15.38}$$

$$\langle\alpha|\widehat{a}^\dagger\widehat{a}|\alpha\rangle = |\alpha|^2 \tag{15.39}$$

$$\langle\alpha|\widehat{a}\widehat{a}^\dagger|\alpha\rangle = \langle\alpha|\left(\widehat{a}^\dagger\widehat{a} + 1\right)|\alpha\rangle = |\alpha|^2 + 1 \tag{15.40}$$

Using these in Eq. (15.36), we obtain

$$\begin{aligned}\left\langle \widehat{E}^2 \right\rangle &= -\frac{\hbar\omega}{2\varepsilon_0 V}\left[2|\alpha|^2\cos 2(kz-\omega t+\phi) - 2|\alpha|^2 - 1\right]\\ &= \frac{2\hbar\omega}{\varepsilon_0 V}|\alpha|^2\sin^2(kz-\omega t+\phi) + \frac{\hbar\omega}{2\varepsilon_0 V}\end{aligned}$$

Hence, the variance in the electric field is given by

$$(\Delta E)^2 = \left\langle \widehat{E}^2 \right\rangle - \left\langle \widehat{E} \right\rangle^2 = \frac{\hbar\omega}{2\varepsilon_0 V} \tag{15.41}$$

which is the same as in the vacuum state. Note that the variance in the electric field of the coherent state is independent of the amplitude α, while the expectation value (which varies sinusoidally with time) increases with α. The variance represents the fluctuations in the electric field associated with the state. A single-mode laser operating much above threshold generates such a coherent state of the electromagnetic field.

Figure 15.3 shows a schematic of the electric field variation in a coherent state given by Eq. (15.35) with the variance given by Eq. (15.41).

We can also evaluate the expectation value and variance in the quadrature operators $\widehat{X}_1$ and $\widehat{X}_2$ (see Eqs. 14.64 and 14.65):

$$\widehat{X}_1 = \frac{\widehat{a}+\widehat{a}^\dagger}{2}; \quad \widehat{X}_2 = \frac{\widehat{a}-\widehat{a}^\dagger}{2i} \tag{15.42}$$

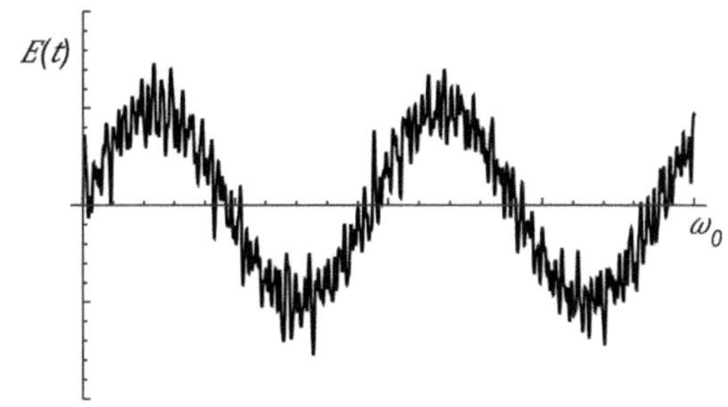

Fig. 15.3 Schematic of the electric field variation of a coherent state

In terms of the quadrature operators, the electric field operator is given by (see Eq. 14.63)

$$\widehat{E}=\sqrt{\frac{2\hbar\omega}{\varepsilon_0 V}}\left[\widehat{X}_1\sin(\omega t-kz)-\widehat{X}_2\cos(\omega t-kz)\right] \tag{15.43}$$

Thus,

$$\left\langle\widehat{X}_1\right\rangle=\langle\alpha|\widehat{X}_1|\alpha\rangle=\langle\alpha|\frac{\left(\widehat{a}+\widehat{a}^{\dagger}\right)}{2}|\alpha\rangle=|\alpha|\cos\phi \tag{15.44}$$

$$\left\langle\widehat{X}_2\right\rangle=\langle\alpha|\widehat{X}_2|\alpha\rangle=\langle\alpha|\frac{\left(\widehat{a}-\widehat{a}^{\dagger}\right)}{2i}|\alpha\rangle=|\alpha|\sin\phi \tag{15.45}$$

$$\left\langle\widehat{X}_1^2\right\rangle=\langle\alpha|\widehat{X}_1^2|\alpha\rangle=\frac{1}{4}+|\alpha|^2\cos^2\phi \tag{15.46}$$

$$\left\langle\widehat{X}_2^2\right\rangle=\langle\alpha|\widehat{X}_2^2|\alpha\rangle=\frac{1}{4}+|\alpha|^2\sin^2\phi \tag{15.47}$$

Thus, the variance in the quadrature components of the electric field is

$$\left(\Delta\widehat{X}_1\right)^2=\left\langle\widehat{X}_1^2\right\rangle-\left\langle\widehat{X}_1\right\rangle^2=\frac{1}{4} \tag{15.48}$$

and

$$\left(\Delta\widehat{X}_2\right)^2=\left\langle\widehat{X}_2^2\right\rangle-\left\langle\widehat{X}_2\right\rangle^2=\frac{1}{4} \tag{15.49}$$

Hence, we obtain

$$\left(\Delta\widehat{X}_1\right)\left(\Delta\widehat{X}_2\right)=\frac{1}{4} \tag{15.50}$$

which is the minimum value of the product of the uncertainties. Thus, the coherent state corresponds to what is referred to as the minimum uncertainty state (MUS).

Since the number states form a complete set of orthogonal states, we can expand the coherent state in terms of the number states. Thus, we write

$$|\alpha\rangle = \sum_{n=0}^{\infty} c_n |n\rangle \tag{15.51}$$

where c_n's are constants to be determined. The coherent state expanded in terms of the number state is given by (see Exercise 15.2)

$$|\alpha\rangle = e^{-\frac{1}{2}|\alpha|^2} \sum_{n=0}^{\infty} \frac{\alpha^n}{\sqrt{n!}} |n\rangle \tag{15.52}$$

We can verify that the state is normalized.

Exercise 15.2 Obtain Eq. (15.52) representing the coherent state as a superposition of number states.

Solution Since $|\alpha\rangle$ satisfies the eigenvalue equation given by Eq. (15.24), we have using Eq. (15.51)

$$\widehat{a}|\alpha\rangle = \widehat{a}\sum_{n=0}^{\infty} c_n|n\rangle = \sum_{n=0}^{\infty} c_n\sqrt{n}|n-1\rangle = \alpha|\alpha\rangle = \alpha\sum_{n=0}^{\infty} c_n|n\rangle$$

Expressing the sum explicitly, we have

$$c_1|0\rangle + c_2\sqrt{2}|1\rangle + c_3\sqrt{3}|2\rangle + \cdots = \alpha(c_0|0\rangle + c_1|1\rangle + c_2|2\rangle + c_3|3\rangle + \cdots)$$

Multiplying by $\langle 0|$, $\langle 1|$, $\langle 2|$, etc. and using the orthonormality condition, we have

$$c_1 = \alpha c_0; \quad c_2 = \frac{\alpha^2}{\sqrt{2!}} c_0; \quad c_3 = \frac{\alpha^3}{\sqrt{3!}} c_0;$$

etc. Thus, we obtain

$$c_n = \frac{\alpha^n}{\sqrt{n!}} c_0$$

Hence,

$$|\alpha\rangle = c_0 \sum_{n=0}^{\infty} \frac{\alpha^n}{\sqrt{n!}} |n\rangle$$

Normalizing $|\alpha\rangle$, we have

$$\langle\alpha|\alpha\rangle = |c_0|^2 \sum_{m=0}^{\infty} \frac{\alpha^{*m}}{\sqrt{m!}} \langle m| \sum_{n=0}^{\infty} \frac{\alpha^n}{\sqrt{n!}} |n\rangle = |c_0|^2 \sum_{m=0}^{\infty} \frac{\alpha^{*m}}{\sqrt{m!}} \sum_{n=0}^{\infty} \frac{\alpha^n}{\sqrt{n!}} \langle m|n\rangle$$

$$= |c_0|^2 \sum_{m=0}^{\infty} \frac{\alpha^{*m}}{\sqrt{m!}} \sum_{n=0}^{\infty} \frac{\alpha^n}{\sqrt{n!}} \delta_{mn}$$

Thus, we have

$$\langle\alpha|\alpha\rangle = |c_0|^2 \sum_{n=0}^{\infty} \frac{\left(|\alpha|^2\right)^n}{n!} = |c_0|^2 e^{|\alpha|^2} = 1$$

Hence,

$$c_0 = e^{-|\alpha|^2/2}$$

Apart from a phase factor. Hence, we have the following expression for the normalized coherent state in terms of the number states:

$$|\alpha\rangle = e^{-|\alpha|^2/2} \sum_{n=0}^{\infty} \frac{\alpha^n}{\sqrt{n!}} |n\rangle$$

Exercise 15.3 Show that two coherent states $|\alpha\rangle$ and $|\beta\rangle$ are not orthogonal to each other. Under what conditions would they be almost orthogonal?

Solution We first express the two coherent states in terms of number states. The coherent state $|\alpha\rangle$ is given by Eq. (15.52) and we have for the coherent state $|\beta\rangle$

$$|\beta\rangle = e^{-|\beta|^2/2} \sum_{m=0}^{\infty} \frac{\beta^m}{\sqrt{m!}} |m\rangle$$

Hence,

$$\langle\beta|\alpha\rangle = e^{-|\alpha|^2/2}e^{-|\beta|^2/2}\sum_{m=0}^{\infty}\frac{\beta^{*m}}{\sqrt{m!}}\sum_{n=0}^{\infty}\frac{\alpha^n}{\sqrt{n!}}\langle m|n\rangle = e^{-|\alpha|^2/2}e^{-|\beta|^2/2}\sum_{m=0}^{\infty}\frac{\beta^{*m}}{\sqrt{m!}} \times\sum_{n=0}^{\infty}\frac{\alpha^n}{\sqrt{n!}}\delta_{mn}$$

which gives us

$$\langle\beta|\alpha\rangle = e^{-|\alpha|^2/2}e^{-|\beta|^2/2}\sum_{n=0}^{\infty}\frac{(\alpha\beta^*)^n}{n!} = e^{-|\alpha|^2/2}e^{-|\beta|^2/2}e^{\alpha\beta^*}$$

and

$$|\langle\beta|\alpha\rangle|^2 = e^{-|\alpha|^2}e^{-|\beta|^2}e^{\alpha\beta^*+\alpha^*\beta}$$

Now

$$|\alpha-\beta|^2 = |\alpha|^2 + |\beta|^2 - \alpha^*\beta - \alpha\beta^*$$

Hence,

$$|\langle\beta|\alpha\rangle|^2 = e^{-|\alpha-\beta|^2}$$

Implying that two coherent states are not orthogonal to each other. They become nearly orthogonal if $|\alpha-\beta| \gg 1$.

The probability of obtaining m photons in the coherent state is given by

$$P_m = |\langle m|\alpha\rangle|^2 = \left|e^{-\frac{1}{2}|\alpha|^2}\sum_{n=0}^{\infty}\frac{\alpha^n}{\sqrt{n!}}\langle m|n\rangle\right|^2 = \left|e^{-\frac{1}{2}|\alpha|^2}\sum_{n=0}^{\infty}\frac{\alpha^n}{\sqrt{n!}}\delta_{mn}\right|^2 = e^{-|\alpha|^2}\frac{|\alpha|^{2m}}{m!}$$

Since $|\alpha|^2 = \langle n\rangle$, the average number of photons in the coherent state we have

$$P_m = e^{-\langle m\rangle}\frac{\langle m\rangle^m}{m!} \tag{15.53}$$

This is nothing but Poisson distribution with an expectation value of $\langle m\rangle$. Figure 15.4 shows the probability distribution for different values of $\langle m\rangle$.

Since

$$|n\rangle = \frac{1}{\sqrt{n}}\hat{a}^\dagger|n-1\rangle = \frac{1}{\sqrt{n(n-1)}}\left(\hat{a}^\dagger\right)^2|n-2\rangle\ldots$$

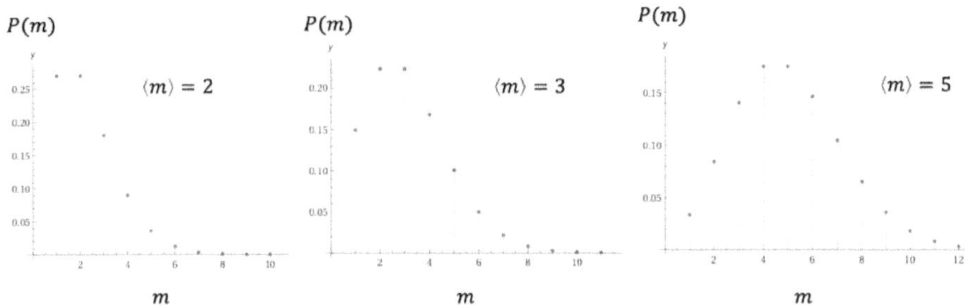

Fig. 15.4 Poisson distribution given by Eq. (15.53) for $\langle m\rangle = 2$, 3 and 5

we have

$$|n\rangle = \frac{1}{\sqrt{n!}}\left(\hat{a}^\dagger\right)^n |0\rangle$$

Hence,

$$|\alpha\rangle = e^{-\frac{1}{2}|\alpha|^2}\sum_{n=0}^{\infty}\frac{\alpha^n}{\sqrt{n!}}|n\rangle = e^{-\frac{1}{2}|\alpha|^2}\sum_{n=0}^{\infty}\frac{\left(\alpha\hat{a}^\dagger\right)^n}{n!}|0\rangle \tag{15.54}$$

Time Evolution of the Coherent State

Since the time variation in the number states $|n\rangle$ is given by $e^{-iE_n t/\hbar}|n\rangle = e^{-i\left(n+\frac{1}{2}\right)\omega t}|n\rangle$ the coherent state evolves with time as

$$\begin{aligned} |\alpha(t)\rangle &= e^{-\frac{1}{2}|\alpha|^2}\sum_{n=0}^{\infty}\frac{\alpha^n}{\sqrt{n!}}e^{-i\left(n+\frac{1}{2}\right)\omega t}|n\rangle \\ &= e^{-\frac{1}{2}|\alpha|^2}e^{-i\omega t/2}\sum_{n=0}^{\infty}\frac{(\alpha)^n}{\sqrt{n!}}e^{-in\omega t}|n\rangle \\ &= e^{-\frac{1}{2}|\alpha|^2}e^{-i\omega t/2}\sum_{n=0}^{\infty}\frac{\left(\alpha e^{-i\omega t}\right)^n}{\sqrt{n!}}|n\rangle \\ &= e^{-i\omega t/2}\left|\alpha e^{-i\omega t}\right\rangle \end{aligned} \tag{15.55}$$

Hence, the coherent state remains a coherent state with an eigenvalue $\alpha e^{-i\omega t}$ which changes with time.

Having obtained the time variation of the coherent state, we can calculate the time variation in the expectation values of the quadrature operators $\widehat{X}_1$ and $\widehat{X}_2$:

$$\left\langle \widehat{X}_1(t) \right\rangle = \langle \alpha(t)|\widehat{X}_1|\alpha(t)\rangle = |\alpha|\cos(\omega t - \phi) \tag{15.56}$$

and

$$\left\langle \widehat{X}_2(t) \right\rangle = \langle \alpha(t)|\widehat{X}_2|\alpha(t)\rangle = -|\alpha|\sin(\omega t - \phi) \tag{15.57}$$

Note that the expectation values of the two quadrature operators have sinusoidal variation with time just like a classical harmonic oscillator.

Exercise 15.4 For a coherent state, obtain the variances in the two quadratures and show that

$$\left(\Delta\widehat{X}_1(t)\right)\left(\Delta\widehat{X}_2(t)\right) = \frac{1}{4} \tag{15.58}$$

independent of time. Assume α to be real.

Solution We have

$$\left\langle \widehat{X}_1^2(t) \right\rangle = \langle \alpha(t)|\frac{\left(\widehat{a}+\widehat{a}^\dagger\right)^2}{4}|\alpha(t)\rangle = \frac{1}{4}\langle \alpha(t)|\left(\widehat{a}^2+\widehat{a}^{\dagger 2}+2\widehat{a}^\dagger\widehat{a}+1\right)|\alpha(t)\rangle$$

which using Eq. (15.55) gives us

$$\left\langle \widehat{X}_1^2(t) \right\rangle = \alpha^2\cos^2\omega t + \frac{1}{4} \tag{15.59}$$

Similarly

$$\left\langle \widehat{X}_2^2(t) \right\rangle = \langle \alpha(t)|\frac{\left(\widehat{a}-\widehat{a}^\dagger\right)^2}{4i^2}|\alpha(t)\rangle = -\frac{1}{4}\langle \alpha(t)|\left(\widehat{a}^2+\widehat{a}^{\dagger 2}-2\widehat{a}^\dagger\widehat{a}-1\right)|\alpha(t)\rangle$$

which gives us

$$\left\langle \widehat{X}_2^2(t) \right\rangle = \alpha^2\sin^2\omega t + \frac{1}{4} \tag{15.60}$$

Hence,

$$\left(\Delta\widehat{X}_1(t)\right) = \sqrt{\left\langle \widehat{X}_1^2(t) \right\rangle - \left\langle \widehat{X}_1(t) \right\rangle^2} = \frac{1}{2} \tag{15.61a}$$

$$\left(\Delta\widehat{X}_2(t)\right) = \sqrt{\left\langle\widehat{X}_2^2(t)\right\rangle - \left\langle\widehat{X}_2(t)\right\rangle^2} = \frac{1}{2} \tag{15.61b}$$

and we obtain

$$\left(\Delta\widehat{X}_1(t)\right)\left(\Delta\widehat{X}_2(t)\right) = \frac{1}{4}$$

Hence, the variances in the two quadratures of a coherent state remain unchanged with time and the state continues to remain as a minimum uncertainty state.

We can define generalized quadrature operators $\widehat{X}_\theta$ and $\widehat{X}_{\theta+\frac{\pi}{2}}$ given by (see Fig. 15.5)

$$\widehat{X}_\theta = \widehat{X}_1\cos\theta + \widehat{X}_2\sin\theta = \frac{\left(\widehat{a}+\widehat{a}^\dagger\right)}{2}\cos\theta + \frac{\left(\widehat{a}-\widehat{a}^\dagger\right)}{2i}\sin\theta = \frac{\widehat{a}}{2}(\cos\theta - i\sin\theta)$$
$$+\frac{\widehat{a}^\dagger}{2}(\cos\theta + i\sin\theta)$$

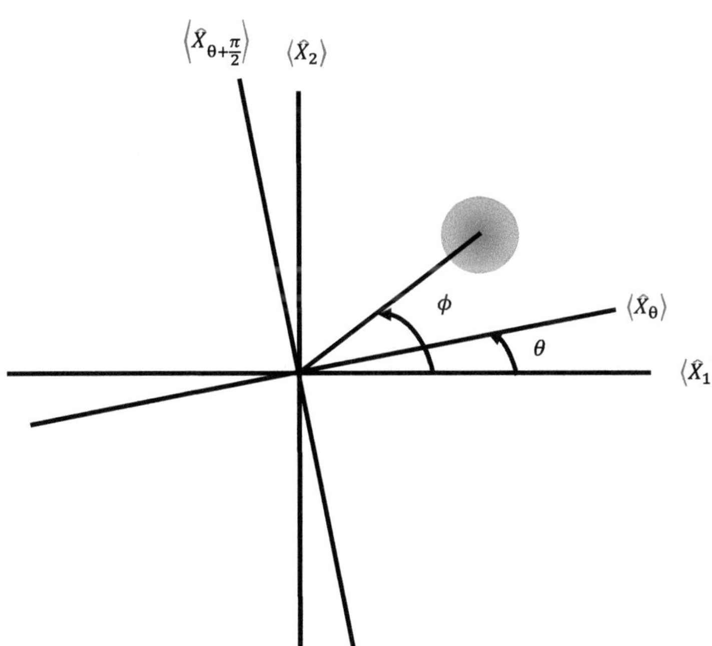

Fig. 15.5 Relationship between the quadrature operators $\widehat{X}_1$ and $\widehat{X}_2$ and the generalized quadrature operators $\widehat{X}_\theta$ and $\widehat{X}_{\theta+\frac{\pi}{2}}$. The circular blob represents a coherent state (see Sec. 15.6)

or

$$\widehat{X}_\theta = \frac{\widehat{a}e^{-i\theta} + \widehat{a}^\dagger e^{i\theta}}{2}$$

Similarly we have

$$\widehat{X}_{\theta+\frac{\pi}{2}} = \frac{\widehat{a}e^{-i\theta} - \widehat{a}^\dagger e^{i\theta}}{2i}$$

Using these operators, it is again possible to evaluate the expectation values and variances in the two generalized quadrature operators for a coherent state.

Thus,

$$\left\langle \widehat{X}_\theta(t) \right\rangle = \langle \alpha(t)|\widehat{X}_\theta|\alpha(t)\rangle = |\alpha|\cos(\omega t + \theta - \phi)$$

and

$$\left\langle \widehat{X}_{\theta+\frac{\pi}{2}}(t) \right\rangle = \langle \alpha(t)|\widehat{X}_{\theta+\frac{\pi}{2}}|\alpha(t)\rangle = -|\alpha|\sin(\omega t + \theta - \phi)$$

We can also evaluate the variances and obtain

$$\left(\Delta\widehat{X}_\theta(t)\right) = \sqrt{\left\langle \widehat{X}_\theta^2(t) \right\rangle - \left\langle \widehat{X}_\theta(t) \right\rangle^2} = \frac{1}{2}$$

and

$$\left(\Delta\widehat{X}_{\theta+\frac{\pi}{2}}(t)\right) = \sqrt{\left\langle \widehat{X}_{\theta+\frac{\pi}{2}}^2(t) \right\rangle - \left\langle \widehat{X}_{\theta+\frac{\pi}{2}}(t) \right\rangle^2} = \frac{1}{2}$$

It is interesting to note that the variances are independent of θ.

15.5.1 Coherent State as a Displaced Vacuum State

Now, we have

$$\sum_{n=0}^{\infty} \frac{\left(\alpha\widehat{a}^\dagger\right)^n}{n!}|0\rangle = e^{\alpha\hat{a}^\dagger}|0\rangle$$

Thus (see Eq. 15.54)

$$|\alpha\rangle = e^{-\frac{1}{2}|\alpha|^2} e^{\alpha\hat{a}^\dagger}|0\rangle = e^{-\frac{1}{2}|\alpha|^2 + \alpha\hat{a}^\dagger}|0\rangle \tag{15.62}$$

We now wish to modify this equation using the following operator identity:

$$e^{\hat{b}} e^{\hat{c}} = e^{\hat{b}+\hat{c}+\frac{1}{2}[\hat{b},\hat{c}]} \tag{15.63}$$

for any pair of operators $\widehat{b}$ and $\widehat{c}$ which commute with their commutators, that is,

$$\left[\widehat{b}, \left[\widehat{b}, \widehat{c}\right]\right] = \left[\widehat{c}, \left[\widehat{b}, \widehat{c}\right]\right] = 0 \tag{15.64}$$

Since

$$\widehat{a}|0\rangle = 0$$

we have

$$e^{-\alpha^* \hat{a}}|0\rangle = |0\rangle$$

Hence, we can write

$$|\alpha\rangle = e^{-\frac{1}{2}|\alpha|^2} e^{\alpha\hat{a}^\dagger} e^{-\alpha^*\hat{a}}|0\rangle \tag{15.65}$$

Now, using Eq. (15.63), we have

$$e^{\alpha\hat{a}^\dagger} e^{-\alpha^*\hat{a}} = e^{\alpha\hat{a}^\dagger - \alpha^*\hat{a} + \frac{1}{2}\left[\alpha\hat{a}^\dagger, -\alpha^*\hat{a}\right]} = e^{\alpha\hat{a}^\dagger - \alpha^*\hat{a} - \frac{[\alpha]^2}{2}\left[\hat{a}^\dagger, \hat{a}\right]} = e^{\frac{1}{2}|\alpha|^2} e^{\alpha\hat{a}^\dagger - \alpha^*\hat{a}}$$

Hence,

$$e^{\alpha\hat{a}^\dagger - \alpha^*\hat{a}} = e^{-\frac{1}{2}|\alpha|^2} e^{\alpha\hat{a}^\dagger} e^{-\alpha^*\hat{a}}$$

Substituting in Eq. (15.65), we obtain

$$|\alpha\rangle = e^{\alpha\hat{a}^\dagger - \alpha^*\hat{a}}|0\rangle \tag{15.66}$$

Thus, the coherent state can be obtained by operating with the operator

$$\widehat{D}(\alpha) = e^{\alpha\hat{a}^\dagger - \alpha^*\hat{a}} \tag{15.67}$$

on the vacuum state $|0\rangle$. The operator $\widehat{D}(\alpha)$ is referred to as the displacement operator.

Coherent states are the closest quantum states to a classical electromagnetic field and the output from a single-mode laser oscillating much above threshold is a coherent state. Note that the coherent state is quite different from a number state; the latter is specified by a definite number of photons in the state while the coherent state does not have a precise number of photons. The probability of finding a certain number of photons is described by a Poisson distribution.

Sometimes a very low intensity laser beam is used to approximate a single photon state. If we wish to have a quantum state in which the probability of finding two or more photons is less than 0.01, then we see from Eq. (15.53) that we must have

$$P_2 = e^{-\langle m\rangle}\frac{\langle m\rangle^2}{2!} < 0.01$$

Solving the above equation gives us the average value of number of photons to be

$$\langle m\rangle \approx 0.1$$

and the probability of detecting one photon is then given by

$$P_1 = e^{-0.1}\frac{(0.1)^1}{1!} \approx 0.09$$

Thus, if we have a very weak laser pulse with an average photon number of 0.1, then the probability of detecting no photon is about 0.9, of detecting one photon will be about 0.09 and of detecting two photons in the pulse will be 0.01. We can also calculate the probability of detecting three photons in the pulse to be about 0.00015. Thus, even with a weak laser source which approximates a coherent state, there is always a finite probability of finding more than a single photon in the pulse and hence the pulse does not represent truly a single photon state.

15.6 Diagrams Representing Quantum States of Radiation

A visualization of the different quantum states of light can be obtained through a diagrammatic representation. Different quantum states can be represented diagrammatically through a phase space diagram to provide a nice qualitative picture. The phasor diagram is a plot in the $X_1 - X_2$ plane depicting the expectation values of the quadratures and their fluctuations.

First we consider a coherent state at $t = 0$. The expectation values of the two quadratures are given by Eqs. (15.56) and (15.57) and their variances are given by Eqs. (15.61a) and (15.61b). Figure 15.6a shows a diagrammatic representation of the coherent state in which the coherent state is represented by a circular patch with the center having coordinates equal to $\left\langle \widehat{X}_1 \right\rangle = |\alpha|\cos\phi$ and $\left\langle \widehat{X}_2 \right\rangle = |\alpha|\sin\phi$. Since the variances in the two quadratures are $\left(\Delta\widehat{X}_1\right)^2 = \frac{1}{4}$ and $\left(\Delta\widehat{X}_2\right)^2 = \frac{1}{4}$ the circular patch

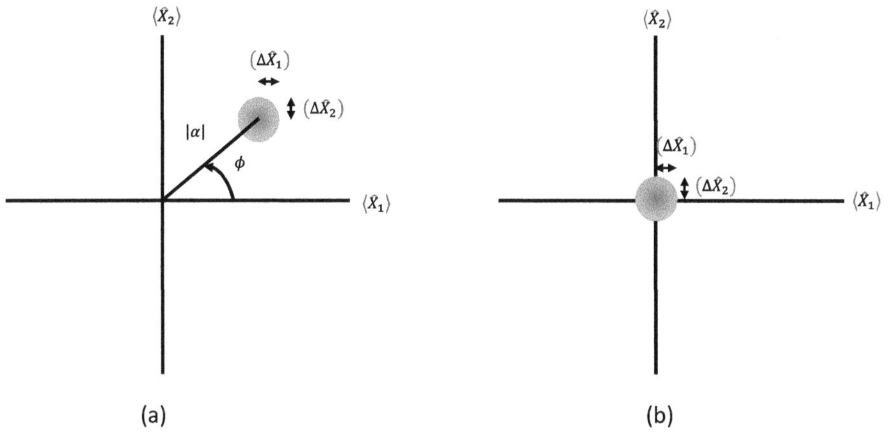

Fig. 15.6 Diagrammatic representations of (**a**) a coherent state and (**b**) a vacuum state. In both cases, $\left(\Delta\widehat{X}_1\right) = \left(\Delta\widehat{X}_2\right) = \frac{1}{2}$

which represents uncertainty in the two quadratures has a radius equal to ½. For coherent states the fluctuations are the same in all directions and hence the shaded regions are drawn as circles.

The length of the arrow from the origin to the center of the circle will be equal to $|\alpha|$. We must remember that the fluctuations depicted in the figure have a quantum origin and must satisfy the Heisenberg uncertainty principle. In a diagram representing a classical state, the area of the circle of fluctuations can be zero since one can in principle have precise simultaneous knowledge of both the quadratures.

If we consider a coherent state with $\alpha = 0$, then the state would be just the vacuum state $|0\rangle$. Figure 15.6b shows the diagram for the vacuum state for which the expectation values of both the quadratures are zero and it has equal fluctuations along all orientations. Thus, a vacuum state would be a circle centered at the origin with a radius of ½.

In Chap. 18, we will present squeezed states and also present a diagrammatic representation of such states. In Chap. 20, we will discuss the technique of balanced optical homodyning to carry out measurements on the various states of the electromagnetic field.

Problems

15.1. Consider a number state $|n\rangle$ with $n = 20$. Calculate the expectation value and the variance in the number of photons in this state.

15.2. Consider a coherent state $|\alpha\rangle$. The value of α is such that the probability of detecting 2 photons is 1% of the probability of detecting one photon. What is the probability of detecting no photons? Assume α to be real.

15.3. Consider a number state $|n\rangle$ with $n = 10$:

(a) Obtain the value of uncertainty in the photon number in this state.
(b) Obtain the variances of the two quadrature operators.
(c) Is this a minimum uncertainty state?

15.4. Consider a superposition of two coherent states $|\alpha\rangle$ and $|-\alpha\rangle$:

$$|\psi\rangle = N(|\alpha\rangle + |-\alpha\rangle)$$

(a) Normalize $|\psi\rangle$.
(b) Show that the state $|\psi\rangle = N(|\alpha\rangle + |-\alpha\rangle)$ is an eigenstate of the operator $\widehat{a}^2$. Obtain the eigenvalue.
(c) Show that the probability of finding an odd number of photons in this state is zero.

This is an example of what is referred to as a Schrödinger cat state.

15.5. Consider an electromagnetic field in a single-mode state $|\psi\rangle$satisfying the following equation:

$$\left(\sqrt{5}\widehat{a} - 2\widehat{a}^{\dagger}\right)|\psi\rangle = 0$$

Obtain the expectation value of photon number in this state.

15.6. A *single-mode* electromagnetic wave is in a state given by the quantum state $|\psi\rangle = \frac{1}{2}|2\rangle + \frac{\sqrt{3}}{2}|3\rangle$. Obtain the expectation value of the energy. What is the expectation value of the quadrature operator $\widehat{X}_1$?

15.7. Many processes involve absorption of single photons from the quantum state. The process of absorption is represented by the action of the annihilation operator $\widehat{a}$.

$$|\psi\rangle = \frac{1}{\sqrt{2}}|2\rangle + \frac{1}{\sqrt{2}}|4\rangle$$

Consider absorption of a photon from the above quantum state. What is the expectation value of the number of photons in the state $\widehat{a}|\psi\rangle$?

Although $\widehat{a}$ is called the annihilation operator, the operator $\widehat{a}$ effectively 'annihilates' the low-energy components of quantum superpositions, increasing the relative weight of the higher-energy components. In the original state, the average number of photons was 3. In the new state, the average number of photons is 7/3 which is greater than 2! Quantum states can behave quite differently from classical expectations.

15.8. Write down a quantum state in which the probabilities of detecting 1, 2, or 3 photons are ½, ¼, and ¼, respectively. Will these probabilities change with time?

15.9. Consider the following state

$$|\psi\rangle = \frac{i}{2}|8\rangle + c|10\rangle$$

(a) What should be the value of c?

(b) What is the expectation value of the number of photons in this state?
(c) What is the expectation value of the quadrature operator $\widehat{X}_1$ in this state?

15.10. Consider a coherent state $|\alpha\rangle$ with $\alpha = 2$. Calculate

(a) The mean number of photons in this state.
(b) The standard deviation of the number of photons.
(c) The probability of detecting 0 and 1 photon in this state.

15.11. Consider a single mode of an electromagnetic field in a number state $|\psi\rangle = |n\rangle$ with $n = 10$ and a coherent state $|\phi\rangle = |\alpha\rangle$ with $\alpha = 10$.

(a) Which of these is an eigenstate of the number operator $\widehat{N} = \widehat{a}^\dagger \widehat{a}$?
(b) Obtain the expectation value of $\widehat{N}$ and $\widehat{N}^2$ in both these states.

15.12. Consider a coherent state $|\alpha\rangle$ with $\alpha = 1$.

(a) What is the expectation value of the number of photons in this state?
(b) What is the probability of finding zero photons, one photon, two photons and three photons in this state? What is the probability of finding more than three photons in this state?

Chapter 16
Multimode Photon States

Abstract This chapter discusses multimode photon states. Examples of two mode states are presented followed by states occupying multiple modes. Concept of single photon temporal wave packet is introduced. This is followed by a brief discussion of the quantum process of detection of photons and formation of interference of single photons.

16.1 Introduction

In Chap. 15, we had discussed single-mode states where only one mode of the radiation field is occupied by photons while all other modes were empty. We have considered different types of single-mode states, namely number states, vacuum states, superposition states, and coherent states.

In this chapter, we will consider multimode states where more than a single mode is occupied. We first discuss the simple case of two mode states to bring out interesting features of such states which is followed by multimode states with more than two modes being occupied. Entangled states that will be discussed in Chap. 17 form a very important class of multimode states and exhibit very interesting nonclassical features that find wide applications in quantum science and technology. For more details on various quantum states of radiation, readers are referred to the excellent text on quantum optics by Grynberg et al. (2010).

16.2 Two Mode States

Let us consider two different modes of the electromagnetic field; these could be two different frequencies, two different polarization states, two different directions of propagation, or two different spatial modes of a waveguide (see Fig. 16.1). We also assume that all modes other than these two are empty.

A general two mode state can be represented by

K. Thyagarajan, *Integrated Quantum Photonics*, Graduate Texts in Physics,
https://doi.org/10.1007/978-3-031-85728-7_16

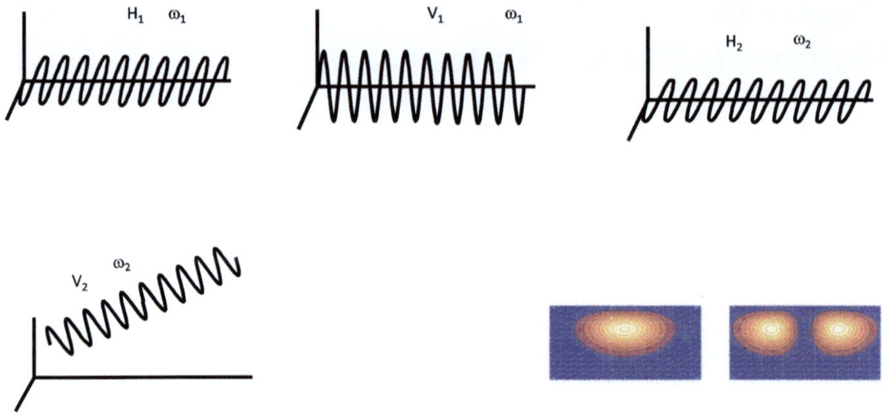

Fig. 16.1 Different modes of the electromagnetic field differing in frequencies or state of polarization or propagation direction or spatial modes of a waveguide. *H* horizontal polarization, *V* vertical polarization

$$|\psi\rangle = \sum_{n_1}\sum_{n_2} c_{n_1,n_2}|n_1, n_2\rangle \tag{16.1}$$

where subscripts 1 and 2 correspond to the two modes under consideration and c_{n_1,n_2} are the expansion coefficients in terms of the number states of the two modes. The above equation represents a superposition state of two mode states with mode # 1 having n_1 photons and mode # 2 having n_2 photons. The coefficients c_{n_1,n_2} satisfy the normalization condition

$$\sum_{n_1}\sum_{n_2} |c_{n_1,n_2}|^2 = 1 \tag{16.2}$$

For example, a two mode state would be

$$|\psi\rangle = |1_1, 1_2, 0, 0, 0, \ldots..\rangle = |1_1, 1_2\rangle \tag{16.3}$$

which represents a state with one photon in mode # 1 and one photon in mode # 2 and is a short form for the following state:

$$|\psi\rangle = |1_1\rangle|1_2\rangle \tag{16.4}$$

which is a product state of the two modes. This state corresponds to $c_{n_1,n_2} = 0$ for $n_1 \neq 1; n_2 \neq 1$ and $c_{1,\,1} = 1$.

Let us try to find out if this state is an eigenstate of the number operator of mode # 1. For this, we calculate

$$\widehat{N}_1|\psi\rangle = \widehat{a}_1^\dagger \widehat{a}_1|1_1, 1_2\rangle = \left(\widehat{a}_1^\dagger \widehat{a}_1|1_1\rangle\right)|1_2\rangle = 1|1_1, 1_2\rangle = 1|\psi\rangle \tag{16.5}$$

Hence, this state is an eigenstate of the number operator of mode # 1 with an eigenvalue 1 which implies that a measurement of the number of photons in mode # 1 will yield a value 1. In a similar fashion, the state is also an eigenstate of the number operator of mode # 2 with eigenvalue 1 implying that there is exactly one photon occupying mode # 2.

We now find out if the given state is also an eigenstate of the total photon number operator. For this, we evaluate

$$\left(\widehat{N}_1 + \widehat{N}_2 + \widehat{N}_3 + \widehat{N}_4 + \cdots\right)|\psi\rangle = \left(\widehat{a}_1^\dagger \widehat{a}_1 + \widehat{a}_2^\dagger \widehat{a}_2 + \widehat{a}_3^\dagger \widehat{a}_3 + \widehat{a}_4^\dagger \widehat{a}_4 + \cdots\right) \times |1_1, 1_2, 0_3, 0_4, \ldots.\rangle$$

where the first bracketed sum is the total number operator which is a sum of the number operator of the different modes.

Let us expand this equation by opening up the brackets. Thus,

$$\left(\widehat{N}_1 + \widehat{N}_2 + \widehat{N}_3 + \widehat{N}_4 + \cdots\right)|\psi\rangle = \widehat{a}_1^\dagger \widehat{a}_1|1_1, 1_2, 0_3, 0_4, \ldots.\rangle + \widehat{a}_2^\dagger \widehat{a}_2|1_1, 1_2, 0_3, 0_4, \ldots.\rangle + \cdots$$

Now the number operator $\widehat{a}_1^\dagger \widehat{a}_1$ operates on the state corresponding to mode 1, the number operator $\widehat{a}_2^\dagger \widehat{a}_2$ operates on the state corresponding to mode 2, etc. Corresponding to a particular mode, operating on the number state corresponding to that mode gives a value equal to the number of photons occupying that number state. Hence, we have

$$\begin{aligned}\left(\widehat{N}_1 + \widehat{N}_2 + \widehat{N}_3 + \widehat{N}_4 + \cdots\right)|\psi\rangle &= \left(\widehat{a}_1^\dagger \widehat{a}_1 + \widehat{a}_2^\dagger \widehat{a}_2 + \widehat{a}_3^\dagger \widehat{a}_3 + \widehat{a}_4^\dagger \widehat{a}_4 + \cdots\right)|1_1, 1_2, 0_3, 0_4, \ldots.\rangle \\ &= (1 + 1 + 0 + 0 + \cdots)|1_1, 1_2, 0_3, 0_4, \ldots.\rangle = 2|1_1, 1_2, 0_3, 0_4, \ldots.\rangle = 2|\psi\rangle\end{aligned} \tag{16.6}$$

Thus, the given state is also an eigenstate of the total photon number operator with an eigenvalue 2. Thus, measurement of the total number of photons in this state will always yield the value 2.

Note that in this state the measurement of the photon number in any of the two individual modes does not affect the photon number measured in the other mode. There is no correlation between the measurements on the two individual modes.

Now let us consider a superposition of two two-mode states. For example, the following state:

$$|\psi\rangle = \frac{1}{\sqrt{2}}(|1_1, 0_2\rangle + |0_1, 1_2\rangle) \tag{16.7}$$

The first term on the right-hand side represents a two-mode state with one photon in mode 1 with no photon in mode 2 and the second term is a two-mode state with one photon in mode 2 and no photon in mode 1. The given state is a superposition of these two states.

Let us now find out if this state is an eigenstate of the photon number operator of mode 1. For this, we evaluate

$$\widehat{N}_1|\psi\rangle = \widehat{a}_1^\dagger\widehat{a}_1\frac{1}{\sqrt{2}}(|1_1, 0_2\rangle + |0_1, 1_2\rangle) = \frac{1}{\sqrt{2}}(1|1_1, 0_2\rangle + 0|0_1, 1_2\rangle) = \frac{1}{\sqrt{2}}|1_1, 0_2\rangle$$

which implies that this state is not an eigenstate of the photon number operator of mode 1 and hence this state does not possess a definite number of photons in mode 1. The expectation value of the number of photons in mode 1 would be

$$\langle\psi|\widehat{N}_1|\psi\rangle = \frac{1}{\sqrt{2}}(\langle 1_1, 0_2| + \langle 0_1, 1_2|)\frac{1}{\sqrt{2}}|1_1, 0_2\rangle = \frac{1}{2} \tag{16.8}$$

Thus, the expectation value of the number of photons in mode 1 is 0.5. Hence, making a measurement of the number of photons in mode 1 will yield a value of 1, 50% of the times and 0, 50% of the times.

Similarly it can be shown that the given state is not an eigenstate of the photon number operator of mode 2 and the expectation value of the number of photons in mode 2 is also 0.5. Hence, in this state a single photon is occupying two modes with equal probability.

We now determine if the given state is an eigenstate of the total photon number operator. Thus, we evaluate

$$\begin{aligned}\left(\widehat{N}_1+\widehat{N}_2+\widehat{N}_3+\widehat{N}_4+\cdots\right)|\psi\rangle &= \left(\widehat{a}_1^\dagger\widehat{a}_1+\widehat{a}_2^\dagger\widehat{a}_2+\widehat{a}_3^\dagger\widehat{a}_3+\widehat{a}_4^\dagger\widehat{a}_4+\cdots\right)\frac{1}{\sqrt{2}}(|1_1,0_2\rangle+|0_1,1_2\rangle)\\ &= \frac{1}{\sqrt{2}}(|1_1,0_2\rangle+|0_1,1_2\rangle) = |\psi\rangle\end{aligned} \tag{16.9}$$

Thus, the given state is an eigenstate of the total photon number operator with an eigenvalue 1. Hence, a measurement of the total number of photons in this state will always give us a definite value and the number of photons occupying this state is 1. Thus, in this state, a single photon is occupying two modes simultaneously and is in a superposition state.

Compared to the state represented by Eq. (16.3) which corresponds to a factorizable state, in the entangled state represented by Eq. (16.7), although photon number measurements on each of the individual modes gives random values of either 0 or 1, a measurement carried out on any one of the modes results in a definite photon number in the other mode. Thus, if a measurement of the photon number in mode 1 gives a value 0, then the measurement of photon number in mode 2 will definitely give a value of 1. Similarly if the measurement of the photon number in mode 1 gives a value 1, then the measurement of photon number in mode 2 will definitely yield 0. Thus, the two measurements are completely correlated.

For example if the two modes corresponded to two different frequencies ω_1 and ω_2, then the photon is occupying both frequencies at the same time. The expectation value of the number of photons occupying frequency ω_1 would be 0.5 and the expectation value of the number of photons occupying mode with the frequency ω_2 would also be 0.5. Hence, in this mode the frequency of the single photon is not defined and a measurement of the frequency would give either ω_1 or ω_2 with equal probabilities.

The most general state with a *maximum* of one photon occupation in either of the two modes will be

$$|\psi\rangle = c_1|0_1, 0_2\rangle + c_2|0_1, 1_2\rangle + c_3|1_1, 0_2\rangle + c_4|1_1, 1_2\rangle \tag{16.10}$$

which can be written as

$$|\psi\rangle = (a_1|0_1\rangle + a_2|1_1\rangle)(b_1|0_2\rangle + b_2|1_2\rangle) \tag{16.11}$$

with $c_1 = a_1b_1$, $c_2 = a_1b_2$, $c_3 = a_2b_1$, and $c_4 = a_2b_2$. Thus, the state given by Eq. (16.10) can be factorized and represented by a product of the states corresponding to the two modes.

A nonfactorizable state would be

$$|\psi\rangle = c_2|0_1, 1_2\rangle + c_3|1_1, 0_2\rangle \tag{16.12}$$

which cannot be written as a product of states corresponding to individual modes and corresponds to what is referred to as an entangled state. To show that this is nonfactorizable, we compare Eq. (16.12) with the factorizable form given by Eq. (16.11) and obtain the following relations:

$$c_1 = a_1b_1 = 0 \tag{16.13a}$$

$$c_2 = a_1b_2 = 1 \tag{16.13b}$$

$$c_3 = a_2b_1 = 1 \tag{16.13c}$$

$$c_4 = a_2b_2 = 0 \tag{16.13d}$$

From Eq. (16.13a), we see that either $a_1 = 0$ or $b_1 = 0$. If $a_1 = 0$, then it is inconsistent with Eq. (16.13b) and if $b_1 = 0$, it is inconsistent with Eq. (16.13c). Thus, the state represented by Eq. (16.12) is nonfactorizable.

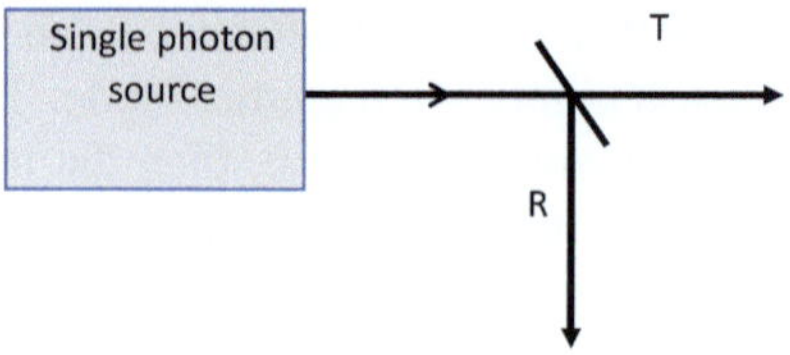

Fig. 16.2 A 50% beam splitter puts an incident photon in a superposition state of being transmitted and reflected

Exercise 16.1 A 50% beam splitter has a probability of 50% of reflecting the incident photon and 50% probability of transmitting the photon. Write down the output state of such a beam splitter.

Solution A beam splitter has two input ports and two output ports (see Fig. 16.2). When a single photon is incident on a 50% beam splitter, at the output it creates a two mode state which is occupied by a single photon. In this case, the two modes correspond to propagation along the transmitted arm and propagation along the reflected arm; these are two different modes of propagation.

If we represent the transmitted and reflected modes by subscripts T and R, the state emerging from the beam splitter would be the following state:

$$|\psi\rangle = \frac{1}{\sqrt{2}}|1_T, 0_R\rangle + \frac{i}{\sqrt{2}}|0_T, 1_R\rangle \quad (16.14)$$

The factor i on the second term on the right hand side is to ensure energy conservation (see Chap. 19). Thus, the probability of detecting the photon in the transmitted arm would be 0.5 and that in the reflected arm would also be 0.5. The beam splitter puts the incident single photon into a superposition state of transmitted mode and the reflected mode. A symmetric waveguide Y-splitter (see Chap. 7) or a 3 dB waveguide directional coupler (see Chap. 5) behaves in a similar fashion and puts a single photon into a superposition state of being in either of the two waveguides.

A beam splitter is a very important device in quantum photonics (see Chap. 19). In Chap. 5, we studied about directional couplers and in Chap. 7, we studied about y-splitters. Both these waveguide devices act as beam splitters and a single photon incident at the input of the device comes out in a superposition state between the two output waveguides.

Exercise 16.2 Consider a single photon having linear polarization with its polarization oriented along an angle θ to the horizontal. The photon is incident on a polarization beam splitter which has a reflectivity of 50% for the horizontal polarization and transmits completely the vertical polarization component. What is the state of the photon in the reflected and transmitted arms?

Solution We will first express the input photon in terms of the horizontal and vertical states $|H\rangle$ and $|V\rangle$. Here $|H\rangle$ represents a single photon with horizontal polarization and $|V\rangle$ represents a single photon with vertical polarization. Thus, we have

$$|\psi\rangle_{in} = \cos\theta|H\rangle + \sin\theta\,|V\rangle \tag{16.15}$$

Given the reflectivity and transmittivity of the beam splitter, the reflected state would be

$$|\psi\rangle_{ref} = \frac{1}{\sqrt{2}}\cos\theta|H\rangle \tag{16.16}$$

and the transmitted state would be

$$|\psi\rangle_{tr} = \frac{1}{\sqrt{2}}\cos\theta|H\rangle + \sin\theta\,|V\rangle \tag{16.17}$$

Exercise 16.3 Consider two modes and assume that there can be a maximum of two photons per mode. Write down the most general state satisfying this condition.

Solution A two mode state with a maximum of two photons per mode is

$$\begin{aligned}|\psi\rangle = c_1|0_1,0_2\rangle + c_2|0_1,1_2\rangle + c_3|0_1,2_2\rangle + c_4|1_1,0_2\rangle + c_5|1_1,1_2\rangle + c_6|1_1,2_2\rangle \\ + c_7|2_1,0_2\rangle + c_8|2_1,1_2\rangle + c_9|2_1,2_2\rangle\end{aligned} \tag{16.18}$$

with the normalization condition

$$\sum_{i=1}^{9}|c_i|^2 = 1 \tag{16.19}$$

These two modes could mean two orthogonal polarization states, or two different frequencies or two different propagation directions.

Referring to Exercise 16.3, if we put a condition that the quantum state is occupied by two photons then the most general state will reduce to

$$|\psi\rangle = c_3|0_1,2_2\rangle + c_5|1_1,1_2\rangle + c_7|2_1,0_2\rangle \tag{16.20}$$

We can show that this state has exactly two photons by showing that this is an eigenstate of the total photon number operator with an eigenvalue of 2:

$$\begin{aligned}\left(\hat{a}_1^\dagger\hat{a}_1+\hat{a}_2^\dagger\hat{a}_2\right)|\psi\rangle &= \hat{a}_1^\dagger\hat{a}_1(c_3|0_1,2_2\rangle+c_5|1_1,1_2\rangle+c_7|2_1,0_2\rangle)\\ &+\hat{a}_2^\dagger\hat{a}_2(c_3|0_1,2_2\rangle+c_5|1_1,1_2\rangle+c_7|2_1,0_2\rangle)=2|\psi\rangle\end{aligned} \tag{16.21}$$

An interesting state out of this is the following state:

$$|\psi\rangle = c_3|0_1,2_2\rangle + c_7|2_1,0_2\rangle \tag{16.22}$$

in which there are two photons but both of them are occupying either mode 1 or mode 2. This state is an eigenstate of the total photon number operator:

$$\begin{aligned}\left(\hat{a}_1^\dagger\hat{a}_1+\hat{a}_2^\dagger\hat{a}_2\right)|\psi\rangle &= \left(\hat{a}_1^\dagger\hat{a}_1+\hat{a}_2^\dagger\hat{a}_2\right)(c_3|0_1,2_2\rangle+c_7|2_1,0_2\rangle)\\ &= 2c_7|2_1,0_2\rangle+2c_3|0_1,2_2\rangle=2|\psi\rangle\end{aligned} \tag{16.23}$$

The state is not an eigenstate of the photon number of either of the two modes. For example, if we calculate

$$\hat{a}_1^\dagger\hat{a}_1|\psi\rangle = \hat{a}_1^\dagger\hat{a}_1(c_3|0_1,2_2\rangle+c_7|2_1,0_2\rangle)=2c_7|2_1,0_2\rangle \tag{16.24}$$

And hence not an eigenstate of $\hat{a}_1^\dagger\hat{a}_1$. Thus, the number of photons in mode # 1 is not defined and measurements on the number of photons in mode # 1 will not give precise values.

Similarly the number of photons in mode # 2 is also not defined. On the other hand, the total number of photons in this state is well defined.

Such states are referred to as NOON states with $N = 2$. A general NOON state would be

$$|\psi\rangle = \frac{1}{\sqrt{2}}|N_1,0_2\rangle + \frac{1}{\sqrt{2}}|0_1,N_2\rangle \tag{16.25}$$

in which there are N photons occupying both the modes with probability of finding all these N photons in mode # 1 being $\frac{1}{2}$ and the probability of finding all the N photons in mode # 2 being $\frac{1}{2}$ with probabilities of all other combinations of photons in the two modes being equal to zero. Such NOON states find applications in quantum metrology and other areas of quantum information processing.

16.3 Multimode States

In this section, we will consider multimode states in which the photons occupy more than two modes. Such states are very important in quantum photonics. An example of this is the polarization entangled state where two photons occupy four modes. Entangled states have very interesting counterintuitive properties and play an extremely important part in quantum key distribution, quantum computing, quantum teleportation, etc. Entangled states are discussed in Chaps. 17 and 23.

Let us begin by considering the following multimode state:

$$\begin{aligned}|\psi\rangle &= c_1|1_1, 0_2, 0_3, 0_4, \ldots\rangle + c_2|0_1, 1_2, 0_3, 0_4, \ldots\rangle + c_3|0_1, 0_2, 1_3, 0_4, \ldots\rangle + \cdots \\ &= \sum\nolimits_l c_l|1_l\rangle\end{aligned} \tag{16.26}$$

where c_l are the expansion coefficients and $|1_l\rangle$ is a short form for the state

$$|0_1, 0_2, 0_3, \ldots 0_{l-1}, 1_l, 0_{l+1}, \ldots\rangle$$

which represents a state with one photon in mode l and all other modes being empty. These different modes could be different frequencies or different directions of propagation.

The given state is a linear combination of states in which only one mode is occupied. For example in the state $|1_1, 0_2, 0_3, 0_4, \ldots\rangle$ mode 1 is occupied by one photon and all other modes are empty, in the state $|0_1, 1_2, 0_3, 0_4, \ldots\rangle$ mode 2 is occupied by a single photon and all other modes are empty, etc.

Let us first find out if this state is an eigenstate of the total photon number operator. For this, we evaluate

$$\begin{aligned}\widehat{N}|\psi\rangle &= \left(\widehat{a}_1^\dagger\widehat{a}_1 + \widehat{a}_2^\dagger\widehat{a}_2 + \widehat{a}_3^\dagger\widehat{a}_3 + \cdots\right)|\psi\rangle \\ &= \left(\widehat{a}_1^\dagger\widehat{a}_1 + \widehat{a}_2^\dagger\widehat{a}_2 + \widehat{a}_3^\dagger\widehat{a}_3 + \cdots\right)(c_1|1_1\rangle + c_2|1_2\rangle + c_3|1_3\rangle + \cdots) = |\psi\rangle\end{aligned} \tag{16.27}$$

where we have used the fact that

$$\widehat{a}_l^\dagger\widehat{a}_l|0_1, 0_2, 0_3, \ldots 0_{l-1}, 1_l, 0_{l+1}, \ldots\rangle = 1|0_1, 0_2, 0_3, \ldots 0_{l-1}, 1_l, 0_{l+1}, \ldots\rangle \tag{16.28}$$

and

$$\widehat{a}_l^\dagger\widehat{a}_l|0_1, 0_2, 0_3, \ldots 1_m, \ldots\ldots\rangle = 0; \quad m \neq l \tag{16.29}$$

Hence, the given state is an eigenstate of the total photon number operator with an eigenvalue of 1. This implies that there is just one photon occupying this state and this single photon is occupying multiple modes at the same time.

Is the given state an eigenstate of the photon number operator of individual modes? For this we determine whether the given state is an eigenstate of the number operator of mode 1:

$$\widehat{N}_1|\psi\rangle = \widehat{a}_1^\dagger\widehat{a}_1(c_1|1_1\rangle + c_2|1_2\rangle + c_3|1_3\rangle + \cdots) = c_1|1_1\rangle \tag{16.30}$$

which shows that the given state is not an eigenstate of the number operator of mode 1. Hence, the number of photons in mode 1 is not well defined. Similarly we can show that the given state is not an eigenstate of the number operator of any of the individual modes. Thus, in the given state a single photon is occupying multiple modes simultaneously.

The probability of finding the photon in mode 1 will be

$$|\langle 1_1, 0_2, 0_3, 0_4, \ldots|\psi\rangle|^2 = |c_1|^2 \tag{16.31}$$

Similarly the probability of finding the photon in mode l would be $|c_l|^2$. The coefficients c_l satisfy the normalization condition

$$\sum |c_l|^2 = 1 \tag{16.32}$$

Exercise 16.4 Consider the following three mode state (the subscripts represent mode numbers):

$$|\psi\rangle = \frac{1}{2}|1_1, 0_2, 0_3\rangle + \alpha|0_1, 1_2, 0_3\rangle + \frac{1}{2}|0_1, 0_2, 1_3\rangle \tag{16.33}$$

(a) Normalize the ket and obtain the value of α.
(b) Check if the state has a well-defined number of photons. If so how many?
(c) What is the expectation value of the number of photons in mode 2?

Solution
(a) Using the normalization, we have

$$\langle\psi|\psi\rangle = 1$$

which gives us $|\alpha| = \frac{1}{\sqrt{2}}$.
(b) To check whether this is an eigenstate of the total number operator, we operate on the state with the total photon number operator (taking $\alpha = \frac{1}{\sqrt{2}}$):

$$\left(\hat{a}_1^\dagger\hat{a}_1 + \hat{a}_2^\dagger\hat{a}_2 + \hat{a}_3^\dagger\hat{a}_3\right)|\psi\rangle = \frac{1}{2}|1_1, 0_2, 0_3\rangle + \frac{1}{\sqrt{2}}|0_1, 1_2, 0_3\rangle + \frac{1}{2}|0_1, 0_2, 1_3\rangle$$
$$= |\psi\rangle$$

Hence, it is an eigenstate of the total photon number operator with an eigenvalue 1. The state has a single photon occupying three discrete modes.
(c) Expectation value of the number of photons in mode 2 is given by

$$\langle\psi|\widehat{N}_2|\psi\rangle = \langle\psi|\widehat{a}_2^\dagger\widehat{a}_2|\psi\rangle = \frac{1}{2}$$

Exercise 16.5 Consider the following state:

$$|\psi\rangle = \frac{1}{\sqrt{2}}(|H_1, H_2, H_3\rangle + |V_1, V_2, V_3\rangle) \tag{16.34}$$

where H and V represent horizontal and vertical polarization respectively and the subscripts refer to three different modes having frequencies ω_1, ω_2 and ω_3. Such a state is referred to as a GHZ state and discussed in Chap. 22.

(a) Is this an eigenstate of the total photon number operator?
(b) How many photons are there in the state?
(c) Is this an eigenstate of the photon number operator of the state represented by H_1?

Solution
(a) Operating with the total photon number operator, we have

$$\widehat{N}|\psi\rangle = \left(\widehat{a}_{H1}^\dagger\widehat{a}_{H1} + \widehat{a}_{H2}^\dagger\widehat{a}_{H2} + \widehat{a}_{H3}^\dagger\widehat{a}_{H3} + \widehat{a}_{V1}^\dagger\widehat{a}_{V1} + \widehat{a}_{V2}^\dagger\widehat{a}_{V2} + \widehat{a}_{V3}^\dagger\widehat{a}_{V3}\right) \left(\frac{1}{\sqrt{2}}|H_1, H_2, H_3\rangle + \frac{1}{\sqrt{2}}|V_1, V_2, V_3\rangle\right) \tag{16.35}$$

Opening the brackets and operating with the number operator of individual mode, we obtain

$$\widehat{N}|\psi\rangle = 3|\psi\rangle \tag{16.36}$$

Yes, this is an eigenstate of the total photon number operator.
(b) The state has three photons.
(c) To check whether the given state is an eigenstate of the photon number operator of the state represented by H_1, we have

$$\widehat{a}_{H1}^\dagger\widehat{a}_{H1}|\psi\rangle = \frac{1}{\sqrt{2}}|H_1, H_2, H_3\rangle \tag{16.37}$$

Hence, the given state is not an eigenstate of the state represented by H_1.

Exercise 16.6 In continuation of Exercise 16.5, consider the following multimode state of photons (such a state is referred to as a W-state)

$$|\psi\rangle = \frac{1}{\sqrt{3}}(|H_1, V_2, V_3\rangle + |V_1, H_2, V_3\rangle + |V_1, V_2, H_3\rangle) \tag{16.38}$$

(a) Show that the state has exactly three photons.
(b) What is the probability of finding the photon having frequency ω_1 to have horizontal polarization or vertical polarization?
(c) What will be the output state when the photon having frequency ω_1 is found to have horizontal polarization and when it is found to have vertical polarization?

Solution

(a) Operating with the total photon number operator, we have

$$\widehat{N}|\psi\rangle = \left(\widehat{a}_{H1}^{\dagger}\widehat{a}_{H1} + \widehat{a}_{H2}^{\dagger}\widehat{a}_{H2} + \widehat{a}_{H3}^{\dagger}\widehat{a}_{H3} + \widehat{a}_{V1}^{\dagger}\widehat{a}_{V1} + \widehat{a}_{V2}^{\dagger}\widehat{a}_{V2} + \widehat{a}_{V3}^{\dagger}\widehat{a}_{V3}\right)|\psi\rangle = 3|\psi\rangle \tag{16.39}$$

Thus, the state has three photons.

(b) Probability of finding the photon having frequency ω_1 to have horizontal polarization is given by

$$|\langle H_1|\psi\rangle|^2 = \frac{1}{3} \tag{16.40}$$

and that of finding it to have vertical polarization is

$$|\langle V_1|\psi\rangle|^2 = \frac{2}{3} \tag{16.41}$$

(c) Output state when the photon having frequency ω_1 is found to have horizontal polarization is

$$|\psi\rangle_{out} = |V_2, V_3\rangle \tag{16.42}$$

and when it is found to have vertical polarization is

$$|\psi\rangle_{out} = \frac{1}{\sqrt{2}}(|H_2, V_3\rangle + |V_2, H_3\rangle) \tag{16.43}$$

Note that in the first case we obtain a separable state while in the second case, the output is a polarization entangled state (see Chap. 17).

16.3.1 Single-Photon Temporal Wave Packet

We consider a multimode state with different modes having different frequencies but all having the same state of polarization and propagating along the same direction. We now form a superposition state of these modes:

$$\begin{aligned}|\psi\rangle &= c_1|1_1, 0_2, 0_3, 0_4, \ldots\rangle + c_2|0_1, 1_2, 0_3, 0_4, \ldots\rangle + c_3|0_1, 0_2, 1_3, 0_4, \ldots\rangle + \cdots \\ &= \sum\nolimits_l c_l|1_l\rangle\end{aligned} \tag{16.44}$$

Here the subscripts 1, 2, 3, etc. correspond to different frequencies. As discussed before this state is an eigenstate of the total photon number operator with an eigenvalue of unity, that is, this state has just a single photon occupying multiple frequencies at the same time. The probability of the photon having frequency ω_l is $|c_l|^2$.

The time evolution of this state in the Schrödinger picture would be given by

$$|\psi(t)\rangle = \sum\nolimits_l c_l e^{-i\omega_l t}|1_l\rangle \tag{16.45}$$

where the energy is measured with respect to the zero point energy. Note that the probability of finding the photon in any specific mode remains independent of time. The state continues to be an eigenstate of the total photon number operator. By an appropriate choice of the expansion coefficients c_l, one can form a single photon wave packet in which the single photon occupies different frequencies.

If we consider a continuum of frequencies, instead of a sum, we can write a single photon wave packet through the following integral:

$$|\psi\rangle = \int c(\omega)|1_\omega\rangle d\omega \tag{16.46}$$

with

$$\int |c(\omega)|^2 d\omega = 1 \tag{16.47}$$

Here $|1_\omega\rangle$ represents a state with a single photon in the mode having a frequency ω.

Equation (16.46) can also be written as

$$|\psi\rangle = \int c(\omega)\hat{a}^\dagger(\omega)|0\rangle d\omega \tag{16.48}$$

where $\hat{a}^\dagger(\omega)$ is the creation operator at frequency ω.

Single photon emission from spontaneous emission can be characterized by the following spectral distribution:

$$c(\omega) = \frac{K}{(\omega - \omega_0) + i\Gamma_{sp}/2} \tag{16.49}$$

where K is a normalization constant satisfying Eq. (16.47), ω_0 is the center frequency, and Γ_{sp} represents the width at half maximum of the spontaneous emission spectrum.

Exercise 16.7 Consider a single photon having a Gaussian spectral amplitude centered at frequency ω_0 and having a spectral width σ. What would be the corresponding $c(\omega)$?

Solution A Gaussian spectral amplitude centered at frequency ω_0 and having a spectral width σ will be given by

$$c(\omega) = Ae^{-(\omega-\omega_0)^2/2\sigma^2} \tag{16.50}$$

where A represents the normalization constant. Using Eq. (16.47), we have

$$\int |c(\omega)|^2 d\omega = A^2 \int e^{-(\omega-\omega_0)^2/\sigma^2} d\omega = A^2\sigma\sqrt{\pi} = 1 \tag{16.51}$$

which gives

$$A = \frac{1}{\sqrt{\sigma\sqrt{\pi}}} \tag{16.52}$$

Thus, we have the spectral amplitude

$$c(\omega) = \frac{1}{\sqrt{\sigma\sqrt{\pi}}} e^{-(\omega-\omega_0)^2/2\sigma^2} \tag{16.53}$$

and the single photon will be described by

$$|\psi\rangle = \frac{1}{\sqrt{\sigma\sqrt{\pi}}} \int e^{-(\omega-\omega_0)^2/2\sigma^2} \hat{a}^\dagger(\omega)|0\rangle d\omega \tag{16.54}$$

Note that due to Fourier transform relationship between frequency and time, a single photon given by Eq. (16.54) will have a temporal duration $\tau \sim \frac{1}{\sigma}$.

16.4 Detection of Photons

Electromagnetic fields at the frequency of light (~10^{14} Hz) are detected by photodetectors which convert the energy of the photon into electron-hole pairs in semiconductors or eject electrons from the cathode of a photomultiplier device. Since the frequencies are very high, the detection process averages out and in classical terms measures the intensity of the radiation which is proportional to E^*E (see Eq. 2.30) where E represents the electric field associated with the electromagnetic wave (assumed here to be a scalar for simplicity).

We now use the Heisenberg picture (see Chap. 21) in which the state is time invariant, while the operators evolve with time. In the quantum picture, the average rate of counting of photons at any given time t for a detector placed at a position $\vec{r}$ is given by (see, e.g., Gerry and Knight 2005)

$$w(\vec{r},t) = s\ \langle\psi|\widehat{E}^{-}(\vec{r},t)\widehat{E}^{+}(\vec{r},t)|\psi\rangle \tag{16.55}$$

where s is a constant of proportionality which depends on the detector characteristics and

$$\widehat{E}^{+}\left(\vec{r},t\right) = \sum i\sqrt{\frac{\hbar\omega_l}{2\epsilon_0 V}}\widehat{a}_l e^{-i\left(\omega_l t - \vec{k}_l.\vec{r}\right)} \tag{16.56}$$

and

$$\widehat{E}^{-}\left(\vec{r},t\right) = -\sum i\sqrt{\frac{\hbar\omega_l}{2\epsilon_0 V}}\widehat{a}_l^{\dagger} e^{+i\left(\omega_l t - \vec{k}_l.\vec{r}\right)} \tag{16.57}$$

represent the positive and negative frequency components of the electromagnetic field in mode l having a frequency ω_l and propagating along the direction given by $\vec{k}_l$. Also $\left|\vec{k}_l\right| = \frac{\omega_l}{c}$; c being the speed of light in free space.

The probability of a photodetector positioned at a point $\vec{r}$ observing a photoionization between t and $(t + dt)$ is proportional to $w\left(\vec{r},t\right)\,dt$.

For a radiation vacuum state given by $|0\rangle$, we will obtain for the photodetection probability

$$\begin{aligned} w(\vec{r},t) &= s\ \langle\psi|\widehat{E}^{-}(\vec{r},t)\widehat{E}^{+}(\vec{r},t)|\psi\rangle \\ &= s\langle 0\left| -\sum i\sqrt{\frac{\hbar\omega_m}{2\epsilon_0 V}}\widehat{a}_m^{\dagger} e^{+i\left(\omega_m t - \vec{k}_m.\vec{r}\right)} \sum i\sqrt{\frac{\hbar\omega_l}{2\epsilon_0 V}}\widehat{a}_l e^{-i\left(\omega_l t - \vec{k}_l.\vec{r}\right)}\right|0\rangle = 0 \end{aligned} \tag{16.58}$$

which is as expected.

As an example let us consider a one photon multimode state as given by Eq. (16.26) and assume that the modes correspond to different frequencies but having the same polarization and propagating along the z-direction. Thus, we will have

$$\widehat{E}^{+}\left(\vec{r},t\right)=\sum i\sqrt{\frac{\hbar\omega_l}{2\epsilon_0 V}}\widehat{a}_l e^{-i(\omega_l t-k_l z)} \tag{16.59}$$

and a similar equation for $\widehat{E}^{-}\left(\vec{r},t\right)$. Hence, we have

$$\begin{aligned} w(\vec{r},t) &= s\,\langle\psi|\widehat{E}^{-}(\vec{r},t)\widehat{E}^{+}(\vec{r},t)|\psi\rangle \\ &= s\sum_q c_q^*\langle 1_q|\sum -i\sqrt{\frac{\hbar\omega_m}{2\epsilon_0 V}}\widehat{a}_m^{\dagger}e^{+i(\omega_m t-k_m z)}\sum i\sqrt{\frac{\hbar\omega_l}{2\epsilon_0 V}}\widehat{a}_l e^{-i(\omega_l t-k_l z)}\sum_p c_p|1_p\rangle \\ &= \sum_q\sum_l\sum_m\sum_p c_q^* c_p\frac{\hbar\sqrt{\omega_l\omega_m}}{2\epsilon_0 V}e^{-i(\omega_l t-k_l z)}e^{+i(\omega_m t-k_m z)}\langle 1_q|\widehat{a}_m^{\dagger}\widehat{a}_l|1_p\rangle \end{aligned} \tag{16.60}$$

Now

$$\langle 1_q|\widehat{a}_m^{\dagger}\widehat{a}_l|1_p\rangle=\langle 1_q|\widehat{a}_m^{\dagger}\delta_{lp}|0\rangle=\langle 0|\delta_{mq}\delta_{lp}|0\rangle=\delta_{mq}\delta_{lp} \tag{16.61}$$

Hence, we have

$$\begin{aligned} w\left(\vec{r},t\right) &= s\sum_l\sum_m c_m^* c_l\frac{\hbar\sqrt{\omega_l\omega_m}}{2\epsilon_0 V}e^{-i(\omega_l t-k_l z)}e^{+i(\omega_m t-k_m z)}=s\left|\sum_l c_l\sqrt{\frac{\hbar\omega_l}{2\epsilon_0 V}}e^{-i(\omega_l t-k_l z)}\right|^2 \\ &= s\left|\sum_l c_l\sqrt{\frac{\hbar\omega_l}{2\epsilon_0 V}}e^{-ik_l(ct-z)}\right|^2 \end{aligned} \tag{16.62}$$

which depends on $(ct - z)$ which implies that in free space, the one photon wave packet travels at a speed c without any distortion.

As an example of single photon interference, we consider a state with a single photon occupying two modes. This corresponds to the following state

$$|\psi\rangle=c_1|1_1,0_2\rangle+c_2|0_1,1_2\rangle$$

where only two modes are mentioned and all other modes are assumed to be empty. As a specific example, we choose $c_1=c_2=\frac{1}{\sqrt{2}}$ and write for the state

$$|\psi\rangle = \frac{1}{\sqrt{2}}|1_1, 0_2\rangle + \frac{1}{\sqrt{2}}|0_1, 1_2\rangle \tag{16.63}$$

with equal probability of occupation of the two modes.

We first consider the case when the two modes represented by subscripts 1 and 2 correspond to two closely lying frequencies $\omega_1 = \omega_0 + \frac{\Delta\omega}{2}$ and $\omega_2 = \omega_0 - \frac{\Delta\omega}{2}$. The propagation constants at these two frequencies are $k_1 = k_0 + \frac{\Delta k}{2}$ and $k_2 = k_0 - \frac{\Delta k}{2}$ respectively with $\frac{\omega_0}{k_0} = c$.

From Eq. (16.62), we obtain for the corresponding number of photons being counted per unit time

$$w(z,t) = s\left|\frac{1}{\sqrt{2}}\sqrt{\frac{\hbar\omega_1}{2\epsilon_0 V}}e^{-i(\omega_1 t - k_1 z)} + \frac{1}{\sqrt{2}}\sqrt{\frac{\hbar\omega_2}{2\epsilon_0 V}}e^{-i(\omega_2 t - k_2 z)}\right|^2$$
$$= s\left|\frac{1}{\sqrt{2}}\sqrt{\frac{\hbar\omega_1}{2\epsilon_0 V}}e^{-i\left(\left[\omega_0 + \frac{\Delta\omega}{2}\right]t - \left[k_0 + \frac{\Delta k}{2}\right]z\right)} + \frac{1}{\sqrt{2}}\sqrt{\frac{\hbar\omega_2}{2\epsilon_0 V}}e^{-i\left(\left[\omega_0 - \frac{\Delta\omega}{2}\right]t - \left[k_0 - \frac{\Delta k}{2}\right]z\right)}\right|^2$$

which simplifies to (assuming $\Delta\omega \ll \omega_0$)

$$w(z,t) = s\frac{\hbar\omega_0}{4\epsilon_0 V}\left|e^{-i\left(\frac{\Delta\omega}{2}t - \frac{\Delta k}{2}z\right)} + e^{i\left(\frac{\Delta\omega}{2}t - \frac{\Delta k}{2}z\right)}\right|^2 = s^2\frac{\hbar\omega_0}{\epsilon_0 V}\cos^2\left(\frac{\Delta\omega t - \Delta k z}{2}\right) \tag{16.64}$$

Thus, for example, a detector placed at a position $z = 0$ will give a signal proportional to $\cos^2\frac{\Delta\omega t}{2}$ which represents the beat between the two frequencies that the single photon is occupying.

The speed of propagation of the peak of the detection probability is given by

$$v_g = \frac{\Delta\omega}{\Delta k} = \frac{d\omega}{dk} \tag{16.65}$$

in the limit $\Delta\omega \to 0$. The speed given by Eq. (16.65) is indeed the group velocity of the single photon wave packet (see Chap. 2).

As another example we consider the single photon in a superposition of two modes assumed to have the same frequency but propagating along two different directions specified by propagation vectors $\vec{k}_1$ and $\vec{k}_2$. In such a case, the positive frequency part of the total electric field operator is given by

$$\hat{E}^+\left(\vec{r},t\right) = \hat{E}_1^+\left(\vec{r},t\right) + \hat{E}_2^+\left(\vec{r},t\right) \tag{16.66}$$

with

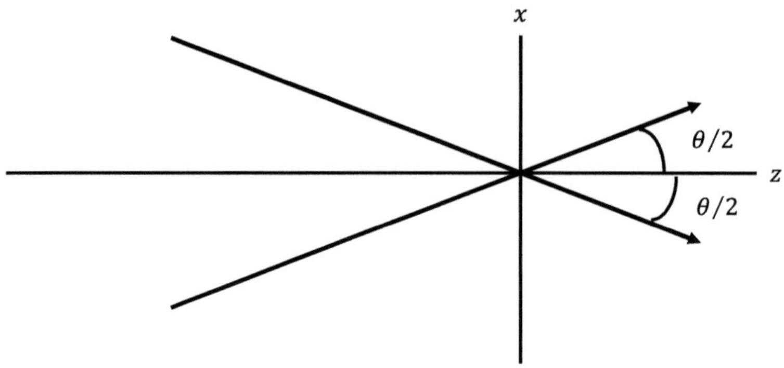

Fig. 16.3 Two plane wave modes have propagation vectors lying in the $x-z$ plane making angles of $+\frac{\theta}{2}$ and $-\frac{\theta}{2}$ with the z-axis

$$\widehat{E}_1^+\left(\vec{r},t\right)=i\sqrt{\frac{\hbar\omega}{2\epsilon_0 V}}\widehat{a}_1 e^{-i\left(\omega t-\vec{k}_1.\vec{r}\right)}=i\frac{E_0}{2}\widehat{a}_1 e^{-i\left(\omega t-\vec{k}_1.\vec{r}\right)} \tag{16.67a}$$

and

$$\widehat{E}_2^+\left(\vec{r},t\right)=i\sqrt{\frac{\hbar\omega}{2\epsilon_0 V}}\widehat{a}_2 e^{-i\left(\omega t-\vec{k}_2.\vec{r}\right)}=i\frac{E_0}{2}\widehat{a}_2 e^{-i\left(\omega t-\vec{k}_2.\vec{r}\right)} \tag{16.67b}$$

where

$$E_0=\sqrt{\frac{2\hbar\omega}{\epsilon_0 V}} \tag{16.68}$$

For specificity, we assume that the two modes have propagation vectors lying in the $x-z$ plane making angles of $+\frac{\theta}{2}$ and $-\frac{\theta}{2}$ with the z-axis (see Fig. 16.3). Thus,

$$\vec{k}_1=\frac{\omega}{c}\left(\widehat{x}\sin\frac{\theta}{2}+\widehat{z}\cos\frac{\theta}{2}\right) \tag{16.69}$$

and

$$\vec{k}_2=\frac{\omega}{c}\left(-\widehat{x}\sin\frac{\theta}{2}+\widehat{z}\cos\frac{\theta}{2}\right) \tag{16.70}$$

If we have a photo detector scan along the x-direction in the plane $z=0$, we have

$$w(z=0,t)=s\,\langle\psi|\widehat{E}^{-}(\vec{r},t)\widehat{E}^{+}(\vec{r},t)|\psi\rangle$$
$$=\frac{sE_0^2}{4\times 2}\left[(\langle 1_1,0_2|+\langle 0_1,1_2|)\left(\widehat{a}_1^{\dagger}e^{i\chi_1}+\widehat{a}_2^{\dagger}e^{i\chi_2}\right)\left(\widehat{a}_1e^{-i\chi_1}+\widehat{a}_2e^{-i\chi_2}\right)(|1_1,0_2\rangle+|0_1,1_2\rangle)\right] \tag{16.71}$$

where

$$\chi_1=\left(\omega t-\vec{k}_1\cdot\vec{r}\right)=\left(\omega t-kx\sin\frac{\theta}{2}\right) \tag{16.72}$$

and

$$\chi_2=\left(\omega t-\vec{k}_2\cdot\vec{r}\right)=\left(\omega t+kx\sin\frac{\theta}{2}\right) \tag{16.73}$$

where we have put $z=0$.

Opening up the brackets in Eq. (16.71), we have

$$w(z=0,t)=\frac{sE_0^2}{8}\Big[(\langle 1_1,0_2|+\langle 0_1,1_2|)\Big(\widehat{a}_1^{\dagger}\widehat{a}_1+\widehat{a}_2^{\dagger}\widehat{a}_2+\widehat{a}_1^{\dagger}\widehat{a}_2e^{i(\chi_1-\chi_2)}$$
$$+\widehat{a}_2^{\dagger}\widehat{a}_1e^{-i(\chi_1-\chi_2)}\Big)(|1_1,0_2\rangle+|0_1,1_2\rangle)\Big]$$

Thus,

$$w(z=0,t)=\frac{sE_0^2}{8}\Big[(\langle 1_1,0_2|+\langle 0_1,1_2|)\Big(|1_1,0_2\rangle+0+0+|0_1,1_2\rangle+e^{i(\chi_1-\chi_2)}\times 0$$
$$+e^{i(\chi_1-\chi_2)}|1_1,0_2\rangle+e^{-i(\chi_1-\chi_2)}|0_1,1_2\rangle+e^{-i(\chi_1-\chi_2)}\times 0\Big)\Big]$$

where we have used the fact that $\widehat{a}_1^{\dagger}\widehat{a}_2|1_1,0_2\rangle=0$ and $\widehat{a}_1^{\dagger}\widehat{a}_2|0_1,1_2\rangle=|1_1,0_2\rangle$. Similarly $\widehat{a}_2^{\dagger}\widehat{a}_1|1_1,0_2\rangle=|0_1,1_2\rangle$ and $\widehat{a}_2^{\dagger}\widehat{a}_1|0_1,1_2\rangle=0$.

Using $\langle 1_1,0_2||1_1,0_2\rangle=1$, $\langle 1_1,0_2||0_1,1_2\rangle=0$, we obtain

$$w(z=0,t)=\frac{sE_0^2}{8}\left(1+e^{i(\chi_1-\chi_2)}+1+e^{-i(\chi_1-\chi_2)}\right)$$

which gives us

$$w(z=0,t)=\frac{sE_0^2}{4}\left[1+\cos\left(2kx\sin\frac{\theta}{2}\right)\right] \tag{16.74}$$

Thus, the detection probability on the screen depends on the position x on the screen clearly showing single photon interference. Note that we have just a single photon occupying the two modes and the detection probability shows the classic interference pattern that would be observed through the detection process. This is just the interference pattern observed in a double-slit experiment performed with single photons.

Exercise 16.8 Consider a state with two photons with each photon occupying a different mode. Such a state would be represented by

$$|\psi\rangle = |1_1\rangle|1_2\rangle = |1_1, 1_2\rangle \tag{16.75}$$

Show that the detection probability does not show any interference effects.

Solution Now as before the positive frequency component of the total electric field operator would be

$$\widehat{E}^+(\vec{r},t) = \widehat{E}_1^+(\vec{r},t) + \widehat{E}_2^+(\vec{r},t) = E_1\widehat{a}_1 e^{-i\chi_1} + E_2\widehat{a}_2 e^{-i\chi_2} \tag{16.76}$$

Hence, the probability of detecting photoionization at a point $\vec{r}$ at time t will be

$$w(\vec{r},t) = s\,\langle\psi|\widehat{E}^-(\vec{r},t)\widehat{E}^+(\vec{r},t)|\psi\rangle$$

Now

$$\widehat{E}^+(\vec{r},t)|\psi\rangle = \left(E_1\widehat{a}_1 e^{-i\chi_1} + E_2\widehat{a}_2 e^{-i\chi_2}\right)|1_1\rangle|1_2\rangle = E_1 e^{-i\chi_1}|0_1\rangle|1_2\rangle + E_2 e^{-i\chi_2}|1_1\rangle|0_2\rangle$$

and

$$\langle\psi|\widehat{E}^-(\vec{r},t) = \langle 1_1, 1_2|E_1\widehat{a}_1^\dagger e^{i\chi_1} + \langle 1_1, 1_2|E_2\widehat{a}_2^\dagger e^{i\chi_2} = \langle 0_1|\langle 1_2|E_1 e^{i\chi_1} + \langle 1_1| \times \langle 0_2|E_2 e^{i\chi_2}$$

Hence,

$$\begin{aligned} w(\vec{r},t) = s\langle\psi|\widehat{E}^-(\vec{r},t)\widehat{E}^+(\vec{r},t)|\psi\rangle = \langle 0_1|\langle 1_2|E_1 e^{i\chi_1}E_1 e^{-i\chi_1}|0_1\rangle|1_2\rangle + \langle 0_1|\langle 1_2|E_1 e^{i\chi_1}E_2 e^{-i\chi_2}|1_1\rangle|0_2\rangle \\ + \langle 1_1|\langle 0_2|E_2 e^{i\chi_2}E_1 e^{-i\chi_1}|0_1\rangle|1_2\rangle + \langle 1_1|\langle 0_2|E_2 e^{i\chi_2}E_2 e^{-i\chi_2}|1_1\rangle|0_2\rangle = \\ |E_1|^2 + 0 + 0 + |E_2|^2 = |E_1|^2 + |E_2|^2 \end{aligned} \tag{16.77}$$

showing that there is no interference between the two photons. Since the two photons belong to two different modes, there is no interference in the detection probability.

Problems

16.1. Write down the most general quantum state of three photons occupying two modes. Show that the state has exactly three photons. Which coefficients need to be zero to make the state in to a NOON state?

16.2. Consider a two mode state given by $|\psi\rangle = \frac{1}{2}\,|0_1, 0_2\rangle + \frac{\sqrt{3}}{2}|1_1, 1_2\rangle$.

(a) Determine if this is an eigenstate of the total photon number operator.
(b) Obtain the expectation value of the number of photons in this state.

16.3. Consider two independent modes i and j of the electromagnetic field. Show that the following two states are eigenstates of the total photon number operator: (a)$|\psi\rangle_1 = |3\rangle_i|3\rangle_j$ and (b) $|\psi\rangle_2 = \frac{1}{\sqrt{2}}\left[|3\rangle_i + |3\rangle_j\right]$ and obtain the number of photons in the two states. Are they also eigenstates of the photon number operators of the individual modes i and j?

16.4. Consider a two mode state given by

$$|\psi\rangle = A(|0_1\rangle + |1_1\rangle)|1_2\rangle$$

(a) Normalize the state and obtain the value of A.
(b) Is this an eigenstate of the photon number operator corresponding to the individual modes 1 and 2? Interpret the result.
(c) What is the expectation value of the total number of photons in this state?

16.5. Write down two expressions corresponding to a three mode single photon state with equal probabilities of finding the photon in any of the three states.

16.6. Consider the following two mode state:

$$|\psi\rangle = \frac{1}{\sqrt{2}}|2_1, 0_2\rangle + \frac{1}{\sqrt{2}}|0_1, 2_2\rangle$$

(a) Check whether the state has a well-defined number of photons.
(b) Show that a measurement of the number of photons in an individual mode will not give me a precise value. Obtain the expectation value of the number of photons in the two modes.

16.7. Consider a two mode state given by $|\psi\rangle = \frac{1}{2}\,|0_1, 0_2\rangle + \frac{\sqrt{3}}{2}|1_1, 1_2\rangle$. Determine if this is an eigenstate of the total photon number operator.

16.8. Write down the most general state of a single photon occupying four different modes.

16.9. Consider three photons at frequencies ω_1, ω_2 and ω_3 which could have either horizontal or vertical polarization states. Write down all possible states of the three photons in the horizontal-vertical basis.

16.10. Consider the following state of photons with H representing horizontal polarization and V representing vertical polarization; the subscripts represent different frequencies:

$$|\psi\rangle = \alpha(|H_1, H_2, H_3, H_4\rangle + |V_1, V_2, V_3, V_4\rangle)$$

(a) Normalize and obtain the value of α.
(b) How many modes are present in the state?
(c) Is this an eigenstate of the total photon number operator? The state contains how many photons?

References

Gerry, C., and P. Knight. 2005. *Introductory quantum optics*. Cambridge: Cambridge University Press.

Grynberg, G., A. Aspect, and C. Fabre. 2010. *Introduction to quantum optics: From the semi-classical approach to quantized light*. Cambridge: Cambridge University Press.

Chapter 17
Entangled Photon States

Abstract This chapter discusses the very important class of states referred to entangled quantum states. Detailed discussion of polarization entangled state, Bell states, spatial mode entangled states, hybrid entangled states, hyperentangled states, and states with higher dimensional entanglement is presented. The very important measure of CHSH (Clauser, Horne, Shinmoy, Holt) inequality is discussed both for entangled photons and distinguishable photons. A brief discussion of GHZ (Greenberger, Horne, Zeilinger) state is also given.

17.1 Introduction

In this chapter, we will consider entangled states which form a very important quantum state finding wide applications in quantum cryptography (see, e.g., Liu et al. 2024), quantum communication and teleportation (see, e.g., Bouwmeester et al. 1997), quantum computing (see, e.g., Thompson et al. 2011), quantum sensing (e.g., Crespi et al. 2012), spectroscopy (Tashima et al. 2024), etc. For an excellent review on quantum entanglement, readers are referred to Horodecki et al. (2009), Zeilinger (2017), Zhang et al. (2024).

In Chap. 12, we discussed difference frequency generation process in which a pump photon at frequency ω_p and a signal photon at frequency ω_s generate a wave at the difference frequency $\omega_i = \omega_p - \omega_s$. In this process, we needed the presence of both the input frequencies. It was also shown in Chap. 13 that with the input at just the pump frequency ω_p no generation of the lower frequency photons takes place.

In Chap. 22, we will discuss about the process of spontaneous parametric down conversion (SPDC) wherein a pump photon at a frequency ω_p down converts to a pair of daughter photons referred to as signal (having frequency ω_s) and idler (having frequency ω_i) with the three frequencies satisfying the energy conservation equation: $\omega_p = \omega_s + \omega_i$. Depending on the phase matching condition and the crystal used, the down converted photons may have the same state of polarization (Type 0 and Type I down conversion) or orthogonal polarization states (Type II down conversion). In Chap. 23, we will discuss different techniques for the generation of entangled states of photons using the SPDC process. In this chapter, we will discuss some of the

K. Thyagarajan, *Integrated Quantum Photonics*, Graduate Texts in Physics,
https://doi.org/10.1007/978-3-031-85728-7_17

properties of entangled states. A nice review of entangled states can be found in Torres et al. (2011).

In order to understand the concept of quantum entanglement, let us consider SPDC process in a z-cut lithium niobate waveguide in which a pump photon that has an ordinary (horizontal) polarization undergoes spontaneous parametric down conversion and generates a pair of orthogonally polarized photons, the signal photon at ω_s frequency having ordinary (horizontal) polarization, and the idler at ω_i frequency having extraordinary (vertical) polarization (see Chap. 22). In this case, the output quantum state of the pair of photons emerging from the crystal can be represented by (see Chap. 22)

$$|\psi\rangle = |H_s\rangle|V_i\rangle = |H_s, V_i\rangle \tag{17.1}$$

where $|H_s\rangle$ corresponds to the state of the signal photon which has horizontal (ordinary) polarization and $|V_i\rangle$ corresponds to the state of the idler photon which has vertical (extraordinary) polarization.

We now perform the following experiment with the emerging pair of photons as shown in Fig. 17.1.

The down converted signal-idler pair of photons is first split into two different paths using a wavelength splitter or a dichroic beam splitter. The beam splitter reflects one wavelength (assumed to be the idler here) while transmits the other wavelength (the signal wavelength). The two photons are then passed through analyzers oriented at appropriate angles and fall on two photodetectors (assumed

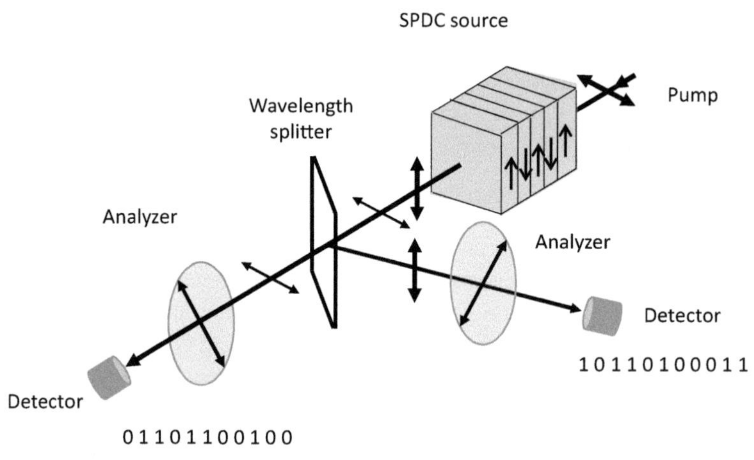

Fig. 17.1 A pump photon having ordinary polarization (horizontal polarization) spontaneously downconverts and generates a pair of photons, a signal photon which has ordinary polarization (horizontal polarization) and an idler photon which has extraordinary polarization (vertical polarization). The photons are separated using a wavelength splitter, pass through two analyzers and are detected

to be perfect) which output a 1 if the photon is detected and a 0 if it is not detected (see Fig. 17.1).

Let us first consider the situation when the signal photon which is horizontally polarized passes through an analyzer whose pass axis is oriented along the horizontal direction and the idler photon which is vertically polarized passes through an analyzer which is oriented along the vertical direction. Since the polarization states of the signal and idler photons coincide with the orientation of the analyzers, both the photons will always pass through the respective analyzers (assuming the analyzers to be lossless) and will generate a sequence of 1s (assuming perfect detectors). On the other hand, if the orientations of the two analyzers are rotated by 90°, then both photons have polarization which are orthogonal to the respective pass axes and thus will not pass through them generating a sequence of 0s. In this experiment, there is no randomness in the output from the two detectors.

Now we orient the two analyzers still oriented with their pass axes orthogonal to each other but at $+\frac{\pi}{4}$ and $-\frac{\pi}{4}$ with the horizontal directions as shown in Fig. 17.1. Since the signal photon which has horizontal polarization is made to pass through an analyzer oriented at $+\frac{\pi}{4}$ to its polarization direction, the probability of the photon passing through the analyzer is $\cos^2\frac{\pi}{4}=\frac{1}{2}$. [This is so since we can express the state $|H_s\rangle$ as a linear combination of the states $\left|+\left(\frac{\pi}{4}\right)_s\right\rangle$ and $\left|-\left(\frac{\pi}{4}\right)_s\right\rangle$ as $|H_s\rangle=\frac{1}{\sqrt{2}}\times\left(\left|+\left(\frac{\pi}{4}\right)_s\right\rangle-\left|-\left(\frac{\pi}{4}\right)_s\right\rangle\right)$. Hence, the probability will be given by $\left|\left\langle\left(\frac{\pi}{4}\right)_s\middle|H_s\right\rangle\right|^2=\frac{1}{2}$]. Thus, the detector will output a series of 1s and 0s at random with almost equal probability. In Fig. 17.1, we have represented the output by a possible sequence of bits.

In an identical fashion, the idler photon passing through another analyzer oriented at −45° to its polarization direction (and orthogonal to the pass axis of the polarizer used in the signal arm) also generates a random bit sequence as shown in Fig. 17.1.

Since the probability of the photon passing through each of the two analyzer is ½, we will get a random sequence of 1s and 0s with almost equal probability. The probability that the signal photon passes through the analyzer *and* the idler photon also passes through the analyzer is $\frac{1}{2}\times\frac{1}{2}=\frac{1}{4}$. Similarly the probability of the other three combinations of passing through and not passing through the respective analyzers will be $\frac{1}{4}$. The signal and idler photons are completely uncorrelated and hence the bit sequence from the signal port and the bit sequence from the idler port will have no correlation. Just like a coin toss, about 50% of the times the outputs from the two detectors will match, while 50% of the times they will not match. The output quantum state emerging from the crystal is a product state of the two photons, horizontal polarization state $|H_s\rangle$ of the signal and vertical polarization state $|V_i\rangle$ of the idler photons. The two photons are independent photons and hence there is no correlation in their measurements. Any polarization measurement carried out on any of the two photons has no influence on the polarization state of the other photon. The two photons are not polarization entangled. This is in contrast to a similar experiment that we will discuss using polarization entangled photon pairs, which will be discussed in Sect. 17.2 wherein the results are very different.

We will first discuss about the polarization entangled state and later discuss an experimental setup which will help us to understand the concept of entanglement in the polarization state between the two photons.

17.2 Polarization Entangled State

In order to introduce the polarization entangled state, we consider an example. We consider four modes of the electromagnetic field with two different frequencies ω_s and ω_i and having two orthogonal polarization states defined through H (Horizontal state of polarization) and V (vertical state of polarization). We assume that all these modes are propagating as fundamental modes along the same direction in the single mode waveguide. The four modes are:

1. Mode # 1: H polarization at frequency ω_s called H_s
2. Mode # 2: V polarization at frequency ω_s called V_s
3. Mode # 3: H polarization at frequency ω_i called H_i
4. Mode # 4: V polarization at frequency ω_i called V_i

We now consider the following quantum state of the four modes:

$$|\psi_1\rangle = |1_1\rangle|0_2\rangle|0_3\rangle|1_4\rangle = |1_1, 0_2, 0_3, 1_4\rangle \tag{17.2}$$

where the state $|1_1\rangle$ corresponds to a single photon occupying Mode # 1, etc. with the subscript corresponds to the mode number. The quantum state given by Eq. (17.2) corresponds to a state with a single photon occupying mode 1 (horizontal polarization at frequency ω_s) and a single photon occupying mode 4 (vertical polarization at frequency ω_i) with the other two modes being empty with no photons. This is the same state as given by Eq. (17.1) except that since we are considering four modes, the complete state with four modes is written with individual photons occupying two modes while the other two modes are unoccupied.

Let us see if this is an eigenstate of the total photon number operator given by (since we are considering only four modes, the number operators corresponding to only these four modes are explicitly written; other number operators operating on the state will give zero as they are all unoccupied):

$$\widehat{N} = \widehat{a}_1^\dagger\widehat{a}_1 + \widehat{a}_2^\dagger\widehat{a}_2 + \widehat{a}_3^\dagger\widehat{a}_3 + \widehat{a}_4^\dagger\widehat{a}_4 \tag{17.3}$$

Now,

$$\begin{aligned}\widehat{N}\big|\psi_1\rangle &= \left(\widehat{a}_1^\dagger\widehat{a}_1 + \widehat{a}_2^\dagger\widehat{a}_2 + \widehat{a}_3^\dagger\widehat{a}_3 + \widehat{a}_4^\dagger\widehat{a}_4\right)|1_1, 0_2, 0_3, 1_4\rangle \\ &= (1+0+0+1)|1_1, 0_2, 0_3, 1_4\rangle = 2|1_1, 0_2, 0_3, 1_4\rangle\end{aligned} \tag{17.4}$$

which implies that the given state is an eigenstate of the total photon number operator with an eigenvalue 2. Hence, this quantum state contains exactly two photons. Any measurement of the number of photons in this state will always yield a precise value of 2.

Let us check if the given state is also an eigenstate of the number operators of individual modes 1, 2, 3, and 4. Thus, to check if this is an eigenstate of mode 1, we operate on the state with the number operator of mode 1:

$$\widehat{a}_1^\dagger \widehat{a}_1 |1_1, 0_2, 0_3, 1_4\rangle = 1|1_1, 0_2, 0_3, 1_4\rangle$$

which implies that this is an eigenstate of the number operator of mode 1 and hence the state has a precise photon number of unity in mode 1. This implies that the state contains a photon at frequency ω_s having horizontal state of polarization. Similarly it is possible to check for other modes and we will find that the given state is an eigenstate of the number operator of each individual mode: with eigenvalues of 1, 0, 0, and 1 of modes 1, 2, 3, and 4, respectively.

Thus, this is a state which contains one photon each in modes 1 (horizontal polarization and frequency ω_s) and 4 (vertical polarization and frequency ω_i) and modes 2 and 3 are empty.

Similarly the following state

$$|\psi_2\rangle = |0_1, 1_2, 1_3, 0_4\rangle \tag{17.5}$$

is an eigenstate of the total photon number operator with an eigenvalue of 2, implying that this state also contains exactly two photons. The state is also an eigenstate of the number operators of modes 1, 2, 3, and 4 with eigenvalues 0, 1, 1, and 0. Thus, this is a state which contains one photon each in modes 2 (vertical polarization with frequency ω_s) and mode 3 (horizontal polarization and frequency ω_i) and modes 1 and 4 are empty.

In the states represented by Eqs. (17.2) and (17.5), there is precisely one photon each in two of the four modes. Thus, any measurement of the number of photons in any of the two states will yield precise values.

Now, let us consider a linear superposition of the two quantum states defined by Eqs. (17.2) and (17.5):

$$|\psi\rangle = \frac{1}{\sqrt{2}}(|1_1, 0_2, 0_3, 1_4\rangle - |0_1, 1_2, 1_3, 0_4\rangle) \tag{17.6}$$

Now operating on this state with the total photon number operator, we find

$$\widehat{N}|\psi\rangle = \left(\widehat{a}_1^\dagger \widehat{a}_1 + \widehat{a}_2^\dagger \widehat{a}_2 + \widehat{a}_3^\dagger \widehat{a}_3 + \widehat{a}_4^\dagger \widehat{a}_4\right)\left\{\frac{1}{\sqrt{2}}(|1_1, 0_2, 0_3, 1_4\rangle - |0_1, 1_2, 1_3, 0_4\rangle)\right\}$$

which gives us

$$\begin{aligned}\widehat{N}|\psi\rangle =& \widehat{a}_1^\dagger\widehat{a}_1\left\{\frac{1}{\sqrt{2}}(|1_1,0_2,0_3,1_4\rangle - |0_1,1_2,1_3,0_4\rangle)\right\}\\ &+\widehat{a}_2^\dagger\widehat{a}_2\left\{\frac{1}{\sqrt{2}}(|1_1,0_2,0_3,1_4\rangle - |0_1,1_2,1_3,0_4\rangle)\right\}\\ &+\widehat{a}_3^\dagger\widehat{a}_3\left\{\frac{1}{\sqrt{2}}(|1_1,0_2,0_3,1_4\rangle - |0_1,1_2,1_3,0_4\rangle)\right\}\\ &+\widehat{a}_4^\dagger\widehat{a}_4\left\{\frac{1}{\sqrt{2}}(|1_1,0_2,0_3,1_4\rangle - |0_1,1_2,1_3,0_4\rangle)\right\}\end{aligned}$$

or

$$\begin{aligned}\widehat{N}|\psi\rangle =& \frac{1}{\sqrt{2}}|1_1,0_2,0_3,1_4\rangle - \frac{1}{\sqrt{2}}|0_1,1_2,1_3,0_4\rangle - \frac{1}{\sqrt{2}}|0_1,1_2,1_3,0_4\rangle\\ &+\frac{1}{\sqrt{2}}|1_1,0_2,0_3,1_4\rangle = 2|\psi\rangle\end{aligned} \tag{17.7}$$

Hence, the superposition state represented by Eq. (17.6) contains exactly two photons. The two photons are in a superposition state occupying the four modes under consideration.

Now let us find out if this state is also an eigenstate of the number operator of individual modes. For example, we operate the number operator of mode 1 on the quantum state and obtain

$$\widehat{a}_1^\dagger\widehat{a}_1|\psi\rangle = \widehat{a}_1^\dagger\widehat{a}_1\left\{\frac{1}{\sqrt{2}}(|1_1,0_2,0_3,1_4\rangle - |0_1,1_2,1_3,0_4\rangle)\right\} = \frac{1}{\sqrt{2}}|1_1,0_2,0_3,1_4\rangle$$

which implies that the given state is *not an eigenstate* of the number operator of mode 1. Thus, a measurement of the number of photons in mode 1 will not give precise values.

Similarly one can show that the given state is not an eigenstate of the number operator of any of the four modes.

This implies that the number of photons in each of the four modes (having frequencies ω_s and ω_i and horizontal and vertical polarization states) is undefined, while the total number of photons in the state is exactly 2.

Let us see how many photons are having a frequency ω_s irrespective of their polarization state. In a measurement setup, this would imply that we pass this quantum state through a filter which passes only frequency ω_s and count the number of photons without bothering about the polarization state of the photon.

Since the modes corresponding to frequency ω_s are numbered 1 and 2, we will check if the given state is an eigenstate of the total photon number operator corresponding to modes 1 and 2:

$$\left(\widehat{a}_1^\dagger\widehat{a}_1 + \widehat{a}_2^\dagger\widehat{a}_2\right)\left\{\frac{1}{\sqrt{2}}(|1_1, 0_2, 0_3, 1_4\rangle - |0_1, 1_2, 1_3, 0_4\rangle)\right\}$$
$$= \left\{\frac{1}{\sqrt{2}}(|1_1, 0_2, 0_3, 1_4\rangle - |0_1, 1_2, 1_3, 0_4\rangle)\right\}$$

Thus, the state is an eigenstate of the sum of the number operators corresponding to modes 1 and 2 and hence the number of photons in this state having frequency ω_s is exactly 1. Thus, although the number of photons in modes 1 or 2 are undefined, the total number of photons in modes 1 and 2 (which are characterized by the same frequency but orthogonal polarization states) is exactly 1.

Similarly one can show that the number of photons with frequency ω_i in this state is exactly 1.

Hence, the state represented by Eq. (17.6) contains one photon at frequency ω_s and one photon at frequency ω_i.

Since the state is not an eigenstate of any of the individual modes but is an eigenstate of the number operator corresponding to each of the individual frequencies, this implies that this state has one photon each corresponding to frequencies ω_s and ω_i, but the individual photons are in an equal amplitude superposition states of the two orthogonal polarizations *H* and *V*. Hence, this state contains one photon at frequency ω_s and one photon at frequency ω_i; at the same time the polarization states of the photons at frequency ω_s and ω_i are not defined.

This implies that a measurement of the polarization state of the photon corresponding to frequency ω_s will not give a precise value and similarly the measurement of the polarization state of the photon at frequency ω_i will not give a precise value. This can also be seen from Exercise 17.1.

Exercise 17.1 Obtain the total probability that the photon at frequency ω_s is horizontally polarized irrespective of the polarization state of the photon at frequency ω_i.

Solution The total probability that the photon at frequency ω_s is horizontally polarized irrespective of the polarization state of the photon at frequency ω_i is the sum of the probability that the photon at frequency ω_s is horizontally polarized and photon at frequency ω_i is horizontally polarized and the probability that the photon at frequency ω_s is horizontally polarized and the photon at frequency ω_i is vertically polarized.

The probability that the photon at frequency ω_s is in the horizontal polarization state (mode # 1) and the photon at frequency ω_i is also in the horizontal polarization state (mode # 3) is given by

$$|\langle 1_1, 1_3|\psi\rangle|^2 = \frac{1}{2}|\langle 1_1, 1_3|1_1, 0_2, 0_3, 1_4\rangle - \langle 1_1, 1_3|0_1, 1_2, 1_3, 0_4\rangle|^2 = 0$$

The probability that the photon at frequency ω_s is in the horizontal polarization state (mode # 1) and the photon at frequency ω_i is in the vertical polarization state (mode # 4) is given by

$$|\langle 1_1, 1_4|\psi\rangle|^2 = \frac{1}{2}|\langle 1_1, 1_4|1_1, 0_2, 0_3, 1_4\rangle - \langle 1_1, 1_4|0_1, 1_2, 1_3, 0_4\rangle|^2 = 1/2$$

Thus, the total probability that photon at frequency ω_s is horizontally polarized and the photon with frequency ω_i is *either* horizontally or vertically polarized is ½.

Similarly it can be shown that the probability that photon at frequency ω_s is vertically polarized (mode # 2) and the photon with frequency ω_i is either horizontally (mode # 3) or vertically polarized (mode # 4) is ½.

Since modes 1 and 2 correspond to the same frequency ω_s but with H and V polarization states and modes 3 and 4 correspond to the same frequency $\omega_{i,}$ but with H and V polarization states, we can write the state given in Eq. (17.6) in a more compact form

$$|\psi\rangle = \frac{1}{\sqrt{2}}(|H_s, V_i\rangle - |V_s, H_i\rangle) \tag{17.8}$$

with the subscripts representing the frequency and the letters H and V corresponding to the polarization states. Here $|H_s, V_i\rangle$ is a shortened form for the state $|1_1, 0_2, 0_3, 1_4\rangle$ and $|V_s, H_i\rangle$ is a shortened form for the state $|0_1, 1_2, 1_3, 0_4\rangle$. In the state defined in Eq. (17.8), only the modes that are occupied are mentioned; all other modes are empty.

First note that the ket represented by Eq. (17.8) cannot be written as a product of kets belonging to signal and idler photons (see Exercise 17.2). Such a state is referred to as a *polarization entangled* state, since the states of polarization of the two photons are quantum correlated. In Chap. 22, we will discuss techniques to generate such polarization entangled states.

Exercise 17.2 Show that the state given by Eq. (17.8) cannot be written as a product of the states of signal and idler photons.

Solution The general polarization state of the signal can be written as

$$|\psi_s\rangle = \alpha_s|H_s\rangle + \beta_s|V_s\rangle$$

and that of the idler can be written as

$$|\psi_i\rangle = \alpha_i|H_i\rangle + \beta_i|V_i\rangle$$

Here $\alpha_s^2 + \beta_s^2 = 1$ and $\alpha_i^2 + \beta_i^2 = 1$. Thus, the most general two state describing the signal and idler will be

$$|\psi\rangle = |\psi_s\rangle|\psi_i\rangle = (\alpha_s|H_s\rangle + \beta_s|V_s\rangle)(\alpha_i|H_i\rangle + \beta_i|V_i\rangle)$$

which can be written as

$$|\psi\rangle = \alpha_s\alpha_i|H_s\rangle|H_i\rangle + \alpha_s\beta_i|H_s\rangle|V_i\rangle + \beta_s\alpha_i|V_s\rangle|H_i\rangle + \beta_s\beta_i|V_s\rangle|V_i\rangle \quad (17.9)$$

Comparing Eqs. (17.9) and (17.8), we obtain the following equations:

$$\alpha_s\alpha_i = 0; \quad \beta_s\beta_i = 0; \quad \alpha_s\beta_i = \frac{1}{\sqrt{2}}; \quad \beta_s\alpha_i = -\frac{1}{\sqrt{2}}$$

These four equations require that $\alpha_s\beta_i\beta_s\alpha_i = \frac{1}{2}$ which is inconsistent with the other two equations. Hence, we cannot represent the state given by Eq. (17.8) as a product state.

We will now discuss the meaning of polarization entanglement.

17.3 Polarization States of Photons at Frequencies ω_s and ω_i

In order to show that the polarization state of each of the individual photons at frequencies ω_s and ω_i are not defined, we assume that we separate the two photons using a dichroic beam splitter which transmits frequency ω_s and reflects frequency ω_i (see Fig. 17.2). We pass the photon at frequency ω_s through an analyzer oriented

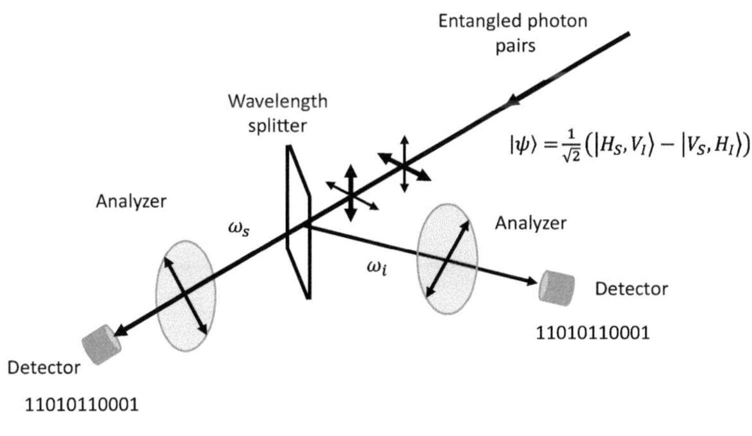

Fig. 17.2 Polarization entangled photon pairs at frequencies corresponding to ω_s and ω_i are incident on a wavelength splitter which reflects photons corresponding to frequency ω_i and passes photons corresponding to frequency ω_s. These photons then pass through appropriately oriented analyzers which are detected and are sent to an electronic coincidence counter

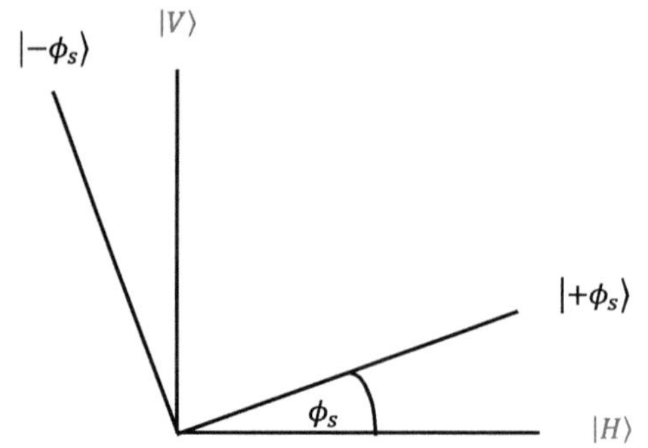

Fig. 17.3 The orientation of the $|+\phi_s\rangle - |-\phi_s\rangle$ basis with respect to $|H\rangle$-$|V$ basis

along a direction making an angle ϕ_s with the horizontal direction and determine the probability of the photon passing through the analyzer as a function of the angle ϕ_s.

In order to perform this calculation, we first represent the polarization making an angle ϕ_s with the horizontal by the ket vector $|+\phi_s\rangle$ where the subscript s refers to frequency ω_s. The corresponding orthogonal state making an angle of $\frac{\pi}{2} + \phi_s$ will be represented by the ket vector $|-\phi_s\rangle$. We express the new basis set in terms of the basis set of H and V orientation as follows (see Fig. 17.3):

$$|H_s\rangle = |+\phi_s\rangle\ \cos\phi_s - |-\phi_s\rangle\ \sin\phi_s \tag{17.10a}$$

$$|V_s\rangle = |+\phi_s\rangle\ \sin\phi_s + |-\phi_s\rangle\ \cos\phi_s \tag{17.10b}$$

Now, the probability that the photon at frequency ω_s passes through the analyzer oriented at an angle ϕ_s with the horizontal and the photon at frequency ω_i will be vertically polarized will be

$$|\langle+\phi_s, V_i|\psi\rangle|^2 = \frac{1}{2}|\langle+\phi_s, V_i|H_s, V_i\rangle - \langle+\phi_s, V_i|V_s, H_i\rangle|^2$$

which can be written as

$$\begin{aligned}|\langle+\phi_s, V_i|\psi\rangle|^2 &= \frac{1}{2}|\langle+\phi_s|H_s\rangle\langle V_i|V_i\rangle - \langle+\phi_s|V_s\rangle\langle V_i|H_i\rangle|^2 \\ &= \frac{1}{2}|\cos\phi_s \times 1 - \sin\phi_s \times 0|^2 = \frac{1}{2}\cos^2\phi_s\end{aligned} \tag{17.11}$$

where we have used Eqs. (17.10a) and (17.10b).

Similarly the probability that the photon at frequency ω_s passes through the analyzer oriented at an angle ϕ_s with the horizontal and the photon at frequency ω_i will be horizontally polarized will be

$$|\langle+\phi_s, H_i|\psi\rangle|^2 = \frac{1}{2}|\langle+\phi_s, H_i|H_s, V_i\rangle - \langle+\phi_s, H_i|V_s, H_i\rangle|^2 = \frac{1}{2}\sin^2\phi_s \tag{17.12}$$

Hence, the total probability that the photon at frequency ω_s passes through the analyzer oriented at an angle ϕ_s with the horizontal irrespective of the polarization state of the photon at frequency ω_i would be

$$|\langle +\phi_s, V_i|\psi\rangle|^2 + |\langle +\phi_s, H_i|\psi\rangle|^2 = \frac{1}{2} \tag{17.13}$$

independent of the angle ϕ_s. Thus, no matter what orientation of the analyzer one chooses, the probability of the photon at frequency ω_s to pass through the analyzer is ½. This implies that the polarization state of the photon at frequency ω_s is completely undefined. Although the quantum state represented by the first and second terms in the entangled state given by Eq. (17.8) have photons with well-defined polarization states, in the superposition state represented by Eq. (17.8) the state of polarization of the two individual photons at the two frequencies is undefined. Note that the photon at frequency ω_s is not in circular polarization state or in an unpolarized state both of which will give the same result as in this experiment.

Similarly one can show that the polarization state of the photon at frequency ω_i is also completely undefined.

At the same time, we will see that although the polarization state of the individual photons are not defined, the two photons are in orthogonal polarization states expressed in any orthogonal basis of polarization.

Note also from the above discussion that if one were to plot the ϕ_s dependence of the probability of the photon at frequency ω_s passing through the analyzer oriented at an angle ϕ_s and the photon at frequency ω_i passing through an analyzer oriented in the vertical direction or in the horizontal direction, the variation will be described, respectively, by Eqs. (17.11) and (17.12) showing a sinusoidal variation.

We now wish to analyze as to what happens if the orientation of the analyzer passing the photons at frequency ω_i is changed from being either horizontal or vertical to an arbitrary orientation.

Consider the situation when the analyzer passing the signal photon makes an angle ϕ_s with the horizontal, while the analyzer passing the idler photon makes an angle θ_i with the horizontal. The probability that the two photons pass through their respective analyzers is given by

$$P_{\phi\theta} = |\langle +\phi_s, +\theta_i|\psi\rangle|^2 = \frac{1}{2}|\langle +\phi_s, +\theta_i|(|H_s, V_i\rangle - |V_s, H_i\rangle)\rangle|^2$$

Now, similar to Eqs. (17.10a) and (17.10b), we can write

$$|H_i\rangle = |+\theta_i\rangle \cos\theta_i - |-\theta_i\rangle \sin\theta_i \tag{17.14a}$$

$$|V_i\rangle = |+\theta_i\rangle \sin\theta_i + |-\theta_i\rangle \cos\theta_i \tag{17.14b}$$

Using Eqs. (17.10a), (17.10b), (17.14a), and (17.14b), we have

$$P_{\phi\theta} = \frac{1}{2}|\langle +\phi_s, +\theta_i|H_s, V_i\rangle - \langle +\phi_s, +\theta_i|V_s, H_i\rangle|^2 = \frac{1}{2}|\cos\phi_s \sin\theta_i - \sin\phi_s \cos\theta_i|^2$$
$$= \frac{1}{2}\sin^2(\phi_s - \theta_i) \tag{17.15}$$

where we have used the fact that (see Eqs. 17.10a, 17.10b, 17.14a, and 17.14b)

$$\langle +\phi_s, +\theta_i|H_s, V_i\rangle = \langle +\phi_s|H_s\rangle\langle +\theta_i|V_i\rangle = \cos\phi_s \sin\theta_i \tag{17.16}$$

and

$$\langle +\phi_s, +\theta_i|V_s, H_i\rangle = \langle +\phi_s|V_s\rangle\langle +\theta_i|H_i\rangle = \sin\phi_s \cos\theta_i \tag{17.17}$$

Note that the probability $P_{\phi\theta}$ depends only on the difference ($\phi_s - \theta_i$). Note also from Eq. (17.15) that if the pass axes of the analyzers passing the signal and idler photons are parallel to each other (no matter what their independent orientations are), then $\phi_s - \theta_i = 0$ and the probability $P_{\phi\theta} = 0$. This implies that written in the horizontal–vertical polarization basis, the two photons are always in orthogonal polarization states. Similar results can also be shown in any other orthogonal basis of polarization such as circular basis or linear basis rotated from horizontal–vertical orientations. Also if the two analyzers have their pass axes orthogonal to each other (no matter what their independent orientations are), then $\phi_s - \theta_i = \frac{\pi}{2}$ and $P_{\phi\theta} = \frac{1}{2}$.

Exercise 17.3 Show that the probability that neither the signal photon passes through nor the idler photon passes through the analyzer (representing it here by $Q_{\phi\theta}$) is again given by $Q_{\phi\theta} = \frac{1}{2}\sin^2(\phi_s - \theta_i)$.

Solution The probability that neither of the two photons pass through the respective analyzers oriented at angles ϕ_s and θ_i with the horizontal is given by

$$Q_{\phi\theta} = |\langle -\phi_s, -\theta_i|\psi\rangle|^2 = \frac{1}{2}|\langle -\phi_s, -\theta_i|(|H_s, V_i\rangle - |V_s, H_i\rangle)\rangle|^2$$

and we obtain

$$Q_{\phi\theta} = \frac{1}{2}|\langle -\phi_s, -\theta_i|H_s, V_i\rangle - \langle -\phi_s, -\theta_i|V_s, H_i\rangle|^2$$
$$= \frac{1}{2}|(-\sin\phi_s)\cos\theta_i - \cos\phi_s(-\sin\theta_i)|^2 = \frac{1}{2}\sin^2(\phi_s - \theta_i) \tag{17.18}$$

From Exercise 17.3, we find that the probability that neither of the two photons pass through their respective analyzers is also $\frac{1}{2}\sin^2(\phi_s - \theta_i)$.

Hence, the probability that either both photons pass through their analyzers or both the photons do not pass through the analyzers is the sum of the two probabilities and corresponds to the coincidence probability given by

$$P_{coinc} = P_{\phi\theta} + Q_{\phi\theta} = \sin^2(\phi_s - \theta_i) \tag{17.19}$$

Figure 17.4 shows the variation in P_{coinc} given by Eq. (17.19) as a function of θ_i for some specific values of ϕ_s. The coincidence probability varies between 0 and 1 as θ_i varies with the curve shifting as the value of ϕ_s is changed. If the two analyzers are oriented with their pass axes orthogonal to each other, then the total coincidence probability will be $P_{coinc} = 1$ showing perfect correlation.

Experimental coincidence measurements on polarization entangled photon pairs show the behavior represented by Eq. (17.19) and are a standard technique used to verify polarization entanglement between photons.

It should be noted that the state given by Eq. (17.8) is very different from the following state:

$$|\psi\rangle = \frac{1}{2}(|H_s\rangle + |V_s\rangle)(|H_i\rangle + |V_i\rangle) = \frac{1}{2}(|H_s, H_i\rangle + |H_s, V_i\rangle + |V_s, H_i\rangle + |V_s, V_i\rangle) \tag{17.20}$$

which is a product state between the signal and the idler unlike the polarization entangled state. The two photons one at ω_s and the other at ω_i are not polarization entangled. A measurement of state of polarization carried out on the photon at ω_s has no influence on the measurement carried out on the state of polarization of the photon at ω_i. This is in contrast to that of the polarization entangled state.

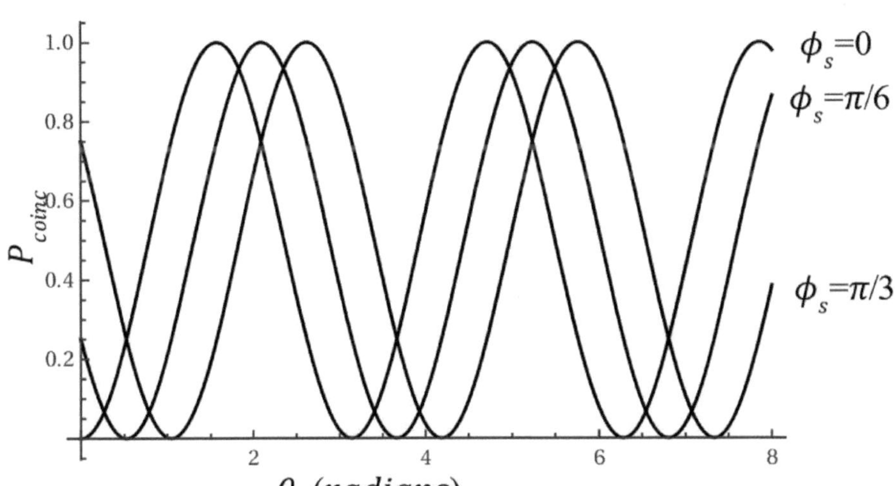

Fig. 17.4 Variation in coincidence probability with the angle θ_i that the analyzer on the path of the idler makes with the horizontal for different values of the angle ϕ_s that the analyzer on the path of the signal photon makes with the horizontal

Exercise 17.4 Consider a state given by

$$|\psi\rangle = |H_s\rangle|V_i\rangle \tag{17.21}$$

which is a product state of a horizontally polarized signal photon and a vertically polarized idler photon. Obtain the coincidence probability if the signal and idler photons are passed through analyzers oriented at angles $+\phi_s$ and $+\theta_i$ with the horizontal.

Solution The state given by Eq. (17.21) is a product state of the signal and idler photons and hence not polarization entangled. The probability that both the signal and idler photons pass through their respective analyzers is given by

$$P_{\phi\theta} = |\langle +\phi_s, +\theta_i|\psi\rangle|^2 = |\langle +\phi_s|H_s\rangle\langle +\theta_i|V_i\rangle|^2 = \cos^2\phi_s \sin^2\theta_i$$

Similarly the probability that both the signal and idler photons do not pass through their respective analyzers is given by

$$Q_{\phi\theta} = |\langle -\phi_s, -\theta_i|\psi\rangle|^2 = |\langle -\phi_s|H_s\rangle\langle -\theta_i|V_i\rangle|^2 = \sin^2\phi_s \cos^2\theta_i$$

Hence, the coincidence probability is given by

$$P_{coinc} = P_{\phi\theta} + Q_{\phi\theta} = \cos^2\phi_s \sin^2\theta_i + \sin^2\phi_s \cos^2\theta_i \tag{17.22}$$

which is very different from Eq. (17.19) when the two photons are polarization entangled.

17.4 Impact of Measuring the State of Polarization of the Signal Photon on the Idler Photon or Vice Versa

Let us assume that the measurement on photon at frequency ω_s is carried out first and assume that the photon passes through an analyzer oriented along the horizontal direction. This implies that the photon at frequency ω_s is projected into the state $|H_s\rangle$. This measurement projects the entire quantum state to the following

$$\langle H_s|\psi\rangle = \frac{1}{\sqrt{2}}\{\langle H_s|H_s, V_i\rangle - \langle H_s|V_s, H_i\rangle\rangle\} = |V_i\rangle \tag{17.23}$$

where the output state has been normalized. This implies that when the photon at frequency ω_s is detected to have horizontal polarization, the photon at frequency ω_i whose state of polarization was undefined before the measurement was carried out on the photon at frequency ω_s becomes definite and the photon at frequency ω_i will be found to have vertical polarization.

Similarly if photon at frequency ω_s does not pass through the analyzer oriented with its pass axis along the horizontal direction, this implies that the photon has been projected on to the state $|V_s\rangle$. This measurement projects the entire quantum state to the following:

$$\langle V_s|\psi\rangle = \frac{1}{\sqrt{2}}\{\langle V_s|H_s, V_i\rangle - \langle V_s|V_s, H_i\rangle\rangle\} = |H_i\rangle \tag{17.24}$$

This implies that when the photon at frequency ω_s is detected to have vertical polarization, the photon at frequency ω_i whose state of polarization was undefined before the measurement on photon at frequency ω_s becomes definite and the photon at frequency ω_i will be found to have horizontal polarization.

What happens if the orientation of both the analyzers is changed to another orientation but still keeping them orthogonal to each other. For example, let us assume that the analyzer for the frequency ω_s corresponds to $|+\theta\rangle$ and the analyzer for the frequency ω_i corresponds to $|-\theta\rangle$ (see Fig. 17.3).

We can use Eqs. (17.10a) and (17.10b) to express the entangled state given by Eq. (17.8) in the $|+\theta\rangle - |-\theta\rangle$ basis. For this, we write

$$|H_s\rangle = (\cos\theta|+\theta_s\rangle - \sin\theta|-\theta_s\rangle) \tag{17.25}$$

$$|V_s\rangle = (\sin\theta|+\theta_s\rangle + \cos\theta|-\theta_s\rangle) \tag{17.26}$$

$$|H_i\rangle = \left(\cos\theta|+\theta_i\rangle - \sin\theta\lfloor-\theta_i\rangle\right) \tag{17.27}$$

$$|V_i\rangle = \left(\sin\theta|+\theta_i\rangle + \cos\theta|-\theta_i\rangle\right) \tag{17.28}$$

Now the given entangled state would be

$$\begin{aligned} |\psi\rangle = \frac{1}{\sqrt{2}}(|H_s, V_i\rangle - |V_s, H_i\rangle) &= \frac{1}{\sqrt{2}}\left\{\begin{matrix}(\cos\theta|+\theta_s\rangle - \sin\theta|-\theta_s\rangle)(\sin\theta|+\theta_i\rangle + \cos\theta|-\theta_i\rangle) - \\ (\sin\theta|+\theta_s\rangle + \cos\theta|-\theta_s\rangle)(\cos\theta|+\theta_i\rangle - \sin\theta|-\theta_i\rangle)\end{matrix}\right\} \\ &= \frac{1}{\sqrt{2}}(|+\theta_s, -\theta_i\rangle - |-\theta_s, +\theta_i\rangle) \end{aligned} \tag{17.29}$$

which is again an entangled state between the $|+\theta_s\rangle$ and $|-\theta_i\rangle$ polarization states.

Thus, if the photon at frequency ω_s passes through the analyzer oriented at $+\theta$ (i.e., it gets projected into the state $|+\theta_s\rangle$), then photon at frequency ω_i is automatically projected on to the state $|-\theta_i\rangle$. Similarly if the photon at frequency ω_s does not pass through the analyzer, then it gets projected in to the state $|-\theta_s\rangle$ and photon at frequency ω_i gets projected to the state $|+\theta_i\rangle$.

One can check that the entangled state given by Eq. (17.8) can be expressed in any orthogonal polarization basis (linear basis, circular basis, elliptical basis, etc.) and the state will be entangled in that basis. This is shown pictorially in Fig. 17.5.

$$|\psi\rangle = \frac{1}{\sqrt{2}}\left(|H_s, V_i\rangle - |V_s, H_i\rangle\right)$$

$$|\psi\rangle = \frac{1}{\sqrt{2}}\left(|+\theta_s, -\theta_i\rangle - |-\theta_s, +\theta_i\rangle\right)$$

$$|\psi\rangle = \frac{1}{\sqrt{2}}\left(|+RCP_s, LCP_i\rangle - |LCP_s, RCP_i\rangle\right)$$

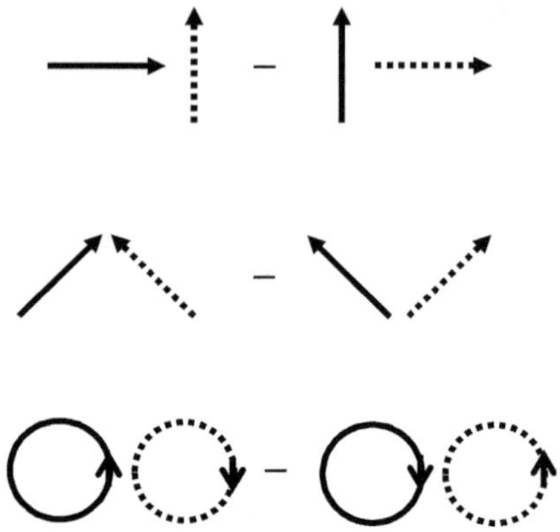

Fig. 17.5 The polarization entangled state in different basis states. Solid curves correspond to the signal photon and dashed to idler photons. RCP and LCP represent right circular polarization and left circular polarization, respectively

Hence, measurement of polarization carried out on any one of the photons instantaneously projects the other photon on to an orthogonal polarization state.

In Chap. 23, we will discuss techniques to generate polarization entangled photon pairs. One of the techniques for direct generation involves satisfying two quasi phase matching conditions in which a pump photon having ordinary polarization generates a signal-idler pair with either of the following combinations: (a) signal photon in horizontal polarization state and idler photon in vertical polarization state and (b) signal photon in vertical polarization state and idler photon in horizontal polarization state. In a classical picture, this will be interpreted as either the output pair corresponds to $|H_s, V_i\rangle$ *or to* $|V_s, H_i\rangle$. In the quantum mechanical picture if the two processes are indistinguishable, the output will be in a superposition state of the two possibilities with the addition of amplitudes. This is similar to the analysis of Young's double slit experiment when single photons are incident on the setup, classically our interpretation would be that the photon *either passes through slit 1 or through slit 2*. In this case, if there is no information as to which slit the photon has passed through, the output state of the photon is a superposition of the photon passing through either of the two slits. Similarly in the case of generation of photon pairs using two equally probable routes which are indistinguishable, the output state would be a superposition of the two states $|H_s, V_i\rangle$ *and* $|V_s, H_i\rangle$ and not $|H_s, V_i\rangle$ *or* $|V_s, H_i\rangle$. If the two paths are fundamentally indistinguishable, the output would be a coherent superposition of the probability amplitudes and will exhibit interference. We now discuss how a measurement of correlation between the measurement on the two photons differentiates between the two possibilities.

Consider a classical picture in which the spontaneous parametric down conversion process for the following two processes is equally probable and is indistinguishable: (a) signal photon is horizontally polarized and idler photon is vertically polarized represented by $|H_s, V_i\rangle$ and (b) signal photon is vertically polarized and idler photon is horizontally polarized represented by $|V_s, H_i\rangle$. Assume that the two photons which have different frequencies are spatially separated and passed through two linear analyzers referred to as A_s and A_i, respectively. We will now evaluate the correlations between the probabilities of the two photons passing through the two analyzers oriented perpendicularly but with different orientations.

If A_s has horizontal orientation and A_i has vertical orientation, then if the pair corresponds to $|H_s, V_i\rangle$ state, then both photons will pass through their respective analyzers. If the pair corresponds to $|V_s, H_i\rangle$, then neither of the photons will pass through the two analyzers and the two measurements would be fully correlated. If the orientations of A_s and A_i correspond to vertical and horizontal orientations, then similar arguments would hold and the measurements would be fully correlated.

We now choose the orientations with A_s along +45° to the horizontal and A_i to be along −45° to the horizontal. Assuming equal probabilities of $\frac{1}{2}$ for the photon pairs to be either in the state $|H_s, V_i\rangle$ or in the state $|V_s, H_i\rangle$, if the pair corresponds to $|H_s, V_i\rangle$ state, then the coincidence probability would be

$$\begin{aligned} P_{H+V-} &= \frac{1}{2}\left[\left|\left\langle +\left(\frac{\pi}{4}\right)_s \middle| H_s\right\rangle\left\langle -\left(\frac{\pi}{4}\right)_i \middle| V_i\right\rangle\right|^2 + \frac{1}{2}\left|\left\langle -\left(\frac{\pi}{4}\right)_s \middle| H_s\right\rangle\left\langle +\left(\frac{\pi}{4}\right)_i \middle| V_i\right\rangle\right|^2\right] \\ &= \frac{1}{2}\left(\cos\frac{\pi}{4}\cos\frac{\pi}{4}\right)^2 + \frac{1}{2}\left(\sin\frac{\pi}{4}\sin\frac{\pi}{4}\right)^2 = \frac{1}{4} \end{aligned} \tag{17.30}$$

Similarly if the pair corresponds to $|V_S, H_I\rangle$, then the coincidence probability would be

$$P_{V+H-} = \frac{1}{2}\left[\left|\left\langle +\left(\frac{\pi}{4}\right)_s \middle| V_s\right\rangle\left\langle -\left(\frac{\pi}{4}\right)_i \middle| H_i\right\rangle\right|^2 + \frac{1}{2}\left|\left\langle -\left(\frac{\pi}{4}\right)_s \middle| V_s\right\rangle\left\langle +\left(\frac{\pi}{4}\right)_i \middle| H_i\right\rangle\right|^2\right] = \frac{1}{4} \tag{17.31}$$

Hence, the total coincidence probability would be $P_{H+V-} + P_{V+H-} = \frac{1}{2}$ as expected just like in two independent coin tosses.

Thus, although the two analyzers are orthogonal to each other, the coincidence probability is not unity but reduces since the two photons are not polarization entangled.

If instead of having the analyzers at +45° and −45°, if the analyzer on the path of the signal photon makes an angle ϕ_s with the horizontal while the analyzer on the path of the idler photon makes an angle θ_i with the horizontal, then it can be shown that the total probability of both photons passing through their respective analyzers is the sum of the two probabilities and there is no interference term:

$$\begin{aligned} P_{\phi\theta} &= \frac{1}{2}\left[|\langle +\phi_s, +\theta_i | H_s, V_i\rangle|^2 + |\langle +\phi_s, +\theta_i | V_s, H_i\rangle|^2\right] \\ &= \frac{1}{2}\left(\cos^2\phi_s \sin^2\theta_i + \sin^2\phi_s \cos^2\theta_i\right) \end{aligned} \tag{17.32a}$$

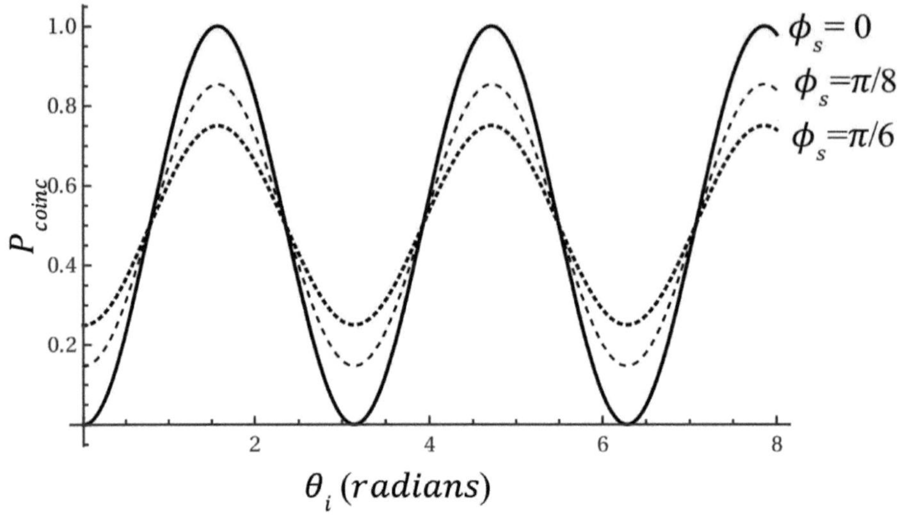

Fig. 17.6 Variation in the coincidence probability P_{coinc} with θ_i for different values of ϕ_s

Similarly the probability that neither of the photons pass through their respective analyzers will be

$$\begin{aligned} Q_{\phi\theta} &= \frac{1}{2}\left[|\langle -\phi_s, -\theta_i|H_s, V_i\rangle|^2 + |\langle -\phi_s, -\theta_i|V_s, H_i\rangle|^2\right] \\ &= \frac{1}{2}\left(\cos^2\phi_s \ \sin^2\theta_i + \sin^2\phi_s \ \cos^2\theta_i\right) \end{aligned} \tag{17.32b}$$

Hence, the total coincidence probability will be given by

$$P_{coinc} = P_{\phi\theta} + Q_{\phi\theta} = \cos^2\phi_s \ \sin^2\theta_i + \sin^2\phi_s \ \cos^2\theta_i \tag{17.33}$$

Figure 17.6 shows the variation in P_{coin} with θ_i for different values of $\phi_s = 0$, $\phi_s = \frac{\pi}{8}$ and $\phi_s = \frac{\pi}{6}$. Note that as ϕ_s changes the correlation drops and the peaks do not shift. Note that in this case the probability depends on the individual values of ϕ_s and θ_i and not on the difference between the two angles as in the case when the photons were entangled in polarization. Compare this variation with the variation in the coincidence probability with polarization entangled photon pairs (see Fig. 17.4).

If the two analyzers are always kept perpendicular to each other, then $\theta_i = \frac{\pi}{2} + \phi_s$ and we will obtain $P_{coinc} = \cos^4\phi_s + \sin^4\phi_s$ which depends on the angle ϕ_s.

In fact, Eq. (17.32a) can be written as

$$P_{\phi\theta} = \langle +\phi_s, +\theta_i | \frac{1}{2} (|H_s, V_i\rangle\langle H_s, V_i| + |V_s, H_i\rangle\langle V_s, H_i|) | +\phi_s, +\theta_i\rangle \\ = \langle +\phi_s, +\theta_i | \widehat{\rho} | +\phi_s, +\theta_i\rangle \tag{17.34}$$

where

$$\widehat{\rho} = \frac{1}{2} (|H_s, V_i\rangle\langle H_s, V_i| + |V_s, H_i\rangle\langle V_s, H_i|) \tag{17.35}$$

is referred to as the density matrix of a mixed state of $|H_s, V_i\rangle$ and $|V_s, H_i\rangle$ with equal probabilities.

Note that in this mixed state the probability of detecting signal photon with the analyzer oriented at an angle ϕ_S with the horizontal is given by

$$P_\phi = \langle +\phi_s, +\theta_i | \widehat{\rho} | +\phi_s, +\theta_i\rangle + \langle +\phi_s, -\theta_i | \widehat{\rho} | +\phi_s, -\theta_i\rangle$$

Evaluating the two terms, we obtain

$$P_\phi = \frac{1}{2} \left(\cos^2\phi_s \, \sin^2\theta_i + \sin^2\phi_s \, \cos^2\theta_i\right) + \frac{1}{2} \left(\cos^2\phi_s \, \cos^2\theta_i + \sin^2\phi_s \, \sin^2\theta_i\right) \\ = \frac{1}{2} \tag{17.36}$$

independent of the angle ϕ_s which implies that the signal photon has undefined polarization like in the entangled state. However, note that $P_{\phi\theta}$ is different for the entangled state and the mixed state.

17.5 Bell States

Bell states (named after John Bell who proposed the states) are maximally entangled superpositions of two qubit states. There are four such orthonormal Bell states and any two-qubit state can be expressed as a linear combination of the four orthogonal Bell states. The four Bell states are given below with the two qubits represented by the subscripts s and i:

$$|\phi^+\rangle = \frac{1}{\sqrt{2}} (|0_s, 0_i\rangle + |1_s, 1_i\rangle) \tag{17.37}$$

$$|\phi^-\rangle = \frac{1}{\sqrt{2}} (|0_s, 0_i\rangle - |1_s, 1_i\rangle) \tag{17.38}$$

$$|\psi^+\rangle = \frac{1}{\sqrt{2}}(|0_s, 1_i\rangle + |1_s, 0_i\rangle) \tag{17.39}$$

$$|\psi^-\rangle = \frac{1}{\sqrt{2}}(|0_s, 1_i\rangle - |1_s, 0_i\rangle) \tag{17.40}$$

Here 0 and 1 could represent orthogonal states of any specific degree of freedom of the qubit. Thus, if we are considering photons, 0 and 1 could correspond to the horizontal (H) and vertical (V) polarization states, respectively, or could correspond to the fundamental mode (F) and first excited spatial mode (E) of the waveguide (see Sect. 17.6), or to two different frequencies, etc.

These four states are extremely important two qubit states and are used extensively in quantum information science, for example, in quantum computing, quantum teleportation, etc.

The four Bell states form a set of orthogonal two bit states and any two bit state can be expressed in terms of the four Bell states. For example, if we consider the following two bit state

$$|\psi\rangle = \alpha|0_s, 1_i\rangle + \beta|1_s, 0_i\rangle \tag{17.41}$$

The form of the state suggests that it must be a linear combination of $|\psi^+\rangle$ and $|\psi^-\rangle$. Writing

$$|\psi\rangle = \alpha|0_s, 1_i\rangle + \beta|1_s, 0_i\rangle = a|\psi^+\rangle + b|\psi^-\rangle \tag{17.42}$$

We can obtain the values of a and b. Thus, we can write

$$|\psi\rangle = \frac{1}{\sqrt{2}}(\alpha+\beta)|\psi^+\rangle + \frac{1}{\sqrt{2}}(\alpha-\beta)|\psi^-\rangle \tag{17.43}$$

Exercise 17.5 Show that the four Bell states are orthogonal to each other.

Solution Consider the following product:

$$\begin{aligned}\langle\phi^-|\phi^+\rangle &= \frac{1}{2}(\langle 0_s, 0_i| - \langle 1_s, 1_i|)(|0_s, 0_i\rangle + |1_s, 1_i\rangle)\\ &= \frac{1}{2}[\langle 0_s, 0_i|0_s, 0_i\rangle + \langle 0_s, 0_i|1_s, 1_i\rangle - \langle 1_s, 1_i|0_s, 0_i\rangle - \langle 1_s, 1_i|1_s, 1_i\rangle]\\ &= \frac{1}{2}(1+0-0-1) = 0\end{aligned}$$

In a similar fashion, it is possible to show that the four Bell states are orthogonal to each other.

17.6 Spatial Mode Entangled State

In the above discussions, we considered a pair of photons having two different frequencies and propagating along the same direction with entanglement in the degree of freedom of polarization. It is also possible to have entanglement in other degrees of freedom.

For example, it is possible that the two photons have the same polarization state but are in two different orthogonal spatial modes, such as the fundamental mode and the first excited mode of an optical waveguide or have two different propagation directions in free space. If we represent the spatial modes by symbols F (corresponding to fundamental spatial mode) and E (corresponding to excited spatial mode), then a spatial mode entangled state will correspond to (corresponding to the $|\psi^-\rangle$ Bell state—see Eq. (17.40))

$$|\psi^-\rangle = \frac{1}{\sqrt{2}}(|F_s, E_i\rangle - |E_s, F_i\rangle) \tag{17.44}$$

which represents a spatial mode entangled state of signal (s) and idler (i) in the fundamental (F) mode or the excited (E) spatial modes of the waveguide. Figure 17.7 shows such an entangled state. In this state, the two photons will always be in orthogonal spatial mode basis. As in the case of polarization entanglement, the state is a coherent superposition of the two quantum states representing spatial modal degree of freedom.

Figure 17.8 shows another combination of spatial mode entangled states represented by (corresponding to the $|\phi^-\rangle$ Bell state)

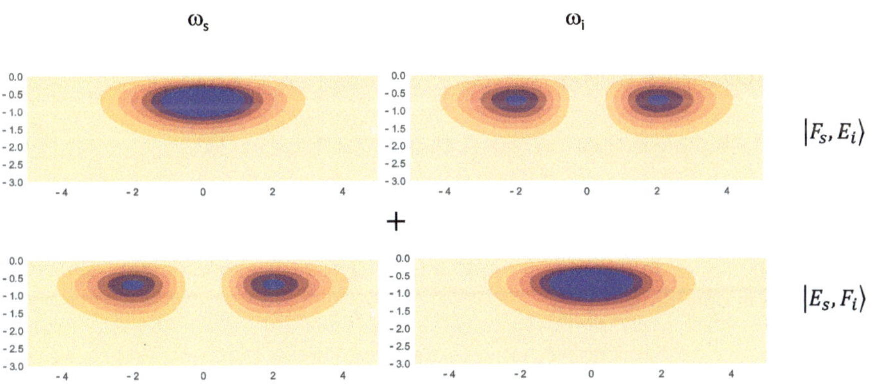

Fig. 17.7 A spatial mode entangled state given by Eq. (17.44) in a waveguide

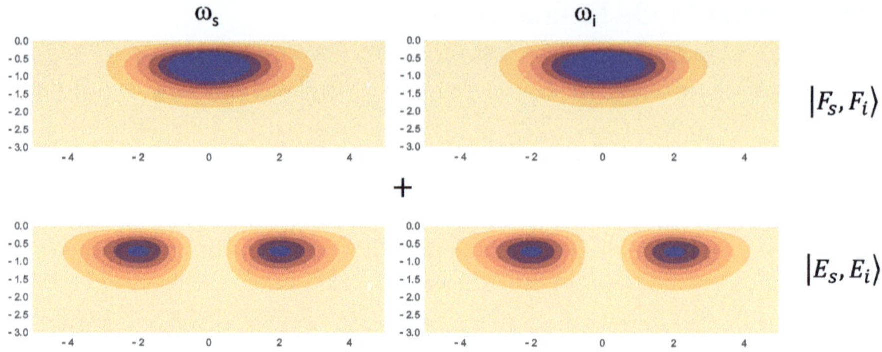

Fig. 17.8 A spatial mode entangled state given by Eq. (17.45) in a waveguide

$$|\phi^-\rangle = \frac{1}{\sqrt{2}}(|F_s, F_i\rangle - |E_s, E_i\rangle) \tag{17.45}$$

By using waveguides supporting multiple modes, it is possible to generate higher dimensional entanglement using the degree of freedom of transverse spatial modes in an optical waveguide.

More discussions on mode entangled states can be found in Ming et al. (2014a, b), Bharadwaj et al. (2015), and Feng et al. (2019).

17.7 Higher Dimensional Entanglement

In the above discussions, we had considered entangled states in which two photons and two modes are involved. It is also possible to entangle two photons in more than two modes leading to what is referred to as higher dimensional entanglement.

For example a state given by

$$|\psi\rangle = C_{00}|H_{s0}, V_{i0}\rangle + C_{11}|H_{s1}, V_{i1}\rangle + C_{22}|H_{s2}, V_{i2}\rangle \tag{17.46}$$

with H_{sn} representing a mode with horizontal polarization with a frequency ω_s in the spatial mode n $(n = 0, 1, 2)$ and V_{in} representing a mode with vertical polarization with a frequency ω_i in the spatial mode n $(n = 0, 1, 2)$. The coefficients satisfy the normalization condition

$$|C_{00}|^2 + |C_{11}|^2 + |C_{22}|^2 = 1 \tag{17.47}$$

In the state given by Eq. (17.46), there are two photons which are entangled in the spatial mode basis among three spatial modes represented by $n = 0$, $n = 1$ and $n = 2$. Such a state is said to have higher dimensional entanglement.

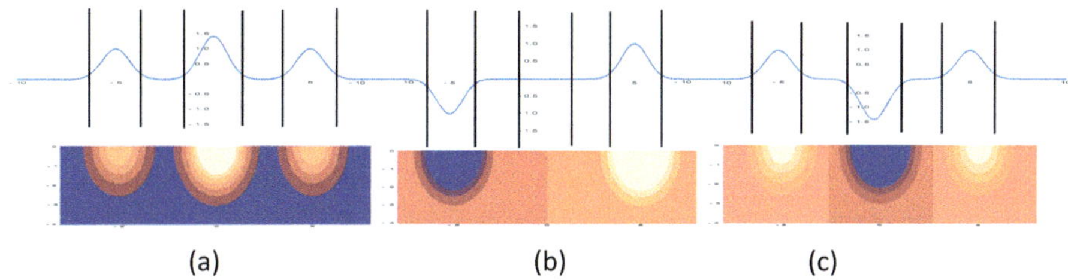

Fig. 17.9 The first three normal modes of a three waveguide directional coupler consisting of equally spaced identical waveguides. The fundamental and the second excited modes (**a**) and (**c**) are symmetric while the first excited mode (**b**) is an antisymmetric mode

It is possible to generate higher dimensional entangled states in waveguides or waveguide systems, such as directional couplers, supporting more than two modes. For example, multimode waveguides which support more than two modes or multiple single mode waveguides supporting more than two normal modes can be designed to generate higher dimensional entanglement. Compared to multimode waveguides, multiple single mode waveguides interacting through evanescent fields (directional couplers-see Chap. 5) provide us with a greater control of the propagation constants and field profiles (Bharadwaj et al. 2015). Figure 17.9 shows the possible three spatial modes in a three waveguide directional coupler consisting of three identical single mode waveguides separated by equal distances. Each of those waveguides is a single mode waveguide and thus the three waveguide directional coupler will support three guided modes, two symmetric (Fig. 17.9a, c), and one antisymmetric (Fig. 17.9b) modes (see Chap. 5).

In order to achieve higher dimensional entanglement in such a structure, notice that the three waveguide directional coupler will support three modes each of the horizontal and vertical polarization states. Depending on the geometry of the device and the input polarization state and the mode in which the pump photon is propagating different combinations of SPDC processes can be achieved and optimized.

Thus, if the pump photon has ordinary polarization (H polarization in a z–cut lithium niobate waveguide) and propagates as a fundamental normal mode in the three waveguide coupler, then using Type II SPDC process, it is possible to achieve any of the following photon pairs depending on the phase matching conditions satisfied: $|H_{s0}, V_{i0}\rangle$, $|H_{s1}, V_{i1}\rangle$, $|H_{s0}, V_{i2}\rangle$, $|H_{s2}, V_{i0}\rangle$ and $|H_{s2}, V_{i2}\rangle$. Note that the remaining combinations of symmetric–antisymmetric pairs of signal and idler are not possible due to parity conservation.

By an appropriate design of the waveguides and their separation, it is possible to choose the following three processes to have almost the same probability: $|H_{s0}, V_{i0}\rangle$ $|H_{s1}, V_{i1}\rangle$ and $|H_{s2}, V_{i2}\rangle$. In such a case, the output quantum state from the device will be a mode entangled state given by

$$|\psi\rangle \sim \int d\omega(C_{00}|H_{s0}, V_{i0}\rangle + C_{11}|H_{s1}, V_{i1}\rangle + C_{22}|H_{s2}, V_{i2}\rangle) \tag{17.48}$$

where C_{ii} ($i = 0, 1, 2$) depends on the overlap integrals of the interacting modes, the nonlinear coefficient that leads to the down conversion process, and the phase matching function. For a given domain reversed grating frequency K, the phase mismatches of the three processes are given by

$$\Delta\beta_{00} = \beta_{p0} - \beta_{s0} - \beta_{i0} - K \tag{17.49}$$

$$\Delta\beta_{11} = \beta_{p0} - \beta_{s1} - \beta_{i1} - K \tag{17.50}$$

$$\Delta\beta_{22} = \beta_{p0} - \beta_{s2} - \beta_{i2} - K \tag{17.51}$$

In Chap. 5, we showed that for a three waveguide directional coupler with identical waveguides and equal spacing the propagation constants of the fundamental, first excited and second excited modes are given by

$$\beta_{m1} = \beta_0 + \sqrt{2}\kappa \tag{17.52}$$

$$\beta_{m2} = \beta_0 \tag{17.53}$$

$$\beta_{m3} = \beta_0 - \sqrt{2}\kappa \tag{17.54}$$

where β_0 is the propagation constant of the fundamental mode of the individual waveguides and κ represents the coupling coefficient among two adjacent waveguides which depends on the spacing between the waveguides as well as the modal fields of the individual waveguides. Eqs. (17.52)–(17.54) can be used to obtain $\Delta\beta_{00}$, $\Delta\beta_{11}$, and $\Delta\beta_{22}$ by using the appropriate propagation constants of the individual waveguide modes at the pump, signal, and idler frequencies. These equations can then be used in designing the three waveguide directional coupler to achieve required characteristics for the generation of photon pairs with higher dimensional entanglement. By using a spectral filter to filter the output photons, one can ensure equality of the coefficients in Eq. (17.48) and also remove any spectral distinguishability between the three processes (Bharadwaj et al. 2015; Schaff et al. 2015). See also Schulz et al. (2024).

17.8 Hybrid Entangled State

It is also possible to entangle two photons in two different degrees of freedom (see, e.g., Ma et al. 2009). For example, it is possible to have two photons entangled between the spatial mode of one photon of frequency ω_s and polarization state of another photon at frequency ω_i. Such a state can be represented by

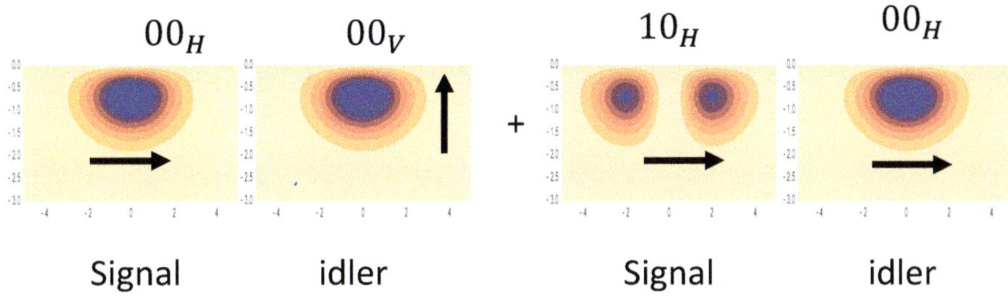

Fig. 17.10 Hybrid entangled state given by Eq. (17.55) in which the photons are entangled between the spatial degree of freedom and polarization degree of freedom. 00 corresponds to the fundamental mode and 10 to the first excited mode of the waveguide

$$|\psi\rangle = \frac{1}{\sqrt{2}}|H_{s0}, V_{i0}\rangle + \frac{1}{\sqrt{2}}|H_{s1}, H_{i0}\rangle \tag{17.55}$$

where H_{s0} corresponds to a horizontally polarized mode at frequency ω_s in the fundamental spatial mode represented by the subscript 0, H_{s1} corresponds to a horizontally polarized mode at frequency ω_s in the first excited spatial mode represented by the subscript 1, V_{i0} corresponds to vertically polarized mode at frequency ω_i in the fundamental spatial mode represented by the subscript 0, and H_{i0} corresponds to horizontally polarized mode at frequency ω_i in the fundamental spatial mode represented by the subscript 0 (Bharadwaj and Thyagarajan 2018).

Figure 17.10 shows a diagrammatic representation of such an interesting quantum state.

It is interesting to find out as to what happens when the idler photon (with polarization degree of freedom) in the hybrid entangled state is passed through an analyzer whose orientation is changed. For this, we refer back to Eq. (17.10a) and (17.10b) and express the state given by Eq. (17.55) in the basis states of $|+\theta_i\rangle$ and $|-\theta_i\rangle$ of the idler. We recall Eqs. (17.10a) and (17.10b) and write for the relationship between the idler states in the $|H\rangle - |V\rangle$ polarization basis and $|+\theta\rangle - |-\theta\rangle$ polarization basis as

$$|H_{i0}\rangle = |+\theta_{i0}\rangle\cos\theta - |-\theta_{i0}\rangle\sin\theta \tag{17.56}$$

$$|V_{i0}\rangle = |+\theta_{i0}\rangle\sin\theta + |-\theta_{i0}\rangle\cos\theta \tag{17.57}$$

Expressed in terms of the new basis states, the hybrid entangled state given by Eq. (17.55) becomes

$$|\psi\rangle = \frac{1}{\sqrt{2}}|H_{s0}\rangle(|+\theta_{i0}\rangle\sin\theta + |-\theta_{i0}\rangle\cos\theta) + \frac{1}{\sqrt{2}}|H_{s1}\rangle \\ \times (|+\theta_{i0}\rangle\cos\theta - |-\theta_{i0}\rangle\sin\theta)$$

which on simplification gives

$$|\psi\rangle = \frac{1}{\sqrt{2}}|+\theta_{i0}\rangle(|H_{s0}\rangle \sin\theta + |H_{s1}\rangle \cos\theta)) + \frac{1}{\sqrt{2}}|-\theta_{i0}\rangle(|H_{s0}\rangle \cos\theta - |H_{s1}\rangle \sin\theta) \tag{17.58}$$

This implies that as the idler photon is passed through an analyzer which is rotated so that its pass axis is rotated leading to change in the angle θ, the transverse field distribution corresponding to the signal photon will change its superposition of the fundamental mode and the first excited mode leading to a change in the transverse intensity pattern. For example, if the idler is passed through an analyzer with its pass axis oriented along $|+\theta_{i0}\rangle$ direction, then when the idler photon passes through the analyzer, the signal photon is projected on to the state $|H_{s0}\rangle$ sin θ + $|H_{s1}\rangle$ cos θ. Hence, if $\theta = 0$, then the signal photon will generate a pattern corresponding to the first excited (10) spatial mode and if $\theta = \frac{\pi}{2}$, then the signal photon will generate a pattern corresponding to the fundamental (00) spatial mode. At other orientations, the pattern generated will be a superposition of the fundamental and first excited spatial modes. Thus, as the orientation of the analyzer passing the idler is changed, the spatial mode pattern that is generated by the signal photon will change. A similar experiment was carried out to achieve real-time imaging of entanglement (see, e.g., Fickler et al. 2013).

17.9 Hyperentangled State

In the above discussion, we have considered entanglement in one degree of freedom between the two entangled photons. It is also possible to simultaneously entangle two photons in multiple degrees of freedom leading to what is referred to as hyperentanglement (see, e.g., Kwiat 1997). For example, it is possible to entangle two photons in both the polarization and spatial modes. Such states find wide applications in quantum photonics (see, e.g., Nemirovsky-Levy et al. 2024)

As an example consider the following state:

$$\begin{aligned} |\psi\rangle &= |0_s, 0_i\rangle|H_s, V_i\rangle + |1_s, 1_i\rangle|H_s, V_i\rangle + |0_s, 0_i\rangle|V_s, H_i\rangle + |1_s, 1_i\rangle|V_s, H_i\rangle \\ &= (|0_s, 0_i\rangle + |1_s, 1_i\rangle) \otimes (|H_s, V_i\rangle + |V_s, H_i\rangle) \end{aligned} \tag{17.59}$$

where 0 and 1 correspond to the fundamental and first excited spatial modes of the waveguide, H and V represent horizontal and vertical polarization states, and subscripts s and i refer to frequencies ω_s and ω_i, respectively.

The above quantum state is a superposition of the following four states:

- Horizontally polarized signal with frequency ω_s in the fundamental mode and vertically polarized idler at frequency ω_i also in the fundamental mode: $|0_s, 0_i\rangle|H_s, V_i\rangle = |H_{s0}, V_{i0}\rangle$.

- Horizontally polarized signal with frequency ω_s in the first excited mode and vertically polarized idler at frequency ω_i also in the first excited mode: $|1_s, 1_i\rangle|H_s, V_i\rangle = |H_{s1}, V_{i1}\rangle$.
- Vertically polarized signal with frequency ω_s in the fundamental mode and horizontally polarized idler at frequency ω_i also in the fundamental mode: $|0_s, 0_i\rangle|V_s, H_i\rangle = |V_{s0}, H_{i0}\rangle$.
- Vertically polarized signal with frequency ω_s in the first excited mode and horizontally polarized idler at frequency ω_i also in the first excited mode: $|1_s, 1_i\rangle|V_s, H_i\rangle = |V_{s1}, H_{i1}\rangle$.

The quantum state represented by Eq. (17.59) corresponds to a hyperentangled state with two degrees of freedom namely the spatial mode of the waveguide and the polarization state of the photons.

If we project the given state onto the fundamental mode at the signal frequency of ω_s, then we obtain

$$\begin{aligned}\langle 0_s|\psi\rangle &= (\langle 0_s|0_s, 0_i\rangle + |\langle 0_s|1_s, 1_i\rangle) \otimes (|H_s, V_i\rangle + |V_s, H_i\rangle) \\ &= |0_i\rangle(|H_s, V_i\rangle + |V_s, H_i\rangle)\end{aligned} \tag{17.60}$$

The projection has resulted in the idler being in the fundamental mode and the photons are still entangled in the polarization degree of freedom.

Similarly if we project the given state on to horizontal polarization at the signal frequency of ω_s, then we obtain

$$\begin{aligned}\langle H_s|\psi\rangle &= (|0_s, 0_i\rangle + |1_s, 1_i\rangle) \otimes (\langle H_s|H_s, V_i\rangle + \langle H_s|V_s, H_i\rangle) \\ &= (|0_s, 0_i\rangle + |1_s, 1_i\rangle)|V_i\rangle\end{aligned} \tag{17.61}$$

The projection has now resulted in the idler being in the vertical polarization state and the photons are still entangled in the spatial modal degree of freedom (Lugani et al. 2013; Huang et al. 2022).

17.10 CHSH Inequality and Polarization Entangled Photons

In this section, we will briefly discuss the CHSH inequality that was proposed by Clauser, Horne, Shinmoy, and Holt (Clauser et al. 1969) as a test of theories based on local hidden variables (i.e., properties possessed by the objects themselves so that distant events cannot have an instantaneous effect on the local ones) and is used to prove Bell's theorem. In local hidden variable theories, the randomness in measured values of observables is explained as a consequence of the presence of some unseen variables (that they possess) of the system. CHSH inequality shows that the predictions of quantum mechanics as regards some measurements are not in agreement with the classical picture of hidden variables.

Here we will very briefly consider the CHSH inequality as regards polarization entangled photon pairs and compare with a classical model in which the generated photons are described through properties that were generated during their birth and which they possess and carry along with them as a local property.

17.10.1 Quantum Picture

We will first consider the quantum picture which describes a polarization entangled state generated through the SPDC process:

$$|\psi^{-}\rangle = \frac{1}{\sqrt{2}}(|H_s, V_i\rangle - |V_s, H_i\rangle) \tag{17.62}$$

where, as before H and V correspond to horizontal and vertical polarizations and s and i to signal and idler. As discussed earlier, the signal and idler photons in the state represented by Eq. (17.62) do not have well-defined polarization states. At the same time, the two photons have orthogonal polarization states in any basis set that may be used to describe them. The polarization state of either of the photons is undefined before any measurement is carried out on either of them and as soon as the state of polarization of either of the photons is measured, the polarization state of the other photon gets projected into a definite state and joint measurements on the two systems show complete correlation. This seems to point towards some instantaneous action by one photon on the other which is not consistent with classical physics. It is also to be pointed out that there is no violation of causality in this as the result of measurement of the first photon is completely random and hence cannot be used for any superluminal communication.

In order to test the inequality, the signal and idler photons are separated through a dichroic mirror and sent to two polarization analyzers which are oriented with their principal axes at angles α and β, respectively, with respect to the horizontal direction (see Fig. 17.11). The signal photon will exit one of the ports of the polarization analyzer. If the signal photon is detected in the state $|+\alpha_s\rangle$, then the output is taken to be +1, while if the signal photon is detected in the state $|-\alpha_s\rangle$, then the output is taken to be −1. Similarly for the idler photon.

We first represent the relation between the two bases $|H\rangle$, $|V\rangle$ and $|+\rangle$, $|-\rangle$ (see Eqs. 17.10a and 17.10b):

$$|H_s\rangle = |+\alpha_s\rangle \cos\alpha - |-\alpha_s\rangle \sin\alpha \tag{17.63a}$$

$$|V_s\rangle = |+\alpha_s\rangle \sin\alpha + |-\alpha_s\rangle \cos\alpha \tag{17.63b}$$

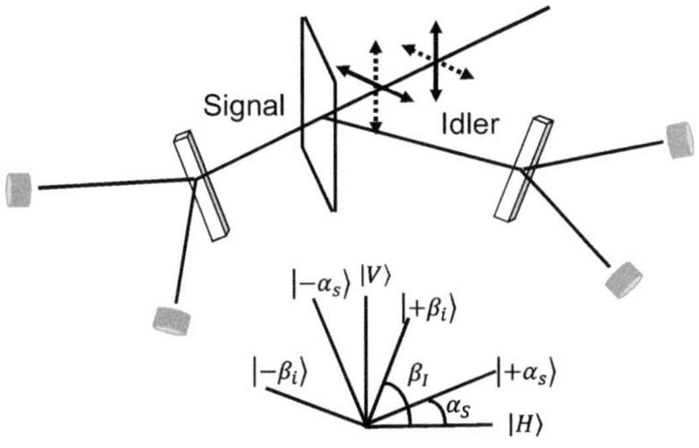

Fig. 17.11 Arrangement to test CHSH inequality of polarization entangled photon pairs

$$|H_i\rangle = |+\beta_i\rangle \cos\beta - |-\beta_i\rangle \sin\beta \tag{17.63c}$$

$$|V_i\rangle = |+\beta_i\rangle \sin\beta + |-\beta_i\rangle \cos\beta \tag{17.63d}$$

The probability that we obtain +1 for the signal photon and +1 for the idler photon is given by

$$P_{++} = |\langle +\alpha_s, +\beta_i|\psi\rangle|^2 = \frac{1}{2}|\langle +\alpha_s, +\beta_i|(|H_s, V_i\rangle - |V_s, H_i\rangle)\rangle|^2 \tag{17.64}$$

Expressing the given state in terms of the rotated basis, we can easily evaluate the probability. For example

$$\langle +\alpha_s, +\beta_i|H_s, V_i\rangle = \langle +\alpha_s|(|+\alpha_s\rangle \cos\alpha - |-\alpha_s\rangle \sin\alpha)\rangle \\ \times \langle +\beta_i|(|+\beta_i\rangle \sin\beta + |-\beta_i\rangle \cos\beta)\rangle = \cos\alpha \sin\beta$$

Similarly

$$\langle +\alpha_s, +\beta_i|V_s, H_i\rangle = \sin\alpha \cos\beta$$

Hence,

$$P_{++} = \frac{1}{2}(\cos\alpha \sin\beta - \sin\alpha \cos\beta)^2 = \frac{1}{2}\sin^2(\alpha - \beta) \tag{17.65}$$

Similarly we can obtain the following relationships:

$$P_{--} = |\langle -\alpha_s, -\beta_i|\psi\rangle|^2 = \frac{1}{2}\sin^2(\alpha - \beta) = P_{++} \tag{17.66}$$

$$P_{\pm} = |\langle +\alpha_s, -\beta_i|\psi\rangle|^2 = \frac{1}{2}\cos^2(\alpha - \beta) = P_{\mp} = |\langle -\alpha_s, +\beta_i|\psi\rangle|^2 \tag{17.67}$$

It is interesting to note that the outcome of measurements on the signal photon (through the orientation angle α) depends on the orientation (through the angle β) of the polarization analyzer placed on the path of the idler photon. In fact the probabilities depend only on the difference in angle $(\alpha - \beta)$ between the orientations of the two analyzers. This shows correlation among the measurements carried out on the two photons.

We define the correlation function through the following average

$$E(\alpha, \beta) = \text{Average } (A(\alpha)B(\beta)) \tag{17.68}$$

where $A(\alpha) = +1$ if the signal photon is found in state $|+\alpha_s\rangle$ and -1 if it is found in the state $|-\alpha_s\rangle$. Similarly $B(\beta) = +1$ if the idler photon is found in the state $|+\beta_i\rangle$ and -1 if it is found in the state $|-\beta_i\rangle$. Hence, the correlation would be

$$\begin{aligned} E(\alpha, \beta) &= (+1) \times P_{++} + (+1) \times P_{--} + (-1)P_{+-} + (-1)P_{-+} \\ &= P_{++} + P_{--} - P_{+-} - P_{-+} \end{aligned}$$

Substituting for the various probabilities, we obtain

$$E(\alpha, \beta) = \sin^2(\alpha - \beta) - \cos^2(\alpha - \beta) = -\cos 2(\alpha - \beta) \tag{17.69}$$

We now calculate the quantity

$$S = E(\alpha, \beta) + E(\alpha, \beta') + E(\alpha', \beta) - E(\alpha', \beta') \tag{17.70}$$

Assuming $\alpha = 0$, $\alpha' = \frac{\pi}{4}$, $\beta = \frac{\pi}{8}$ and $\beta' = -\frac{\pi}{8}$, we obtain

$$S = -\cos\frac{\pi}{4} - \cos\frac{\pi}{4} - \sin\frac{\pi}{4} + \cos\frac{3\pi}{4}$$

which gives us

$$|S| = 2\sqrt{2} \tag{17.71}$$

Notice that the angle differences between the settings (α, β), (α, β'), (α', β) are all equal to $\frac{\pi}{8}$ which have negative correlations. On the other hand, the angle difference between the settings (α', β') is $\frac{3\pi}{8}$, which has positive correlation. Thus, in the quantity S, having a negative sign on the last term ensures that the four terms all add giving the possibility of maximizing the value of S.

We can see that the quantity S is easily amenable to experimental measurements using entangled photon pairs generated by SPDC processes.

We now consider the classical picture of the SPDC process and obtain the quantity S.

17.10.2 Distinguishable Photon Pairs

In a classical picture, when the pump photon undergoes SPDC through *either* of the two possible down conversion processes, the classical interpretation would be that the pump photon splits into *either* the pair $|H_s, V_i\rangle$ or into the pair $|V_s, H_i\rangle$. Thus, in the classical picture each of the two photons has well-defined polarization state which they carry as they propagate and which are not influenced by any measurement carried out on the sister photon that was created at the same time. In contrast, in the quantum picture, the output state is a coherent superposition of these two states since we need to add the probability amplitudes of both the two indistinguishable processes, the state of the generated photon pair represented by Eq. (17.62).

We will now evaluate the same quantity S as given by Eq. (17.70) assuming that the pump photon has equal probability of generating *either* the pair $|H_s, V_i\rangle$ or the pair $|V_s, H_i\rangle$.

If the output is the pair $|H_s, V_i\rangle$, then we have

$$P_{++1} = |\langle +\alpha_s, +\beta_i|\psi\rangle|^2 = |\langle +\alpha_s, +\beta_i|H_s, V_i\rangle|^2 = \cos^2\alpha\sin^2\beta \tag{17.72}$$

Similarly if the output is the pair $|V_s, H_i\rangle$, then we would obtain

$$P_{++2} = |\langle +\alpha_s, +\beta_i|\psi\rangle|^2 = |\langle +\alpha_s, +\beta_i|V_s, H_i\rangle|^2 = \sin^2\alpha\cos^2\beta \tag{17.73}$$

Hence, assuming equal probability of obtaining either $|H_s, V_i\rangle$ or $|V_s, H_i\rangle$, we have for the total probability

$$P_{++} = \frac{1}{2}(P_{++1} + P_{++2}) = \frac{1}{2}\left(\cos^2\alpha\sin^2\beta + \sin^2\alpha\cos^2\beta\right) \tag{17.74}$$

In fact, we can rewrite P_{++} in a slightly different form:

$$\begin{aligned} P_{++} &= \frac{1}{2}|\langle +\alpha_s, +\beta_i|H_s, V_i\rangle|^2 + |\langle +\alpha_s, +\beta_i|V_s, H_i\rangle|^2 \\ &= \langle +\alpha_s, +\beta_i|\frac{1}{2}(|H_s, V_i\rangle\langle H_s, V_i| + |V_s, H_i\rangle\langle V_s, H_i|)|+\alpha_s, +\beta_i\rangle \end{aligned} \tag{17.75}$$

Thus, P_{++} is nothing but the expectation value of the following quantity referred to as the density operator of the state:

$$\widehat{\rho} = \frac{1}{2}(|H_s, V_i\rangle\langle H_s, V_i| + |V_s, H_i\rangle\langle V_s, H_i|) \tag{17.76}$$

Equation (17.76) represents a mixed state with 50% probability of the photons being in the $|H_s, V_i\rangle$ state and 50% probability of the photons being in the $|V_s, H_i\rangle$ state. Note that this is a very different state compared to the entangled state and there is no coherence between the two output states unlike in the entangled state.

Similarly, we can obtain the following relations:

$$P_{--}=\frac{1}{2}|\langle-\alpha_s,-\beta_i|H_s,V_i\rangle|^2+|\langle-\alpha_s,-\beta_i|V_s,H_i\rangle|^2=\frac{1}{2}\left(\sin^2\alpha\cos^2\beta+\cos^2\alpha\sin^2\beta\right) \tag{17.77}$$

$$P_{+-}=\frac{1}{2}|\langle+\alpha_s,-\beta_i|H_s,V_i\rangle|^2+|\langle+\alpha_s,-\beta_i|V_s,H_i\rangle|^2=\frac{1}{2}\left(\cos^2\alpha\cos^2\beta+\sin^2\alpha\sin^2\beta\right) \tag{17.78}$$

$$P_{-+}=\frac{1}{2}|\langle-\alpha_s,+\beta_i|H_s,V_i\rangle|^2+|\langle-\alpha_s,+\beta_i|V_s,H_i\rangle|^2=\frac{1}{2}\left(\sin^2\alpha\sin^2\beta+\cos^2\alpha\cos^2\beta\right) \tag{17.79}$$

Hence, we have

$$E(\alpha,\beta)=P_{++}+P_{--}-P_{+-}-P_{-+}=-\cos 2\alpha\cos 2\beta \tag{17.80}$$

We now calculate the quantity S:

$$S=E(\alpha,\beta)+E(\alpha,\beta')+E(\alpha',\beta)-E(\alpha',\beta') \tag{17.81}$$

Assuming $\alpha=0$, $\alpha'=\frac{\pi}{4}$, $\beta=\frac{\pi}{8}$ and $\beta'=-\frac{\pi}{8}$, we obtain

$$S=-\cos 0\cos\frac{\pi}{4}-\cos 0\cos\left(-\frac{\pi}{4}\right)-\cos\frac{\pi}{2}\cos\frac{\pi}{4}+\cos\frac{\pi}{2}\cos\left(-\frac{\pi}{4}\right)=-\sqrt{2}$$

which is contrary to the prediction of the quantum picture (see Eq. 17.71).

Although this is a specific example, it can be shown that if we assume that there are hidden variables, then the following inequality is satisfied:

$$|S|\leq 2 \tag{17.82}$$

Actual experiments carried out on polarization entangled photon pairs and other entangled systems give $|S|\geq 2$ and thus are consistent with the quantum picture rather than with the picture of hidden variables.

Exercise 17.6 Consider a state given by

$$|\psi\rangle=(|H_s,V_i\rangle-\sigma|V_s,H_i\rangle)/\sqrt{1+\sigma^2} \tag{17.83}$$

which represents a nonmaximally entangled state for $\sigma\neq 1$. Obtain P_{++} and discuss the variation in P_{++} as a function of the angle β for different values of σ. Compare with the result when $\sigma=1$.

Solution We have

$$P_{++} = |\langle +\alpha_s, +\beta_i|\psi\rangle|^2 = \frac{1}{(1+\sigma^2)} |\langle +\alpha_s, +\beta_i|(|H_s, V_i\rangle - \sigma|V_s, H_i\rangle)\rangle|^2$$

Using Eqs. (17.63a)–(17.63d), the terms can be easily evaluated and we obtain

$$P_{++} = \frac{(\cos\alpha\sin\beta - \sigma\sin\alpha\cos\beta)^2}{(1+\sigma^2)} \tag{17.84}$$

For $\sigma = 1$, which corresponds to maximally entangled state, we have

$$P_{++} = \frac{1}{2}\sin^2(\alpha - \beta)$$

In contrast to the case with $\sigma \neq 1$, for $\sigma = 1$ (corresponding to maximally entangled pair), P_{++} depends only on the difference in angle $(\alpha - \beta)$ of the analyzer settings. Nonmaximally entangled states find applications in quantum information (see, e.g., White et al. 1999; Grossman and Reznik 2005; Modlawska and Grudka 2008; Xue et al. 2001).

Exercise 17.7 For the following nonmaximally entangled state

$$|\psi\rangle = (|H_S, V_I\rangle - \sigma|V_S, H_I\rangle)/\sqrt{1+\sigma^2}$$

Assuming $\sigma = 0.9$, obtain

$$S = E(\alpha,\beta) + E(\alpha,\beta') + E(\alpha',\beta) - E(\alpha',\beta')$$

for $\alpha = 0$, $\alpha' = \frac{\pi}{4}$, $\beta = \frac{\pi}{8}$ and $\beta' = -\frac{\pi}{8}$.

Solution For the given state, we obtain the following:

$$E(\alpha,\beta) = \frac{1}{1+\sigma^2}\Big[(\cos\alpha\sin\beta - \sigma\sin\alpha\cos\beta)^2 + (\sin\alpha\cos\beta - \sigma\cos\alpha\sin\beta)^2 - (\cos\alpha\cos\beta + \sigma\sin\alpha\sin\beta)^2 - (\sin\alpha\sin\beta + \sigma\cos\alpha\cos\beta)^2\Big]$$

Substituting the values $\alpha = 0$, $\alpha' = \frac{\pi}{4}$, $\beta = \frac{\pi}{8}$ and $\beta' = -\frac{\pi}{8}$ and $\sigma = 0.9$ we obtain

$$S = E(\alpha,\beta) + E(\alpha,\beta') + E(\alpha',\beta) - E(\alpha',\beta')$$
$$\approx (-0.707 - 0.707 - 0.703 - 0.703) \approx -2.82$$

which still violates the classical limit of the CHSH inequality.

Note that if $\sigma = 0$, the state is no more an entangled state and $E(\alpha,\beta) = -\cos 2\alpha \cos 2\beta$ which gives $S = -\sqrt{2}$. If $\sigma = 1$, the state becomes maximally entangled and we have $E(\alpha,\beta) = -\cos 2(\alpha - \beta)$ as obtained earlier.

17.11 Three Photon Entangled State: The GHZ State

Until now we have discussed entangled states with two photons. It is also possible to have entanglement among more than two photons leading to what is referred to as multiphoton entanglement.

One of the most important multiphoton entangled state is referred to as the GHZ state (Greenberger, Horne and Zeilinger) which is a maximally entangled state among three photons (Greenberger et al. 1990). The GHZ state exhibits nonlocal correlations wherein measurements on one qubit affect the outcome of measurements on the other qubits.

An example of such a GHZ state is the following polarization entangled state:

$$|\psi\rangle_{GHZ} = \frac{1}{\sqrt{2}}(|H_1\rangle|H_2\rangle|H_3\rangle + |V_1\rangle|V_2\rangle|V_3\rangle) \tag{17.85}$$

where H and V refer to horizontal and vertical polarization states and the subscripts 1, 2, and 3 refer to the three individual photons. The above state is a quantum superposition of the state in which all the three photons are horizontally polarized and the state in which all the three photons are vertically polarized. Similar to the two photon entangled state, in such a quantum state none of the photons have any well-defined polarization state of their own. Such a three photon entangled state enables various tests of the foundations of quantum mechanics (see, e.g., Bouwmeester et al. 1999; Pan et al. 2000) and is also expected to play a significant role in quantum information processing.

Just like in the case of two photon entanglement, here too we can write the GHZ state in a different basis. For example, we consider the following diagonal/antidiagonal bases:

$$|D\rangle = \frac{1}{\sqrt{2}}(|H\rangle + |V\rangle) \tag{17.86}$$

$$|A\rangle = \frac{1}{\sqrt{2}}(|H\rangle - |V\rangle) \tag{17.87}$$

where $|D\rangle$ and $|A\rangle$ correspond to the diagonal $\left(\left|+\frac{\pi}{4}\right\rangle\right)$ and antidiagonal $\left(\left(\left|-\frac{\pi}{4}\right\rangle\right)\right)$ bases (see Fig. 17.3).

We can indeed express the GHZ state given in Eq. (17.85) in terms of the new bases. For example, we can express the state in terms of the states $|D\rangle$ and $|A\rangle$ and obtain the following:

$$|\psi\rangle = \frac{1}{2}(|D_1\rangle|D_2\rangle|D_3\rangle + |D_1\rangle|A_2\rangle|A_3\rangle + |A_1\rangle|D_2\rangle|A_3\rangle + |A_1\rangle|A_2\rangle|D_3\rangle) \tag{17.88}$$

The above equation implies that measuring the state of polarization of photon 1 in the *D*/*A* basis will result in *D* or *A* with equal probabilities of 50%. Measuring the state of polarization of photons 1 and 2 in the *D*/*A* bases will yield any of the combinations $|D_1\rangle|D_2\rangle$, $|D_1\rangle|A_2\rangle$, $|A_1\rangle|D_2\rangle$ or $|A_1\rangle|A_2\rangle$ with equal probability of 25%. At the same time, the result of the measurement carried out on the two photons will immediately specify the state of polarization of the third photon. Thus, if the photons 1 and 2 are found to be in the state $|D_1\rangle$ and $|A_2\rangle$, respectively, then the photon 3 will definitely be found to be in the state $|A_3\rangle$.

GHZ states have been realized experimentally both in bulk and integrated photonic configurations using different degrees of freedom of the photons (see, e.g., Pont et al. 2022; Chen et al. 2024; Cao et al. 2024).

Exercise 17.7 Express the GHZ state given by Eq. (17.85) expressing photons 1 and 2 in the right-left circular bases and photon 3 in the diagonal–antidiagonal basis.

Solution The right and left circular polarization are related to the horizontal/vertical bases through the following transformations:

$$|R\rangle = \frac{1}{\sqrt{2}}(|H\rangle + i|V\rangle) \tag{17.89}$$

$$|L\rangle = \frac{1}{\sqrt{2}}(|H\rangle - i|V\rangle) \tag{17.90}$$

where and $|R\rangle$ and $|L\rangle$ correspond to the right circular and left circular bases. Substituting from the above equations in Eq. (17.85) after a bit of algebra, we obtain:

$$|\psi\rangle = \frac{1}{2}(|R_1\rangle|R_2\rangle|A_3\rangle + |R_1\rangle|L_2\rangle|D_3\rangle + |L_1\rangle|R_2\rangle|D_3\rangle + |L_1\rangle|L_2\rangle|A_3\rangle) \tag{17.91}$$

The above equation implies that measuring the state of polarization of photon 1 in the circular basis will yield R or L with equal probabilities of

50%. Measuring the state of polarization of photons 1 and 2 in the circular polarization bases will yield any of the combinations *RR*, *RL*, *LR*, or *LL* with equal probability of 25%. At the same time, the result of the measurement carried out on the two photons will immediately specify the state of polarization of the third photon. Thus, if the photons 1 and 2 are found to be in the state $|R_1\rangle$ and $|L_2\rangle$, respectively, then from the second term in Eq. (17.91) we can conclude that the photon 3 will be found to be in the state $|D_3\rangle$.

Exercise 17.8 Consider the GHZ state given in Eq. (17.85). Assume that photon 1 is passed through a linear analyzer oriented at an angle θ with the horizontal. Show that the probability of the photon passing through the analyzer is always $\frac{1}{2}$ irrespective of the angle θ. Thus, the state of polarization of the photon is undefined like in the two photon polarization entangled state.

Solution If we represent the state with a polarization at an angle θ with the horizontal by $|+\theta\rangle$ and the corresponding orthogonal state by $|-\theta\rangle$, then we have from Eqs. (17.10a) and (17.10b), the probability that photon 1 passes through the analyzer and the other photons are horizontally polarized is

$$P_1 = \left[\langle +\theta_1|\langle H_2|\langle H_3|\psi\rangle\right]^2 = \frac{1}{2}\cos^2\theta \tag{17.92}$$

Similarly the probability that photon 1 passes through the analyzer and the other photons are vertically polarized is

$$P_2 = \left[\langle +\theta_1|\langle V_2|\langle V_3|\psi\rangle\right]^2 = \frac{1}{2}\sin^2\theta \tag{17.93}$$

Hence, the total probability that the photon 1 passes through the analyzer irrespective of the state of polarization of the other two photons is

$$P = P_1 + P_2 = \frac{1}{2} \tag{17.94}$$

which is independent of the angle θ made by the analyzer with the horizontal axis. Hence, the polarization state of photon 1 is undefined. Similarly one can show that the polarization states of the other two photons are also undefined.

Problems

17.1. Is the following state a polarization entangled state?

$$|\psi\rangle = \frac{1}{\sqrt{2}}\left(|H_s, V_i\rangle + |H_s, H_i\rangle\right)$$

17.2. Express the following polarization entangled state in the basis set of right circular polarization (RCP) and left circular polarization (LCP)

$$|\psi\rangle = \frac{1}{\sqrt{2}}(|H_s, V_i\rangle - |V_s, H_i\rangle)$$

17.3. Consider a general two qubit state given by

$$|\psi\rangle = \alpha|0_A, 0_B\rangle + \beta|0_A, 1_B\rangle + \gamma|1_A, 0_B\rangle + \delta|1_A, 1_B\rangle$$

Express this state in terms of the Bell states. Obtain the probabilities of finding the state in the four Bell states.

17.4. Show that the following state is a mixture of two Bell states:

$$|\psi\rangle = \frac{1}{2}|H_1, V_2\rangle + \frac{\sqrt{3}}{2}|V_1, H_2\rangle$$

17.5. Consider a symmetric Y-splitter in z-cut lithium niobate with single mode waveguides with the two arms of the Y far enough not to have any interactions between them. The two waveguide section has periodic poling of an appropriate period. Pump photons are incident on the Y splitter and undergo SPDC in the two waveguide region. Considering the spatial modes in the waveguide, which modal entanglement state can be generated in the process?

17.6. Write down the quantum state of a photon in a circularly polarized state in terms of horizontal and vertical polarization states. Such a state is passed through a linear analyzer oriented at an angle θ with the horizontal. Show that the probability of the photon passing through a linear analyzer is independent θ.

17.7. Consider the four Bell states given in Eqs. (17.37)–(17.40). Assuming the 0 corresponds to H polarization and 1 corresponds to V polarization, express the states in the (+, −) basis where $|+\rangle$ is a polarization state rotated by +45° from the horizontal axis and $|-\rangle$ is its orthogonal state.

17.8. Referring to Sect. 17.7 instead of the pump photon being in the fundamental mode, if the pump photon is chosen to be in the first excited mode, which mode combinations of signal and idler will be allowed by parity conservation? Assume Type II SPDC process. Write down the corresponding QPM conditions required to be satisfied for efficient SPDC.

17.9. Given that an SPDC process generates the following Bell state in the polarization basis:

$$|\psi^+\rangle = \frac{1}{\sqrt{2}}(|H_s, V_i\rangle + |V_s, H_i\rangle)$$

(a) Show that if the signal photon is passed through a half wave plate corresponding to the signal frequency ω_S with an appropriate orientation, the state $|\psi^+\rangle$ is converted to the state $|\psi^-\rangle$.
(b) For what orientation of the half wave plate will the state $|\psi^+\rangle$ get transformed to the state $|\phi^+\rangle$?
(c) Show that using two half wave plates with appropriate orientations, the state $|\psi^+\rangle$ can be transformed to the state $|\phi^-\rangle$ (Mishra and Singh 2024).

17.10. Consider the following Bell state:

$$|\phi^+\rangle = \frac{1}{\sqrt{2}}(|H_S, H_I\rangle + |V_S, V_I\rangle)$$

Obtain the probability $P_{\phi\theta}$. Also consider the case when the emitted photon pair is interpreted as either both are horizontally polarized or both are vertically polarized. Compare the two cases.

17.11. Consider the following entangled state

$$|\phi^+\rangle = \frac{1}{\sqrt{2}}(|H_s, H_i\rangle + |V_s, V_i\rangle)$$

Following the same procedure discussed in Sect. 17.10, obtain the corresponding expression S for this state. Check for violation of CHSH inequality.

17.12. Consider the following state:

$$|\psi\rangle - \frac{1}{\sqrt{2}}(|H_s\rangle + |V_s\rangle)|V_i\rangle$$

Obtain the probability P_{++} and interpret the result.

17.13. Consider the following state

$$|\psi\rangle = \frac{1}{2}|H_S, V_I\rangle - \frac{\sqrt{3}}{2}|V_S, H_I\rangle$$

Obtain S as given by Eq. (17.81) and check the value for $\alpha = 0$, $\alpha' = \frac{\pi}{4}$, $\beta = \frac{\pi}{8}$ and $\beta' = -\frac{\pi}{8}$.

17.14. Write the entangled state given by Eq. (17.8) in an elliptical polarization basis expressed in linear polarization basis as given by

$$|E_+\rangle = \frac{1}{\sqrt{5}}(|H\rangle + 2i|V\rangle); \quad |E_-\rangle = \frac{1}{\sqrt{5}}(2|H\rangle - i|V\rangle)$$

Obtain the probability of detecting (a) both the signal photon and the idler photons in the $|E_+\rangle$ state and (b) the signal photon in the $|E_+\rangle$ state and the idler photon in the $|E_-\rangle$ state.

17.15. Show that the state represented by Eq. (17.46) is an eigenstate of the total photon number operator. What is the number of photons in the state?

References

Bharadwaj, D., and K. Thyagarajan. 2018. Direct generation of hybrid entangled photon pairs in waveguides. *J Opt Soc Am* 35:790–796.

Bharadwaj, D., K. Thyagarajan, M. Karpinski, and K. Banaszek. 2015. Generation of higher-dimensional modal entanglement using a three-waveguide directional coupler. *Phys Rev A* 91: 033824.

Bouwmeester, D., J.-W. Pan, K. Mattle, M. Eibi, H. Weinfurter, and A. Zeilinger. 1997. Experimental quantum teleportation. *Nature* 390:575–579.

Bouwmeester, D., J.-W. Pan, M. Daniell, H. Weinfurter, and A. Zeilinger. 1999. Observation of three photon Greenberger-Horne-Zeilinger entanglement. *Phys Rev Lett* 82:1345–1349.

Cao, H., L. M. Hansen, F. Giorgino, L. Carosini, P. Zahálka, F. Zilk, J. C. Loredo, and P. Walther. 2024. Photonic source of heralded Greenberger-Horne-Zeilinger states. *Phys Rev Lett* 132: 130604.

Chen, C., Y. Chen, Z. Fang, R. Ge, J. Wu, and X. Chen. 2024. Hybrid material integration for active photonic applications. *APL Photonics* 9:030903.

Clauser, J. F., M. A. Horne, and A. Shimony. 1969. Proposed experiment to test local hidden variable theories. *Phys Rev Lett* 23:880–884.

Crespi, A., M. Lobino, J. C. F. Mathews, A. Politi, C. R. Neal, R. Ramponi, R. Osellame, and J. L. O'Brien. 2012. Measuring protein concentration with entangled photons. *Appl Phys Letts* 100:233704

Feng, L.-T., M. Zhang, X. Xiong, Y. Chen, H. W. Ming, G.-P. Guo, G.-C. Guo, D.-X. Dai, and X.-F. Ren. 2019. On chip transverse mode entangled photon pair source. *npj Quant Inform* 5:2. https://doi.org/10.1038/s41534-018-0121-z.

Fickler, R., M. Krenn, R. Lapkiewicz, S. Ramelow, and A. Zeilinger. 2013. Real time imaging of quantum entanglement. *Sci Rep* 3:1914.

Greenberger, D. M., M. A. Horne, A. Shinmoy, and A. Zeilinger. 1990. Bell's theorem without inequalities. *Am J Phys* 58:1131–1143.

Grossman, B., and B. Reznik. 2005. Implementing nonlocal gates with non maximally entangled states. *Phys Rev A* 71:032322.

Horodecki, R., P. Horodecki, M. Horodecki, and K. Horodecki. 2009. Quantum entanglement. *Rev Mod Phys* 81:865–942.

Huang, Y., J. Feng, Y. Li, Z. Qi, C. Lu, Y. Zheng, and X. Chen. 2022. High performance hyperentanglement generation and manipulation on lithium niobate waveguides. *Phys Rev Appl* 17:054002.

Kwiat, P. G. 1997. Hyper-entangled states. *J Modern Opt* 44:2173–2184.

Liu, J., Z. Lin, D. Liu, X. Feng, F. Liu, K. Cui, Y. Huang, and W. Zhang. 2024. High-dimensional quantum key distribution using energy-time entanglement over 242 km partially deployed fiber. *Quant Sci Technol* 9:015003.

Lugani, J., S. Ghosh, and K. Thyagarajan. 2013. Switchable hyperentangled photon pairs from an integrated optic waveguide device. *J Opt Soc Am* 30:795–805.

Ma, X., A. Qarry, J. Kofler, T. Jennewein, and A. Zeilinger. 2009. Experimental violation of a Bell inequality with two different degrees of freedom of entangled particle pairs. *Phys Rev A* 79: 042101.

Ming, Y., Z. J. Wu, G. Cui, A. Tan, and Y. Lu. 2014a. Integrated source of tunable non maximally mode-entangled photons in a domain engineered lithium niobate waveguide. *Appl Phys Lett* 104:171110.

Ming, Y., A.-H. Tan, Z.-J. Wu, Z.-X. Chen, F. Xu, and Y.-Q. Lu. 2014b. Tailoring entanglement through domain engineering in a lithium niobate waveguide. *Sci Rep* 4:4812. https://doi.org/10.1038/srep04812.

Mishra, S., and R. P. Singh. 2024. Transformation of Bell states using linear optics. *Phys Open* 18: 100199.

Modlawska, J., and A. Grudka. 2008. Non maximally entangled states can be better for multiple linear operations. *Phys Rev Lett* 100:11503.

Nemirovsky-Levy, L., U. Pereg, and M. Segev. 2024. Increasing quantum communication rates using hyperentangled photonic states. *Opt Quant* 2:165–172.

Pan, J.-W., D. Bouwmeester, M. Daniell, H. Weinfuter, and A. Zeilinger. 2000. Experimental test of quantum nonlocality in three photon Greenberger-Horne-Zeilinger entanglement. *Nature* 403: 515–519.

Pont, M., G. Corrielli, A. Fyrillas, I. Agresti, G. Carvacho, N. Maring, P.-E. Emeriau, F. Ceccarelli, R. Albiero, P. H. D. Ferreira, N. Somaschi, J. Senellart, I. Sagnes, M. Morassi, A. Lemaître, P. Senellart, F. Sciarrino, M. Liscidini, N. Belabas, and R. Osellame. 2022. High fidelity generation of four photon GHZ states on a chip. *arXiv*. 2211.1562v1 [quant-ph].

Schaff, C., R. Polster, M. Huber, S. Ramelow, and A. Zeiliinger. 2015. Experimental access to higher dimensional entangled quantum systems using integrated optics. *Optica* 2:523–529.

Schulz, E. M., J. D. Elliot, M. van Nierkerk, D. Procotor, M. Ciminelli, T. Palone, C. C. Tison, M. L. Fanto, S. F. Preble, and G. A. Howland. 2024. Generating high-dimensional entanglement using a foundry-fabricated photonic integrated circuit. *Opt Quantum* 2:35–40.

Tashima, T., Y. Mukai, M. Arahata, N. Oda, M. Hiramitsu, K. Tokuda, R. Okamoto, and S. Takeguchi. 2024. Ultra-broadband quantum infrared spectroscopy. *Optica* 11:81–87.

Thompson, M. G., A. Politi, J. C. F. Mathews, and J. L. O'Brien. 2011. Integrated waveguide circuits for optical quantum computing. *IET Circuits Devices Syst* 5:94–102.

Torres, J. P., K. Banaszek, and I. A. Walmsley. 2011. Engineering nonlinear optic sources of photonic entanglement. *Prog Optics* 56:227–331.

White, A. G., D. F. V. James, P. H. Eberhard, and P. G. Kwiat. 1999. Nonmaximally entangled states: Production, characterization and utilization. *Phys Rev Lett* 83:3103–3107.

Xue, P., C.-F. Li, and G.-C. Guo. 2001. Efficient quantum-key-distribution scheme with nonmaximally entangled states. *Phys Rev A* 64:032305.

Zeilinger, A. 2017. Light for the quantum. Entangled photons and their applications: a very personal perspective. *Phys Scr* 92:072501. (33pp).

Zhang, Z., C. You, O. S. Magaña-Loaiza, R. Fickler, R. de J. León-Montiel, J. P. Torres, T. S. Humble, S. Liu, Y. Xia, and Q. Zhuang. 2024. Entanglement-based quantum information technology: a tutorial. *Adv Opt Photonics* 16:60–162.

Chapter 18
Squeezed Photon States

Abstract Squeezed states form a very important concept in quantum photonics. This chapter discusses squeezed photon states and their properties. Amplitude and phase squeezed states are discussed with diagrammatic representation of these states. Also discussed are squeezed vacuum state and two mode squeezed states including generation of squeezed vacuum state.

18.1 Introduction

In Chap. 15, we discussed about number states and coherent states. In this chapter, we will discuss squeezed states and some of their properties. Squeezed states are extremely interesting quantum states of light and find applications in quantum metrology, quantum sensing, etc. Squeezed states have reduced noise in one of the quadrature at the expense of the other quadrature and hence provide for high precision in optical measurements (see e.g., Stokowski et al. 2023; Kashiwazaki et al. 2021; Kashiwazaki et al. 2023). In Chap. 22, we will discuss a technique for generating such quantum states using nonlinear interactions. For a nice review of squeezed states, see Lvovsky (2014). For more discussions on properties of squeezed states, refer to Walls (1983), Henry and Glotzer (1988), Teich and Saleh (1989), Breitenbach et al. (1997), Breitenbach and Schiller (1997), Andersen et al. (2016), and Mehmet et al. (2010). Squeezed states have also been generated using lithium niobate on insulator platform, by integrating multiple photonic components on a single chip (Park et al. 2024).

18.2 Defining Squeezed States

In Chap. 14, we introduced the two quadrature operators $\widehat{X}_1$ and $\widehat{X}_2$ (see Eqs. 14.64 and 14.65) and in Chap. 15, we discussed coherent states which are minimum uncertainty states with the product of uncertainties in the two quadratures to be

K. Thyagarajan, *Integrated Quantum Photonics*, Graduate Texts in Physics,
https://doi.org/10.1007/978-3-031-85728-7_18

equal to $\frac{1}{4}$ with equal uncertainties in both $\widehat{X}_1$ and $\widehat{X}_2$ quadratures each being equal to $\frac{1}{2}$:

$$\langle \Delta \hat{X}_1 \rangle = \frac{1}{2}; \quad \langle \Delta \hat{X}_2 \rangle = \frac{1}{2}$$
$$\langle \Delta \hat{X}_1 \rangle \langle \Delta \hat{X}_2 \rangle = \frac{1}{4} \tag{18.1}$$

This is the minimum value of the uncertainty product as determined by the uncertainty relation.

We now define squeezed states as eigenstates of the following operator which is a linear superposition of the annihilation and creation operators:

$$\widehat{b} = \mu \widehat{a} + \nu \widehat{a}^{\dagger} \tag{18.2}$$

where μ and ν are, in general, complex quantities satisfying the following equation:

$$|\mu|^2 - |\nu|^2 = 1 \tag{18.3}$$

In view of Eq. (18.3), we can write $\mu = \cosh \sigma$ and $\nu = e^{i\phi} \sinh \sigma$ and hence the operator $\widehat{b}$ becomes

$$\widehat{b} = \cosh \sigma \, \widehat{a} + e^{i\phi} \sinh \sigma \, \widehat{a}^{\dagger} \tag{18.4}$$

where σ and ϕ are two parameters specifying the squeezed state. Note that for $\sigma = 0$, the state $|\beta\rangle$ reduces to the coherent state $|\alpha\rangle$.

The eigenvalue equation satisfied by the operator $\widehat{b}$ is given by

$$\widehat{b}|\beta\rangle = (\mu \widehat{a} + \nu \widehat{a}^{\dagger})|\beta\rangle = \beta|\beta\rangle \tag{18.5}$$

with β is the eigenvalue and $|\beta\rangle$ is the eigenket. The operator $\widehat{b}$ is also referred to as a generalized annihilation operator since as we will see it satisfies the same commutation relation as the annihilation operator $\widehat{a}$. Note that when $\nu = 0$ then $\mu = 1$ and $\widehat{b} = \widehat{a}$. Thus, the generalized annihilation operator becomes equal to the annihilation operator when $\nu = 0$.

Note also that $|\beta\rangle$ is not an eigenstate of the number operator $\widehat{N} = \widehat{a}^{\dagger}\widehat{a}$ and hence the state does not have a well-defined number of photons. In fact, the squeezed state can be expressed as a linear combination of number states (see Exercise 18.3).

From Eq. (18.5), we also have

$$\langle \beta | \widehat{b}^{\dagger} = \beta^* \langle \beta | \tag{18.6}$$

From Eq. (18.2), we note that

$$\widehat{b}^{\dagger}=\mu^{*}\widehat{a}^{\dagger}+\nu^{*}\widehat{a} \tag{18.7}$$

To simplify our algebra and the discussion, we will assume that μ and ν are real quantities. Thus,

$$\widehat{b}^{\dagger}=\mu\widehat{a}^{\dagger}+\nu\widehat{a} \tag{18.8}$$

We can invert the equations for $\widehat{b}$ and $\widehat{b}^{\dagger}$ and obtain the following equations:

$$\widehat{a}=\mu\widehat{b}-\nu\widehat{b}^{\dagger} \tag{18.9}$$

and

$$\widehat{a}^{\dagger}=\mu\widehat{b}^{\dagger}-\nu\widehat{b} \tag{18.10}$$

It can also be shown that (see Exercise 18.1)

$$\left[\widehat{b},\widehat{b}^{\dagger}\right]=1 \tag{18.11}$$

Thus, $\widehat{b}$ and $\widehat{b}^{\dagger}$ satisfy the same commutation relations as $\widehat{a}$ and $\widehat{a}^{\dagger}$ and hence share similar properties.

Exercise 18.1 Show that $\widehat{b}$ and $\widehat{b}^{\dagger}$ satisfy the commutation relation given by Eq. (18.11).

Solution We have

$$\left[\widehat{b},\widehat{b}^{\dagger}\right]=\left[\left(\mu\widehat{a}+\nu\widehat{a}^{\dagger}\right),\left(\mu\widehat{a}^{\dagger}+\nu^{*}\widehat{a}\right)\right]=\mu^{2}\left[\widehat{a},\widehat{a}^{\dagger}\right]+|\nu|^{2}\left[\widehat{a}^{\dagger},\widehat{a}\right]=\mu^{2}-|\nu|^{2}=1$$

where we have used the fact that $\left[\widehat{a},\widehat{a}^{\dagger}\right]=1$ and $\left[\widehat{a}^{\dagger},\widehat{a}\right]=-1$.

18.3 Expectation Values

In order to better understand squeezed states, we will now calculate the expectation values of various quantities in the squeezed state and show some very interesting properties of these states.

Referring to Chap. 14, the electric field operator for a mode corresponding to a plane wave propagating along the z-direction is given by (see Eq. 14.62)

$$\widehat{E} = \frac{i}{2}E_0\left[\widehat{a}e^{-i\chi} - \widehat{a}^{\dagger}e^{i\chi}\right] \tag{18.12}$$

where

$$\chi = \omega t - kz \tag{18.13}$$

and

$$E_0 = \sqrt{\frac{2\hbar\omega}{\varepsilon_0 V}} \tag{18.14}$$

We are using scalar relation for simplicity.

We use the relations given by Eq. (18.9) and (18.10) and write the electric field operator in terms of $\widehat{b}$ and $\widehat{b}^{\dagger}$:

$$\widehat{E} = \frac{i}{2}E_0\left[\left(\mu\widehat{b} - \nu\widehat{b}^{\dagger}\right)e^{-i\chi} - \left(\mu\widehat{b}^{\dagger} - \nu\widehat{b}\right)e^{i\chi}\right] \tag{18.15}$$

The expectation value of the electric field then is given by

$$\langle\widehat{E}\rangle = \langle\beta|\widehat{E}|\beta\rangle = \frac{i}{2}E_0\left[\langle\beta|\left(\mu\widehat{b} - \nu\widehat{b}^{\dagger}\right)|\beta\rangle e^{-i\chi} - \langle\beta|\left(\mu\widehat{b}^{\dagger} - \nu\widehat{b}\right)|\beta\rangle e^{i\chi}\right] \tag{18.16}$$

To simplify the analysis, we will also assume β to be real. The analysis can be easily extended to the case when β is complex.

Hence, we have

$$\left\langle\widehat{E}\right\rangle = \frac{i}{2}E_0\left[\beta(\mu - \nu)e^{-i\chi} - \beta(\mu - \nu)e^{i\chi}\right] = E_0\beta(\mu - \nu)\sin\chi$$

Using $\mu = \cosh\sigma$ and $\nu = \sinh\sigma$, we have

$$\left\langle\widehat{E}\right\rangle = E_0\beta e^{-\sigma}\sin\chi \tag{18.17}$$

Notice that if $\nu = 0$, then $\mu = 1$ and $\sigma = 0$ and the expectation value $\left\langle\widehat{E}\right\rangle$ reduces to that of a coherent state of amplitude β.

In order to describe the squeezed state, we also need to evaluate the variance in the electric field. For this, we need to evaluate $\left\langle\widehat{E}^2\right\rangle$. After some straightforward algebra, we will find the following expression:

$$\left\langle\widehat{E}^2\right\rangle = \frac{E_0^2}{4}e^{2\sigma}\cos^2\chi + E_0^2\left(\beta^2 + \frac{1}{4}\right)e^{-2\sigma}\sin^2\chi \tag{18.18}$$

Thus, we have for the variance in the electric field of the squeezed state

$$(\Delta E)^2 = \left\langle \widehat{E}^2 \right\rangle - \left\langle \widehat{E} \right\rangle^2 = \frac{E_0^2}{4} e^{-2\sigma} \sin^2\chi + \frac{E_0^2}{4} e^{2\sigma} \cos^2\chi \tag{18.19}$$

If we choose $z = 0$, then $\chi = \omega t$ and we have

$$(\Delta E)^2 = \frac{E_0^2}{4} e^{-2\sigma} \sin^2\omega t + \frac{E_0^2}{4} e^{2\sigma} \cos^2\omega t \tag{18.20}$$

and

$$\left\langle \widehat{E} \right\rangle = E_0 \beta e^{-\sigma} \sin \omega t \tag{18.21}$$

We see from Eq. (18.21) that the expectation value of the electric field oscillates sinusoidally at the frequency ω and from Eq. (18.20) that the variance changes periodically with time. This is unlike the coherent state where the variance was constant and independent of time (see Chap. 15). Indeed, if we put $\sigma = 0$, the state becomes a coherent state and $(\Delta E)^2 = \frac{E_0^2}{4}$.

The expectation value of the photon number in the squeezed state is given by

$$\langle \beta | \widehat{N} | \beta \rangle = \langle \beta | \widehat{a}^\dagger \widehat{a} | \beta \rangle = \langle \beta | \left(\mu \widehat{b}^\dagger - \nu \widehat{b} \right) \left(\mu \widehat{b} - \nu \widehat{b}^\dagger \right) | \beta \rangle$$

Opening the brackets and using the commutation relation given by Eq. (18.11) give us

$$\langle \beta | \widehat{N} | \beta \rangle = \langle \beta | \left(\mu^2 \widehat{b}^\dagger \widehat{b} - \mu\nu \widehat{b}^{\dagger 2} - \mu\nu \widehat{b}^2 + \nu^2 \widehat{b}^\dagger \widehat{b} + \nu^2 \right) | \beta \rangle$$

Using Eqs. (18.5) and (18.6), we have

$$\left\langle \widehat{N} \right\rangle = (\mu - \nu)^2 \beta^2 + \nu^2 = e^{-2\sigma} \beta^2 + \nu^2 \tag{18.22}$$

Note that a coherent state would correspond to $\nu = 0$ and thus $\sigma = 0$ and we obtain $\left\langle \widehat{N} \right\rangle = \beta^2$.

Exercise 18.2 Consider a squeezed state with $\widehat{b} = \mu \widehat{a} - \sqrt{15} \widehat{a}^\dagger$. (a) What should be the value of μ? (b) Show that this state cannot have less than 15 photons on an average.

Solution

(a) Using Eq. (18.3) we have $\mu^2 - 15 = 1$, which gives us $|\mu| = 4$. Hence, $\widehat{b} = 4\widehat{a} - \sqrt{15}\widehat{a}^\dagger$.

(b) The eigenvalue equation of the operator $\widehat{b}$ is given by Eq. (18.5) where β is a complex quantity. The expectation value of the photon number in the state is then

$$\langle \widehat{N} \rangle = \langle \beta | \widehat{N} | \beta \rangle = \langle \beta | \widehat{a}^\dagger \widehat{a} | \beta \rangle$$

Recalling Eq. (18.9) we have $\widehat{a} = 4\widehat{b} + \sqrt{15}\widehat{b}^\dagger$ and $\widehat{a}^\dagger = 4\widehat{b}^\dagger + \sqrt{15}\widehat{b}$. Hence

$$\widehat{a}^\dagger \widehat{a} = \left(4\widehat{b}^\dagger + \sqrt{15}\widehat{b}\right)\left(4\widehat{b} + \sqrt{15}\widehat{b}^\dagger\right) = 31\widehat{b}^\dagger \widehat{b} + 4\sqrt{15}\widehat{b}^{\dagger 2} + 4\sqrt{15}\widehat{b}^2 + 15$$

Thus, the expectation value of the photon number in this state is

$$\left\langle \widehat{N} \right\rangle = 31|\beta|^2 + 4\sqrt{15}\left(\beta^{*2} + \beta^2\right) + 15$$

For a squeezed vacuum state $|\beta| = 0$ and we have $\left\langle \widehat{N} \right\rangle = 15$. For any other squeezed state with $\widehat{b} = 4\widehat{a} - \sqrt{15}\widehat{a}^\dagger$, the expectation value of photon number $\left\langle \widehat{N} \right\rangle > 15$.

18.4 Amplitude Squeezed State

We refer to Eqs. (18.20) and (18.21) and first assume $\sigma > 0$. Then at times given by

$$t = \frac{\pi}{2\omega}, \frac{3\pi}{2\omega} \cdots \tag{18.23}$$

the variance is

$$(\Delta E)^2 = \frac{E_0^2}{4} e^{-2\sigma} \tag{18.24}$$

which is less than that of a coherent state for which $(\Delta E)^2 = \frac{E_0^2}{4}$ (see Eq. 15.41). At these times $\left\langle \widehat{E} \right\rangle$ has maximum values of $E_0 \beta e^{-\sigma}$.

Similarly at

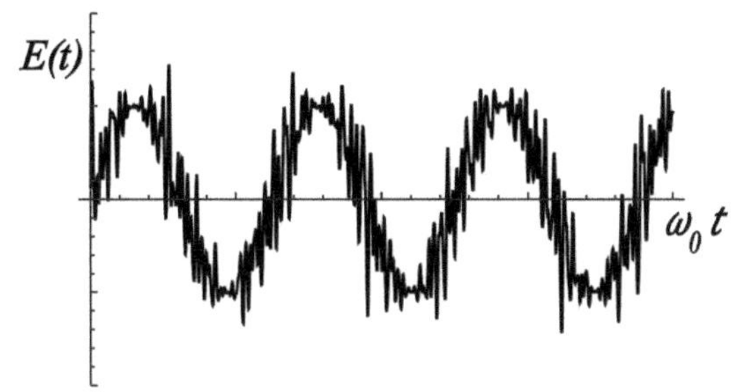

Fig. 18.1 Schematic of the electric field variation in an amplitude squeezed light. Notice the reduced noise at the times when the amplitude is maximum. At the same time, the noise is much higher as the field passes through zero

$$t = 0, \frac{\pi}{\omega}, \frac{2\pi}{\omega} \ldots \tag{18.25}$$

the variance is

$$(\Delta E)^2 = \frac{E_0^2}{4} e^{2\sigma} \tag{18.26}$$

which is greater than that of a coherent state. At these instants $\left\langle \widehat{E} \right\rangle = 0$.

Such a state is referred to as amplitude squeezed state since the noise is less than that of a coherent state at the peaks of the electric field and thus the amplitude of the field has less uncertainty at those instants of time. Notice that the noise is larger than a coherent state at times when the field passes through the origin, that is, at times when $\left\langle \widehat{E} \right\rangle = 0$. Thus, in contrast to a coherent state, such a squeezed state has lower noise in amplitude and additional noise in phase.

Figure 18.1 represents the time dependence of an amplitude squeezed light and Fig. 18.2 shows a schematic representation of amplitude squeezing. The projection on the vertical axis (representing the expectation value and noise in the electric field) changes its width with time and at times corresponding to Eq. (18.23) has minimum widths and at times corresponding to Eq. (18.25) has maximum widths.

18.5 Phase Squeezed State

If we now assume $\sigma < 0$, then at

$$t = \frac{\pi}{2\omega}, \frac{3\pi}{2\omega} \ldots \tag{18.27}$$

the variance is

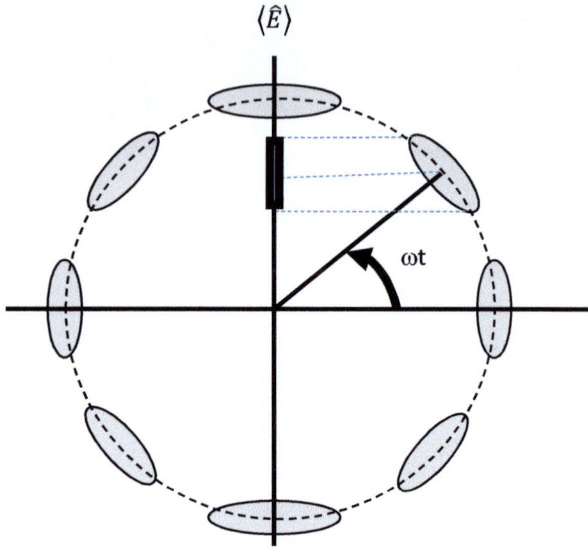

Fig. 18.2 Schematic representation of an amplitude squeezed state by a rotating ellipse with its major axis tangential to the rotating circle representing the time variation in the expectation value of the electric field. As the ellipse rotates, the projection on the vertical axis changes its width and has minimum at times given by Eq. (18.23) and maximum at times given by Eq. (18.25)

$$(\Delta E)^2 = \frac{E_0^2}{4} e^{2\sigma} \tag{18.28}$$

which is greater than that of a coherent state for which $(\Delta E)^2 = \frac{E_0^2}{4}$. At these times $\left\langle \widehat{E} \right\rangle$ has maximum values of $E_0 \beta e^{-\sigma}$.

Similarly at

$$t = 0, \frac{\pi}{\omega}, \frac{2\pi}{\omega} \ldots \tag{18.29}$$

the variance is

$$(\Delta E)^2 = \frac{E_0^2}{4} e^{-2\sigma} \tag{18.30}$$

which is less than that of a coherent state. At these instants $\left\langle \widehat{E} \right\rangle = 0$.

This corresponds to what is referred to as phase squeezing since the phase of the wave is more precisely defined as compared to a coherent state. Figure 18.3 shows a phase squeezed light and Fig. 18.4 shows a schematic representation of phase squeezing. The projection on the vertical axis changes its width with time and at times corresponding to Eq. (18.27) has maximum widths and at times corresponding to Eq. (18.29) has minimum widths.

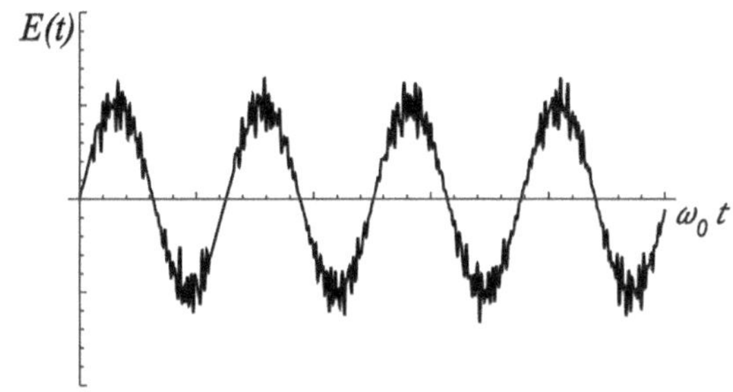

Fig. 18.3 Electric field variation in a phase squeezed state. Note that the noise is very low at times when the field passes through a zero and much higher at the times when the amplitude is maximum

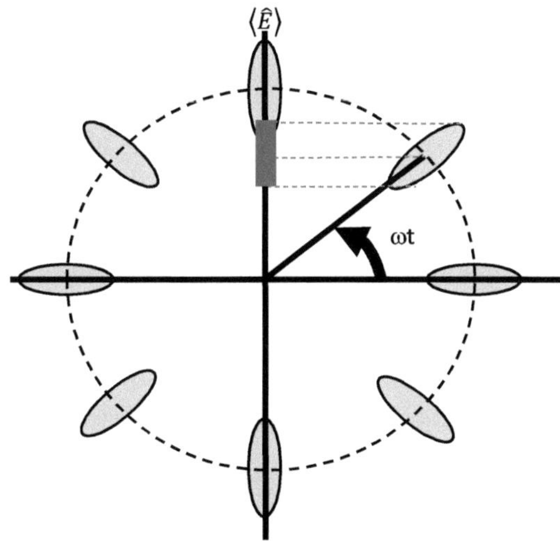

Fig. 18.4 Schematic representation of the phase squeezed state by a rotating ellipse with its minor axis tangential to the rotating circle representing the time variation in the expectation value of the electric field. As the ellipse rotates, the projection on the vertical axis changes its width and has maximum at times given by Eq. (18.27) and minimum at times given by Eq. (18.29)

18.6 Variances in the Quadratures

We will show that the variances in the two individual quadratures can be below ¼ with of course the other quadrature having a variance larger than ¼ so that Heisenberg uncertainty relation is satisfied. In order to evaluate the time dependence of the variance of the two quadratures, we use the Heisenberg picture (see Chap. 21) in which the operators evolve in time and the state remains fixed.

In the following discussion for simplicity, we will assume $\phi = 0$.

We first note that the annihilation and creation operators have the following time variations (see Chap. 15):

$$\hat{a}(t) = \hat{a}(0)e^{-i\omega t} \tag{18.31}$$

Hence,

$$\hat{a}^{\dagger}(t) = \hat{a}^{\dagger}(0)e^{i\omega t} \tag{18.32}$$

Thus, we have

$$\widehat{b}(0) = \mu\widehat{a}(0) + \nu\widehat{a}^{\dagger}(0) \tag{18.33}$$

and

$$\widehat{b}(t) = \mu\widehat{a}(t) + \nu\widehat{a}^{\dagger}(t) = \mu\widehat{a}(0)e^{-i\omega t} + \nu\widehat{a}^{\dagger}(0)e^{i\omega t} \tag{18.34}$$

We can invert Eq. (18.33) and the equation for $\widehat{b}^{\dagger}(0)$ to obtain

$$\widehat{a}(0) = \mu\widehat{b}(0) - \nu\widehat{b}^{\dagger}(0) \tag{18.35}$$

and

$$\widehat{a}^{\dagger}(0) = \mu\widehat{b}^{\dagger}(0) - \nu\widehat{b}(0) \tag{18.36}$$

where for simplicity we have assumed μ and ν to be real quantities.

Now

$$\left\langle \widehat{X}_1(t) \right\rangle = \langle\beta|\widehat{X}_1(t)|\beta\rangle = \langle\beta|\frac{\widehat{a}(t) + \widehat{a}^{\dagger}(t)}{2}|\beta\rangle$$

Substituting from Eqs. (18.31) and (18.32), we obtain

$$\begin{aligned}\left\langle \widehat{X}_1(t) \right\rangle &= \langle\beta|\frac{\widehat{a}(0)e^{-i\omega t} + \widehat{a}^{\dagger}(0)e^{i\omega t}}{2}|\beta\rangle \\ &= \langle\beta|\frac{\left(\mu\widehat{b}(0) - \nu\widehat{b}^{\dagger}(0)\right)e^{-i\omega t} + \left(\mu\widehat{b}^{\dagger}(0) - \nu\widehat{b}(0)\right)e^{i\omega t}}{2}|\beta\rangle\end{aligned}$$

which using the eigenvalue equation (Eq. 18.5) simplifies to

$$\left\langle \widehat{X}_1(t) \right\rangle = (\mu - \nu)\beta\cos\omega t \tag{18.37}$$

Similarly

$$\left\langle \widehat{X}_2(t) \right\rangle = -(\mu - \nu)\beta\sin\omega t \tag{18.38}$$

We can also evaluate

$$\left\langle \widehat{X}_1^2(t) \right\rangle = \langle\beta|\widehat{X}_1^2(t)|\beta\rangle = \langle\beta|\frac{\left(\widehat{a}(t) + \widehat{a}^{\dagger}(t)\right)\left(\widehat{a}(t) + \widehat{a}^{\dagger}(t)\right)}{4}|\beta\rangle$$

which after some mathematical steps comes out to be

$$\left\langle \widehat{X}_1^2(t) \right\rangle = \frac{1}{4}(\mu-\nu)^2 + (\mu-\nu)^2\beta^2\cos^2\omega t + \mu\nu\sin^2\omega t \tag{18.39}$$

Thus, variance in the quadrature $\widehat{X}_1$ is given by

$$\left(\Delta\widehat{X}_1(t)\right)^2 = \left\langle \widehat{X}_1^2(t) \right\rangle - \left\langle \widehat{X}_1(t) \right\rangle^2 = \frac{1}{4}(\mu-\nu)^2 + \mu\nu\sin^2\omega t \tag{18.40}$$

We can similarly evaluate $\left\langle \widehat{X}_2(t) \right\rangle$ and $\left\langle \widehat{X}_2^2(t) \right\rangle$ and obtain the variance in the quadrature $\widehat{X}_2$ as

$$\left(\Delta\widehat{X}_2(t)\right)^2 = \left\langle \widehat{X}_2^2(t) \right\rangle - \left\langle \widehat{X}_2(t) \right\rangle^2 = \frac{1}{4}(\mu-\nu)^2 + \mu\nu\cos^2\omega t \tag{18.41}$$

We can see from Eqs. (18.40) and (18.41) that the variances in the two quadratures vary with time. The variances in both quadratures vary from $\frac{1}{4}(\mu-\nu)^2$ to $\frac{1}{4}(\mu+\nu)^2$. Since $\mu = \cosh\sigma$ and $\nu = \sinh\sigma$, the variance oscillates between $\frac{1}{4}e^{-2\sigma}$ to $\frac{1}{4}e^{+2\sigma}$. Thus, the variances in the two quadratures have values lower than that in the coherent state (which was ¼) at certain instants of time.

The quadrature $\widehat{X}_1$ has minimum variance (assuming $\sigma > 0$) of $\frac{1}{4}e^{-2\sigma}$ at times $t = 0, \frac{\pi}{\omega}, \frac{2\pi}{\omega}, \ldots$. Similarly the quadrature $\widehat{X}_2$ has minimum variance of $\frac{1}{4}e^{-2\sigma}$ at times $t = \frac{\pi}{2\omega}, \frac{3\pi}{2\omega}, \ldots$. Hence, the variances vary periodically with time with minimum values lower than the coherent states. Such a low variance in the quadrature can be used for enhancing the sensitivity of detectors and find applications in quantum metrology, gravitational wave detection, etc.

Figure 18.5 shows a diagrammatic representation of squeezed state at $t = 0$ with $\sigma > 0$ and $\sigma < 0$ referred to as amplitude squeezed state and phase squeezed states, respectively.

A squeezed vacuum state (see Fig. 18.5c) corresponds to $\beta = 0$ and has $\left\langle \widehat{X}_1(t) \right\rangle = 0$ and $\left\langle \widehat{X}_2(t) \right\rangle = 0$. The variances are still given by Eqs. (18.40) and (18.41) and vary with time.

At times when $\sin\omega t = 0$, the two variances are

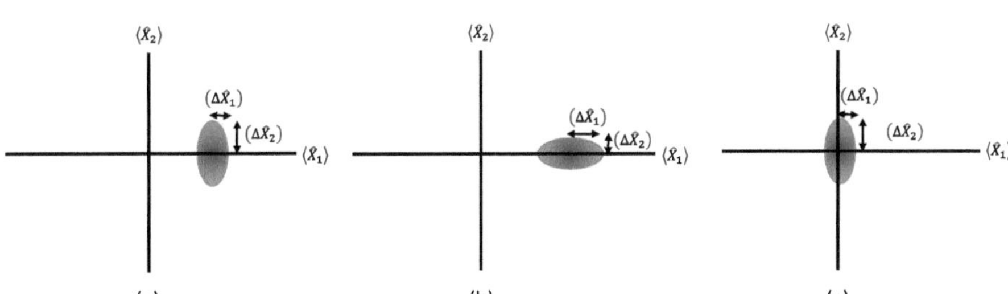

Fig. 18.5 Diagrammatic representations of squeezed state at $t = 0$ with (**a**) $\sigma > 0$ and (**b**) $\sigma < 0$. (**c**) A squeezed vacuum state

$$\left(\Delta\widehat{X}_1\right)^2 = \frac{1}{4}(\mu - \nu)^2 = \frac{e^{-2\sigma}}{4} \tag{18.42}$$

and

$$\left(\Delta\widehat{X}_2\right)^2 = \frac{1}{4}(\mu + \nu)^2 = \frac{e^{2\sigma}}{4} \tag{18.43}$$

Thus, we have

$$\left(\Delta\widehat{X}_1\right)^2\left(\Delta\widehat{X}_2\right)^2 = \frac{1}{16} \tag{18.44}$$

giving us

$$\left\langle\Delta\widehat{X}_1\right\rangle\left\langle\Delta\widehat{X}_2\right\rangle = \frac{1}{4} \tag{18.45}$$

showing that the state is a minimum uncertainty state.

At times when $\cos\omega t = 0$, the two variances are

$$\left(\Delta\widehat{X}_1\right)^2 = \frac{1}{4}(\mu + \nu)^2 = \frac{e^{2\sigma}}{4} \tag{18.46}$$

and

$$\left(\Delta\widehat{X}_2\right)^2 = \frac{1}{4}(\mu - \nu)^2 = \frac{e^{-2\sigma}}{4} \tag{18.47}$$

For $\sigma > 0$, we note that the uncertainty (ΔX_1) in the sine quadrature is less than that of a coherent state, while the uncertainty (ΔX_2) for the cosine quadrature is greater than that of a coherent state; recall that for a coherent state the uncertainties in both the quadratures are equal to $\frac{1}{2}$.

For $\sigma < 0$, we note that the uncertainty (ΔX_1) in the sine quadrature is greater than for a coherent state, while the uncertainty (ΔX_2) for the cosine quadrature is less than that of a coherent state.

18.7 Squeezed Vacuum State

A squeezed vacuum state is a state that satisfies the following eigenvalue equation:

$$\widehat{b}|0_S\rangle = 0 \tag{18.48}$$

where the subscript S refers to squeezed state. This is similar to the vacuum state that satisfies the equation

$$\hat{a}|0\rangle = 0 \tag{18.49}$$

The expectation value of the photon number in a vacuum state is

$$\langle 0|\hat{N}|0\rangle = \langle 0|\hat{a}^{\dagger}\hat{a}|0\rangle = 0 \tag{18.50}$$

For the squeezed vacuum state, we have

$$\langle 0_s|\hat{N}|0_s\rangle = \langle 0_s|\hat{a}^{\dagger}\hat{a}|0_s\rangle = \langle 0_s|\left(\mu\hat{b}^{\dagger} - \nu\hat{b}\right)\left(\mu\hat{b} - \nu\hat{b}^{\dagger}\right)|0_s\rangle = \nu^2 \tag{18.51}$$

where we have used the commutation relation given by Eq. (18.11) and assumed the state to be normalized. Thus, unlike the vacuum state, the squeezed vacuum state has finite number of photons. If we choose $\nu = 0$, the state becomes a normal vacuum state with an expectation value of zero.

For a squeezed vacuum state, we can also get expressions for other quantities such as expectation value of electric field and its variance, the expectation value, and variance of the two quadrature operators by just putting $\beta = 0$ in the expressions derived earlier.

Thus, we have for a squeezed vacuum state

$$\left\langle \hat{E} \right\rangle = E_0 \beta e^{-\sigma} \sin \omega t = 0 \tag{18.52}$$

$$(\Delta E)^2 = \frac{E_0^2}{4} e^{-2\sigma} \sin^2 \omega t + \frac{E_0^2}{4} e^{2\sigma} \cos^2 \omega t \tag{18.53}$$

$$\left\langle \hat{X}_1 \right\rangle = (\mu - \nu)\beta = 0; \quad \left\langle \hat{X}_2 \right\rangle = 0 \tag{18.54}$$

$$(\Delta X_1)^2 = \frac{1}{4}\left(e^{2\sigma} \sin^2 \omega t + e^{-2\sigma} \cos^2 \omega t\right) \tag{18.55}$$

$$(\Delta X_2)^2 = \frac{1}{4}\left(e^{2\sigma} \cos^2 \omega t + e^{-2\sigma} \sin^2 \omega t\right) \tag{18.56}$$

Thus, although the expectation value of the electric field is zero in the squeezed vacuum state (just like the vacuum state), the variance is finite and fluctuates with time. For $\sigma > 0$, the fluctuations in the electric field at $\omega t = (2n + 1)\frac{\pi}{2}$ will be less than that for a vacuum state and for $\omega t = n\pi$, the fluctuations will be greater than that for a vacuum state.

Figure 18.6 shows a schematic of the electric field variation in squeezed vacuum. Chapter 20 presents balanced homodyne technique for the measurement of different quantum states of light.

Fig. 18.6 The squeezed vacuum state with an expectation value of zero and a variance which varies with time. At certain instants, the variance is less than that in a vacuum state and this aspect finds applications in interferometry

Exercise 18.3 Express the squeezed vacuum state in terms of the number states and show that the squeezed vacuum state is a superposition of even number photon states.

Solution The squeezed vacuum state is defined by Eq. (18.48). Expanding the squeezed vacuum state in terms of photon number states, which form an orthonormal set, we write

$$|0_S\rangle = \sum c_n|n\rangle \tag{18.57}$$

Using Eq. (18.48), we have

$$\widehat{b}|0_S\rangle = \left(\mu\widehat{a} + \nu\widehat{a}^\dagger\right)\sum c_n|n\rangle = 0$$

which gives us

$$\sum c_n\left(\mu\widehat{a} + \nu\widehat{a}^\dagger\right)|n\rangle = \sum c_n(\mu\sqrt{n}|n-1\rangle + \nu\sqrt{n+1}|n+1\rangle) = 0$$

Opening the sum, we have

$$\begin{aligned} c_0\nu|1\rangle + c_1\left(\mu|0\rangle + \nu\sqrt{2}|2\rangle\right) + c_2\left(\mu\sqrt{2}|1\rangle + \nu\sqrt{3}|3\rangle\right) \\ + c_3\left(\mu\sqrt{3}|2\rangle + \nu\sqrt{4}|4\rangle\right) + c_4\left(\mu\sqrt{4}|3\rangle + \nu\sqrt{5}|5\rangle\right)\cdots = 0 \end{aligned}$$

which gives us

$$\begin{aligned} c_1\mu &= 0 \\ c_0\nu + c_2\mu\sqrt{2} &= 0 \\ c_1\nu\sqrt{2} + c_3\mu\sqrt{3} &= 0 \end{aligned}$$

$$c_2\nu\sqrt{3} + c_4\mu\sqrt{4} = 0$$

and so on. These give us

$$c_1 = c_3 = \cdots = c_{(2n+1)} = 0$$
$$c_2 = -c_0\frac{\nu}{\mu\sqrt{2}}$$
$$c_4 = -c_2\frac{\nu\sqrt{3}}{\mu\sqrt{4}}$$

that is, all odd coefficients are zero. The even coefficients are given by

$$c_{2m} = -c_{2m-2}\frac{\nu}{\mu}\frac{\sqrt{2m-1}}{\sqrt{2m}} = -c_{2m-2}(\tanh\sigma)\frac{\sqrt{2m-1}}{\sqrt{2m}}; \quad m = 1, 2, \ldots$$

where we have used $\mu = \cosh\sigma$ and $\nu = \sinh\sigma$. Using this relation, we obtain

$$c_{2m} = (-1)^m(\tanh\sigma)^m\frac{\sqrt{(2m)!}}{2^m m!}c_0 \tag{18.58}$$

The normalized squeezed vacuum state is thus given by

$$|0_S\rangle = \frac{1}{\sqrt{\cosh\sigma}}\sum(-1)^m(\tanh\sigma)^m\frac{\sqrt{(2m)!}}{2^m m!}|2m\rangle$$

Thus, the squeezed vacuum state has only even photon number states in its expansion.

For a squeezed vacuum state, the probability of finding zero photons

$$P_0 = |\langle 0|0_S\rangle|^2 = \frac{1}{\cosh\sigma}$$

For $\sigma = 1.0$, the probability is $P_0 = 0.65$.

Exercise 18.4 Obtain the variance in the photon number in a squeezed vacuum state.

Solution The variance in photon number is given by

$$(\Delta N)^2 = \left\langle \widehat{N}^2 \right\rangle - \left\langle \widehat{N} \right\rangle^2 \tag{18.59}$$

Now,

$$\left\langle \hat{N}^2 \right\rangle = \langle 0_s | \hat{N}^2 | 0_s \rangle = \langle 0_s | \hat{a}^\dagger \hat{a} \hat{a}^\dagger \hat{a} | 0_s \rangle = \langle 0_s | \hat{a}^\dagger \left(\hat{a}^\dagger \hat{a} + 1 \right) \hat{a} | 0_s \rangle = \langle 0_s | \hat{a}^{\dagger 2} \hat{a}^2 + \hat{a}^\dagger \hat{a} | 0_s \rangle$$

Substituting for $\hat{a}$ and $\hat{a}^\dagger$ in terms of $\hat{b}$ and $\hat{b}^\dagger$ from Eqs. (18.9) and (18.10), we have

$$\left\langle \hat{N}^2 \right\rangle = \langle 0_s | \left(\mu \hat{b}^\dagger - \nu \hat{b} \right)^2 \left(\mu \hat{b} - \nu \hat{b}^\dagger \right)^2 | 0_s \rangle + \langle 0_s | \left(\mu \hat{b}^\dagger - \nu \hat{b} \right) \left(\mu \hat{b} - \nu \hat{b}^\dagger \right) | 0_s \rangle$$

We open up the brackets and using the commutation relation given by Eq. (18.11) and the fact that $\hat{b}|0_s\rangle = 0$ and $\langle 0_s | \hat{b}^\dagger = 0$, we obtain

$$\langle 0_s | \hat{N}^2 | 0_s \rangle = 3\nu^4 + 2\nu^2$$

Hence, the variance in the photon number in the squeezed vacuum state is

$$(\Delta N)^2 = \left\langle \hat{N}^2 \right\rangle - \left\langle \hat{N} \right\rangle^2 = 2\left(\nu^4 + \nu^2\right) \tag{18.60}$$

Note that compared to a squeezed vacuum state, a vacuum state $|0\rangle$ is an eigenstate of the number operator and hence $\left\langle \hat{N} \right\rangle = 0$ and $(\Delta N)^2 = 0$.

18.8 Two-Mode Squeezed State

Consider two annihilation operators $\hat{b}_s$ and $\hat{b}_i$ corresponding to two different modes of the electromagnetic field satisfying the following equations:

$$\hat{b}_s = \mu \hat{a}_s + \nu \hat{a}_i^\dagger \tag{18.61}$$

$$\hat{b}_i = \nu \hat{a}_s^\dagger + \mu \hat{a}_i \tag{18.62}$$

And μ and ν satisfy the following equation:

$$|\mu|^2 - |\nu|^2 = 1 \tag{18.63}$$

Here $\hat{a}_s$ and $\hat{a}_i$ are annihilation operators of two modes of the electromagnetic field. The subscripts s and i correspond to signal and idler which are two different modes of the electromagnetic field; we will see that such squeezed states are produced through the process of spontaneous parametric down conversion (SPDC) in a nonlinear medium (see Chap. 22).

From Eqs. (18.61) and (18.62), we also have

$$\widehat{b}_s^\dagger = \mu \widehat{a}_s^\dagger + \nu \widehat{a}_i \tag{18.64}$$

$$\widehat{b}_i^\dagger = \nu \widehat{a}_s + \mu \widehat{a}_i^\dagger \tag{18.65}$$

where for simplicity we have assumed μ and ν to be real.

Using the commutation relations $\left[\widehat{a}_s, \widehat{a}_s^\dagger\right] = 1$ and $\left[\widehat{a}_i, \widehat{a}_i^\dagger\right] = 1$, we can show that the operator $\widehat{b}_S$ and $\widehat{b}_i$ satisfy the required commutation relations:

$$\left[\widehat{b}_s, \widehat{b}_s^\dagger\right] = 1 \tag{18.66}$$

$$\left[\widehat{b}_i, \widehat{b}_i^\dagger\right] = 1 \tag{18.67}$$

We can invert Eqs. (18.61) and (18.62) and obtain

$$\widehat{a}_s = \mu \widehat{b}_s + \nu b_i^\dagger \tag{18.68}$$

$$\widehat{a}_i = \mu \widehat{b}_i + \nu b_s^\dagger \tag{18.69}$$

We now define a two-mode squeezed vacuum state through the following equations:

$$\widehat{b}_s |0_s, 0_i\rangle = 0 \tag{18.70}$$

$$\widehat{b}_i |0_s, 0_i\rangle = 0 \tag{18.71}$$

We first calculate the mean photon number and the fluctuations in the photon number in the two modes. Thus,

$$\langle \widehat{N}_s \rangle = \langle 0_s, 0_i | \widehat{N}_s | 0_s, 0_i \rangle = \langle 0_s, 0_i | \widehat{a}_s^\dagger \widehat{a}_s | 0_s, 0_i \rangle$$

Substituting from Eq. (18.68), we have

$$\langle \widehat{N}_s \rangle = \langle 0_s, 0_i | \left(\mu b_s^\dagger + \nu \widehat{b}_i\right) \left(\mu \widehat{b}_s + \nu b_i^\dagger\right) |0_s, 0_i\rangle$$

Opening the brackets and using Eqs. (18.70) and (18.71), we have

$$\langle N_s \rangle = \nu^2 \tag{18.72}$$

Similarly we have

$$\langle N_i \rangle = \langle 0_s, 0_i | \hat{a}_i^\dagger \hat{a}_i | 0_s, 0_i \rangle = \nu^2 \tag{18.73}$$

Thus, the mean photon numbers in the two modes are the same and equal to ν^2.

We also calculate the variance in the photon-number difference between the two modes.

$$\left\langle \Delta\left(\hat{N}_s - \hat{N}_i\right)^2 \right\rangle = \left\langle \left(\hat{N}_s - \hat{N}_i\right)^2 \right\rangle - \left\langle \left(\hat{N}_s - \hat{N}_i\right) \right\rangle^2 \tag{18.74}$$

Now

$$\hat{N}_s - \hat{N}_i = \hat{a}_s^\dagger \hat{a}_s - \hat{a}_i^\dagger \hat{a}_i \tag{18.75}$$

Substituting from Eqs. (18.68) and (18.69), we obtain

$$\hat{N}_s - \hat{N}_i = \hat{b}_s^\dagger \hat{b}_s - \hat{b}_i^\dagger \hat{b}_i \tag{18.76}$$

and

$$\left(\hat{N}_s - \hat{N}_i\right)^2 = \left(\hat{b}_s^\dagger \hat{b}_s - \hat{b}_i^\dagger \hat{b}_i\right)^2 = \hat{b}_s^{\dagger 2} \hat{b}_s^2 + \hat{b}_i^{\dagger 2} \hat{b}_i^2 - 2\hat{b}_s^\dagger \hat{b}_s \hat{b}_i^\dagger \hat{b}_i + \hat{b}_s^\dagger \hat{b}_s + \hat{b}_i^\dagger \hat{b}_i$$

Hence,

$$\left\langle \hat{N}_s - \hat{N}_i \right\rangle = \langle 0_s, 0_i | \left(\hat{N}_s - \hat{N}_i\right) | 0_s, 0_i \rangle = 0 \tag{18.77}$$

and

$$\left\langle \left(\hat{N}_s - \hat{N}_i\right)^2 \right\rangle = \langle 0_s, 0_i | \left(\hat{N}_s - \hat{N}_i\right)^2 | 0_s, 0_i \rangle = 0 \tag{18.78}$$

Hence,

$$\left\langle \Delta\left(\hat{N}_s - \hat{N}_i\right)^2 \right\rangle = 0 \tag{18.79}$$

showing that in this state, the variance of the difference in the photon-number of the two modes has no fluctuations.

It is interesting to compare this two-mode squeezed state with a two-mode state with two independent modes, for example, two coherent states $|\alpha_s\rangle$ and $|\alpha_i\rangle$. We write the composite state as $|\alpha_s, \alpha_i\rangle$. We refer to Chap. 15 for a discussion on single-mode coherent states.

The difference in the expectation value of the photon numbers in the two states is

$$\langle (N_s - N_i) \rangle = \langle \alpha_s, \alpha_i | \left(\widehat{a}_s^\dagger \widehat{a}_s - \widehat{a}_i^\dagger \widehat{a}_i \right) | \alpha_s, \alpha_i \rangle = \alpha_s^2 - \alpha_i^2 \tag{18.80}$$

assuming α_s and α_i to be real.

We also have

$$\left\langle (N_s - N_i)^2 \right\rangle = \alpha_s^4 + \alpha_s^2 + \alpha_i^4 + \alpha_i^2 - 2\alpha_s^2 \alpha_i^2 \tag{18.81}$$

Hence,

$$\left\langle \Delta \left(\widehat{N}_s - \widehat{N}_i \right)^2 \right\rangle = \left\langle \left(\widehat{N}_s - \widehat{N}_i \right)^2 \right\rangle - \left\langle \left(\widehat{N}_s - \widehat{N}_i \right) \right\rangle^2 = \alpha_s^2 + \alpha_i^2 \tag{18.82}$$

Since $\langle N_s \rangle = \alpha_s^2$ and $\langle N_i \rangle = \alpha_i^2$, we obtain the variance as

$$\left\langle \Delta \left(\widehat{N}_s - \widehat{N}_i \right)^2 \right\rangle = \langle N_s \rangle + \langle N_i \rangle \tag{18.83}$$

Since the variances in the two coherent states are simply $\langle N_s \rangle$ and $\langle N_i \rangle$ (see Chap. 15), it can be seen that the variance in the difference in the photon number between the two modes is nothing but the sum of the variances of the two modes. This is quite contrary to the two-mode squeezed state defined earlier.

We define the quadrature operators

$$\widehat{X}_{s1} = \frac{\widehat{a}_S + \widehat{a}_s^\dagger}{2}; \quad \widehat{X}_{s2} = \frac{\widehat{a}_S - \widehat{a}_s^\dagger}{2i} \tag{18.84}$$

$$\widehat{X}_{i1} = \frac{\widehat{a}_i + \widehat{a}_i^\dagger}{2}; \quad \widehat{X}_{i2} = \frac{\widehat{a}_i - \widehat{a}_i^\dagger}{2i} \tag{18.85}$$

Substituting from Eqs. (18.68) and (18.69), we have

$$\widehat{X}_{s1} = \frac{\left(\mu \widehat{b}_s + \nu b_i^\dagger \right) + \left(\mu b_s^\dagger + \nu \widehat{b}_i \right)}{2} = \mu \widehat{Y}_{s1} + \nu \widehat{Y}_{i1} \tag{18.86}$$

where

$$\widehat{Y}_{s1} = \frac{\widehat{b}_S + \widehat{b}_s^\dagger}{2}; \quad \widehat{Y}_{i1} = \frac{\widehat{b}_i + \widehat{b}_i^\dagger}{2} \tag{18.87}$$

Thus,

$$\left\langle \widehat{X}_{s1} \right\rangle = \mu \left\langle \widehat{Y}_{s1} \right\rangle + \nu \left\langle \widehat{Y}_{i1} \right\rangle \tag{18.88}$$

and

$$\left\langle \widehat{X}_{s1}^2 \right\rangle = \mu^2 \left\langle \widehat{Y}_{s1}^2 \right\rangle + \nu^2 \left\langle \widehat{Y}_{i1}^2 \right\rangle + 2\mu\nu \left\langle \widehat{Y}_{s1} \widehat{Y}_{i1} \right\rangle \tag{18.89}$$

Hence, the variance in the quadrature is given by

$$(\Delta X_{s1})^2 = \left\langle \widehat{X}_{s1}^2 \right\rangle - \left\langle \widehat{X}_{s1} \right\rangle^2 = \mu^2 (\Delta Y_{s1})^2 + \nu^2 (\Delta Y_{i1})^2 + 2\mu\nu \left(\left\langle \widehat{Y}_{s1} \widehat{Y}_{i1} \right\rangle - \left\langle \widehat{Y}_{s1} \right\rangle \left\langle \widehat{Y}_{i1} \right\rangle \right) \tag{18.90}$$

As an example let us consider a case when $\widehat{Y}_{s1} = \widehat{Y}_{i1}$. In such a case

$$\left(\Delta X_{s1}^2 \right) = \left(\mu^2 + \nu^2 + 2\mu\nu \right) (\Delta Y_{s1})^2 \tag{18.91}$$

If we write $\mu = \cosh \sigma$, $\nu = \sinh \sigma$, we have

$$\left\langle \widehat{X}_{s1} \right\rangle = e^{\sigma} \left\langle \widehat{Y}_{s1} \right\rangle \tag{18.92}$$

$$\left(\Delta X_{s1}^2 \right) = e^{2\sigma} (\Delta Y_{s1})^2 \tag{18.93}$$

We note that

$$\frac{\left\langle \widehat{X}_{s1} \right\rangle^2}{\left(\Delta X_{s1}^2 \right)} = \frac{\left\langle \widehat{Y}_{s1} \right\rangle^2}{(\Delta Y_{s1})^2} \tag{18.94}$$

We will use these discussions in Chap. 22 where we will discuss about parametric amplifier.

Exercise 18.5 Show that $\widehat{b}_s$ satisfies the commutation relation given by Eq. (18.56).

Solution Using Eq. (18.61) we obtain

$$\left[\widehat{b}_s, \widehat{b}_s^\dagger \right] = \left[\left(\mu \widehat{a}_s + \nu \widehat{a}_i^\dagger \right), \left(\mu \widehat{a}_s^\dagger + \nu \widehat{a}_i \right) \right] = \mu^2 \left[\widehat{a}_s, \widehat{a}_s^\dagger \right] + \mu\nu \left[\widehat{a}_i^\dagger, \widehat{a}_s^\dagger \right]$$
$$+ \mu\nu [\widehat{a}_s, \widehat{a}_i] + \nu^2 \left[\widehat{a}_i^\dagger, \widehat{a}_i \right] = \mu^2 - \nu^2 = 1$$

18.9 Generation of Squeezed Vacuum State

In Chap. 13, we discussed the parametric amplification process in which a pump at a frequency 2ω can amplify a signal at a frequency ω which is the case of degenerate parametric amplification. We also showed that in the absence of the signal input at frequency ω, the classical equations cannot predict the generation of the frequency ω. However, it was mentioned that experimentally one does observe the generation of frequency ω from an input pump at frequency 2ω provided phase matching condition is satisfied. This is called spontaneous parametric down conversion and has quantum mechanical origin. In Chap. 22, we will discuss the SPDC process in greater detail; here we will employ a quantum mechanical formalism and show that the output from such an interaction is a squeezed vacuum state that shows lower noise in one quadrature at the expense of higher noise in the other quadrature.

In order to describe the parametric process quantum mechanically, we first recall that in a degenerate parametric amplifier the equation governing the amplification process is given by (see Eq. 13.12)

$$E_s(z) = E_{s0}e^{-i\phi_{s0}}\left[\cosh\gamma z + ie^{-i\Delta\phi}\sinh\gamma z\right] \tag{18.95}$$

where

$$\Delta\phi = \phi_{p0} - 2\phi_{s0} \tag{18.96}$$

with ϕ_{p0} being the initial phase of the pump and ϕ_{s0} the initial phase of the signal, E_{s0} is the amplitude of the input signal and γ is defined in Eq. (13.7). Note that if $E_{s0} = 0$, that is, there is no input at the frequency ω, then the signal will never get generated as $E_s(z) = 0$ for all z. The generation of the wave at ω with no input at the frequency ω is a purely quantum mechanical phenomenon.

In Chap. 19, we will discuss the quantum mechanical treatment of a beam splitter which is described through a transfer matrix between the input and output annihilation operators. The quantum picture was obtained from the classical description of the beam splitter and then using that to describe the beam splitter in the quantum mechanical picture.

Here we will adopt the same procedure and use Eq. (18.95) to describe the parametric process through an equation connecting the input and output annihilation and creation operators.

Opening the bracket in Eq. (18.95), we obtain,

$$E_s(L) = E_{s0}e^{-i\phi_{s0}}\cosh\gamma L + E_{s0}e^{i\phi_{s0}}e^{i\chi}\sinh\gamma L \tag{18.97}$$

where we have replaced z by the length of interaction L and $\chi = \left(\frac{\pi}{2} - \phi_{p0}\right)$.

In order to get the quantum picture, we replace the complex field amplitude of the input electric field $E_{s0}e^{-i\phi_{s0}}$ and $E_{s0}e^{i\phi_{s0}}$ by annihilation and creation operators, respectively, and the output electric field $E_s(L)$ by annihilation operator to obtain the following relation between the annihilation ($\widehat{a}$) and creation ($\widehat{a}^\dagger$) operators of the input modes and annihilation operator ($\widehat{b}$) of the output mode:

$$\widehat{b} = \widehat{a} \cosh \gamma L + e^{i\chi} \widehat{a}^\dagger \sinh \gamma L \tag{18.98}$$

Writing

$$\mu = \cosh \gamma L; \quad \nu = e^{i\chi} \sinh \gamma L \tag{18.99}$$

we obtain

$$\widehat{b} = \mu \widehat{a} + \nu \widehat{a}^\dagger \tag{18.100}$$

which is nothing but the squeezing operator (see Eq. 18.2). Thus, the parametric process leads to squeezing of the electromagnetic field.

In the case of nondegenerate parametric amplifier, the equations connecting the output and input fields at the signal and idler are given by (see Eqs. 13.48 and 13.49)

$$\tilde{A}_s(L) = \tilde{A}_{s0} \cosh gL + \tilde{A}_{i0}^* e^{i\chi} \sinh gL \tag{18.101}$$

$$\tilde{A}_i^*(L) = \tilde{A}_{s0} e^{-i\chi} \sinh gL + \tilde{A}_{i0}^* \cosh gL \tag{18.102}$$

As before to move to the quantum picture, we replace the complex field amplitudes of the input and output electric fields of the two fields by the corresponding annihilation and creation operators of the signal and idler fields and obtain

$$\widehat{b}_s = \cosh gL \widehat{a}_s + e^{i\chi} \sinh gL \widehat{a}_i^\dagger \tag{18.103}$$

$$\widehat{b}_i^\dagger = e^{-i\chi} \sinh gL \widehat{a}_s + \cosh gL\, \widehat{a}_i^\dagger \tag{18.104}$$

Comparing with Eqs. (18.61) and (18.62), we see that if we define $\mu = \cosh gL$ and $\nu = e^{i\chi} \sinh gL$ with vacuum input the parametric process generates two-mode squeezed vacuum state. Thus, the nondegenerate parametric interaction leads to two-mode squeezed state.

Problems

18.1. Consider an electromagnetic field in a single-mode state $|\psi\rangle$ satisfying the following equation:

$$\left(\sqrt{5}\widehat{a} - 2\widehat{a}^\dagger\right)|\psi\rangle = 0$$

(a) Obtain the expectation value of the photon number in this state.
(b) Obtain the expectation value and variance of $\widehat{X}_1$ and $\widehat{X}_2$ quadratures and draw the corresponding phasor diagram depicting the state.
(c) Is this a minimum uncertainty state?

18.2. Obtain the expectation value and variance of photon number in the squeezed vacuum state with $\widehat{b} = 2\widehat{a} - \sqrt{3}\widehat{a}^{\dagger}$.

18.3. If $|\beta\rangle$ is an eigenstate of the operator $\widehat{b} = \sqrt{2}\widehat{a} - \widehat{a}^{\dagger}$ with eigenvalue β, obtain the variances in the two quadrature operators and plot the corresponding uncertainty area graphically in a figure. Assume β to be real.

18.4. Consider a single-mode quantum state given by

$$|\psi\rangle = \alpha|0\rangle + \beta|1\rangle$$

(a) Normalize $|\psi\rangle)$ and obtain the relationship between α and β.
(b) Obtain the values of α and β for which the quantum state will exhibit squeezing in the quadrature represented by $\widehat{X}_1$.

18.5. An single-mode state is described by

$$|\psi\rangle = \frac{1}{\sqrt{2}}(|n\rangle + |n+2\rangle)$$

Obtain the variances in the two quadratures and check whether the state can be quadrature squeezed.

18.6. Obtain the variance in the photon numbers in a squeezed state. Check that the expression reduces to the correct expression for a coherent state.

18.7. Consider a squeezed vacuum state with $\mu = 1.5$. Obtain an expression for the probability of finding 2 photons in the state.

18.8. Consider a single-mode quantum state given by (assume α and β to be real quantities) the superposition of two number states

$$|\psi\rangle = \alpha|0\rangle + \beta|2\rangle$$

(a) Obtain the variances in both the quadratures $\widehat{X}_1$ and $\widehat{X}_2$.
(b) Assuming $\beta = \epsilon \ll 1$, make appropriate approximations and show that there is squeezing in one of the quadratures.

References

Andersen, U. L., T. Gehring, C. Marquardt, and G. Leuchs. 2016. 30 years of squeezed light generation. *Phys Scr* 91:053001.

Breitenbach, G., and S. Schiller. 1997. Homodyne tomography of classical and non-classical light. *J Modern Opt* 44:2207–2225.

Breitenbach, G., S. Schiller, and J. Mlynek. 1997. Measurement of the quantum states of squeezed light. *Nature* 387:471–475.

Henry, R. W., and S. C. Glotzer. 1988. A squeezed-state primer. *Am J Phys* 56:318–328.

Kashiwazaki, T., T. Yamashima, N. Takanashi, A. Inoue, T. Umeki, and A. Furusawa. 2021. Fabrication of low-loss quasi-single-mode PPLN waveguide and its application to a modularized broadband high-level squeezer. *Appl Phys Lett* 119:251104.

Kashiwazaki, T., T. Yamashima, K. Enbutsu, T. Kazama, A. Inoue, K. Fukui, M. Endo, T. Umeki, and A. Furusawa. 2023. Over 8-dB squeezed light generation by a broadband waveguide optical parametric amplifier towards fault tolerant ultra-fast quantum computers. *Appl Phys Lett* 122: 234003. https://doi.org/10.1063/5.0144385.

Lvovsky, A. I. 2014. Squeezed light. *A guide to experiments in quantum optics*. https://doi.org/10.1002/9783527695805.ch9.

Mehmet, M., H. Vahlbruch, N. Lastzka, K. Danzmann, and R. Schnabel. 2010. Observation of squeezed states with strong photon-number oscillations. *Phys Rev A* 81:013814.

Park, T., H. Stokowsky, V. Ansari, S. Gyger, K. K. S. Multani, O. T. Celik, A. Y. Huwang, D. J. Dean, F. Mayor, T. P. McKenna, M. M. Fejer, and A. Safavi-Naeni. 2024. Single-mode squeezed light generation and tomography with an integrated optical parametric oscillator. *Sci Adv* 10 (11): eadl1814. https://doi.org/10.1126/sciadv.adl1814.

Stokowski, H. S., T. P. McKenna, T. Park, A. Y. Hwang, D. J. Dean, O. T. Celik, V. Ansari, M. M. Fejer, and A. H. Safavi-Naeini. 2023. Integrated quantum optical phase sensor in thin film lithium niobate. *Nat Commun* 14 (1): 3355. https://doi.org/10.1038/s41467-023-38246-6.

Teich, M. C., and B. E. A. Saleh. 1989. Squeezed states of light. *Quant Opt* 1:153–191.

Walls, D. F. 1983. Squeezed states of light. *Nature* 306:141–146.

Chapter 19
Beam Splitter

Abstract Beam splitters form very important components of quantum photonic devices and this chapter presents a quantum description of the beam splitter. Output states from beam splitters under different inputs such as single photons entering through one port, two photons entering through the two input ports, single photon in a multimode state, and entangled photons are discussed. Optical directional couplers as beam splitters are also presented. Mach-Zehnder interferometer which uses two beam splitters is presented showing single photon interference. Interaction-free measurement is also presented.

19.1 Introduction

A beam splitter is a device with two inputs and two outputs and forms a very important component in many optical setups. It is also a very important component in quantum optics and quantum photonics setups where the input could be in a nonclassical state.

A bulk beam splitter consists of a substrate coated with a dielectric film which partially reflects and partially transmits an incident beam. The most common beam splitter is a 50% splitter which reflects 50% and transmits 50% of the incident light, assuming no scattering or absorption in the beam splitter. Such a splitter is also referred to as a 3dB splitter since 3 dB corresponds to 50%. Losses in a device can also be treated in the form of a beam splitter with a very small percentage of reflection corresponding to the loss and a very high percentage of transmission. In integrated optics, waveguide directional couplers behave as beam splitters (see Chap. 5), have similar properties, and are an important component in many integrated quantum photonic devices.

In this chapter, we will obtain some general relations between the amplitude reflectivity and transmittivity of a 50% beam splitter through energy conservation principles. We will then extend the discussion to the representation of the beam splitter in a quantum mechanical system as an operator operating on the annihilation operators of the input states of light. We will study the quantum mechanical analysis of how the beam splitter behaves under different input conditions such as pairs of photons incident on the two input arms which leads to two photon interference.

K. Thyagarajan, *Integrated Quantum Photonics*, Graduate Texts in Physics,
https://doi.org/10.1007/978-3-031-85728-7_19

In order to keep the analysis simple, we will consider a symmetric beam splitter which has the same property for waves incident from either of the input ports.

19.2 Classical Beam Splitter

Figure 19.1 shows a symmetric beam splitter represented by a line and having the same medium on both sides of the beam splitter. The input ports are labeled as 1 and 2 and the corresponding two output ports are labeled 3 and 4. Let the amplitude of the electric fields of the waves input from ports 1 and 2 be represented by E_1 and E_2 and the corresponding electric fields at the two output ports be represented by E_3 and E_4. Let the complex amplitude reflectivity and transmittivity of the beam splitter be represented by r and t, respectively. Thus, from Fig. 19.1, we have the following relations:

$$E_3 = rE_1 + tE_2 \tag{19.1}$$

and

$$E_4 = tE_1 + rE_2 \tag{19.2}$$

If there is no absorption or scattering loss in the beam splitter, then energy conservation requires

$$|E_3|^2 + |E_4|^2 = |E_1|^2 + |E_2|^2 \tag{19.3}$$

Now,

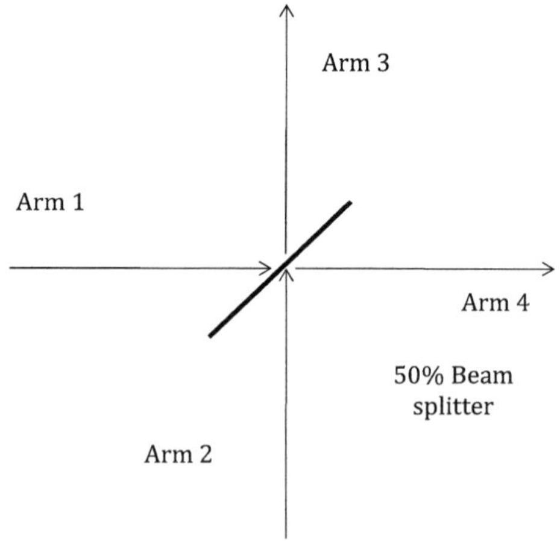

Fig. 19.1 A classical symmetric bulk beam splitter

$$|E_3|^2 + |E_4|^2 = |rE_1 + tE_2|^2 + |tE_1 + rE_2|^2 \tag{19.4}$$

which on simplification gives

$$|E_3|^2 + |E_4|^2 = \left(|r|^2 + |t|^2\right)|E_1|^2 + \left(|r|^2 + |t|^2\right)|E_2|^2 + (r^*t + t^*r)\left(E_1^*E_2 + E_2^*E_1\right)$$

Since we need to satisfy Eq. (19.3) for arbitrary values of E_1 and E_2, we obtain the following relationships:

$$|r|^2 + |t|^2 = 1 \tag{19.5}$$

and

$$r^*t + t^*r = 0 \tag{19.6}$$

For a symmetric 3 dB beam splitter, we have

$$|r| = |t|$$

and hence we obtain

$$|r| = |t| = \frac{1}{\sqrt{2}} \tag{19.7}$$

We write

$$r = |r|e^{i\phi_r} \tag{19.8a}$$

and

$$t = |t|e^{i\phi_t} \tag{19.8b}$$

where ϕ_r and ϕ_t represent the phase shifts occurring at the beam splitter corresponding to the reflected wave and transmitted waves, respectively. Thus, Eq. (19.6) gives

$$|r||t|e^{-i(\phi_r - \phi_t)} + |r||t|\, e^{i(\phi_r - \phi_t)} = 0$$

which implies

$$\cos(\phi_r - \phi_t) = 0$$

Hence, we obtain

$$(\phi_r - \phi_t) = \pm \frac{\pi}{2} \tag{19.9}$$

If we assume that there is no phase shift in transmission through the beam splitter, that is, $\phi_t = 0$, then we have

$$\phi_r = \frac{\pi}{2}$$

where we have taken the positive sign. Thus, we have for the 50% beam splitter

$$r = \frac{i}{\sqrt{2}} \tag{19.10a}$$

$$t = \frac{1}{\sqrt{2}} \tag{19.10b}$$

Energy consideration at the beam splitter gives us the fact that for a symmetric 3 dB lossless beam splitter, the phase shifts at reflection and transmission have to satisfy a relationship (Eq. 19.9). Thus, all lossless beam splitters, bulk or integrated optic with two input ports and two output ports have to satisfy specific phase relationships between the input and output ports as required by energy conservation considerations.

Using the above relationship, we obtain the following equations connecting the output and input electric field amplitudes at a 3 dB lossless beam splitter:

$$E_3 = \frac{i}{\sqrt{2}} E_1 + \frac{1}{\sqrt{2}} E_2 \tag{19.11a}$$

$$E_4 = \frac{1}{\sqrt{2}} E_1 + \frac{i}{\sqrt{2}} E_2 \tag{19.11b}$$

which can be written in a matrix form as

$$\begin{pmatrix} E_3 \\ E_4 \end{pmatrix} = \frac{1}{\sqrt{2}} \begin{pmatrix} i & 1 \\ 1 & i \end{pmatrix} \begin{pmatrix} E_1 \\ E_2 \end{pmatrix} \tag{19.12}$$

Figure 19.2 shows an integrated optic directional coupler of an appropriate length which behaves like a 3 dB beam splitter (see Chap. 5). Note that in Chap. 5, we had shown that in a directional coupler consisting of two identical waveguides acting like a 3 dB splitter where the input power is equally divided among the two output ports, there is a $\frac{\pi}{2}$ phase difference between the two ouput arms exactly as we have shown in the case of bulk beam splitter (see Eqs. 5.34 and 5.35). The transfer matrix of a 3 dB directional coupler consisting of two identical waveguides is given by (see Eq. 5.60 with $\kappa z = \frac{\pi}{4}$)

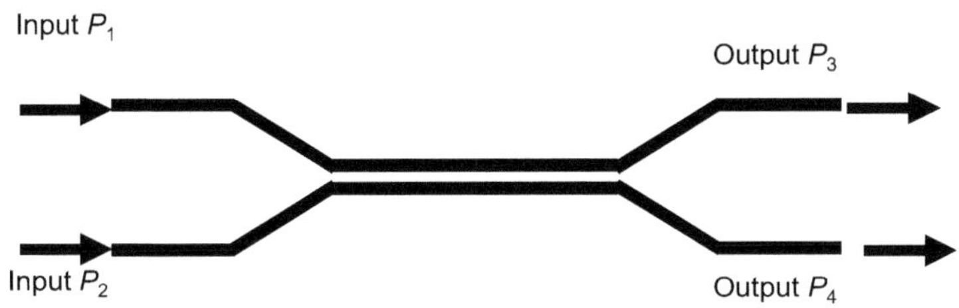

Fig. 19.2 An integrated optic directional coupler of an appropriate length of interaction behaves like a 3 dB beam splitter

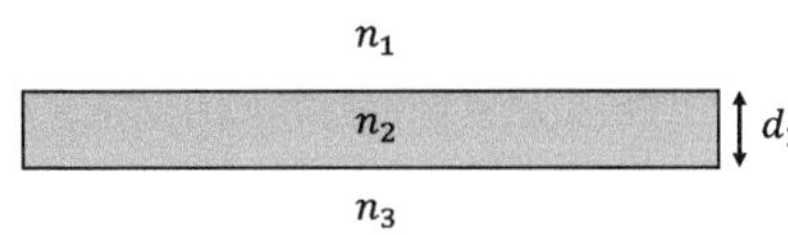

Fig. 19.3 Figure for Exercise 19.1

$$T = \frac{1}{\sqrt{2}}\begin{pmatrix} 1 & i \\ i & 1 \end{pmatrix} \tag{19.13}$$

It may be mentioned that some texts and references use the following transformation for the beam splitter which also satisfies the energy conservation principle (see Problem 19.6):

$$\begin{pmatrix} E_3 \\ E_4 \end{pmatrix} = \frac{1}{\sqrt{2}}\begin{pmatrix} 1 & 1 \\ 1 & -1 \end{pmatrix}\begin{pmatrix} E_1 \\ E_2 \end{pmatrix} \tag{19.14}$$

In this book, we will consistently use Eq. (19.12) to represent a bulk beam splitter and Eq. (19.13) to represent an optical waveguide directional coupler.

Exercise 19.1 Consider reflection of light waves at normal incidence from a parallel film of refractive index n_2 and thickness d_2 which is surrounded by media of refractive indices n_1 and n_3 on either side (see Fig. 19.3). The Fresnel relations for the amplitude reflection and transmission coefficients are well known and for normal incidence are given by (see, e.g., Ghatak and Thyagarajan 1989):

$$r = \frac{r_1 + r_2}{1 + r_1 r_2} \tag{19.15}$$

where

$$r_1 = \frac{n_1 - n_2}{n_1 + n_2}; \quad r_2 = \frac{n_2 - n_3}{n_2 + n_3} \tag{19.16}$$

and the incidence on the film is from the medium of refractive index n_1. Here for simplicity we have assumed the case when $e^{i\delta_2} = 1$ where $\delta_2 = k_0 n_2 d_2$, $k_0 = \frac{2\pi}{\lambda_0}$. Compare the phase of the amplitude reflection coefficients when the wave is incident from region with refractive index n_1 and when the wave is incident from the region with refractive index n_3.

Solution When the incidence is from the medium of refractive index n_3, then the amplitude reflection coefficient can be obtained from Eq. (19.16) by interchanging n_1 and n_3.Thus, to obtain the reflection coefficient r' we replace r_1 by $-r_2$ and r_2 by $-r_1$. Hence, we have (see Eq. 19.15)

$$r' = \frac{-r_2 - r_1}{1 + r_1 r_2} = -\frac{r_1 + r_2}{1 + r_1 r_2} = -r \tag{19.17}$$

Thus, the reflection coefficients have a π phase difference between reflection from the two sides of the film. The reader can check that the amplitude transmission coefficient does not change sign in this process.

Exercise 19.2 Show that the matrix connecting the input and output states of the beam splitter is a unitary matrix.

Solution A matrix M is unitary if it satisfies the following property:

$$M^{\dagger}M = MM^{\dagger} = I \tag{19.18}$$

where $M^{\dagger}$ is the conjugate transpose of M and I is the identity matrix.

Now, from Eq. (19.12), we have

$$M = \frac{1}{\sqrt{2}}\begin{pmatrix} i & 1 \\ 1 & i \end{pmatrix} \tag{19.19}$$

Thus,

$$M^{\dagger} = \frac{1}{\sqrt{2}}\begin{pmatrix} -i & 1 \\ 1 & -i \end{pmatrix} \tag{19.20}$$

Hence,

$$MM^{\dagger} = \frac{1}{2}\begin{pmatrix} i & 1 \\ 1 & i \end{pmatrix}\begin{pmatrix} -i & 1 \\ 1 & -i \end{pmatrix} = I = M^{\dagger}M \tag{19.21}$$

Similarly from Eq. (19.14), we have

$$M = \frac{1}{\sqrt{2}}\begin{pmatrix} 1 & 1 \\ 1 & -1 \end{pmatrix} \tag{19.22}$$

and

$$MM^{\dagger} = \frac{1}{2}\begin{pmatrix} 1 & 1 \\ 1 & -1 \end{pmatrix}\begin{pmatrix} 1 & 1 \\ 1 & -1 \end{pmatrix} = I = M^{\dagger}M \tag{19.23}$$

Exercise 19.3 Consider a directional coupler of length L made up of two nonidentical waveguides. The relationship between the output and input ports is given by Eq. (5.59). Show that the transfer matrix is unitary.

Solution The transfer matrix is given by

$$T = \begin{pmatrix} P & iQ \\ iQ & P^{*} \end{pmatrix} \tag{19.24}$$

where

$$P = \cos \Gamma L + i\frac{\delta}{\Gamma}\sin \Gamma L \tag{19.25}$$

and

$$Q = \frac{\kappa}{\Gamma}\sin \Gamma L; \Gamma = \sqrt{(\kappa^2 + \delta^2)}; \delta = \frac{\Delta\beta}{2} \tag{19.26}$$

Now,

$$T^{\dagger}T = \begin{pmatrix} P^{*} & -iQ \\ -iQ & P \end{pmatrix}\begin{pmatrix} P & iQ \\ iQ & P^{*} \end{pmatrix} = \begin{pmatrix} |P|^2 + |Q|^2 & iP^{*}Q - iP^{*}Q \\ iPQ - iPQ & |P|^2 + |Q|^2 \end{pmatrix} = \begin{pmatrix} 1 & 0 \\ 0 & 1 \end{pmatrix} = I \tag{19.27}$$

Similarly, $TT^{\dagger} = I$.

Exercise 19.4 Consider a 3 dB directional coupler made up of two nonidentical waveguides. Obtain the transfer matrix for such a coupler.

Solution For the coupler to be a 3 dB coupler, we must have

$$Q = \frac{\kappa}{\Gamma}\sin \Gamma L = \frac{1}{\sqrt{2}} \tag{19.28}$$

In this case

$$P = |P|e^{i\tan^{-1}\left(\frac{\delta}{\Gamma}\tan \Gamma L\right)} = |P|e^{i\phi} \tag{19.29}$$

where $\phi = \tan^{-1}\left(\frac{\delta}{\Gamma}\tan \Gamma L\right)$ and from Eq. (19.25), we have

$$|P| = \sqrt{\cos^2\Gamma L + \left(\frac{\delta}{\Gamma}\right)^2 \sin^2\Gamma L} = \frac{1}{\sqrt{2}} \tag{19.30}$$

as expected. In obtaining Eq. (19.30), we have used Eq. (19.28).

We also have

$$\phi = \tan^{-1}\left(\frac{\delta}{\Gamma}\tan\Gamma L\right) = \tan^{-1}\left(\frac{\delta}{\Gamma}\frac{\sin\Gamma L}{\sqrt{1-\sin^2\Gamma L}}\right) = \tan^{-1}\left(\frac{1}{\sqrt{\frac{\kappa^2}{\delta^2}-1}}\right) \tag{19.31}$$

The transfer matrix is then given by

$$T = \frac{1}{\sqrt{2}}\begin{pmatrix} e^{i\phi} & i \\ i & e^{-i\phi} \end{pmatrix} \tag{19.32}$$

Hence, in this 3 dB coupler made up of two nonidentical waveguides (which behaves like a 50% beam splitter) the phases of the output arms depend on the coupler parameters. It is left to the reader to show that the matrix T as given by Eq. (19.32) is unitary. Also check that if the waveguides are identical, then the transfer matrix given by Eq. (19.32) reduces to that given in Eq. (19.13).

Exercise 19.5 It is interesting to note that a waveguide with two guided modes with coupling induced between them using periodic perturbation (see Chap. 9) is also a device with two inputs and two outputs. Referring to the two modes as mode a and mode b, the coupled mode equations with $\Delta\beta - K = 0$ are described by Eqs. (9.34) and (9.35). Consider an input condition given by $A(0) = 0 = A_0$, $B(0) = B_0$. Obtain the relation between the amplitudes of the output modes and input modes.

Solution With the given initial conditions, we obtain

$$\begin{pmatrix} A(z) \\ B(z) \end{pmatrix} = \begin{pmatrix} \cos\kappa z & \sin\kappa z \\ -\sin\kappa z & \cos\kappa z \end{pmatrix}\begin{pmatrix} A(0) \\ B(0) \end{pmatrix}$$

If we assume a length $z = L$ satisfying $\kappa L = \frac{\pi}{2}$, we obtain

$$\begin{pmatrix} A(L) \\ B(L) \end{pmatrix} = \frac{1}{\sqrt{2}}\begin{pmatrix} 1 & 1 \\ -1 & 1 \end{pmatrix}\begin{pmatrix} A(0) \\ B(0) \end{pmatrix}$$

which gives another transfer matrix between the input and output ports of the 2×2 device. It is left to the reader to show that the transfer matrix is unitary.

19.3 Quantum Treatment of Beam Splitter

In order to treat the beam splitter quantum mechanically, we replace the complex field amplitudes E_i by annihilation operators $\widehat{a}_i$ for the ith ports.

19.3.1 Neglecting Vacuum Input

Let us first consider a case when there is only one input E_1 on the beam splitter. In such a case, the outputs will be given by

$$E_3 = rE_1 \tag{19.33a}$$

$$E_4 = tE_1 \tag{19.33b}$$

where for a 3 dB beam splitter, $r = \frac{i}{\sqrt{2}}$ and $t = \frac{1}{\sqrt{2}}$. If we replace the complex electric field amplitudes by the annihilation operators, we obtain the following equations:

$$\widehat{a}_3 = r\widehat{a}_1 \tag{19.34a}$$

$$\widehat{a}_4 = t\widehat{a}_1 \tag{19.34b}$$

However, the operators of each field are supposed to satisfy the commutation relations:

$$\left[\widehat{a}_i, \widehat{a}_j^\dagger\right] = \delta_{ij} \tag{19.35a}$$

$$\left[\widehat{a}_i, \widehat{a}_j\right] = 0 \tag{19.35b}$$

$$\left[\widehat{a}_i^\dagger, \widehat{a}_j^\dagger\right] = 0 \tag{19.35c}$$

Now,

$$\left[\widehat{a}_3, \widehat{a}_3^\dagger\right] = \left[r\widehat{a}_1, r^*\widehat{a}_1^\dagger\right] = |r|^2 \tag{19.36a}$$

$$\left[\widehat{a}_4, \widehat{a}_4^\dagger\right] = \left[t\widehat{a}_1, t^*\widehat{a}_1^\dagger\right] = |t|^2 \tag{19.36b}$$

$$\left[\widehat{a}_3, \widehat{a}_4^\dagger\right] = \left[r\widehat{a}_1, t^*\widehat{a}_1^\dagger\right] = rt^* \neq 0 \tag{19.36c}$$

Thus, the output annihilation and creation operators given by Eqs. (19.34a) and (19.34b) do not satisfy the correct commutation relations and hence **cannot provide the correct quantum description of the beam splitter.**

19.3.2 Correct Quantum Description

In the above discussion, we assumed that there is only one input to the beam splitter and that on the other input port there is no incident wave. In the quantum picture, the unused port has an input corresponding to the vacuum state and in a quantum description it is important to take inputs into both ports into consideration.

Thus, in a quantum description of the beam splitter the relationships between the input annihilation operators and output annihilation operators are given by

$$\widehat{a}_3 = r\widehat{a}_1 + t\widehat{a}_2 \tag{19.37a}$$

$$\widehat{a}_4 = t\widehat{a}_1 + r\widehat{a}_2 \tag{19.37b}$$

With these relations between the input and output operators, we have

$$\begin{aligned}\left[\widehat{a}_3, \widehat{a}_3^\dagger\right] &= \left[(r\widehat{a}_1 + t\widehat{a}_2), \left(r^*\widehat{a}_1^\dagger + t^*\widehat{a}_2^\dagger\right]\right. \\ &= \left[r\widehat{a}_1, r^*\widehat{a}_1^\dagger\right] + \left[r\widehat{a}_1, t^*\widehat{a}_2^\dagger\right] + \left[t\widehat{a}_2, r^*\widehat{a}_1^\dagger\right] + \left[t\widehat{a}_2, t^*\widehat{a}_2^\dagger\right] = |r|^2 + 0 + 0 + |t|^2 = 1\end{aligned} \tag{19.38}$$

which is the correct commutation relation for the output port 3. Similarly it can be shown that

$$\left[\widehat{a}_4, \widehat{a}_4^\dagger\right] = 1 \tag{19.39}$$

$$\left[\widehat{a}_3, \widehat{a}_4^\dagger\right] = 0 \tag{19.40}$$

Thus, following Eqs. (19.11a) and (19.11b), we have the following correct relationship between the input and output annihilation operators of the beam splitter:

$$\widehat{a}_3 = \frac{i}{\sqrt{2}}\widehat{a}_1 + \frac{1}{\sqrt{2}}\widehat{a}_2 \tag{19.41a}$$

$$\widehat{a}_4 = \frac{1}{\sqrt{2}}\widehat{a}_1 + \frac{i}{\sqrt{2}}\widehat{a}_2 \tag{19.41b}$$

Equations (19.41a) and (19.41b) can be combined into the following equation connecting the annihilation operators of the output states and input states of the beam splitter:

$$\begin{pmatrix} \widehat{a}_3 \\ \widehat{a}_4 \end{pmatrix} = \frac{1}{\sqrt{2}} \begin{pmatrix} i & 1 \\ 1 & i \end{pmatrix} \begin{pmatrix} \widehat{a}_1 \\ \widehat{a}_2 \end{pmatrix} \tag{19.42}$$

The transfer matrix relating the output and input annihilation operator is unitary and hence the inverse of the transfer matrix is just its Hermitian conjugate. Thus,

$$\begin{pmatrix} \widehat{a}_1 \\ \widehat{a}_2 \end{pmatrix} = \frac{1}{\sqrt{2}} \begin{pmatrix} -i & 1 \\ 1 & -i \end{pmatrix} \begin{pmatrix} \widehat{a}_3 \\ \widehat{a}_4 \end{pmatrix} \tag{19.43}$$

We describe the input quantum state to the beam splitter by the following:

$$|\psi_{in}\rangle = |\psi_{1,2}\rangle \tag{19.44}$$

corresponding to the input ports 1 and 2. The beam splitter coverts this input quantum state into an output quantum state, which we can be described by

$$|\psi_{out}\rangle = |\psi_{3,4}\rangle \tag{19.45}$$

corresponding to the ports 3 and 4 of the beam splitter. Since the beam splitter converts the state $|\psi_{1,\,2}\rangle$ to the state $|\psi_{3,\,4}\rangle$, we describe the operation of the beam splitter by a unitary operator $\widehat{U}$ such that

$$|\psi_{3,4}\rangle = \widehat{U}|\psi_{1,2}\rangle \tag{19.46}$$

Now consider the problem of obtaining the expectation value of an operator $\widehat{O}_{out}$ at the output of the beam splitter. For this, we need to evaluate the following quantity:

$$\left\langle \widehat{O}_{out} \right\rangle = \langle \psi_{3,4}|\widehat{O}_{out}|\psi_{3,4}\rangle \tag{19.47}$$

As an example, we may need to obtain the expectation value of the number of photons in each of the output ports 3 and 4. In this case $\widehat{O}_{out} = \widehat{a}_3^\dagger \widehat{a}_3$ for output port 3 and $\widehat{O}_{out} = \widehat{a}_4^\dagger \widehat{a}_4$ for the output port 4.

We now substitute for $|\psi_{3,4}\rangle$ and $\langle\psi_{3,4}|$ from Eq. (19.46) in Eq. (19.47) and obtain

$$\left\langle \widehat{O}_{out} \right\rangle = \langle \psi_{1,2}|\widehat{U}^\dagger \widehat{O}_{out} \widehat{U}|\psi_{1,2}\rangle \tag{19.48}$$

which implies that the expectation value of the operator $\widehat{O}_{out}$ between the output state $|\psi_{3,\,4}\rangle$ is the same as the expectation value of the operator $\widehat{U}^\dagger \widehat{O}_{out} \widehat{U}$ taken between the input state $|\psi_{1,\,2}\rangle$.

It should be noted that the same unitary operator $\widehat{U}$ is the operator that transforms between the input and output operators in a change of basis. Thus, we can write

$$\widehat{O}_{in} = \widehat{U}^{\dagger}\widehat{O}_{out}\widehat{U} \tag{19.49}$$

and we obtain

$$\left\langle \widehat{O}_{out} \right\rangle = \langle \psi_{1,2} | \widehat{O}_{in} | \psi_{1,2} \rangle \tag{19.50}$$

which is the expectation value of the observable $\widehat{O}_{out}$ written in the input space as $\widehat{O}_{in}$ and taken with respect to the input state $|\psi_{1,\ 2}\rangle$.

The observable $\widehat{O}_{out}$ will contain annihilation and creation operators corresponding to output ports 3 and 4. To convert $\widehat{O}_{out}$ to $\widehat{O}_{in}$, we have to express $\widehat{a}_3$ and $\widehat{a}_4$ and their Hermitian conjugates in terms of the annihilation and creation operators $\widehat{a}_1$ and $\widehat{a}_2$ corresponding to the input ports 1 and 2. This is nothing but the relation given by Eq. (19.42).

The above discussion follows the excellent lecture by Alain Aspect on beam splitters; the link is given below:

https://www.coursera.org/lecture/quantum-optics-single-photon/3-1-beam-splitter-in-quantum-optics-8yYR1

We now discuss some interesting examples:

19.3.2.1 Single Photon Incident on a Beam Splitter

Let us consider the incidence of a single photon in arm 1 of the beam splitter and with no input in arm 2 which in quantum mechanical terms implies input of vacuum state in arm 2. The input state is then given by

$$|\psi_{in}\rangle = |1\rangle_1 |0\rangle_2 = |1_1, 0_2\rangle \tag{19.51}$$

The subscripts refer to the arm number and the numbers represent the number of photons in that arm.

The expectation value of the number of photons in port 3 will be given by

$$\left\langle \widehat{a}_3^{\dagger}\widehat{a}_3 \right\rangle = \langle \psi_{3,4} | \widehat{a}_3^{\dagger}\widehat{a}_3 | \psi_{3,4} \rangle = \langle \psi_{1,2} | \widehat{O}_{in} | \psi_{1,2} \rangle \tag{19.52}$$

with $\widehat{a}_3$ expressed in terms of $\widehat{a}_1$ and $\widehat{a}_2$:

$$\begin{aligned} \widehat{O}_{in} &= \left(-\frac{i}{\sqrt{2}}\widehat{a}_1^{\dagger} + \frac{1}{\sqrt{2}}\widehat{a}_2^{\dagger} \right) \left(\frac{i}{\sqrt{2}}\widehat{a}_1 + \frac{1}{\sqrt{2}}\widehat{a}_2 \right) \\ &= \frac{1}{2} \left(\widehat{a}_1^{\dagger}\widehat{a}_1 - i\widehat{a}_1^{\dagger}\widehat{a}_2 + i\widehat{a}_2^{\dagger}\widehat{a}_1 + \widehat{a}_2^{\dagger}\widehat{a}_2 \right) \end{aligned} \tag{19.53}$$

Thus, we have

$$\begin{aligned}\left\langle \hat{a}_3^\dagger \hat{a}_3 \right\rangle &= \langle 1_1, 0_2| \frac{1}{2}\left(\hat{a}_1^\dagger \hat{a}_1 - i\hat{a}_1^\dagger \hat{a}_2 + i\hat{a}_2^\dagger \hat{a}_1 + \hat{a}_2^\dagger \hat{a}_2\right)|1_1, 0_2\rangle \\ &= \frac{1}{2}(1+0+0+0) = \frac{1}{2}\end{aligned} \tag{19.54}$$

Similarly the expectation value of the number of photons in port 4 will be

$$\left\langle \hat{a}_4^\dagger \hat{a}_4 \right\rangle = \frac{1}{2} \tag{19.55}$$

Thus, the input photon goes into a superposition state of being in both output arms of the beam splitter. Here, since we are considering a 50% beam splitter, the superposition is with equal probabilities.

19.3.2.2 Output State from the Beam Splitter

In the above discussion, we were able to calculate the expectation value of an observable at the output expressed in terms of the input operators and input state. In order to write the output quantum state for a given input quantum state, we give heuristic arguments below.

If we have a single photon incident on port 1 with vacuum in port 2, then the input state is given by

$$|\psi_{in}\rangle = |1\rangle_1|0\rangle_2 = |1_1, 0_2\rangle = \hat{a}_1^\dagger|0_1, 0_2\rangle = \hat{a}_1^\dagger|0\rangle \tag{19.56}$$

where $|0\rangle = |0_1, 0_2\rangle$ is the radiation vacuum state with no photons in *any* mode of the field.

Now the photon in port 1 has equal probability of exiting from port 3 or port 4. Thus, the output state can be represented by (remembering that the output arm 3 has an extra phase difference of $\pi/2$)

$$|\psi_{out}\rangle = \frac{1}{\sqrt{2}}(i|1_3, 0_4\rangle + |0_3, 1_4\rangle) \tag{19.57}$$

which can be written as follows:

$$\begin{aligned}|\psi_{out}\rangle &= \frac{1}{\sqrt{2}}\left(i\hat{a}_3^\dagger|0_3, 0_4\rangle + \hat{a}_4^\dagger|0_3, 0_4\rangle\right) = \frac{1}{\sqrt{2}}\left(i\hat{a}_3^\dagger + \hat{a}_4^\dagger\right)|0_3, 0_4\rangle \\ &= \frac{1}{\sqrt{2}}\left(i\hat{a}_3^\dagger + \hat{a}_4^\dagger\right)|0\rangle\end{aligned} \tag{19.58}$$

where again $|0\rangle = |0_3, 0_4\rangle$ is the radiation vacuum state with no photons in *any* mode of the field.

Hence, we see that an input state written as $\widehat{a}_1^\dagger|0\rangle$ is transformed to an output state given by $\frac{1}{\sqrt{2}}\left(i\widehat{a}_3^\dagger + \widehat{a}_4^\dagger\right)|0\rangle$. Comparing the input and output states, we see that we can get the output state by replacing $\widehat{a}_1^\dagger$ acting on the radiation vacuum state by $\frac{1}{\sqrt{2}}\left(i\widehat{a}_3^\dagger + \widehat{a}_4^\dagger\right)$ acting on the radiation vacuum state. We notice that the relationship between the annihilation operators of the input arms and those of the output arms are given by Eq. (19.43) and the beam splitter actually transforms $\widehat{a}_1^\dagger$ to $\frac{1}{\sqrt{2}}\left(i\widehat{a}_3^\dagger + \widehat{a}_4^\dagger\right)$.

This suggests that to obtain the output quantum state for a given input quantum state, we expand the input state in terms of creation and annihilation operators of ports 1 and 2 operating on radiation vacuum state. Then we expand the creation and annihilation operators of the input ports in terms of the creation and annihilation operators of the output ports operating on the radiation vacuum state as given by the equations relating the input and output annihilation and creation operators. Operating on the radiation vacuum state with the creation and annihilation operators of ports 3 and 4, we then obtain the output quantum state from the beam splitter.

19.3.2.3 Output State with Single Photon Incident on a Beam Splitter

Let us again consider the incidence of a single photon on arm 1 of the beam splitter and vacuum state in arm 2. The input state is then given by

$$|\psi_{in}\rangle = |1_1, 0_2\rangle = \widehat{a}_1^\dagger|0_1, 0_2\rangle \tag{19.59}$$

As discussed earlier, in order to obtain the output quantum state we express the operator $\widehat{a}_1^\dagger$ in terms of $\widehat{a}_3^\dagger$ and $\widehat{a}_4^\dagger$ using Eq. (19.43):

$$\begin{aligned} |1_1, 0_2\rangle = \widehat{a}_1^\dagger|0_1, 0_2\rangle &\to \left(\frac{i}{\sqrt{2}}\widehat{a}_3^\dagger + \frac{1}{\sqrt{2}}\widehat{a}_4^\dagger\right)|0_3, 0_4\rangle \\ &= \frac{i}{\sqrt{2}}|1_3, 0_4\rangle + \frac{1}{\sqrt{2}}|0_3, 1_4\rangle \end{aligned}$$

Thus, the output state is given by

$$|\psi_{out}\rangle = \frac{i}{\sqrt{2}}|1_3, 0_4\rangle + \frac{1}{\sqrt{2}}|0_3, 1_4\rangle \tag{19.60}$$

Hence, the output state is a superposition state of finding either one photon in arm 3 and no photon in arm 4 represented by $|1_3, 0_4\rangle$ or finding one photon in arm 4 and no photon in arm 3 represented by $|0_3, 1_4\rangle$. Since we are considering a 50% beam splitter, both these have equal probabilities of ½. Thus, the incident single photon goes into a superposition state of being in either of the two arms. Please note that the photon is not in the state of being *either* in arm 3 or in arm 4 but in a superposition state of being in both arms. Of course, any detection performed at the output will result in the photon being detected either in arm 3 or in arm 4 with equal probabilities.

19.3.2.4 Output State with Input of One Photon Each in Both of the Input Ports

Consider now a situation in which one photon each is simultaneously incident in each of the input arms. We assume that the two photons are indistinguishable having the same frequency and same state of polarization. Thus, the incident state can be written as

$$|\psi_{in}\rangle = |1_1, 1_2\rangle = \hat{a}_1^\dagger \hat{a}_2^\dagger |0_1, 0_2\rangle \tag{19.61}$$

Using Eq. (19.43), we can introduce the effect of the beam splitter and obtain

$$\begin{aligned} |\psi_{out}\rangle &= \left(\frac{i}{\sqrt{2}}\hat{a}_3^\dagger + \frac{1}{\sqrt{2}}\hat{a}_4^\dagger\right)\left(\frac{1}{\sqrt{2}}\hat{a}_3^\dagger + \frac{i}{\sqrt{2}}\hat{a}_4^\dagger\right)|0_3, 0_4\rangle \\ &= \frac{1}{2}\left(i\hat{a}_3^\dagger \hat{a}_3^\dagger - \hat{a}_3^\dagger \hat{a}_4^\dagger + \hat{a}_4^\dagger \hat{a}_3^\dagger + i\hat{a}_4^\dagger \hat{a}_4^\dagger\right)|0_3, 0_4\rangle \\ &= \frac{i}{\sqrt{2}}(|2_3, 0_4\rangle + |0_3, 2_4\rangle) \end{aligned} \tag{19.62}$$

where we have used the fact that $\hat{a}_3^\dagger \hat{a}_3^\dagger |0_3\rangle = \sqrt{2}|2_3\rangle$ and $\hat{a}_4^\dagger \hat{a}_4^\dagger |0_4\rangle = \sqrt{2}|2_4\rangle$.

The output state given by Eq. (19.62) implies that both incident photons exit from the same port, either port 3 or port 4 with equal probabilities of $\frac{1}{2}$. The probability of one photon exiting from each of the two arms is zero. Thus, detectors placed at the two output ports of the beam splitter will never register simultaneous detection events.

In fact, the coincidence probability of detecting one photon in port 3 and one photon in port 4 is

$$P_{coin} = |\langle 1_3, 1_4 || \psi_{out}\rangle|^2 = 0$$

An experimental demonstration of this was first carried out by Hong et al. (1987). In the experiment as the time of arrival of the photon on one of the arms of the beam is delayed, the rate of coincidence counting measured after the beam splitter would show a dip when the two photons arrive simultaneously at the beam splitter (see Sect. 19.3.2.3). This is referred to as Hong Ou Mandel (HOM) dip and is a purely quantum phenomenon with no classical explanation. HOM interference has also been demonstrated using integrated photonic directional couplers (see, e.g., Babel et al. 2023; Mohanty et al. 2017). Readers are referred Mandel (1999) which is an excellent review on one photon and two photon interference.

Exercise 19.6 Consider the HOM dip experiment carried out using two incident photons that are distinguishable, for example, having two different polarization states or two different frequencies. Obtain the output of the beam splitter and show that there will be no two-photon interference like in the case with indistinguishable photons. Assume the beam splitter to be independent of frequency and polarization.

Solution Let us assume that the two incident photons have orthogonal polarizations referred to as Horizontal (H) and Vertical (V). In such a case we write the input state as

$$|\psi_{in}\rangle = |1_{1H}, 1_{2V}\rangle = \widehat{a}_{1H}^{\dagger}\widehat{a}_{2V}^{\dagger}|0_1, 0_2\rangle \tag{19.63}$$

We introduce the effect of the beam splitter through the following transformation:

$$|1_{1H}, 1_{2V}\rangle = \widehat{a}_{1H}^{\dagger}\widehat{a}_{2V}^{\dagger}|0_1, 0_2\rangle = \left(\frac{i}{\sqrt{2}}\widehat{a}_{3H}^{\dagger} + \frac{1}{\sqrt{2}}\widehat{a}_{4H}^{\dagger}\right)\left(\frac{1}{\sqrt{2}}\widehat{a}_{3V}^{\dagger} + \frac{i}{\sqrt{2}}\widehat{a}_{4V}^{\dagger}\right)|0_3, 0_4\rangle \tag{19.64}$$

which gives the output state as

$$|\psi_{out}\rangle = \frac{i}{\sqrt{2}}\widehat{a}_{3H}^{\dagger}\frac{1}{\sqrt{2}}\widehat{a}_{3V}^{\dagger}|0_3, 0_4\rangle + \frac{i}{\sqrt{2}}\widehat{a}_{3H}^{\dagger}\frac{i}{\sqrt{2}}\widehat{a}_{4V}^{\dagger}|0_3, 0_4\rangle$$
$$+ \frac{1}{\sqrt{2}}\widehat{a}_{4H}^{\dagger}\frac{1}{\sqrt{2}}\widehat{a}_{3V}^{\dagger}|0_3, 0_4\rangle + \frac{1}{\sqrt{2}}\widehat{a}_{4H}^{\dagger}\frac{i}{\sqrt{2}}\widehat{a}_{4V}^{\dagger}|0_3, 0_4\rangle$$

which simplifies to

$$|\psi_{out}\rangle = \frac{i}{2}|1_{3H}, 1_{3V}\rangle - \frac{1}{2}|1_{3H}, 1_{4V}\rangle + \frac{1}{2}|1_{3V}, 1_{4H}\rangle + \frac{i}{2}|1_{4H}, 1_{4V}\rangle \tag{19.65}$$

The first term corresponds to both photons in the output arm 3, the last term to both the photons in the output arm 4 and the second and third terms to one photon each from arms 3 and 4. Note that in the case two photons were in identical modes and indistinguishable, the terms with one photon in each arm canceled. However, in the case when the photons are distinguishable no such cancellation happens. Thus, we see that in this case with the photons being distinguishable through their polarization states, there is no two-photon interference.

The coincidence probability will be given by

$$P_{coin} = |\langle 1_{3H}, 1_{4V}\, ||\psi_{out}\rangle|^2 + |\langle 1_{3V}, 1_{4H}\, ||\psi_{out}\rangle|^2 = \frac{1}{2} \tag{19.66}$$

The same result can be obtained using the projection operators. Thus, the coincidence probability of detecting one photon in each of the output ports 3 and 4 is given by

$$P_{coin} = \langle\psi_{out}\,|(|1_{3H}, 1_{4V}\rangle\langle 1_{3H}, 1_{4V}\,| + |1_{3V}, 1_{4H}\rangle\langle 1_{3V}, 1_{4H}\,||\psi_{out}\rangle) \tag{19.67}$$

where

$$\widehat{P} = |1_{3H}, 1_{4V}\rangle\langle 1_{3H}, 1_{4V}\,| + |1_{3V}, 1_{4H}\rangle\langle 1_{3V}, 1_{4H}\,| \tag{19.68}$$

is the projection operator corresponding to detection of one photon each in arms 3 and 4 irrespective of the polarization state of the photon.

Substituting from Eq. (19.65), we obtain the same result as Eq. (19.66).

Exercise 19.7 Assume that the photon incident in arm 1 is horizontally polarized, while the photon incident in arm 2 is linearly polarized making an angle θ with the horizontal, that is, the input state is given by

$$|\psi_{in}\rangle = |1_{1H}\rangle(\cos\theta|1_{2H}\rangle + \sin\theta|1_{2V}\rangle) \tag{19.69}$$

Obtain the coincidence probability of getting one photon each in output arms 3 and 4. In this case, the two photons are partially indistinguishable. Check if the result gives correct values for $\theta = 0$ and $\theta = \frac{\pi}{2}$.

Solution The input state can be written as

$$|\psi_{in}\rangle = \cos\theta\widehat{a}^{\dagger}_{1H}\widehat{a}^{\dagger}_{2H}|0_{1H}, 0_{2H}\rangle + \sin\theta\widehat{a}^{\dagger}_{1H}\widehat{a}^{\dagger}_{2V}|0_{1H}, 0_{2V}\rangle \tag{19.70}$$

Hence, the output state is given by

$$\begin{aligned}|\psi_{out}\rangle = {}& \cos\theta\left(\frac{i}{\sqrt{2}}\widehat{a}^{\dagger}_{3H} + \frac{1}{\sqrt{2}}\widehat{a}^{\dagger}_{4H}\right)\left(\frac{1}{\sqrt{2}}\widehat{a}^{\dagger}_{3H} + \frac{i}{\sqrt{2}}\widehat{a}^{\dagger}_{4H}\right)|0\rangle \\ & + \sin\theta\left(\frac{i}{\sqrt{2}}\widehat{a}^{\dagger}_{3H} + \frac{1}{\sqrt{2}}\widehat{a}^{\dagger}_{4H}\right)\left(\frac{1}{\sqrt{2}}\widehat{a}^{\dagger}_{3V} + \frac{i}{\sqrt{2}}\widehat{a}^{\dagger}_{4V}\right)|0\rangle\end{aligned}$$

where $|0\rangle$ is the radiation vacuum state with no photon occupying any of the states. Simplifying, we obtain the output state as

$$\begin{aligned}|\psi_{out}\rangle = {}& \cos\theta\left(\frac{i}{2}|2_{3H}, 0_{4H}\rangle + \frac{i}{2}|0_{3H}, 2_{4H}\rangle\right) \\ & + \sin\theta\left(\frac{i}{2}|1_{3H}, 1_{3V}\rangle - \frac{1}{2}|1_{3H}, 1_{4V}\rangle + \frac{1}{2}|1_{3V}, 1_{4H}\rangle + \frac{i}{2}|1_{4H}, 1_{4V}\rangle\right)\end{aligned} \tag{19.71}$$

For $\theta = 0$, the output shows two photon interference with both photons appearing in either of the arms. For $\theta = \frac{\pi}{2}$, the output shows no two-photon interference with 0.25 probability of both photons appearing in either of the arms and 0.5 probability of one photon each in either of the arms. In the general case. The indistinguishability between the two photons is determined by the value of θ; for $\theta = 0$, the photons are indistinguishable, while for $\theta = \frac{\pi}{2}$ the two photons are fully distinguishable. The probability of coincidence detection of photons in arms 3 and 4 is given by

$$P_{coin} = +|\langle 1_{3H}, 1_{4V} ||\psi_{out}\rangle|^2 + |\langle 1_{3V}, 1_{4H} ||\psi_{out}\rangle|^2 \tag{19.72}$$

which on substitution from Eq. (19.71) gives us

$$P_{coin} = \frac{1}{2} \sin^2\theta \tag{19.73}$$

This shows that when the photons are perfectly indistinguishable, the coincidence dip goes to zero, while if the photons are fully distinguishable, there will be no dip in coincidence.

19.3.3 Incident Single Photon in Multimode State

In the above sections, we considered the incidence of photons in which the photons were occupying a single mode. In this section, we consider the case when each of the two input photons is in a multimode superposed state with various frequencies forming single photon wave packets.

We can write the state of the input photons in arms 1 and 2 as (see Chap. 16)

$$|\psi_1\rangle = \int \phi(\omega)\hat{a}_1^\dagger(\omega)|0\rangle d\omega \tag{19.74}$$

and

$$|\psi_2\rangle = \int \chi(\omega)\hat{a}_2^\dagger(\omega)|0\rangle d\omega \tag{19.75}$$

where $\phi(\omega)$ and $\chi(\omega)$ are the spectral amplitudes of the two photons. Note that in each of the two arms a single photon is incident; the single photon is in a superposition state of being in multiple frequencies (see Chap. 16).

Hence, we can write the input state as

$$|\psi_{in}\rangle = \int \phi(\omega_1)\hat{a}_1^\dagger(\omega_1)|0\rangle d\omega_1 \int \chi(\omega_2)\hat{a}_2^\dagger(\omega_2)|0\rangle d\omega_2 \tag{19.76}$$

Using the relation for the beam splitter (see Eq. 19.43), the output state of the beam splitter will be given by

$$|\psi_{out}\rangle = \int \phi(\omega_1)\left(\frac{i}{\sqrt{2}}\hat{a}_3^\dagger(\omega_1) + \frac{1}{\sqrt{2}}\hat{a}_4^\dagger(\omega_1)\right)|0\rangle d\omega_1 \int \chi(\omega_2) \times \left(\frac{1}{\sqrt{2}}\hat{a}_3^\dagger(\omega_2) + \frac{i}{\sqrt{2}}\hat{a}_4^\dagger(\omega_2)\right)|0\rangle d\omega_2$$

which simplifies to

$$|\psi_{out}\rangle = \int d\omega_1 \phi(\omega_1) \int d\omega_2 \chi(\omega_2) \left[\frac{i}{2}\hat{a}_3^\dagger(\omega_1)\hat{a}_3^\dagger(\omega_2) - \frac{1}{2}\hat{a}_3^\dagger(\omega_1)\hat{a}_4^\dagger(\omega_2) + \frac{1}{2}\hat{a}_3^\dagger(\omega_2)\hat{a}_4^\dagger(\omega_1) + \frac{i}{2}\hat{a}_4^\dagger(\omega_1)\hat{a}_4^\dagger(\omega_2)\right]|0_3, 0_4\rangle \tag{19.77}$$

The above equation has four terms on the right hand side. The first term $\hat{a}_3^\dagger(\omega_1)\hat{a}_3^\dagger(\omega_2)$ corresponds to both the photons exiting arm 3 and the last term $\hat{a}_4^\dagger(\omega_1)\hat{a}_4^\dagger(\omega_2)$ corresponds to both the photons exiting arm 4.

The two middle terms $\hat{a}_3^\dagger(\omega_1)\hat{a}_4^\dagger(\omega_2)$ and $\hat{a}_3^\dagger(\omega_2)\hat{a}_4^\dagger(\omega_1)$ correspond to the two photons exiting two different ports. The first of these corresponds to the photon with the spectral amplitude $\phi(\omega)$ exiting arm 3 and the photon with spectral amplitude $\chi(\omega)$ exiting arm 4. The second term corresponds to the output where the photon with the spectral amplitude $\phi(\omega)$ exiting arm 4 and the photon with spectral amplitude $\chi(\omega)$ exiting arm 3. These are the two terms that will account for coincidence measurements.

Now, the projection operator corresponding to detection in port 3 is given by

$$\hat{P}_3 = \int d\omega \hat{a}_3^\dagger(\omega)|0_3\rangle\langle 0_3|\hat{a}_3(\omega) \tag{19.78}$$

Similarly, the projection operator corresponding to detection in port 4 is given by

$$\hat{P}_4 = \int d\omega \hat{a}_4^\dagger(\omega)|0_4\rangle\langle 0_4|\hat{a}_4(\omega) \tag{19.79}$$

The coincidence probability will be given by

$$P_{coin} = \langle \psi_{out} |\hat{P}_3 \otimes \hat{P}_4| \psi_{out}\rangle \tag{19.80}$$

Substituting for the various terms in Eq. (19.80), we obtain

$$P_{coin}=\frac{1}{2}\int d\omega_1\phi^*(\omega_1)\int d\omega_2\chi^*(\omega_2)\langle 0_3,0_4|$$

$$\left[-i\hat{a}_3(\omega_1)\hat{a}_3(\omega_2)-\hat{a}_3(\omega_1)\hat{a}_4(\omega_2)+\hat{a}_4(\omega_1)\hat{a}_3(\omega_2)-i\hat{a}_4(\omega_1)\hat{a}_4(\omega_2)\right]$$

$$\int d\omega_3\hat{a}_3^\dagger(\omega_3)|0_3\rangle\langle 0_3|\hat{a}_3(\omega_3)\int d\omega_4\hat{a}_4^\dagger(\omega_4)|0_4\rangle\langle 0_4|\hat{a}_4(\omega_4)\left[\frac{1}{2}\int d\omega_5\phi(\omega_5)\int d\omega_6\chi(\omega_6)\right.$$

$$\left.\left[i\hat{a}_3^\dagger(\omega_5)\hat{a}_3^\dagger(\omega_6)-\hat{a}_3^\dagger(\omega_5)\hat{a}_4^\dagger(\omega_6)+\hat{a}_4^\dagger(\omega_5)\hat{a}_3^\dagger(\omega_6)+i\hat{a}_4^\dagger(\omega_5)\hat{a}_4^\dagger(\omega_6)\right]|0_3,0_4\rangle\right]$$

which simplifies to

$$\begin{aligned}P_{coin}=&\frac{1}{4}\int d\omega_1\int d\omega_2\int d\omega_3\int d\omega_4\int d\omega_5\int d\omega_6\phi^*(\omega_1)\chi^*(\omega_2)\phi(\omega_5)\chi(\omega_6)\langle 0_3,0_4|\\&[-i\hat{a}_3(\omega_1)\hat{a}_3(\omega_2)-\hat{a}_3(\omega_1)\hat{a}_4(\omega_2)+\hat{a}_4(\omega_1)\hat{a}_3(\omega_2)-i\hat{a}_4(\omega_1)\hat{a}_4(\omega_2)]\hat{a}_3^\dagger(\omega_3)\hat{a}_4^\dagger(\omega_4)|0_3,0_4\rangle\\&\langle 0_3,0_4|\hat{a}_3(\omega_3)\hat{a}_4(\omega_4)\left[i\hat{a}_3^\dagger(\omega_5)\hat{a}_3^\dagger(\omega_6)-\hat{a}_3^\dagger(\omega_5)\hat{a}_4^\dagger(\omega_6)+\hat{a}_4^\dagger(\omega_5)\hat{a}_3^\dagger(\omega_6)+i\hat{a}_4^\dagger(\omega_5)\hat{a}_4^\dagger(\omega_6)\right]|0_3,0_4\rangle\end{aligned} \tag{19.81}$$

Terms of the type

$$\langle 0_3,0_4\,|\hat{a}_3(\omega_1)\hat{a}_3(\omega_2)\hat{a}_3^\dagger(\omega_3)\hat{a}_4^\dagger(\omega_4)|0_3,0_4\rangle$$

having an odd number of operators corresponding to one of the outputs go to zero. On the other hand, terms of the type

$$\langle 0_3,0_4\,|\hat{a}_3(\omega_1)\hat{a}_4(\omega_2)\hat{a}_3^\dagger(\omega_3)\hat{a}_4^\dagger(\omega_4)|0_3,0_4\rangle$$

give Dirac delta functions of the form $\delta(\omega_1-\omega_3)\delta(\omega_2-\omega_4)$. Thus, Eq. (19.81) simplifies to

$$\begin{aligned}P_{coin}=\frac{1}{4}\int d\omega_1\int d\omega_2\int d\omega_3\int d\omega_4\int d\omega_5\int d\omega_6\phi^*(\omega_1)\chi^*(\omega_2)\phi(\omega_5)\chi(\omega_6)\\ [-\delta(\omega_1-\omega_3)\delta(\omega_2-\omega_4)+\delta(\omega_1-\omega_4)\delta(\omega_2-\omega_3)]\\ [-\delta(\omega_3-\omega_5)\delta(\omega_4-\omega_6)+\delta(\omega_3-\omega_6)\delta(\omega_4-\omega_5)]\end{aligned} \tag{19.82}$$

Carrying out the integration using the property of Dirac delta functions, we obtain

$$
\begin{aligned}
P_{coin} = \frac{1}{4}\Bigg[& \int d\omega_1 \int d\omega_2 \phi^*(\omega_1)\chi^*(\omega_2)\phi(\omega_1)\chi(\omega_2) \\
& - \int d\omega_1 \int d\omega_2 \phi^*(\omega_1)\chi^*(\omega_2)\phi(\omega_2)\chi(\omega_1) \\
& - \int d\omega_1 \int d\omega_2 \phi^*(\omega_1)\chi^*(\omega_2)\phi(\omega_2)\chi(\omega_1) \\
& + \int d\omega_1 \int d\omega_2 \phi^*(\omega_1)\chi^*(\omega_2)\phi(\omega_1)\chi(\omega_2)\Bigg]
\end{aligned}
$$

or

$$
P_{coin} = \frac{1}{2}\left[\int d\omega_1 \phi^*(\omega_1)\phi(\omega_1) \int d\omega_2 \chi^*(\omega_2)\chi(\omega_2) - \int d\omega_1 \phi^*(\omega_1)\chi(\omega_1) \int d\omega_2 \chi^*(\omega_2)\phi(\omega_2)\right] \tag{19.83}
$$

Assuming the following normalization conditions

$$
\int d\omega_1 \phi^*(\omega_1)\phi(\omega_1) = 1 \tag{19.84a}
$$

$$
\int d\omega_2 \chi^*(\omega_2)\chi(\omega_2) = 1 \tag{19.84b}
$$

we obtain

$$
P_{coin} = \frac{1}{2} - \frac{1}{2}\int d\omega_1 \phi^*(\omega_1)\chi(\omega_1) \int d\omega_2 \chi^*(\omega_2)\phi(\omega_2) \tag{19.85}
$$

which gives us the coincidence probability with incident single photons which are in a multimode superposition state.

- If the two spectra of the single photons are identical, that is, $\phi(\omega) = \chi(\omega)$, then $P_{coin} = 0$.
- If there is no overlap between the two spectra of the single photons, that is, they are fully distinguishable, then $\int d\omega_1 \phi^*(\omega_1)\chi(\omega_1) = 0$ and we have $P_{coin} = \frac{1}{2}$.

The HOM experiment is carried out by changing the delay between the arrival of the two photons at the beam splitter and measuring the coincidence counts as a function of the delay. We can easily extend the above analysis for the case of delay between the two photons.

Let us assume that a time delay of τ is given on one of the photons (say in input arm 2). In this case instead of Eq. (19.75), we would have

$$|\psi_2\rangle = \int \chi(\omega)\hat{a}_2^\dagger(\omega)e^{i\omega\tau}|0\rangle d\omega \tag{19.86}$$

A delay by τ in the time domain leads to a phase shift of $e^{i\omega\tau}$ in the Fourier space, that is, $\chi(\omega)$ is replaced by $\chi(\omega)e^{i\omega\tau}$. The probability of coincidence can be obtained from Eq. (19.85) simply by replacing $\chi(\omega)$ by $\chi(\omega)e^{i\omega\tau}$. Thus, we would obtain the following expression:

$$P_{coin} = \frac{1}{2} - \frac{1}{2}\int d\omega_1 \phi^*(\omega_1)\chi(\omega_1)e^{i\omega_1\tau} \int d\omega_2 \chi^*(\omega_2)\phi(\omega_2)e^{-i\omega_2\tau} \tag{19.87}$$

If the two photons are characterized by the same spectra, then,

$$P_{coin} = \frac{1}{2} - \frac{1}{2}\int d\omega_1 |\phi(\omega_1)|^2 e^{i\omega_1\tau} \int d\omega_2 |\phi(\omega_2)|^2 e^{-i\omega_2\tau} \tag{19.88}$$

As an example if we assume the two single photons to have Gaussian spectra, then we have the following normalized spectrum:

$$\phi(\omega) = \frac{1}{\sqrt{\sigma\sqrt{\pi}}} e^{-(\omega-\omega_0)^2/2\sigma^2} \tag{19.89}$$

and we obtain for the coincidence probability

$$P_{coin} = \frac{1}{2} - \frac{1}{2}e^{-\sigma^2\tau^2/2} \tag{19.90}$$

which gives an expression for the probability of coincidence as a function of delay of one of the photons as it arrives on the beam splitter. Figure 19.4 shows a plot of P_{coin} versus the normalized delay $\sigma\tau$ showing the dip in coincidence when the delay is zero and the two photons characterized by the same spectra enter the beam splitter together.

Many experiments to demonstrate HOM dip have been reported using various single photon sources. Since optical waveguide directional couplers behave as beam splitters, HOM dip can also be observed using waveguide directional couplers with the two input photons coupled to the two input waveguides of the directional coupler. Babel et al. (2023) report such measurements using directional coupler fabricated in lithium niobate on insulator (LNOI) platform showing excellent characteristics. HOM interference between photons occupying different spatial waveguide modes has also been demonstrated (Mohanty et al. 2017).

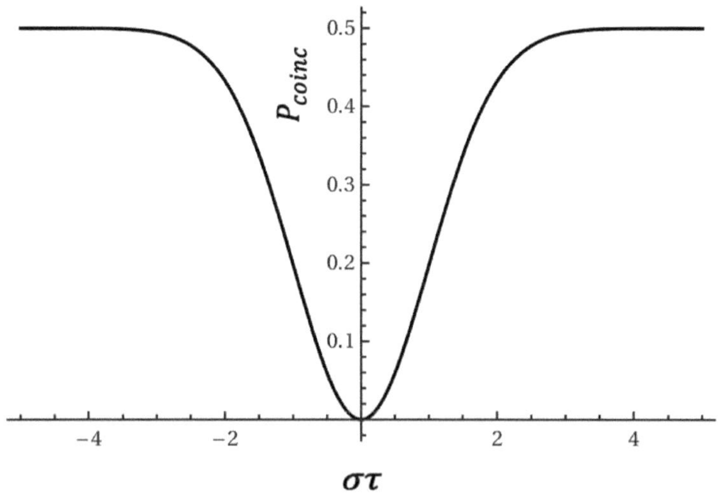

Fig. 19.4 Variation in coincidence probability with the normalized delay between two single photons in identical multimode states. The dip is the famous Hong-Ou-Mandel (HOM) dip

19.4 Two Photons Through a Single Input Port

If instead of having one photon each enter each of the two input ports, let us consider the case when two indistinguishable photons enter through one single port, say for example port 1. Hence,

$$|\psi_{in}\rangle = |2_1, 0_2\rangle = \frac{1}{\sqrt{2}}\left(\hat{a}_1^\dagger\right)^2 |0_1, 0_2\rangle \tag{19.91}$$

The beam splitter converts this input state to the following output state:

$$|\psi_{out}\rangle = \frac{1}{\sqrt{2}}\left(\frac{i}{\sqrt{2}}\hat{a}_3^\dagger + \frac{1}{\sqrt{2}}\hat{a}_4^\dagger\right)^2 |0_3, 0_4\rangle = \frac{1}{\sqrt{2}}\left(-\frac{1}{2}\left(\hat{a}_3^\dagger\right)^2 + \frac{1}{2}\left(\hat{a}_4^\dagger\right)^2 + \frac{i}{2}\hat{a}_3^\dagger\hat{a}_4^\dagger + \frac{i}{2}\hat{a}_4^\dagger\hat{a}_3^\dagger\right)|0_3, 0_4\rangle$$

which gives

$$|\psi_{out}\rangle = -\frac{1}{2}|2_3, 0_4\rangle + \frac{1}{2}|0_3, 2_4\rangle + \frac{i}{\sqrt{2}}|1_3, 1_4\rangle \tag{19.92}$$

which implies that there are three possibilities, both photons exiting port 3 or both photons exiting port 4 with equal probabilities of 0.25 and one photon exiting port 3 and the other photon exiting port 4 with the probability of 0.5. This case is quite different from the earlier example wherein the two photons were incident from the two different ports leading to interference among the paths and both photons exiting from either of the two exit ports with zero probability of one photon exiting from each individual output port.

19.5 Entangled State on a Beam Splitter

In this example, we consider the incidence of a polarization entangled state on the beam splitter and assume that the beam splitter is polarization insensitive. The input state is assumed to be given by (see Chap. 17 for a discussion on polarization entangled state)

$$|\psi_{in}\rangle = \frac{1}{\sqrt{2}}(|H_1, V_2\rangle + |V_1, H_2\rangle) \tag{19.93}$$

This state can be written as

$$|\psi_{in}\rangle = \frac{1}{\sqrt{2}}\left(\widehat{a}_{H1}^{\dagger}\widehat{a}_{V2}^{\dagger}|0\rangle + \widehat{a}_{V1}^{\dagger}\widehat{a}_{H2}^{\dagger}|0\rangle\right) = \frac{1}{\sqrt{2}}\left(\widehat{a}_{H1}^{\dagger}\widehat{a}_{V2}^{\dagger} + \widehat{a}_{V1}^{\dagger}\widehat{a}_{H2}^{\dagger}\right)|0\rangle \tag{19.94}$$

Now as before we have

$$\widehat{a}_{H1}^{\dagger} = \left(\frac{i}{\sqrt{2}}\widehat{a}_{H3}^{\dagger} + \frac{1}{\sqrt{2}}\widehat{a}_{H4}^{\dagger}\right) \tag{19.95a}$$

$$\widehat{a}_{H2}^{\dagger} = \left(\frac{1}{\sqrt{2}}\widehat{a}_{H3}^{\dagger} + \frac{i}{\sqrt{2}}\widehat{a}_{H4}^{\dagger}\right) \tag{19.95b}$$

$$\widehat{a}_{V1}^{\dagger} = \left(\frac{i}{\sqrt{2}}\widehat{a}_{V3}^{\dagger} + \frac{1}{\sqrt{2}}\widehat{a}_{V4}^{\dagger}\right) \tag{19.95c}$$

$$\widehat{a}_{V2}^{\dagger} = \left(\frac{1}{\sqrt{2}}\widehat{a}_{V3}^{\dagger} + \frac{i}{\sqrt{2}}\widehat{a}_{V4}^{\dagger}\right) \tag{19.95d}$$

Using the procedure outlined above, we obtain for the output state

$$\begin{aligned}|\psi_{out}\rangle &= \frac{1}{\sqrt{2}}\left(\frac{i}{\sqrt{2}}\widehat{a}_{H3}^{\dagger} + \frac{1}{\sqrt{2}}\widehat{a}_{H4}^{\dagger}\right)\left(\frac{1}{\sqrt{2}}\widehat{a}_{V3}^{\dagger} + \frac{i}{\sqrt{2}}\widehat{a}_{V4}^{\dagger}\right)|0\rangle \\ &+ \frac{1}{\sqrt{2}}\left(\frac{1}{\sqrt{2}}\widehat{a}_{H3}^{\dagger} + \frac{i}{\sqrt{2}}\widehat{a}_{H4}^{\dagger}\right)\left(\frac{i}{\sqrt{2}}\widehat{a}_{V3}^{\dagger} + \frac{1}{\sqrt{2}}\widehat{a}_{V4}^{\dagger}\right)|0\rangle\end{aligned} \tag{19.96}$$

Expanding the brackets, we obtain

$$\begin{aligned}|\psi_{out}\rangle &= \frac{1}{\sqrt{2}}\left(\frac{i}{2}\widehat{a}_{H3}^{\dagger}\widehat{a}_{V3}^{\dagger} - \frac{1}{2}\widehat{a}_{H3}^{\dagger}\widehat{a}_{V4}^{\dagger} + \frac{1}{2}\widehat{a}_{H4}^{\dagger}\widehat{a}_{V3}^{\dagger} + \frac{i}{2}\widehat{a}_{H4}^{\dagger}\widehat{a}_{V4}^{\dagger}\right)|0\rangle + \frac{1}{\sqrt{2}} \\ &\times \left(\frac{i}{2}\widehat{a}_{H3}^{\dagger}\widehat{a}_{V3}^{\dagger} + \frac{1}{2}\widehat{a}_{H3}^{\dagger}\widehat{a}_{V4}^{\dagger} - \frac{1}{2}\widehat{a}_{H4}^{\dagger}\widehat{a}_{V3}^{\dagger} + \frac{i}{2}\widehat{a}_{H4}^{\dagger}\widehat{a}_{V4}^{\dagger}\right)|0\rangle\end{aligned}$$

which gives

$$
\begin{aligned}
|\psi_{out}\rangle = & \frac{1}{\sqrt{2}}\left(\frac{i}{2}|1_{H3}\rangle|1_{V3}\rangle - \frac{1}{2}|1_{H3}\rangle|1_{V4}\rangle + \frac{1}{2}|1_{H4}\rangle|1_{V3}\rangle + \frac{i}{2}|1_{H4}\rangle|1_{V4}\rangle\right) \\
& + \frac{1}{\sqrt{2}}\left(\frac{i}{2}|1_{H3}\rangle|1_{V3}\rangle + \frac{1}{2}|1_{H3}\rangle|1_{V4}\rangle - \frac{1}{2}|1_{H4}\rangle|1_{V3}\rangle + \frac{i}{2}|1_{H4}\rangle|1_{V4}\rangle\right) \\
& = \frac{i}{\sqrt{2}}\left(|1_{H3}\rangle|1_{V3}\rangle + |1_{H4}\rangle|1_{V4}\rangle\right)
\end{aligned}
\tag{19.97}
$$

At the output both photons now appear in the same port but the state is still an entangled state.

19.6 Coherent State Input on a Beam Splitter

In this section, we consider the incidence of a coherent state on a beam splitter and determine the state of the output from the beam splitter. The input state is now given by

$$
\begin{aligned}
|\psi_{in}\rangle = |\alpha_1\rangle|0_2\rangle = e^{-|\alpha|^2/2}\sum\frac{\alpha^n}{\sqrt{n!}}|n_1\rangle|0_2\rangle = e^{-|\alpha|^2/2}\sum\frac{\alpha^n}{n!}\left(a_1^\dagger\right)^n|0_1\rangle|0_2\rangle \\
= e^{-|\alpha|^2/2}\sum\frac{\alpha^n}{n!}\left(a_1^\dagger\right)^n|0\rangle
\end{aligned}
\tag{19.98}
$$

where again $|0\rangle$ is the radiation vacuum state. Since (see Eq. (19.43))

$$
a_1^\dagger = \frac{1}{\sqrt{2}}\left(ia_3^\dagger + a_4^\dagger\right)
$$

we have

$$
|\psi_{out}\rangle = e^{-|\alpha|^2/2}\sum\frac{\alpha^n}{2^{n/2}n!}\left(ia_3^\dagger + a_4^\dagger\right)^n|0\rangle \tag{19.99}
$$

Now, since $a_3^\dagger$ and $a_4^\dagger$ commute with each other, we have

$$
\left(ia_3^\dagger + a_4^\dagger\right)^n = \sum_m \frac{n!i^m}{m!(n-m)!}\left(a_3^\dagger\right)^m\left(a_4^\dagger\right)^{n-m} \tag{19.100}
$$

Hence, we can write for an arbitrary state $|l_3\rangle$

$$\langle l_3|\psi_{out}\rangle = e^{-|\alpha|^2/2}\sum\sum\frac{\alpha^n i^m n!}{2^{n/2}m!n!(n-m)!}\langle l_3|\left(a_3^\dagger\right)^m\left(a_4^\dagger\right)^{n-m}|0\rangle$$
$$= e^{-|\alpha|^2/2}\sum\sum\frac{\alpha^n i^m n!}{2^{n/2}m!n!(n-m)!}\langle l_3|\left(a_3^\dagger\right)^m|0_3\rangle\left(a_4^\dagger\right)^{n-m}|0_4\rangle$$

Since

$$\left(a_3^\dagger\right)^m|0_3\rangle = \sqrt{m!}|m_3\rangle \tag{19.101}$$

and

$$\langle l_3|m_3\rangle = \delta_{lm} \tag{19.102}$$

we have

$$\langle l_3|\psi_{out}\rangle = e^{-|\alpha|^2/2}\sum\frac{\alpha^n i^l n!}{2^{n/2}n!l!(n-l)!}\sqrt{l!}\sqrt{(n-l)!}\left|(n-l)_4\right\rangle$$

which can be rewritten as

$$\langle l_3|\psi_{out}\rangle = e^{-|\alpha|^2/4}\sum\frac{\left(\frac{\alpha}{\sqrt{2}}\right)^{n-l}}{\sqrt{(n-l)!}}\left|(n-l)_4\right\rangle e^{-|\alpha|^2/4}\frac{\left(\frac{i\alpha}{\sqrt{2}}\right)^l}{\sqrt{l!}} \tag{19.103}$$

Since

$$\left|\frac{\alpha}{\sqrt{2}}\right\rangle = e^{-|\alpha|^2/4}\sum\frac{(\alpha/\sqrt{2})^p}{p!}|p\rangle \tag{19.104}$$

we have

$$\langle l_3|\psi_{out}\rangle = e^{-|\alpha|^2/4}\frac{\left(\frac{i\alpha}{\sqrt{2}}\right)^l}{\sqrt{l!}}\left|\left(\frac{\alpha}{\sqrt{2}}\right)_4\right\rangle = \langle l_3|e^{-|\alpha|^2/4}\sum_n\frac{\left(\frac{i\alpha}{\sqrt{2}}\right)^{n_3}}{\sqrt{n_3!}}|n_3\rangle\left|\left(\frac{\alpha}{\sqrt{2}}\right)_4\right\rangle$$
$$= \left\langle l_3\middle|\left(\frac{i\alpha}{\sqrt{2}}\right)_3\right\rangle\left|\left(\frac{\alpha}{\sqrt{2}}\right)_4\right\rangle$$

Since this is valid for any $\langle l_3|$, we find that the output state must be

$$|\psi_{out}\rangle = \left|\left(\frac{i\alpha}{\sqrt{2}}\right)_3\right\rangle\left|\left(\frac{\alpha}{\sqrt{2}}\right)_4\right\rangle \tag{19.105}$$

which represents a product state of the two output ports. Hence, when a coherent state is incident on one of the input ports of the beam splitter, the output consists of two coherent states. This is similar to a classical case when the beam splitter splits the intensity evenly between the two outputs. The average photon number in the input coherent state is $|\alpha|^2$ and the average photon numbers in either of the arms is $|\alpha|^2/2$. Note that the case of a single photon incident on the beam splitter cannot be obtained as a limiting case of a coherent state input with small value of α.

19.7 Mach-Zehnder Interferometer

Let us now consider the incidence of a photon on one of the arms of a Mach-Zehnder interferometer as shown in Fig. 19.5. Figure 19.6 shows an equivalent Mach-Zehnder interferometer based on waveguides in which two 3 dB directional couplers behave as the beam splitters (see Chap. 7).

Let us assume that in one of the arms (arm 4), there is a phase shifter which shifts the phase by φ. For a single photon incident in arm 1, we write for the input quantum state of the interferometer as

$$|\psi_{in}\rangle = |1_1, 0_2\rangle = \hat{a}_1^\dagger |0_1, 0_2\rangle \tag{19.106}$$

After the first beam splitter BS_1, the state becomes:

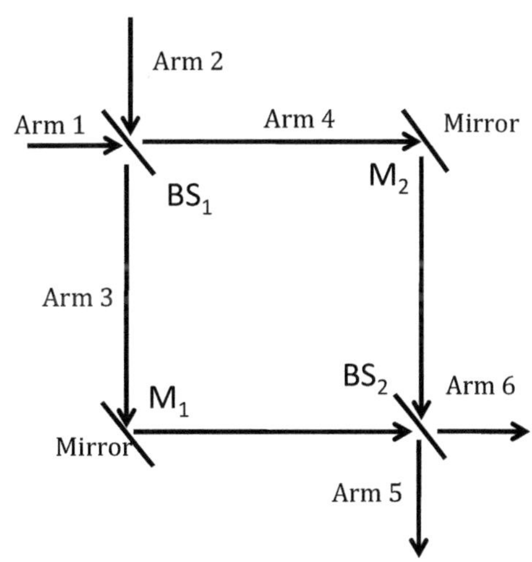

Fig. 19.5 A bulk optics Mach-Zehnder interferometer

Fig. 19.6 An integrated optic Mach-Zehnder interferometer consisting of two 3 dB directional couplers behaving as the beam splitters and two waveguides behaving as arms of the interferometer

$$|\psi_{BS1}\rangle = \left(\frac{i}{\sqrt{2}}\hat{a}_3^\dagger + \frac{1}{\sqrt{2}}\hat{a}_4^\dagger\right)|0_3, 0_4\rangle = \frac{i}{\sqrt{2}}|1_3, 0_4\rangle + \frac{1}{\sqrt{2}}|0_3, 1_4\rangle \tag{19.107}$$

Let us assume that mirrors M_1 and M_2 introduce a phase shift of π/2 each. Thus, the state of the field just after the two mirrors would be

$$-\frac{1}{\sqrt{2}}|1_3, 0_4\rangle + \frac{i}{\sqrt{2}}|0_3, 1_4\rangle$$

After the phase shifter in arm 4, the state evolves to

$$-\frac{1}{\sqrt{2}}|1_3, 0_4\rangle + \frac{i}{\sqrt{2}}|0_3, 1_4\rangle e^{i\varphi}$$

giving a phase difference between the two quantum states passing through the two arms of the interferometer.

In order to take account of the effect of beam splitter BS_2, we use the same procedure as employed for the first beam splitter and write the state incident on beam splitter 2 as

$$-\frac{1}{\sqrt{2}}|1_3, 0_4\rangle + \frac{i}{\sqrt{2}}|0_3, 1_4\rangle e^{i\varphi} = -\frac{1}{\sqrt{2}}\hat{a}_3^\dagger|0_3, 0_4\rangle + \frac{i}{\sqrt{2}}\hat{a}_4^\dagger|0_3, 0_4\rangle e^{i\varphi}$$

Since BS_2 will convert $|0_3, 0_4\rangle$ to $|0_5, 0_6\rangle$, we write for the state after BS_2 as

$$|\psi_{BS2}\rangle = -\frac{1}{\sqrt{2}}\left(\frac{i}{\sqrt{2}}\hat{a}_5^\dagger + \frac{1}{\sqrt{2}}\hat{a}_6^\dagger\right)|0_5, 0_6\rangle + \frac{i}{\sqrt{2}}\left(\frac{1}{\sqrt{2}}\hat{a}_5^\dagger + \frac{i}{\sqrt{2}}\hat{a}_6^\dagger\right)|0_5, 0_6\rangle e^{i\varphi} \tag{19.108}$$

which on simplification gives us

$$|\psi_{out}\rangle = -\frac{i}{2}(1 - e^{i\varphi})|1_5, 0_6\rangle - \frac{1}{2}(1 + e^{i\varphi})|0_5, 1_6\rangle \tag{19.109}$$

Thus, the probability of finding a photon in arm 5 will be

$$|\langle 1_5, 0_6|\psi_{out}\rangle|^2 = \left|-\frac{i}{2}(1 - e^{i\varphi})\right|^2 = \frac{1}{2}(1 - \cos\varphi) \tag{19.110}$$

Similarly the probability of finding a photon in arm 6 would be

$$|\langle 0_5, 1_6|\psi_{out}\rangle|^2 = \left|-\frac{1}{2}(1 + e^{i\varphi})\right|^2 = \frac{1}{2}(1 + \cos\varphi) \tag{19.111}$$

This clearly shows interference effects between the two possible paths of the single photon to the output of the interferometer. Note that it is a single photon which is interfering with itself; actually, the probability amplitudes of the two indistinguishable paths are interfering to produce the interference effects in the probability. If we choose $\varphi = 0$, then the probability of detecting a photon in arm 5 is zero; the photon will always exit from the arm 6. Thus, in this setup it is incorrect to state that the photon chooses one of the arms when it arrives on BS1; it chooses both the paths of the interferometer. Integrated optic platforms using lithium niobate on insulator for tunable quantum interference have been demonstrated for applications in integrated quantum photonic processing (Maeder et al. 2024).

Exercise 19.8 Consider a waveguide-based Mach-Zehnder interferometer with two 3 dB directional couplers (see Fig. 19.6) described by the transfer matrix given by Eq. (19.32). Assume that the upper arm of the interferometer has a phase shifter which introduces a phase shift of χ on the path. Obtain the transfer matrix of the Mach-Zehnder interferometer and find the condition under which a photon incident on Port 1 exits the interferometer from Port 4.

Solution The transfer matrix of the two directional couplers is given in Eq. (19.32) and the transfer matrix of the phase shifter in the upper arm of the interferometer is given by

$$T_\chi = \begin{pmatrix} e^{i\chi} & 0 \\ 0 & 1 \end{pmatrix} \tag{19.112}$$

Hence, the transfer matrix of the Mach-Zehnder interferometer would be

$$\begin{aligned} T_{MZI} &= \frac{1}{\sqrt{2}}\begin{pmatrix} e^{i\phi} & i \\ i & e^{-i\phi} \end{pmatrix}\begin{pmatrix} e^{i\chi} & 0 \\ 0 & 1 \end{pmatrix}\frac{1}{\sqrt{2}} \\ &\times \begin{pmatrix} e^{i\phi} & i \\ i & e^{-i\phi} \end{pmatrix} = \frac{1}{2}\begin{pmatrix} e^{i(2\phi+\chi)} - 1 & ie^{i(\phi+\chi))} + ie^{-i\phi} \\ ie^{i(\phi+\chi))} + ie^{-i\phi} & -e^{i\chi} + e^{-2i\phi} \end{pmatrix} \end{aligned} \tag{19.113}$$

If the input photon enters Port 1, then the condition under which the photon will always exit from Port 4 will require

$$e^{i(2\phi+\chi)} - 1 = 0$$

or

$$\chi = -2\phi \tag{19.114}$$

Under this condition, the transfer matrix becomes

$$T_{MZI} = ie^{-i\phi}\begin{pmatrix} 0 & 1 \\ 1 & 0 \end{pmatrix}$$

It is left to the reader to show that the transfer matrix T_{MZI} given in Eq. (19.113) is unitary.

19.7.1 Unstable Mirror in One Arm of the Interferometer

In the above discussion note that the two paths that the photon could take are indistinguishable and there is no possible way to determine as to which path the photon followed. It is primarily due to the absence of this information anywhere in the experiment, whether we care to determine or not, that there was interference among the two paths.

We now consider a slight modification of the above interferometer. We assume that mirror M_2 is not fixed but can move. We assume that the mirror is so sensitive that even if one photon hits it, it will vibrate and thus such an interferometer will contain the "which path" information about the photon. We will show that this information will erase the interference pattern from the interferometer.

Let us denote the two states of the mirror namely stationary and vibrating mirror by the following two orthogonal states:

$$|\psi\rangle_S \text{ and } |\psi\rangle_V$$

Now let us again assume the incidence of a single photon in arm 1 of the interferometer. The analysis remains the same as before until the state passes through the mirrors 1 and 2. Thus, to repeat, the state after BS_1 is

$$|\psi_{BS1}\rangle = \frac{i}{\sqrt{2}}|1\rangle_3|0\rangle_4 + \frac{1}{\sqrt{2}}|0\rangle_3|1\rangle_4 \tag{19.115}$$

After mirrors M_1 and M_2, the state becomes

$$\frac{i}{\sqrt{2}}|1\rangle_3|0\rangle_4|\psi\rangle_S + \frac{1}{\sqrt{2}}|0\rangle_3|1\rangle_4|\psi\rangle_V$$

where we have used the fact that for the portion corresponding to the photon being found in path 3 the mirror will remain stationary and for the path in which the photon may be found in arm 4 the mirror would vibrate.

Continuing as before the state exiting the interferometer after BS_2 would be

$$
\begin{aligned}
|\psi_{out}\rangle &= -\frac{1}{\sqrt{2}}\left(\frac{i}{\sqrt{2}}\widehat{a}_5^+ + \frac{1}{\sqrt{2}}\widehat{a}_6^+\right)|0\rangle_5|0\rangle_6|\psi\rangle_S + \frac{i}{\sqrt{2}}\left(\frac{1}{\sqrt{2}}\widehat{a}_5^+ + \frac{i}{\sqrt{2}}\widehat{a}_6^+\right)|0\rangle_5|0\rangle_6|\psi\rangle_V \\
&= \left(-\frac{i}{2}|1\rangle_5|0\rangle_6|\psi\rangle_S - \frac{1}{2}|0\rangle_5|1\rangle_6|\psi\rangle_S\right) + \left(\frac{i}{2}|1\rangle_5|0\rangle_6|\psi\rangle_V - \frac{1}{2}|0\rangle_5|1\rangle_6|\psi\rangle_V\right) \\
&= -\frac{i}{2}|1\rangle_5|0\rangle_6\left(|\psi\rangle_S - |\psi\rangle_V\right) - \frac{1}{2}|0\rangle_5|1\rangle_6\left(|\psi\rangle_S + |\psi\rangle_V\right)
\end{aligned}
\tag{19.116}
$$

Thus, the probability of finding the photon in arm 5 of the interferometer would be

$$
|\langle 1_5, 0_6, \psi_S|\psi_{out}\rangle|^2 + |\langle 1_5, 0_6, \psi_V|\psi_{out}\rangle|^2 = \frac{1}{2} \tag{19.117}
$$

Notice that now the probability of detecting a photon on arm 5 is ½. In the case discussed earlier when there was no information available about the path of the photon, the probability of finding the photon in arm 5 was zero (for $\phi = 0$). By having a vibrating mirror in one of the arms, we have a device in the interferometer which can tell us about the path of the photon in the interferometer and immediately the interference term vanishes from the output.

In fact if we had an absorbing object in arm 4 of the interferometer, then the probability of detecting a photon in arm 5 is still half. Thus, appearance of a photon in arm 5 in a balanced interferometer wherein no photon is supposed to have exited arm 5 signifies the presence of an object in arm 4. Thus, the exiting photon is able to show the presence of an object in arm 4 without ever having interacted with it. Such effects are referred to as interaction free measurements and are very interesting from both a philosophical perspective and for practical applications.

Problems

19.1. 100 photons are individually incident on a 3 dB beam splitter. It is found that the first 20 incident photons get reflected. What is the probability that the 21st photon will be reflected?

19.2. The output state from a beam splitter is given by

$$
|\psi\rangle = \alpha|1_R, 0_T\rangle + 0.9\ |0_R, 1_T\rangle
$$

If 1000 photons are incident on the beam splitter, obtain the mean number of photons in the reflected arm

19.3. Show that the following matrix is unitary

$$
U = \frac{1}{\sqrt{2}}\begin{pmatrix} i & 1 \\ 1 & i \end{pmatrix}
$$

19.4. When 1 mW of power is incident on Port 1 of a directional coupler made up of two identical waveguides, it is found that 0.2 mW of power exits Port 3 and 0.8 mW of power exits Port 4. (a) Write down the matrix relation connecting the input and output annihilation operators. (b) What will be the output state when a single photon is incident on Port 1?

19.5. Optical waveguide directional couplers may have polarization dependence. Consider a directional coupler which has a transmission coefficient of 30% for horizontal polarization and 50% for vertical polarization. Write down the output state when a single photon incident on Port 1 has (a) vertical polarization (b) horizontal polarization and (c) is linearly polarized and making an angle θ with the horizontal direction (d) What are the probabilities of detecting the photon in Port 3 and Port 4 of the coupler?

19.6. Show that Eqs. (19.12) and (19.14) satisfy energy conservation.

19.7. Consider a Sagnac interferometer arrangement as shown in Fig. 19.7. Assuming a single photon incident in Port 1 of the device, (a) follow through the path of the photon through the interferometer and obtain the output state. (b) What is the probability of detecting the photon in Port 1 and Port 2? (c) If a nonreciprocal effect is introduced so that only the clockwise path suffers a phase shift ϕ, what will be the output state? What will be the corresponding probabilities of detecting the photon in Port 1 and Port 2? Note that both the input ports 1 and 2 are also the output ports of the device. Such an interferometer is used as an optical gyroscope.

19.8. Consider a symmetric directional coupler consisting of three identical waveguides spaced equally. Obtain the relationship between the three output ports and the three input ports in matrix form. Check if the result satisfies energy conservation (see Sect. 25.3.3).

19.9. Obtain thc probability of detecting 2 photons in arm 3 of a Mach-Zehnder interferometer if the input state to the beam splitter is given by $|2_1, 0_2\rangle$.

Fig. 19.7 For Problem 19.7

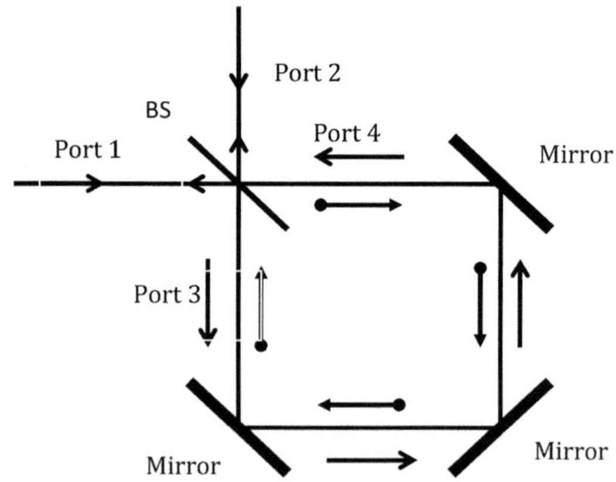

19.10. Consider a 3 dB directional coupler made of identical waveguides when two identical photons are incident in a superposition state of both being in arm 1 or both being in arm 2 as described by the following quantum state:

$$|\psi_{in}\rangle = \frac{1}{\sqrt{2}}\left(|2_1, 0_2\rangle + e^{2i\phi}|0_1, 2_2\rangle\right)$$

where 2ϕ is the phase difference. Obtain the output state from the directional coupler and interpret the result.

19.11. Consider the incidence of the following entangled state on a beam splitter:

$$|\psi_{in}\rangle = \frac{1}{\sqrt{2}}(|H_1, V_2\rangle - |V_1, H_2\rangle)$$

Obtain the output state from the beam splitter which is assumed to be independent of polarization. Compare your results with the result obtained in Sect. 19.5.

19.12. Consider two single photons with the same central wavelength but having Gaussian spectra of widths σ_1 and σ_2. Show that, the coincidence probability or detecting one photon in each of the output arms is

$$P_{coin} = \frac{1}{2} - \frac{\sigma_1\sigma_2}{\sigma_1^2 + \sigma_2^2}\exp\left[-\frac{\sigma_1^2\sigma_2^2}{\sigma_1^2 + \sigma_2^2}\tau^2\right]$$

Note that when the spectra are identical, then $\sigma_1 = \sigma_2$ and for $\tau = 0$ we obtain $P_{coin} = 0$. Also if the two photons have very different spectral widths, for example, if $\sigma_1 \gg \sigma_2$, then

$$P_{coin}(\tau = 0) \approx \frac{1}{2} - \frac{1}{2}\sqrt{\frac{2\sigma_2}{\sigma_1}}$$

Thus, if $\sigma_1 = 10\sigma_2$, then $P_{coin} \approx 0.28$ showing that the HOM dip does not reach zero value which it should have if the two photons were indistinguishable.

References

Babel, S., L. Bollmers, M. Massaro, K. H. Luo, M. Stefeszky, F. Pegoraro, P. Held, H. Herrmann, C. Eigner, B. Brecht, L. Padberg, and C. Silberhorn. 2023. Demonstration of Hong-Ou-Mandel interference in an LNOI directional coupler. *Opt Express* 31:23140–23148.

Ghatak, A., and K. Thyagarajan. 1989. *Optical electronics*. New York: Plenum Publishing Corporation.

Hong, C. K., Z. Y. Ou, and L. Mandel. 1987. Measurement of subpicosecond time intervals between two photons by interference. *Phys Rev Lett* 59:2044–2046.

Maeder, A., G. Finco, F. Kaufmann, A. Sabatti, J. Kellner, R. J. Chapman, and R. Grange. 2024. On-chip tunable quantum interference in a lithium niobate on insulator photonic integrated circuit. *Quant Sci Tech* 9:035040.

Mandel, L. 1999. Quantum effects in one-photon and two-photon interference. *Rev Mod Phys* 71: S274–S282.

Mohanty, A., M. Zhang, A. Dutt, S. Ramelow, P. Nussenzveig, and M. Lipson. 2017. Quantum interference between transverse spatial waveguide modes. *Nat Commun* 8:14010. https://doi.org/10.1038/ncomms14010.

Chapter 20
Optical Homodyning

Abstract Optical homodyning is very important technique in quantum photonics for carrying out measurements on different quantum states of light. This chapter discusses the principle behind balanced optical homodyning and its quantum description. Optical homodyning techniques applied on incident coherent state, squeezed state, vacuum state are given.

20.1 Introduction

Homodyne detection is a process in which a weak modulated signal is mixed with a strong wave (called the local oscillator) having the same frequency which results in the amplification of the signal and extraction of the information contained in the modulation. Optical homodyne detection is very important in quantum state measurements and allows us to carry out measurements on the different quadratures of the signal. Using homodyning techniques, it is possible to measure noise in the vacuum state and squeezed states.

In this chapter, we will first discuss the classical analysis of homodyne detection and then present the quantum analysis.

20.2 Classical Picture

Consider an optical signal with an electric field variation given by

$$E_s = E_{s0} \sin (\omega t - \phi_s) \tag{20.1}$$

where E_{s0} represents the amplitude and ϕ_s the phase of the signal field. Instead of expressing the field in terms of amplitude and phase, we can express them in terms of the two quadrature components:

K. Thyagarajan, *Integrated Quantum Photonics*, Graduate Texts in Physics,
https://doi.org/10.1007/978-3-031-85728-7_20

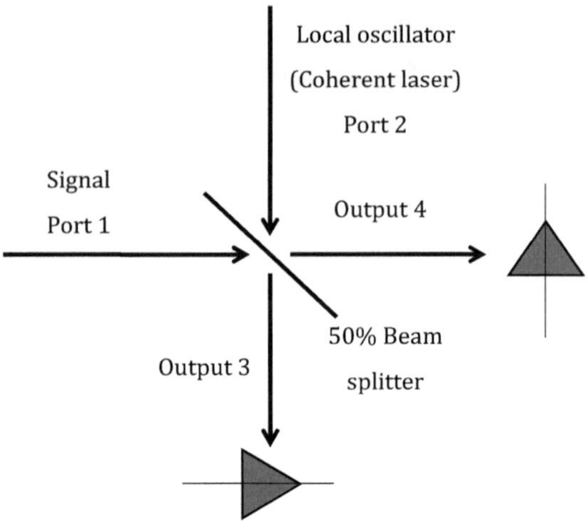

Fig. 20.1 Optical homodyning technique in which the signal to be measured is mixed with a local oscillator at the same frequency using a 50% beam splitter

$$E_s = E_{s0} \cos \phi_s \sin \omega t - E_{s0} \sin \phi_s \cos \omega t \tag{20.2}$$

The time variation in the two terms are of the form sinωt and cosωt and are the two quadrature components. Their amplitudes are given by $E_{s0} \cos \phi_s$ and $E_{s0} \sin \phi_s$. Using the homodyning technique, we can either measure the sine quadrature or the cosine quadrature by appropriate choice of the phase of the local oscillator.

Figure 20.1 shows the technique of balanced optical homodyning in which the signal to be measured is mixed with a local oscillator having the same frequency through a 50% beam splitter. Each of the two incident waves is partially reflected and partially transmitted and interfere with each other in the two output arms of the beam splitter.

We represent the signal and local oscillator fields in the complex representation:

$$E_s = E_{s0} e^{i(\phi_s - \omega t)} \tag{20.3}$$

$$E_L = E_{L0} e^{i(\phi_L - \omega t)} \tag{20.4}$$

where E_{L0} is the amplitude and ϕ_L is the phase of the local oscillator field, respectively.

We assume that the signal field is incident on Port 1 and the local oscillator on Port 2 of the beam splitter (see Fig. 20.1).

Now following the analysis of beam splitter given in Chap. 19, we obtain the following expressions for the fields in the output ports 3 and 4 of the beam splitter:

$$E_3 = \frac{1}{\sqrt{2}}(iE_s + E_L) = \frac{i}{\sqrt{2}} E_{s0} e^{i(\phi_s - \omega t)} + \frac{1}{\sqrt{2}} E_{L0} e^{i(\phi_L - \omega t)} \tag{20.5}$$

$$E_4 = \frac{1}{\sqrt{2}}(E_s + iE_L) = \frac{1}{\sqrt{2}}E_{s0}e^{i(\phi_s - \omega t)} + \frac{i}{\sqrt{2}}E_{L0}e^{i(\phi_L - \omega t)} \tag{20.6}$$

The outputs from the two arms of the beam splitter are incident on photodetectors which convert the optical signal into current signals. The currents generated by the photodetectors from ports 3 and 4 will be proportional to $|E_3|^2$ and $|E_4|^2$, respectively.

If I_3 and I_4 represent the two currents and the proportionality constant is represented by K, then we have

$$I_3 = K|E_3|^2 \tag{20.7}$$

and

$$I_4 = K|E_4|^2 \tag{20.8}$$

Substituting from Eqs. (20.5) and (20.6) for E_3 and E_4 in the above two equations gives us

$$I_3 = \frac{K}{2}\left[E_{s0}^2 + E_{L0}^2 - 2E_{s0}E_{L0}\sin(\phi_s - \phi_L)\right] \tag{20.9}$$

$$I_4 = \frac{K}{2}\left[E_{s0}^2 + E_{L0}^2 + 2E_{s0}E_{L0}\sin(\phi_s - \phi_L)\right] \tag{20.10}$$

An electronic circuit measures the difference in the two currents and we obtain

$$I = I_4 - I_3 = 2KE_{s0}E_{L0}\ \sin(\phi_s - \phi_L) \tag{20.11}$$

which is proportional to E_{s0} sin (ϕ_s − ϕ_L), that is, to the signal electric field amplitude E_{s0} and the phase difference (ϕ_s − ϕ_L) between the signal field and the local oscillator field.

If we choose $\phi_L = 0$, then the current is proportional to E_{s0} sin ϕ_s which is the amplitude of the cosine quadrature of the signal field. Similarly if we choose $\phi_L = \frac{\pi}{2}$, then the current is proportional to E_{s0} cos ϕ_s which is the amplitude of the sine quadrature of the signal. Thus, by appropriately choosing the phase of the local oscillator with respect to the signal field, it is possible to measure either of the two quadratures of the signal.

Note that since I is proportional to E_{L0}, which is the amplitude of the local oscillator, this measurement technique leads to amplification of the signal current that is being measured. This is indeed a standard technique in classical signal processing.

20.3 Quantum Analysis of Balanced Homodyne Detection

We now analyze balanced homodyne detection in the quantum picture. As discussed in Chap. 19, we can relate the annihilation operators of the output ports 3 and 4 in terms of the annihilation operators of the input ports 1 and 2 as follows (see Eqs. 19.41a and 19.41b):

$$\widehat{a}_3 = \frac{i}{\sqrt{2}}\widehat{a}_1 + \frac{1}{\sqrt{2}}\widehat{a}_2 \tag{20.12}$$

$$\widehat{a}_4 = \frac{1}{\sqrt{2}}\widehat{a}_1 + \frac{i}{\sqrt{2}}\widehat{a}_2 \tag{20.13}$$

We now assume that the quantum state that needs to be measured (represented by $|\psi\rangle$) is incident on port 1 and the local oscillator that is incident on port 2 is a coherent state $|\alpha\rangle$ having an amplitude α. The local oscillator is assumed to be a strong field. Thus, the input to the beam splitter is given by the quantum state

$$|\psi_{in}\rangle = |\psi_{1,2}\rangle = |\psi_1\rangle|\alpha_2\rangle \tag{20.14}$$

Now the optical intensities measured in ports 3 and 4 will be proportional to the expectation values of the number operators $\widehat{a}_3^\dagger\widehat{a}_3$ and $\widehat{a}_4^\dagger\widehat{a}_4$ of ports 3 and 4, respectively. The expectation value of the number of photons exiting port 3 is given by

$$\begin{aligned}\langle\widehat{n}_3\rangle = \langle\widehat{a}_3^\dagger\widehat{a}_3\rangle = \langle\psi_{3,4}|\widehat{a}_3^\dagger\widehat{a}_3|\psi_{3,4}\rangle &= \langle\psi_{1,2}\Big|\left(-\frac{i}{\sqrt{2}}\widehat{a}_1^\dagger + \frac{1}{\sqrt{2}}\widehat{a}_2^\dagger\right)\left(\frac{i}{\sqrt{2}}\widehat{a}_1 + \frac{1}{\sqrt{2}}\widehat{a}_2\right)\Big|\psi_{1,2}\rangle \\ &= \frac{1}{2}\left[\langle\psi_{1,2}|\left(\widehat{a}_1^\dagger\widehat{a}_1 - i\widehat{a}_1^\dagger\widehat{a}_2 + i\widehat{a}_2^\dagger\widehat{a}_1 + \widehat{a}_2^\dagger\widehat{a}_2\right)|\psi_{1,2}\rangle\right]\end{aligned} \tag{20.15}$$

Similarly the expectation value of the number of photons exiting port 4 is given by

$$\begin{aligned}\langle\widehat{n}_4\rangle = \langle\widehat{a}_4^\dagger\widehat{a}_4\rangle = \langle\psi_{3,4}|\widehat{a}_4^\dagger\widehat{a}_4|\psi_{3,4}\rangle &= \langle\psi_{1,2}\Big|\left(\frac{1}{\sqrt{2}}\widehat{a}_1^\dagger - \frac{i}{\sqrt{2}}\widehat{a}_2^\dagger\right)\left(\frac{1}{\sqrt{2}}\widehat{a}_1 + \frac{i}{\sqrt{2}}\widehat{a}_2\right)\Big|\psi_{1,2}\rangle \\ &= \frac{1}{2}\left[\langle\psi_{1,2}|\left(\widehat{a}_1^\dagger\widehat{a}_1 + i\widehat{a}_1^\dagger\widehat{a}_2 - i\widehat{a}_2^\dagger\widehat{a}_1 + \widehat{a}_2^\dagger\widehat{a}_2\right)|\psi_{1,2}\rangle\right]\end{aligned} \tag{20.16}$$

Hence, the difference in the photon counts from ports 3 and 4 will be

$$\langle\widehat{n}_d\rangle = \langle\widehat{n}_4\rangle - \langle\widehat{n}_3\rangle = \langle\psi_{1,2}|\left(i\widehat{a}_1^\dagger\widehat{a}_2 - i\widehat{a}_2^\dagger\widehat{a}_1\right)|\psi_{1,2}\rangle \tag{20.17}$$

It may be noted that since the signal and local oscillator have the same time dependence, $\widehat{a}_1^{\dagger}\widehat{a}_2$ and $\widehat{a}_2^{\dagger}\widehat{a}_1$ will be time independent. Since $|\alpha\rangle$ is a coherent state in port 2, we have

$$\widehat{a}_2|\alpha\rangle = \alpha|\alpha\rangle \tag{20.18}$$

Thus,

$$\langle\widehat{n}_d\rangle = \langle\psi_1|\left(i\widehat{a}_1^{\dagger}\alpha - i\alpha^*\widehat{a}_1\right)|\psi_1\rangle \tag{20.19}$$

If the coherent state has an amplitude $|\alpha|$ and phase ϕ, that is,

$$\alpha = |\alpha|e^{i\phi} \tag{20.20}$$

then we have

$$\langle\widehat{n}_d\rangle = i|\alpha|\langle\psi_1|\left(\widehat{a}_1^{\dagger}e^{i\phi} - \widehat{a}_1e^{-i\phi}\right)|\psi_1\rangle$$

which can be written as

$$\langle\widehat{n}_d\rangle = 2|\alpha|\langle\psi_1|\frac{\left(\widehat{a}_1e^{-i\phi} - \widehat{a}_1^{\dagger}e^{i\phi}\right)}{2i}|\psi_1\rangle \tag{20.21}$$

If we represent the general quadrature operator by

$$\widehat{X}_\theta = \frac{\left(\widehat{a}_1e^{-i\theta} + \widehat{a}_1^{\dagger}e^{i\theta}\right)}{2} \tag{20.22}$$

then

$$\frac{\left(\widehat{a}_1e^{-i\phi} - \widehat{a}_1^{\dagger}e^{i\phi}\right)}{2i} = \frac{\left(-i\widehat{a}_1e^{-i\phi} + i\widehat{a}_1^{\dagger}e^{i\phi}\right)}{2} = \frac{\left(\widehat{a}_1e^{-i\theta} + \widehat{a}_1^{\dagger}e^{i\theta}\right)}{2}$$

where $\theta = \phi + \frac{\pi}{2}$.

Hence,

$$\langle\widehat{n}_d\rangle = 2|\alpha|\left\langle\widehat{X}_\theta\right\rangle \tag{20.23}$$

where $\left\langle\widehat{X}_\theta\right\rangle = \langle\psi_1|\widehat{X}_\theta|\psi_1\rangle$.

Thus, we have for $\theta = 0$, $\widehat{X}_0 = \widehat{X}_1$ and for $\theta = \frac{\pi}{2}$ we have $\widehat{X}_{\frac{\pi}{2}} = \widehat{X}_2$ which are the two quadratures of the electromagnetic field. Hence, depending on the phase of the local oscillator, namely ϕ, it is possible to measure any one of the two quadratures of the signal field input in port 1.

If we choose $\phi = -\frac{\pi}{2}$, we have $\theta = 0$ and

$$\langle \widehat{n}_d \rangle = 2|\alpha| \langle \psi_1 | \frac{\left(\widehat{a}_1 + \widehat{a}_1^\dagger\right)}{2} | \psi_1 \rangle \tag{20.24}$$

which implies that the measured current is proportional to the expectation value of the quadrature $X_1 = \frac{\left(\widehat{a}_1 + \widehat{a}_1^\dagger\right)}{2}$ in the input unknown quantum state $|\psi\rangle$.

Similarly if we choose $\phi = 0$, then $\theta = \frac{\pi}{2}$ and

$$\langle \widehat{n}_d \rangle = 2|\alpha| \langle \psi_1 | \frac{\left(\widehat{a}_1 - \widehat{a}_1^\dagger\right)}{2i} | \psi_1 \rangle \tag{20.25}$$

and the measured current is proportional to the expectation value of the quadrature $X_2 = \frac{\left(\widehat{a}_1 - \widehat{a}_1^\dagger\right)}{2i}$.

The fluctuations in the difference in the number of photon counts from ports 3 and 4 are given by

$$\left(\Delta n_d^2\right) = \left\langle \widehat{n}_d^2 \right\rangle - \langle \widehat{n}_d \rangle^2 \tag{20.26}$$

Now

$$\begin{aligned}\left\langle \widehat{n}_d^2 \right\rangle &= \left\langle \left(i\widehat{a}_1^\dagger \widehat{a}_2 - i\widehat{a}_2^\dagger \widehat{a}_1 \right)^2 \right\rangle = -\left\langle \left(\widehat{a}_1^\dagger \widehat{a}_2 - \widehat{a}_2^\dagger \widehat{a}_1 \right) \left(\widehat{a}_1^\dagger \widehat{a}_2 - \widehat{a}_2^\dagger \widehat{a}_1 \right) \right\rangle \\ &= -\left\langle \left(\widehat{a}_1^{\dagger 2} \widehat{a}_2^{\,2} + \widehat{a}_1^2 \widehat{a}_2^{\dagger 2} - 2\widehat{a}_1^\dagger \widehat{a}_1 \widehat{a}_2^\dagger \widehat{a}_2 - \widehat{a}_1^\dagger \widehat{a}_1 - \widehat{a}_2^\dagger \widehat{a}_2 \right) \right\rangle\end{aligned}$$

where we have used the commutation relations satisfied by the operators $\widehat{a}_1$, $\widehat{a}_1^\dagger$ and $\widehat{a}_2$, $\widehat{a}_2^\dagger$.

Now,

$$\begin{aligned}\left\langle \widehat{a}_1^{\dagger 2} \widehat{a}_2^{\,2} \right\rangle &= \left\langle \widehat{a}_1^{\dagger 2} \right\rangle \left\langle \widehat{a}_2^{\,2} \right\rangle = |\alpha|^2 e^{2i\phi} \left\langle \widehat{a}_1^{\dagger 2} \right\rangle \\ \left\langle \widehat{a}_1^2 \widehat{a}_2^{\dagger 2} \right\rangle &= \left\langle \widehat{a}_1^2 \right\rangle \left\langle \widehat{a}_2^{\dagger 2} \right\rangle = |\alpha|^2 e^{-2i\phi} \left\langle \widehat{a}_1^2 \right\rangle \\ \left\langle \widehat{a}_1^\dagger \widehat{a}_1 \widehat{a}_2^\dagger \widehat{a}_2 \right\rangle &= \left\langle \widehat{a}_1^\dagger \widehat{a}_1 \right\rangle \left\langle \widehat{a}_2^\dagger \widehat{a}_2 \right\rangle = |\alpha|^2 \left\langle \widehat{a}_1^\dagger \widehat{a}_1 \right\rangle \\ \left\langle \widehat{a}_2^\dagger \widehat{a}_2 \right\rangle &= |\alpha|^2\end{aligned}$$

Hence, we have

$$\begin{aligned}\left\langle\widehat{n}_d^2\right\rangle &= |\alpha|^2\left(-e^{2i\phi}\left\langle\widehat{a}_1^{\dagger 2}\right\rangle - e^{-2i\phi}\left\langle\widehat{a}_1{}^2\right\rangle + 2\left\langle\widehat{a}_1^{\dagger}\widehat{a}_1\right\rangle + 1\right)\\ &= |\alpha|^2\left(e^{2i\theta}\left\langle\widehat{a}_1^{\dagger 2}\right\rangle + e^{-2i\theta}\left\langle\widehat{a}_1{}^2\right\rangle + 2\left\langle\widehat{a}_1^{\dagger}\widehat{a}_1\right\rangle + 1\right)\end{aligned} \tag{20.27}$$

where we have neglected $\left\langle\widehat{a}_1^{\dagger}\widehat{a}_1\right\rangle$ assuming that $|\alpha|^2 \gg \left\langle\widehat{a}_1^{\dagger}\widehat{a}_1\right\rangle$ and have substituted $\theta = \phi + \frac{\pi}{2}$.

Now,

$$\left\langle\widehat{X}_\theta{}^2\right\rangle = \frac{\left\langle\left(\widehat{a}_1 e^{-i\theta} + \widehat{a}_1^{\dagger} e^{i\theta}\right)^2\right\rangle}{4} = \frac{1}{4}\left(e^{-2i\theta}\left\langle\widehat{a}_1{}^2\right\rangle + e^{2i\theta}\left\langle\widehat{a}_1^{\dagger 2}\right\rangle + 2\left\langle\widehat{a}_1^{\dagger}\widehat{a}_1\right\rangle + 1\right) \tag{20.28}$$

Comparing Eqs. (20.27) and (20.28), we obtain

$$\left\langle\widehat{n}_d^2\right\rangle = 4|\alpha|^2\left\langle\widehat{X}_\theta{}^2\right\rangle \tag{20.29}$$

Hence,

$$\begin{aligned}\left(\Delta n_d^2\right) &= \left\langle\widehat{n}_d^2\right\rangle - \left\langle\widehat{n}_d\right\rangle^2 = 4|\alpha|^2\left\langle\widehat{X}_\theta{}^2\right\rangle - 4|\alpha|^2\left\langle\widehat{X}_\theta\right\rangle^2 = 4|\alpha|^2\left(\left\langle\widehat{X}_\theta{}^2\right\rangle - \left\langle\widehat{X}_\theta\right\rangle^2\right)\\ &= 4|\alpha|^2\left(\Delta X_\theta^2\right)\end{aligned} \tag{20.30}$$

Thus, the fluctuations in the photon count from the balanced homodyne measurement are proportional to the fluctuations in the appropriate quadrature of the incident quantum state which enters port 1 of the beam splitter. Measurement of the photon number fluctuations directly gives us the fluctuations in the appropriate quadrature of the input signal state.

Thus, by appropriately choosing the phase of the coherent state (the local oscillator which is a classical laser beam), it is possible to make measurements on the quadrature components of the unknown quantum state. In fact, by choosing arbitrary values of ϕ, one can measure the signal amplitude in different quadrature bases.

For example, if the unknown state is a squeezed state, then as we discussed in Chap. 18, we will observe squeezing in the appropriate quadrature, amplitude, or phase squeezing.

Exercise 20.1 Express $\widehat{X}_\theta$ in terms of the cosine and sine quadratures.

Solution From Eq. (20.22), we have

$$\widehat{X}_\theta = \frac{\left(\widehat{a}_1 e^{-i\theta} + \widehat{a}_1^\dagger e^{i\theta}\right)}{2} = \frac{\left[\widehat{a}_1(\cos\theta - i\sin\theta\,) + \widehat{a}_1^\dagger(\cos\theta + i\sin\theta)\right]}{2}$$

which can be simplified to give

$$\widehat{X}_\theta = \cos\theta\frac{\left(\widehat{a}_1 + \widehat{a}_1^\dagger\right)}{2} + \sin\theta\frac{\left(\widehat{a}_1 - \widehat{a}_1^\dagger\right)}{2i} = \widehat{X}_1\cos\theta + \widehat{X}_2\sin\theta.$$

The corresponding expectation value is given by

$$\left\langle \widehat{X}_\theta \right\rangle = \left\langle \widehat{X}_1 \right\rangle \cos\theta + \left\langle \widehat{X}_2 \right\rangle \sin\theta \tag{20.31}$$

Thus, depending on the value of θ, the technique provides us with the signal corresponding to the two quadratures.

In Exercise 20.1, we showed that the expectation value $\left\langle \widehat{X}_\theta \right\rangle$ depends on θ. If the time variation of the phase θ is given by $\theta = \Omega t$, then from Eq. (20.31) we have

$$\left\langle \widehat{X}_\theta(t) \right\rangle = \left\langle \widehat{X}_1 \right\rangle \cos\Omega t + \left\langle \widehat{X}_2 \right\rangle \sin\Omega t \tag{20.32}$$

and we have

$$\langle \widehat{n}_d \rangle = 2|\alpha|\left\langle \widehat{X}_\theta \right\rangle = 2|\alpha|\left(\left\langle \widehat{X}_1 \right\rangle \cos\Omega t + \left\langle \widehat{X}_2 \right\rangle \sin\Omega t\right) \tag{20.33}$$

showing a sinusoidal time variation of the detected mean photon number.

We also have

$$\begin{aligned}\left\langle \widehat{X}_\theta^2 \right\rangle &= \left\langle \left(\widehat{X}_1\cos\theta + \widehat{X}_2\sin\theta\right)^2 \right\rangle \\ &= \left\langle \widehat{X}_1^2 \right\rangle \cos^2\theta + \left\langle \widehat{X}_2^2 \right\rangle \sin^2\theta + \left\langle \widehat{X}_1\widehat{X}_2 + \widehat{X}_2\widehat{X}_1 \right\rangle \sin\theta\cos\theta\end{aligned}$$

and hence

$$\begin{aligned}\left(\Delta n_d^2\right) &= 4|\alpha|^2\left(\Delta X_\theta^2\right)\\ &= 4|\alpha|^2\left(\left(\Delta X_1^2\right)\cos^2\Omega t + \left(\Delta X_2^2\right)\sin^2\Omega t + \left\{\left\langle \widehat{X}_1\widehat{X}_2 + \widehat{X}_2\widehat{X}_1\right\rangle - 2\left\langle \widehat{X}_1\right\rangle\left\langle \widehat{X}_2\right\rangle\right\}\sin\Omega t\cos\Omega t\right)\end{aligned} \tag{20.34}$$

Hence, in general, the fluctuations in the measured signal of the homodyne detector also vary with time at a frequency determined by Ω.

We now consider some examples.

20.3.1 Coherent State as the Input Signal

Let us assume that the input signal in port 1 to be measured is a coherent state $|\tilde{\alpha}\rangle$ with an eigenvalue $\tilde{\alpha} = |\tilde{\alpha}|e^{i\chi}$. Thus,

$$\langle \widehat{a}_1\rangle = \langle \tilde{\alpha}|\widehat{a}_1|\tilde{\alpha}\rangle = |\tilde{\alpha}|e^{i\chi} \tag{20.35a}$$

and

$$\left\langle \widehat{a}_1^\dagger\right\rangle = \langle \tilde{\alpha}|\widehat{a}_1^\dagger|\tilde{\alpha}\rangle = |\tilde{\alpha}|e^{-i\chi} \tag{20.35b}$$

Hence,

$$\left\langle \widehat{X}_1\right\rangle = |\tilde{\alpha}|\cos\chi \tag{20.36}$$

and

$$\left\langle \widehat{X}_2\right\rangle = |\tilde{\alpha}|\sin\chi \tag{20.37}$$

Hence,

$$\langle \widehat{n}_d\rangle = 2|\alpha|\left(\left\langle \widehat{X}_1\right\rangle\cos\Omega t + \left\langle \widehat{X}_2\right\rangle\sin\Omega t\right) = 2|\alpha||\tilde{\alpha}|\cos(\Omega t - \chi) \tag{20.38}$$

showing that the mean photon number varies sinusoidally with time.

Also, we have that for a coherent state (see Chap. 15)

$$\left(\Delta X_1^2\right) = \left(\Delta X_2^2\right) = \frac{1}{4} \tag{20.39}$$

We also see that

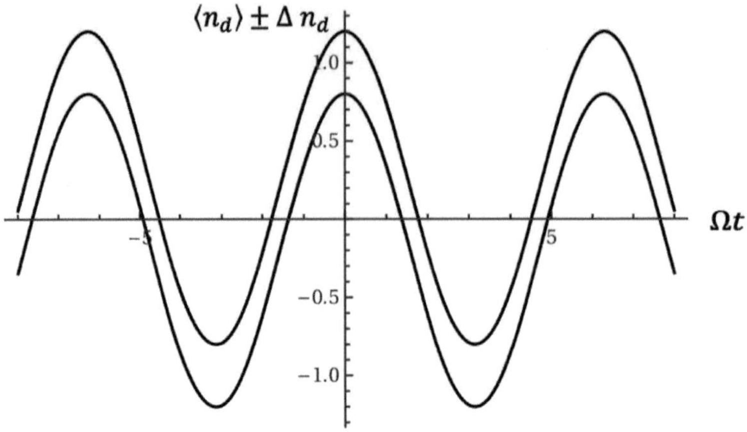

Fig. 20.2 Plot of mean photon number with the fluctuation for a coherent state. Note that the noise is independent of time

$$\left\langle \widehat{X}_1\widehat{X}_2 + \widehat{X}_2\widehat{X}_1 \right\rangle = \frac{1}{4i}\left\langle \left(\left(\widehat{a}_1 + \widehat{a}_1^\dagger \right)\left(\widehat{a}_1 - \widehat{a}_1^\dagger \right) + \left(\widehat{a}_1 - \widehat{a}_1^\dagger \right)\left(\widehat{a}_1 + \widehat{a}_1^\dagger \right) \right) \right\rangle$$

On opening the brackets and using the commutation relation between $\widehat{a}_1$ and $\widehat{a}_1^\dagger$, we obtain

$$\left\langle \widehat{X}_1\widehat{X}_2 + \widehat{X}_2\widehat{X}_1 \right\rangle = \frac{1}{2i}\left\langle \left(\widehat{a}_1^2 - a_1^{\dagger 2} \right) \right\rangle = |\tilde{\alpha}|^2 \sin 2\chi \tag{20.40}$$

Hence, we have

$$\left(\Delta n_d^2\right) = 4|\alpha|^2\left[\left(\frac{1}{4}\right)\cos^2\Omega t + \left(\frac{1}{4}\right)\sin^2\Omega t + \left(|\tilde{\alpha}|^2 \sin 2\chi - 2|\tilde{\alpha}|^2 \sin\chi\cos\chi\right)\sin\Omega t\cos\Omega t\right]$$

which on simplification gives us

$$\left(\Delta n_d^2\right) = |\alpha|^2 \tag{20.41}$$

which is indeed independent of time.

Thus, when the input field is a coherent state, the expectation value of the mean photon number varies sinusoidally with time with a constant uncertainty independent of time. Figure 20.2 shows the plot of mean photon number with the fluctuation for a coherent state.

Note that $\tilde{\alpha} = 0$ corresponds to an input vacuum state which has the same noise as above with $\langle \widehat{n}_d \rangle = 0$.

20.3.2 Squeezed State as the Input Signal

We now consider the input to be a squeezed state $|\beta\rangle$ which is an eigenstate of the operator $\widehat{b} = \mu\widehat{a}_1 + \nu\widehat{a}_1^\dagger$ with $|\mu|^2 - |\nu|^2 = 1$ (see Chap. 18):

$$\widehat{b}|\beta\rangle = \left(\mu\widehat{a}_1 + \nu\widehat{a}_1^\dagger\right)|\beta\rangle = \beta|\beta\rangle \tag{20.42}$$

For simplicity, we will assume that both μ and ν are real and the eigenvalue β also to be real. We recall the following from Chap. 18 (see Eqs. 18.55 and 18.56 with $t = 0$):

$$(\Delta X_1)^2 = \frac{e^{-2\sigma}}{4} \tag{20.43}$$

$$(\Delta X_2)^2 = \frac{e^{+2\sigma}}{4} \tag{20.44}$$

We also have

$$\widehat{X}_1 = \frac{(\widehat{a} + \widehat{a}^\dagger)}{2} = (\mu - \nu)\frac{(\widehat{b} + \widehat{b}^\dagger)}{2} \tag{20.45}$$

and

$$\widehat{X}_2 = \frac{(\widehat{a} - \widehat{a}^\dagger)}{2i} = (\mu + \nu)\frac{(\widehat{b} - \widehat{b}^\dagger)}{2i} \tag{20.46}$$

Thus,

$$\left\langle \widehat{X}_1 \right\rangle = (\mu - \nu)\beta \tag{20.47}$$

and

$$\left\langle \widehat{X}_2 \right\rangle = 0 \tag{20.48}$$

Also from Eqs. (20.45) and (20.46) and using the commutation relation between $\widehat{b}$ and $\widehat{b}^\dagger$, we obtain

$$\widehat{X}_1\widehat{X}_2 = \frac{1}{4i}\left(\widehat{b}^2 - \widehat{b}^{\dagger 2} - 1\right) \tag{20.49}$$

$$\widehat{X}_2\widehat{X}_1 = \frac{1}{4i}\left(\widehat{b}^2 - \widehat{b}^{\dagger 2} + 1\right) \tag{20.50}$$

Thus,

$$\widehat{X}_1\widehat{X}_2 + \widehat{X}_2\widehat{X}_1 = \frac{1}{2i}\left(\widehat{b}^2 - \widehat{b}^{\dagger 2}\right) \tag{20.51}$$

Hence, we have (see Eq. 20.33)

$$\langle \widehat{n}_d \rangle = 2|\alpha|\left(\left\langle \widehat{X}_1 \right\rangle \cos \Omega t + \left\langle \widehat{X}_2 \right\rangle \sin \Omega t\right) = 2|\alpha|(\mu - \nu)\beta \cos \Omega t = 2|\alpha| e^{-\sigma}\beta \cos \Omega t \tag{20.52}$$

and (see Eq. 20.34)

$$\left(\Delta n_d^2\right) = 4|\alpha|^2\left(\Delta X_\theta^2\right) = |\alpha|^2\left(e^{-2\sigma}\cos^2\Omega t + e^{+2\sigma}\sin^2\Omega t\right) \tag{20.53}$$

Here $\mu = \cosh \sigma$ and $\nu = \sinh \sigma$.

Thus, in the squeezed state we are considering, the mean value of the difference in photon counts varies sinusoidally with time and the fluctuation in the photon number difference varies periodically with time and varies between $|\alpha|^2 e^{-2\sigma}$ and $|\alpha|^2 e^{+2\sigma}$.

Thus if $\sigma > 0$, the fluctuations in the photon number difference are minimum at times $t = m\frac{\pi}{\Omega}; m = 0, 1, 2, \ldots$ and maximum when $t = (2m + 1)\frac{\pi}{2\Omega}; m = 0, 1, 2, \ldots$ From Eq. (20.52), it can be seen that at $t = m\frac{\pi}{\Omega}; m = 0, 1, 2, \ldots$ the mean photon number difference is maximum and at $t = (2m + 1)\frac{\pi}{2\Omega}; m = 0, 1, 2, \ldots$, the mean photon number difference is zero. This corresponds to amplitude squeezed state.

On the other hand if $\sigma < 0$, then the fluctuations in the photon number difference are minimum at times $t = (2m + 1)\frac{\pi}{2\Omega}; m = 0, 1, 2, \ldots$ and maxima when $t = m\frac{\pi}{\Omega}; m = 0, 1, 2, \ldots$. From Eq. (20.52), it can be seen that the fluctuations are minimum when the difference in the mean number of photons is zero and maximum when the difference in mean photon number is maximum. This corresponds to phase squeezed state.

Note that $\sigma = 0$ corresponds to a coherent state and the noise is independent of time as discussed earlier.

Figures 20.3 and 20.4 show simulated plots of the difference in mean photon number variation with the fluctuations for an amplitude and a phase squeezed state (see also Figs. 18.1 and 18.3). Figure 20.5 shows simulated plot of the difference in mean photon number and the fluctuation in the photon number difference for a squeezed vacuum state.

For very nice measurements using balanced homodyne detection of different quantum states of light including coherent states, squeezed states, squeezed vacuum states, etc., readers are referred to Breitenbach et al. (1997) and Breitenbach and Schiller (1997). Detection of squeezed light using integrated optic technology and homodyning can be found, for example, in Porto et al. (2018) and Raffaelli et al. (2018). See also Leonhardt (1997) and Protte et al. (2024).

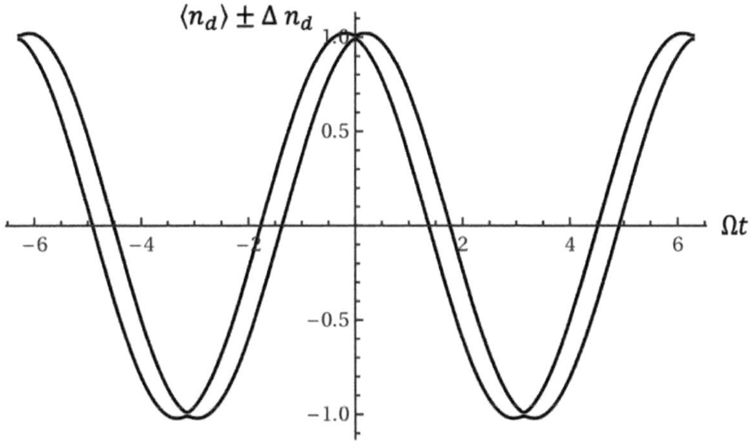

Fig. 20.3 Plot of difference in mean photon number with the fluctuation for an amplitude squeezed state ($\sigma > 0$). Note the reduction in noise at the peaks of the signal

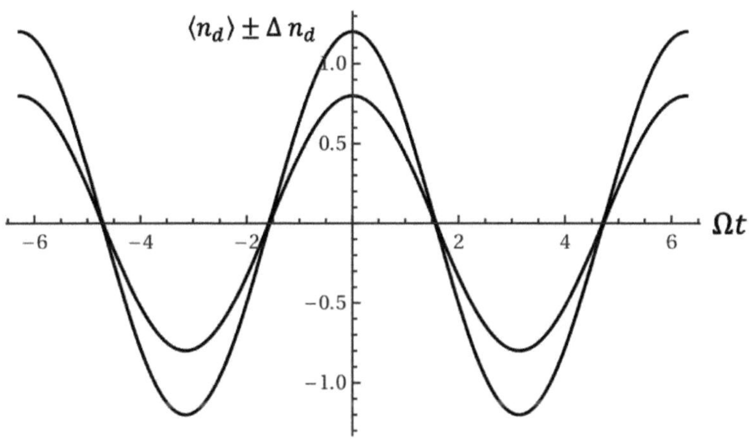

Fig. 20.4 Plot of difference in mean photon number with the fluctuation for a phase squeezed state ($\sigma < 0$)

Fig. 20.5 Plot of difference in mean photon number with the fluctuation for a squeezed vacuum state

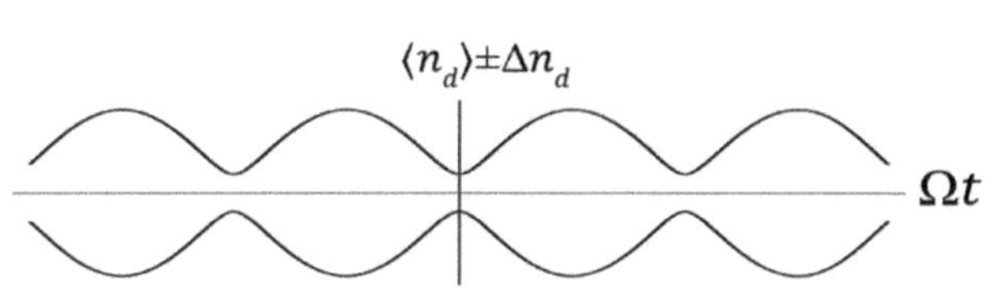

Exercise 20.2 Consider the incidence of a number state in the signal arm of the homodyne measurement setup. Obtain the measured signal.

Solution An input number state is given by $|\psi_{in}\rangle = |n\rangle$. For this state, we have

$$\left\langle \widehat{X}_1 \right\rangle = \frac{1}{2}\left\langle \left(\widehat{a}_1 + \widehat{a}_1^\dagger\right) \right\rangle = \frac{1}{2}\langle n|\left(\widehat{a}_1 + \widehat{a}_1^\dagger\right)|n\rangle = 0$$

and

$$\left\langle \widehat{X}_1^{\,2} \right\rangle = \frac{1}{4}\left\langle \left(\widehat{a}_1 + \widehat{a}_1^\dagger\right)^2 \right\rangle = \frac{1}{4}\langle n|\left(\widehat{a}_1^{\,2} + \widehat{a}_1^{\dagger 2} + 2\widehat{a}_1^\dagger\widehat{a}_1 + 1\right)|n\rangle = \frac{1}{4}(2n+1)$$

Also

$$\left\langle \widehat{X}_2 \right\rangle = \frac{1}{2i}\left\langle \left(\widehat{a}_1 - \widehat{a}_1^\dagger\right) \right\rangle = \frac{1}{2i}\langle n|\left(\widehat{a}_1 - \widehat{a}_1^\dagger\right)|n\rangle = 0$$

$$\left\langle \widehat{X}_2^2 \right\rangle = -\frac{1}{4}\left\langle \left(\widehat{a}_1 - \widehat{a}_1^\dagger\right)^2 \right\rangle = -\frac{1}{4}\langle n|\left(\widehat{a}_1^{\,2} + \widehat{a}_1^{\dagger 2} - 2\widehat{a}_1^\dagger\widehat{a}_1 - 1\right)|n\rangle = \frac{1}{4} \times (2n+1)$$

$$\left\langle \widehat{X}_1\widehat{X}_2 + \widehat{X}_2\widehat{X}_1 \right\rangle = \frac{1}{4i}\left\langle \left(\left(\widehat{a}_1 + \widehat{a}_1^\dagger\right)\left(\widehat{a}_1 - \widehat{a}_1^\dagger\right) + \left(\widehat{a}_1 - \widehat{a}_1^\dagger\right)\left(\widehat{a}_1 + \widehat{a}_1^\dagger\right)\right) \right\rangle = \frac{1}{2i}\left\langle \widehat{a}_1^2 - \widehat{a}_1^{\dagger 2} \right\rangle = 0$$

Hence,

$$\langle \widehat{n}_d \rangle = 2|\alpha|\left(\left\langle \widehat{X}_1 \right\rangle \cos\Omega t + \left\langle \widehat{X}_2 \right\rangle \sin\Omega t\right) = 0$$

$$\left(\Delta n_d^2\right) = 4|\alpha|^2\left(\frac{1}{4}(2n+1)\cos^2\Omega t + \frac{1}{4}(2n+1)\sin^2\Omega t\right) = (2n+1)|\alpha|^2$$

Problems

20.1. Consider the following state incident on port 1 of a balanced homodyne measurement setup:

$$|\psi\rangle = \frac{1}{\sqrt{2}}(|n\rangle + |n+1\rangle)$$

Obtain the difference in the mean photon number and the fluctuations in the photon number difference.

20.2. Consider the incidence of a squeezed state with $\mu = \sqrt{2}$ and $\nu = -1$ in a homodyne measurement setup. Obtain the measured mean number of photons and the fluctuations in the photon number difference at the output.

References

Breitenbach, G., and S. Schiller. 1997. Homodyne tomography of classical and non-classical light. *J Modern Opt* 44:2207–2225.

Breitenbach, G., S. Schiller, and J. Mlynek. 1997. Measurement of the quantum states of squeezed light. *Nature* 387:471–475.

Leonhardt, U. 1997. *Measuring the quantum state of light*. Cambridge: Cambridge University Press.

Porto, C., D. Rusca, S. Cialdi, A. Crespi, R. Osselame, D. Tamascelli, S. Olivares, and M. G. A. Paris. 2018. Detection of squeezed light with glass integrated technology embedded into a homodyne detector setup. *J Opt Soc Am B* 35:1596–1602.

Protte, M., T. Schapeler, J. Sperling, and T. J. Bartley. 2024. Low-noise balanced homodyne detection with superconducting nanowire single-photon detectors. *Opt Quantum* 2 (1): 1–6. https://doi.org/10.1364/OPTICAQ.502201.

Raffaelli, F., G. Ferranti, D. H. Mahler, P. Sibson, J. E. Kennard, A. Santamato, G. Sinclair, D. Bonneau, M. G. Thompson, and J. C. F. Matthews. 2018. A homodyne detector integrated onto a photonic chip for measuring quantum states and generating random numbers. *Quantum Sci Technol* 3:025003.

Chapter 21
Schrödinger, Heisenberg, and Interaction Pictures

Abstract This chapter discusses three important pictures in quantum mechanics: these include the Schrödinger picture, Heisenberg picture, and the interaction picture. Also discussed is time-dependent perturbation theory and application of the interaction picture to parametric down conversion.

21.1 Introduction

In this chapter, we will briefly discuss the three primary pictures of quantum mechanics; these are the Schrödinger picture, the Heisenberg picture, and the interaction picture. All these pictures are equivalent pictures and depending on the problem at hand one of them may be more convenient than the others.

As discussed in Chap. 3, in quantum mechanics the state of a system is represented by a wave function or a ket vector which contains all the information regarding the state. The observables are represented by Hermitian operators which act on the state of the system and the time evolution of the state is determined by the Schrödinger equation.

The time evolution of a dynamical system can be described by infinitely many quantum mechanical pictures out of which two are very important; these are the Schrödinger and Heisenberg pictures. In the Schrödinger picture, the state vector changes with time as per Schrödinger equation, while the observables are independent of time. In the Heisenberg picture which is diametrically opposite to Schrödinger picture, the observables change with time, while the state vector remains constant and independent of time. Both these pictures are completely equivalent and give the same results. They are just two different pictures for analyzing the time evolution of a quantum system. In the interaction picture, we write the total Hamiltonian as a sum of two parts; the first part represents the Hamiltonian of several isolated parts of the system and the second part represents the interaction among them. In this picture, the interaction part of the Hamiltonian is used to obtain the time variation of the quantum state. We will use the interaction picture in Chap. 22, when we study spontaneous parametric down conversion process.

K. Thyagarajan, *Integrated Quantum Photonics*, Graduate Texts in Physics,
https://doi.org/10.1007/978-3-031-85728-7_21

It is important to mention that representations that we discussed in Chap. 3 and the pictures that we are discussing in this chapter are quite different. Representations correspond to the choice of the complete orthonormal set of functions with respect to which the quantum states and the operators are described. Thus, we have coordinate representation which is related to the standard wave function, and the momentum representation which is related to the coordinate representation through a Fourier transform and so on. On the other hand, the pictures are quite different in which the time dependence is either associated with the quantum state itself while the operators are independent of time (Schrödinger picture) or the Heisenberg picture where the states are independent of time while the operators change with time. In the interaction picture, both the quantum state and the observables are time dependent.

One of the most important quantities is the expectation value of an operator $\widehat{O}$ which is given by (see Chap. 3)

$$\left\langle \widehat{O} \right\rangle = \langle \psi | \widehat{O} | \psi \rangle \tag{21.1}$$

where $|\psi\rangle$ represents the quantum state of the system. The expectation value of any operator must be the same in all pictures irrespective of whether the time dependence is accounted for in the state vector or in the observable. We will see that the different quantum mechanical pictures follow from each other via unitary transformations.

21.2 Schrödinger Picture

In the Schrödinger picture, the observables are represented by time-independent Hermitian operators, while the quantum state is represented by time-dependent kets. The time-dependent quantum state of the system satisfies the Schrödinger equation:

$$i\hbar \frac{d}{dt} \left| \psi^{(S)}(t) \right\rangle = \widehat{H}^{(S)} \left| \psi^{(S)}(t) \right\rangle \tag{21.2}$$

where the superscript S is to indicate that it is in the Schrödinger picture. The Hamiltonian $\widehat{H}^{(S)}$ is the Hamiltonian operator and independent of time, except for any explicit time dependence, which we are not considering here.

The formal integral of Eq. (21.2) can be written as

$$\left| \psi^{(S)}(t) \right\rangle = e^{-\frac{i}{\hbar} \widehat{H}^{(S)} t} \left| \psi^{(S)}(0) \right\rangle \tag{21.3}$$

where $e^{-\frac{i}{\hbar} \widehat{H}^{(S)} t}$ is the time evolution operator.

If $|\psi^{(S)}(0)\rangle$ is an eigenstate of the Hamiltonian operator with an eigenvalue E, then we obtain

$$\left|\psi^{(S)}(t)\right\rangle = e^{-\frac{i}{\hbar}Et}\left|\psi^{(S)}(0)\right\rangle \tag{21.4}$$

and the time evolution is just a time-dependent phase factor. In general, since the initial state $|\psi^{(S)}(0)\rangle$ will not be an eigenstate of the Hamiltonian operator, the time dependence is not just a phase factor.

In the Schrödinger picture, the state vector evolves with time, while the operators are time independent except for some explicit time dependence that may be present. Hence, the expectation value of an operator will be given by

$$\left\langle \widehat{O}^{(S)} \right\rangle = \left\langle \psi^{(S)}(t) \middle| \widehat{O}^{(S)} \middle| \psi^{(S)}(t) \right\rangle \tag{21.5}$$

where the time dependence is shown explicitly in the state vector and the operators are assumed to be time independent.

Chapter 3 covered this picture in more detail.

Exercise 21.1 Consider a simple harmonic oscillator in the initial state given by a superposition of energy eigenstates (see Chap. 3):

$$\left|\psi^{(S)}(0)\right\rangle = \frac{1}{2}|n\rangle + \frac{\sqrt{3}}{2}|n+1\rangle \tag{21.6}$$

Obtain the time evolution $|\psi^{(S)}(t)\rangle$ of the state and the expectation value of the energy in the state using the Schrödinger picture.

Solution The time evolution of the state is given by Eq. (21.3). Thus, we have

$$\left|\psi^{(S)}(t)\right\rangle = e^{-\frac{i}{\hbar}\widehat{H}^{(S)}t}\left(\frac{1}{2}|n\rangle + \frac{\sqrt{3}}{2}|n+1\rangle\right) = \frac{1}{2}e^{-\frac{i}{\hbar}\widehat{H}^{(S)}t}|n\rangle + \frac{\sqrt{3}}{2}e^{-\frac{i}{\hbar}\widehat{H}^{(S)}t}|n+1\rangle$$

which simplifies to

$$\left|\psi^{(S)}(t)\right\rangle = \frac{1}{2}e^{-\frac{i}{\hbar}E_n t}|n\rangle + \frac{\sqrt{3}}{2}e^{-\frac{i}{\hbar}E_{n+1}t}|n+1\rangle \tag{21.7}$$

where E_n represents the energy of the nth eigenstate. The Hamiltonian operator is $\widehat{H} = \hbar\omega\left(\widehat{a}^{\dagger}\widehat{a} + \frac{1}{2}\right)$ and is independent of time in the Schrödinger picture. The expectation value of the energy is given by

$$\langle E\rangle = \left\langle \psi^{(S)}(t)\middle|\hbar\omega\left(\widehat{a}^{\dagger}\widehat{a} + \frac{1}{2}\right)\middle|\psi^{(S)}(t)\right\rangle$$

which becomes

$$\langle E\rangle = \frac{1}{2}\hbar\omega$$

$$+\hbar\omega\left(\frac{1}{2}e^{\frac{i}{\hbar}E_n t}\langle n| + \frac{\sqrt{3}}{2}e^{\frac{i}{\hbar}E_{n+1}t}\langle n+1|\right)\hat{a}^{\dagger}\hat{a}\left(\frac{1}{2}e^{-\frac{i}{\hbar}E_n t}|n\rangle + \frac{\sqrt{3}}{2}e^{-\frac{i}{\hbar}E_{n+1}t}|n+1\rangle\right)$$

which simplifies to

$$\langle E\rangle = \frac{1}{2}\hbar\omega + \hbar\omega\left(\frac{1}{4}n + \frac{3}{4}[n+1]\right) = \left(n + \frac{5}{4}\right)\hbar\omega \tag{21.8}$$

where we have used the orthogonality among the eigenstates.

21.3 Heisenberg Picture

In the Heisenberg picture, the states are time independent, while the operators are time dependent. The Schrödinger and Heisenberg pictures are related through unitary transformations. We write the relation between the quantum state in the Heisenberg picture and Schrödinger picture as

$$\left|\psi^{(S)}(t)\right\rangle = \hat{U}(t)\left|\psi^{(H)}\right\rangle \tag{21.9}$$

where the superscript H stands for Heisenberg picture and $\hat{U}(t)$ represents a unitary operator that connects the Schrödinger and Heisenberg pictures. Here $|\psi^{(H)}\rangle$ is the ket representing the state in the Heisenberg picture and is time independent. Also if we demand that the two states be identical at some time $t = 0$, we must have $\hat{U}(t=0) = 1$:

$$\left|\psi^{(S)}(t=0)\right\rangle = \hat{U}(t=0)\left|\psi^{(H)}\right\rangle = \left|\psi^{(H)}\right\rangle \tag{21.10}$$

Substituting from Eq. (21.9) in the Schrödinger equation, Eq. (21.2), we obtain

$$i\hbar\frac{d}{dt}\left(\hat{U}(t)\left|\psi^{(H)}\right\rangle\right) = \hat{H}^{(S)}\hat{U}(t)\left|\psi^{(H)}\right\rangle$$

Since $|\psi^{(H)}\rangle$ is independent of time, we have

$$i\hbar\frac{d\hat{U}(t)}{dt}\left|\psi^{(H)}\right\rangle = \hat{H}^{(S)}\hat{U}(t)\left|\psi^{(H)}\right\rangle$$

Hence,

$$i\hbar \frac{d\widehat{U}(t)}{dt} = \widehat{H}^{(S)} \widehat{U}(t); \quad \widehat{U}(t=0) = 1 \tag{21.11}$$

For conservative systems, the Hamiltonian is time independent and the formal solution of the above equation is

$$\widehat{U}(t) = \mathrm{e}^{-\frac{i}{\hbar}\widehat{H}^{(S)} t} \tag{21.12}$$

where the exponential of an operator is defined through the power series expansion:

$$e^{\widehat{O}} = 1 + \widehat{O} + \frac{1}{2!}\widehat{O}\widehat{O} + \frac{1}{3!}\widehat{O}\widehat{O}\widehat{O} + \dots . \tag{21.13}$$

Note that

$$\widehat{U}^{\dagger}(t) = \widehat{U}^{-1}(t) \tag{21.14}$$

and hence $\widehat{U}(t)$ is a unitary operator.

Since the expectation value of any operator must be independent of the picture, using Eq. (21.9) we must have

$$\langle \psi^{(S)}(t) | \widehat{O}^{(S)} | \psi^{(S)}(t) \rangle = \langle \psi^{(H)} | \widehat{U}^{\dagger}(t) \widehat{O}^{(S)} \widehat{U}(t) | \psi^{(H)} \rangle = \langle \psi^{(H)} | \widehat{O}^{(H)}(t) | \psi^{(H)} \rangle \tag{21.15}$$

Thus, the relationship between the operators in the Schrödinger and Heisenberg pictures is given by

$$\widehat{O}^{(H)}(t) = \widehat{U}^{\dagger}(t) \widehat{O}^{(S)} \widehat{U}(t) \tag{21.16}$$

The operators in the Schrödinger and Heisenberg pictures coincide at $t = 0$ since $\widehat{U}(t=0) = 1$.

Since the operators in the Heisenberg pictures are time dependent, we can obtain an equation for the operators by differentiating Eq. (21.16) to obtain

$$\frac{d\widehat{O}^{(H)}(t)}{dt} = \frac{d\widehat{U}^{\dagger}(t)}{dt} \widehat{O}^{(S)} \widehat{U}(t) + \widehat{U}^{\dagger}(t) \widehat{O}^{(S)} \frac{d\widehat{U}(t)}{dt}$$

We have used the fact that the operators in the Schrödinger picture are time independent. Using Eq. (21.11) this becomes

$$\begin{aligned} \frac{d\widehat{O}^{(H)}(t)}{dt} &= \frac{i}{\hbar} \widehat{U}^{\dagger}(t) \widehat{H}^{(S)} \widehat{O}^{(S)} \widehat{U}(t) - \frac{i}{\hbar} \widehat{U}^{\dagger}(t) \widehat{O}^{(S)} \widehat{H}^{(S)} \widehat{U}(t) \\ &= \frac{i}{\hbar} \widehat{U}^{\dagger}(t) \widehat{H}^{(S)} \widehat{U}(t) \widehat{U}^{\dagger}(t) \widehat{O}^{(S)} \widehat{U}(t) - \frac{i}{\hbar} \widehat{U}^{\dagger}(t) \widehat{O}^{(S)} \widehat{U}(t) \widehat{U}^{\dagger}(t) \widehat{H}^{(S)} \widehat{U}(t) \end{aligned}$$

where we have used the fact that $\widehat{U}(t)\widehat{U}^{\dagger}(t)=\widehat{U}^{\dagger}(t)\widehat{U}(t)=I$.

Since the operators in the Schrödinger picture and Heisenberg picture are related through the following relation,

$$\widehat{U}^{\dagger}(t)\widehat{O}^{(S)}\widehat{U}(t)=\widehat{O}^{(H)}(t) \tag{21.17}$$

Putting $\widehat{O}^{(S)}=\widehat{H}^{(S)}$ the Hamiltonian operator, we have

$$\widehat{U}^{\dagger}(t)\widehat{H}^{(S)}\widehat{U}(t)=\widehat{H}^{(H)}(t)=\widehat{H}=\widehat{H}^{(S)} \tag{21.18}$$

The last step follows since $\widehat{H}^{(S)}$ and $\widehat{U}(t)$ commute and that $\widehat{U}(t)$ is a unitary operator.

Hence, the equation of motion for the operator in the Heisenberg picture is

$$i\hbar\frac{d\widehat{O}^{(H)}(t)}{dt}=\left[\widehat{O}^{(H)}(t),\widehat{H}\right] \tag{21.19}$$

Equation (21.19) defines the time evolution of the operator in the Heisenberg picture.

Usually the superscripts are suppressed and when Heisenberg picture is used the operators will be written with time as an argument and in Schrödinger picture they will be written without any time argument.

One can show that the commutation relations have the same form in both Schrödinger and Heisenberg pictures (see Problem 21.1).

The Heisenberg picture is formally analogous to classical mechanics since the equations of motion of the operators resemble the corresponding classical equations of motion (see, e.g., Dirac 1982; Marcuse 1980).

Exercise 21.2 Solve Exercise 21.1 in the Heisenberg picture.

Solution In the Heisenberg picture, the state does not evolve with time and is given by

$$\left|\psi^{(H)}\right\rangle=\left|\psi^{(S)}(0)\right\rangle=\frac{1}{2}|n\rangle+\frac{\sqrt{3}}{2}|n+1\rangle$$

The Hamiltonian operator is given by $\widehat{H}(t)=\hbar\omega\left(\widehat{a}^{\dagger}(t)\widehat{a}(t)+\frac{1}{2}\right)$ where we are not specifically putting a superscript to refer to Heisenberg picture but showing it by mentioning the time dependence of the operator. Now in the Heisenberg picture [see Eq. (21.26)]

$$\widehat{a}(t) = \widehat{a}(0)e^{-i\omega t}$$

and

$$\widehat{a}^{\dagger}(t) = \widehat{a}^{\dagger}(0)e^{i\omega t}$$

Thus, in the Heisenberg picture

$$\widehat{H}(t) = \hbar\omega\left(\widehat{a}^{\dagger}(t)\widehat{a}(t) + \frac{1}{2}\right) = \hbar\omega\left(\widehat{a}^{\dagger}(0)\widehat{a}(0) + \frac{1}{2}\right)$$

and independent of time. The expectation value of energy is then given by

$$\begin{aligned}\langle E\rangle &= \langle\psi^{(H)}|\hbar\omega\left(\widehat{a}^{\dagger}(t)\widehat{a}(t) + \frac{1}{2}\right)|\psi^{(H)}\rangle \\ &= \frac{1}{2}\hbar\omega + \hbar\omega\left(\frac{1}{2}\langle n| + \frac{\sqrt{3}}{2}\langle n+1|\right)\widehat{a}^{\dagger}(0)\widehat{a}(0)\left(\frac{1}{2}|n\rangle + \frac{\sqrt{3}}{2}|n+1\rangle\right)\end{aligned}$$

which gives us

$$\langle E\rangle = \left(n + \frac{5}{4}\right)\hbar\omega$$

The same result as obtained in Exercise 21.1 where we had used the Schrödinger picture.

21.4 Time Evolution of the Coherent State

Since in the Heisenberg picture the operators satisfy the equation of motion given in Eq. (21.19), the annihilation operator of the electromagnetic field satisfies the following equation:

$$i\hbar\frac{d\widehat{a}(t)}{dt} = \left[\widehat{a}(t), \widehat{H}\right] \tag{21.20}$$

For the electromagnetic field

$$\widehat{H} = \hbar\omega\left(\widehat{a}^{\dagger}\widehat{a} + \frac{1}{2}\right) \tag{21.21}$$

Thus,

$$i\hbar\frac{d\widehat{a}(t)}{dt} = \left[\widehat{a}(t), \hbar\omega\left(\widehat{a}^\dagger(t)\widehat{a}(t) + \frac{1}{2}\right)\right] \tag{21.22}$$

Now

$$\left[\widehat{a}(t), \hbar\omega\left(\widehat{a}^\dagger(t)\widehat{a}(t) + \frac{1}{2}\right)\right] = \left[\widehat{a}(t), \hbar\omega\widehat{a}^\dagger(t)\widehat{a}(t)\right] + \left[\widehat{a}(t), \hbar\omega\frac{1}{2}\right]$$
$$= \hbar\omega\left[\widehat{a}(t), \widehat{a}^\dagger(t)\widehat{a}(t)\right]$$

since the second commutator on the right hand side is zero. We also have

$$\left[\widehat{a}(t), \widehat{a}^\dagger(t)\widehat{a}(t)\right] = \widehat{a}(t)\widehat{a}^\dagger(t)\widehat{a}(t) - \widehat{a}^\dagger(t)\widehat{a}(t)\widehat{a}(t) \tag{21.23}$$

Using the commutation relation satisfied by $\widehat{a}(t)$ and $\widehat{a}^\dagger(t)$, we have

$$\left[\widehat{a}(t), \widehat{a}^\dagger(t)\widehat{a}(t)\right] = \left(\widehat{a}^\dagger(t)\widehat{a}(t) + 1\right)\widehat{a}(t) - \widehat{a}^\dagger(t)\widehat{a}(t)\widehat{a}(t) = \widehat{a}(t) \tag{21.24}$$

Hence, we have

$$i\hbar\frac{d\widehat{a}(t)}{dt} = \hbar\omega\widehat{a}(t) \tag{21.25}$$

whose solution is

$$\widehat{a}(t) = \widehat{a}(t-0)e^{-i\omega t} = \widehat{a}e^{-i\omega t} \tag{21.26}$$

which gives the time evolution of the annihilation operator in the Heisenberg picture.

We now consider coherent states which are eigenstates of the annihilation operator $\widehat{a}$ (see Chap. 15):

$$\widehat{a}|\alpha\rangle = \alpha|\alpha\rangle \tag{21.27}$$

Thus, we can evaluate the expectation value of the electric field in this state to obtain

$$\begin{aligned}\left\langle\widehat{E}\right\rangle &= \langle\alpha|\widehat{E}(t)|\alpha\rangle = i\sqrt{\frac{\hbar\omega}{2\varepsilon_0 V}}\left[\langle\alpha|\widehat{a}|\alpha\rangle e^{i(kz-\omega t)} - \langle\alpha|\widehat{a}^\dagger|\alpha\rangle e^{-i(kz-\omega t)}\right]\\ &= i\sqrt{\frac{\hbar\omega}{2\varepsilon_0 V}}\left[\alpha e^{i(kz-\omega t)} - \alpha^* e^{-i(kz-\omega t)}\right]\end{aligned} \tag{21.28}$$

In the Schrödinger picture, the electric field operator will be time independent and the operator in these two representations is identical at $t = 0$. Hence, in the Schrödinger picture the electric field operator will be

$$\widehat{E} = i\sqrt{\frac{\hbar\omega}{2\varepsilon_0 V}}\left[\widehat{a}e^{ikz} - \widehat{a}^{\dagger}e^{-ikz}\right] \tag{21.29}$$

and the quantum state vector will be time dependent and should be written as $|\alpha(t)\rangle$.

In order to obtain the time dependence of the state, we express the state at $t = 0$ in terms of the number state (see Chap. 15):

$$|\alpha(t=0)\rangle = e^{-\frac{1}{2}|\alpha|^2}\sum_{n=0}^{\infty}\frac{\alpha^n}{\sqrt{n!}}|n\rangle \tag{21.30}$$

Since we know that the number state evolves as described by

$$|n(t)\rangle = e^{-\frac{iE_n t}{\hbar}}|n\rangle = e^{-i\left(n+\frac{1}{2}\right)\omega t}|n\rangle \tag{21.31}$$

we have

$$\begin{aligned}|\alpha(t)\rangle &= e^{-\frac{1}{2}|\alpha|^2}\sum_{n=0}^{\infty}\frac{\alpha^n}{\sqrt{n!}}e^{-i\left(n+\frac{1}{2}\right)\omega t}|n\rangle\\ &= e^{-\frac{1}{2}|\alpha|^2}e^{-i\omega t/2}\sum_{n=0}^{\infty}\frac{\left(\alpha e^{-i\omega t}\right)^n}{\sqrt{n!}}|n\rangle = e^{-i\omega t/2}|\alpha e^{-i\omega t}\rangle\end{aligned} \tag{21.32}$$

Hence, as the coherent state evolves with time, apart from the phase factor $e^{-i\omega t/2}$, it remains a coherent state with a complex eigenvalue $\alpha e^{-i\omega t}$. The state evolves periodically with a period given by $T = \frac{2\pi}{\omega}$.

Thus, the expectation value of the electric field will be

$$\begin{aligned}\left\langle\widehat{E}\right\rangle = \langle\alpha(t)|\widehat{E}|\alpha(t)\rangle &= i\sqrt{\frac{\hbar\omega}{2\varepsilon_0 V}}\left[\left\langle\alpha e^{-i\omega t}\left|\widehat{a}\right|\alpha e^{-i\omega t}\right\rangle e^{ikz} - \left\langle\alpha e^{-i\omega t}\left|\widehat{a}^{\dagger}\right|\alpha e^{-i\omega t}\right\rangle e^{-ikz}\right]\\ &= i\sqrt{\frac{\hbar\omega}{2\varepsilon_0 V}}\left[\alpha e^{i(kz-\omega t)} - \alpha^* e^{-i(kz-\omega t)}\right]\end{aligned} \tag{21.33}$$

which, as expected, is the same as obtained in the Heisenberg picture as it should indeed be.

21.5 Interaction Picture

Let us consider a quantum system that is perturbed in such a way that the Hamiltonian of the system is given by

$$\widehat{H}^{(S)}(t) = \widehat{H}_0^{(S)}(t) + \widehat{W}^{(S)}(t) \tag{21.34}$$

where $\widehat{W}^{(S)}(t)$ is the perturbation and $\widehat{H}_0^{(S)}(t)$ is the Hamiltonian in the absence of the perturbation. In most cases, the Schrödinger equation cannot be solved using the entire Hamiltonian $\widehat{H}^{(S)}(t)$ and one resorts to approximate solutions assuming that the perturbation defined by $\widehat{W}^{(S)}(t)$ to be weak. In order to do so we separate the fast time evolution due to $\widehat{H}_0^{(S)}(t)$ from the much slower time evolution determined by the perturbation $\widehat{W}^{(S)}(t)$. Usually the unperturbed Hamiltonian $\widehat{H}_0^{(S)}(t)$ does not depend explicitly on time, while the perturbation could be time dependent.

In order to do so, we define the quantum state in the interaction picture (represented by superscript I) by the following unitary transformation

$$|\psi^{(I)}(t)\rangle = U_0^{\dagger}(t)|\psi^{(S)}(t)\rangle \tag{21.35}$$

where $\widehat{U}_0(t)$ is a unitary operator. Note that in this picture both the operator and the state vector are time dependent with the state vector having a time dependence due to the interaction part of the Hamiltonian and the time dependence of the operators is determined by the unperturbed Hamiltonian $\widehat{H}_0^{(S)}(t)$.

Differentiating Eq. (21.35) with respect to time, we have

$$\begin{aligned} i\hbar\frac{d}{dt}|\psi^{(I)}(t)\rangle &= i\hbar\frac{dU_0^{\dagger}(t)}{dt}|\psi^{(S)}(t)\rangle + i\hbar U_0^{\dagger}(t)\frac{d|\psi^{(S)}(t)\rangle}{dt} \\ &= i\hbar\frac{dU_0^{\dagger}(t)}{dt}|\psi^{(S)}(t)\rangle + U_0^{\dagger}(t)\widehat{H}^{(S)}(t)|\psi^{(S)}(t)\rangle \end{aligned} \tag{21.36}$$

where we have used the Schrödinger equation satisfied by $|\psi^{(S)}(t)\rangle$. Substituting from Eq. (21.34), we obtain

$$\begin{aligned} i\hbar\frac{d}{dt}|\psi^{(I)}(t)\rangle = i\hbar\frac{d\widehat{U}_0^{\dagger}(t)}{dt}|\psi^{(S)}(t)\rangle + \widehat{U}_0^{\dagger}(t)\widehat{H}_0^{(S)}(t)|\psi^{(S)}(t)\rangle \\ +\widehat{U}_0^{\dagger}(t)\widehat{W}^{(S)}(t)|\psi^{(S)}(t)\rangle \end{aligned} \tag{21.37}$$

If the operator $\widehat{U}_0(t)$ satisfies the following equation (see Eq. (21.11)),

$$i\hbar\frac{d\widehat{U}_0(t)}{dt}=\widehat{H}_0^{(S)}(t)\widehat{U}_0(t);\quad \widehat{U}_0(t=0)=1 \tag{21.38}$$

we have

$$-i\hbar\frac{d\widehat{U}_0^{\dagger}(t)}{dt}=\widehat{U}_0^{\dagger}(t)\widehat{H}_0^{(S)}(t) \tag{21.39}$$

Using Eq. (21.39) in Eq. (21.37), then $|\psi^{(I)}(t)\rangle$ satisfies the following equation

$$i\hbar\frac{d}{dt}|\psi^{(I)}(t)\rangle=\widehat{U}_0^{\dagger}(t)\widehat{W}^{(S)}(t)|\psi^{(S)}(t)\rangle=\widehat{U}_0^{\dagger}(t)\widehat{W}^{(S)}(t)\widehat{U}_0(t)\widehat{U}_0^{\dagger}(t)|\psi^{(S)}(t)\rangle$$

where we have used $\widehat{U}_0(t)\widehat{U}_0^{\dagger}(t)=1$. Thus,

$$i\hbar\frac{d}{dt}|\psi^{(I)}(t)\rangle=\widehat{W}^{(I)}(t)|\psi^{(I)}(t)\rangle \tag{21.40}$$

where we have used Eq. (21.35) and we have

$$\widehat{W}^{(I)}(t)=\widehat{U}_0^{\dagger}(t)\widehat{W}^{(S)}(t)\widehat{U}_0(t) \tag{21.41}$$

represents the perturbation operator in the interaction picture.

Equation (21.40) shows that the evolution of the state in the interaction picture (represented by $|\psi^{(I)}(t)\rangle$) is the same form as Schrödinger equation except for the fact that $\widehat{H}^{(S)}(t)$ is replaced by the interaction part $\widehat{W}^{(I)}(t)$.

We can formally integrate Eq. (21.40) to obtain

$$|\psi^{(I)}(t)\rangle=|\psi^{(I)}(0)\rangle+\frac{1}{i\hbar}\int_0^t\widehat{W}^{(I)}(t')|\psi^{(I)}(t')\rangle dt' \tag{21.42}$$

Note that in Eq. (21.42), $|\psi^{(I)}(t)\rangle$ appears both on the left hand side and within the integral on the right hand side and rarely has analytical solutions.

Usually the perturbation $\widehat{W}^{(S)}(t)$ is quite weak and we can make a first order approximation to Eq. (21.42) by replacing $|\psi^{(I)}(t')\rangle$ within the integral by $|\psi^{(I)}(0)\rangle$ and obtain an approximate solution (to first order)

$$|\psi^{(I)}(t)\rangle\approx|\psi^{(I)}(0)\rangle+\left[\frac{1}{i\hbar}\int_0^t\widehat{W}^{(I)}(t')dt'\right]|\psi^{(I)}(0)\rangle \tag{21.43}$$

If $H_0^{(S)}(t)$ is independent of time, then the formal solution of Eq. (21.38) is

$$\widehat{U}_0(t) = e^{-\frac{i}{\hbar}\widehat{H}_0^{(S)}t}\widehat{U}_0(t=0) = e^{-\frac{i}{\hbar}\widehat{H}_0^{(S)}t} \tag{21.44}$$

Thus,

$$\widehat{W}^{(I)}(t) = e^{\frac{i}{\hbar}\widehat{H}_0^{(S)}t}\widehat{W}^{(S)}(t)e^{-\frac{i}{\hbar}\widehat{H}_0^{(S)}t} \tag{21.45}$$

and

$$\left|\psi^{(I)}(t)\right\rangle = e^{\frac{i}{\hbar}\widehat{H}_0^{(S)}t}\left|\psi^{(S)}(0)\right\rangle \tag{21.46}$$

We will use this interaction picture in Chap. 22 to discuss parametric down conversion process in which a pump photon spontaneously generates a pair of lower frequency photons due to nonlinear interaction within the medium.

If the interaction is time independent, then $\widehat{W}^{(I)}$ is time independent and then we can write the formal solution of Eq. (21.40) as

$$\left|\psi^{(I)}(t)\right\rangle = e^{-\frac{i}{\hbar}\widehat{W}^{(I)}t}\left|\psi^{(I)}(t=0)\right\rangle \tag{21.47}$$

As an example, we recall that in Chap. 14 we discussed the quantization of electromagnetic fields and in the quantized picture the electric field of a particular mode is represented by an operator as defined below:

$$\widehat{E}(t) = i\sqrt{\frac{\hbar\omega}{2\varepsilon_0 V}}\left[\widehat{a}e^{i(kz-\omega t)} - \widehat{a}^{\dagger}e^{-i(kz-\omega t)}\right] \tag{21.48}$$

This form of the electric field operator is in the Heisenberg picture since the operator is time dependent. In this picture, the quantum state of the electromagnetic field is time independent.

21.6 Time-Dependent Perturbation Theory and Periodic Transitions Between Eigenstates

As an example of the application of the interaction picture, we consider a time-dependent perturbation on a time-independent Hamiltonian (see Eq. (21.34)). Thus, $\widehat{H}_0^{(S)}$ will be independent of time, while the perturbation $\widehat{W}^{(S)}(t)$ will be time dependent. Referring to Eq. (21.38), we have

$$i\hbar\frac{d\widehat{U}_0(t)}{dt}=\widehat{H}_0^{(S)}\widehat{U}_0(t) \tag{21.49}$$

Since $\widehat{H}_0^{(S)}$ is independent of time, the solution of Eq. (21.49) is

$$\widehat{U}_0(t)=e^{-i\frac{\widehat{H}_0^{(S)}t}{\hbar}} \tag{21.50}$$

We also have (see Eq. (21.41))

$$\widehat{W}^{(I)}(t)=e^{i\frac{\widehat{H}_0^{(S)}t}{\hbar}}\widehat{W}^{(S)}(t)e^{-i\frac{\widehat{H}_0^{(S)}t}{\hbar}} \tag{21.51}$$

As an example let us consider quantum system with a set of discrete eigenstates and eigenvalues such as a one dimensional harmonic oscillator. In this case, the Hamiltonian is independent of time and we represent the eigenfunctions by kets $|n\rangle$ having energy eigenvalue E_n. Since they form a complete set, we expand $|\psi^{(I)}(t)\rangle$ in terms of the eigenkets $|n\rangle$:

$$|\psi^{(I)}(t)\rangle=\sum_n c_n(t)|n\rangle \tag{21.52}$$

Thus, from Eq. (21.35), we have

$$|\psi^{(S)}(t)\rangle=\widehat{U}_0(t)|\psi^{(I)}(t)\rangle=e^{-i\frac{\widehat{H}_0^{(S)}t}{\hbar}}\sum_n c_n(t)|n\rangle=\sum_n c_n(t)e^{-i\frac{E_nt}{\hbar}}|n\rangle \tag{21.53}$$

where we have used the fact that $\widehat{H}_0^{(S)}|n\rangle=E_n|n\rangle$.

Substituting from Eq. (21.52) in Eq. (21.40), we have

$$i\hbar\frac{d}{dt}\sum_n c_n(t)|n\rangle=\widehat{W}^{(I)}(t)\sum_n c_n(t)|n\rangle$$

which gives us

$$\sum_n i\hbar\frac{dc_n(t)}{dt}|n\rangle=\sum_n c_n(t)\widehat{W}^{(I)}(t)|n\rangle=\sum_n\sum_m|m\rangle\langle m|c_n(t)\widehat{W}^{(I)}(t)|n\rangle \tag{21.54}$$

where we have replaced unity by $\sum_m|m\rangle\langle m|$ which satisfies the completeness condition $\sum_m|m\rangle\langle m|=1$. Hence, we have

$$\sum_n i\hbar \frac{dc_n(t)}{dt}|n\rangle = \sum_n \sum_m c_n(t)\langle m|\widehat{W}^{(I)}(t)|n\rangle|m\rangle \tag{21.55}$$

If we define the matrix element

$$W_{mn}^{(I)}(t) = \langle m|\widehat{W}^{(I)}(t)|n\rangle \tag{21.56}$$

we have

$$\sum_n i\hbar \frac{dc_n(t)}{dt}|n\rangle = \sum_n \sum_m c_n(t) W_{mn}^{(I)}(t)|m\rangle \tag{21.57}$$

Multiplying from the left by $\langle m^{'}|$ and using the fact that $\langle m|m'\rangle = \delta_{mm'}$, we obtain (then replacing $m^{'}$ by m)

$$i\hbar \frac{dc_m(t)}{dt} = \sum_n W_{mn}^{(I)}(t) c_n(t) \tag{21.58}$$

We also have

$$\begin{aligned} W_{mn}^{(I)}(t) &= \langle m|\widehat{W}^{(I)}(t)|n\rangle = \langle m|e^{i\frac{\widehat{H}_0^{(S)}t}{\hbar}}\widehat{W}^{(S)}(t)e^{-i\frac{\widehat{H}_0^{(S)}t}{\hbar}}|n\rangle \\ &= e^{i(E_m-E_n)t/\hbar}\langle m|\widehat{W}^{(S)}(t)|n\rangle \end{aligned} \tag{21.59}$$

Let us consider a harmonic perturbation at frequency Ω:

$$\widehat{W}^{(S)}(t) = \widehat{W}^{(0)} \cos \Omega t \tag{21.60}$$

where $\widehat{W}^{(0)}$ is a constant. In such a case

$$W_{mn}^{(I)}(t) = e^{i(E_m-E_n)t/\hbar} \cos \Omega t \langle m|\widehat{W}^{(0)}|n\rangle = e^{i(E_m-E_n)t/\hbar} \cos \Omega t \widehat{W}_{mn}^{(0)} \tag{21.61}$$

Thus, we have

$$i\hbar \frac{dc_m(t)}{dt} = \cos \Omega t \sum_n e^{i(E_m-E_n)t/\hbar} \widehat{W}_{mn}^{(0)} c_n(t) \tag{21.62}$$

Let us consider two states labeled 1 and 2 and assume that the perturbation defined by $\widehat{W}^{(S)}(t)$ induces coupling among these two states. We now write down the equations that couple the two states:

$$i\hbar\frac{dc_1(t)}{dt} = \widehat{W}_{11}^{(0)}\cos\Omega t c_1(t) + e^{-i(E_2-E_1)t/\hbar}\cos\Omega t\widehat{W}_{12}^{(0)}c_2(t) \tag{21.63}$$

$$i\hbar\frac{dc_2(t)}{dt} = \widehat{W}_{21}^{(0)}e^{i(E_2-E_1)t/\hbar}\cos\Omega t c_1(t) + \widehat{W}_{22}^{(0)}\cos\Omega t c_2(t) \tag{21.64}$$

These equations are very similar to the equations determining periodic coupling among modes in a waveguide (see Chap. 9). We expand the cosine function and write

$$\begin{aligned} i\hbar\frac{dc_1(t)}{dt} &= \frac{1}{2}\widehat{W}_{11}^{(0)}\left(e^{i\Omega t}+e^{-i\Omega t}\right)c_1(t) \\ &+\frac{1}{2}\widehat{W}_{12}^{(0)}c_2(t)\left(e^{-i[(E_2-E_1)-\hbar\Omega]t/\hbar}+e^{-i[(E_2-E_1)+\hbar\Omega]t/\hbar}\right) \end{aligned} \tag{21.65}$$

$$\begin{aligned} i\hbar\frac{dc_2(t)}{dt} &= \frac{1}{2}\widehat{W}_{22}^{(0)}\left(e^{i\Omega t}+e^{-i\Omega t}\right)c_2(t) \\ &+\frac{1}{2}\widehat{W}_{21}^{(0)}c_1(t)\left(e^{i[(E_2-E_1)+\hbar\Omega]t/\hbar}+e^{i[(E_2-E_1)-\hbar\Omega]t/\hbar}\right) \end{aligned} \tag{21.66}$$

Very similar to the arguments given in Chap. 9, only the terms with $(E_2 - E_1) \approx \hbar\Omega$ on the right hand side (assuming $E_2 > E_1$) will have significant contributions when $\Omega \approx \frac{(E_2-E_1)}{\hbar}$ and neglect the term that varies with time as $e^{\pm i\Omega t}$ and $e^{-i[(E_1-E_2)-\hbar\Omega]t/\hbar}$ (leading to what is referred to as rotating wave approximation); we can then approximate Eqs. (21.65) and (21.66) by

$$i\hbar\frac{dc_1(t)}{dt} \approx \frac{1}{2}\widehat{W}_{12}^{(0)}c_2(t)e^{-i[(E_2-E_1)-\hbar\Omega]t/\hbar} \tag{21.67}$$

$$i\hbar\frac{dc_2(t)}{dt} \approx \frac{1}{2}\widehat{W}_{21}^{(0)}c_1(t)e^{i[(E_2-E_1)-\hbar\Omega]t/\hbar} \tag{21.68}$$

Note that the matrix element $\widehat{W}_{12}^{(0)}$ must be nonzero for the transitions between levels 1 and 2 to be possible. Also note that $\widehat{W}_{12}^{(0)} = \langle 1|\widehat{W}^{(0)}|2\rangle$ and $\widehat{W}_{21}^{(0)} = \langle 2|\widehat{W}^{(0)}|1\rangle$ and we will assume that $\widehat{W}_{12}^{(0)} = \widehat{W}_{21}^{(0)}$.

As a specific example, we consider the case with $\Omega = \frac{(E_2-E_1)}{\hbar}$ (see also Problem 21.6) and obtain

$$\frac{dc_1(t)}{dt} = -i\tau_{12}c_2(t) \tag{21.69a}$$

$$\frac{dc_2(t)}{dt} = -i\tau_{12}c_1(t) \tag{21.69b}$$

where

$$\tau_{12} = \frac{1}{2\hbar}\widehat{W}_{12}^{(0)} \tag{21.70}$$

The solutions of Eqs. (21.69a) and (21.69b) with initial conditions $c_1(t=0)=1$ and $c_2(t=0)=0$ are

$$c_1(t) = \cos \tau_{12} t \tag{21.71}$$

$$c_2(t) = -i \sin \tau_{12} t \tag{21.72}$$

Substituting the solution in Eq. (21.53), we obtain the time evolution of the quantum state to be

$$\left|\psi^{(S)}(t)\right\rangle = e^{-i\frac{E_1 t}{\hbar}} \cos(\tau_{12} t)|1\rangle - ie^{-i\frac{E_2 t}{\hbar}} \sin(\tau_{12} t)|2\rangle \tag{21.73}$$

In the presence of the appropriate perturbation, the quantum system transitions periodically between the states $|1\rangle$ and $|2\rangle$ and with a time period of $T = \frac{\pi}{2\tau_{12}}$; the corresponding frequency is referred to as Rabi frequency. This is very similar to the periodic exchange of energy between two modes in a waveguide due to spatially periodic perturbation in the waveguide (see Chap. 9).

Exercise 21.3 Obtain the commutation relation between $\widehat{a}(t)$ and $\widehat{a}^\dagger(t)$ where the operators are expressed in the Heisenberg picture.

Solution We first note that the time-independent operator satisfies the commutation relation

$$\left[\widehat{a}, \widehat{a}^\dagger\right] = \widehat{a}\widehat{a}^\dagger - \widehat{a}^\dagger\widehat{a} = 1 \tag{21.74}$$

Multiplying by $\widehat{U}^\dagger(t)$ from the left and $\widehat{U}(t)$ from the right and using the fact that $\widehat{U}(t)\widehat{U}^\dagger(t) = \widehat{I}$, we have

$$\widehat{U}^\dagger(t)\widehat{a}\widehat{U}(t)\widehat{U}^\dagger(t)\widehat{a}^\dagger\widehat{U}(t) - \widehat{U}^\dagger(t)\widehat{a}^\dagger\widehat{U}(t)\widehat{U}^\dagger(t)\widehat{a}\widehat{U}(t) = 1$$

Using the fact that $\widehat{a}(t) = \widehat{U}^\dagger(t)\widehat{a}\widehat{U}(t)$ and $\widehat{a}^\dagger(t) = \widehat{U}^\dagger(t)\widehat{a}^\dagger\widehat{U}(t)$, we obtain

$$\left[\widehat{a}(t), \widehat{a}^\dagger(t)\right] = 1 \tag{21.75}$$

Exercise 21.4 Obtain the time variation of the expectation value of $\widehat{a}(t)$ in the coherent state in the Heisenberg picture.

Solution In the Heisenberg picture, the state remains independent of time and is given by $|\alpha(t)\rangle = |\alpha\rangle$, while the operator $\widehat{a}(t)$ evolves with time described by

$\widehat{a}(t) = \widehat{a}(t=0)e^{-i\omega t} = \widehat{a}e^{-i\omega t}$. Hence, the expectation value of the operator $\widehat{a}(t)$ in the Heisenberg picture is given by

$$\langle \widehat{a}(t) \rangle = \langle \alpha | \widehat{a}(t) | \alpha \rangle = \langle \alpha | \widehat{a} e^{-i\omega t} | \alpha \rangle = \alpha e^{-i\omega t} = \langle \widehat{a}(0) \rangle e^{-i\omega t} \tag{21.76}$$

which is the same as obtained using the Schrödinger picture.

21.7 Parametric Down Conversion

As an example of the interaction picture, we will consider degenerate parametric down conversion in which a photon at a frequency 2ω spontaneously down converts to a pair of photons at frequency ω (see Chap. 22). In this case, the interaction Hamiltonian is time independent and given by

$$\widehat{W}^{(I)} = i\hbar \frac{\zeta}{2} \left(\widehat{a}^{\dagger 2} - \widehat{a}^2 \right) \tag{21.77}$$

where ζ is a constant depending on the nonlinear interaction process. The perturbation is time independent and the evolution of the quantum state is given by (see Eq. (21.47))

$$\left| \psi^{(I)}(t) \right\rangle = e^{-\frac{i}{\hbar} \widehat{W}^{(I)} t} \left| \psi^{(I)}(t=0) \right\rangle \tag{21.78}$$

If the input state is a coherent state $|\alpha\rangle$, then

$$\left| \psi^{(I)}(t) \right\rangle = e^{-\frac{i}{\hbar} \widehat{W}^{(I)} t} |\alpha\rangle = e^{\frac{\zeta}{2} \left(\widehat{a}^{\dagger 2} - \widehat{a}^2 \right)} |\alpha\rangle = \widehat{S} |\alpha\rangle \tag{21.79}$$

where

$$\widehat{S} = e^{\frac{\zeta}{2} \left(\widehat{a}^{\dagger 2} - \widehat{a}^2 \right)} \tag{21.80}$$

It can be shown that (see Exercise 21.6)

$$\widehat{S} \widehat{a} \widehat{S}^{-1} = \widehat{a} \cosh \zeta + \widehat{a}^{\dagger} \sinh \zeta \tag{21.81}$$

Now, coherent state is defined by

$$\widehat{a} |\alpha\rangle = \alpha |\alpha\rangle \tag{21.82}$$

Using the fact that $\widehat{S}^{-1} \widehat{S} = \widehat{I}$, we can write

$$\widehat{S}^{-1}\widehat{S}\widehat{a}|\alpha\rangle = \alpha|\alpha\rangle \tag{21.83}$$

Multiplying both sides by $\widehat{S}$, we obtain the following equation:

$$\widehat{S}\widehat{a}|\alpha\rangle = \alpha\left(\widehat{S}|\alpha\rangle\right) \tag{21.84}$$

Also

$$\widehat{S}\widehat{a}|\alpha\rangle = \widehat{S}\widehat{a}\widehat{S}^{-1}\left(\widehat{S}|\alpha\rangle\right) = \left(\widehat{a}\cosh\zeta + \widehat{a}^{\dagger}\sinh\zeta\right)\left(\widehat{S}|\alpha\rangle\right) \tag{21.85}$$

Hence, using Eq. (21.84), we obtain

$$\left(\widehat{a}\cosh\zeta + \widehat{a}^{\dagger}\sinh\zeta\right)\left(\widehat{S}|\alpha\rangle\right) = \alpha\left(\widehat{S}|\alpha\rangle\right) \tag{21.86}$$

Thus, $\widehat{S}|\alpha\rangle$ is an eigenstate of the operator $\widehat{S}\widehat{a}\widehat{S}^{-1} = \left(\widehat{a}\cosh\zeta + \widehat{a}^{\dagger}\sinh\zeta\right)$.

We showed in Chap. 18 that the eigenstate of the operator of the form $\left(\widehat{a}\cosh\zeta + \widehat{a}^{\dagger}\sinh\zeta\right)$ are nothing but squeezed states. Thus, the eigenstate of $\widehat{S}|\alpha\rangle = e^{\frac{\zeta}{2}\left(\widehat{a}^{\dagger 2} - \widehat{a}^{2}\right)}|\alpha\rangle$ is nothing but a squeezed state.

Thus, the state $|\psi^{(I)}(t)\rangle$ generated by the nonlinear interaction process described by the interaction Hamiltonian given by Eq. (21.77) is a squeezed state. If the input state is a vacuum state, then the output is a squeezed vacuum state. Squeezed states are discussed in more detail in Chap. 15.

Exercise 21.5 Derive Eq. (21.81).

Solution Now, for any pair of operators $\widehat{A}$ and $\widehat{B}$, we start with the following identity (Baker Hausdorf lemma):

$$e^{\widehat{B}}\widehat{A}e^{-\widehat{B}} = \widehat{A} + \frac{1}{1!}\left[\widehat{B},\widehat{A}\right] + \frac{1}{2!}\left[\widehat{B},\left[\widehat{B},\widehat{A}\right]\right] + \frac{1}{3!}\left[\widehat{B},\left[\widehat{B},\left[\widehat{B},\widehat{A}\right]\right]\right] + \ldots. \tag{21.87}$$

If we consider $\widehat{A} = \widehat{a}$ and $\widehat{B} = \frac{\zeta}{2}\left(\widehat{a}^{2} - \widehat{a}^{\dagger 2}\right)$, then we have

$$\left[\widehat{B},\widehat{A}\right] = \zeta\widehat{a}^{\dagger}$$

$$\left[\widehat{B},\widehat{a}^{\dagger}\right]=\zeta\widehat{a}$$

and

$$\widehat{S}\widehat{a}\widehat{S}^{-1}=e^{\widehat{B}}\widehat{A}e^{-\widehat{B}}=\widehat{a}+\zeta\widehat{a}^{\dagger}+\frac{1}{2!}\zeta^{2}\widehat{a}+\frac{1}{3!}\zeta^{3}\widehat{a}^{\dagger}+\ldots.$$

$$=\widehat{a}\left(1+\frac{\zeta^{2}}{2!}+\frac{\zeta^{4}}{4!}+\ldots.\right)+\widehat{a}^{\dagger}\left(\zeta+\frac{\zeta^{3}}{3!}+\frac{\zeta^{5}}{5!}+\ldots\right)=\widehat{a}\cosh\zeta+\widehat{a}^{\dagger}\sinh\zeta$$

Exercise 21.6 The commutation relation between the operators $\widehat{x}$ and $\widehat{p}$ in the Schrödinger picture is given by $[\widehat{x},\widehat{p}]=i\hbar$. Obtain the commutation relation in the Heisenberg picture.

Solution The operators in the Heisenberg picture and Schrödinger picture are related through Eq. (21.16). Thus, we have the operators $\widehat{x}(t)$ and $\widehat{p}(t)$ in the Heisenberg picture are given by

$$\widehat{x}(t)=\widehat{U}^{\dagger}(t)\widehat{x}\widehat{U}(t)$$
$$\widehat{p}(t)=\widehat{U}^{\dagger}(t)\widehat{p}\widehat{U}(t)$$

Hence,

$$[\widehat{x},\widehat{p}]=\widehat{x}\widehat{p}-\widehat{p}\widehat{x}=\widehat{U}^{\dagger}(t)\widehat{x}\widehat{U}(t)\widehat{U}^{\dagger}(t)\widehat{p}\widehat{U}(t)-\widehat{U}^{\dagger}(t)\widehat{p}\widehat{U}(t)\widehat{U}^{\dagger}(t)\widehat{x}\widehat{U}(t)$$

where we have multiplied by $\widehat{U}^{\dagger}(t)$ from the left and $\widehat{U}(t)$ from the right and used the fact that $\widehat{U}(t)\widehat{U}^{\dagger}(t)=\widehat{I}$. Thus,

$$[\widehat{x},\widehat{p}]=\widehat{x}(t)\widehat{p}(t)-\widehat{p}(t)\widehat{x}(t)=i\hbar$$

implying that the commutator relation remains the same.

Problems

21.1. Consider a linear harmonic oscillator and write down the expressions for the position operator $\widehat{x}$ in the Schrödinger picture and the Heisenberg picture.

21.2. Show that the commutation relations have the same form in both Schrödinger and Heisenberg pictures.

21.3. Consider the following number state at $t=0$:

$$|\psi(t=0)\rangle = |100\rangle$$

(a) What will be $|\psi(t)\rangle$ in (a) the Schrödinger picture and (b) the Heisenberg picture?

(b) Obtain the expectation value of the number operator $\widehat{N} = \widehat{a}^{\dagger}\widehat{a}$ in both the pictures.

21.4. Consider a linear harmonic oscillator in a state described by the following equation at $t = 0$:

$$|\psi(t=0)\rangle = \frac{1}{\sqrt{2}}(|0\rangle + |1\rangle)$$

Obtain the time evolution of the expectation value of position and momentum for the above state using the Schrödinger and Heisenberg pictures and show that both of them lead to the same result. Show that $\frac{d\langle\widehat{x}\rangle}{dt} = \frac{\langle\widehat{p}\rangle}{m}$.

21.5. Show that if the Schrödinger operator commutes with the Hamiltonian, then the corresponding Heisenberg operator is a constant of motion.

21.6. Solve Eqs. (21.67) and (21.68) when $\Omega \neq \frac{(E_2 - E_1)}{\hbar}$ with the initial condition $c_1(t = 0) = 1$ and $c_2(t = 0) = 0$ and obtain the corresponding solution. Compare with a similar solution obtained in Chap. 9 while discussing periodic interaction in waveguides.

References

Dirac, P. A. M. 1982. *The principles of quantum mechanics*. Oxford: Clarendon Press.

Marcuse, D. 1980. *Principles of quantum electronics*. New York: Academic Press.

Chapter 22
Quantum Treatment of Parametric Processes

Abstract Quantum mechanical treatment of parametric processes including spontaneous parametric down conversion (SPDC) and parametric amplifier are discussed in this chapter. Expressions for the output quantum state of the SPDC process is derived, and it is shown that at degeneracy the output state is a squeezed state. Also discussed are techniques to achieve spectral separability of the output photon pair. Phase-sensitive and phase-insensitive parametric amplifier is analyzed showing how phase-sensitive parametric amplifiers can provide us with noise-free amplification.

22.1 Introduction

We saw in Chap. 11 that a wave at frequency ω propagating in a medium possessing second-order nonlinearity can under certain conditions generate a wave at 2ω resulting in second harmonic generation. The question arises as to whether input of a wave at 2ω through the same medium can generate the frequency ω leading to what is referred to as subharmonic generation.

In Chap. 13, we discussed degenerate parametric amplification and showed that with a strong input at frequency 2ω, an input signal at frequency ω gets amplified or attenuated depending on the relative phase of the two inputs resulting in phase-sensitive amplification. It was also shown that with input at frequency 2ω, in the absence of an input at frequency ω, classical coupled mode equations do not predict the generation of the frequency ω. This is primarily because launching a wave at 2ω does not produce any nonlinear polarization at ω and thus no generation at the frequency ω. Thus, subharmonic generation is prohibited by classical equations. On the other hand, if at the input we have both 2ω and ω, then there will be a finite nonlinear polarization at ω leading to interaction between the frequencies 2ω and ω. This leads to parametric amplification or attenuation depending on the phases of the two inputs (see Chap. 13).

Experiments show that subharmonic at frequency ω does get generated with the incidence of a wave at 2ω and the explanation for this process is purely quantum mechanical. It is actually the quantum zero-point fluctuations at the frequency ω (see Chap. 14) which mix with the wave at frequency 2ω to generate nonlinear

K. Thyagarajan, *Integrated Quantum Photonics*, Graduate Texts in Physics,
https://doi.org/10.1007/978-3-031-85728-7_22

polarization at ω which then leads to generation of waves at frequency ω if appropriate phase matching conditions are satisfied. This is referred to as spontaneous parametric down conversion (SPDC).

In fact, for an input wave at frequency ω_p if phase matching conditions are satisfied for a pair of lower frequencies ω_s (referred to as the signal) and ω_i (referred to as the idler) such that $\omega_p = \omega_s + \omega_i$, then SPDC will generate waves at the two lower frequencies without any input at the two lower frequencies. Thus, although, for a given pump frequency ω_p there are an infinite number of combinations of frequencies satisfying the equation $\omega_p = \omega_s + \omega_i$, the specific pair of frequencies that will get generated will be those that satisfy the phase matching condition or the quasi-phase matching condition.

If the pump photon, signal photon, and the idler photons have the same polarization state, then the process is termed Type 0 process. If the signal and idler pair has the same polarization state which is orthogonal to the polarization state of the pump photon, then the process is termed Type I process. Finally, if the signal and idler have orthogonal polarizations, then it is termed Type II process.

In Chap. 12, we discussed parametric amplifiers in which an incident signal can get amplified through nonlinear interaction provided some conditions are satisfied. In this chapter, we will study the phenomenon of spontaneous parametric down conversion (SPDC) in which photon pairs are produced through a spontaneous emission process when only the pump mode is excited. The classical picture prohibits the generation of the pair of photons with no input in either of the down converted modes. The generation of signal–idler pair requires using the quantum picture, which is presented in the chapter.

22.2 Interaction Hamiltonian for Second-Order Parametric Process

In order to get the output quantum state under SPDC, we will use the interaction picture (see Chap. 21). We first evaluate the interaction Hamiltonian which is then used to obtain the output quantum state. To obtain the interaction Hamiltonian, we first calculate the nonlinear interaction energy density and assume the pump to be a classical field (as the pump powers are quite high in such processes) and consider the down converted signal and idler to be described by quantum fields.

22.2.1 Effective Electromagnetic Energy

Here we present an approximate heuristic analysis to obtain the effective electromagnetic energy.

The effective electromagnetic energy in a certain volume is given by

$$U = \int d\tau \int_0^{\vec{D}} \vec{E}.d\vec{D} = \varepsilon_0 \int d\tau \int_0^{\vec{E}} \vec{E}.d\vec{E} + \int d\tau \int_0^{\vec{P}} \vec{E}.d\vec{P} \tag{22.1}$$

where we have used $\vec{D} = \varepsilon_0 \vec{E} + \vec{P}$. Here $\vec{E}$ represents the total electric field within the medium, $\vec{P}$ represents electric polarization, and $\vec{D}$ is the electric displacement. Now, in the presence of second-order nonlinearity, we have

$$P_j = \epsilon_0 \chi_{jk} E_k + \epsilon_0 \chi_{jkl}^{(2)} E_k E_l \tag{22.2}$$

where, as per convention, we assume summation of repeated indices and hence not writing the summation sign explicitly. The first term on the right hand side of Eq. (22.2) is the linear polarization term and the second term on the right hand side is the nonlinear polarization term. We are reserving the index i for representing idler and hence using other indices as subscripts.

Hence,

$$\vec{E}.d\vec{P} = E_j dP_j = \epsilon_0 \chi_{jk} E_j dE_k + \epsilon_0 \chi_{jkl}^{(2)} E_j d(E_k E_l) \tag{22.3}$$

Now

$$\begin{aligned} \epsilon_0 \chi_{jkl}^{(2)} d(E_k E_l) &= \epsilon_0 \chi_{jkl}^{(2)} E_k dE_l + \epsilon_0 \chi_{jkl}^{(2)} E_l dE_k = \epsilon_0 \chi_{jkl}^{(2)} E_k dE_l + \epsilon_0 \chi_{jlk}^{(2)} E_k dE_l \\ &= 2\epsilon_0 \chi_{jkl}^{(2)} E_k dE_l \end{aligned} \tag{22.4}$$

assuming symmetry of the nonlinear susceptibility χ_{jkl}, that is, $\chi_{jkl} = \chi_{jlk}$.

Similarly

$$\begin{aligned} \epsilon_0 \chi_{jkl}^{(2)} d(E_j E_k E_l) &= \epsilon_0 \chi_{jkl}^{(2)} E_j E_k d(E_l) + \epsilon_0 \chi_{jkl}^{(2)} E_k E_l d(E_j) + \epsilon_0 \chi_{jkl}^{(2)} E_j E_l d(E_k) \\ &= 3\epsilon_0 \chi_{jkl}^{(2)} E_k E_l d(E_j) \end{aligned} \tag{22.5}$$

again using complete symmetry (Kleinman's symmetry) between all the indices of $\chi_{jkl}^{(2)}$. Also

$$\epsilon_0 \chi_{jkl}^{(2)} d(E_j E_k E_l) = \epsilon_0 \chi_{jkl}^{(2)} E_k E_l d(E_j) + \epsilon_0 \chi_{jkl}^{(2)} E_j d(E_k E_l) \tag{22.6}$$

From Eqs. (22.5) and (22.6), we obtain

$$\begin{aligned} \epsilon_0 \chi_{jkl}^{(2)} E_j d(E_k E_l) &= \epsilon_0 \chi_{jkl}^{(2)} d(E_j E_k E_l) - \epsilon_0 \chi_{jkl}^{(2)} E_k E_l d(E_j) \\ &= \epsilon_0 \chi_{jkl}^{(2)} d(E_j E_k E_l) - \frac{1}{3} \epsilon_0 \chi_{jkl}^{(2)} d(E_j E_k E_l) \end{aligned}$$

Thus,

$$\epsilon_0 \chi^{(2)}_{jkl} E_j d(E_k E_l) = \frac{2}{3} \epsilon_0 \chi^{(2)}_{jkl} d(E_j E_k E_l) \tag{22.7}$$

where again we have used the symmetry of $\chi^{(2)}_{jkl}$.

Hence, the linear and nonlinear terms in the energy associated with the polarization will be given, respectively, by

$$U_l = \int d\tau \int_0^{\vec{P}} \epsilon_0 \chi_{jk} E_j dE_k = \frac{\epsilon_0}{2} \int d\tau \int_0^{\vec{D}} \chi_{jk} d(E_j E_k) = \frac{\epsilon_0}{2} \int d\tau \chi_{jk} E_j E_k \tag{22.8}$$

and

$$U_{nl} = \int d\tau \int_0^{\vec{P}} \epsilon_0 \chi^{(2)}_{jkl} E_j d(E_k E_l) = \frac{2\epsilon_0}{3} \int d\tau \int_0^{\vec{D}} \chi^{(2)}_{jkl} d(E_j E_k E_l) = \frac{2\epsilon_0}{3} \int d\tau \chi^{(2)}_{jkl} E_j E_k E_l \tag{22.9}$$

We now consider the three-wave interaction between a pump (p) with frequency ω_p, signal (s) with frequency ω_s, and idler (i) with frequency ω_i and assume that all the interacting waves are propagating along the $z-$ direction. The jth component of the total electric field is given by the sum of the jth components of the pump, signal, and idler waves:

$$E_j = \frac{1}{2} \left[E_{pj} e^{i(k_p z - \omega_p t)} + cc + E_{sj} e^{i(k_s z - \omega_s t)} + cc + E_{ij} e^{i(k_i z - \omega_i t)} + cc \right] \tag{22.10}$$

where the subscripts p, s, and i correspond to the pump, signal, and idler, respectively, and k_p, k_s, and k_i are the propagation constants at the pump, signal, and idler frequencies, respectively. We will have similar expressions for the kth and l^{th} components. Here for simplicity we are assuming the same propagation constants for the different components of the fields.

We substitute the expressions for the electric field components in Eq. (22.9) and collect terms that satisfy the energy conservation equation:

$$\omega_p = \omega_s + \omega_i \tag{22.11}$$

and from the product $E_j E_k E_l$ obtain the following six terms and their complex conjugates:

$$\frac{1}{8} E^*_{pl} E_{sj} E_{ik} e^{-i(k_p - k_s - k_i)z}, \frac{1}{8} E^*_{pl} E_{sk} E_{ij} e^{-i(k_p - k_s - k_i)z}, \frac{1}{8} E^*_{pk} E_{sj} E_{il} e^{-i(k_p - k_s - k_i)z},$$
$$\frac{1}{8} E^*_{pk} E_{sl} E_{ij} e^{-i(k_p - k_s - k_i)z}, \frac{1}{8} E^*_{pj} E_{sk} E_{il} e^{-i(k_p - k_s - k_i)z},$$
$$\frac{1}{8} E^*_{pj} E_{sl} E_{ik} e^{-i(k_p - k_s - k_i)z},$$

Using the symmetry of $\chi_{jkl}^{(2)}$, we obtain finally

$$\begin{aligned} U_{nl} &= \frac{2}{3}\frac{1}{8}\int d\tau 6\epsilon_0 \chi_{jkl}^{(2)} E_{sj} E_{ik} E_{pl}^{*} e^{-i\left(k_p - k_s - k_i\right)z} + c.c \\ &= \frac{1}{2}\left[\int d\tau \epsilon_0 \chi_{jkl}^{(2)} E_{sj} E_{ik} E_{pl}^{*} e^{-i\left(k_p - k_s - k_i\right)z} + c.c\right] \end{aligned} \tag{22.12}$$

If we write $\chi_{jkl}^{(2)} = 2d_{jkl}$, then we obtain the following equation for the electromagnetic energy

$$U_{nl} = \frac{1}{2}\left[2\int d\tau \epsilon_0 d_{jkl} E_{sj} E_{ik} E_{pl}^{*} e^{-i\left(k_p - k_s - k_i\right)z} + c.c\right] \tag{22.13}$$

22.3 SPDC with Quantized Radiation Field

In this section, we will study the process of spontaneous parametric down conversion (SPDC) using quantized fields. Since the pump field is reasonably strong, we will treat the pump as a classical electromagnetic wave and use quantized fields for the signal and idler fields.

As a specific example, we will consider SPDC in a lithium niobate waveguide with the propagation direction to be along the $\widehat{x}$ direction, the input pump to have an ordinary polarization along the $\widehat{y}$ direction and the signal to be an extraordinary wave polarized along the $\widehat{z}$ direction and the idler to be an ordinary wave polarized along the $\widehat{y}$ direction. This is referred to as Type II interaction. A similar analysis can be also carried out for other types of phase matching.

Although the pump is assumed to be monochromatic, the down conversion will take place over a range of signal and idler frequencies close to the phase matching pair of signal and idler frequencies. This is because as we saw in Chap. 12 even if the phase matching condition is not exactly satisfied, exchange of power between the interacting modes can still take place over a range of frequencies around the phase matching pair. Of course, every pair of signal and idler frequencies within the bandwidth will have to satisfy the energy conservation condition $\omega_p = \omega_s + \omega_i$. In Sect. 22.6, we will discuss SPDC using pulsed pump with the pump field having a finite bandwidth and hence in such a case, the input pump would have a range of frequencies.

We express the quantized electromagnetic field at the signal frequency ω_s and the idler frequency ω_i as an integral over a range of frequencies and given by (see Chap. 14)

$$\widehat{E}_{se} = i \int d\omega_s e_{se}(y,z) \sqrt{\frac{\hbar\omega_s}{2\epsilon_{se} L}} \left(\widehat{a}_{se} e^{i(\beta_{se} x - \omega_s t)} - \widehat{a}_{se}^{\dagger} e^{-i(\beta_{se} x - \omega_s t)} \right) \widehat{z} \tag{22.14}$$

and

$$\widehat{E}_{io} = i \int d\omega_i e_{io}(y,z) \sqrt{\frac{\hbar\omega_i}{2\epsilon_{io} L}} \left(\widehat{a}_{io} e^{i(\beta_{io} x - \omega_i t)} - \widehat{a}_{io}^{\dagger} e^{-i(\beta_{io} x - \omega_i t)} \right) \widehat{y} \tag{22.15}$$

The subscripts e and o correspond to extraordinary and ordinary polarizations. In the above equations, $e_{se}(y, z)$ and $e_{io}(y, z)$ represent the normalized transverse modal field distributions of the interacting waveguide modes corresponding to the extraordinary polarization at the signal frequency and the ordinary polarizations at the idler frequency, respectively, β_{se} and β_{io} represent the corresponding modal propagation constants, $\widehat{a}_{se}$ and $\widehat{a}_{io}$ are the annihilation operators corresponding to the extraordinary signal polarization and ordinary idler polarization, ϵ_{se} and ϵ_{io} are the permittivity of the medium corresponding to the respective frequencies and polarizations, and L is the length of the quantization volume.

The field at the pump is written as a classical electromagnetic field given by

$$\vec{E}_{po} = \frac{1}{2} e_{po}(y,z) E_{p0} \left(e^{i(\beta_{po} x - \omega_p t)} + cc \right) \widehat{y} \tag{22.16}$$

where $e_{po}(y, z)$ represents the normalized transverse modal field distribution at the pump frequency, β_{po} is the propagation constant of the mode at the pump frequency, and E_{p0} is the amplitude of the pump field. We will use the no pump depletion approximation and hence assume E_{p0} to be independent of propagation distance x.

The fields $e_{se}(y, z)$, $e_{io}(y, z)$, and $e_{po}(y, z)$ are assumed to satisfy the following normalization conditions:

$$\iint e_{se}^2(y,z) dy dz = 1$$

$$\iint e_{io}^2(y,z) dy dz = 1$$

$$\iint e_{po}^2(y,z) dy dz = 1$$

For the chosen set of polarization states of pump, signal, and idler, the effective nonlinear coefficient that is responsible for the down conversion is $d_{eff} = d_{24}$ (see Chap. 10). If first-order QPM is used, then $d_{eff} = \frac{2}{\pi} d_{24}$.

Hence, the interaction energy expression becomes

$$U_{nl} = \frac{1}{2}\left[2\int d\tau\epsilon_0 d_{eff} E_{se} E_{io} E_{po}^{*} e^{-i(\beta_{po}-\beta_{se}-\beta_{io})x} + c.c\right] \tag{22.17}$$

In order to obtain an expression for the interaction Hamiltonian, we replace the various expressions as given below (comparing Eqs. 22.10 and 22.14–22.16) in Eq. (22.17):

$$E_{se} e^{i(\beta_{se}x-\omega_s t)} \rightarrow i\int d\omega_s e_{se}(y,z)\sqrt{\frac{2\hbar\omega_s}{\epsilon_{se}L}}\hat{a}_{se} e^{i(\beta_{se}x-\omega_s t)}$$

$$E_{io} e^{i(\beta_{io}x-\omega_i t)} \rightarrow i\int d\omega_i e_{io}(y,z)\sqrt{\frac{2\hbar\omega_i}{\epsilon_{io}L}}\hat{a}_{io} e^{i(\beta_{io}x-\omega_i t)}$$

$$E_{po}^{*} e^{-i(\beta_{po}x-\omega_p t)} \rightarrow e_{po}(y,z)E_{po}^{*} e^{-i(\beta_{po}x-\omega_p t)}$$

This gives us

$$\hat{H}_{int} = \int d\omega_s \int d\omega_i \left(\frac{2\hbar E_{po}\sqrt{\omega_s\omega_i}}{L}\right)\int_0^{L_{int}} d_{eff}(x)\frac{I_{eo}}{n_{se}n_{io}} \left\{\hat{a}_{se}^{\dagger}\hat{a}_{io}^{\dagger} e^{i[\beta_{po}-\beta_{se}-\beta_{io}]x} e^{-i(\omega_p-\omega_s-\omega_i)t} + \hat{a}_{se}\hat{a}_{io} e^{-i[\beta_{po}-\beta_{se}-\beta_{io}]x} e^{i(\omega_p-\omega_s-\omega_i)t}\right\}dx \tag{22.18}$$

where

$$I_{eo} = \iint e_{po}(y,z)e_{se}(y,z)e_{io}(y,z)dydz \tag{22.19}$$

represents an overlap integral between the pump, signal, and idler modal fields. The quantity $(I_{eo})^{-2}$ represents the effective area of interaction between pump, signal, and idler modes. We have also assumed E_{po} to be real. We have used $\epsilon_{se} = \epsilon_0 n_{se}^2$ and $\epsilon_{io} = \epsilon_0 n_{io}^2$ with n_{se} and n_{io} representing the effective indices of the respective modes.

Exercise 22.1 Show that the quantity $(I_{eo})^{-2}$ has dimensions of area.

Solution The fields $e_{po}(y,z)$, $e_{io}(y,z)$, and $e_{se}(y,z)$ are normalized fields and satisfy the following equation: $\iint |e_{po}(y,z)|^2 dydz = 1$,etc. Thus, the quantities $e_{po}(y,z)$, $e_{io}(y,z)$, and $e_{se}(y,z)$ have dimensions of L^{-1}. Thus, I_{eo} has dimesions of L^{-1} implying that $(I_{eo})^{-2}$ has dimensions of area.

Using quasi phase matching, the nonlinear coefficient $d_{24}(x)$ has different spatial frequency components and we consider the relevant spatial frequency terms using first-order quasi phase matching and the effective d coefficient is then given by, apart from a phase factor (see Chap. 11)

$$d_{eff} = \frac{2}{\pi} d_{24}\left(e^{iKx} + e^{-iKx}\right) + \text{terms with other spatial frequencies} \tag{22.20}$$

Here

$$K = \frac{2\pi}{\Lambda} \tag{22.21}$$

where Λ is the spatial period of the QPM grating. Thus, keeping terms satisfying $\beta_{po} - \beta_{se} - \beta_{io} - K \approx 0$, we have

$$\begin{aligned}\widehat{H}_{int} = &\frac{2}{\pi} d_{24} \int d\omega_s \int d\omega_i \left(\frac{2E_{p0}\hbar\sqrt{\omega_s\omega_i}}{L}\right) \frac{I_{eo}}{n_{se}n_{io}}\\ &\left[\widehat{a}_{se}^{\dagger}\widehat{a}_{io}^{\dagger} e^{-i(\omega_p-\omega_s-\omega_i)t} \int_0^{L_{int}} e^{i[\beta_{po}-\beta_{se}-\beta_{io}-K]x} dx + \widehat{a}_{se}\widehat{a}_{io} e^{i(\omega_p-\omega_s-\omega_i)t} \int_0^{L_{int}} e^{-i[\beta_{po}-\beta_{se}-\beta_{io}-K]x} dx\right]\end{aligned} \tag{22.22}$$

where L_{int} is the interaction length. Integrating, we obtain

$$\begin{aligned}\widehat{H}_{int} = &\frac{2}{\pi} d_{24} \int d\omega_s \int d\omega_i \left(\frac{2E_{p0}\hbar\sqrt{\omega_s\omega_i}}{L}\right) \frac{I_{eo}}{n_{se}n_{io}} \left[\widehat{a}_{se}^{\dagger}\widehat{a}_{io}^{\dagger} e^{-i(\omega_p-\omega_s-\omega_i)t} e^{i\Delta\beta_{eo}L_{int}/2} L_{int} \left[\frac{\sin\left(\frac{\Delta\beta_{eo}L_{int}}{2}\right)}{\left(\frac{\Delta\beta_{eo}L_{int}}{2}\right)}\right]\right.\\ &\left. + \widehat{a}_{se}\widehat{a}_{io} e^{i(\omega_p-\omega_s-\omega_i)t} e^{-i\Delta\beta_{eo}L_{int}/2} L_{int} \left[\frac{\sin\left(\frac{\Delta\beta_{eo}L_{int}}{2}\right)}{\left(\frac{\Delta\beta_{eo}L_{int}}{2}\right)}\right]\right]\end{aligned}$$

Here

$$\Delta\beta_{eo} = \beta_{po} - \beta_{se} - \beta_{io} - K \tag{22.23}$$

Defining

$$\begin{aligned}C_{eo} &= \frac{2}{\pi} d_{24} \left(\frac{2E_{p0}\hbar\sqrt{\omega_s\omega_i} I_{eo}}{n_{se}n_{io}L}\right) e^{\frac{i\Delta\beta_{eo}L_{int}}{2}} L_{int} \left[\frac{\sin\left(\frac{\Delta\beta_{eo}L_{int}}{2}\right)}{\left(\frac{\Delta\beta_{eo}L_{int}}{2}\right)}\right]\\ &= \frac{2}{\pi} d_{24} \left(\frac{2E_{p0}\hbar\sqrt{\omega_s\omega_i} I_{eo}}{n_{se}n_{io}L}\right) e^{\frac{i\Delta\beta_{eo}L_{int}}{2}} L_{int} sinc\left(\frac{\Delta\beta_{eo}L_{int}}{2}\right)\end{aligned} \tag{22.24}$$

we have

$$\widehat{H}_{int} = \int d\omega_s \int d\omega_i \left[C_{eo} \widehat{a}_{se}^{\dagger} \widehat{a}_{io}^{\dagger} e^{-i(\omega_p - \omega_s - \omega_i)t} + C_{eo}^{*} \widehat{a}_{se} \widehat{a}_{io} e^{i(\omega_p - \omega_s - \omega_i)t} \right] \tag{22.25}$$

The quantity C_{eo} can be expressed in terms of pump power by noting that the incident pump power is given by (see Chap. 4)

$$P_{p0} = \frac{\beta_{po}}{2\omega_p \mu_0} |E_{p0}|^2 = \frac{c\epsilon_0 n_{po}}{2} |E_{p0}|^2 \tag{22.26}$$

where n_{po} is the effective index of the pump mode. Hence, we have

$$C_{eo} = \frac{2}{\pi} d_{24} \hbar \left(\frac{8P_{p0}\omega_s\omega_i}{c\epsilon_0 n_{se}^2 n_{io}^2 n_{po} L^2} \right)^{1/2} L_{int} I_{eo} e^{\frac{i\Delta\beta_{eo}L_{int}}{2}} sinc \left(\frac{\Delta\beta_{eo} L_{int}}{2} \right) \tag{22.27}$$

We now use the interaction picture (see Chap. 21) and since the perturbation due to nonlinearity is assumed to be very small, we use Eq. (21.43).

Now,

$$\frac{1}{i\hbar} \int \widehat{H}_{int} dt = -\frac{i}{\hbar} \left(\int d\omega_s \int d\omega_i \left[C_{eo} \widehat{a}_{se}^{\dagger} \widehat{a}_{io}^{\dagger} \int_0^t e^{-i(\omega_p - \omega_s - \omega_i)t} dt + C_{eo}^{*} \widehat{a}_{se} \widehat{a}_{io} \int_0^t e^{i(\omega_p - \omega_s - \omega_i)t} dt \right] \right)$$

Since at steady state, the system has no memory of the initial and final times, we can extend the integral over time from $-\infty$ to $+\infty$ and use the fact that

$$\int_{-\infty}^{+\infty} e^{\pm i(\omega_p - \omega_s - \omega_i)t} dt = 2\pi\delta(\omega_p - \omega_s - \omega_i) \tag{22.28}$$

to obtain the following equation:

$$\frac{1}{i\hbar} \int \widehat{H}_{int} dt = -\frac{i}{\hbar} \left(2\pi \int d\omega_s \int d\omega_i \left[C_{eo} \widehat{a}_{se}^{\dagger} \widehat{a}_{io}^{\dagger} \delta(\omega_p - \omega_s - \omega_i) + C_{eo}^{*} \widehat{a}_{se} \widehat{a}_{io} \delta(\omega_p - \omega_s - \omega_i) \right] \right)$$

which gives us

$$\frac{1}{i\hbar} \int \widehat{H}_{int} dt = -\frac{2\pi i}{\hbar} \int d\omega_s \left[C_{eo} \widehat{a}_{se}^{\dagger} \widehat{a}_{io}^{\dagger} + C_{eo}^{*} \widehat{a}_{se} \widehat{a}_{io} \right]$$

Thus, the quantum state at the output will be (see Eq. 21.40)

$$|\psi^{(I)}(t)\rangle = e^{-\frac{i}{\hbar}\int \widehat{H}_{int}dt}|\psi^{(I)}(t=0)\rangle \approx \left(1 - \frac{2\pi i}{\hbar}\int d\omega_s\left[C_{eo}\widehat{a}^{\dagger}_{se}\widehat{a}^{\dagger}_{io} + C^{*}_{eo}\widehat{a}_{se}\widehat{a}_{io}\right]\right)|0_{se}, 0_{io}\rangle \tag{22.29}$$

Hence, we obtain

$$\begin{aligned} |\psi^{(I)}(t)\rangle &\approx |0_{se}, 0_{io}\rangle - \frac{2\pi i}{\hbar}\int\left[C_{eo}\widehat{a}^{\dagger}_{se}\widehat{a}^{\dagger}_{io} + C^{*}_{eo}\widehat{a}_{se}\widehat{a}_{io}\right]|0_{se}, 0_{io}\rangle d\omega_s \\ &= |0_{se}, 0_{io}\rangle - \frac{2\pi i}{\hbar}\int C_{eo}|1_{se}, 1_{io}\rangle d\omega_s \end{aligned} \tag{22.30}$$

Thus, under the approximation of weak interaction (low pump power), the ouput state consists of a superposition of the vacuum state and a two photon state with one photon in the mode corresponding to the signal frequency and one photon in the mode corresponding to the idler frequency. The amplitude of the state with signal frequency ω_s and idler frequency ω_i is given by $\frac{2\pi i}{\hbar}C_{eo}$. Since C_{eo} is proportional to $sinc\left(\frac{\Delta\beta_{eo}L_{int}}{2}\right)$ (see Eq. 22.27) which has the maximum amplitude at $\Delta\beta_{eo} = 0$, the maximum probability of finding the pair will be at the frequency corresponding to the quasi phase matching condition:

$$\Delta\beta_{eo} = \beta_{po} - \beta_{se} - \beta_{io} - K = 0 \tag{22.31}$$

Since the amplitude of the photon pair generation rate is defined by C_{eo}, the spectra corresponding to the signal and idler photons will be proportional to $|C_{eo}|^2$. Thus, the spectra will be defined by the phase matching function $sinc^2\left(\frac{\Delta\beta_{eo}L_{int}}{2}\right)$ (see Fig. 22.1). Notice that even though the pump is monochromatic, signal idler pairs are generated over a certain bandwidth and are characterized by a spectrum.

For a very narrow bandwidth of down conversion, we can approximate the output state by

$$|\psi_{out}\rangle \approx |0_{se}, 0_{io}\rangle - \frac{2\pi i}{\hbar}C_{eo}|1_{se}, 1_{io}\rangle \tag{22.32}$$

with the signal and idler frequencies to correspond to the quasi phase matching condition. Hence, at low efficiencies, the spontaneous parametric down conversion generates a signal–idler pair of photons. In this example, they have orthogonal polarizations; however, one may use other processes to generate other polarization combinations. One of these photons can be used to herald the arrival of the other photon and is referred to as heralded photon. This is very useful in conducting experiments using single photons.

In order to obtain the spectra of signal and idler, we note that the phase mismatch is given by Eq. (22.23):

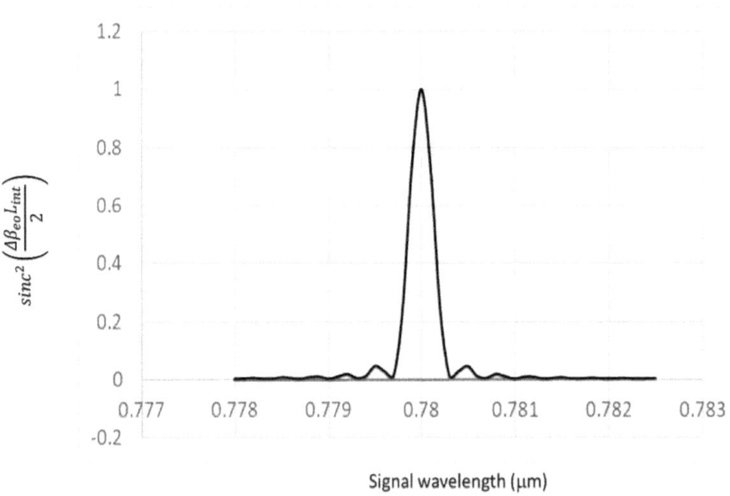

Fig. 22.1 Typical variation in $sinc^2\left(\Delta\beta_{eo}L_{int}/2\right)$ with signal wavelength for an interaction length of 10 mm

$$\Delta\beta_{eo} = \beta_{po} - \beta_{se} - \beta_{io} - K = \frac{\omega_p}{c}n_{po} - \frac{\omega_s}{c}n_{se} - \frac{\omega_i}{c}n_{io} - K \tag{22.33}$$

with n_{po}, n_{se} and n_{io} representing the effective indices of the corresponding modes at the pump, signal, and idler frequencies, respectively.

The pump, signal, and idler frequencies are related through Eq. (22.11). Assuming the pump to be almost monochromatic, we have

$$\Delta\omega_s = -\Delta\omega_i \tag{22.34}$$

We assume that the phase matching condition is exactly satisfied at the signal and idler frequencies given by ω_{s0} and ω_{i0}. Hence, we have

$$\Delta\beta_{eo} = \frac{\omega_p}{c}n_{po} - \frac{\omega_{s0}}{c}n_{se}(\omega_{s0}) - \frac{\omega_{i0}}{c}n_{io}(\omega_{i0}) - K = 0 \tag{22.35}$$

Also corresponding to a signal frequency $\omega_{S0} + \Delta\omega_S$, we have corresponding to this frequency pair

$$\Delta\beta_{eo} = \frac{\omega_p}{c}n_{po} - \frac{(\omega_{s0} + \Delta\omega_s)}{c}n_{se}(\omega_{s0} + \Delta\omega_s) - \frac{(\omega_{i0} + \Delta\omega_i)}{c}n_{io}(\omega_{i0} + \Delta\omega_i) - K \tag{22.36}$$

Assuming $\Delta\omega_s \ll \omega_{s0}$ and $\Delta\omega_i \ll \omega_{i0}$, we have

$$n_{se}(\omega_{s0}+\Delta\omega_s) \approx n_{se}(\omega_{s0}) + \Delta\omega_s \frac{dn_{se}}{d\omega_s}\bigg|_{\omega_s=\omega_{s0}} \tag{22.37a}$$

and

$$n_{io}(\omega_{i0}+\Delta\omega_i) \approx n_{io}(\omega_{i0}) + \Delta\omega_i \frac{dn_{io}}{d\omega_i}\bigg|_{\omega_i=\omega_{i0}} \tag{22.37b}$$

Substituting in Eq. (22.36), keeping terms up to order $\Delta\omega_s$ and $\Delta\omega_i$, and using Eqs. (22.34) and (22.35), we have

$$\Delta\beta_{eo} \approx -\frac{\Delta\omega_s}{c}N_{se} - \frac{\Delta\omega_i}{c}N_{io} = \frac{\Delta\omega_s}{c}(N_{io}-N_{se}) \tag{22.38}$$

where

$$N_{se} = n_{se} + \omega_s \frac{dn_{se}}{d\omega_s}\bigg|_{\omega_s=\omega_{s0}} \tag{22.39a}$$

and

$$N_{io} = n_{io} + \omega_i \frac{dn_{io}}{d\omega_i}\bigg|_{\omega_i=\omega_{i0}} \tag{22.39b}$$

are the group effective indices at the signal and idler wavelengths $\omega_s = \omega_{s0}$ and $\omega_i = \omega_{i0}$, respectively, (see Chap. 4).

Since the amplitude of the down conversion process is proportional to C_{eo} and $sinc^2(x) = 0.5$ for $x \approx \pm 1.39$, the values of $\Delta\beta_{eo}$ when the efficiency of the SPDC process will drop by a factor of 0.5 will correspond to

$$\frac{\Delta\beta_{eo}L_{int}}{2} \approx \pm 1.39 \tag{22.40}$$

Substituting for $\Delta\beta_{eo}$ from Eq. (22.38), we obtain for the full spectral width at half maximum (FWHM)

$$\Delta\omega_{s,FWHM} = \Delta\omega_{i,FWHM} \approx \frac{5.56\,c}{|N_{io}-N_{se}|L_{int}} \tag{22.41}$$

which in terms of wavelength becomes

$$\Delta\lambda_{s,FWHM} \approx \frac{\lambda_s^2}{2\pi c}\Delta\omega_{s,FWHM} = \frac{\lambda_s^2}{2\pi}\frac{5.56}{|N_{io} - N_{se}|L_{int}} \tag{22.42}$$

Since $\Delta\omega_s = -\Delta\omega_i$, we have $\Delta\lambda_i = -\frac{\lambda_i^2}{\lambda_s^2}\Delta\lambda_s$ and thus

$$\Delta\lambda_{i,FWHM} \approx \frac{\lambda_i^2}{2\pi}\frac{5.56}{|N_{io} - N_{se}|L_{int}} \tag{22.43}$$

It is interesting to note that classically with just the input at the pump frequency, there can be no down conversion. In contrast, in the quantum mechanical analysis, with vacuum input we do find the generation of signal–idler photon pairs from the process of spontaneous parametric down conversion. It is the zero point fluctuations in the electromagnetic field that leads to the generation of down converted photons.

For an electromagnetic wave having a finite bandwidth $\Delta\lambda$ at a free space wavelength of λ_0, the coherence length is defined by

$$L_c = \frac{\lambda_0^2}{\Delta\lambda} \tag{22.44}$$

From Eqs. (22.42) and (22.43), we note that the coherence length of the output signal and idler photons are the same although their spectral widths are different. Coherence length is an important property which defines the maximum path difference that can lead to interference effects. In Chap. 23, we will see the importance of coherence length in the generation of time-bin entangled photon states for different applications.

The spatial frequency of the periodic reversal of nonlinear coefficient determines the frequencies of the signal–idler pair of wavelengths that are generated and the bandwidth of the process is also determined by the length of interaction and the difference in group indices at the signal and idler wavelengths. It is possible to appropriately design the spatial frequency profile of the periodic nonlinear coefficient to design a required down conversion bandwidth. This is termed domain engineering and has been used extensively [see, e.g., Branczyk et al. (2010), Fraine et al. (2012), Ming et al. (2014)].

Compared to interaction in the bulk in the case of waveguides, the group indices at the signal and idler wavelengths will also depend on the waveguide characteristics such as the width, the depth, and the refractive index difference between the guiding region and the surrounding (see Chap. 4). In such a case apart from material dispersion, we would also have contributions from waveguide dispersion which can be used to achieve appropriate down conversion characteristics.

As an example we consider bulk interaction in lithium niobate and assume a pump wavelength of 519 nm, a signal wavelength of 780 nm giving an idler wavelength of 1551.03 nm from energy conservation. The corresponding refractive indices are given by

$$n_o(519 \text{ nm}) = 2.3299$$
$$n_e(780 \text{ nm}) = 2.1778$$
$$n_o(1551.03 \text{ nm}) = 2.2111$$

The corresponding QPM grating period Λ satisfies the following equation:

$$\frac{1}{\Lambda} = \frac{n_{po}}{\lambda_p} - \frac{n_{se}}{\lambda_s} - \frac{n_{io}}{\lambda_i}$$

Substituting the values, we obtain

$$\Lambda \approx 3.68 \ \mu\text{m}$$

The group refractive indices of lithium niobate at the three wavelengths are given below [Ref: https://refractiveindex.info/?shelf=main&book=LiNbO3&page=Zelmon-o]:

$$N_o(519 \text{ nm}) = 2.6091$$
$$N_e(780 \text{ nm}) = 2.2710$$
$$N_o(1551.03 \text{ nm}) = 2.2636$$

For an interaction length of 10 mm, we obtain the signal and idler bandwidths to be approximately given by 7.3 and 28.8 nm, respectively (see Fig. 22.2). Note that the signal and idler bandwidths are different.

The propagation constants β_m for pump, signal, and idler would depend on the waveguide geometry and values of parameters such as the width and depth of the waveguide, the index difference between the guiding and cladding regions of the waveguide as well as the polarization state of the interacting waves. Thus, it is possible to optimize the interaction through waveguide engineering for a given material and thus achieve functionalities that may not be possible in bulk using the same materials.

Figure 22.2 shows the plot of the spectrum of the signal photon that emerges from an SPDC process from a bulk lithium niobate crystal over a length of interaction of 10 mm with the pump wavelength of 519 nm for two processes in which a pump photon having ordinary polarization undergoes parametric down conversion to a pair of signal–idler photons with the signal having an extraordinary polarization and the idler having an ordinary polarization and the other process in which a pump photon having ordinary polarization undergoes parametric down conversion to a pair of signal–idler photons with the signal having an ordinary polarization and the idler having an extraordinary polarization. The QPM condition is exactly satisfied for a signal wavelength of 780 nm and a corresponding idler wavelength of 1551.03

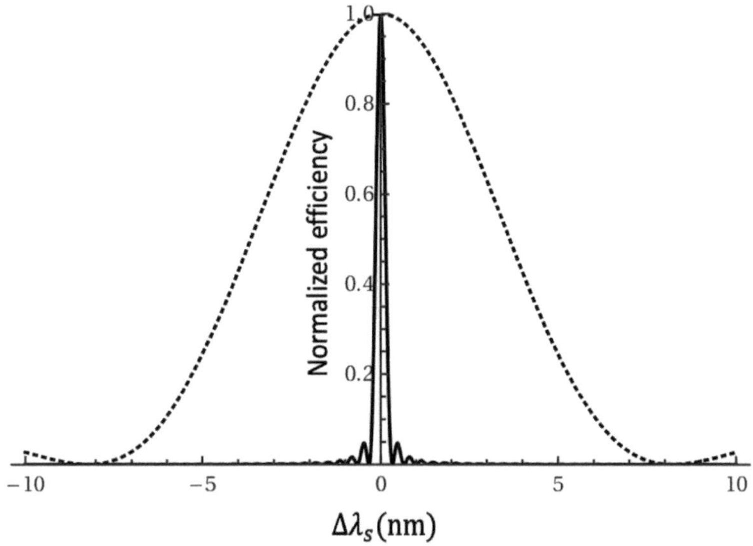

Fig. 22.2 Spectrum of signal photons emerging from SPDC process with an extraordinarily polarized signal and ordinarily polarized idler (dashed curve) and ordinarily polarized signal and extraordinarily polarized idler (solid curve). The quantity $\Delta\lambda_s = \lambda_s - \lambda_{s0}$ is the deviation from the phase matching wavelength λ_{s0}. Note the large difference in the spectra of the two processes

nm. Note that due to material dispersion, the spectra of the signal are different in the two cases.

It is also interesting to note that the output photons are entangled in frequency. The frequencies of the signal and idler photons exiting the process are undefined and the photons are in superposition states of a continuum of frequencies around a specific signal frequency and a corresponding specific idler frequency. If we separate the signal and idler pairs into two different paths through, for example, a wavelength selective beam splitter and pass the signal photon through a frequency filter having a specific frequency ω_{s1} and if the signal photon passes through the filter, the frequency of the idler photon is determined to be equal to $\omega_{i1} = \omega_p - \omega_{s1}$.

Let us assume that the signal and idler photons are passed through filters having multiple passbands. For example, we consider the filter for the signal to transmit only frequencies $\omega_{s1}, \omega_{s2}, \omega_{s3}, \ldots \omega_{sn}$ and the filter for the idler to transmit only the frequencies $\omega_{i1}, \omega_{i2}, \omega_{i3}, \ldots \omega_{in}$ with $\omega_p = \omega_{s1} + \omega_{i1} = \omega_{s2} + \omega_{i2} + \ldots = \omega_{sn} + \omega_{in}$. We assume that all these frequencies lie within the band of frequencies of the bandwidth of the signal and idler photons as specified by the sinc function. Then depending on the frequency port from which the signal photon exits, the idler photon will exit from the corresponding port satisfying the condition mentioned above.

It must be noted that although in the analysis we have assumed the pump to be monochromatic, the generated signal–idler pair is not monochromatic but gets generated over a band of frequencies around the phase matching wavelengths.

Exercise 22.2 Consider a z-cut-$x-$propagating lithium niobate crystal with channel waveguides which have a symmetric refractive index profile along the $y-$ axis and an asymmetric profile along the $z-$axis. Is it possible to achieve SPDC with the pump, signal, and idler all having extraordinary polarization in the following spatial modes: $E_{11}^{(z)}(\omega_p)$, $E_{11}^{(z)}(\omega_s)$, and $E_{12}^{(z)}(\omega_i)$ with the first subscript corresponding to variation along the $z-$direction and the second subscript corresponding to variation along the $y-$direction.

Solution Note that since the waveguide is given to have a symmetric refractive index profile along the $y-$direction, the modes $E_{11}^{(z)}(\omega_p)$ and $E_{11}^{(z)}(\omega_s)$ will be symmetric along the $y-$direction while the mode $E_{12}^{(z)}(\omega_i)$ will be antisymmetric along the $y-$direction. Hence, in this case the integral given in Eq. (22.19) which for the case under consideration is given by

$$I_{ee} = \iint e_{pe}(y,z)e_{se}(y,z)e_{ie}(y,z)dydz \tag{22.45}$$

will be zero since in this case the integrand will be an antisymmetric function of y. Thus, there will no photon pair generation in this case. On the other hand if the signal mode is $E_{12}^{(z)}(\omega_s)$ which will also be antisymmetric along the $y-$direction, the integral will be nonzero and photon pair generation is possible provided appropriate phase matching condition is satisfied.

Exercise 22.3 In Chap. 4, we discussed the variational technique to obtain modes of a graded index channel waveguide in which the modal distributions are represented by a trial field which is given by the following for the pump, signal, and idler fields:

$$e_p(x,y) = N_p x e^{-x^2/w_{px}^2} e^{-y^2/w_{py}^2}; \quad x < 0 \tag{22.46a}$$

$$e_s(x,y) = N_s x e^{-x^2/w_{sx}^2} e^{-y^2/w_{sy}^2}; \quad x < 0 \tag{22.46b}$$

$$e_i(x,y) = N_i x e^{-x^2/w_{ix}^2} e^{-y^2/w_{iy}^2}; \quad x < 0 \tag{22.46c}$$

with all fields being equal to zero for $x > 0$. Here, N_j; $j = p, s, i$ are the normalization constants and w_{jx} and w_{jy}; $j = p, s, i$ represent the widths of the modes along the $x-$ and $y-$directions. Here, the fields are written in the laboratory coordinate system with the x—direction chosen to be perpendicular to the substrate and the y—direction parallel to the substrate and the propagation direction is assumed to be along the z—direction

(a) Normalize the modal fields and obtain expressions for the amplitudes N_j; $j = p, s, i$.
(b) Substitute the fields in the expression for $I = \iint e_p(x,y)e_s(x,y)e_i(x,y)dxdy$ and obtain an expression for the effective interaction area $A_{eff} = (I)^{-2}$.

Solution Normalizing the fields gives us

$$\iint\limits_{-\infty\,-\infty}^{\infty\;\infty} \left|e_j(x,y)\right|^2 dx\,dy = 1 \tag{22.47}$$

(a) On substituting and evaluating the integrals, we obtain

$$N_j = \frac{4}{\left(\pi w_{jx}^3 w_{jy}\right)^{1/2}}; \quad j = p, s, i \tag{22.48}$$

(b) Substituting the expressions for the fields in the integral for the effective interaction area, we have

$$I = N_p N_s N_i \int_{-\infty}^{0} x^3 e^{-x^2\left(\frac{1}{w_{px}^2}+\frac{1}{w_{sx}^2}+\frac{1}{w_{ix}^2}\right)} dx \int_{-\infty}^{+\infty} e^{-y^2\left(\frac{1}{w_{py}^2}+\frac{1}{w_{sy}^2}+\frac{1}{w_{iy}^2}\right)} dy$$

Evaluating the integrals, we have

$$I = \frac{32}{\left(\pi^2 w_{px}^3 w_{sx}^3 w_{ix}^3 w_{py} w_{sy} w_{iy}\right)^{1/2}} \left(\frac{1}{w_{px}^2}+\frac{1}{w_{sx}^2}+\frac{1}{w_{ix}^2}\right)^{-2} \left(\frac{1}{w_{py}^2}+\frac{1}{w_{sy}^2}+\frac{1}{w_{iy}^2}\right)^{-1/2} \tag{22.49}$$

Check that the dimension of I is inverse of length dimension.

22.3.1 Signal Power Spectral Density

In this section, we will obtain an expression for the signal power generated in the Type II SPDC process in the waveguide. For this, we use Fermi's golden rule and follow the approach given in Koch et al. (1995) and Fiorentino (2007).

We first note that SPDC process occurs over a continuous range of signal and corresponding idler frequencies satisfying the energy conservation condition $\omega_p = \omega_s + \omega_i$; the efficiency of the generation depends on the phase mismatch corresponding to the specific frequency. Taking a quantization length of L, we first note that the values of k_s and k_i of the signal and idler modes are restricted to discrete values (see Chap. 14) with the allowed values of k_s and k_i being spaced by $\frac{2\pi}{L}$. Hence, the number of possible states in $dk_s dk_i$ is given by

$$dN = \frac{dk_s dk_i}{\left(\frac{2\pi}{L}\right)^2} = \frac{L^2}{4\pi^2} dk_s dk_i \tag{22.50}$$

Using $k_s = \frac{\omega_s}{c} n_s$ where n_s is the effective index of the mode at the signal frequency and $k_i = \frac{\omega_i}{c} n_i$ where n_i is the effective index of the mode at the idler frequency, we have

$$dN = \frac{L^2}{4\pi^2} \frac{n_s n_i}{\hbar c^2} d\omega_s dE \tag{22.51}$$

where $dE = \hbar d\omega_i$ is the element of change in the energy of the idler photon. Hence, the number of states per unit energy interval or the density of states into which the signal can get generated in a frequency interval $d\omega_s$ is given by

$$\rho = \frac{L^2}{4\pi^2} \frac{n_s n_i}{\hbar c^2} d\omega_s \tag{22.52}$$

According to Fermi's golden rule [see, e.g., Schiff (1968)], the rate of transition of the pump photon into pairs of signal–idler photons is given by

$$W = \frac{2\pi}{\hbar} \left| \langle 1_s, 1_i | \widehat{H}_{int} | 0_s, 0_i \rangle \right|^2 \rho \tag{22.53}$$

where the initial state is given by $|0_s, 0_i\rangle$ with no photons either in the signal mode or in the idler mode and the final state is given by $\langle 1_s, 1_i|$ with one photon each in the signal mode and the idler mode.

The generated signal power in the frequency interval $d\omega_s$ is given by

$$dP_s = \hbar \omega_s W \tag{22.54}$$

Substituting for $\widehat{H}_{int}$ from Eq. (22.25), ρ from Eq. (22.52) and W from Eq. (22.53), we obtain

$$dP_s = \frac{2\pi}{\hbar} (C_{oe})^2 \frac{L^2}{4\pi^2} \frac{n_s n_i}{\hbar c^2} \hbar \omega_s d\omega_s$$

which simplifies to

$$dP_s = \frac{16\hbar d_{24}^2 \omega_s^2 \omega_i}{\pi^3 \epsilon_0 c^3 n_{se} n_{io} n_p} P_{p0} L_{int}^2 I_{eo}^2 \, sinc^2 \left(\frac{\Delta\beta_{eo} L_{int}}{2} \right) d\omega_s \tag{22.55}$$

We can write Eq. (22.55) in terms of free space pump, signal, and idler wavelengths and obtain

$$dP_s = \frac{256\pi\hbar c d_{24}^2}{\epsilon_0 n_{se} n_{io} n_p \lambda_s^4 \lambda_i} P_{p0} L_{int}^2 I_{eo}^2 \, sinc^2\left(\frac{\Delta\beta_{eo} L_{int}}{2}\right) d\lambda_s \tag{22.56}$$

We take typical values of various parameters given earlier for a pump wavelength of 519 nm, a signal wavelength of 780 nm and an idler wavelength of 1551.03 nm, an interaction length of 10 mm, a pump power of 1 mW, using $d_{24} \approx 3.9 \times 10^{-12}$ m/V, an effective interaction area of $\frac{1}{I_{eo}^2} = 20 \times 10^{-12}$ m^2 for a spectral bandwidth $d\lambda_s = 1$ nm, at the phase matching wavelengths, we obtain $dP_s \approx 33$ pW. The signal wavelength of 780 nm corresponds to a photon energy $\hbar\omega_s \approx 2.55 \times 10^{-19}$ J and thus the number of signal photons generated (which is equal to the number of idler photons generated) over a signal bandwidth of 1 nm is approximately given by 1.3×10^8 signal photons/second which is a typical number that is achieved in experiments. The same number of idler photons will also be produced in the process over a corresponding idler bandwidth.

Compared to the SPDC process in waveguides, the generation rate when the interaction takes place in bulk is much smaller showing one of the advantages used in generating photon pairs using waveguides (Fiorentino et al. 2007).

Exercise 22.4 Consider Type 0 SPDC in a channel waveguide in $Y-$cut $X-$propagating lithium niobate with the refractive index profile defined by Eq. (4.148) and an input pump which has extraordinary polarization. It is given that $\Delta n_s = 0.03$ at the pump (800 nm), signal (1530 nm) and idler wavelengths. Assuming $h = 4$ μm and $w = 6$ μm, using the variational technique discussed in Chap. 4, (a) obtain the transverse field distributions of the $z-$polarized modes at the three wavelengths, (b) obtain the interaction area of the SPDC process, and (c) Obtain the generated signal power in units of pW/nm. Assume the pump power to be 1 mW and an interaction length of 1 cm.

Solution For a pump wavelength of 800 nm and a signal at 1530 nm, the idler wavelength is ~1677 nm. The extraordinary refractive indices of the substrate at the pump, signal, and idler wavelengths are 2.1755, 2.1381, and 2.134, respectively. Using the variational technique given in Chap. 4, we obtain at the pump wavelength $w_x \approx 1.8$ μm and $w_y \approx 2.2$ μm. At the signal wavelength $w_x \approx 2.7$ μm and $w_y \approx 3.24$ μm and at the idler wavelength $w_x \approx 2.88$ μm and $w_y \approx 3.45$ μm. Using these values in Eq. (22.49), we obtain $I \approx 0.2$ μm^{-1} giving the interaction area of $A_{\text{int}} \approx 25$ μm^2. Substituting all the values in Eq. (22.56), we get a signal power of about 36 pW over a bandwidth of 1 nm.

22.4 Ordinary Pump Down Converting to Ordinary Signal and Extraordinary Idler

In Sect. 22.3, we had discussed the case when the incident pump was having ordinary polarization and the signal had extraordinary polarization, while the idler had ordinary polarization. It is also possible to use the same nonlinearity for an SPDC process with a pump having ordinary polarization to generate a signal having ordinary polarization and an idler having extraordinary polarization. In this case, the phase matching condition will be determined by

$$\Delta\beta_{oe} = \beta_p - \beta_{so} - \beta_{ie} - K = 0 \tag{22.57}$$

and the down converted two photon state would be given by

$$\begin{aligned} \left|\psi^{(I)}(t)\right\rangle &\approx |0_s, 0_i\rangle - \frac{2\pi i}{\hbar}\int \left[C_{oe}\widehat{a}_{so}^{\dagger}\widehat{a}_{ie}^{\dagger} + C_{oe}^{*}\widehat{a}_{so}\widehat{a}_{ie}\right]|0_{so}, 0_{ie}\rangle d\omega_s \\ &= |0_s, 0_i\rangle - \frac{2\pi i}{\hbar}\int C_{oe}|1_{so}, 1_{ie}\rangle d\omega_s \end{aligned} \tag{22.58}$$

with

$$C_{oe} = \frac{2}{\pi}d_{24}\left(\frac{2E_{p0}\hbar\sqrt{\omega_s\omega_i}I_{oe}}{n_{so}n_{ie}L}\right)L_{int}e^{\frac{i\Delta\beta_{oe}L_{int}}{2}}sinc\left(\frac{\Delta\beta_{oe}L_{int}}{2}\right) \tag{22.59}$$

and

$$I_{oe} = \iint e_{po}(y,z)e_{so}(y,z)e_{ie}(y,z)dydz \tag{22.60}$$

As an example we consider lithium niobate and take the same pump wavelength of 519 nm, a signal wavelength of 780 nm giving an idler wavelength of 1551.03 nm. The corresponding refractive indices are given by n_o(519 nm) = 2.3299, n_o(780 nm) = 2.2580, and n_e(1551.03 nm) = 2.1375.

The corresponding QPM grating period is given by

$$\Lambda \approx 4.63\ \mu\text{m}$$

which is different from the earlier case when the signal was having extraordinary polarization and idler was in the ordinary polarization.

In order to estimate the bandwidth of this process, we need the group refractive indices at the ordinary pump, ordinary signal, and extraordinary idler wavelengths which are given by N_o(519 nm) = 2.6091, N_o(780 nm) = 2.3665, and N_e(1551.03) nm = 2.1823. The corresponding signal and idler bandwidths can be obtained by using Eqs. (22.42) and (22.43) and are given by 0.29 and 1.15 nm, respectively, showing much smaller bandwidths than the case in which the signal had extraordinary polarization and idler had ordinary polarization. Figure 22.2 shows the corresponding normalized efficiency as a function of signal wavelength showing much smaller bandwidth.

Note that in both the above cases, the generated photon pairs are in polarization separable states and are not polarization entangled. In Chap. 23, we will study a technique for the generation of polarization entangled photon pairs by allowing for both the above two SPDC processes to take place simultaneously.

22.5 Degenerate Case

We consider the degenerate case when pump photons at frequency 2ω undergo spontaneous down conversion to a pair of photons both having the same frequeny ω. In this case, the signal and idler frequencies are equal and this corresponds to degeneracy.

To be specific, we consider Type 0 down conversion in lithium niobate wherein an extraordinarily polarized pump photon at frequency 2ω down converts to a pair of extraordinarily polarized photons at frequency ω. The nonlinear coefficient that is responsible for the down conversion will be d_{33} (see Chap. 10). We also assume perfect quasi phase matching and a narrow bandwidth process. Following a procedure similar to the one presented in Sect. 22.3, we can show that the interaction Hamiltonian in this case will be given by

$$\widehat{H}_{int} = \int d\omega_s \int d\omega_i \left[C\widehat{a}_{1e}^\dagger \widehat{a}_{1e}^\dagger e^{-i\omega t} + C^* \widehat{a}_{1e} \widehat{a}_{1e} e^{i\omega t} \right] \tag{22.61}$$

where $\widehat{a}_{1e}$ and $\widehat{a}_{1e}^\dagger$ represent the annihilation and creation operators corresponding to the extraordinary polarization at frequency ω and

$$C = \frac{2}{\pi} d_{33} \left(\frac{2E_{pe}\hbar\omega I_{ee}}{n_{1e}^2} \right) e^{\frac{i\Delta\beta L_{int}}{2}} \text{sinc} \left(\frac{\Delta\beta L_{int}}{2} \right) \tag{22.62}$$

and

$$I_{ee} = \iint e_{pe}(y, z) e_{1e}^2(y, z) dy dz \tag{22.63}$$

is the overlap integral between the field profiles corresponding to the mode at frquency 2ω and the down converted frequency ω. Also $\Delta\beta = \beta_{2\omega} - 2\beta_\omega - K$.

As before we use the interaction picture and follow the same procedure followed in Sect. 22.3 to obtain the output state as

$$|\psi_{out}\rangle \approx |0\rangle - \frac{2\pi i}{\hbar} C|2_{1e}\rangle \tag{22.64}$$

where we have assumed that the interaction bandwidth to be very narrow and the state represented by $|2_{1e}\rangle$ corresponds to two photons at frequency ω both having extraordinary polarization. Thus, under the approximation, the output state is a superposition of vacuum state and a two photon state.

We note that the coefficient of the second term in Eq. (22.64) is very small compared to unity. Hence, if we define $\alpha = \frac{2\pi}{\hbar} C\ (\ll 1)$, then the output state is given by

$$|\psi_{out}\rangle \approx (|0\rangle - i\alpha|2\rangle) \tag{22.65}$$

where for compactness we have removed the subscripts. Since α is small we neglect terms of order α^2 and higher.

We will now show that the output state given by Eq. (22.65) is a squeezed state (see Chap. 18).

For this, we start with our generalized quadrature operator (see Chap. 15, Sect. 15.5):

$$\widehat{X}_\theta = \frac{\widehat{a}_1 e^{-i\theta} + \widehat{a}_1^\dagger e^{i\theta}}{2} \tag{22.66}$$

Note that for $\theta = 0$, $\widehat{X}_0 = \widehat{X}_1$ and for $\theta = \frac{\pi}{2}$ we have $\widehat{X}_{\frac{\pi}{2}} = \widehat{X}_2$.

Now,

$$\begin{aligned}\left\langle \widehat{X}_\theta \right\rangle &= \langle \psi_{out}|\widehat{X}_\theta|\psi_{out}\rangle = \langle \psi_{out}| \frac{\widehat{a}_1 e^{-i\theta} + \widehat{a}_1^\dagger e^{i\theta}}{2} |\psi_{out}\rangle \\ &= (\langle 0| + i\alpha\langle 2|)\left(\frac{\widehat{a}_1 e^{-i\theta} + \widehat{a}_1^\dagger e^{i\theta}}{2}\right)(|0\rangle - i\alpha|2\rangle) = 0\end{aligned} \tag{22.67}$$

Also

$$\begin{aligned}\left\langle \widehat{X}_\theta^2 \right\rangle &= \langle \psi_{out}| \left(\frac{\widehat{a}_1 e^{-i\theta} + \widehat{a}_1^\dagger e^{i\theta}}{2}\right)^2 |\psi_{out}\rangle \\ &= \frac{1}{4}\langle \psi_{out}| \left(\widehat{a}_1^2 e^{-2i\theta} + \widehat{a}_1^{\dagger 2} e^{+2i\theta} + 2\widehat{a}_1^\dagger \widehat{a}_1 + 1\right)|\psi_{out}\rangle\end{aligned} \tag{22.68}$$

or

$$\left\langle \widehat{X}_\theta^2 \right\rangle = \frac{1}{4}(\langle 0| + i\alpha\langle 2|)\left(\widehat{a}_1^2 e^{-2i\theta} + \widehat{a}_1^{\dagger 2} e^{+2i\theta} + 2\widehat{a}_1^\dagger \widehat{a}_1 + 1\right)(|0\rangle - i\alpha|2\rangle)$$

which can be evaluated to give

$$\left\langle \widehat{X}_\theta^2 \right\rangle = \frac{1}{4}\left(1 - 2\sqrt{2}\alpha \sin 2\theta\right) \tag{22.69}$$

where we have neglected terms of the order α^2.

Hence, the variance is given by

$$\left\langle \Delta\widehat{X}_\theta^2 \right\rangle = \left(\left\langle \widehat{X}_\theta^{\ 2} \right\rangle - \left\langle \widehat{X}_\theta \right\rangle^2\right) = \frac{1}{4}\left(1 - 2\sqrt{2}\alpha \sin 2\theta\right) \tag{22.70}$$

For a vacuum state $|\psi\rangle_{vac} = |0\rangle$ and we have

$$\left\langle \Delta\widehat{X}_\theta^2 \right\rangle = \frac{1}{4} \tag{22.71}$$

Hence, the output state has a variance which varies with the phase θ and goes below the variance for vacuum state for $\sin 2\theta > 0$.

Figure 22.3 shows the variation in $\left\langle \Delta\widehat{X}_\theta^2 \right\rangle$ with θ. As we can see from Eq. (22.70), the variance $\left\langle \Delta\widehat{X}_\theta^2 \right\rangle$ varies with θ and goes from a maximum of $\frac{1}{4} \times \left(1 + 2\sqrt{2}\alpha\right) > \frac{1}{4}$ for $\theta = \frac{3\pi}{4}, \frac{7\pi}{4} \ldots$ to a minimum of $\frac{1}{4}\left(1 - 2\sqrt{2}\alpha\right) < \frac{1}{4}$ for $\theta = \frac{\pi}{4}, \frac{5\pi}{4} \ldots$ Thus, depending on the phase, the variance of the quadrature operator is below that of vacuum and thus the output state is a squeezed state and is referred to as squeezed vacuum state.

Lithium niobate waveguides offer an efficient method to generate squeezed light (see, e.g., Kashiwazaki et al. 2022; Park et al. 2024)

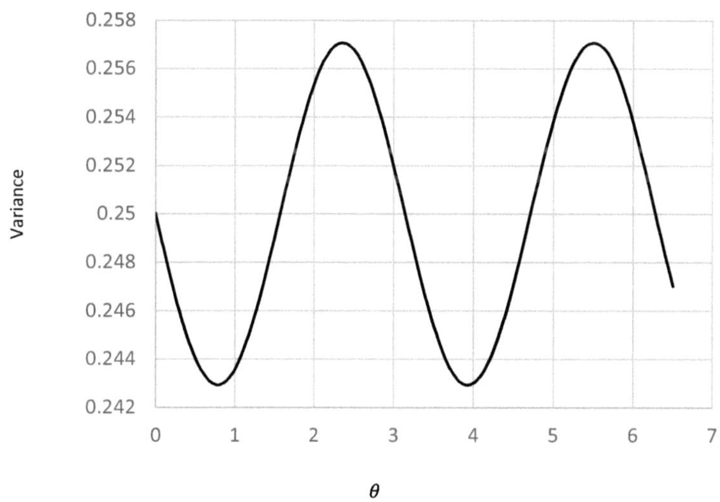

Fig. 22.3 Plot of variance given by Eq. (22.70) with respect to θ (in radians) for =0.01. Note that the variance goes below 0.25 which is the value for vacuum state

Exercise 22.5 Assuming an almost monochromatic pump, obtain the bandwidth of the signal photons generated in the Type-II SPDC process using bulk lithium niobate with an input ordinary pump photon down converting to an extraordinary signal photon and an ordinary idler photon given the wavelength of the pump is 800 nm and the signal photon to have a wavelength of 1530 nm and the length of interaction to be 10 mm.

Solution The idler wavelength is given by 1676.7 nm. The bandwidth of the generated signal photons is given by Eq. (22.42).

$$\Delta\lambda_{s,FWHM} \approx \frac{\lambda_s^2}{2\pi}\frac{5.56}{|N_{io} - N_{se}|L_{int}}$$

The group refractive indices at the signal and idler wavelengths are given by $N_{se}(\omega_s) = 2.2642$ and $N_{io}(\omega_i) = 2.2611$, which gives us

$$\Delta\lambda_{s,FWHM} \approx \frac{\lambda_s^2}{2\pi}\frac{5.56}{|N_{io} - N_{se}|L_{int}} \approx 0.67 \text{ nm}$$

$$\Delta\lambda_i = -\frac{\lambda_i^2}{\lambda_s^2}\Delta\lambda_s \approx 0.8 \text{ nm}$$

Exercise 22.6 Consider an SPDC process in KTP in which a pump with a wavelength of 800 nm having polarization parallel to the $z-$axis generates a pair of signal (having a wavelength of 1400 nm) and idler photons both of which are having polarization parallel to the $z-$axis, leading to Type-0-conversion. (a) Assuming all the three interacting waves to propagate along the same direction obtain the required spatial period for first-order QPM. (b) If the pump and signal propagate along the $x-$direction and the idler propagates along the $-x$-direction, what is the required period for first-order QPM?

Solution For a pump wavelength of 800 nm and signal wavelength of 1400 nm, the idler wavelength would be $\lambda_i \approx 1.867$ μm.

For the case when all the three interacting waves are propagating parallel to each other, the required QPM period satisfies the following equation:

$$\frac{1}{\Lambda_1} = \frac{n_{pz}}{\lambda_p} - \frac{n_{sz}}{\lambda_s} - \frac{n_{iz}}{\lambda_i}$$

and for the case when the idler is propagating in an opposite direction to the pump and signal wavelengths, the required grating period is given by

$$\frac{1}{\Lambda_2} = \frac{n_{pz}}{\lambda_p} - \frac{n_{sz}}{\lambda_s} + \frac{n_{iz}}{\lambda_i}$$

The values of n_z of KTP at the pump, signal, and idler wavelengths are given by 1.8446, 1.8194, and 1.8086, respectively. Using these values, we obtain the following QPM periods:

$$\Lambda_1 \approx 26.7\ \mu\text{m}$$

$$\Lambda_2 \approx 506\ \text{nm}$$

Note that for backward conversion, the required periods are very small and require advanced techniques for fabrication. Recently such QPM periods have been fabricated and backward SPDC process has been obtained [Kuo et al. (2023)]

Exercise 22.7 Obtain the bandwidth of the backward SPDC process in which the pump and signal travel in the same direction and the idler propagates in the reverse direction and all the interacting modes correspond to extraordinary polarization in lithium niobate. Refractive index values and group refractive index values can be obtained from the following URL: https://refractiveindex.info/?shelf=main&book=LiNbO3&page=Zelmon-e

Solution For a backward SPDC process with the idler propagating in the reverse direction, we have

$$\Delta\beta_{ee} = \beta_{pe} - \beta_{se} + \beta_{ie} - K = \frac{\omega_p}{c} n_{pe} - \frac{\omega_s}{c} n_{se} + \frac{\omega_i}{c} n_{ie} - K$$

Thus, if the phase matching wavelengths correspond to ω_{s0} and ω_{i0} and if $\omega_s = \omega_{s0} + \Delta\omega_s$ and $\omega_i = \omega_{i0} + \Delta\omega_i$, then we have following the same procedure as in obtaining Eq. (22.38) we obtain

$$\Delta\beta_{ee} \approx -\frac{\Delta\omega_s}{c} N_{se} + \frac{\Delta\omega_i}{c} N_{io} = -\frac{\Delta\omega_s}{c} (N_{io} + N_{se})$$

The bandwidth of the interaction will be given by

$$\Delta\omega_{s,FWHM} = \Delta\omega_{i,FWHM} \approx \frac{5.56\ c}{|N_{io} + N_{se}| L_{int}}$$

The corresponding bandwidths in terms of wavelengths will be

$$\Delta\lambda_{s,FWHM} \approx \frac{\lambda_s^2}{2\pi} \frac{5.56}{|N_{io} + N_{se}| L_{int}}$$

and

$$\Delta\lambda_{i,FWHM} \approx \frac{\lambda_i^2}{2\pi} \frac{5.56}{|N_{io} + N_{se}| L_{int}}$$

Assuming a pump wavelength of 765 nm and a signal wavelength of 1550 nm, we obtain the idler wavelength of 1510.5 nm. Using $N_{ie}(1510.5\,\text{nm}) = 2.1833$ and $N_{se}(1550\,\text{nm}) = 2.1823$, we obtain the bandwidth for an interaction length of 5 mm of about 0.1 nm for the signal and idler.

Exercise 22.8 Consider a waveguide fabricated on a $Y-$cut, $X-$propagating lithium niobate substrate. (For the principal axes directions, we are using upper case letters X, Y, Z). At the input $Z-$polarized mode (TE mode corresponding to extraordinary polarization) at frequency ω_p is launched. (a) Show that it is possible to generate a pair of $Y-$polarized (TM mode corresponding to ordinary polarization) photons at frequencies ω_s and ω_i with $\omega_p = \omega_s + \omega_i$. (b) Which nonlinear coefficient is involved in the down conversion process. (c) Write down an expression for the QPM period required in terms of the effective indices of the modes.

Solution The nonlinear tensor of lithium niobate is given in Chap. 10. The nonlinear polarization is given by

$$P_i^{(NL)} = 2\epsilon_0 d_{ijk} E_j E_k$$

It is interesting to note that nonlinear coefficient involved in the down conversion process is the same as that for the upconversion process with the same states of polarization and direction of propagation. Here, we will obtain the nonlinear polarization at ω_p generated by the electric fields at ω_s and ω_i. It is given that the electric fields at ω_s and ω_i have $Y-$components only. This corresponds to $j = 2$; $k = 2$. Hence, we have for the nonlinear polarization generated along the $Z-$direction to be

$$P_Z^{(NL)} = 2\epsilon_0 d_{322} E_Y E_Y = 2\epsilon_0 d_{32} E_Y E_Y$$

where the $Y-$component of the total electric field consisting of the signal and idler is given by

$$E_Y = \frac{1}{2}\left[\left(E_Y^{(\omega_s)} e^{i(\beta_{so}x - \omega_s t)} + c.c\right) + \left(E_Y^{(\omega_i)} e^{i(\beta_{io}x - \omega_i t)} + c.c\right)\right]$$

where β_{so} and β_{io} correspond to the propagation constants of the TM modes (ordinary polarization) of the waveguide. Substituting the expression for E_Y and picking terms corresponding to the frequency $\omega_p = \omega_s + \omega_i$, we have

$$P_Z^{(\omega_p)} = \frac{1}{2}\left[2\epsilon_0 d_{32} E_Y^{(\omega_s)} E_Y^{(\omega_i)} e^{i(\{\beta_{so}+\beta_{io}\}x - \omega_p t)} + c.c.\right]$$

Hence, the nonlinear coefficient responsible for the generation of ω_p from ω_s and ω_i or generation of the pair of frequencies ω_s and ω_i from the frequency ω_p is the coefficient $d_{32} = d_{31}$.

The QPM condition would be given by

$$K = \beta_{pe} - \beta_{so} - \beta_{io}$$

where β_{pe} is the propagation constant of the TE mode (extraordinary polarization) at the frequency ω_p and β_{so} and β_{io} are the propagation constants of the ordinary polarization (TM modes) at the signal and idler frequencies, respectively.

To get an idea of the required period we consider bulk lithium niobate and assume a pump wavelength of 800 nm, a signal wavelength of 1550 nm and an idler wavelength of 1653.3 nm. The extraordinary refractive index at 800 nm is given by $n_{pe} = 2.1755$ and the ordinary refractive indices at 1550 nm and 1653.3 nm are given by $n_{so} = 2.2111$ and $n_{io} = 2.2077$, respectively. Hence, the required period QPM grating period is given by

$$\frac{1}{\Lambda} = \left| \frac{n_{pe}}{\lambda_p} - \frac{n_{so}}{\lambda_s} - \frac{n_{io}}{\lambda_i} \right|$$

Substituting the values we obtain 23.5 μm.

22.6 Pulsed Pump

In the above analysis, we assumed the pump to be a monochromatic wave at frequency ω_p. If the pump is in the form of a pulse or assumed to have a certain bandwidth, then we write for the pump field (expressed as a classical field) as

$$\vec{E}_{po} = \frac{1}{2} e_{po}(y, z) E_{p0} \left(\int \tilde{\alpha}_p(\omega_p) e^{i(\beta_{po} x - \omega_p t)} d\omega_p + cc \right) \hat{y} \tag{22.72}$$

where $\tilde{\alpha}_p(\omega_p)$ represents the spectrum of the pump pulse assumed to be peaked around a central pump frequency ω_{p0}. For example, if the input pump pulse is Gaussian of pulse width τ_p, then the corresponding frequency spectrum would also be Gaussian with a width $\frac{2}{\tau_p}$:

$$\tilde{\alpha}_p(\omega_p) = \alpha_0 \sqrt{\pi} \tau_p e^{-\frac{(\omega_p - \omega_{p0})^2 \tau_p^2}{4}} \tag{22.73}$$

where α_0 is a constant. Since the pump is in the form of a pulse, we may note that the pump contains a spectrum of frequencies around the central frequency ω_{p0}. The width of the pump spectrum depends inversely on the pump pulse width τ_p.

We again consider an SPDC process with pump having ordinary polarization generating a signal with extraordinary polarization and an idler having ordinary polarization.

For a monochromatic pump, we have $\tilde{\alpha}_p(\omega_p) = \alpha_0 \delta(\omega_p - \omega_{p0})$ and Eq. (22.72) reduces to Eq. (22.16) with $\alpha_0 = 1$. Note that since the pump has a continuum of frequencies, each of the pump frequency can lead to down conversion depending on the phase matching condition.

In such a case Eq. (22.25) becomes

$$\widehat{H}_{int} = \int d\omega_p \int d\omega_s \int d\omega_i \left[C_{eo} \tilde{\alpha}_p(\omega_p) \widehat{a}_{se}^\dagger \widehat{a}_{io}^\dagger e^{-i(\omega_p - \omega_s - \omega_i)t} + C_{eo}^* \tilde{\alpha}_p^*(\omega_p) \widehat{a}_{se} \widehat{a}_{io} e^{i(\omega_p - \omega_s - \omega_i)t} \right] \tag{22.74}$$

Thus, we have

$$\begin{aligned} \frac{1}{i\hbar} \int_0^t \widehat{H}_{int} dt &\approx -\frac{i}{\hbar} \left(2\pi \int d\omega_s \int d\omega_i \left[C_{eo} \widehat{a}_{se}^\dagger \widehat{a}_{io}^\dagger \tilde{\alpha}_p(\omega_s + \omega_i) + C_{eo}^* \widehat{a}_{se} \widehat{a}_{io} \tilde{\alpha}_p^*(\omega_s + \omega_i) \right] \right) \\ &= -\frac{2\pi i}{\hbar} \int d\omega_s \int d\omega_i \left[\tilde{C}_{eo} \widehat{a}_{se}^\dagger \widehat{a}_{io}^\dagger + \tilde{C}_{eo}^* \widehat{a}_{se} \widehat{a}_{io} \right] \end{aligned} \tag{22.75}$$

where we have used the fact that

$$\int e^{-i(\omega_p - \omega_s - \omega_i)t} d\omega_p = 2\pi \delta(\omega_p - \omega_s - \omega_i) \tag{22.76}$$

and

$$\tilde{C}_{eo} = C_{eo} \tilde{\alpha}_p(\omega_s + \omega_i) = \frac{2}{\pi} d_{24} \left(\frac{2E_{p0} \hbar \sqrt{\omega_s \omega_i} I_{eo}}{n_{se} n_{io} L} \right) e^{\frac{i\Delta\beta_{eo} L_{int}}{2}} L_{int} \tilde{\alpha}_p(\omega_s + \omega_i) \operatorname{sinc}\left(\frac{\Delta\beta_{eo} L_{int}}{2} \right) \tag{22.77}$$

For example for a Gaussian pump pulse centered around the pump frequency ω_{p0}, we would have

$$\tilde{\alpha}_p(\omega_s + \omega_i) = \alpha_0 \sqrt{\pi} \tau_p e^{-\frac{(\omega_s + \omega_i - \omega_{p0})^2 \tau_p^2}{4}} \tag{22.78}$$

Thus, the output two photon state can be written as

$$\left|\psi^{(I)}(t)\right\rangle \approx \left(1 - \frac{2\pi i}{\hbar}\int d\omega_s \int d\omega_i \left[\tilde{C}_{eo}\hat{a}^{\dagger}_{se}\hat{a}^{\dagger}_{io} + \tilde{C}^{*}_{eo}\hat{a}_{se}\hat{a}_{io}\right]\right)|0_s, 0_i\rangle$$

which becomes

$$\begin{aligned}\left|\psi^{(I)}(t)\right\rangle &\approx |0_s, 0_i\rangle - \frac{2\pi i}{\hbar}\int d\omega_s \int d\omega_i \left[\tilde{C}_{eo}\hat{a}^{\dagger}_{se}\hat{a}^{\dagger}_{io} + \tilde{C}^{*}_{eo}\hat{a}_{se}\hat{a}_{io}\right]|0_{se}, 0_{io}\rangle \\ &= |0_s, 0_i\rangle - \frac{2\pi i}{\hbar}\int d\omega_s \int d\omega_i \tilde{C}_{eo}|1_{se}, 1_{io}\rangle\end{aligned} \tag{22.79}$$

Note that since the pump is pulsed, it is no more monochromatic and has a spectrum defined by $\tilde{\alpha}_p(\omega_p)$ implying that the input pump frequency is not precisely defined and now covers a range of frequencies. Since the SPDC process is a random process, the down conversion can take place from a pump photon of any specific frequency within the spectrum of the pump. Since the pump frequency is not precisely defined, there is a range of signal and idler photon pairs that get generated with the sum of their frequencies being equal to the pump frequency photon that resulted in the generation of this pair. Thus, if the down conversion takes place from a pump photon of frequency ω_{p0}, then the signal and idler pair frequencies will satisfy the equation $\omega_{p0} = \omega_{s0} + \omega_{i0}$. Similarly if the pump photon has a frequency $\omega_{p0} + \delta\omega_{p0}$, then the generated signal and idler pair would satisfy the equation $\omega_{p0} + \delta\omega_{p0} = \omega_{s0} + \delta\omega_{s0} + \omega_{i0} + (\delta\omega_{p0} - \delta\omega_{s0})$. The double integral over both signal and idler frequencies is a result of the pump frequency occupying a range of frequencies.

In the case of a monochromatic pump, for a given signal frequency, the corresponding idler frequency is well defined. On the other hand, when the pump is not monochromatic but is characterized by a spectrum, then a given signal frequency does not define the corresponding idler frequency since the pump frequencies form a band and it is not known which pump photon frequency was responsible for the generation of the signal–idler pair.

The joint spectral distribution $g(\omega_s, \omega_i)$ of the signal–idler photon pair is defined as the product of the pump spectral distribution and the phase matching function:

$$g(\omega_s, \omega_i) = \tilde{\alpha}_p(\omega_s + \omega_i)\text{sinc}\left(\frac{\Delta\beta_{eo}L_{int}}{2}\right) \tag{22.80}$$

which determines the spectral distributions of the signal and idler from the pulsed pump. Thus,

$$\tilde{C}_{eo} = C_{eo}\tilde{\alpha}_p(\omega_s + \omega_i) = \frac{2}{\pi}d_{24}\left(\frac{2E_{p0}\hbar\sqrt{\omega_s\omega_i}I_{eo}}{n_{se}n_{io}L}\right)L_{int}e^{\frac{i\Delta\beta_{eo}L_{int}}{2}}g(\omega_s, \omega_i) \tag{22.81}$$

and we obtain

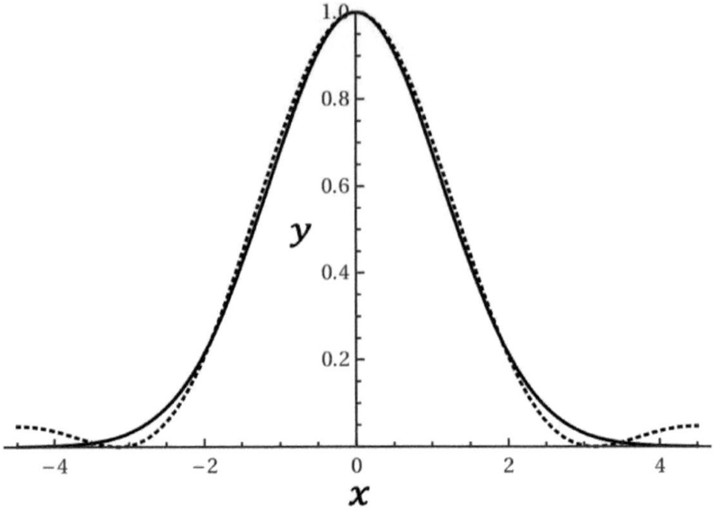

Fig. 22.4 Comparison between $y = sinc^2\,(x)$ (dashed curve) and $y = e^{-0.386x^2}$ (solid curve)

$$
\begin{aligned}
\left|\psi^{(I)}(t)\right\rangle &\approx |0_s, 0_i\rangle - \frac{2\pi i}{\hbar}\int d\omega_s \int d\omega_i \left[\tilde{C}_{eo}\hat{a}^{\dagger}_{se}\hat{a}^{\dagger}_{io} + \tilde{C}^{*}_{eo}\hat{a}_{se}\hat{a}_{io}\right]|0_{se}, 0_{io}\rangle \\
&= |0_s, 0_i\rangle - \frac{4i}{\hbar}d_{24}\left(\frac{2E_{p0}\hbar\sqrt{\omega_s\omega_i}I_{eo}}{n_{se}n_{io}L}\right)L_{int}\int d\omega_s \int d\omega_i e^{\frac{i\Delta\beta_{eo}L_{int}}{2}} g(\omega_s, \omega_i)\hat{a}^{\dagger}_{se}\hat{a}^{\dagger}_{io}|0_{se}, 0_{io}\rangle
\end{aligned}
\tag{22.82}
$$

where we have taken the factor $\sqrt{\omega_s\omega_i}$ outside the integral as the bandwidths are quite small and the variation in this quantity over the integral can be neglected. Thus, the joint spectral distribution $g(\omega_s, \omega_i)$ specifies the spectra of the signal and idler photons that are generated in the process of SPDC using a pulsed pump.

It is also interesting to observe that the phase matching function described by the sinc function can be approximated by a Gaussian function [Brecht et al. (2011)]. Figure 22.4 shows a plot showing the approximation of the sinc function by an appropriate Gaussian function.

$$
\text{sinc}\left(\frac{\Delta\beta_{eo}L_{int}}{2}\right) \approx e^{-0.193\left(\frac{\Delta\beta_{eo}L_{int}}{2}\right)^2}
\tag{22.83}
$$

An interesting way to understand the signal–idler pair generation process is through measurements of the joint spectral distribution corresponding to the down converted photon pairs. The first experiments in joint spectral distribution measurements were carried out by Wasilewski et al. (2006).

In Sect. 22.7, we will derive an explicit expression for the joint spectral distribution $g(\omega_s, \omega_i)$ in terms of group effective indices of the medium at the pump, signal, and idler frequencies and deviation from the phase matching wavelengths given by ω_{s0}, ω_{i0}; see Eq. (22.106). Here we present simulations corresponding to lithium

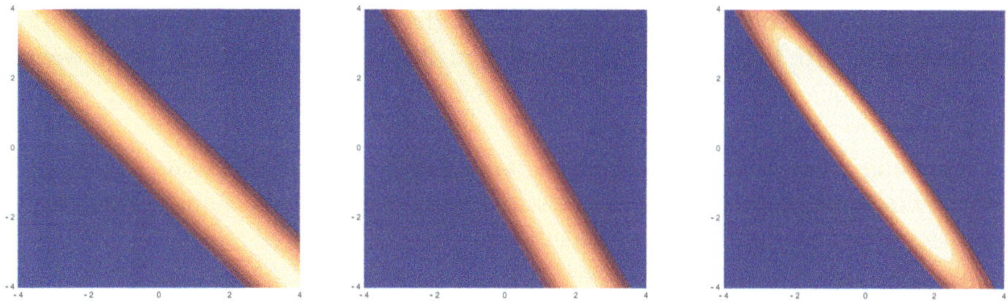

a b c

Fig. 22.5 (**a**) Pump spectrum, (**b**) Phase matching function and (**c**) joint spectral distribution given by Eq. (22.80)

niobate in Type-II interaction (pump with ordinary polarization down converting to signal with extraordinary polarization and idler with ordinary polarization) for the following values of various parameters: $\lambda_{p0} = 800$ nm; $\lambda_{s0} = 1500$ nm; $\lambda_{i0} = 1714.3$ nm. Also, the group indices are given by (with the subscript representing ordinary (o) or extraordinary e polarization states) $N_o(\omega_{p0}) = 2.3585$; $N_e(\omega_{s0}) = 2.1836$; $N_o(\omega_{i0}) = 2.2606$. The length of interaction is assumed to be 1 mm and $\tau_p = 0.2$ ps.

Figure 22.5a shows the density plot corresponding to the power spectrum of the pump $\tilde{\alpha}_p^2(\omega_s + \omega_i)$ as a function of the deviation of the signal frequency ω_s and idler frequency ω_i from the phase matching frequencies. The central line along the plot gives the central frequency of the pump. Figure 22.5b shows the density plot of the phase matching function $sinc^2\left(\frac{\Delta\beta_{eo}L_{int}}{2}\right)$. Along the central line of the plot the phase matching is exactly satisfied and $\Delta\beta_{eo} = 0$. Figure 22.5c shows the complete joint spectral distribution which is a product of the pump power spectrum and the phase matching function [see also Shimizu and Edamatsu (2009)].

22.7 Spectral Factorizability of Photon Pairs Generated by SPDC

Spontaneous parametric down conversion is one of the most important techniques to generate photon pairs. Due to the requirement of energy conservation, the signal–idler photon pairs are usually correlated in frequency. Depending on the device, the photons can be generated in polarization entangled states, mode entangled states, hybrid entangled states, etc. In many applications, it is required that the emerging photons are uncorrelated. For example, in the application to generate heralded single photons, one of the photons is detected and used as the heralding photon to announce the arrival of the other photon. In this application in order to have pure photon states, it is important that there be no frequency correlation between the signal and idler photons. In order to realize such pure spectrally factorizable photons, many techniques have been developed (see, e.g., Grice et al. 2001; U'Ren et al. 2005; Laudenbach 2016; Liu et al. 2021; Xinhui et al. 2021; Xin et al. 2022). Here, we will discuss the idea behind some of the important techniques to achieve spectral factorizability of the photon pair.

Recalling from Sect. 22.6, using a broadband or a pulsed pump, the Type II spontaneous parametric down conversion process generates signal–idler photon pairs described by the following equation (see Eq. 22.79):

$$\left|\psi^{(I)}(t)\right\rangle \approx |0_s, 0_i\rangle - \frac{2\pi i}{\hbar}\int d\omega_s \int d\omega_i \tilde{C}_{eo}|1_{se}, 1_{io}\rangle \tag{22.84}$$

where, for an SPDC process using the d_{24} coefficient of lithium niobate,

$$\tilde{C}_{eo} = \frac{2}{\pi} d_{24}\left(\frac{2E_{p0}\hbar\sqrt{\omega_s\omega_i}I_{eo}}{n_{se}n_{io}}\right) e^{\frac{i\Delta\beta_{eo}L}{2}} g(\omega_s, \omega_i) \tag{22.85}$$

with L being the interaction length and

$$g(\omega_s, \omega_i) = \tilde{\alpha}_p(\omega_s + \omega_i)\ \mathrm{sinc}\left(\frac{\Delta\beta_{eo}L}{2}\right) \tag{22.86}$$

representing the joint spectral amplitude (JSA). Here $\tilde{\alpha}_p$ is the pump spectral distribution, assumed to be a Gaussian spectral distribution with width σ_P, centered at the frequency ω_{p0}:

$$\tilde{\alpha}_p(\omega_s + \omega_i) = \alpha_0 e^{-\frac{(\omega_p - \omega_{p0})^2}{\sigma_p^2}} = \alpha_0 e^{-\frac{(\omega_s + \omega_i - \omega_{p0})^2}{\sigma_p^2}} \tag{22.87}$$

and $\mathrm{sinc}\left(\frac{\Delta\beta_{eo}L}{2}\right)$ represents the phase matching function with

$$\Delta\beta_{eo} = \beta_{po} - \beta_{se} - \beta_{io} - K \tag{22.88}$$

Here β_{po} is the propagation constant of the pump mode having ordinary polarization, β_{se} is the propagation constant of the signal mode having extraordinary polarization and β_{io} is the propagation constant of the idler mode having ordinary polarization and K is the spatial frequency of the QPM grating used for quasi phase matching.

From Eq. (22.84), we notice that the generated photon pairs are frequency entangled as the state cannot be expressed as a product state of signal and idler photons. In order to achieve pure single photon states, we need to be able to factor out the signal and idler photon states. This is possible if the JSA can be factored into two terms, one depending on the signal frequency and the other on the idler frequency:

$$g(\omega_s, \omega_i) = f_s(\omega_s) f_i(\omega_i) \tag{22.89}$$

which then gives (apart from the vacuum state)

$$|\psi^{(I)}(t)\rangle \approx \frac{2\pi i}{\hbar} \int d\omega_s f_s(\omega_s) \hat{a}^\dagger_{so} |0_{so}\rangle \int d\omega_i f_i(\omega_i) \hat{a}^\dagger_{ie} |0_{ie}\rangle \tag{22.90}$$

which is a separable state as it is a product state of a signal photon and an idler photon.

In order to investigate the possibility of generating pure single photon states, we first note that we can approximate the *sinc* function as given by Eq. (22.83) (see Fig. 22.4)

Thus, substituting for $\tilde{\alpha}_p$ and the *sinc* function, the JSA becomes

$$g(\omega_s, \omega_i) \approx \alpha_0 e^{-\frac{(\omega_s + \omega_i - \omega_{p0})^2}{\sigma_p^2}} e^{-0.193\left(\frac{\Delta\beta_{eo} L}{2}\right)^2} \tag{22.91}$$

To investigate under what conditions is $g(\omega_s, \omega_i)$ is factorable, we first note that

$$\Delta\beta_{eo} = \beta_{po} - \beta_{se} - \beta_{io} - K = \beta_o(\omega_p) - \beta_e(\omega_s) - \beta_o(\omega_i) - K \tag{22.92}$$

If ω_{p0}, ω_{s0}, and ω_{i0} represent the central pump, signal, and idler frequencies satisfying the QPM condition, we have

$$\beta_o(\omega_{p0}) - \beta_e(\omega_{s0}) - \beta_o(\omega_{i0}) - K = 0 \tag{22.93}$$

with

$$\omega_{p0} = \omega_{s0} + \omega_{i0} \tag{22.94}$$

Now for a pump frequency $\omega_{p0} + \Delta\omega_p$ with the corresponding signal and idler frequencies $\omega_{s0} + \Delta\omega_s$ and $\omega_{i0} + \Delta\omega_i$, respectively, we have

$$\begin{aligned} \Delta\beta_{eo} &= \beta_o(\omega_p) - \beta_e(\omega_s) - \beta_o(\omega_i) - K \\ &= \beta_o(\omega_{p0} + \Delta\omega_p) - \beta_e(\omega_{s0} + \Delta\omega_s) - \beta_o(\omega_{i0} + \Delta\omega_i) - K \end{aligned} \tag{22.95}$$

We make a Taylor series expansion assuming $\Delta\omega_p \ll \omega_{p0}$ and obtain

$$\beta_o(\omega_{p0} + \Delta\omega_p) = \beta_o(\omega_{p0}) + \Delta\omega_p \frac{d\beta_o}{d\omega}\bigg|_{\omega_{p0}} = \beta_o(\omega_{p0}) + \Delta\omega_p \frac{N_{eff,o}(\omega_{p0})}{c} \tag{22.96}$$

where

$$N_{eff}(\omega)=n_{eff}(\omega)+\omega\frac{dn_{eff}}{d\omega} \tag{22.97}$$

is the group effective index of the mode (see Chap. 4) which is related to the group velocity through $v_{gr}(\omega)=\frac{c}{N_{\mathrm{eff}}(\omega)}$.

Making a similar expansion of $\beta_e(\omega_{s0}+\Delta\omega_s)$ and $\beta_o(\omega_{i0}+\Delta\omega_i)$ assuming $\Delta\omega_s \ll \omega_{s0}$ and $\Delta\omega_i \ll \omega_{i0}$ and using Eq. (22.95), we obtain

$$\Delta\beta_{eo}=\Delta\omega_p\frac{N_{eff,o}(\omega_{p0})}{c}-\Delta\omega_s\frac{N_{eff,e}(\omega_{s0})}{c}-\Delta\omega_i\frac{N_{eff,o}(\omega_{i0})}{c} \tag{22.98}$$

Since

$$\omega_p=\omega_s+\omega_i \tag{22.99}$$

we have

$$\Delta\omega_p=\Delta\omega_s+\Delta\omega_i \tag{22.100}$$

Using this in Eq. (22.98), we obtain

$$\Delta\beta_{eo}=\frac{\Delta\omega_s}{c}\left(N_{eff,o}(\omega_{p0})-N_{eff,e}(\omega_{s0})\right)+\frac{\Delta\omega_i}{c}\left(N_{eff,o}(\omega_{p0})-N_{eff,e}(\omega_{i0})\right) \tag{22.101}$$

Writing

$$\Delta N_{ps}=N_{eff,o}(\omega_{p0})-N_{eff,e}(\omega_{s0}) \tag{22.102}$$

and

$$\Delta N_{pi}=N_{eff,o}(\omega_{p0})-N_{eff,o}(\omega_{i0}) \tag{22.103}$$

we can write Eq. (22.101) as

$$\Delta\beta_{eo}=\frac{1}{c}\left(\Delta\omega_s\Delta N_{ps}+\Delta\omega_i\Delta N_{pi}\right) \tag{22.104}$$

We also have

$$\omega_s+\omega_i-\omega_{p0}=\omega_{s0}+\Delta\omega_s+\omega_{i0}+\Delta\omega_i-\omega_{p0}=\Delta\omega_s+\Delta\omega_i \tag{22.105}$$

where we have used Eq. (22.94). Thus, we have

$$g(\omega_s, \omega_i) \approx \alpha_0 e^{-\frac{(\Delta\omega_s + \Delta\omega_i)^2}{\sigma_p^2}} e^{-0.193\frac{L^2}{4c^2}\left(\Delta\omega_s \Delta N_{ps} + \Delta\omega_i \Delta N_{pi}\right)^2} \tag{22.106}$$

We first note that the function $g(\omega_s, \omega_i)$ cannot be written as a product of factors depending on $\Delta\omega_s$ and $\Delta\omega_i$ and thus the generated photon pairs are entangled in frequency.

An interesting picture of the JSA is through a density plot of the function with $\Delta\omega_s$ and $\Delta\omega_i$ as the axes. In such a figure, the slope of the first exponential factor in Eq. (22.106) will be at -45° to the horizontal (see Fig. 22.6a), while that of the second exponential factor will be given by

$$\theta = \tan^{-1}\left(-\frac{\Delta N_{ps}}{\Delta N_{pi}}\right) \tag{22.107}$$

Figure 22.6 gives plots of the first exponential factor and second exponential factor in in Eq. (22.106). Figure 22.7 gives the JSA which is a product of the two exponential factors.

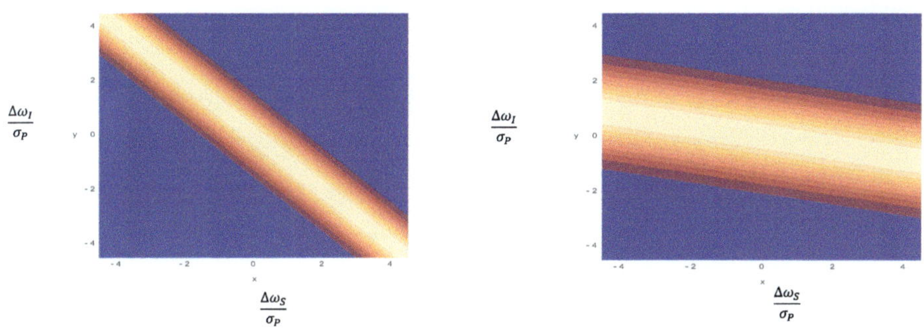

Fig. 22.6 Plot of (**a**) the first exponential factor and (**b**) the second exponential factor given in Eq. (22.106)

Fig. 22.7 Simulated plot of a typical JSA

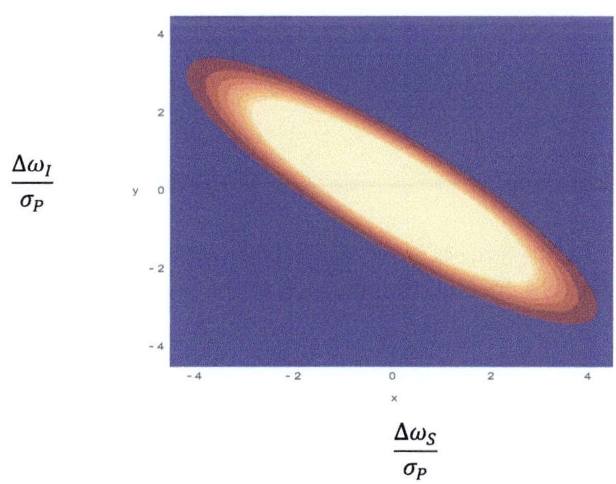

The plots correspond to the following values of various parameters: $L = 1\ \text{mm}, \sigma_P = 10^{13}\,\frac{\text{rad}}{\text{s}}, \Delta N_{pi} = 0.1, \frac{\Delta N_{ps}}{\Delta N_{pi}} = 0.5$

22.7.1 Bandwidth of the Generated Photons

As discussed in Sect. 22.3 when $\frac{\Delta\beta_{eo}L}{2} = \pm 1.39$ the generation efficiency will drop to 3 dB from the maximum. Hence, the FWHM is now given by

$$\frac{\Delta\beta_{eo}L}{2} = \frac{L}{c}\left(\Delta\omega_s\Delta N_{ps} + \Delta\omega_i\Delta N_{pi}\right) = \pm 1.39 \tag{22.108}$$

where ΔN_{ps} is the group index difference between the pump wavelength and signal wavelength and ΔN_{pi} is the group index difference between the pump and the idler frequency.

Measuring the idler spectrum at a signal frequency $\omega_s = \omega_{s0}$, we obtain the idler bandwidth as

$$\Delta\omega_{i,\text{FWHM}} = \frac{c}{L}\frac{5.56}{\Delta N_{pi}} \tag{22.109}$$

or in terms of wavelength

$$\Delta\lambda_{i,\text{FWHM}} \approx \frac{\lambda_i^2}{2\pi L}\frac{5.56}{\Delta N_{pi}} \tag{22.110}$$

Similarly measuring the signal spectrum at the idler frequency $\omega_i = \omega_{i0}$, we will obtain the signal bandwidth as

$$\Delta\omega_{s,\text{FWHM}} = \frac{c}{L}\frac{5.56}{\Delta N_{ps}} \tag{22.111}$$

or in terms of wavelength as

$$\Delta\lambda_{s,\text{FWHM}} \approx \frac{\lambda_s^2}{2\pi L}\frac{5.56}{\Delta N_{ps}} \tag{22.112}$$

If the process is a backward interaction with the pump and signal traveling in the forward direction and the idler in the reverse direction, then

$$\Delta N_{pi} = N_{eff}\left(\omega_p\right) + N_{eff}\left(\omega_i\right) \tag{22.113}$$

$$\Delta N_{ps} = N_{eff}(\omega_p) - N_{eff}(\omega_s) \tag{22.114}$$

Note that since $\Delta N_{pi} = N_{eff}(\omega_p) + N_{eff}(\omega_i)$, the backward propagating idler bandwidth is expected to be much smaller than the forward propagating signal bandwidth.

As an example, we consider a Type 0 backward SPDC process in lithium niobate with all the waves having extraordinary polarization and with the pump at 765 nm, signal at 1550 nm, and idler at 1510 nm. We have $N_{eff}(\omega_p) \approx 2.2765$, $N_{eff}(\omega_s) \approx 2.1823$, $N_{eff}(\omega_i) \approx 2.1833$. For these values, we obtain $\Delta N_{pi} = N_{eff}(\omega_p) + N_{eff}(\omega_i) = 4.4598$ and $\Delta N_{ps} = N_{eff}(\omega_p) - N_{eff}(\omega_s) = 0.0942$.

For an interaction length of 40 mm, the corresponding bandwidths of the idler and signal come out to be 11.3 pm and 0.54 nm, respectively. Note the extremely small bandwidth of the backward propagating idler as compared to the forward propagating signal.

For the backward interaction process, it should be noted that to use first-order QPM, the required grating period will be about 0.35 μm which is indeed very small. In order to reduce fabrication issues, it is possible to use higher order QPM process at the expense of efficiency as the effective nonlinear coefficient would be smaller. The required period for 3rd order and 5th order QPM would be approximately 1.04 and 1.73 μm, respectively.

Exercise 22.9 Consider Type II SPDC in KTP with a pump wavelength of 775 nm and QPM to satisfy the phase matching condition corresponding to a signal wavelength of 1510 nm and an idler wavelength of 1592 nm. The pump is polarized along the x−direction, signal is polarized along the z−direction and idler polarized along the x−direction. (a) Which nonlinear coefficient will be responsible for the SPDC process? The group indices at these wavelengths are given in Table 22.1. Obtain ΔN_{ps} and ΔN_{pi}.

Solution

(a) For the polarization states given, the nonlinear coefficient responsible for the SPDC process would be $d_{xzx} = d_{15}$.

(b) $\Delta N_{ps} = N_x(\lambda_p) - N_z(\lambda_s) = -0.0538; \Delta N_{pi} = N_x(\lambda_p) - N_x(\lambda_i) = +0.0447$

Note that the ratio $\frac{\Delta N_{ps}}{\Delta N_{pi}}$ is negative which gives an anticorrelation between the signal and idler photons.

Table 22.1 Group effective indices of KTP at 775 nm (pump), 1510 nm (signal) and 1592 nm (idler) wavelengths

	λ_p	λ_s	λ_i
N_x	1.7983	1.7542	1.7536
N_y	1.8102	1.7632	1.7626
N_z	1.9128	1.8521	1.8510

Exercise 22.10 In the backward SPDC process, show that when the pump frequency changes, the change in the generated idler frequency is very small compared to the change in the signal frequency.

Solution We first note that for a backward SPDC process, assuming that the frequency ω_i corresponds to the backward wave, we will have (see Eq. (22.98))

$$\Delta\beta = \Delta\omega_p \frac{N_{eff}(\omega_{p0})}{c} - \Delta\omega_s \frac{N_{eff}(\omega_{s0})}{c} + \Delta\omega_i \frac{N_{eff}(\omega_{i0})}{c} \tag{22.115}$$

And we also have

$$\Delta\omega_p = \Delta\omega_s + \Delta\omega_i$$

The maximum efficiency will be when $\Delta\beta = 0$. Substituting for $\Delta\omega_s$ in terms of $\Delta\omega_p$ and $\Delta\omega_i$ in Eq. (22.115), we obtain

$$\Delta\omega_i = \frac{N_{eff}(\omega_{s0}) - N_{eff}(\omega_{p0})}{N_{eff}(\omega_{s0}) + N_{eff}(\omega_{i0})} \Delta\omega_p \tag{22.116}$$

If we assume all the interacting waves to be extraordinarily polarized and the interaction takes place in lithium niobate with the $\lambda_p = 800$ nm, $\lambda_s = 1550$ nm, $\lambda_i = 1653.3$ nm, we have typical values $N_{eff}(\omega_{p0}) \approx 2.2642$, $N_{eff}(\omega_{s0}) \approx 2.1823$, $N_{eff}(\omega_{i0}) \approx 2.1804$. This gives us $\Delta\omega_i \approx -1.9 \times 10^{-2} \Delta\omega_p$ or equivalently in wavelength terms, $\Delta\lambda_i = -1.9 \times 10^{-2} \frac{\lambda_i^2}{\lambda_p^2} \Delta\lambda_p$. As an example, if the pump wavelength changes by 10 nm from 800 nm to 810 nm, then the idler wavelength will shift by about 0.8 nm to 1652.5 nm and the signal wavelength will change by about 38 nm to 1588 nm.

Exercise 22.11 We now consider SPDC process with all the interacting waves propagating along the same direction in lithium niobate with the pump wavelength being 780 nm, the signal wavelength being 1300 nm and the corresponding idler wavelength of 1950 nm. At these wavelengths $n_e(\omega_p) = 2.1778$, $n_e(\omega_s) = 2.1455$ and $n_e(\omega_i) = 2.1267$ and $N_e(\omega_p) = 2.2710$, $N_e(\omega_s) = 2.1917$ and $N_e(\omega_i) = 2.1786$. Estimate the QPM period required and also the shift in idler wavelength caused due to changes in the pump wavelength.

Solution In the case when all the interacting modes are propagating along the same direction, the required QPM period satisfies the following equation:

$$\frac{1}{\Lambda} = \frac{n_e(\omega_p)}{\lambda_p} - \frac{n_e(\omega_s)}{\lambda_s} - \frac{n_e(\omega_i)}{\lambda_i}$$

Substituting the values, we obtain $\Lambda \approx 19.59$ μm which is much larger than that required for backward interaction.

In this case, we also have

$$\Delta\omega_i = \frac{N_{eff}(\omega_{s0}) - N_{eff}(\omega_{p0})}{N_{eff}(\omega_{s0}) - N_{eff}(\omega_{i0})}\Delta\omega_p \tag{22.117}$$

Hence, if the pump wavelength shifts from 780 nm to 780.1 nm, then the shift in idler frequency will be $\Delta\omega_i \approx -6.05\Delta\omega_p$ and correspondingly the shift in the idler wavelength will be $\Delta\lambda_i = -6.05\frac{\lambda_i^2}{\lambda_p^2}\Delta\lambda_p$. Using $\Delta\lambda_p = 0.1$ nm, the idler wavelength changes by about -3.78 nm to 1946.22 nm and the signal wavelength changes by about 1.958 nm to 1301.96 nm. This shows significant changes in both the idler and the signal wavelengths even for a very small change in the pump wavelength contrary to the case of backward SPDC process.

For further details on backward interaction, readers may look up the following references: Luo et al. (2020), Booth et al. (2002), Christ et al. (2009)

22.7.2 Narrow Band Filtering

In order to generate spectrally separable photon states, one can employ different techniques. For example, by using a narrow band filter for the signal photons or the idler photons or both, it is possible to generate almost spectrally separable states. This technique of course results in reduced efficiency of photon pair generation.

Thus, for example, if a very narrow band spectral filter for signal wavelength centered at $\omega_s = \omega_{s0}$ is used, and if we describe the filter through a Gaussian function centered at $\Delta\omega_s = 0$, given by

$$F(\omega_s) = Ae^{-\frac{\Delta\omega_s{}^2}{\sigma_s^2}} \tag{22.118}$$

where σ_s is the spectral width of the filter which is assumed to be very small to lead to narrow band filtering, then the JSA after the filter would be described by (see Eqs. 22.91 and 22.104)

$$\tilde{g}(\omega_s,\omega_i) = g(\omega_s,\omega_i)Ae^{-\Delta\omega_s{}^2/\sigma_s^2} \approx A\alpha_0 e^{-\frac{(\Delta\omega_s+\Delta\omega_i)^2}{\sigma_p^2}} e^{-0.193\frac{L^2}{4c^2}\left(\Delta\omega_s\Delta N_{ps}+\Delta\omega_i\Delta N_{pi}\right)^2} e^{-\Delta\omega_s{}^2/\sigma_s^2}$$

If

$$\frac{1}{\sigma_s} \gg \frac{1}{\sigma_p}, \sqrt{0.193}\frac{L}{2c}\Delta N_{ps} \tag{22.119}$$

then we obtain

$$\tilde{g}(\omega_s, \omega_i) \approx A\alpha_0 e^{-\Delta\omega_i^2\left(\frac{1}{\sigma_p^2}+0.193\frac{L^2}{4c^2}\Delta N_{pi}^2\right)} e^{-\frac{\Delta\omega_s^2}{\sigma_s^2}} \tag{22.120}$$

thus leading to factorization of the JSA and consequently spectral separation of the signal and idler photons.

Similar arguments can be given for a filter for the idler.

22.7.3 Direct Generation of Separable States

It is also possible to employ techniques in which the generated photon pairs are in spectrally separable states so that no filtering is required which results in higher generation efficiency. For example, by appropriate choice of the dispersion characteristics of the optical waveguide, it is possible to ensure that the signal photons are generated in a narrow bandwidth which will automatically result in separable photon states. Since normal waveguides do not exhibit strong dispersion, with appropriate design special kinds of waveguides referred to as Bragg reflection waveguides have been shown to generate spectrally separable states (Thyagarajan et al. 2008; Svozilik et al. 2011) (see also Exercise 22.12).

Exercise 22.12 Assuming that the pump, signal, and idler propagate along the same direction, show that if the group indices at the pump and idler wavelengths are almost equal, then the generated signal would have extremely small bandwidth.

Solution Similar to Eq. (22.117) which relates the shift in idler frequency corresponding to a shift in the pump frequency, we can obtain the following equation describing a relationship between the change in signal frequency for a given shift in the pump frequency:

$$\Delta\omega_s = \frac{N_{eff}(\omega_{i0}) - N_{eff}(\omega_{p0})}{N_{eff}(\omega_{s0}) - N_{eff}(\omega_{i0})} \Delta\omega_p \tag{22.121}$$

Since the group indices of the propagating modes in a waveguide depend on the material as well as on the waveguide design, if it is possible to design a waveguide to have $N_{eff}(\omega_{p0}) \approx N_{eff}(\omega_{i0})$, then with a change in the pump wavelength, the signal wavelength will have negligible change while the idler frequency will change appreciably. In order to achieve this condition, with appropriate design, Bragg reflection waveguides in which Bragg waveguide modes exhibit appreciable dispersion have been shown to lead to this condition. This results in appreciable pump acceptance bandwidth as well as the generation of separable states (Thyagarajan et al. 2008; Svozilik et al. 2011).

Another possibility exists for generating factorizable states by an appropriate choice of waveguide with the following condition satisfied:

$$\Delta N_{ps} = N_{eff}(\omega_{p0}) - N_{eff}(\omega_{s0}) = 0 \qquad (22.122)$$

and choosing a pump bandwidth to be much larger than the phase matching bandwidth as given by

$$\sigma_p \gg \frac{2c}{0.44L\Delta N_{pi}} \qquad (22.123)$$

In this case, we would have

$$\tilde{g}(\omega_s, \omega_i) \approx A\alpha_0 e^{-\frac{\Delta\omega_i^2}{\sigma_i^2}} e^{-\frac{\Delta\omega_s^2}{\sigma_p^2}} \qquad (22.124)$$

where $\sigma_i^2 = \frac{4c^2}{0.193L^2\Delta N_{pi}^2}$. Equation (22.124) shows that the state is a factorized state of signal and idler photons.

Figure 22.8 shows a plot of $\tilde{g}(\omega_s, \omega_i)$ with the following parameters: $\sigma_p = 5 \times 10^{12}$ rad/s, $\frac{2c}{\sqrt{0.193}L\Delta N_{pi}} = 0.4 \times 10^{12}$ rad/s and $\Delta N_{ps} = 0$ clearly showing the generation of separable photon state.

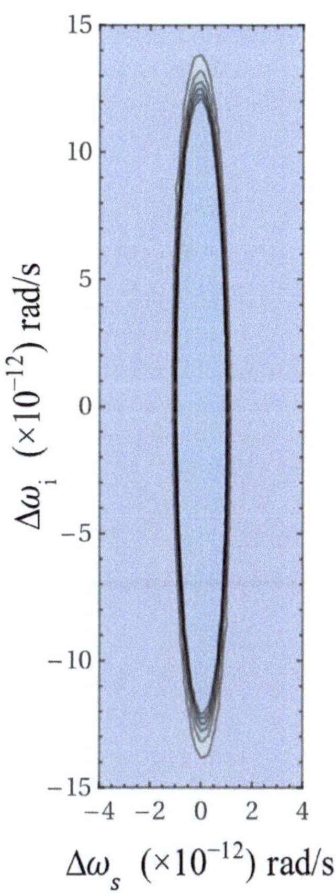

Fig. 22.8 JSA plot showing generation of separable state under conditions given by Eqs. (22.122) and (22.123)

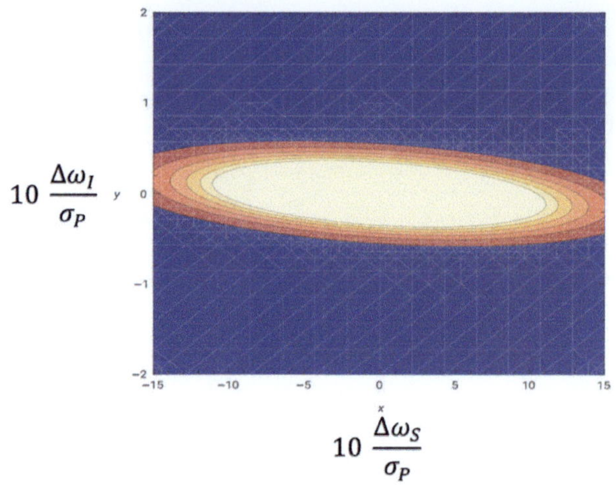

Fig. 22.9 JSA plot for a backward spontaneous parametric generation process showing the generation of an almost separable quantum state

In Sect. 22.7.1, we discussed an example of backward spontaneous parametric generation and showed that due to the backward propagation of one of the generated photons, the spectral bandwidth of that photon is extremely small. In such a case the output state of the photon pair is automatically in a separable state.

Figure 22.9 shows the joint spectral amplitude for $\sigma_p = 10^{13}$ rad/s, $L = 1$ mm, $\Delta N_{ps} = 0.04$, $\Delta N_{pi} = 4.3$. Note that the plot is almost horizontal showing an almost separable state generation.

22.7.4 Group Index/Group Velocity Matching

It is also possible to achieve separable photon state generation by appropriate combination of the nonlinear material and the waveguide geometry. In order to understand this we first note that if we choose a medium and a waveguide satisfying the following condition:

$$\Delta N_{ps} = -\Delta N_{pi} \tag{22.125}$$

That is,

$$N_{eff}(\omega_p) - N_{eff}(\omega_s) = N_{eff}(\omega_i) - N_{eff}(\omega_p)$$

or

$$N_{eff}(\omega_p) = \frac{1}{2}\left(N_{eff}(\omega_s) + N_{eff}(\omega_i)\right) \tag{22.126}$$

Then Eq. (22.106) becomes

$$g(\omega_s, \omega_i) \approx \alpha_0 e^{-\frac{(\Delta\omega_s+\Delta\omega_i)^2}{\sigma_p^2}} e^{-0.193\frac{L^2}{4c^2}\Delta N_{ps}^2(\Delta\omega_s-\Delta\omega_i)^2} \tag{22.127}$$

Equation (22.126) implies that the group index of the waveguide at the pump frequency must be the average of the group indices at the signal and idler frequencies. If in addition, we choose the pump duration to satisfy the following equation:

$$\frac{1}{\sigma_p^2} = \frac{0.193}{4}\frac{L^2}{c^2}\Delta N_{ps}^2 \tag{22.128}$$

then Eq. (22.127) becomes

$$g(\omega_s, \omega_i) \approx \alpha_0 e^{-\frac{1}{\sigma_p^2}\left[(\Delta\omega_s+\Delta\omega_i)^2+(\Delta\omega_s-\Delta\omega_i)^2\right]} = \alpha_0 e^{-2\frac{\Delta\omega_s{}^2}{\sigma_p^2}} e^{-2\frac{\Delta\omega_i{}^2}{\sigma_p^2}}$$

or

$$g(\omega_s, \omega_i) = \alpha_0 e^{-\frac{2}{\sigma_p^2}(\omega_s-\omega_{s0})^2} e^{-\frac{2}{\sigma_p^2}(\omega_i-\omega_{i0})^2} \tag{22.129}$$

which is now factored into a product of a term depending only on ω_s and a term depending only on ω_i. In such a situation, the SPDC process would generate pure photon states without any frequency entanglement between the signal and idler photons. Figure 22.10 gives the corresponding JSA; the circular nature of the plot indicates factorized signal–idler photon states. The corresponding joint spectral intensity (JSI) is given by

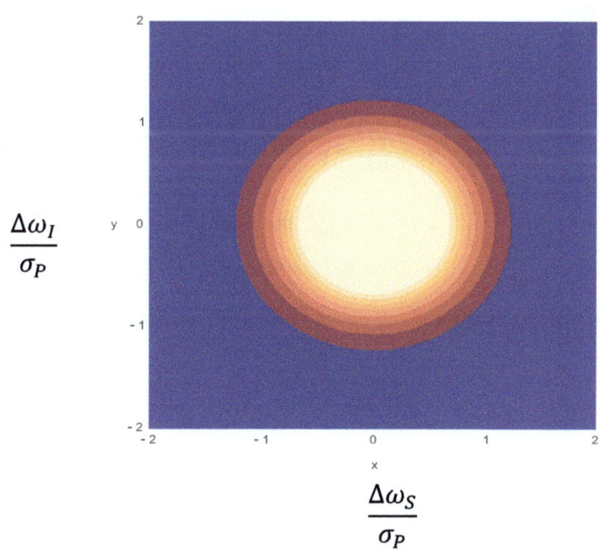

Fig. 22.10 JSA corresponding to Eq. (22.129) showing a factorized signal–idler photon state

$$g^2(\omega_s, \omega_i) = \alpha_0^2 e^{-\frac{4}{\sigma_p^2}(\omega_s - \omega_{s0})^2} e^{-\frac{4}{\sigma_p^2}(\omega_i - \omega_{i0})^2} \tag{22.130}$$

The condition given by Eq. (22.126) depends on the material used in the SPDC process as well as the wavelengths of the pump, signal, and idler and the waveguide characteristics.

To estimate an order of magnitude of the required pump pulse width, we assume $L = 1$ mm, $\Delta N_{ps} = 0.1$ giving us $\sigma_p \approx 1.3 \times 10^{13}$ rad/s.

Note from Eq. (22.128) that longer interaction lengths and larger differences in group indices require smaller pump bandwidths. Increased interaction length or increased value of group index difference, requires smaller bandwidth pulses to generate pure photon states.

22.8 Quantum Treatment of Parametric Amplifiers

In Chap. 13, we had discussed optical parametric amplification using classical equations. In this section, we will discuss the quantum treatment of optical parametric amplifiers.

We first discuss the quantum treatment of a degenerate parametric amplifier, and in Sect. 22.8.2 we will discuss a nondegenerate phase-insensitive and a phase-sensitive parametric amplifier.

22.8.1 Degenerate Parametric Amplifier

Let us first consider a degenerate parametric amplifier with the pump having a frequency 2ω and the signal with a frequency ω. The equations describing the input–output relation of the amplitude of the signal assuming no-pump depletion approximation and perfect phase matching (or quasi phase matching) are given by (see Chap. 13, Eq. 13.12)

$$\begin{aligned} A_s(z) &= A_{s0} e^{-i\phi_{s0}} \cosh \gamma z + i A_{s0} e^{i\phi_{s0}} e^{-i\phi_{p0}} \sinh \gamma z \\ &= A_s(z=0) \cosh \gamma z + i A_S^*(z=0) e^{-i\phi_{p0}} \sinh \gamma z \end{aligned} \tag{22.131}$$

where (see Eq. 13.10)

$$A_s(z=0) = A_{s0} e^{-i\phi_{s0}} \tag{22.132}$$

is the complex field amplitude of the signal at the input to the amplifier and γ is defined in Eq. (13.7).

In order to have a quantum description of the process, we replace the electric fields by annihilation and creation operators (see Chap. 19):

$$A_s(z=0) \to \widehat{a}_s \tag{22.133a}$$

$$A_s^*(z=0) \to \widehat{a}_s^\dagger \tag{22.133b}$$

$$A_s(z) \to \widehat{b}_s \tag{22.133c}$$

$$A_s^*(z) \to \widehat{b}_s^\dagger \tag{22.133d}$$

and obtain

$$\widehat{b}_s = \cosh \gamma z \, \widehat{a}_s + ie^{-i\phi_{p0}} \sinh \gamma z \, \widehat{a}_s^\dagger \tag{22.134}$$

Since $\left[\widehat{a}_s, \widehat{a}_s^\dagger\right] = 1$, the operators $\widehat{b}_s$ and $\widehat{b}_s^\dagger$ satisfy the necessary commutation relation:

$$\left[\widehat{b}_s, \widehat{b}_s^\dagger\right] = 1 \tag{22.135}$$

as it indeed should.

If the input state is $|n\rangle$, then the expectation value of the photon number in the input is given by

$$\langle N_{in} \rangle = \langle n | \widehat{a}_s^\dagger \widehat{a}_s | n \rangle = n \tag{22.136}$$

The expectation value of the photon number at the output of the parametric amplifier after a length of interaction L is given by (see Chap. 19)

$$\langle N_{out} \rangle = \langle n | \widehat{b}_s^\dagger \widehat{b}_s | n \rangle \tag{22.137}$$

which using Eqs. (22.134) becomes

$$\langle N_{out} \rangle = \langle n | \left(\widehat{a}_s^\dagger \cosh \gamma L - ie^{i\phi_{p0}} \widehat{a}_s \sinh \gamma L \right) \left(\widehat{a}_s \cosh \gamma L + ie^{-i\phi_{p0}} \widehat{a}_s^\dagger \sinh \gamma L \right) | n \rangle$$

Opening the brackets, we have

$$\begin{aligned} \langle N_{out} \rangle = \langle n | \Big(& \widehat{a}_s^\dagger \widehat{a}_s \cosh^2 \gamma L + ie^{-i\phi_{p0}} \widehat{a}_s^{\dagger 2} \cosh \gamma L \sinh \gamma L \\ & - ie^{i\phi_{p0}} \widehat{a}_s^{\,2} \sinh \gamma L \cosh \gamma L + \widehat{a}_s \widehat{a}_s^\dagger \sinh^2 \gamma L \Big) | n \rangle \end{aligned}$$

On simplification, we have

$$\langle N_{out} \rangle = n \cosh^2 \gamma L + \sinh^2 \gamma L \tag{22.138}$$

Thus, the parametric amplifier amplifies the signal by a factor $\cosh^2 \gamma L$ and also adds noise given by the term $\sinh^2 \gamma L$. This noise is due to spontaneous parametric emission into the signal mode. Note also that if there is no input at the signal, then $n = 0$, and the output in this case is nonzero and is given by

$$\langle N_{out} \rangle = \sinh^2 \gamma L \tag{22.139}$$

This output is spontaneous parametric down conversion into the mode at frequency ω. We shall now consider the quadrature operators. The input quadrature operators are given by

$$\widehat{X}_{s,in} = \frac{\widehat{a}_s + \widehat{a}_S^\dagger}{2} \tag{22.140a}$$

$$\widehat{Y}_{s,in} = \frac{\widehat{a}_s - \widehat{a}_S^\dagger}{2i} \tag{22.140b}$$

Here for brevity we are representing the quadrature operators by $\widehat{X}$ and $\widehat{Y}$ instead of by $\widehat{X}_1$ and $\widehat{X}_2$. Similarly, the output quadrature operators are given by

$$\widehat{X}_{s,out} = \frac{\widehat{b}_s + \widehat{b}_s^\dagger}{2} \tag{22.141a}$$

$$\widehat{Y}_{s,out} = \frac{\widehat{b}_s - \widehat{b}_s^\dagger}{2i} \tag{22.141b}$$

Substituting from Eq. (22.134) in Eq. (22.141a) and (22.141b), we find

$$\widehat{X}_{s,out} = \frac{\cosh \gamma L \widehat{a}_s + ie^{-i\phi_{p0}} \sinh \gamma L \widehat{a}_s^\dagger + \cosh \gamma L \widehat{a}_s^\dagger - ie^{i\phi_{p0}} \sinh \gamma L \widehat{a}_s}{2}$$

or

$$\widehat{X}_{s,out} = \frac{\widehat{a}_s \left(\cosh \gamma L - ie^{i\phi_{p0}} \sinh \gamma L \right) + \widehat{a}_s^\dagger \left(\cosh \gamma L + ie^{-i\phi_{p0}} \sinh \gamma L \right)}{2} \tag{22.142}$$

Similarly,

$$\widehat{Y}_{s,out} = \frac{\widehat{a}_s \left(\cosh \gamma L + ie^{i\phi_{p0}} \sinh \gamma L \right) - \widehat{a}_s^\dagger \left(\cosh \gamma L - ie^{-i\phi_{p0}} \sinh \gamma L \right)}{2i} \tag{22.143}$$

If we choose $\phi_{p0} = \frac{\pi}{2}$, we have

$$\widehat{X}_{s,out} = \frac{\widehat{a}_s(\cosh\gamma L + \sinh\gamma L) + \widehat{a}_s^\dagger(\cosh\gamma L + \sinh\gamma L)}{2} = e^{\gamma L}\widehat{X}_{s,in} \quad (22.144a)$$

$$\widehat{Y}_{s,out} = \frac{\widehat{a}_s(\cosh\gamma L - \sinh\gamma L) - \widehat{a}_s^\dagger(\cosh\gamma L - \sinh\gamma L)}{2i} = e^{-\gamma L}\widehat{Y}_{s,in} \quad (22.144b)$$

which shows that for this value of ϕ_{p0} the $\widehat{X}$ quadrature gets amplified by a factor $e^{\gamma L}$, while the $\widehat{Y}$ quadrature gets attenuated by the factor $e^{-\gamma L}$.

On the other hand if we choose $\phi_{p0} = -\frac{\pi}{2}$, then we have

$$\widehat{X}_{s,out} = e^{-\gamma L}\widehat{X}_{s,in} \quad (22.145)$$

and

$$\widehat{Y}_{s,out} = e^{+\gamma L}\widehat{Y}_{s,in} \quad (22.146)$$

implying amplification of the $\widehat{Y}$ quadrature and attenuation of the $\widehat{X}$ quadrature. This shows the phase-sensitive nature of the amplification process.

If we consider the case with $\phi_{p0} = \frac{\pi}{2}$, then we have $\widehat{b}_s = \cosh\gamma L\widehat{a}_s + \sinh\gamma L\widehat{a}_s^\dagger$ using which we obtain

$$\left\langle \widehat{X}_{s,out} \right\rangle = e^{\gamma L}\left\langle \widehat{X}_{s,in} \right\rangle \quad (22.147)$$

and

$$\left\langle \widehat{X}^2_{s,out} \right\rangle = e^{2\gamma L}\left\langle \widehat{X}^2_{s,in} \right\rangle \quad (22.148)$$

Hence, we have

$$\Delta X_{s,out} = \sqrt{\left\langle \widehat{X}^2_{s,out} \right\rangle - \left\langle \widehat{X}_{s,out} \right\rangle^2} = e^{\gamma L}\Delta X_{s,in} \quad (22.149)$$

Thus, the signal to noise ratio (SNR) of the output and input is related through the following relation:

$$\frac{\left\langle \widehat{X}_{s,out} \right\rangle}{\Delta X_{s,out}} = \frac{\left\langle \widehat{X}_{s,in} \right\rangle}{\Delta X_{s,in}} \quad (22.150)$$

Implying that the signal to noise ratio at the output is the same as the signal to noise ratio at the input. Thus, the degenerate parametric amplifier does not add any noise in the amplification process. The noise figure (NF) is the ratio of the input SNR to the output SNR:

$$NF = \frac{(SNR)_{in}}{(SNR)_{out}} \tag{22.151}$$

The NF of the degenerate parametric amplifier is thus unity or in logarithmic units, 0 dB. This is in contrast to standard phase-insensitive amplifiers which have a minimum NF of 3 dB for high gain (see Sect. 22.8.2). This noise characteristic of parametric amplifiers is very interesting for applications involving optical communication (see, e.g., Umeki et al. 2011).

22.8.2 Nondegenerate Phase-Insensitive Parametric Amplifier

We next consider a nondegenerate phase-insensitive parametric amplifier with a pump at frequency ω_p, a signal at frequency ω_s, and an idler at frequency ω_i with $\omega_p = \omega_s + \omega_i$. In Chap. 13, we had discussed parametric amplification and under no-pump depletion approximation and perfect phase matching or perfect quasi phase matching and had obtained the following relations between the input signal and idler and the corresponding outputs, assuming inputs at pump and signal frequencies only (see Eqs. 12.57 and 12.58):

$$A_s^*(z) = A_{s0}^* \cosh gz \tag{22.152}$$

$$A_i(z) = i\sqrt{\frac{\omega_i n_s}{\omega_s n_i}} A_{s0}^* \sinh gz \tag{22.153}$$

where g is defined in Eq. (13.28).

We recall Eqs. (13.44) and (13.45):

$$\tilde{A}_s(z) = \sqrt{\frac{n_s}{\omega_s}} A_s(z); \tilde{A}_i(z) = \sqrt{\frac{n_i}{\omega_i}} A_i(z) \tag{22.154}$$

Using these new variables, for a length of interaction equal to L, Eqs. (22.152) and (22.153) become

$$\tilde{A}_s^*(L) = \tilde{A}_{s0}^* \cosh gL \tag{22.155}$$

and

$$\tilde{A}_i(L) = i\tilde{A}_{i0}^* \sinh gL \tag{22.156}$$

In order to move to the quantum domain, we replace the amplitudes by the corresponding annihilation operators. Thus, we write Eq. (22.155) as

$$\widehat{b}_s^\dagger = \widehat{a}_s^\dagger \cosh gL \tag{22.157}$$

where, as before $\widehat{a}_s$ and $\widehat{b}_s$ correspond to annihilation operators of the input field and the output field, respectively.

Now, it is imperative that the annihilation and creation operators obey commutation relations. Thus, for the input we have

$$\left[\widehat{a}_s, \widehat{a}_s^\dagger\right] = 1 \tag{22.158}$$

On the other hand, we have

$$\left[\widehat{b}_s, \widehat{b}_s^\dagger\right] = \left[\widehat{a}_s \cosh gL, \widehat{a}_s^\dagger \cosh gL\right] = \cosh^2 gL\left[\widehat{a}_s, \widehat{a}_s^\dagger\right] = \cosh^2 gL \tag{22.159}$$

which violates the requirement of the commutation relation. In this context, it is interesting to recall our discussion on beam splitters (see Chap. 19) where a simple extension of the classical beam splitting equations gave us incorrect commutation relation since we were not including the vacuum noise entering from the other port of the beam splitter. Here again the issue is the same as in obtaining Eq. (22.157) for the signal evolution, we have not included the vacuum entering the idler port of the amplifier. This requires us to correct Eq. (22.157) by adding a term corresponding to the noise added by the idler port. Hence, we replace Eq. (22.157) by the following equation:

$$\widehat{b}_s^\dagger = \widehat{a}_s^\dagger \cosh gL + \alpha\widehat{c}_i \tag{22.160}$$

where $\widehat{c}_i$ is the annihilation operator corresponding to the idler port with vacuum input and α is a constant. The operators $\widehat{a}_s$ and $\widehat{c}_i$ commute with each other as they belong to two different modes.

Now we have

$$\begin{aligned}\left[\widehat{b}_s, \widehat{b}_s^\dagger\right] &= \left[\left(\widehat{a}_s \cosh gL + \alpha\widehat{c}_i^\dagger\right), \left(\widehat{a}_s^\dagger \cosh gL + \alpha\widehat{c}_i\right)\right] \\ &= \left[\widehat{a}_s, \widehat{a}_s^\dagger\right] \cosh^2 gL + \alpha^2\left[\widehat{c}_i^\dagger, \widehat{c}_i\right] = 1\end{aligned}$$

Using $\left[\widehat{a}_s, \widehat{a}_s^\dagger\right] = 1$, if $\left[\widehat{b}_s, \widehat{b}_s^\dagger\right] = 1$, then we have

$$\alpha^2\left[\widehat{c}_i, \widehat{c}_i^\dagger\right] = \left(\cosh^2 gL - 1\right)$$

If we assume $\left[\widehat{c}_i, \widehat{c}_i^\dagger\right] = 1$, then we have

$$\alpha = \sqrt{\left(\cosh^2 gL - 1\right)} = \sinh gL \tag{22.161}$$

We write Eq. (22.160) in one possible form given below:

$$\widehat{b}_s^\dagger = \widehat{a}_s^\dagger \cosh gL + \sinh gL\widehat{c}_i \tag{22.162}$$

It is easy to check that the above operator satisfies the required commutation relation.

Using the quadrature operators as given in Eqs. (22.140a), (22.140b) and (22.141a), (22.141b) we have

$$\widehat{X}_{s,out} = \frac{\cosh gL\widehat{a}_s + \sinh gL\widehat{c}_i^\dagger + \cosh gL\,\widehat{a}_s^\dagger + \sinh gL\widehat{c}_i}{2}$$

which can be written as

$$\widehat{X}_{s,out} = \frac{\cosh gL\left(\widehat{a}_s + \widehat{a}_s^\dagger\right) + \sinh gL\left(\widehat{c}_i + \widehat{c}_i^\dagger\right)}{2} = \cosh gL\,\widehat{X}_{s,in} + \sinh gL\widehat{X}_{c,in}$$

where $\widehat{X}_{c,in} = \frac{\widehat{c_i} + \widehat{c_i^\dagger}}{2}$. Hence,

$$\left\langle \widehat{X}_{s,out} \right\rangle = \cosh gL \left\langle \widehat{X}_{s,in} \right\rangle \tag{22.163}$$

where we have used the fact that $\left\langle \widehat{X}_{c,in} \right\rangle = \left\langle \frac{\left(\widehat{c_i} + \widehat{c_i^\dagger}\right)}{2} \right\rangle = 0$ since this represents vacuum noise. Thus, the amplitude gain of the amplifier is given by $\cosh\gamma L$ and the power gain will be $\cosh^2\gamma L$.

Similarly we have

$$\left\langle \widehat{X}_{s,out}^2 \right\rangle = \cosh^2 gL \left\langle \widehat{X}_{s,in}^2 \right\rangle + \sinh^2 gL \left\langle \widehat{X}_{c,in}^2 \right\rangle \tag{22.164}$$

where $\left\langle \widehat{X}_{c,in}^2 \right\rangle$ corresponds to contribution from the noise term from the idler port.

Hence, the variance in the output signal is given by

$$(\Delta X_{s,out})^2 = \left\langle \widehat{X}_{s,out}^2 \right\rangle - \left\langle \widehat{X}_{s,out} \right\rangle^2 = \cosh^2 gL(\Delta X_{s,in})^2 + \sinh^2 gL(\Delta X_{c,in})^2 \tag{22.165}$$

The first term on the right-hand side represents the amplification of the variance of the input signal and the second term is the noise in variance added by the vacuum input into the idler. Compared to phase-sensitive amplification, we note that there is an additional noise term coming from the unused input port (idler port) of the amplifier.

If the input corresponds to coherent state, then $\Delta X_{s,in} = \Delta X_{c,in}$ (see Chap. 15) and we have

$$(\Delta X_{s,out})^2 = (2G - 1)(\Delta X_{s,in})^2 \tag{22.166}$$

where $G = \cosh^2 gL$ represents the amplifier gain. Thus, the signal gets amplified by the factor G, while the noise gets amplified by $(2G - 1)$. Hence, the NF of the amplifier is given by

$$NF = \frac{(SNR)_{in}}{(SNR)_{out}} = \frac{(Signal)_{in}}{(Signal)_{out}}\frac{(Noise)_{out}}{(Noise)_{in}} = \frac{2G-1}{G} = 2 - \frac{1}{G} \tag{22.167}$$

Hence, for large gain $G \gg 1$, the factor for noise figure is equal to a factor of 2 or equivalently in logarithmic scale, 3 dB.

22.8.3 Nondegenerate Phase-Sensitive Parametric Amplifier

We now consider a nondegenerate phase-sensitive amplifier described by the following classical equations (see Chap. 13, Eqs. 13.48 and 13.49):

$$\tilde{A}_s(L) = \tilde{A}_{s0} \cosh gL + i\tilde{A}_{i0}^* e^{-i\phi_{p0}} \sinh gL \tag{22.168}$$

$$\tilde{A}_i^*(L) = -i\tilde{A}_{s0} e^{i\phi_{P0}} \sinh gL + \tilde{A}_{i0}^* \cosh gL \tag{22.169}$$

We replace and field amplitudes by the corresponding annihilation and creation operators of the signal and idler field:

$$\tilde{A}_{s0} \to \widehat{a}_s; \quad \tilde{A}_{s0}^* \to \widehat{a}_s^\dagger; \quad \tilde{A}_{i0} \to \widehat{a}_i; \quad \tilde{A}_{i0}^* \to \widehat{a}_i^\dagger \tag{22.170a}$$

$$\tilde{A}_s(L) \to \widehat{b}_s; \quad \tilde{A}_s^*(L) \to \widehat{b}_s^\dagger; \quad \tilde{A}_i(L) \to \widehat{b}_i; \quad \tilde{A}_i^*(L) \to \widehat{b}_i^\dagger \tag{22.170b}$$

and obtain

$$\widehat{b}_s = \widehat{a}_s \cosh gL + ie^{-i\phi_{p0}} \widehat{a}_i^\dagger \sinh gL \tag{22.171}$$

which gives us

$$\widehat{b}_s^\dagger = \widehat{a}_s^\dagger \cosh gL - ie^{i\phi_{p0}} \widehat{a}_i \sinh gL \tag{22.172}$$

Now the output operator $\widehat{b}_s$ satisfies the required commutation relation $\left[\widehat{b}_s, \widehat{b}_s^\dagger\right] = 1$. In this case unlike the phase-insensitive case with only signal input, there is an additional term coming from the idler port which ensures that the required commutation relations are satisfied.

We can similarly write for the output idler creation and annihilation operators

$$\widehat{b}_i^\dagger = -ie^{i\phi_{p0}}\widehat{a}_s \sinh gL + \widehat{a}_i^\dagger \cosh gL \tag{22.173}$$

and hence

$$\widehat{b}_i = ie^{-i\phi_{p0}}\widehat{a}_s^\dagger \sinh gL + \widehat{a}_i \cosh gL \tag{22.174}$$

These operators also satisfy the required commutation relation $\left[\widehat{b}_i, \widehat{b}_i^\dagger\right] = 1$.

Using the definition of the quadrature operator $\widehat{X}_{s,out}$, we have

$$\widehat{X}_{s,out} = \frac{\widehat{a}_s \cosh gL + ie^{-i\phi_{p0}}\widehat{a}_i^\dagger \sinh gL + \widehat{a}_s^\dagger \cosh gL - ie^{i\phi_{p0}}\widehat{a}_i \sinh gL}{2}$$

which is simplified to

$$\widehat{X}_{s,out} = \frac{\cosh gL\left(\widehat{a}_s + \widehat{a}_s^\dagger\right) - i\sinh gL\left(\widehat{a}_i e^{i\phi_{p0}} - \widehat{a}_i^\dagger e^{-i\phi_{p0}}\right)}{2} \tag{22.175}$$

Let us consider the case when $\phi_{p0} = \frac{\pi}{2}$. In such a case, we have

$$\widehat{X}_{s,out} = \frac{\cosh gL\left(\widehat{a}_s + \widehat{a}_s^\dagger\right) + \sinh gL\left(\widehat{a}_i + \widehat{a}_i^\dagger\right)}{2} = \cosh gL\widehat{X}_{s,in} + \sinh gL\widehat{X}_{i,in}$$

and we have

$$\left\langle \widehat{X}_{s,out}\right\rangle = \cosh gL\left\langle \widehat{X}_{s,in}\right\rangle + \sinh gL\left\langle \widehat{X}_{i,in}\right\rangle \tag{22.176}$$

Also

$$\left\langle \widehat{X}_{s,out}^2\right\rangle = \cosh^2 gL\left\langle \widehat{X}_{s,in}^2\right\rangle + \sinh^2 gL\left\langle \widehat{X}_{i,in}^2\right\rangle + 2\cosh gL \sinh gL\left\langle \widehat{X}_{s,in}\widehat{X}_{i,in}\right\rangle \tag{22.177}$$

Hence, the variance is given by

$$(\Delta X_{s,out})^2 = \left\langle \widehat{X}_{s,out}^2 \right\rangle - \left\langle \widehat{X}_{s,out} \right\rangle^2 = \cosh^2 gL (\Delta X_{s,in})^2 + \sinh^2 gL (\Delta X_{i,in})^2$$
$$+2\cosh gL \sinh gL \left(\left\langle \widehat{X}_{s,in} \widehat{X}_{i,in} \right\rangle - \left\langle \widehat{X}_{s,in} \right\rangle \left\langle \widehat{X}_{i,in} \right\rangle \right) \tag{22.178}$$

In Chap. 13, we discussed about a phase-matched nondegenerate parametric amplifier with the input polarization oriented at 45° to the optic axis resulting in one component (parallel or perpendicular) acting like the signal and the other component (perpendicular or parallel) acting like the idler. This situation leads to a phase-sensitive parametric amplifier.

Considering this case, we write

$$\widehat{X}_{i,in} = \widehat{X}_{s,in} \tag{22.179}$$

and we have

$$\left\langle \widehat{X}_{s,out} \right\rangle = (\cosh gL + \sinh gL) \left\langle \widehat{X}_{s,in} \right\rangle = e^{gL} \left\langle \widehat{X}_{s,in} \right\rangle \tag{22.180}$$

and

$$(\Delta X_{s,out})^2 = \left(\cosh^2 gL + \sinh^2 gL + 2\cosh gL \sinh gL\right) (\Delta X_{s,in})^2 = e^{2gL} (\Delta X_{s,in})^2 \tag{22.181}$$

Using Eqs. (22.180) and (22.181), we have the following relationship between the output and input signal to noise ratios:

$$\frac{\left\langle \widehat{X}_{s,out} \right\rangle^2}{(\Delta X_{s,out})^2} = \frac{\left\langle \widehat{X}_{s,in} \right\rangle^2}{(\Delta X_{s,in})^2} \tag{22.182}$$

Thus, the phase-sensitive nondegenerate parametric amplifier does not degrade the signal to noise ratio during amplification and its noise figure is 0 dB. Compare this with a phase-insensitive amplifier which has a NF of 3 dB for large gain. For a discussion on reduction of quantum noise in parametric amplifiers, see Levenson et al. (1993).

Exercise 22.13 In a degenerate parametric amplifier, instead of a vacuum state consider the incidence of a coherent state $|\psi_{in}\rangle = |\alpha\rangle$. Assume $\alpha = \alpha_0 \, e^{-i\phi}$ and obtain the expectation value of the number operator at the output of the amplifier.

Solution The expectation value of the output number operator will be

$$\langle\alpha|\widehat{N}_{out}|\alpha\rangle = \langle\alpha|\widehat{b}^{\dagger}\widehat{b}|\alpha\rangle = \langle\alpha|\left(\mu^{*}\widehat{a}^{\dagger} + \nu^{*}\widehat{a}\right)\left(\mu\widehat{a} + \nu\widehat{a}^{\dagger}\right)|\alpha\rangle$$

where $\mu = \cosh gL$ and $\nu = ie^{-i\phi_{p0}} \sinh gL$. Using the fact that $\widehat{a}|\alpha\rangle = \alpha|\alpha\rangle$ and $\langle\alpha|\widehat{a}^{\dagger} = \alpha^{*}\langle\alpha|$, we obtain

$$\langle\alpha|\widehat{N}_{out}|\alpha\rangle = \alpha_0^2\left(\cosh^2 gL + \sinh^2 gL + 2\cosh gL \sinh gL \sin\left(\phi_{p0} - 2\phi\right)\right) + \sinh^2 gL$$

The expectation value of the photon number at the input is α_0^2.

The expectation value of the output photon number depends on the phase of the input coherent state namely. Thus, if $\phi_{p0} - 2\phi = \frac{\pi}{2}$, then the output expectation value would be

$$\langle\alpha|\widehat{N}_{out}|\alpha\rangle = \alpha_0^2 e^{2\gamma L} + \sinh^2 gL$$

showing amplification with added noise.

For $\phi_{p0} - 2\phi = -\frac{\pi}{2}$,

$$\langle\alpha|\widehat{N}_{out}|\alpha\rangle = \alpha_0^2 e^{-2\gamma L} + \sinh^2 gL$$

showing attenuation with added noise.

Hence, the parametric amplification is phase-sensitive (see Chap. 13).

Problems

22.1. State whether the following statement is true: A single frequency pump undergoes SPDC. The emitted signal and idler photons will also be emitted at precisely defined frequencies determined by the phase matching conditions.

22.2. Consider the incidence of a pump wave with a frequency ω_p in lithium niobate and assume that the quasi phase matching condition for down conversion to a pair of signal and idler frequencies ω_s and $\omega_i = \omega_p - \omega_s$ is satisfied; all the modes are supposed be extraordinarily polarized. Show that if there is no input at the signal and idler frequencies, classical coupled mode equations do not predict the generation of the signal idler pair. Assume no pump depletion approximation.

22.3. Carry out the analytical calculations for the case when an ordinary polarization pump photon down converts to an ordinary polarization signal and an extraordinary polarization idler to obtain Eq. (22.58) and obtain expressions for the corresponding signal and idler bandwidths.

22.4. Consider a diffused channel waveguide in z-cut x-propagating lithium niobate in which SPDC is obtained with all interacting modes being z-polarized. The refractive index profile is given as (see Eq. 4.81)

$$n^2(y,z) = n_b^2 + 2n_b \Delta n_b e^{-y^2/w^2} e^{-z^2/h^2}; \quad z < 0$$
$$= n_c^2; \quad z > 0$$

The modal fields of the interacting modes to be fundamental modes and assumed to be described by the following equation (see Eq. 4.87):

$$E(y,z) = \sqrt{\frac{16\alpha_{yq}\alpha_{zq}}{\pi wh}} \alpha_{zq} \left(\frac{z}{h}\right) e^{-\alpha_{yq}^2 y^2/w^2} e^{-\alpha_{zq}^2 z^2/h^2}; \quad z < 0$$
$$= 0; \quad z > 0$$

with α_{yq}; $q = p, s, i$ and α_{zq}; $q = p, s, i$ representing the optimized variational parameters corresponding to the pump, signal, and idler, respectively.

(a) Substitute the expressions for the fields and show that the overlap integral given by Eq. (22.19) comes out to be

$$I = \frac{2}{\pi\sqrt{wh}} \frac{(\alpha_{zp}\alpha_{zs}\alpha_{zi})^{3/2}\sqrt{\alpha_{yp}\alpha_{ys}\alpha_{yi}}}{\sqrt{\alpha_{yp}^2 + \alpha_{ys}^2 + \alpha_{yi}^2}\left(\alpha_{zp}^2 + \alpha_{zs}^2 + \alpha_{zi}^2\right)^2}$$

(b) For a pump wavelength of 519 nm, a signal wavelength of 780 nm, obtain the idler wavelength.

(c) Given that $w = h = 10$ μm, $\Delta n_e = 0.0037$,0.0030 and 0.0025 at the pump, signal, and idler wavelengths, obtain the optimum values of α_{yq} and α_{zq} at the pump, signal, and idler wavelengths. Assume n_{be}(519 nm) = 2.2393; n_{be}(780 nm) = 2.1778 and n_{be}(1551 nm) = 2.1375. From these values obtain the value of I.

22.5. Consider Type 0 SPDC process in lithium niobate with a pump at a wavelength of 520 nm and an idler wavelength of 1550 nm. (a) Obtain the wavelength of the signal photon. (b) If the group extraordinary refractive indices at pump, signal, and idler are given by 2.4752, 2.2701 and 2.1823, respectively, and the interaction length is 10 mm obtain the bandwidth of the generated signal and idler photons. (c) What will be the required first order QPM period for efficient generation of the signal–idler pair? Given the refractive indices at the pump, signal and idler wavelengths to be 2.2388, 2.1775, and 2.1376, respectively.

References

Booth, M. C., M. Atature, G. Di Giuseppe, B. E. A. Saleh, A. V. Sergienko, and M. C. Teich. 2002. Counter propagating entangled photons from a waveguide with periodic nonlinearity. *Phys Rev A* 66:023815.

Branczyk, A. M., A. Fedrizzi, T. M. Stade, T. D. Ralph, and A. G. White. 2010. Engineered optical nonlinearity for quantum light sources. *Opt Express* 19:55–65.

Brecht, B., A. Eckstein, A. Christ, H. Suche, and C. Silberhorn. 2011. From quantum pulse gate to quantum pulse shaper—engineered frequency conversion in nonlinear optical waveguides. *N J Phys* 13:065029.

Christ, A., A. Eckstein, P. J. Mosley, and C. Silberhorn. 2009. Pure single photon generation by type I PDC with backward-wave amplification. *Opt Express* 17:3441–3446.

Fiorentino, M., S. M. Spillane, R. G. Beausoleil, T. D. Roberts, P. Battle, and M. W. Munro. 2007. Spontaneous parametric down-conversion in periodically poled KTP waveguides and bulk crystals. *Opt Express* 15:7479–7488.

Fraine, A., O. Minaeva, D. S. Simon, R. Egorov, and A. V. Sergienko. 2012. Broadband source of polarization entangled photons. *Opt Lett* 37:1910–1912.

Grice, W. P., A. B. U'Ren, and I. A. Walmsley. 2001. Eliminating frequency and space-time correlations in multiphoton states. *Phys Rev A* 64:063815.

Kashiwazaki, T., A. Inoue, and T. Umeki. 2022. Broadband continuous-wave optical quadrature squeezer for ultra-fast optical quantum computers. *NTT Tech Rev* 20:47–51.

Koch, K., E. C. Cheung, Gerald T. Moore, Stephen H. Chakmakjian, and J. M. Liu. 1995. Hot spots in parametric fluorescence with a pump beam of finite cross section. *IEEE J Quant Electron* 31: 769–781.

Kuo, P. S., D. V. Reddy, V. Verma, S. W. Nam, A. Zukauskas, and C. Canalias. 2023. Photon pair production and frequency translation using backward-wave spontaneous parametric downconversion. *Opt Quant* 1:43–48.

Laudenbach, F., H. Hübel, M. Hentschel, P. Walther, and A. Poppe. 2016. Modelling parametric down-conversion yielding spectrally pure photon states. *Opt Express* 24:2712–2727.

Levenson, J. A., I. Abram, Th. Riviera, and Ph. Grangier. 1993. Reduction of quantum noise in optical parametric amplification. *J Opt Soc Am B* 10:2233–2238.

Liu, Y.-C., D-. J. Guo, K-. Q. Ren, R. Yang, M. Shang, W. Zhou, X. Li, C-. W. Sun, P. Xu, Z. Xie, Y-. X. Gong, and S-. N. Zhu. 2021. Observation of frequency-uncorrelated photon pairs generated by counter-propagating spontaneous parametric down conversion. *Sci Rep* 11: 12628. https://doi.org/10.1038/s41598-021-92141-y.

Luo, K., V. Ansari, M. Massaro, M. Santandrea, C. Eigner, R. Ricken, H. Herrmann, and C. Silberhorn. 2020. Counter-propagating photon pair generation in a nonlinear waveguide. *Opt Express* 28:3215–3225.

Ming, Y., Z. J. Wu, G. Cui, A. Tan, and Y. Lu. 2014. Integrated source of tunable non maximally mode-entangled photons in a domain engineered lithium niobate waveguide. *Appl Phys Lett* 104:171110.

Park, T., H. Stokowsky, V. Ansari, S. Gyger, K. K. S. Multani, O. T. Celik, A. Y. Huwang, D. J. Dean, F. Mayor, T. P. McKenna, M. M. Fejer, and A. Safavi-Naeni. 2024. Single-mode squeezed light generation and tomography with an integrated optical parametric oscillator. *Sci Adv* 10 (11): eadl1814. https://doi.org/10.1126/sciadv.adl1814.

Schiff, L. 1968. *Quantum mechanics*. USA: McGraw Hill.

Shimizu, R., and K. Edamatsu. 2009. High-flux and broadband biphoton sources with controlled frequency entanglement. *Opt Express* 17:16385–16393.

Svozilík, J., M. Hendrych, A. S. Helmy, and J. P. Torres. 2011. Generation of paired photons in a quantum separable state in Bragg reflection waveguide. *Opt Express* 19:3115–3123.

Thyagarajan, K., R. Das, O. Alibart, M. de Micheli, D. B. Ostrowsky, and S. Tanzilli. 2008. Increased pump acceptance bandwidth in spontaneous parametric down conversion process using Bragg reflection waveguides. *Opt Express* 16:3577–3582.

U'Ren, A. B., C. Silberhorn, K. Banaszek, I. A. Walmsley, R. Erdmann, W. P. Grice, and M. G. Raymer. 2005. Generation of pure state single photon wavepackets by conditional preparation based on spontaneous parametric downconversion. *Laser Phys* 15:146–161.

Umeki, T., O. Tadanaga, A. Takada, and Masaki Asobe. 2011. Phase sensitive degenerate parametric amplification using directly-bonded PPLN ridge waveguides. *Opt Express* 19:6326–6332.

Wasilewski, W., P. Wasylczyk, P. Kolenderski, K. Banaszek, and C. Radzewicz. 2006. Joint spectrum of photon pairs measured by coincidence spectroscopy. *Opt Lett* 31:1130–1132.

Xin, C. J., J. Mishra, C. Chen, D. Zhu, A. Shams-Ansari, C. Langrock, N. Sinclair, F. N. C. Wong, M. M. Fejer, and M. Lončar. 2022. Spectrally separable photon-pair generation in dispersion engineered thin-film lithium niobate. *Opt Lett* 47:2830–2833.

Xinhui, L., S. Chang-Wei, X. Ping, X. Zhenda, G. Yan-Xiao, and Z. Shi-Ning. 2021. Observation of frequency uncorrelated photon pairs generated by counter propagating spontaneous parametric down conversion. *Sci Rep* 11:12628.

Chapter 23
Entangled Photon Pair Generation

Abstract This chapter presents different techniques for the generation of entangled photon states including polarization entangled states, spatial mode entangled states, frequency entangled states, and energy-time entangled states.

23.1 Introduction

In Chap. 22, we discussed the process of spontaneous parametric down conversion (SPDC) in which an incident pump photon at a frequency ω_p splits into a pair of daughter photons referred to as the signal and idler photons with frequencies ω_s and ω_i, respectively, such that $\omega_p = \omega_s + \omega_i$ through the nonlinearity in the medium. Efficient generation of the pair also requires one to satisfy the phase matching condition $\beta_p = \beta_s + \beta_i$, where β_m, $m = p, s, i$ correspond to the propagation constants of the mode of the waveguide at the pump, signal, and idler frequencies. Efficient generation is also possible by satisfying the quasi phase matching condition $\beta_p = \beta_s + \beta_i + K$ where K is the spatial frequency of the nonlinear grating in the waveguide. Since the propagation constants of the modes depend on the interacting frequencies, polarization states, and spatial modes which depend on the waveguide, the required spatial frequency for efficient SPDC is usually unique for a given set of interacting modes with specific polarization, spatial mode, and frequencies. Thus, usually the SPDC process takes place and generates a pair of photons occupying well-defined modes in terms of frequency, polarization, and spatial mode.

We have seen in Chap. 17 that entangled photon pairs occupy multiple modes and thus generating entangled photon pairs requires some special designs. In this chapter, we will consider a few techniques for the generation of entangled photon pairs in different degrees of freedom such as polarization, spatial mode, frequency, or time-bin. For a recent review, readers are referred to Chen et al. (2021).

K. Thyagarajan, *Integrated Quantum Photonics*, Graduate Texts in Physics,
https://doi.org/10.1007/978-3-031-85728-7_23

23.2 Polarization Entangled Photon Pairs

In Chap. 22, we discussed the SPDC process in $x-$propagating, $z-$cut lithium niobate waveguide with TE like mode (having horizontal polarization) corresponding to ordinary polarization and TM like mode (having vertical polarization) corresponding to extraordinary polarization. We discussed how a pump photon with horizontal polarization (or H polarization) generates a pair of signal and idler photons with the signal being V polarized (extraordinary polarization) and the idler being H polarized (horizontal polarization or ordinary polarization) if the QPM grating is chosen with a specific spatial frequency K_{eo} given by

$$\Delta\beta_{eo} = \beta_{po} - \beta_{se} - \beta_{io} - K_{eo} = 0 \tag{23.1}$$

where β_{po} represents the propagation constant of the ordinary polarization (horizontal polarization) at the pump frequency, β_{se} and β_{io} correspond, respectively, to the propagation constants of the signal having extraordinary polarization (vertical polarization) and idler having ordinary polarization (horizontal polarization).

For a pump photon with H polarization to generate a pair of signal and idler photons with signal being H polarized (ordinary polarization) and idler being V polarized (extraordinary polarization), we need a QPM grating with a spatial frequency K_{oe} given by

$$\Delta\beta_{oe} = \beta_{po} - \beta_{so} - \beta_{ie} - K_{oe} = 0 \tag{23.2}$$

where β_{so} and β_{ie} represent propagation constants of the ordinary (o) polarization at the signal frequency and extraordinary (e) polarization at the idler frequency. In general due to dispersion and anisotropy of the waveguide, the two spatial frequencies K_{eo} and K_{oe} are different (see Exercise 23.1) and depending on the spatial frequency of the QPM grating, either of the above two processes takes place. In such a situation, the output state is proportional to (see Chap. 22)

$$\int C_{eo}|1_{se}, 1_{io}\rangle d\omega_s$$

if the spatial frequency of the QPM grating is K_{eo} and is proportional to

$$\int C_{oe}|1_{so}, 1_{ie}\rangle d\omega_s$$

if the spatial frequency of the QPM grating is K_{oe}. Here $|1_{se}, 1_{io}\rangle$ corresponds to a state with one signal photon having extraordinary polarization and one idler photon having ordinary polarization. Similarly the state $|1_{so}, 1_{ie}\rangle$ corresponds to one signal photon having ordinary polarization and one idler photon having extraordinary polarization.

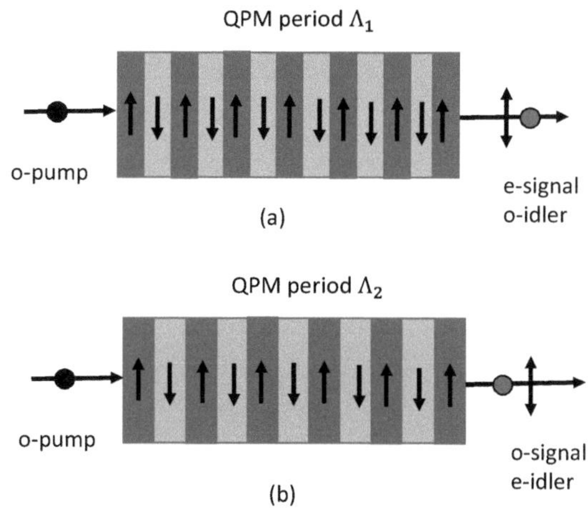

Fig. 23.1 A pump with ordinary polarization (H-polarized) generates (**a**) an extraordinary signal (V-polarized) and an ordinary idler (H-polarized) if the QPM grating satisfies Eq. (23.1) and (**b**) an ordinary signal (H-polarized) and an extraordinary idler (V-polarized) if the QPM grating satisfies Eq. (23.2)

In either case the polarization states of the signal and idler photons are well defined and any measurement of the polarization state of the signal or idler will yield a definite value. Thus, such a state is not a polarization entangled state.

Figure 23.1 shows schematics of the two processes.

Exercise 23.1 Consider SPDC of a pump photon having ordinary polarization at a wavelength of 800 nm in to (a) a signal photon having extraordinary polarization at a wavelength of 1550 nm and an idler photon having ordinary polarization and (b) a signal photon having ordinary polarization at the wavelength of 1550 nm and an idler photon having extraordinary polarization in bulk lithium niobate crystal. Obtain the spatial periods of the QPM process required for efficient SPDC the two processes.

Solution For a pump photon at 800 nm and a signal photon at 1550 nm, the idler wavelength is 1653.3 nm. The refractive indices of lithium niobate at these wavelengths are given by

$$n_o(\lambda_p) = 2.2553; \quad n_o(\lambda_s) = 2.2111; \quad n_o(\lambda_i) = 2.2077$$
$$n_e(\lambda_s) = 2.1376; \quad n_e(\lambda_i) = 2.1346$$

The required periods for the two processes are given by

$$\Lambda_{eo} = \left(\frac{n_o(\lambda_p)}{\lambda_p} - \frac{n_e(\lambda_s)}{\lambda_s} - \frac{n_o(\lambda_i)}{\lambda_i}\right)^{-1} \approx 9.55\ \mu\text{m}$$

$$\Lambda_{oe} = \left(\frac{n_o(\lambda_p)}{\lambda_p} - \frac{n_o(\lambda_s)}{\lambda_s} - \frac{n_e(\lambda_i)}{\lambda_i}\right)^{-1} \approx 9.85\ \mu\text{m}$$

Note that the two periods are different.

A polarization entangled state would correspond to (see Chap. 17)

$$|\psi\rangle \sim \int (C_{eo}|1_{se}, 1_{io}\rangle + C_{oe}|1_{so}, 1_{ie}\rangle)d\omega_s \tag{23.3}$$

For maximal entanglement, we must have $C_{eo} = C_{oe}$.

In the state given by Eq. (23.3), the polarization states of the signal and idler are not defined but as discussed in Chap. 17, the signal and idler photon polarization states are always orthogonal to each other no matter which polarization basis is chosen.

In order to directly generate photons in the state defined by Eq. (23.3), we must ensure that the QPM nonlinear grating contains both the spatial frequencies K_{eo} and K_{oe} so that both the above mentioned SPDC processes will take place with equal efficiency. This can be achieved by designing a substate which is poled with dual periodicities.

The spatial variation in the nonlinear coefficient d along the propagation direction decides the spatial frequencies available for satisfying the quasi phase matching condition (see Chap. 11). It is indeed possible to have both QPM conditions given by Eqs. (23.1) and (23.2) to be satisfied simultaneously by noting first that if a periodic function with fundamental spatial frequency K_0 is amplitude or phase modulated by another periodic function with spatial frequency K_p ($<K_0$), then the modulated function would have spatial frequency components at $nK_0 + mK_p$ where $n = \pm 1, \pm 2, \ldots$ and $m = \pm 1, \pm 2, \ldots$ etc. Since the amplitude of the resulting Fourier coefficients at various frequencies determine the strength of the nonlinear interaction, for significant efficiency we must choose lower order terms with small values of n and m.

As discussed in Chap. 11, periodic poling in the crystal leads to alternating regions of positive or negative value of the appropriate nonlinear coefficient. Figure 23.2 shows a schematic of a domain engineered crystal with multiple periodicities.

For specificity, we consider the pump to be polarized along the y—direction and the signal and idler to have polarizations along the y—and z—directions which lead to the nonlinear coefficient d_{24} of lithium niobate.

Recalling our discussion in Chap. 11, we write for the x—dependence of the nonlinear coefficient (propagation direction is along the x—direction) for a dual periodic QPM grating as

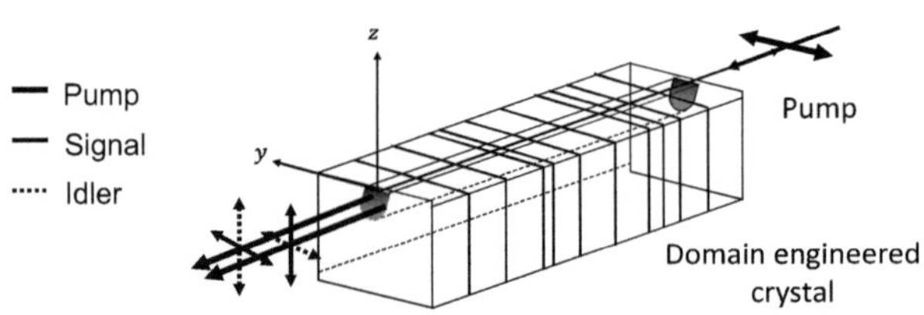

Fig. 23.2 A domain engineered crystal (with multiple periodicities) can be designed to satisfy both the QPM conditions given by Eqs. (23.1) and (23.2) and the output will correspond to an entangled state

$$\overline{d} = d_{24} f_1(x) f_2(x) \tag{23.4}$$

where

$$\begin{aligned} f_1(x) &= +1; \quad 0 < x < \frac{\Lambda_1}{2} \\ &= -1; \quad \frac{\Lambda_1}{2} < x < \Lambda_1 \end{aligned} \tag{23.5}$$

and

$$f_1(x + \Lambda_1) = f_1(x) \tag{23.6}$$

Similarly

$$\begin{aligned} f_2(x) &= +1; \quad 0 < x < \frac{\Lambda_2}{2} \\ &= -1; \quad \frac{\Lambda_2}{2} < x < \Lambda_2 \end{aligned} \tag{23.7}$$

and

$$f_2(x + \Lambda_2) = f_2(x) \tag{23.8}$$

Here Λ_1 and Λ_2 are the spatial periods of the two periodic variations. We are assuming the two periodic variations to have 50% duty cycles.

Fourier expansion of each of the functions $f_1(x)$ and $f_2(x)$ gives us an infinite series with the first two terms written down explicitly (see Chap. 11):

$$f_1(x) = i\frac{2}{\pi} e^{-iK_1 x} - i\frac{2}{\pi} e^{iK_1 x} + \ldots \tag{23.9}$$

where $K_1 = \frac{2\pi}{\Lambda_1}$ is the fundamental spatial frequency of $f_1(x)$.

Similarly

$$f_2(x) = i\frac{2}{\pi} e^{-iK_2 x} - i\frac{2}{\pi} e^{iK_2 x} + \ldots \tag{23.10}$$

where $K_2 = \frac{2\pi}{\Lambda_2}$. Thus, we have

$$\overline{d} = d_{24} f_1(x) f_2(x) = \frac{4d_{24}}{\pi^2} \left(e^{iK_{oe}x} + e^{-iK_{oe}x} - e^{iK_{eo}x} - e^{-iK_{eo}x} + \ldots \right) \tag{23.11}$$

where we have assumed K_1 and K_2 such that

$$K_{oe} = K_1 - K_2 \quad (23.12)$$

$$K_{eo} = K_1 + K_2 \quad (23.13)$$

By choosing appropriate values of K_1 and K_2, the nonlinear coefficient will have terms that will satisfy the QPM conditions for both the two SPDC processes of converting of pump photon with ordinary polarization to (a) signal photon with ordinary polarization and idler photon with extraordinary polarization and (b) signal photon with extraordinary polarization and idler photon with ordinary polarization. In such a case, the output state will then be proportional to

$$\int (C_{eo}|1_{se}, 1_{io}\rangle + C_{oe}|1_{so}, 1_{ie}\rangle) d\omega_s$$

which corresponds to polarization entangled photon state.

Figure 23.3 shows a schematic of the resulting dual periodically poled lithium niobate crystal.

In the case of a waveguide, due to both material and waveguide dispersions, the spectra of the signal photons of the two down conversion processes can have very different bandwidths. As an example considering a lithium niobate waveguide, corresponding to a pump wavelength of 519 nm and signal wavelength of 780 nm with a corresponding idler wavelength, SPDC leading to ordinary signal and extraordinary idler gives a bandwidth of 0.29 nm, while the one leading to extraordinary signal and ordinary idler gives a bandwidth of 6.35 nm (Thyagarajan et al. 2009).

Since the two bandwidths are different, in order to obtain a maximally entangled state one can use a narrow bandpass filter to select only the region where the amplitudes of the two spectra are identical. The use of a filter having a bandpass shorter than the shortest bandwidth can erase the bandwidth distinguishability and let

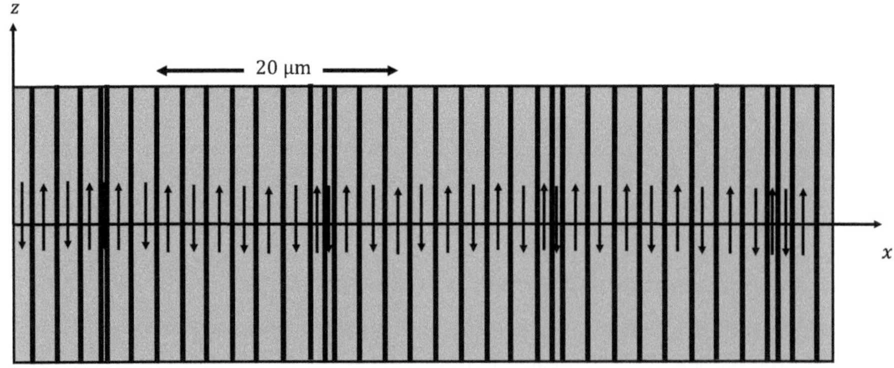

Fig. 23.3 A domain engineered lithium niobate crystal to satisfy QPM conditions for two SPDC processes simultaneously

the coefficients C_{eo} and C_{oe} be the only relevant parameters concerning entanglement.

Note that these two coefficients depend on the overlap integral between the interacting modes. Thus, by appropriate waveguide design, it is possible to ensure equality of the two coefficients leading to maximally polarization entangled state.

It is interesting to note that with dual periodic poling, there are two indistinguishable possible paths for the SPDC process and thus from the principles of quantum mechanics the output will be a superposition of the two possible output states. This leads to polarization entanglement at the output of the device.

Experimental results generating polarization entangled photon pairs using double period quasi phase matching have been reported in multiple references: e.g., Herrmann et al. (2013), Ueno et al. (2012), and Suhara et al. (2009).

Instead of the above technique of dual periodic poling, it is also possible to devise domain engineered substrates using a phase modulated QPM grating with a single period which through Fourier transform provides us with multiple periods for QPM for both polarization combination of signal and idler pairs. This again can lead to the generation of polarization entangled photon pairs (Asobe et al. 2003).

There are other techniques for the generation of polarization entangled photon pairs. For example, Fig. 23.4 shows such a device in which there are two equal length sections of domain reversed titanium in-diffused lithium niobate waveguides with two different spatial frequencies (Suhara et al. 2009; Ueno et al. 2012). The two spatial frequencies correspond to the generation of H polarized signal at wavelength λ_s and V polarized idler at a wavelength λ_i or V polarized signal at wavelength λ_s and H polarized idler at the wavelength λ_i from a single pump wavelength. This results in the generation of polarization entangled signal–idler photon pairs. Experiments reported in Suhara et al. (2009) correspond to a pump wavelength of 773.49 nm and signal wavelength of 1546 nm and idler wavelength of 1551 nm which are obtained by using wavelength filters at the output of the device. Similar results can be obtained by choosing the same domain reversal period in both the sections but modifying the effective indices of one of the sections by using a suitable superstrate on top of the waveguide. Nonlinear directional couplers have also been used to generate spatially entangled photon pairs (Setzpfandt et al. 2016). See Suhara (2009) for a review of QPM based waveguide nonlinear optics for the generation of entangled photon pairs.

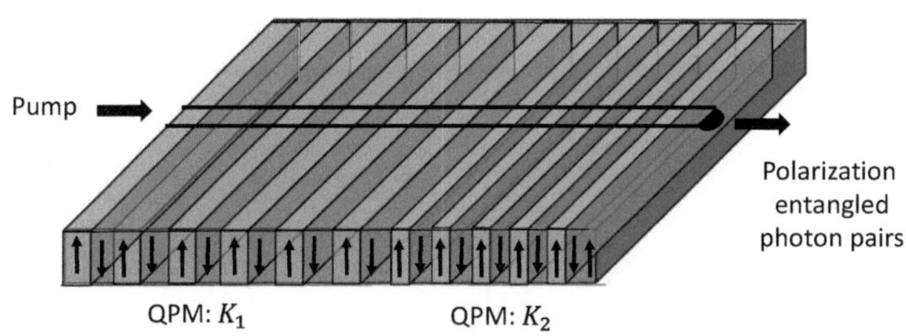

Fig. 23.4 Domain reversed crystal having two different regions for the generation of polarization entangled photon pairs. (Adapted from Suhara et al. (2009))

It is also possible to generate polarization entangled photon pairs using a *single* QPM period provided we satisfy certain conditions. For example, consider a Type II SPDC process with the pump frequency to be ω_p, the signal frequency to be ω_s, and the idler frequency to be ω_i with the input pump having ordinary polarization, then two spatial frequencies K_{eo} and K_{oe} (see Eqs. 23.1 and 23.2) will be required for the generation of the polarization entangled state. If we require both the above processes to be satisfied using a single QPM grating, then we must have $K_{oe} = K_{eo}$. Using Eqs. (23.1) and (23.2), we obtain

$$\beta_{se} + \beta_{io} = \beta_{so} + \beta_{ie}$$

which gives us the following condition:

$$\omega_s(n_{se} - n_{so}) = \omega_i(n_{ie} - n_{io}) = (\omega_p - \omega_s)(n_{ie} - n_{io})$$

where n_{se}, n_{so}, n_{ie}, and n_{io} are the effective indices of the modes with extraordinary and ordinary polarization at the signal and idler frequencies, respectively. Such a device generates the following polarization entangled state:

$$|\psi\rangle = \frac{1}{\sqrt{2}}\left(|H_s, V_i\rangle + e^{i\theta}|V_s, H_i\rangle\right)$$

where θ represents the phase difference between the two quantum states. Here $|H_s, V_i\rangle = |1_{so}, 1_{ie}\rangle$ and $|V_s, H_i\rangle = |1_{se}, 1_{io}\rangle$.

Depending on the geometry of the SPDC source generating entangled photon pairs, the two processes generating the (H, V) pair and the (V, H) pair may be distinguishable at the output due to their different delays in propagating from being generated to being detected. For them to be entangled, this distinguishability needs to be erased. This is usually carried out by using a compensator which cancels out this information (Shih 2003; Suhara 2009).

The ability to satisfy this condition depends on the nonlinear crystal, the waveguide, the pump wavelength, and the signal–idler pair of wavelengths. Using the Sellmeier equations for the refractive indices of the crystal, it is possible to verify the possibility of satisfying such a condition. Interestingly such a possibility exists with certain periodically poled crystals such as stoichiometric lithium tantalate crystals (PPSLT) (Hojo and Tanaka 2024). See also Exercise 23.2

Waveguides fabricated in lithium niobate on insulator (LNOI) have also been used to generate high-quality entangled photon pairs (see e.g., Zhao et al. 2020; Ebers et al. 2022).

Exercise 23.2 Consider Type II SPDC process in lithium tantalate ($LiTaO_3$) which is a positive uniaxial crystal with a pump wavelength of 475 nm and a signal wavelength of 725 nm and idler wavelength of 1377.5 nm. Given the following refractive index values of lithium tantalate: $n_{po} = 2.229$, $n_{so} = 2.1623$,

$n_{io} = 2.1244$; $n_{pe} = 2.2336$, $n_{se} = 2.1666$, $n_{ie} = 2.1281$, obtain the QPM periods required for the two possible Type-II down conversion processes.

Solution For Type II process, the input pump has ordinary polarization and the signal–idler pair has orthogonal polarizations.

Hence, for ordinary signal and extraordinary idler combination, the required QPM period would be

$$\frac{1}{\Lambda_1} = \frac{n_{po}}{\lambda_p} - \frac{n_{so}}{\lambda_s} - \frac{n_{ie}}{\lambda_i} = \frac{2.229}{0.475} - \frac{2.1623}{0.725} - \frac{2.1281}{1.3775}$$

which gives $\Lambda_1 \approx 6.05$ μm.

For the case of extraordinary signal and ordinary idler combination, the required QPM period would be

$$\frac{1}{\Lambda_2} = \frac{n_{po}}{\lambda_p} - \frac{n_{se}}{\lambda_s} - \frac{n_{io}}{\lambda_i} = \frac{2.229}{0.475} - \frac{2.1666}{0.725} - \frac{2.1244}{1.3775}$$

which gives $\Lambda_2 \approx 6.17$ μm. Note that the required periods are close but different for the two cases.

It has been shown that it is possible to obtain the same QPM period for both the above processes by an appropriate choice of the pump wavelength and the operating temperature of the device on which the principal refractive indices of the crystal depends. Thus, under this condition with a single grating period, one can generate polarization entangled pair of photons (Hojo and Tanaka 2024).

Exercise 23.3 Consider SPDC in a buried step index planar waveguide in *Y*—cut, *X*—propagating lithium niobate substrate with a waveguide width of 12 μm and $\Delta n_e = 0.003$. The extraordinary refractive indices of the substrate at the pump wavelength of 775 nm, the signal wavelength of 1506 nm, and the idler wavelength of 1596.6 nm are $n_{be}(\lambda_p) = 2.1784$, $n_{be}(\lambda_s) = 2.1388$, and $n_{be}(\lambda_i) = 2.1362$. The subscript b corresponds to bulk.

(a) Obtain the effective indices of the TE_0 and TE_1 modes of the waveguide at the three wavelengths.
(b) What is the spatial period required for first order QPM for conversion from the TE_0 mode at pump to the TE_0 mode at signal and idler?
(c) Obtain the corresponding spatial period required for first order QPM from the TE_0 mode at pump to the TE_1 mode at signal and TE_1 mode at idler.
(d) Will these processes take place even if QPM condition is satisfied? What about the process $TE_0 \rightarrow TE_0 + TE_1$ or the processes $TE_1 \rightarrow TE_0 + TE_1$; $TE_1 \rightarrow TE_1 + TE_0$? What would be the corresponding QPM periods required?

Solution

(a) The effective indices of the TE_0 modes at pump, signal, and idler can be obtained by using the method discussed in Chap. 4 by solving appropriate eigenvalue equations and are given by

$$n_{\text{eff}}(\lambda_p, \text{TE}_0) = 2.18126$$

$$n_{\text{eff}}(\lambda_p, \text{TE}_1) = 2.18084$$

$$n_{\text{eff}}(\lambda_s, \text{TE}_0) = 2.14124$$

$$n_{\text{eff}}(\lambda_s, \text{TE}_1) = 2.13993$$

$$n_{\text{eff}}(\lambda_i, \text{TE}_0) = 2.13867$$

$$n_{\text{eff}}(\lambda_i, \text{TE}_1) = 2.1372$$

(b) The required QPM period for $\text{TE}_0 \rightarrow \text{TE}_0 + \text{TE}_0$ process is given by

$$\frac{1}{\Lambda_1} = \frac{n_{\text{eff}}(\lambda_p, \text{TE}_0)}{\lambda_p} - \frac{n_{\text{eff}}(\lambda_s, \text{TE}_0)}{\lambda_s} - \frac{n_{\text{eff}}(\lambda_i, \text{TE}_0)}{\lambda_i}$$

which gives $\Lambda_1 \approx 18.79$ μm.

(c) The required QPM period for $\text{TE}_0 \rightarrow \text{TE}_1 + \text{TE}_1$ process is given by

$$\frac{1}{\Lambda_2} = \frac{n_{\text{eff}}(\lambda_p, \text{TE}_0)}{\lambda_p} - \frac{n_{\text{eff}}(\lambda_s, \text{TE}_1)}{\lambda_s} - \frac{n_{\text{eff}}(\lambda_i, \text{TE}_1)}{\lambda_i}$$

which gives $\Lambda_2 \approx 18.18$ μm.

(d) SPDC process can generate photon pairs only if the following integral is nonzero (see Chap. 22):

$$I_{eo} = \iint e_{po}(y,z)e_{se}(y,z)e_{io}(y,z)dydz$$

which is an overlap of the modal fields of the pump, signal, and idler. The given waveguide is a symmetric waveguide and hence the TE_0 mode will be symmetric and the TE_1 mode will be antisymmetric.

In the case of $\text{TE}_0 \rightarrow \text{TE}_0 + \text{TE}_0$ process, all the fields are symmetric and the integral is nonzero.

In the case of for $\text{TE}_0 \rightarrow \text{TE}_1 + \text{TE}_1$ process, the pump field is symmetric, while the signal and idler fields are antisymmetric which imply that the integrand is a symmetric function and hence the integral is nonzero. Hence, this process is also possible.

The process $TE_0 \rightarrow TE_0 + TE_1$ implies that the integrand is antisymmetric and hence the integral vanishes implying that even if the QPM condition is satisfied, the photon pair generation cannot take place.

The process $TE_1 \rightarrow TE_0 + TE_1$ can take place since in this case the integrand is symmetric and the integral is nonvanishing. The corresponding QPM period can be calculated.

$$\frac{1}{\Lambda_3} = \frac{n_{\text{eff}}(\lambda_p, \text{TE}_1)}{\lambda_p} - \frac{n_{\text{eff}}(\lambda_s, \text{TE}_0)}{\lambda_s} - \frac{n_{\text{eff}}(\lambda_i, \text{TE}_1)}{\lambda_i}$$

which on substitution of the values gives us $\Lambda_3 \approx 18.47$ μm.

The process $TE_1 \rightarrow TE_1 + TE_0$ can take place since in this case the integrand is symmetric and the integral is nonvanishing. The corresponding QPM period can be calculated:

$$\frac{1}{\Lambda_4} = \frac{n_{\text{eff}}(\lambda_p, \text{TE}_1)}{\lambda_p} - \frac{n_{\text{eff}}(\lambda_s, \text{TE}_1)}{\lambda_s} - \frac{n_{\text{eff}}(\lambda_i, \text{TE}_0)}{\lambda_i}$$

which on substitution of the values gives us $\Lambda_4 \approx 18.49$ μm.

The calculations imply that the two processes $TE_1 \rightarrow TE_0 + TE_1$ and $TE_1 \rightarrow TE_1 + TE_0$ require almost the same QPM period and hence if the waveguide has a domain reversal at this period, then both these processes can take place and will be able to generate a spatial mode entangled state (see Sect. 23.3).

23.3 Spatial Mode Entangled States

Apart from polarization degree of freedom, spatial modes also form a degree of freedom of photons propagating in a waveguide. Thus, it is also possible to have spatial mode entangled photon pairs generated by nonlinear interactions in waveguide supporting more than one mode (see e.g., Lugani et al. 2013; Ming et al. 2014; Feng et al. 2019; Saleh et al. 2009, 2010). Photons can also be entangled in multiple degrees of freedom leading to what is referred to as hyper-entanglement (see e.g., Barreiro et al. 2005; Bharadwaj and Thyagarajan 2018).

We first consider a geometry to generate spatial mode entangled photon pairs using a single QPM grating (Lugani et al. 2013).

Figure 23.5 shows the device similar to a Mach-Zehnder interferometer except for the output waveguide being a two-mode waveguide rather than a single-mode waveguide as in the conventional Mach-Zehnder interferometer (see Chap. 7). In the device, a single-mode waveguide splits into two single-mode waveguides using an adiabatic Y-splitter and after a certain length the two single-mode waveguides combine through a Y-combiner into a two-mode waveguide supporting both the fundamental symmetric and the first excited anti symmetric mode.

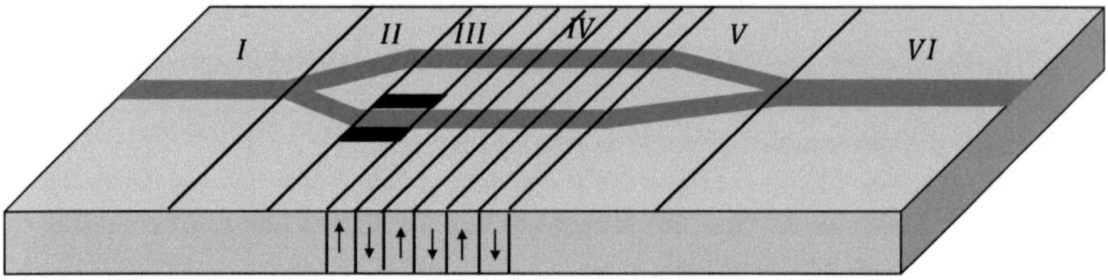

Fig. 23.5 Modified Mach-Zehnder interferometer with the output waveguide being a two-mode waveguide. The structure leads to spatial mode entanglement and the electro optic device can be used to switch between two Bell states

It is important to note the following:

1. Since the Y-splitter is a symmetric splitter and the input waveguide is a single-mode waveguide, the input field excites the symmetric mode of the two waveguide region of the interferometer (see Chap. 7). Since the two waveguides in region IV are assumed to be identical and far apart without any evanescent field interaction (i.e., coupling coefficient $\kappa = 0$), the propagation constants of the symmetric fundamental and antisymmetric first excited mode of the two waveguide region will be equal to the propagation constant of the fundamental mode of individual waveguides (see Chap. 5).
2. Since the output waveguide after the Y-combiner is a two-mode device and the two waveguide region supports two modes, the power in the fundamental mode in the two waveguide region (region IV) will propagate as the fundamental mode in the output waveguide and the power in the first excited mode in the two waveguide region (region IV) will propagate as the first excited mode of the two-mode region. If the output waveguide is a single-mode waveguide as in conventional Mach-Zehnder interferometer, then the power in the first excited mode in the two waveguide region (region IV) will be radiated into the substrate and will not exit the output waveguide as a guided mode (see Chap. 7).
3. In the device, the region IV with the two noninteracting waveguides is periodically poled with a spatial frequency K. In the two-mode region (Region IV), we assume that the propagation constant of the fundamental mode at the pump frequency is β_{ps}, that of the fundamental and first excited modes at the signal frequency are given by β_{ss} and β_{sa} and that of the fundamental and first excited mode at the idler frequency are given by β_{is} and β_{ia},respectively. The first subscript corresponds to pump (p), signal (s), and idler (i) frequencies and the second subscript to symmetric (s) and antisymmetric (a) modes. Note that since the two waveguides in the two waveguide region are far apart, we have $\beta_{ps} = \beta_{pa}$, $\beta_{ss} = \beta_{sa}$ and $\beta_{is} = \beta_{ia}$ (see Chap. 5).

   The spatial frequency of the QPM grating is chosen to satisfy the following equation:

$$K = \beta_{ps} - \beta_{ss} - \beta_{is} = \beta_{ps} - \beta_{sa} - \beta_{ia} \tag{23.14}$$

4. With the appropriate QPM grating given by Eq. (23.14) and polarization states to use an effective nonlinear coefficient, an incident pump photon at frequency ω_p in the symmetric fundamental mode can undergo SPDC process and generate signal–idler photon pairs both in the fundamental symmetric mode or signal–idler photon pairs both in the first excited antisymmetric mode. It may be pointed out here that the symmetric pump mode can only down convert to a signal–idler pair with either both the photons in the symmetric spatial mode or both in the antisymmetric spatial mode; due to parity nonconservation, it cannot generate a symmetric (0) and antisymmetric (1) combination of signal and idler photon pairs as the overlap integral will vanish (see Exercise 23.2). This follows from the fact that the efficiency of the down conversion process will depend on the following overlap integral (see Chap. 22):

$$I = \iint e_p(y,z)e_s(y,z)e_i(y,z)dydz$$

with $e_p(y, z)$, $e_s(y, z)$ and $e_i(y, z)$ representing the transverse electric fields of the interacting modes at the pump, signal, and idler frequencies, respectively. Since the pump field $e_p(y, z)$ is a symmetric function of the $y-$ coordinate (parallel to the substrate surface), for a finite value of the integral, the product $e_s(y, z)e_i(y, z)$ of the transverse fields corresponding to the signal and idler fields must also be symmetric in $y-$ coordinate. Note that both the signal field $e_s(y, z)$ and idler field $e_i(y, z)$ are either symmetric or antisymmetric functions of y due to the geometry of the waveguide device we are considering. This essentially rules out the generation of pairs of signal and idler which do not occupy the same transverse spatial mode.

5. Again since both the SPDC processes are allowed in the domain reversed region and hence the output from the device will be a spatial mode entangled state given by

$$|\psi\rangle \sim (|1_{ss}, 1_{is}\rangle + |1_{sa}, 1_{ia}\rangle) \tag{23.15}$$

where, as before, the subscripts ss, is, sa, ia correspond to symmetric modes at signal and idler and antisymmetric modes at signal and idler frequencies, respectively. The state represented by Eq. (23.15) corresponds to a spatial mode entangled state with both signal and idler occupying the fundamental mode or both signal and idler occupying the first excited mode.

6. The two waveguides are recombined into a two-mode waveguide and if the transition is adiabatic then the photons generated in the fundamental mode will propagate in the fundamental mode of the two-mode waveguide and the photons generated in the first excited mode will propagate as the first excited mode in the two-mode waveguide region. This leads to spatial mode entangled states in the output of the device (see Fig. 23.6).

7. It is interesting to note that if a phase difference of π is introduced between the pump fields in the two arms of the Mach-Zehnder interferometer through the

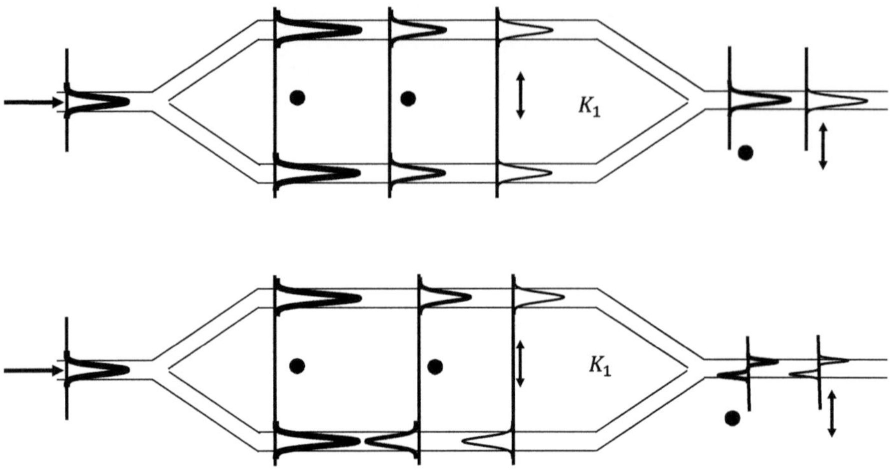

Fig. 23.6 Generating spatial mode entangled state (see Eq. 23.15) using a modified waveguide Mach-Zehnder interferometer. The signal and idler have orthogonal polarization states

electro optic effect (see Chap. 6) before entering the periodically poled region, then after passing through the electro optic modulator, the pump mode would correspond to the first excited mode. Again due to parity conservation the SPDC process would ensure the generation of the signal–idler pair in the following two combinations: (a) symmetric signal-antisymmetric idler and (b) antisymmetric signal and symmetric idler. These processes will require the following spatial frequencies for efficient generation:

$$K' = \beta_{pa} - \beta_{ss} - \beta_{ia} = \beta_{ps} - \beta_{ss} - \beta_{is} = K$$

and

$$K'' = \beta_{pa} - \beta_{sa} - \beta_{is} = \beta_{ps} - \beta_{ss} - \beta_{is} = K$$

Since the required spatial frequencies are the same as in the earlier down conversion processes, in this case, the following spatial mode entangled state will be generated:

$$|\psi\rangle \sim (|1_{ss}, 1_{ia}\rangle + |1_{sa}, 1_{is}\rangle) \tag{23.16}$$

in which the entanglement is between states with signal in the symmetric mode and idler in the antisymmetric mode and signal in the antisymmetric mode and idler in the symmetric mode (see Fig. 23.7). The other spatial mode combinations of the signal idler pair will have zero efficiency.

In fact by introducing an arbitrary phase in the pump field in one of the arms of the interferometer leading to a superposition of fundamental and first excited mode

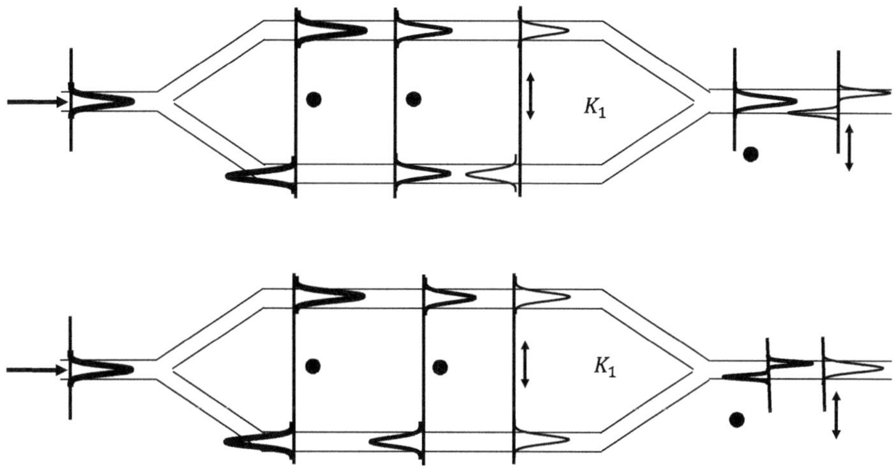

Fig. 23.7 Due to π phase shift introduced by the electro optic modulator, the output state is now given by Eq. (23.16) instead of Eq. (23.15)

of the pump, it is possible to generate more complex spatial mode entangled states (see Problem 23.5).

> ***Exercise 23.4*** Assume that the device shown in Fig. 23.6 is fabricated in $y-$ cut, $x-$ propagating lithium niobate. What input mode would one choose to generate the photon pairs and assuming Type 0 down conversion and what will be the corresponding nonlinear coefficient responsible for the conversion?
>
> ***Solution*** For Type 0 interaction in lithium niobate we choose the pump, signal, and idler to have extraordinary polarizations. Thus, the modes would correspond to TE modes of the waveguide. The nonlinear coefficient responsible for the down conversion would be d_{33}.

If instead of a single spatial frequency QPM grating a dual periodic poled structure is created with spatial frequencies K_1 and K_2 where

$$K_1 = \beta_{po} - \beta_{so} - \beta_{ie}$$

and

$$K_2 = \beta_{po} - \beta_{se} - \beta_{io}$$

then the spatial frequency K_1 will lead to the generation of signal photon in the ordinary polarization and idler photon in the extraordinary polarization. Similarly the spatial frequency K_2 will lead to the generation of signal photon in the extraordinary polarization and idler photon in the ordinary polarization. Also in each of these two processes, the pump photon in the ordinary polarization can either generate signal and idler photons both in the fundamental mode or both in the first excited mode. This leads to the generation of the following (non-normalized) hyperentangled state (entanglement in multiple degrees of freedom, here in polarization and spatial mode):

$$|\psi\rangle \sim (|H_s, V_i\rangle + |V_s, H_i\rangle) \otimes (|1_{ss}, 1_{ia}\rangle + |1_{sa}, 1_{is}\rangle) \tag{23.17}$$

Such hyperentangled states are very interesting for applications in quantum computing.

Three waveguide directional couplers have also been employed for the generation of higher dimensional mode entangled states (Bharadwaj et al. 2015).

Exercise 23.5 Assuming waveguides in $y-$ cut, $x-$ propagating lithium niobate, what input polarization of the pump would be required for the generation of the hyperentangled state given by Eq. (23.17)?

Solution Since the process corresponds to Type II interaction, the nonlinear coefficient responsible for the process in lithium niobate is d_{24}. For this the pump needs to be having ordinary polarization which will correspond to TM mode in the waveguide and the signal idler pairs will be in combinations of TE and TM modes.

23.4 Path Entangled Photon Pairs

An important class of entangled states are the NOON states or path-entangled states defined by the following equation:

$$|\psi\rangle = \frac{1}{\sqrt{2}} \left(|N\rangle_a |0\rangle_b + |0\rangle_a |N\rangle_b \right) \tag{23.18}$$

where a and b represent two different waveguides/paths. The state defined by Eq. (23.18) is a coherent superposition of N photons in two different waveguides/paths.

If the output waveguides shown in Figs. 23.6 or 23.7 are followed by an asymmetric Y-splitter that can separate the symmetric and the antisymmetric modes of the two-mode waveguide (see Chap. 7), then the power corresponding to the two modes will exit from two different waveguides. If the two output waveguides are numbered 1 and 2, then the output state corresponding to Eq. (23.15) and Fig. 23.6 will be given by

$$|\psi\rangle \sim (|1_{s1}, 1_{i1}\rangle + |1_{s2}, 1_{i2}\rangle) \tag{23.19}$$

This path entangled state corresponds to both photons exiting from waveguide #1 or both photons exiting from waveguide #2 which is a NOON state with $N = 2$.

Similarly, the output state corresponding to Eq. (23.16) or Fig. 23.7 will be given by

$$|\psi\rangle \sim (|1_{s1}, 1_{i2}\rangle + |1_{s2}, 1_{i1}\rangle) \tag{23.20}$$

The states corresponding to Eqs. (23.19) and (23.20) correspond to path entangled states. In this case, the state consists of a superposition of the following two states: (i) signal photon exiting waveguide #1 and idler photon exiting waveguide #2 and (ii) signal photon exiting waveguide #2 and the idler photon exiting waveguide #1 (see e.g., Lugani et al. 2011; Solntsev and Sukhorukov 2017; Kruse et al. 2015; Belsley et al. 2020)

23.5 Frequency Entangled State

Let us consider the following quantum state:

$$|\psi\rangle = \int C_1 |\omega_{s1}, \omega_{i1}\rangle d\omega_{s1} + \int C_2 |\omega_{s2}, \omega_{i2}\rangle d\omega_{s2} \tag{23.21}$$

The state given by Eq. (23.21) corresponds to a frequency entangled state.
If the entangled frequencies satisfy the following relation

$$\omega_{s1} + \omega_{i1} = \omega_{s2} + \omega_{i2} = \omega_P \tag{23.22}$$

where the pairs of frequencies ω_{s1}: ω_{i1} and ω_{s2}; ω_{i2} sum to the same value of pump frequency, thus satisfying the energy conservation principle, then the frequency entangled state can be generated via SPDC processes using a single pump frequency.

Generation of the state $|\omega_{s1}, \omega_{i1}\rangle$ using QPM would require a spatial frequency given by

$$K_1 = \frac{2\pi}{\Lambda_1} = 2\pi\left(\frac{n_p}{\lambda_p} - \frac{n_{s1}}{\lambda_{s1}} - \frac{n_{i1}}{\lambda_{i1}}\right) \tag{23.23a}$$

while generation of the state $|\omega_{s2}, \omega_{i2}\rangle$ would require a spatial frequency given by

$$K_2 = \frac{2\pi}{\Lambda_2} = 2\pi\left(\frac{n_p}{\lambda_p} - \frac{n_{s2}}{\lambda_{s2}} - \frac{n_{i2}}{\lambda_{i2}}\right) \tag{23.23b}$$

Here, n_p, n_{s1}, n_{i1}, n_{s2}, and n_{i2} correspond to the effective indices of the fundamental modes corresponding to the frequencies ω_p, ω_{s1}, ω_{i1}, ω_{s2}, and ω_{i2}, respectively.

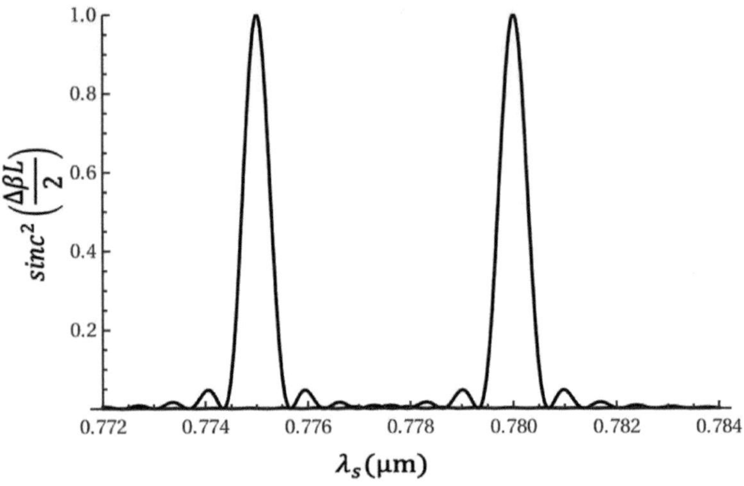

Fig. 23.8 Signal spectra of a frequency entangled state that can be generated using dual periodic poling

In order to satisfy the required QPM conditions, one can use the dual periodic poling to have a doubly periodic QPM grating that has both spatial frequencies of K_1 and K_2(see Sect. 23.2) to generate the frequency entangled state (see, e.g., Warke and Thyagarajan 2021).

Figure 23.8 shows the spectra corresponding to Type 0 SPDC process in lithium niobate of the frequency entangled state given by Eq. (23.21). The two signal wavelengths are 780 nm and 775 nm for a length of interaction given by 10 mm.

It is worth mentioning that as compared to bulk crystals, using these processes in waveguide geometry can provide additional degree of freedom using the waveguide dispersion aspects of the waveguides which is controlled by the dimensions and the index step provided by the waveguide. A number of designs have been investigated for the generation of frequency entangled photon pairs (Reimer et al. 2016; Hojo and Tanaka 2024; Francesconi et al. 2023; Morrison et al. 2022; Jin et al. 2016)

Exercise 23.6 Consider bulk lithium niobate and Type 0 SPDC process with the pump, signal, and idler all having extraordinary polarizations. Assume a pump wavelength of 519 nm and signal wavelengths of 775 nm and 780 nm. (a) Obtain the corresponding idler wavelengths. (b) From the refractive index values of lithium niobate, obtain the QPM periods required to achieve the two processes.

Solution

(a) The idler wavelength is given by

$$\frac{1}{\lambda_i} = \frac{1}{\lambda_p} - \frac{1}{\lambda_s}$$

The idler wavelength for a signal wavelength of 775 nm is 1571.19 nm and for a signal wavelength of 780 nm is 1551.03 nm, respectively.

(b) To obtain the QPM period for the two processes, we first determine the extraordinary refractive index of lithium niobate at the various pump, signal, and idler wavelengths. These are given by

$$n_{pe}(519\text{ nm}) = 2.2393$$
$$n_{se}(775\text{ nm}) = 2.1784$$
$$n_{ie}(1571.19\text{ nm}) = 2.1370$$
$$n_{se}(780\text{ nm}) = 2.1778$$
$$n_{ie}(1551.03\text{ nm}) = 2.1375$$

The spatial period of the QPM grating required or the signal wavelength of 775 nm is

$$\frac{1}{\Lambda_1} = \frac{n_{pe}}{\lambda_p} - \frac{n_{se}}{\lambda_s} - \frac{n_{ie}}{\lambda_i} = \frac{2.2393}{0.519} - \frac{2.1784}{0.775} - \frac{2.1370}{1.57119}$$

giving $\Lambda_1 \approx 6.96\ \mu$m.

The spatial period of the QPM grating required for the signal wavelength of 780 nm is

$$\frac{1}{\Lambda_2} = \frac{n_{pe}}{\lambda_p} - \frac{n_{se}}{\lambda_s} - \frac{n_{ie}}{\lambda_i} = \frac{2.2393}{0.519} - \frac{2.1778}{0.780} - \frac{2.1375}{1.55103}$$

giving $\Lambda_2 \approx 6.92\ \mu$m.

Exercise 23.7 Consider SPDC process in a lithium niobate crystal with the interacting waves propagating along the x-direction. For a pump wavelength with ordinary polarization at a wavelength of 775 nm, (a) obtain the first order QPM period required for a Type 0 SPDC process to generate a signal in the extraordinary polarization with an idler in the ordinary polarization. (b) What would be the QPM period required if the signal has ordinary polarization and the idler has extraordinary polarization. Assume the signal and idler wavelengths to be given by 1506 nm and 1596.6 nm.

Solution The ordinary refractive index of lithium niobate at 775 nm is $n_{po} = 2.2587$. The ordinary and extraordinary refractive indices at 1506 nm and 1596.6 nm are given by $n_{so} = 2.2126$, $n_{se} = 2.1388$ and $n_{io} = 2.2095$, $n_{ie} = 2.1363$.

Thus, the QPM period for the signal having ordinary polarization and idler having extraordinary polarization will be

$$\frac{1}{\Lambda_1} = \frac{n_{po}}{\lambda_p} - \frac{n_{so}}{\lambda_s} - \frac{n_{ie}}{\lambda_i} = \frac{2.2587}{0.775} - \frac{2.2126}{1.506} - \frac{2.1363}{1.5966}$$

giving $\Lambda_1 \approx 9.33\ \mu\text{m}$.

Similarly the QPM period for the signal having extraordinary polarization and idler having ordinary polarization will be

$$\frac{1}{\Lambda_2} = \frac{n_{po}}{\lambda_p} - \frac{n_{se}}{\lambda_s} - \frac{n_{io}}{\lambda_i} = \frac{2.2587}{0.775} - \frac{2.1388}{1.506} - \frac{2.2095}{1.5966}$$

giving $\Lambda_2 \approx 9.06\ \mu\text{m}$.

Thus, a crystal with both these spatial periods can generate the following entangled state

$$|\psi\rangle = \frac{1}{\sqrt{2}}\left(|H_{\omega_s}, V_{\omega_i}\rangle + e^{i\theta}|V_{\omega_s}, H_{\omega_i}\rangle\right)$$

If this state is passed through a polarization beam splitter that separates the H (along path A) and V polarization (along path B), then we will obtain the following state:

$$|\psi\rangle = \frac{1}{\sqrt{2}}\left(|\omega_s\rangle_{A,H}|\omega_i\rangle_{B,V} + e^{i\theta}|\omega_s\rangle_{B,V}|\omega_i\rangle_{A,H}\right)$$

which is nothing but a frequency-bin entangled state.

On the other hand, if the generated photons are passed through a dichroic mirror that separates the two frequencies, with signal along path A and idler along path B, then we will have the following state:

$$|\psi\rangle = \frac{1}{\sqrt{2}}\left(|H\rangle_{A,\omega_s}|V\rangle_{B,\omega_i} + e^{i\theta}|V\rangle_{A,\omega_s}|H\rangle_{B,\omega_i}\right)$$

which is nothing but a polarization entangled state (Kaneda et al. 2019).

Exercise 23.8 In Exercise 23.7 we had considered the interaction to take place in bulk. In this Exercise, we consider a $y-$ cut, $x-$ propagating planar waveguide in lithium niobate and consider Type 0 interaction with signal and idler at 1506 and 1594 nm, respectively. The thickness of the waveguide is assumed to be 5 μm and $\Delta n_o = \Delta n_e = 0.003$. The effective indices of the corresponding modes are given below:

$$n_{\text{eff},po} = 2.26082;\ \ n_{\text{eff},so} = 2.21338;\ \ n_{\text{eff},io} = 2.21023;\ n_{\text{eff},se} = 2.13966;\ \ n_{\text{eff},ie} = 2.13723$$

Obtain the corresponding QPM periods for the two different SPDC processes.

Solution The QPM period for the signal having ordinary polarization (TM mode) and idler having extraordinary polarization (TE mode) will be

$$\frac{1}{\Lambda_1} = \frac{n_{po}}{\lambda_p} - \frac{n_{so}}{\lambda_s} - \frac{n_{ie}}{\lambda_i} = \frac{2.26082}{0.7744} - \frac{2.21338}{1.506} - \frac{2.13723}{1.594}$$

giving $\Lambda_1 \approx 9.18$ μm.

Similarly the QPM period for the signal having extraordinary polarization and idler having ordinary polarization will be

$$\frac{1}{\Lambda_2} = \frac{n_{po}}{\lambda_p} - \frac{n_{se}}{\lambda_s} - \frac{n_{io}}{\lambda_i} = \frac{2.26082}{0.7744} - \frac{2.13966}{1.506} - \frac{2.21023}{1.594}$$

giving $\Lambda_2 \approx 8.92$ μm.

These QPM periods are different from bulk due to the presence of waveguide dispersion.

Exercise 23.9 Consider backward phase matching in KTP with the pump and signal propagating along the x–direction and idler propagating in the $-x$ direction. What are the QPM spatial frequencies required for Type 0 and Type I phase matching? If the signal–idler pair is degenerate, then what are the spatial frequencies required?

Solution In view of the nonlinear tensor of KTP (see Chap. 10), the pump is assumed to be polarized along the z—direction and in Type 0 conversion the signal and idler will also be polarized along the z—direction. Hence, if the effective indices of the waveguide at the pump, signal, and idler wavelengths of the z—polarized modes are represented by n_{pz}, n_{sz}, and n_{iz}, respectively, then for mth—order QPM the required spatial frequency mK_1 where $K_1 = \frac{2\pi}{\Lambda_1}$, with Λ_1 being the fundamental spatial period, is given by

$$mK_1 = \frac{\omega_p}{c} n_{pz} - \frac{\omega_s}{c} n_{sz} + \frac{\omega_i}{c} n_{iz}$$

For Type I interaction, the signal and idler will be polarized along the y–direction and if the corresponding effective indices of the signal and idler are represented by n_{sy} and n_{iy}, then the required spatial frequency mK_2 where $K_2 = \frac{2\pi}{\Lambda_2}$, with Λ_2 being the fundamental spatial period, is given by

$$mK_2 = \frac{\omega_p}{c} n_{pz} - \frac{\omega_s}{c} n_{sy} + \frac{\omega_i}{c} n_{iy}$$

which is different from that required for Type 0 process. In case of degeneracy, we will have $\omega_s = \omega_i$ and $n_{sz} = n_{iz}$ and $n_{sy} = n_{iy}$ and the spatial frequencies become

$$K_1 = K_2 = \frac{\omega_p}{mc} n_{pz}$$

The corresponding spatial period for mth–order QPM is given by

$$\Lambda_m = m \frac{\lambda_p}{n_{pz}}$$

Note that the nonlinear coefficients involved in the two conversion processes are d_{33} and d_{32} which are not equal. Assuming both the processes to take place, the generated signal–idler photon pair will be nonmaximally entangled.

As an example we assume a pump wavelength of 775 nm and the degenerate pair of signal idler photons, then have a wavelength of 1550 nm. Assuming an effective index $n_{pz} = 1.85$, the required period for mth–order QPM is given by $\Lambda_m = 0.419\ m\ \mu$m. Recently extremely short period periodic poling have been reported in lithium niobate on insulator (LNOI) platform (see e.g., Ma et al. 2023b). Higher order QPM process will require larger periods but at the expense of reduced effective nonlinear coefficient. Thus, the 7th order QPM will require a spatial period of 2.93 μm.

For an interesting comparison between counter propagating and co propagating twin photon generation, readers are referred to Gatti and Brambilla (2018). Counter propagating entangled photon pair generation from periodically poled lithium niobate is also discussed in Booth et al. (2002).

23.6 Energy-Time Entangled States

Until now we have discussed polarization entanglement, spatial mode entanglement, and frequency entanglement. There is another very important form of entanglement that is referred to as energy-time entanglement (Franson 1989; Brendel et al. 1999; Honjo et al. 2007; Maclean et al. 2018) which can also be used for testing Bell's inequalities and also for generating quantum keys in quantum key distribution applications (see, e.g., Tittel et al. 2000; Tanzilli et al. 2002; Marcikic et al. 2004; Kaiser et al. 2016; Chen et al. 2018; Kim et al. 2022; Inagaki and Takesue 2014).

Consider a quasi monochromatic pump that generates a signal–idler pair of photons through the process of spontaneous parametric down conversion. Since the pump is almost monochromatic, its frequency is well defined. Although the pump is almost monochromatic, the generated signal and idler photons have a finite bandwidth which depends on the parameters of the device generating the pair (see Chap. 22). This implies an uncertainty in the frequencies of the generated signal and idler photons. At the same time, due to energy conservation, the sum of the energy of the signal and idler photons must be equal to that of the pump. Thus, although there is uncertainty in the individual frequencies of the signal and idler, the sum of the frequencies of any generated pair of photons is well defined. At the same time, since

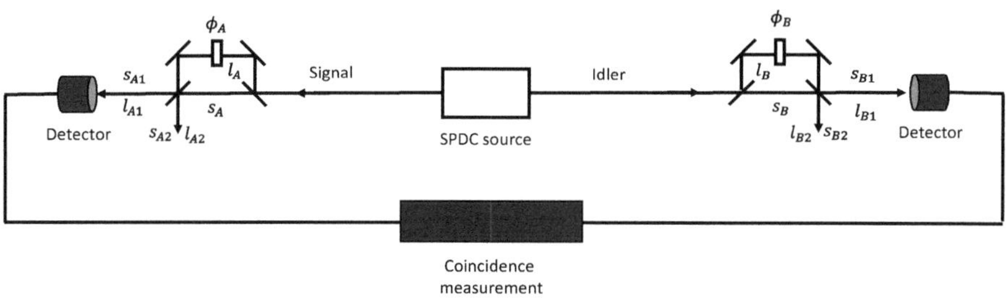

Fig. 23.9 Franson interferometer with two unbalanced Mach-Zehnder interferometers and an SPDC source of signal-idler pair of photons

the signal–idler pair is simultaneously generated, the time difference between the time of birth of the two photons is equal to zero. In this situation, we do not know the exact time of birth and energy of the individual photons but know precisely the sum of their energies and the difference in their birth times. Such an entanglement is referred to as energy-time entanglement.

Energy-time entanglement can be observed using what is referred to as Franson interferometer shown in Fig. 23.9 (Franson 1989). The setup consists of two identical unbalanced Mach-Zehnder interferometers (see Chap. 7) in which one of the arms of the interferometer is longer than the other arm. We refer to the interferometers as A and B and the lengths of the short arms as s_A and s_B and the lengths of the long arms as l_A and l_B. An SPDC source pumped by a continuous wave quasi monochromatic pump produces a signal-idler pair of photons and each of these photons is sent to each of the two interferometers.

The optical path length difference between the arms of the two interferometers is made to satisfy the following conditions:

(a) Although the pump is quasi monochromatic, we have seen in Chap. 22 that the generated signal–idler pair of photons have a finite bandwidth determined by the phase matching condition. Since the coherence length is inversely related to the spectral bandwidth of the photons, large bandwidth implies short coherence lengths and small bandwidths implies long coherence lengths. Calling the coherence length of the signal and idler photons as L_c^s and L_c^i, respectively, we impose the following condition:

$$\Delta L_A = (l_A - s_A) \text{ and } \Delta L_B = (l_B - s_B) \gg L_c^{s,i} \tag{23.24}$$

That is, the optical path length difference between the two arms of the interferometer is larger than the coherence length of the generated photons. If this condition is satisfied, then the short and long paths through the interferometers will not lead to any single photon interference.

Thus, the output state of the signal photon exiting from the signal interferometer (with subscript A) shown in Fig. 23.9 would be given by the following state:

$$|\psi\rangle_A = \frac{1}{2}\left(|s_{A1}\rangle + i|s_{A2}\rangle - e^{i\phi_A}|l_{A1}\rangle + ie^{i\phi_A}|l_{A2}\rangle\right) \tag{23.25}$$

Here $|s\rangle_{A1}$ is the output state produced by the photon propagating through the short arm of the interferometer and exiting towards the detector, $|s\rangle_{A2}$ is the output state produced by the photon propagating through the short arm and exiting towards the arm without the detector (see Fig. 23.9), $|l\rangle_{A1}$ is the output state produced by the photon propagating through the long arm and exiting towards the detector, and $|l\rangle_{A2}$ is the output state produced by the photon propagating through the long arm of the interferometer and exiting towards the arm without the detector. The phase ϕ_A is the controllable phase change on the long arm of the interferometer. Note that two of the terms in Eq. (23.25) lead to the photon exiting the arm with subscript $A1$ (where the photon falls on the detector shown in Fig. 23.9) and two of the terms lead to the photon exiting the arm with subscript $A2$.

The component of the state that corresponds to the photon impinging on the detector will be given by

$$|\psi\rangle_A = \frac{1}{2}\left(|s_{A1}\rangle - e^{i\phi_A}|l_{A1}\rangle\right) \tag{23.26}$$

Similarly the output state of the idler photon in arm B that is exiting towards the detector will be given by

$$|\psi\rangle_B = \frac{1}{2}\left(|s_{B1}\rangle - e^{i\phi_B}|l_{B1}\rangle\right) \tag{23.27}$$

where $|s_{B1}\rangle$ is the output state produced by the idler photon propagating through the short arm and $|l_{B1}\rangle$ is the output state produced by the idler photon propagating through the long arm of the interferometer. Here ϕ_B is the controllable phase change on the long arm of the interferometer.

Note that in view of the condition given by Eq. (23.24), the two paths are distinguishable and hence there is no interference between the two paths.

We also note that if one of the photons (say the signal photon) passes through the short arm while the other photon (idler photon) passes through the long arm, they are distinguishable at the output of each interferometer as one of them arrives before the other. On the other hand, if both photons (signal as well as idler) pass through the short arms or both of them pass through the long arms of the two interferometers, these two situations are indistinguishable at the output since the time of birth of the signal–idler pair is unknown. Due to this indistinguishability, one can observe interference between the coincidence detection of the two photons exiting each of the interferometer for the case when both of them travel via the short arms or both of them travel through the long arms. We will come back to this point later in the section.

(b) The pump is assumed to be quasi monochromatic and thus is characterized by a coherence length L_c^p. We also need to satisfy the condition

$$\Delta L_A, \Delta L_B \ll L_c^p \tag{23.28}$$

so that when both photons take the short path or when both photons take the long path they remain coherent, leading to interference between the two probabilities which can be observed using coincidence detection. This condition will ensure that the signal–idler pair that we are using in the measurement is born from coherent pump photons.

(c) The two interferometers are also chosen to have identical optical path differences between the two arms, that is,

$$\Delta L_A = \Delta L_B \tag{23.29}$$

We can write for the two photon state at the outputs arms with the detectors of the two interferometers as

$$\begin{aligned} |\psi\rangle_{A,B} &= |\psi\rangle_A |\psi\rangle_B \\ &= \frac{1}{4}\left(|s_{A1}\rangle|s_{B1}\rangle - e^{i\phi_A}|l_{A1}\rangle|s_{B1}\rangle - e^{i\phi_B}|s_{A1}\rangle|l_{B1}\rangle + e^{i(\phi_A+\phi_B)}|l_{A1}\rangle|l_{B1}\rangle\right) \end{aligned} \tag{23.30}$$

Each of the outputs is detected by detectors which are assumed to be perfect.

Now, if we look for coincidence among the two detected signals at the outputs of the two interferometers, then we can see that if the signal and idler photons both take the short path or both take the long path, we will observe coincidence. On the other hand, if either of the photons takes the short path and the other photon takes a long path, then this will not result in any coincidence. Since the two situations when both the photons take the short path or both the photons take the long path are indistinguishable since the time of birth of the signal–idler pair is unknown, we will see interference among the two probability amplitudes.

Hence, we obtain for the probability of coincidence to be the absolute square of the sum of the probability amplitudes of the two photons both taking the short paths or both taking the long paths. Hence, we obtain the coincidence probability as

$$P_{\text{coin}} = \left|\left(\langle s_{A1}|\langle s_{B1}||\psi\rangle_{A,B} + \langle l_{A1}|\langle l_{B1}||\psi\rangle_{A,B}\right)\right|^2 \tag{23.31}$$

Substituting for $|\psi\rangle_{A,\,B}$ and using the fact that the states are normalized, we obtain

$$P_{\text{coin}} = \frac{1}{16}\left|1 + e^{i(\phi_A+\phi_B)}\right|^2 = \frac{1}{8}[1 + \cos(\phi_A + \phi_B)] \tag{23.32}$$

Note that in the coincidence measurement only the two parts $|s\rangle_{A1}|s\rangle_{B1}$ and $|l\rangle_{A1}|l\rangle_{B1}$ take part, and the other projection terms reduce to zero.

We can also write Eq. (23.31) as

$$\begin{aligned} P_{\text{coin}} &= \left| (\langle s_{A1}|\langle s_{B1}| + \langle l_{A1}|\langle l_{B1}|) \left[\frac{1}{4} \left(|s\rangle_{A1}|s\rangle_{B1} + e^{i(\phi_A+\phi_B)}|l\rangle_{A1}|l\rangle_{B1} \right) \right] \right|^2 \\ &= \left| (\langle s_{A1}|\langle s_{B1}| + \langle l_{A1}|\langle l_{B1}|)|\psi\rangle_{\text{Ent}} \right|^2 \end{aligned} \tag{23.33}$$

The state

$$|\psi\rangle_{\text{Ent}} = \frac{1}{4} \left(|s_{A1}\rangle|s_{B1}\rangle + e^{i(\phi_A+\phi_B)}|l_{A1}\rangle|l_{B1}\rangle \right) \tag{23.34}$$

represents an energy-time entangled state.

It is interesting to note that the coincidence probability as given by Eq. (23.32) depends on the phase (ϕ_A + ϕ_B). Thus, as this phase changes, the coincidence probability will vary between $\frac{1}{4}$ and zero. As an example, if $\phi_A = 0$, then the maxima of coincidence occurs when $\phi_B = 2m\pi$; $m = 0, 1, 2\ldots$ and the coincidence becomes zero when $\phi_B = (2m + 1)\pi$; $m = 0, 1, 2\ldots$

If now $\phi_A = \frac{\pi}{3}$, then the maxima of coincidence occurs when $\phi_B = 2m\pi - \frac{\pi}{3}; m = 0, 1, 2\ldots$ and the coincidence becomes zero when $\phi_B = (2m + 1)\pi - \frac{\pi}{3}; m = 0, 1, 2\ldots$

Referring to Sect. 23.2 we see that the behavior is very similar to that of polarization entangled photon pairs when the photons are detected through linear analyzers whose orientations are changed and we obtain sinusoidal variation in coincidence.

The entangled state given by Eq. (23.34) has indeed been used to test Bell's inequalities (see, e.g., Tittel et al. 1998).

Exercise 23.10 Using a quasi monochromatic pump at 780 nm, signal–idler pair of wavelengths with a signal at a wavelength of 1550 nm is generated using Type-0 SPDC process in a lithium niobate crystal of length 10 mm with all interacting modes having extraordinary polarization. The bandwidth of the signal photon is about 1.3 nm and that of the idler photon is approximately 5.3 nm. What condition should ΔL_A and ΔL_B satisfy for generating time-bin entangled photons using this source?

Solution For the given pump and signal wavelengths, the idler wavelength is 1570 nm. The coherence length is related to the spectral width through the following equation:

$$L_c = \frac{\lambda_0^2}{\Delta\lambda}$$

Hence, the coherence lengths of the signal photon would be about 1.85 mm and that of the idler photons will be about 0.47 mm. Hence, the optical path length difference between the two arms of the unbalanced Mach-Zehnder interferometer must satisfy the following condition:

$$\Delta L_A, \Delta L_B \gg 2 \text{ mm}$$

If the pump pulse has a duration of $\Delta\tau = 1$ ns, then the coherence length of the pump would be $L_c^p = c\Delta\tau = 30$ cm. Hence, as per conditions given by Eqs. (23.28) and (23.24), we must have 2 mm $\ll \Delta L_A, \Delta L_B \ll$ 30 cm.

23.6.1 Time-Bin Entanglement

In order to realize a practical source which can utilize the energy-time entanglement, a modified design is used in which we add a third unbalanced Mach-Zehnder interferometer as shown in Fig. 23.10. In this setup, the pump is taken to be pulsed source and the pump pulse is first passed through an unbalanced Mach-Zehnder interferometer referred to as MZ1. Assuming that the optical path length difference between the two arms is longer than the coherence length of the pump pulse, the output of the first interferometer will consist of a superposition of pump modes given by

$$|\psi\rangle_p = \frac{1}{2}\left(|s\rangle_p - e^{i\phi_P}|l\rangle_p\right) \tag{23.35}$$

where the subscript p corresponds to pump. Note that we are writing only the part of the state that corresponds to the pump photon moving towards the interferometers, the total probability of which is $\frac{1}{2}$.

This superposition state of the pump pulse then enters the nonlinear crystal which is phase matched (or quasi phase matched) to generate a signal idler pair. Thus, after the nonlinear crystal the state becomes

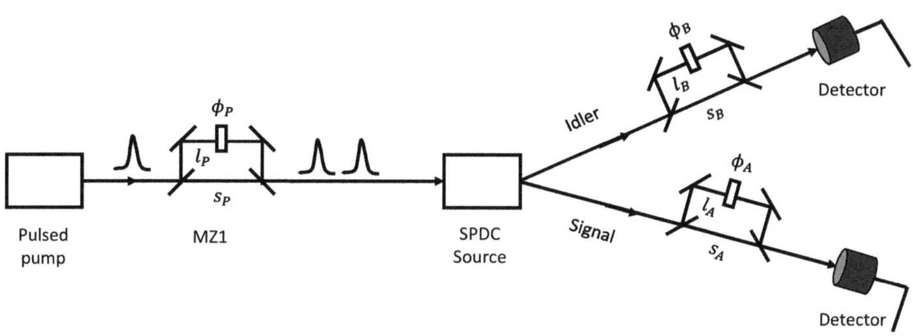

Fig. 23.10 Setup to generate time-bin entangled photons pairs for quantum key distribution application

$$|\psi\rangle_{cr} = \frac{1}{2}\left(|s\rangle_{ss}|s\rangle_{si} - e^{i\phi_P}|l\rangle_{ls}|l\rangle_{li}\right) \tag{23.36}$$

The first subscript s denotes the fact that the signal–idler pair is generated by the pump pulse that took the short path in MZ1, while the subscript l denotes the fact that the signal–idler pair took the long path in MZ1. The second subscripts correspond to signal and idler.

Like before, the signal and idler photons are then sent through two unbalanced Mach-Zehnder interferometers (see Fig. 23.10) satisfying the same conditions stated before. Thus, the signal photons will get modified to the following state (the component of the state responsible for detection at the two detectors):

$$|s\rangle_{ss} = \frac{1}{2}\left(|s\rangle_{sA} - e^{i\phi_A}|l\rangle_{sA}\right) \tag{23.37}$$

and

$$|l\rangle_{ls} = \frac{1}{2}\left(|s\rangle_{lA} - e^{i\phi_A}|l\rangle_{lA}\right) \tag{23.38}$$

Similarly the idler photons will be described by the states

$$|s\rangle_{si} = \frac{1}{2}\left(|s\rangle_{sB} - e^{i\phi_B}|l\rangle_{sB}\right) \tag{23.39}$$

$$|l\rangle_{li} = \frac{1}{2}\left(|s\rangle_{lB} - e^{i\phi_B}|l\rangle_{lB}\right) \tag{23.40}$$

Thus, the complete state of the two photons after the two MZ interferometers will be given by

$$\begin{aligned}|\psi\rangle_t = \frac{1}{8}\Big[&|s\rangle_{sA}|s\rangle_{sB} - e^{i\phi_B}|s\rangle_{sA}|l\rangle_{sB} - e^{i\phi_A}|l\rangle_{sA}|s\rangle_{sB} + e^{i(\phi_A+\phi_B)}|l\rangle_{sA}|l\rangle_{sB}\\ &- e^{i\phi_P}\left\{|s\rangle_{lA}|s\rangle_{lB} - e^{i\phi_B}|s\rangle_{lA}|l\rangle_{lB} - e^{i\phi_A}|l\rangle_{lA}|s\rangle_{lB} + e^{i(\phi_A+\phi_B)}|l\rangle_{lA}|l\rangle_{lB}\right\}\Big]\end{aligned}$$

Out of all the eight possible paths, the following paths are indistinguishable:

- Short path in the pump interferometer and short paths in both the other interferometers or long paths in both the other interferometers.
- Long path in the pump interferometer and short paths in both the other interferometers or long paths in both the other interferometers.

Note that the paths (i) short path in pump interferometer and short paths in both the other interferometers and (ii) long path in pump interferometer and long paths in both the other interferometers are distinguishable due to their different travel times and thus do not interfere. The only paths that are indistinguishable are the following two paths:

- Short path in the pump interferometer and long paths in both the other interferometers
- Long path in the pump interferometer and short paths in both the other interferometers

Since these are coherent, the state that is responsible for coincidence counting of the outputs from the two interferometers will be

$$|\psi\rangle_{\text{coinc}} = \frac{1}{8}\left(e^{i(\phi_A+\phi_B)}|l\rangle_{sA}|l\rangle_{sB} - e^{i\phi_P}|s\rangle_{lA}|s\rangle_{lB}\right) \tag{23.41}$$

which is an entangled state. Hence the coincidence probability will be given by

$$P_{\text{coin}} = \frac{1}{64}\left|\left(\langle s_A|\langle s_B| + \langle l_A|\langle l_B|\right)\left(e^{i(\phi_A+\phi_B)}|l\rangle_{sA}|l\rangle_{sB} - e^{i\phi_P}|s\rangle_{lA}|s\rangle_{lB}\right)\right|^2$$

which simplifies to

$$P_{\text{coin}} = \frac{1}{64}\left|e^{i(\phi_A+\phi_B)} - e^{i\phi_P}\right|^2 = \frac{1}{32}\left(1 - \cos[\phi_A + \phi_B - \phi_P]\right) \tag{23.42}$$

Hence, the coincidence probability depends on all the three phases namely ϕ_P, ϕ_A, and ϕ_B. Interferometer configurations can be used to generate quantum keys for quantum key distribution applications (see Chap. 25) (Tittel et al. 2000; Tanzilli 2002; Rossi et al. 2008; Mueller et al. 2024).

Problems

23.1. Consider a channel waveguide in Y—cut X—propagating lithium niobate having a depth of 8 μm and a width of 10 μm. It is proposed to generate photon pairs using the Type 0 process with an extraordinary pump generating a signal idler photon pair both having extraordinary polarization. The pump wavelength is 800 nm and the signal wavelength is 1550 nm. (a) Obtain the effective indices of the fundamental extraordinary modes at the pump, signal, and idler wavelengths. (b) Obtain the required first order QPM period for this process. (c) Are the generated photons entangled in polarization?

23.2. Consider a directional coupler made of two identical symmetric waveguides. The propagation constants of the two individual waveguides are β_{p0} , β_{s0}, and β_{i0} at the pump, signal, and idler frequencies. The corresponding coupling coefficients are κ_p , κ_s, and κ_i, respectively. If the input pump excites the fundamental mode of the directional coupler, which of the following processes can take place? For those processes, obtain an expression for the spatial frequency required for first-order QPM for efficient conversion: (a) symmetric pump to a combination of symmetric signal and idler modes, (b) symmetric pump mode to a pair of antisymmetric signal and idler modes,

(c) symmetric pump mode to a symmetric signal and an antisymmetric idler mode, (d) antisymmetric pump to a pair of symmetric signal and idler modes, and (e) antisymmetric pump mode to a symmetric signal and an antisymmetric idler modes.

23.3. It is proposed to generate signal–idler photon pairs using SPDC in a directional coupler made up of identical waveguides in a *z*—cut *x*—propagating lithium niobate waveguide. The propagation constants of the extraordinary modes at the pump, signal, and idler wavelengths of the individual waveguides are β_p, β_s, and β_i, respectively, and the coupling coefficients are κ_p, κ_s, and κ_i, respectively. It is desired to generate a mode entangled state using a pump propagating in the symmetric mode of the coupler. What QPM periods would be needed to generate a spatial mode entangled state and what will be the output entangled state?

23.4. Consider a *Z*—cut *X*—propagating lithium niobate waveguide with an effective index represented by $n_{\text{eff}}(\lambda_j, TE)$ and $n_{\text{eff}}(\lambda_j, TM)$ with $j = p, s, i$ corresponding to the pump, signal, and idler frequencies and to TE and TM modes. (a) Show that an incident pump photon in the TM mode can generate a signal idler pair both having TM polarization. What will be the corresponding nonlinear coefficient responsible for the process? Write down an expression for the corresponding required first order QPM period. (b) Show that an incident photon in the TM mode can also generate a signal idler pair both having TE polarization. What will be the corresponding nonlinear coefficient responsible for the process? Write down an expression for the corresponding first order QPM period. (c) which one of these two processes will be having higher efficiency of down conversion? (d) If the waveguide has dual periodic poling with both the obtained periods, will it generate a maximally entangled photon pair?

23.5. Referring to the device shown in Fig. 23.5, if the electro optic modulator introduces a phase difference of $\frac{\pi}{2}$ between the modes propagating in the upper and lower waveguide, what will be the output state produced by the device?

23.6. SPDC process takes place in a *Z*—cut *X*—propagating lithium niobate channel waveguide with a refractive index that is symmetric along the *Y*-direction. The waveguide supports the following two spatial modes $E_{11}^{(z)}$ and $E_{21}^{(z)}$ at the pump, signal and idler frequencies; the first subscript represents the mode number along the *Y*-direction. State all the possible pairs of spatial modes in which the signal and idler photons be generated if the pump is in the $E_{11}^{(z)}$ or the $E_{21}^{(z)}$ mode.

23.7. Consider SPDC in bulk lithium niobate whose wavelength dependence of the ordinary and extraordinary refractive indices are given in Table 23.1. Consider a lithium niobate crystal with a pump at a wavelength of 800 nm propagating as an extraordinary wave along the *x*-direction in a periodically poled lithium niobate crystal with a poling period of 20.4 μm and undergoing Type 0 SPDC. What pair of signal idler wavelengths will have the highest efficiency of generation? Use the Sellmeier equation for calculating the refractive indices.

Table 23.1 Sellmeier coefficients for congruently grown lithium niobate (Zelmon et al. 1997); see also Fig. 23.11: $n^2(\lambda_0) = 1 + \frac{A\lambda_0^2}{(\lambda_0^2 - B)} + \frac{C\lambda_0^2}{(\lambda_0^2 - D)} + \frac{E\lambda_0^2}{(\lambda_0^2 - F)}$

Coefficient	n_e	n_o
A	2.9804	2.6734
B	0.02047	0.01764
C	0.5981	1.2290
D	0.0666	0.05914
E	8.9543	12.614
F	416.08	474.6

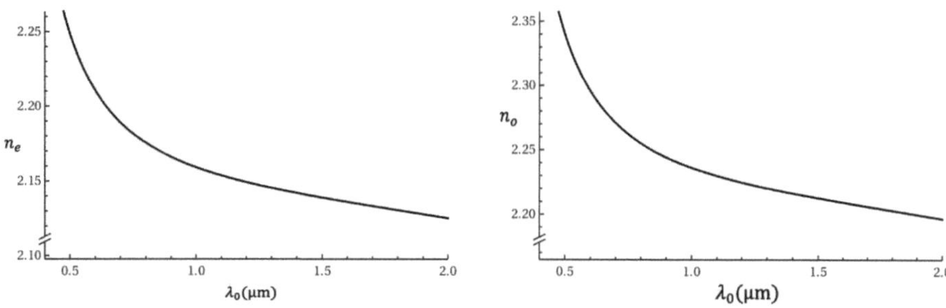

Fig. 23.11 Wavelength variation in the extraordinary and ordinary refractive indices of congruently grown lithium niobate using Table 23.1

23.8. Consider Type I SPDC in bulk lithium niobate crystal with the input pump at 750 nm propagating as an ordinary wave and the signal at 1300 nm being an extraordinary wave. (a) What will be the wavelength of the generated idler? (b) What QPM period would have to be used for maximum efficiency? (c) What would be the required period if the signal is an ordinary wave?

23.9. Consider lithium niobate and assume an input pump which has extraordinary polarization.

(a) Show that such a pump photon can spontaneously down convert to a signal idler pair of photons both having extraordinary polarizations.
(b) Which nonlinear coefficient is involved in this process?
(c) Assuming the pump to have a wavelength of 800 nm and the signal photon to have a wavelength of 1500 nm, obtain the idler wavelength.
(d) Taking bulk refractive indices of congruently grown lithium niobate (given in Table 23.1 and Fig. 23.11) obtain the required QPM period for phase matching.

References

Asobe, M., O. Tadanaga, H. Miyazawa, Y. Nishida, and H. Suzuki. 2003. Multiple quasi-phase-matched $LiNbO_3$ wavelength converter with a continuously phase-modulated domain structure. *Opt Lett* 28:558–560.

Barreiro, J. T., N. K. Langford, N. A. Peters, and P. G. Kwiat. 2005. Generation of hyperentangled photon pairs. *Phys Rev Lett* 95:260501.

Belsley, A., T. Pertsch, and F. Setzpfandt. 2020. Generating path entangled states in waveguide systems with second order nonlinearity. *Opt Express* 28:28792–28809.

Bharadwaj, D., and K. Thyagarajan. 2018. Direct generation of hybrid entangled photon pairs in waveguides. *J Opt Soc Am* 35:790–796.

Bharadwaj, D., K. Thyagarajan, M. Karpinski, and K. Banaszek. 2015. Generation of higher-dimensional modal entanglement using a three-waveguide directional coupler. *Phys Rev A* 91: 033824.

Booth, M. C., M. Atature, G. Di Giuseppe, B. E. A. Saleh, A. V. Sergienko, and M. C. Teich. 2002. Counter propagating entangled photons from a waveguide with periodic nonlinearity. *Phys Rev A* 66:023815.

Brendel, J., N. Gisin, W. Tittel, and H. Zbinden. 1999. Pulsed energy-time entangled twin photon source for quantum communication. *Phys Rev Lett* 82:2594–2597.

Chen, H., S. Auchter, M. Prilmuller, A. Schlager, T. Kauten, K. Laiho, B. Pressl, H. Suchomel, M. Kamp, S. Hofling, C. Schneiderand, and G. Weihs. 2018. Time-bin entangled photon pairs from Bragg reflection waveguides. *APL Photonics* 3:080804.

Chen, X., Z. Fu, Q. Gong, and J. Wang. 2021. Quantum entanglement on photonic chips: a review. *Adv Photonics* 3 (6): 064002.

Ebers, L., A. Ferreri, M. Hammer, M. Albert, C. Meier, J. Forstner, and P. R. Sharapova. 2022. Flexible source of correlated photons based on LNOI rib waveguides. *J Phys Photonics* 4: 025001.

Feng, L.-T., M. Zhang, X. Xiong, Y. Chen, H. W. Ming, G.-P. Guo, G.-C. Guo, D.-X. Dai, and X.-F. Ren. 2019. On chip transverse mode entangled photon pair source. *npj Quant Inform* 5:2. https://doi.org/10.1038/s41534-018-0121-z.

Francesconi, S., A. Raymond, R. Duhamel, P. Filloux, A. Lemaître, P. Milman, M. I. Amanti, F. Baboux, and S. Ducci. 2023. On-chip generation of hybrid polarization-frequency entangled biphoton state. *Photonics Res* 11:270–278.

Franson, J. D. 1989. Bell inequality for position and time. *Phys Rev Lett* 62:2205–2208.

Gatti, A., and E. Brambilla. 2018. Heralding pure single photons: A comparison between counterpropagating and copropagating twin photons. *Phys Rev A* 97:013838.

Herrman, H., X. Yang, A. Thomas, A. Poppe, W. Sohler, and C. Silberhorn. 2013. Post selection free, integrated optical source of non-degenerate, polarization entangled photon pairs. *Opt Express* 21:27981–27991.

Hojo, M., and K. Tanaka. 2024. Single pass generation of widely tunable frequency domain entangled photon pairs. *Opt Express* 32:1902–1913.

Honjo, T., H. Takesue, and K. Inoue. 2007. Generation of energy-time entangled photon pairs at 1.5 μm band with periodically poled lithium niobate waveguide. *Opt Express* 15:1679–1683.

Inagaki, T., and H. Takesue. 2014. Distribution of entangled photon pairs over 300 km of optical fiber. *NTT Tech Rev* 12 (9): 14–18.

Jin, R.-B., R. Shimizu, M. Fujiwara, M. Takeoka, R. Wakabayashi, T. Yamashita, S. Miki, H. Terai, T. Gerrits, and M. Sasaki. 2016. Simple method of generating and distributing frequency-entangled qudits. *Quant Sci Technol* 1:015004.

Kaiser, F., D. Aktas, B. Fedrici, T. Lunghi, L. Labonte, and S. Tanzilli. 2016. Optimal analysis for ultra broadband energy-time entanglement for high bit rate dense wavelength division multiplexed quantum networks. *Appl Phys Lett* 108:231108.

Kaneda, F., H. Suzuki, R. Shimizu, and K. Edamatsu. 2019. Direct generation of frequency-bin entangled photons via two-period quasi-phase-matched parametric downconversion. *Opt Express* 27:1416–1424.

Kim, J.-H., J.-W. Chae, Y.-C. Jeong, and Y.-H. Kim. 2022. Quantum communication with time-bin entanglement over a wavelength multiplexed fiber network. *APL Photonics* 7:016106. https://doi.org/10.1063/5.0073040.

Kruse, R., L. Sansoni, S. Brauner, R. Ricken, C. S. Hamilton, I. Jex, and C. Silberhorn. 2015. Dual-path source engineering in integrated quantum optics. *Phys Rev A* 92:053841.

Lugani, J., S. Ghosh, and K. Thyagarajan. 2011. Generation of modal-path entangled photons using a domain engineered integrated optical waveguide device. *Phys Rev A* 83:062333.

———. 2013. Switchable hyperentangled photon pairs from an integrated optic waveguide device. *J Opt Soc Am* 30:795–805.

Ma, Z., J.-Y. Chen, M. Garikapati, Z. Li, C. Tang, Y. M. Sua, and Y.-P. Huang. 2023a. Highly efficient and pure few-photon source on a chip. *Phys Rev Lett* 20:044033.

Ma, J., X. Cheng, N. Zheng, P. Chen, X. Xu, T. Wang, D. Wei, Y. Nie, S. Zhu, M. Xiao, and Y. Zhang. 2023b. Fabrication of 100-nm-period domain structure in lithium niobate on insulator. *Opt Express* 31:37464–37471.

Maclean, J.-P. W., J. M. Donohue, and K. J. Resch. 2018. Direct characterization of ultrafast energy-time entangled photon pairs. *Phys Rev Lett* 120:053601.

Marcikic, I., H. de Reidmatten, W. Tittel, H. Zbinden, M. Legre, and N. Gisin. 2004. Distribution of time-bin entangled qubits over 50 km of optical fiber. *Phys Rev Lett* 93:180502.

Ming, Y., Z. J. Wu, G. Cui, A. Tan, and Y. Lu. 2014. Integrated source of tunable non maximally mode-entangled photons in a domain engineered lithium niobate waveguide. *Appl Phys Lett* 104:171110.

Morrison, C. L., F. Graffitti, P. Barrow, A. Pickston, J. Ho, and A. Fedrizzi. 2022. Frequency-bin entanglement from domain-engineered down conversion. *APL Photonics* 7:066102.

Mueller, A., S. I. Davis, B. Korzh, R. Valivarthi, A. D. Beyer, R. Youssef, N. Sinclair, C. Pena, M. D. Shaw, and M. Spiropulu. 2024. High-rate multiplexed entanglement source based on time-bin qubits for advanced quantum networks. *Opt Quant* 2:64–71.

Reimer, C., M. Kues, P. Roztocki, B. Wetzel, F. Grazioso, B. E. Little, S. T. Chu, T. Johnston, Y. Bromberg, L. Caspani, D. J. Moss, and R. Morandotti. 2016. Generation of multiphoton entangled quantum states by means of integrated frequency combs. *Science* 351:1176–1180.

Rossi, A., G. Vallone, F. De Martini, and P. Mataloni. 2008. Generation of time-bin entangled photons without temporal postselection. *Phys Rev A* 78:012345.

Saleh, M. F., B. E. A. Saleh, and M. C. Teich. 2009. Modal, spectral and polarization entanglement in guided-wave parametric down conversion. *Phys Rev A* 79:053842.

Saleh, M. F., G. Di Giuseppe, B. E. A. Saleh, and M. C. Teich. 2010. Photonic circuits for generating modal, spectral and polarization entanglement. *IEEE Photonics J* 2:736–752.

Setzpfandt, F., A. S. Solntsev, J. Titchener, C. W. Wu, C. Xiong, R. Schiek, T. Pertsch, D. N. Neshev, and A. A. Sukhorukov. 2016. Tunable generation of entangled photons in a nonlinear directional coupler. *Laser Photonics* 10:131–136.

Shih, Y. 2003. Entangled photons. *IEEE J Sel Topics Quant Electron* 9:1455–1467.

Solntsev, A. S., and A. A. Sukhorukov. 2017. Path-entangled photon sources on nonlinear chips. *Rev Phys* 2:19–31.

Suhara, T. 2009. Generation of quantum-entangled twin photons by waveguide nonlinear-optic devices. *Laser Photon Rev* 3:370–393.

Suhara, T., G. Nayaka, J. Kawashima, and M. Fujimura. 2009. Quasi-phase matched waveguide devices for generation of postselection-free polarization entangled twin photons. *IEEE Photon Tech Lett* 21:1096–1098.

Tanzilli, J-S. 2002. *Optique Integree pour les communication quantiques*. Ph D thesis, Universite de Nice Sophia Antipolis, France.

Tanzilli, S., W. Tittel, H. de Reidmatten, H. Zbinden, P. Baldi, M. de Micheli, D. B. Ostrowsky, and N. Gisin. 2002. PPLN waveguide for quantum communication. *Eur J Phys D* 18:155–160.

Thyagarajan, K., J. Lugani, S. Ghosh, K. Sinha, A. Martin, D. B. Ostrowsky, O. Alibart, and S. Tanzilli. 2009. Generation of polarization-entangled photons using type-II doubly periodically poled lithium niobate waveguides. *Phys Rev A* 80:052321.

Tittel, W., J. Brendel, H. Zbinden, and N. Gisin. 1998. Violation of Bell inequalities by photons more than 10 km apart. *Phys Rev Lett* 81:3563–3566.

———. 2000. Quantum cryptography using entangled photons in Energy-Time Bell states. *Phys Rev Lett* 84:4737–4740.

Ueno, W., F. Kaneda, H. Suzuki, S. Nagano, A. Syouji, R. Shimizu, K. Suizu, and K. Edamatsu. 2012. Entangled photon generation in two-period quasi phase matched parametric down conversion. *Opt Express* 20:5508–5517.

Warke, A., and K. Thyagarajan. 2021. Direct generation of two-pair frequency entanglement via dual periodic poling in lithium niobate waveguides. *Eur Phys J Plus* 137:697ra.

Zelmon, D. E., D. L. Small, and D. Jundt. 1997. Infrared corrected Sellemeier coefficients for congruently grown lithium niobate and 5 mol.% magnesium oxide-doped lithium niobate. *J Opt Soc Am* 14:3319–3322.

Zhao, J., C. Ma, M. Rusing, and S. Mookherjea. 2020. High quality entangled photon pair generation in periodically poled thin film lithium niobate waveguides. *Phys Rev Lett* 124: 163603.

Chapter 24
Some Interesting Experiments in Quantum Photonics

Abstract This chapter discusses some interesting experiments in quantum photonics demonstrating indistinguishability and interference, quantum imaging with entangled photons, delayed choice and wave particle duality and provides a brief mathematical explanation of the observed counter-intuitive results.

24.1 Introduction

In this chapter, we will briefly discuss some very interesting and counter-intuitive experiments using photons. We will also show that all the observations of the experiments are explainable using the quantum formulation.

24.2 Indistinguishability and Optical Interference

In an interesting experiment (Zou et al. 1991) using two nonlinear crystals, pump photons at ultraviolet wavelength are incident on a beam splitter BS_p (see Fig. 24.1). The incident pump photon is put into a superposition state (see Chap. 19) with one path going through the nonlinear crystal NL1 and the other path going through an identical nonlinear crystal NL2. The pump photon undergoes SPDC in the two crystals. In view of very low down-conversion probabilities, there is expected to be only one signal–idler pair present in the experiment at any time. The paths of the idlers from NL1 and NL2 are perfectly aligned so that when an idler photon is detected in detector D_i, it is impossible to determine as to whether the idler came from NL1 or from NL2. Consequently, it is impossible to know whether the signal photon that is present in the experiment came from NL1 or from NL2. Thus, the signal photon can arrive through two indistinguishable paths, namely via path s_1 or via path s_2. Since the two paths are indistinguishable for the signal photon arriving on detector D_s, as per Feynman's interference rule, when a signal photon is detected in D_s in coincidence with an idler arriving in detector D_i, we need to add the probability amplitudes corresponding to the signal propagating along the two

K. Thyagarajan, *Integrated Quantum Photonics*, Graduate Texts in Physics,
https://doi.org/10.1007/978-3-031-85728-7_24

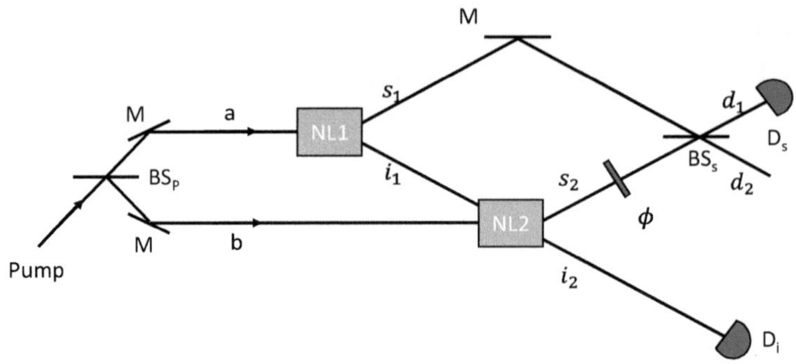

Fig. 24.1 Experimental arrangement to observe interference of signal photons through coincidence counting between the signals from detectors D_s and D_i. *M* Mirrors, *BSp, BSs* beam splitters, *NL1* and *NL2* nonlinear crystals. (Adapted from Zou et al. (1991))

indistinguishable paths. Now, if a phase change is introduced in any one of the signal arms, the phase difference between the two paths of the signal photon changes which results in change in the coincidence count between the detectors D_s and D_i.

An interesting aspect of the experiment is that interference effect is seen only as long as the two idler paths are exactly aligned so that it is *in principle impossible* to know as to which crystal generated the idler photon and consequently the signal photon whose interference is being observed. If the path i_1 is blocked in front of the crystal NL2, then when an idler photon is detected in the detector D_i, it must have been generated from NL2 and thus there is knowledge of the source of the idler photon and consequently the signal photon. In this case, experiment shows that there is no interference effect! Thus, the mere knowledge of the path of the idler is enough to wash off the interference (Zou et al. 1991).

We will now give a simplified analysis of the entire process. The pump photon incident on the beam splitter BS_p generates the following pump state after the beam splitter:

$$|\psi\rangle = \frac{1}{\sqrt{2}}(|1_{Pa}\rangle|0_{Pb}\rangle + i|0_{Pa}\rangle|1_{Pb}\rangle) \tag{24.1}$$

where the subscript P corresponds to pump and a and b correspond to the paths after the beam splitter. After passing through the nonlinear crystals NL1 and NL2, the states $|1_{Pa}\rangle$ and $|1_{Pb}\rangle$ undergo spontaneous down conversion processes with the outputs:

$$|1_{Pa}\rangle \rightarrow |0_{Pa}\rangle|1_{s1}\rangle|1_{i1}\rangle$$

and

$$|1_{Pb}\rangle \rightarrow |0_{Pb}\rangle|1_{s2}\rangle|1_{i2}\rangle$$

Here subscripts s and i correspond to signal and idler, respectively. Hence, the state after the nonlinear crystals NL1 and NL2 is given by

$$
\begin{aligned}
|\chi\rangle &= \frac{1}{\sqrt{2}}\left(|0_{Pa}\rangle|0_{Pb}\rangle|1_{s1}\rangle|1_{i1}\rangle + i|0_{Pa}\rangle|0_{Pb}\rangle e^{i\phi}|1_{s2}\rangle|1_{i2}\rangle\right) \\
&= \frac{1}{\sqrt{2}}|0_{Pa}\rangle|0_{Pb}\rangle\left(|1_{s1}\rangle|1_{i1}\rangle + ie^{i\phi}|1_{s2}\rangle|1_{i2}\rangle\right)
\end{aligned}
\tag{24.2}
$$

where we have introduced a phase change ϕ in the path of the signal photon. Here subscripts $s1$ and $s2$ correspond to the two signal photons and $i1$ and $i2$ correspond to the two idler photons.

Exercise 24.1 From Eq. (24.2) obtain the total number of photons present in the state.

Solution The state given by Eq. (24.2) contains six modes corresponding to Pa, Pb, $s1$, $i1$, $s2$, and $i2$. The total number operator corresponding to these relevant modes is given by

$$
\widehat{N} = \widehat{a}^{\dagger}_{Pa}\widehat{a}_{Pa} + \widehat{a}^{\dagger}_{Pb}\widehat{a}_{Pb} + \widehat{a}^{\dagger}_{s1}\widehat{a}_{s1} + \widehat{a}^{\dagger}_{i1}\widehat{a}_{i1} + \widehat{a}^{\dagger}_{s2}\widehat{a}_{s2} + \widehat{a}^{\dagger}_{i2}\widehat{a}_{i2} \tag{24.3}
$$

We operate on the state $|\chi\rangle$ by the total number operator and obtain

$$
\widehat{N}|\chi\rangle = \left(\widehat{a}^{\dagger}_{Pa}\widehat{a}_{Pa} + \widehat{a}^{\dagger}_{Pb}\widehat{a}_{Pb} + \widehat{a}^{\dagger}_{s1}\widehat{a}_{s1} + \widehat{a}^{\dagger}_{i1}\widehat{a}_{i1} + \widehat{a}^{\dagger}_{s2}\widehat{a}_{s2} + \widehat{a}^{\dagger}_{i2}\widehat{a}_{i2}\right)|\chi\rangle \tag{24.4}
$$

The first term in $|\chi\rangle$ gives us

$$
\begin{aligned}
|\rho_1\rangle = &\left(\widehat{a}^{\dagger}_{Pa}\widehat{a}_{Pa} + \widehat{a}^{\dagger}_{Pb}\widehat{a}_{Pb} + \widehat{a}^{\dagger}_{s1}\widehat{a}_{s1} + \widehat{a}^{\dagger}_{i1}\widehat{a}_{i1} + \widehat{a}^{\dagger}_{s2}\widehat{a}_{s2} + \widehat{a}^{\dagger}_{i2}\widehat{a}_{i2}\right) \\
&\frac{1}{\sqrt{2}}|0_{Pa}\rangle|0_{Pb}\rangle|1_{s1}\rangle|1_{i1}\rangle|0_{s2}\rangle|0_{i2}\rangle
\end{aligned}
$$

where we have specifically included the term $|0_{s2}\rangle|0_{i2}\rangle$ to indicate that in this term there are no photons in the state corresponding to $s2$ and $i2$. Operating on the kets by the individual number operators, we obtain

$$
|\rho_1\rangle = \frac{1}{\sqrt{2}}2|0_{Pa}\rangle|0_{Pb}\rangle|1_{s1}\rangle|1_{i1}\rangle|0_{s2}\rangle|0_{i2}\rangle
$$

Here we have used the fact that $\widehat{a}^{\dagger}_{s1}\widehat{a}_{s1}|1_{s1}\rangle = |1_{s1}\rangle$, $\widehat{a}^{\dagger}_{i1}\widehat{a}_{i1}|1_{i1}\rangle = |1_{i1}\rangle$, $\widehat{a}^{\dagger}_{s2}\widehat{a}_{s2}|0_{s2}\rangle = 0$, etc. Similarly the second term gives us

$$
|\rho_2\rangle = \frac{1}{\sqrt{2}}2|0_{Pa}\rangle|0_{Pb}\rangle|0_{s1}\rangle|0_{i1}\rangle|1_{s2}\rangle|1_{i2}\rangle
$$

Hence, we obtain

$$\widehat{N}|\chi\rangle = 2\frac{1}{\sqrt{2}}|0_{Pa}\rangle|0_{Pb}\rangle\left(|1_{s1}\rangle|1_{i1}\rangle + ie^{i\phi}|1_{s2}\rangle|1_{i2}\rangle\right) = 2|\chi\rangle \tag{24.5}$$

Thus, the state has exactly two photons: one signal photon and one idler photon.

Now at the beam splitter BS_s, the signals s_1 and s_2 transform as follows:

$$|1_{s1}\rangle \to \frac{1}{\sqrt{2}}(i|1_{d1}\rangle|0_{d2}\rangle + |0_{d1}\rangle|1_{d2}\rangle) \tag{24.6}$$

and

$$|1_{s2}\rangle \to \frac{1}{\sqrt{2}}(|1_{d1}\rangle|0_{d2}\rangle + i|0_{d1}\rangle|1_{d2}\rangle) \tag{24.7}$$

Hence, the output state after the beam splitter $\mathrm{BS_s}$ is given by

$$|\psi_{\text{out}}\rangle = \frac{1}{2}|0_{Pa}\rangle|0_{Pb}\rangle\left[(i|1_{d1}\rangle|0_{d2}\rangle + |0_{d1}\rangle|1_{d2}\rangle)|1_{i1}\rangle + ie^{i\phi}(|1_{d1}\rangle|0_{d2}\rangle + i|0_{d1}\rangle|1_{d2}\rangle)|1_{i2}\rangle\right] \tag{24.8}$$

If the idlers i_1 and i_2 are identical and perfectly aligned, then the modes corresponding to i_1 and i_2 represent the same mode and we can write $|1_{i1}\rangle = |1_{i2}\rangle$. Using this in Eq. (24.8), we obtain

$$|\psi_{\text{out}}\rangle = \frac{1}{2}|0_{Pa}\rangle|0_{Pb}\rangle|1_{i1}\rangle\left[(i|1_{d1}\rangle|0_{d2}\rangle + |0_{d1}\rangle|1_{d2}\rangle) + ie^{i\phi}(|1_{d1}\rangle|0_{d2}\rangle + i|0_{d1}\rangle|1_{d2}\rangle)\right] \tag{24.9}$$

which can be simplified to

$$|\psi_{\text{out}}\rangle = \frac{1}{2}|0_{Pa}\rangle|0_{Pb}\rangle|1_{i1}\rangle\left[i(1 + e^{i\phi})|1_{d1}\rangle|0_{d2}\rangle + (1 - e^{i\phi})|0_{d1}\rangle|1_{d2}\rangle\right] \tag{24.10}$$

It is important to note that we have obtained the above equation under the condition that the modes corresponding to i_1 and i_2 are identical.

The probability of detecting an idler photon in arm i_2 and simultaneously a signal photon in arm d_1 is given by

$$P_{i_1,d_1} = |\langle 1_{i1}|\langle 1_{d1}|\psi_{\text{out}}\rangle|^2 = \frac{1}{2}(1 + \cos\phi) \tag{24.11}$$

Similarly the probability of detecting an idler photon and simultaneously a signal photon in arm d_2 is given by

$$P_{i_1,d_2} = |\langle 1_{i1}|\langle 1_{d2}|\psi_{\text{out}}\rangle|^2 = \frac{1}{2}(1 - \cos\phi) \tag{24.12}$$

These expressions clearly show interference effects as explained earlier with the probabilities depending on the phase shift ϕ introduced in the signal arm. It is interesting to note that the single signal photon is either generated in NL1 or in NL2 and in this case, these are indistinguishable thus leading to interference due to absence of information of the path of the signal photon.

In fact, since the idler state $|1_{i1}\rangle$ is factored out of the expression, the probability of detecting a signal photon in d_1 without any consideration for the idler photon is still given by Eq. (24.11).

Here comes an interesting part. If the idler i_1 is blocked from propagating to the nonlinear crystal NL2 and further, then the output state becomes

$$|\psi_{\text{out}}\rangle = \frac{1}{2}|0_{Pa}\rangle|0_{Pb}\rangle\left[(i|1_{d1}\rangle|0_{d2}\rangle + |0_{d1}\rangle|1_{d2}\rangle)|0_{i1}\rangle + ie^{i\phi}(|1_{d1}\rangle|0_{d2}\rangle + i|0_{d1}\rangle|1_{d2}\rangle)|1_{i2}\rangle\right] \tag{24.13}$$

and the idler state cannot now be factored out. The probability of detecting an idler photon and a signal photon in arm d_1 is now given by

$$P_{i_1,d_1} = |\langle 1_{i2}|\langle 1_{d1}|\psi_{\text{out}}\rangle|^2 = \frac{1}{4} \tag{24.14}$$

and we do not observe any interference effect at the detector. This is due to the fact that when the idler i_1 is blocked, then the two signal photon paths become distinguishable. If an idler photon is detected, then it implies that it was generated in the crystal NL2 which also implies that the signal photon came from NL2, thus giving away the path information of the signal photon. Also if no idler photon is detected but a signal photon is detected, then it implies that the signal photon was generated by the crystal NL1, again giving away the path information. The information of the path of the signal photon is coming from the entangled idler photon and not the signal photon itself.

The question that arises is that if there is no detector to detect the idler photon would we be able to observe interference pattern? The probability of detecting a signal photon in the path d_1 is given by

$$P_{d_1} = |\langle 1_{d1}|\psi_{\text{out}}\rangle|^2 = \langle\psi_{\text{out}}|1_{d1}\rangle\langle 1_{d1}|\psi_{\text{out}}\rangle = \frac{1}{2} \tag{24.15}$$

again showing no interference effect.

This very interesting experiment shows that the mere possibility of existence of information about the path taken, and not necessarily having information about the path, destroys interference effects. The two interfering paths of the signal photon are revealed by the idler photon which is of course entangled to the signal photon.

Experiments clearly show interference effect of the coincidence counting rate between detectors D_s and D_i as the transverse position of the beam splitter BS_s is changed introducing a phase difference between the two paths when the two idler paths are indistinguishable (Zou et al. 1991). At the same time when the idler i_1 is blocked, the coincidence measurements do not show any interference effect. The experiment shows a strange behavior since the two signal paths correspond to signal photons generated by two different crystals and between which interference is not expected.

A brief review of quantum effects in interference of photons can be found in Mandel (1999).

Exercise 24.2 Consider a Mach-Zehnder interferometer in which horizontally polarized single photons are incident. On one of the arms is a half wave plate with its principal axis oriented at an angle θ with the horizontal. A phase plate is inserted in one of the paths so that it introduces a phase difference ϕ between the two arms of the interferometer. In such a case obtain the probability of detecting the photon in one of the output arms. What is the effect of the angle θ on the probability?

Solution The state of the photon after the first beam splitter is given by (see Chap. 19)

$$|\psi\rangle = \frac{1}{\sqrt{2}}(|H_a\rangle|0_b\rangle + i|0_a\rangle|H_b\rangle) = \frac{1}{\sqrt{2}}(|H_a, 0_b\rangle + i|0_a, H_b\rangle) \quad (24.16)$$

where a corresponds to the transmitted path and b to the reflected path (see Fig. 24.2) and H represents horizontal polarization. Let us assume that the half wave plate is placed in the transmitted path a. The half wave plate rotates the polarization state by an angle 2θ and hence the half wave plate transforms the state $|H_a\rangle$ to the state $\frac{1}{\sqrt{2}}(|H_a\rangle \cos 2\theta + |V_a\rangle \sin 2\theta)$. Hence, the state after the half wave plate will be given by

$$|\psi\rangle = \frac{1}{\sqrt{2}}(|H_a, 0_b\rangle \cos 2\theta + |V_a, 0_b\rangle \sin 2\theta) + \frac{1}{\sqrt{2}} i|0_a, H_b\rangle \quad (24.17)$$

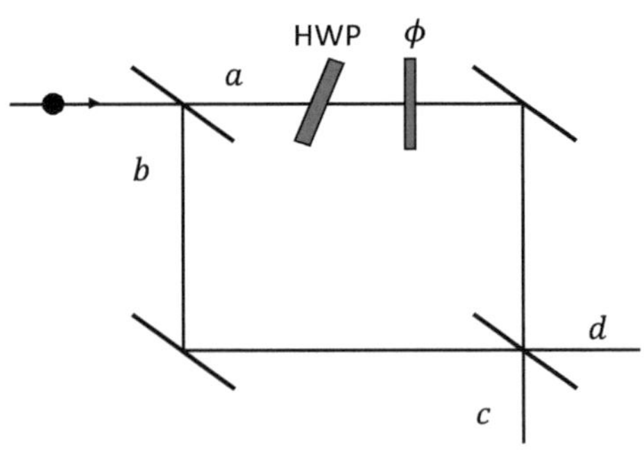

Fig. 24.2 A Mach-Zehnder interferometer in which in one of the arms, a half wave plate, and a phase shifter are introduced

Let us assume that there is a polarization independent phase shift of ϕ in arm a with respect to arm b. Then the state before entering the second beam splitter is given by

$$|\psi\rangle = \frac{1}{\sqrt{2}}(|H_a, 0_b\rangle \cos 2\theta + |V_a, 0_b\rangle \sin 2\theta)e^{i\phi} + \frac{1}{\sqrt{2}} i|0_a, H_b\rangle \quad (24.18)$$

At the second beam splitter, the following transformation takes place:

$$|H_a\rangle \rightarrow \frac{1}{\sqrt{2}}(|H_c\rangle + i|H_d\rangle) \quad (24.19)$$

$$|V_a\rangle \rightarrow \frac{1}{\sqrt{2}}(|V_c\rangle + i|V_d\rangle) \quad (24.20)$$

and

$$|H_b\rangle \rightarrow \frac{1}{\sqrt{2}}(i|H_c\rangle + |H_d\rangle) \quad (24.21)$$

where c and d correspond to the two paths after the beam splitter. Using these expressions in Eq. (24.18), we obtain

$$|\psi\rangle = \frac{1}{2}\cos 2\theta(|H_c, 0_d\rangle + i|0_c, H_d\rangle)e^{i\phi} + \frac{1}{2}\sin 2\theta(|V_c, 0_d\rangle + i|0_c, V_d\rangle)e^{i\phi} + \frac{i}{2}(i|H_c, 0_d\rangle + |0_c, H_d\rangle)$$

which can be simplified to

$$|\psi\rangle = \left(\frac{1}{2}\cos 2\theta e^{i\phi} - \frac{1}{2}\right)|H_c, 0_d\rangle + \left(\frac{i}{2}\cos 2\theta e^{i\phi} + \frac{i}{2}\right)|0_c, H_d\rangle + \frac{1}{2}\sin 2\theta e^{i\phi}|V_c, 0_d\rangle + \frac{i}{2}\sin 2\theta e^{i\phi}|0_c, V_d\rangle \quad (24.22)$$

Hence, the probability of detecting a photon in arm c of the interferometer irrespective of the state of polarization is

$$P_c = |\langle H_c, 0_d|\psi\rangle|^2 + |\langle V_c, 0_d|\psi\rangle|^2 = \frac{1}{2}(1 - \cos 2\theta \cos \phi) \quad (24.23)$$

If $\theta = 0$, then the half wave plate has no effect on the polarization state of the photon in path a and the probability varies with the phase difference ϕ as

$$P_c = \frac{1}{2}(1 - \cos \phi) \tag{24.24}$$

showing single photon interference between the two paths. The two paths become completely indistinguishable and the probability in this case varies from 0 to 1. Similarly for the case when $\theta = \frac{\pi}{2}$.

If $\theta = \frac{\pi}{4}$, then along path a the half wave plate transforms the horizontal polarization to vertical polarization with the result that the two paths a and b through the interferometer become fully distinguishable. In this case, we have

$$P_c = \frac{1}{2} \tag{24.25}$$

and the no interference is observed. Thus, complete path information erases the interference effects completely.

As θ increases from 0 to $\frac{\pi}{4}$, the indistinguishability between the two paths decreases resulting in lower contrast in the probability variation with ϕ. The maximum probability is $P_{c,\max} = \frac{1}{2}(1 + \cos 2\theta)$ and the minimum probability is $P_{c,\min} = \frac{1}{2}(1 - \cos 2\theta)$. We can define visibility as $V = \frac{P_{c,\max} - P_{c,\min}}{P_{c,\max} + P_{c,\min}} = \cos 2\theta$. Also note that since in one of the arms the state of polarization is horizontal, the component $|V_a\rangle \sin 2\theta$ in the other arm of the interferometer is responsible for the distinguishability of the two paths. As we make the two paths more and more distinguishable by increasing θ from 0 to $\frac{\pi}{4}$, the contrast or the visibility in the interference effect decreases. The more indistinguishable the two paths are, the greater is the contrast of the probability variation with change of phase. There is a close connection between distinguishability and visibility of the interference pattern in the experiment.

24.3 Quantum Imaging with Entangled Photons

In this section, we consider another very interesting experiment which shows that knowledge can be extracted by and about a photon that is never detected (Lemos et al. 2014). Figure 24.3 shows the experimental arrangement similar to the one considered in Sect. 24.2 except that here there is no coincidence counting. Similar to the experiment detailed in Sect. 24.2, a pump photon passes through a beam splitter BS_p and is put in a superposition state with one path going through a nonlinear crystal NL1 and another path through a nonlinear crystal NL2 where SPDC takes place and a signal–idler pair is generated. Following this there is a dichroic beam splitter which reflects the idler photon (shown as dashed line) and transmits the signal (shown as solid line) photon. The generated idler photon passes through the object to be imaged and is then combined with the other path of the pump photon in another dichroic beam splitter. The two beams (pump and idler from NL1) then pass

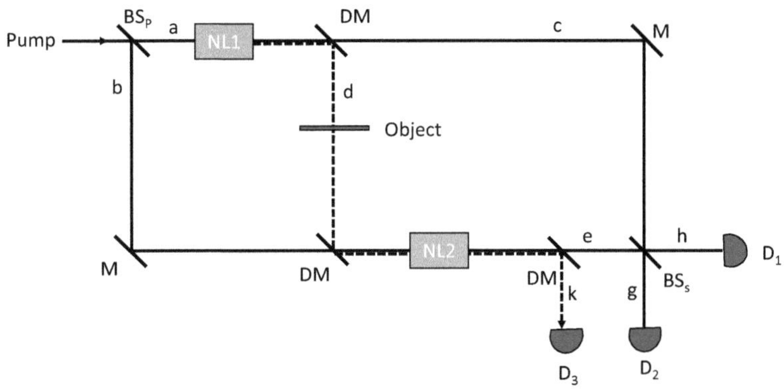

Fig. 24.3 Experimental arrangement to image objects using entangled photon pairs. *BSp* Pump beam splitter; *BSs* Signal beam splitter, *M* Mirror, *DM* Dichroic mirror, *D* Detector. *NL1, NL2* Nonlinear crystals. (Adapted from Lemos et al. (2014))

through another identical nonlinear crystal NL2 through which the idler photon passes unchanged, while the pump photon undergoes an SPDC process to generate an identical signal–idler pair of photons. The idler photon is reflected and the two paths of the signal photon one from NL1 and the other from NL2 are combined at a beam splitter BS$_s$. The signal photon is then detected and no coincidence measurements are made. The single photon counts at either of the detectors D_1 or D_2 show interference effects.

If the detectors are replaced with cameras sensitive to single photons, an image of an object placed on the path of the idler photon generated by NL1 can be generated (see Lemos et al. 2014).

The same experimental arrangement also shows the concept of interaction-free measurement. If no object is placed on the path of the idler photon from NL1, then there is no signal at the detector D_1. However, if an object is placed which blocks the photon in the path of the idler from NL1, then coincidence counts in the detector D_1 and detector D_3 reveal the presence of the object, although no photon has interacted with the object.

Here we give a simple mathematical basis of the experiment. We use subscripts P, s, and i for the pump, the signal, and the idler, respectively.

The state of pump photon after the beam splitter BS$_P$ is (see Chap. 19)

$$|\psi_{\text{in}}\rangle = \frac{1}{\sqrt{2}}(|1_{Pa}, 0_{Pb}\rangle + i|0_{Pa}, 1_{Pb}\rangle) \tag{24.26}$$

where the subscripts a and b refer to the paths marked a and b in Fig. 24.3. After NL1 the state $|1_{Pa}\rangle$ becomes (before interacting with the object)

$$|1_{Pa}\rangle \rightarrow |1_{sc}\rangle|1_{id}\rangle$$

After interacting with the object which is represented by a transmission $Te^{i\phi}$ the state is

$$|1_{sc}\rangle|1_{id}\rangle Te^{i\phi}$$

Similarly, the pump photon passing through NL2 becomes

$$|1_{Pb}\rangle \rightarrow |1_{se}\rangle|1_{ik}\rangle$$

where is it assumed that the idler photon is reflected to the reflected arm (represented by k) of the dichroic mirror DM.

When the signal photon in arm c interacts with the beam splitter BSs, the state evolves to

$$|1_{sc}\rangle \rightarrow \frac{1}{\sqrt{2}}\left(i|1_{sh}, 0_{sg}\rangle + |0_{sh}, 1_{sg}\rangle\right)$$

Similarly, after passing through the beam splitter BSs, the signal photon on arm e becomes

$$|1_{se}\rangle \rightarrow \frac{1}{\sqrt{2}}\left(|1_{sh}, 0_{sg}\rangle + i|0_{sh}, 1_{sg}\rangle\right)$$

Hence, the state of the signal photons after beam splitter BSs will be

$$|\psi_{\text{out}}\rangle = \frac{1}{2}\left(i|1_{sh}, 0_{sg}, 1_{ik}\rangle Te^{i\phi} + |0_{sh}, 1_{sg}, 1_{ik}\rangle Te^{i\phi} + i|1_{sh}, 0_{sg}, 1_{ik}\rangle - |0_{sh}, 1_{sg}, 1_{ik}\rangle\right) \tag{24.27}$$

The probability of detecting a signal photon in arm h will be given by

$$P_h = |\langle 1_{sh}|\psi_{\text{out}}\rangle|^2 = \langle \psi_{\text{out}}|1_{sh}\rangle\langle 1_{sh}|\psi_{\text{out}}\rangle = \frac{1}{4}\left(1 + 2T\cos\phi + T^2\right) \tag{24.28}$$

and that of detecting a signal photon in arm g will be given by

$$P_g = \langle \psi_{\text{out}}|1_{sg}\rangle\langle 1_{sg}|\psi_{\text{out}}\rangle = \frac{1}{4}\left(1 - 2T\cos\phi + T^2\right) \tag{24.29}$$

One can see that the probabilities of detecting a signal in the signal output arms depend on the transmittance of the object and show interference in spite of the fact that the signal photons never interacted with the object!

Also it is interesting to observe that if there is no object present, then $T = 1$, $\phi = 0$ and we get $P_g = 0$. On the other hand, if an object is present, then $T\cos\phi$ is finite and $P_g \neq 0$ which reveals the presence of the object in the arm although the signal photon being detected never interacted with the object. This is referred to as interaction-free measurement and is very interesting from the point of view of its applications. Since the two photons can be of different wavelengths, this technique can be used for probing delicate biological samples by sending low energy photons through the sample and using visible range photons to form the image (Lemos et al. 2014). See also Pearce et al. (2023).

Exercise 24.3 What would happen if the object placed in the path d is a pure phase object which only changes the phase without affecting the transmission?

Solution For a pure phase object, $T = 1$ and the phase ϕ will depend on the transverse coordinates. Thus, in this case the probability P_h will be given by

$$P_h = \frac{1}{4}(1 + 2\cos\phi + 1) = \frac{1}{2}(1 + \cos\phi)$$

implying interference and that the phase change due to the object on the idler photon is revealed by the probability of detecting a signal photon.

24.4 Detecting an Object Without Interacting with It

We now discuss another experiment in which it is shown that using single photons it is possible to detect the presence of an object using a photon that has never interacted with it. The experiment is essentially a symmetric Mach-Zehnder interferometer with two 50% beam splitters and two detectors which are placed at the output (see Fig. 24.4).

We again use a similar analysis to that given in Sect. 24.3.

We assume single photons to be incident on a Mach-Zehnder interferometer. After the beam splitter BS1, we have the following state:

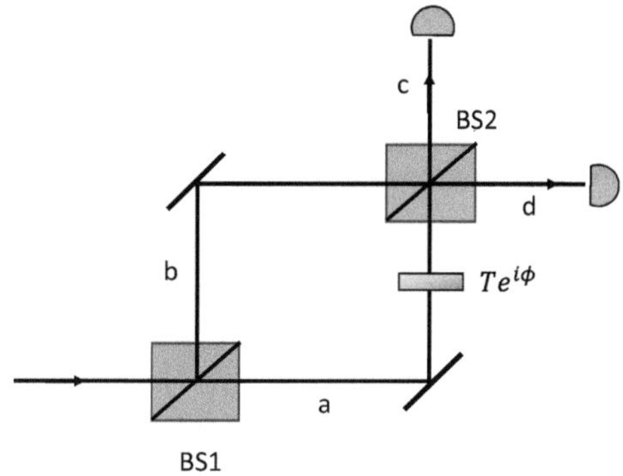

Fig. 24.4 Setup for interaction-free measurement using a Mach-Zehnder interferometer with an object in one of the arms of the interferometer

$$|\psi_{\text{in}}\rangle = \frac{1}{\sqrt{2}}(|1_a, 0_b\rangle + i|0_a, 1_b\rangle) \tag{24.30}$$

That is, the photon exits in a superposition state. Path a intersects an object with a transmittance given by $Te^{i\phi}$. Hence, the state after the path a crosses the beam splitter BS2 becomes

$$|1_a\rangle \rightarrow \frac{1}{\sqrt{2}}(|1_c, 0_d\rangle + i|0_c, 1_d\rangle)\mathrm{T}e^{i\phi}$$

and path b when it crosses beam splitter BS2 becomes

$$|1_b\rangle \rightarrow \frac{1}{\sqrt{2}}(i|1_c, 0_d\rangle + |0_c, 1_d\rangle)$$

Thus, the overall state after the beam splitter BS2 can be written as

$$|\psi_{\text{out}}\rangle = \frac{1}{2}(|1_c, 0_d\rangle + i|0_c, 1_d\rangle)Te^{i\phi} + \frac{1}{2}i(i|1_c, 0_d\rangle + |0_c, 1_d\rangle)$$

which can be written as

$$|\psi_{\text{out}}\rangle = \frac{1}{2}|1_c, 0_d\rangle\left(Te^{i\phi} - 1\right) + \frac{1}{2}|0_c, 1_d\rangle\left(iTe^{i\phi} + i\right) \tag{24.31}$$

The probability of detecting a photon in arms c and d are, respectively, given by

$$P_c = |\langle 1_c, 0_d|\psi_{\text{out}}\rangle|^2 = \frac{1}{4}\left(1 - 2T\cos\phi + T^2\right) \tag{24.32}$$

and

$$P_d = |\langle 0_c, 1_d|\psi_{\text{out}}\rangle|^2 = \frac{1}{4}\left(1 + 2T\cos\phi + T^2\right) \tag{24.33}$$

If there is no object in path a, then $T = 1$ and $\phi = 0$ and we have $P_d = 1$ and $P_c = 0$. Thus, all the incident photons arrive at the port d and none of the photons arrive in port c.

Now if an opaque object is present in path a, then $T = 0$ and $P_c = 0.25$. Thus, when we send single photons through the interferometer, there are three possibilities:

(a) None of the two detectors click. This may be because the photon has been scattered or absorbed by the object. The probability of this occurring is 0.5.
(b) Detector on path d clicks. The probability of this occurring is 0.25. This does not confirm the presence or the absence of an object on the path a.

(c) Detector at port *c* clicks. The probability of this occurring is 0.25. In this case, we are sure that there is an object in the path *a* thus revealing the presence of the object.

Note that since the object in path *a* is opaque, the photon that we are detecting has passed through path *b* and thus has not even interacted with the object!

This is the well-known Elitzur-Vaidman experiment that the two scientists had suggested (Elitzur and Vaidman 1993). This again corresponds to an interaction-free measurement. For interesting experimental results using such an arrangement, see White et al. (1998).

24.5 Delayed Choice Experiment and Wave Particle Duality

We consider the two experimental setups shown in Fig. 24.5. The first one has only one beam splitter, while the second one has two beam splitters in the paths of the beams. It is obvious that the first setup corresponds to photons incident on just a single beam splitter and the incident photon has equal probability of being found in any of the two output arms. In the second experimental setup, the second beam splitter remixes the two paths and corresponds to a Mach-Zehnder interferometer. In this setup any incident photon can be made to exit only through one of the arms by adjusting the phase along the paths of the beams. In the first experiment, the photon seems to exhibit particle behavior, while in the second experiment it seems to exhibit wave behavior.

John Archibald Wheeler proposed what is referred to as delayed choice experiment to address the issue as to whether the photon entering the setup somehow "senses" the experimental setup and behaves accordingly either as a wave or as a particle.

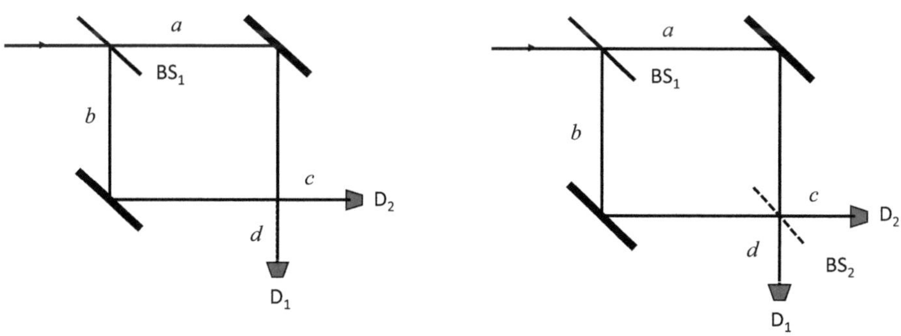

Fig. 24.5 Without the beam splitter BS_2 the incident photon acts like a particle and has equal probability of appearing in either of the output arms. With BS_2 in position, the photon acts like a wave. The question is what happens if the choice of introducing BS_2 is implemented after the photon passes BS_1

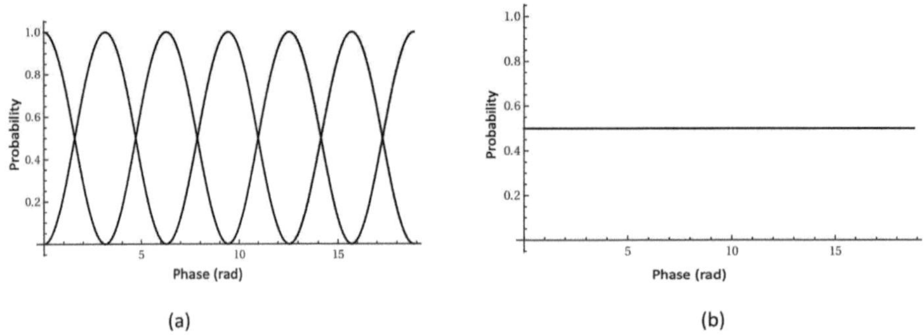

Fig. 24.6 Schematic of the results of the delayed choice experiment. (**a**) Shows interference pattern when both paths are made to interfere with the beam splitter BS2 in position and (**b**) when BS2 is absent corresponding to particle behavior. For actual experimental results see Jacques et al. (2007)

In order to check this possibility an experiment is carried out in which the beam splitter BS2 is introduced in the setup after the photon has entered the experimental setup via beam splitter BS1 at which point the photon needs to "decide" to either exhibit particle behavior by proceeding along one of the two paths or wave behavior. This experiment was carried out by Jacques et al. (2007). In the experiment, the decision to insert or not the beam splitter BS2 is left to a quantum random number generator so that it is completely random.

Figure 24.6 shows a schematic representing the experimental results of the output with and without the beam splitter BS2. As can be seen the behavior of the photon in the interferometer depends on the choice of the observable that is measured, irrespective of when and where the choice is made.

Exercise 24.4 Referring to Fig. 24.5b, if the beam splitter BS_1 has energy reflectivity and transmittivity of R_1 and T_1, respectively, and the beam splitter BS_2 has energy reflectivity and transmittivity of R_2 and T_2, respectively, obtain the probability of detecting a photon in the two output arms of the interferometer.

Solution Assuming a single photon to be incident on the interferometer, the state after BS_1 will be given by

$$|\psi\rangle = \sqrt{T_1}|1_a, 0_b\rangle + i\sqrt{R_1}|0_a, 1_b\rangle$$

Assuming a phase change of ϕ in arm a of the interferometer, the state before reaching the beam splitter BS_2 will be

$$|\psi\rangle = \sqrt{T_1}|1_a, 0_b\rangle e^{i\phi} + i\sqrt{R_1}|0_a, 1_b\rangle$$

At the beam splitter BS2, the following transformation takes place:

$$|1_a, 0_b\rangle \rightarrow i\sqrt{R_2}|1_c, 0_d\rangle + \sqrt{T_2}|0_c, 1_d\rangle$$

$$|0_a, 1_b\rangle \rightarrow \sqrt{T_2}|1_c, 0_d\rangle + i\sqrt{R_2}|0_c, 1_d\rangle$$

Thus, the output state will be given by

$$|\psi\rangle = \sqrt{T_1}\left(i\sqrt{R_2}|1_c, 0_d\rangle + \sqrt{T_2}|0_c, 1_d\rangle\right)e^{i\phi} + i\sqrt{R_1} \times \left(\sqrt{T_2}|1_c, 0_d\rangle + i\sqrt{R_2}|0_c, 1_d\rangle\right)$$

which can be simplified to give

$$|\psi\rangle = |1_c, 0_d\rangle\left(i\sqrt{R_2T_1}e^{i\phi} + i\sqrt{R_1T_2}\right) + |0_c, 1_d\rangle\left(\sqrt{T_1T_2}e^{i\phi} - \sqrt{R_1R_2}\right)$$

Hence, the probabilities of detecting the photons in the output arms c and d will be given by

$$P_c = R_1T_2 + R_2T_1 + 2\sqrt{R_1R_2T_1T_2}\cos\phi$$

and

$$P_d = R_1R_2 + T_1T_2 - 2\sqrt{R_1R_2T_1T_2}\cos\phi$$

Note that $P_c + P_d = 1$ as it should be. The maximum and minimum probabilities are given by $P_{c,max} = \left(\sqrt{R_1T_2} + \sqrt{R_2T_1}\right)^2$ and $P_{c,\min} = \left(\sqrt{R_1T_2} - \sqrt{R_2T_1}\right)^2$which gives a visibility of

$$V = \frac{P_{c,\max} - P_{c,\min}}{P_{c,\max} + P_{c,\min}} = \frac{2\sqrt{R_1R_2T_1T_2}}{(R_1T_2 + R_2T_1)}$$

24.5.1 Quantum Beam Splitter

Consider an experimental arrangement shown in Fig. 24.7. It consists of a beam splitter BS1 which is a polarization-independent 50% beam splitter and a beam splitter BS2 which is a polarization-dependent beam splitter that reflects horizontal polarization while behaving as a 50% beam splitter for the vertical polarization. If the incident photon is horizontally polarized, then since the beam splitter BS2 completely reflects the horizontal polarization, the photon behavior will be that of a particle with 50% chance of appearing in the output path c and 50% chance of

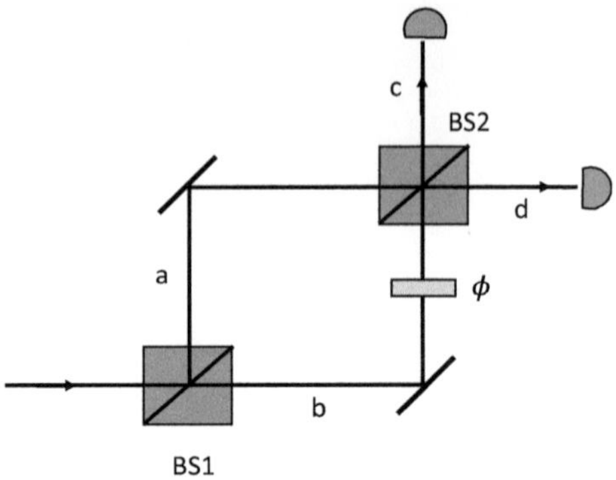

Fig. 24.7 A quantum beam splitter. BS1 is a 50% beam splitter independent of polarization, while the beam splitter BS2 has 100 % reflectance for horizontal polarization and is a 50% beam splitter for the vertical polarization. ϕ represents the variable phase that can be introduced in path b of the photon

appearing in the output path d and is independent of the phase ϕ introduced in the arm b.

On the other hand, if the incident photon is vertically polarized, then since the beam splitter BS2 is a 50% reflector for the vertical polarization, the photon will behave like a wave and lead to interference at the output arms c and d with the probability depending on the phase shift ϕ introduced in arm b.

We now consider a situation when the incident photon is in a superposition state of polarization and given by

$$|\psi\rangle_{\text{in}} = \alpha|H\rangle + \beta|V\rangle \tag{24.34}$$

with $|\alpha|^2 + |\beta|^2 = 1$.

Since the beam splitter BS1 is a polarization independent 50% beam splitter, the state after BS1 can be written as

$$\alpha\frac{1}{\sqrt{2}}(i|H_a, 0_b\rangle + |0_a, H_b\rangle) + \beta\frac{1}{\sqrt{2}}(i|V_a, 0_b\rangle + |0_a, V_b\rangle)$$

Just before entering the beam splitter BS2, the state becomes

$$\alpha\frac{1}{\sqrt{2}}\left(i|H_a, 0_b\rangle + e^{i\phi}|0_a, H_b\rangle\right) + \beta\frac{1}{\sqrt{2}}\left(i|V_a, 0_b\rangle + e^{i\phi}|0_a, V_b\rangle\right)$$

where we assume that the phase change ϕ is independent of the state of polarization.

In view of the polarization dependence of the beam splitter BS2, at the beam splitter BS2, the following transformation takes place of each of the above components:

$$|H_a, 0_b\rangle \to i|H_c, 0_d\rangle$$

$$|0_a, H_b\rangle \to i|0_c, H_d\rangle$$

$$|V_a, 0_b\rangle \rightarrow \frac{1}{\sqrt{2}}(i|V_c, 0_d\rangle + |0_c, V_d\rangle)$$

$$|0_a, V_b\rangle \rightarrow \frac{1}{\sqrt{2}}(|V_c, 0_d\rangle + i|0_c, V_d\rangle)$$

Hence, the complete state after the beam splitter BS2 will be given by

$$\begin{aligned}|\psi\rangle_{\text{out}} = &\alpha\frac{1}{\sqrt{2}}(-|H_c, 0_d\rangle + ie^{i\phi}|0_c, H_d\rangle) \\ &+\frac{\beta}{2}[(-|V_c, 0_d\rangle + i|0_c, V_d\rangle) + e^{i\phi}(|V_c, 0_d\rangle + i|0_c, V_d\rangle)]\end{aligned} \tag{24.35}$$

which can be simplified to

$$|\psi\rangle_{\text{out}} = \alpha\frac{1}{\sqrt{2}}(-|H_c, 0_d\rangle + ie^{i\phi}|0_c, H_d\rangle) + \frac{\beta}{2}[|V_c, 0_d\rangle(-1 + e^{i\phi}) + i|0_c, V_d\rangle(1 + e^{i\phi})]$$

or

$$\begin{aligned}|\psi\rangle_{out} = &\alpha\frac{1}{\sqrt{2}}(-|H_c, 0_d\rangle + ie^{i\phi}|0_c, H_d\rangle) \\ &+i\beta e^{i\phi/2}\left[|V_c, 0_d\rangle \sin\left(\frac{\phi}{2}\right) + |0_c, V_d\rangle \cos\left(\frac{\phi}{2}\right)\right]\end{aligned} \tag{24.36}$$

Note that the output has both the particle aspect and the wave aspect; the particle aspect is associated with the horizontal component of the polarization and the wave aspect is associated with the vertical component of the polarization with the corresponding output depending on the phase ϕ. Thus, as α changes from the value 1 to 0, the behavior changes from particle nature to a wave nature demonstrating interesting situation for intermediate values. In this context, it may be mentioned that the wave nature of the photon is associated with the ability to produce interference and the particle nature to the inability to produce an interference effect.

A very interesting extension of the delayed choice experiment was carried out by Kaiser et al. (2012) using polarization entangled photon pairs wherein the photon entering the beam splitter BS1 of Fig. 24.7 is in a polarization entangled state:

$$|\phi^+\rangle = \frac{1}{\sqrt{2}}(|H_s, H_i\rangle + |V_s, V_i\rangle) \tag{24.37}$$

with the photon with subscript s (signal) entering the quantum beam splitter shown in Fig. 24.8 and the other photon with subscript i (idler) is sent to another beam splitter after passing through an electro-optic modulator (see Fig. 24.8).

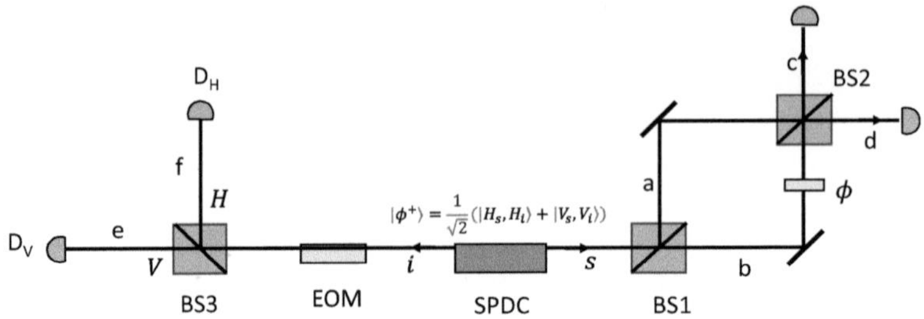

Fig. 24.8 Schematic of the experimental setup described in Kaiser et al. (2012). The SPDC source sends polarization entangled photon pairs. *EOM* Electro-optic modulator

The electro-optic modulator is used to rotate the polarization state of the photon and if the angle of rotation is given by θ, then after the electro-optic modulator, the state of the idler photon gets changed as represented below:

$$|H_i\rangle \rightarrow \cos\theta|H_i\rangle + \sin\theta|V_i\rangle$$

and

$$|V_i\rangle \rightarrow -\sin\theta|H_i\rangle + \cos\theta|V_i\rangle$$

Following an analysis similar to that presented earlier, we can obtain the output state of the system as

$$\begin{aligned}|\psi\rangle_{\text{out}} = &\frac{1}{2}(-|H_{sc},0_{sd}\rangle + ie^{i\phi}|0_{sc},H_{sd}\rangle)(\cos\theta|H_i\rangle + \sin\theta|V_i\rangle)\\ &+i\frac{1}{\sqrt{2}}e^{i\phi/2}\left[|V_{sc},0_{sd}\rangle\sin\left(\frac{\phi}{2}\right) + |0_{sc},V_{sd}\rangle\cos\left(\frac{\phi}{2}\right)\right](-\sin\theta|H_i\rangle + \cos\theta|V_i\rangle)\end{aligned} \tag{24.38}$$

Let us first assume that $\theta = 0$. Then the output state will be

$$\begin{aligned}|\psi\rangle_{out} = &\frac{1}{2}(-|H_{sc},0_{sd}\rangle + ie^{i\phi}|0_{sc},H_{sd}\rangle)|H_i\rangle\\ &+i\frac{1}{\sqrt{2}}e^{i\phi/2}\left[|V_{sc},0_{sd}\rangle\sin\left(\frac{\phi}{2}\right) + |0_{sc},V_{sd}\rangle\cos\left(\frac{\phi}{2}\right)\right]|V_i\rangle\end{aligned} \tag{24.39}$$

If now the detector D_{H} clicks, it implies that the idler photon is projected in to the state $|H_i\rangle$ projecting the signal photon to the state (apart from a normalization constant)

$$(-|H_{sc},0_{sd}\rangle + ie^{i\phi}|0_{sc},H_{sd}\rangle)$$

which shows *particle behavior* of the photon with equal probabilities of 50% chance of being detected by the detector in path c and by the detector in path d with no effect of the phase ϕ on the detection.

On the other hand, if the detector D_V clicks, it implies that the idler photon is projected in to the state $|V_i\rangle$ projecting the signal photon to the following state (apart from a normalization constant)

$$|V_{sc}, 0_{sd}\rangle \sin\left(\frac{\phi}{2}\right) + |0_{sc}, V_{sd}\rangle \cos\left(\frac{\phi}{2}\right)$$

which shows the *wave behavior* of the signal photon; the probability being detected at the output arms c and d now depending on the phase ϕ. The above discussion implies that the behavior of the signal photon is now consequent on the idler photon.

If $\theta \neq 0$, then the detection of the idler photon by the detector D_H implies that the signal photon is projected into the following state:

$$\frac{1}{2}\cos\theta(-|H_{sc}, 0_{sd}\rangle + ie^{i\phi}|0_{sc}, H_{sd}\rangle)$$
$$-i\frac{1}{\sqrt{2}}\sin\theta e^{i\phi/2}\left[|V_{sc}, 0_{sd}\rangle \sin\left(\frac{\phi}{2}\right) + |0_{sc}, V_{sd}\rangle \cos\left(\frac{\phi}{2}\right)\right]$$

which is indeed a mixture of both particle and wave aspects.

If the idler photon is detected by the detector D_V, then the signal photon is projected to the following state:

$$\frac{1}{2}\sin\theta(-|H_{sc}, 0_{sd}\rangle + ie^{i\phi}|0_{sc}, H_{sd}\rangle)$$
$$+i\frac{1}{\sqrt{2}}e^{i\phi/2}\cos\theta\left[|V_{sc}, 0_{sd}\rangle \sin\left(\frac{\phi}{2}\right) + |0_{sc}, V_{sd}\rangle \cos\left(\frac{\phi}{2}\right)\right]$$

which is again a mixture of both particle and wave aspects.

The coincidence probability that the detector in the arm c and the detector D_H click is given by

$$P(c, D_H) = \frac{1}{4}\cos^2\theta + \frac{1}{2}\sin^2\theta \sin^2\left(\frac{\phi}{2}\right) \tag{24.40}$$

Note that when $\theta = 0$, then the coincidence probability is independent of the phase ϕ showing the particle behavior. When $\theta = \frac{\pi}{2}$, then the coincidence varies sinusoidally with the phase ϕ, showing wave like behavior. With $0 < \theta < \frac{\pi}{2}$, the behavior of the signal photon continuously varies between the particle and wave aspects.

For more details of the experiment, readers are referred to Kaiser et al. (2012). For more interesting delayed choice experiments, see Ionicioiu and Terno (2011), Tang et al. (2012), Kaiser et al. (2012), Peruzzo et al. (2012), and Rab et al. (2017).

Problems

24.1. Referring to Fig. 24.1 show that if the modes corresponding to i_1 and i_2 are not identical and are distinguishable, no interference effect would be observed in the detector D_s. Can there be distinguishability even when the two paths of the idlers are exactly overlapping?

24.2. Referring to Chap. 5 on directional couplers, consider a directional coupler made of two identical single mode channel waveguides with coupling lengths corresponding to the horizontal (H) and vertical (V) polarizations as $L_c(H) = 1.25$ mm and $L_c(V) = 1.5$ mm. (a) After how much propagation distance will the two polarizations exit from different waveguides? (b) After how much minimum distance will the horizontal polarization get completely coupled and at this distance what fraction of the vertical polarization would be coupled to the other waveguide?

24.3. In the delayed choice experiment, suggest a possible way to be able to obtain information as to when the photon has entered the Mach-Zehnder interferometer shown in Fig. 24.4?

24.4. Referring to Fig. 24.5b, in which arm will an incident photon appear at the exit if the two paths through the interferometer are identical in optical path length? How would you make the photon appear in the other arm?

24.5. Derive Eq. (24.38).

24.6. From Eq. (24.38) show that the total probability of coincidence detection between detectors c and D_H and D_V and between d and D_H and D_V is unity.

References

Elitzur, A. C., and L. Vaidman. 1993. Quantum mechanical interaction-free measurements. *Found Phys* 23:987–997.

Ionicioiu, R., and R. Terno. 2011. Proposal for a quantum delayed-choice experiment. *Phys Rev Lett* 107:230406.

Jacques, V., E. Wu, F. Grosshans, F. Treussart, P. Grangier, A. Aspect, and J.-F. Roch. 2007. Experimental realization of Wheeler's delayed-choice Gadenken experiment. *Science* 315:966–968.

Kaiser, F., T. Coudreau, P. Milman, D. B. Ostrowsky, and S. Tanzilli. 2012. Entangled enabled delayed choice experiment. *Science* 338:637–640.

Lemos, B. G., V. Borish, G. D. Cole, S. Ramelow, R. Lapkiewicz, and A. Zeilinger. 2014. Quantum imaging with undetected photons. *Nature* 512:409–412.

Mandel, L. 1999. Quantum effects in one-photon and two-photon interference. *Rev Mod Phys* 71: S274–S282.

Pearce, E., N. R. Gemmel, J. Florez, J. Ding, R. F. Oulton, A. S. Clark, and C. C. Philips. 2023. Practical quantum imaging with undetected photons. *Optics Continuum* 2:2386–2397.

Peruzzo, A., P. Shadbolt, N. Brunner, S. Popescu, and J. L. O'Brien. 2012. A quantum delayed choice experiment. *Science* 338:634–637.

Rab, A. S., E. Polino, Z.-X. Man, N. B. An, Y.-J. Xia, N. Spagnolo, R. L. Franco, and F. Sciarrino. 2017. Entanglement of photons in their dual wave-particle nature. *Nat Commun* 8:915. https://doi.org/10.1038/s41467-017-01058-6.

Tang, J.-S., Y.-L. Li, X.-Y. Xu, G.-Y. Xiang, C.-F. Li, and G.-C. Guo. 2012. Realization of quantum Wheeler's delayed choice experiment. *Nat Photonics* 6:600–604. https://doi.org/10.1038/NPHOTON.2012.179.

White, A. G., J. R. Mitchell, O. Nairz, and P. G. Kwiat. 1998. "Interaction free" imaging. *Phys Rev A* 58:605–613.

Zou, X. Y., L. J. Wang, and L. Mandel. 1991. Induced coherence and indistinguishability in optical interference. *Phys Rev Lett* 67:318–321.

Chapter 25
Some Applications of Quantum Photonic Technology

Abstract This chapter gives brief description of some applications of quantum photonic technology. The applications discussed include quantum key distribution using BB84 and E91 protocols, photonic qubits and photonic quantum computation, photonic C-NOT gates, Boson sampling, and quantum teleportation.

25.1 Introduction

In this chapter, we will briefly discuss the principles behind three important applications of integrated quantum photonic technology. In Sect. 25.2, we discuss quantum key distribution (QKD) which ensures exchange of unbreakable cryptographic keys among parties who wish to exchange secret information. In Sect. 25.3, we will discuss the basic principles of photonic quantum computation and in Sect. 25.4, we will discuss briefly application in quantum teleportation. The discussions given here are with the aim to introduce the basic concept in the application of integrated quantum photonics and are very brief; for more details, readers are referred to excellent texts dealing in detail with these applications.

25.2 Quantum Key Distribution

Quantum key distribution (QKD) is one of the first practical applications of quantum photonics and commercial devices are already available and being implemented (see, e.g., ID Quantique, Switzerland; Toshiba, Japan; QuantumXchange; USA, LuxQuanta; Spain, etc.).

In order to send a secret message that is very difficult to be decoded by any unauthorized person, the actual message is encoded using a key that is used to encrypt the secret message using an algorithm referred to as the cipher. In classical cryptographic techniques, the only completely secure method is the use of one-time pad, that is, use of a key only once. Although this is completely secure, the issue is how to share the secret key between the two parties. This problem is addressed in

K. Thyagarajan, *Integrated Quantum Photonics*, Graduate Texts in Physics,
https://doi.org/10.1007/978-3-031-85728-7_25

quantum key distribution where the common key can be shared securely by using the basic principles of quantum mechanics of superposition and measurement.

Photons, which are also referred to as flying qubits, are the best resource for quantum key distribution. In QKD, the quantum properties of photons are used to generate the secret key required to encode information. For this application, various protocols have been proposed for implementation. Here we will briefly discuss the basic principle of two important protocols, namely BB84 and E91.

The techniques we will discuss use a one-time pad and are absolutely secure techniques for encrypting messages. Using the techniques, the two parties who wish to exchange a secret message generate an identical random sequence of bits 0 and 1 which can then be used to code a secret message. For security, the key should be at least as long as the message to be sent. Once the coded message based on the secret key is sent, the two parties must renew the key as the one-time pad can only be used once to achieve absolute secrecy.

Consider a single photon which has a polarization state defined by the following state written in the (H, V) basis (horizontal-vertical basis):

$$|\psi\rangle = c_0|H\rangle + c_1|V\rangle \tag{25.1}$$

where $|H\rangle$ and $|V\rangle$ correspond to horizontal and vertical polarization states of the photon and c_0 and c_1 are constants satisfying the normalization condition $|c_0|^2 + |c_1|^2 = 1$. If the coefficients c_0 and c_1 are unknown, it is impossible to determine the polarization state of the photon with a single measurement. For example, if the polarizer is oriented along the horizontal direction, a single measurement in which the photon passes through the polarizer cannot distinguish between the states $|H\rangle, \frac{1}{\sqrt{2}}(|H\rangle + |V\rangle), \frac{1}{2}|H\rangle + \frac{\sqrt{3}}{2}|V\rangle$ and so on. This is an important feature that is used in the protocols used in QKD using polarization degree of freedom since the key distribution is accomplished through communication through single photons.

25.2.1 BB84 Protocol

In 1984, Bennett and Brassard proposed a method to create a secret key using single photons. In this scheme, we assume that in the (H, V) basis, the state H corresponds to the bit 0, while the V state corresponds to the bit 1. Similarly in the (D, A) basis, the state D corresponds to the bit 0 and the state A corresponds to the bit 1. Here D (for diagonal) corresponds to $+45^\circ$ orientation of the analyzer and A (anti diagonal) corresponds to -45° orientation of the analyzer with respect to the horizontal state. Note that (D, A) basis can be written as a superposition of the (H, V) basis and vice versa.

As mentioned earlier, it is not possible to find the polarization state of a single photon unless one has an a priori knowledge of the orthogonal basis in which the single photon is prepared. Thus, if the single photon is polarized in the (H, V) basis, while measurement is carried out in the (D, A) basis, the photon has equal probability of exiting either in the polarization state corresponding to D or corresponding to A irrespective of whether the incident photon had H polarization or V polarization. Thus, there is no way of knowing as to whether the photon was having H polarization or V polarization before measurement. Note that the (H, V) basis and the (D, A) basis are bases which are nonorthogonal to each other. On the other hand, if it is known that the basis in which the photon is prepared is the (H, V) basis, a single measurement of photon polarization state using polarization anayzers in the (H, V) basis can provide us with the information on the state of the polarization of the photon.

In order to generate the key, the two parties who wish to exchange a secret message will be referred to here as Alice and Bob. In order to generate the key for encrypting the message, Alice sends to Bob, single photons which are linearly polarized either in the (H, V) basis or in the (D, A) basis randomly changing the basis about half of the times. Bob who receives these photons then passes them through linear polarization analyzers and determines the state of polarization of the single photons that were sent but again choosing randomly either the (H, V) basis or the (D, A) basis for the measurement. After sending a certain number of photons, Alice and Bob check on a *public communication system* as to the *bases used* by Alice in sending the different photons. Note that the information being exchanged is only as to the *bases* used and not the result of the measurement. Whenever the bases of Alice and Bob are the same, both of them will obtain identical bits. On the other hand, when the bases do not coincide, the results will, in general, be different and those bits are discarded. This procedure is referred to as key sifting. After sifting, Alice and Bob have two identical sequences of bits which will be approximately half the number of bits exchanged.

In order to check whether there has been any eavesdropping, Alice and Bob compare the values of the bit for a certain fraction (about 10%) of the key that was generated. If there was any eavesdropping with Eve having measured in an arbitrary basis (since the basis used by Alice in sending the bit is unknown to both Eve and Bob) and sending to Bob a photon having the polarization that she has measured, there would be errors when Alice and Bob compare their bits. This will indicate the presence of a spy on the link and the key is not used.

As an example, Table 25.1 gives the bases used by Alice and Bob for preparing the key.

In the example, out of ten cases, the bases used by Alice and Bob match six times and in those six times, both of them obtain the same bit. So when Alice and Bob communicate with each other, they will know that only in the S. Nos. 1, 2, 4, 6, 8, and 9 the two bases match and hence both of them have identical bits. Out of these six bits they can compare say two of the cases and check whether they obtained the same bit. In this way, they can easily check for errors and get information as to whether any eavesdropper is present. The sifted key is then used to encrypt the secret message that Alice wishes to send to Bob which knowing the key, Bob can decipher.

Table 25.1 A sample of results in BB84 protocol

S. No.	Polarization state sent by Alice	Bit obtained	Basis chosen by Bob	Bit obtained	Select
1	H	0	(H, V)	0	Y
2	D	0	(D, A)	0	Y
3	V	1	(D, A)	1	
4	H	0	(H, V)	0	Y
5	A	1	(H, V)	0	
6	V	1	(H, V)	1	Y
7	D	0	(H, V)	0	
8	D	0	(D, A)	0	Y
9	H	0	(H, V)	0	Y
10	A	1	(H, V)	1	

25.2.2 E91 Protocol

In this section, we will describe how entangled photon pairs can be used for generating a secure key for cryptographic application. Specifically, this method was invented by Artur Eckert in 1991 and is referred to as E91 protocol.

In this technique, the two parties referred to here as Alice and Bob have one photon each of a polarization entangled pair. Both Alice and Bob pass their photons through a linear polarization analyzer such that depending on the port through which the photon exits, the bit is referred to as 0 or 1. For example if the polarization analyzer has an orientation corresponding to the (H, V) basis, then the output H may correspond to the bit 0 and the output V to a bit 1. As we discussed in Chap. 17, the polarization states of the two photons of the polarization entangled pair are undefined before any measurement is carried out and hence both Alice and Bob will each generate a random sequence of bits of 0 and 1. At the same time, since the two photons are entangled in polarization, if both of them use the same bases for measurement, they will obtain results which will be perfectly correlated to each other thus providing them a technique to generate the key.

As a specific example, let us assume that the polarization entangled state is given by the following Bell state:

$$|\phi^+\rangle = \frac{1}{\sqrt{2}}(|H_s, H_i\rangle + |V_s, V_i\rangle) \tag{25.2}$$

where, as before the subscripts s and i refer to signal and idler which are the two entangled photons. Since the two photons are polarization entangled, if the orientation of the analyzers of Alice and Bob is identical, then both of them will get identical bits, either both will be 0 or both will be 1, since with polarization entangled photon pairs, there is perfect correlation between the measurements of the two parties. Hence, with the polarization entangled photon pairs, Alice and Bob will

Table 25.2 A sample of measurements carried out using entangled state given by Eq. (25.2)

S. No.	Basis chosen by Alice	Bit obtained	Basis chosen by Bob	Bit obtained	Select
1	(H, V)	1	(H, V)	1	Y
2	(H, V)	0	(D, A)	0	
3	(D, A)	0	(H, V)	0	
4	(H, V)	1	(H, V)	1	Y
5	(D, A)	0	(D, A)	0	Y
6	(D, A)	1	(H, V)	1	
7	(H, V)	0	(H, V)	0	Y
8	(H, V)	0	(D, A)	1	
9	(D, A)	1	(H, V)	0	
10	(H, V)	1	(H, V)	1	Y

have two identical but random sequence of bits and which can be used as a one-time pad for encoding.

If there is an eavesdropper who wishes to obtain the key and if she knows the orientation of the analyzer chosen by Alice and Bob, then she can choose the same orientation and after getting the result to be either 0 or 1, she can send another photon with an appropriate polarization state to Bob. In this process, Eve can get to know the one-time pad that Alice and Bob are trying to decide for encoding their message without becoming visible to either Alice or Bob. With such an interruption, the security of the key is completely lost.

In order to beat the eavesdropper Alice and Bob can decide on changing the orientations of the polarization analyzer randomly along for example two different orientations, for example, (H, V) basis or (D, A) where D (for diagonal) corresponds to $+45^{\circ}$ orientation of the analyzer and A (anti diagonal) corresponds to -45° orientation of the analyzer. Now, when the entangled photons pass through the respective analyzers of Alice and Bob, if their orientations match, then they will obtain exactly the same bit irrespective of whether they use the (H, V) basis or the (D, A) basis. If the bases do not match, then their results will not be correlated.

As an example, Table 25.2 gives a sequence of ten measurements carried out by Alice and Bob on the entangled photons with the basis shown.

From the table we see that in the cases at S. Nos. 1, 4, 5, 7, 10, the orientations of the two analyzers match and both Alice and Bob obtain the same bit. The other results when the two bases did not match and where there is no correlation between the two outputs are discarded.

Hence, both Alice and Bob have the following identical copies of the key: 11001. This bit sequence can be used as a one-time pad for encrypting secret messages.

To find out whether there is a spy on the line, Alice and Bob can check for correlations between the entangled photon pair with other angles of $\left(\frac{\pi}{8}, \frac{\pi}{4}, \frac{3\pi}{8}\right)$ using Bell's inequality or CHSH inequality (see Chap. 17). Results of correlation measurement can immediately warn Alice and Bob about the presence of a spy on the line and who will then refrain from using the key that has been obtained. This is so since in order to spy, Eve would have to make a measurement on one of the photons before sending it back in the communication link. The moment a measurement is carried out the two-photon state will become a factorizable state and the correlation between the measurements of Alice and Bob will satisfy Bell's inequality. This way the presence of the eavesdropper becomes visible to the parties.

Integrated quantum photonics offers a unique platform to realize practical QKD systems offering compactness, stability, and low loss systems. QKD chips to process information have been under development and is expected to revolutionize QKD (see, e.g., Labonte et al. 2024; Sax et al. 2023; Wei et al. 2023; Sibson et al. 2017). For a nice review on quantum repeaters, readers are referred to Azuma et al. (2023).

25.3 Photonic Quantum Computation

Conventional computers can handle most problems with ease. There are some computational problems that a conventional computer will find it difficult to solve. For example, multiplying two very large prime numbers can be performed very easily by a classical computer, while if the product of the two very large prime numbers is given with the problem to find out the factors, then a conventional computer will take an exponentially large time to factorize. The number of operations required to factorize increases exponentially with the size of the integer to be factorized. It is in such difficult and intractable problems that quantum computers can play a significant role.

In classical computers, information is coded in the form of bits 0 and 1, while in quantum computers, information is stored as qubits which are superpositions of 0 and 1. Unlike classical bits where the bit is either 0 or 1, a qubit is in a superposition state. A general state of qubit can be written as

$$|\psi\rangle = c_0|0\rangle + c_1|1\rangle \tag{25.3}$$

with the normalization $|c_0|^2 + |c_1|^2 = 1$. The quantum information is stored in the complex quantities c_0 and c_1. A physical qubit can be in the form of a photon which is in a superposition state of two orthonormal polarization states say H and V, or it could be two orthogonal spatial modes of an optical waveguide, or it could be in the form of spin of electrons being in a superposition of spin up and spin down states, or it could be a two level system with the system being in a superposition state of the ground and first excited state and so on. Note that the qubit is in a superposition state of 0 and 1 and measurement of the value of the qubit is probabilistic with probability of $|c_0|^2$of being in the state $|0\rangle$ and a probability of $|c_1|^2$of being in the state $|1\rangle$.

It is interesting to note that if we have a register with two qubits, then the general state of the system is a linear superposition of four states:

$$|\psi\rangle = c_{00}|0\rangle_1|0\rangle_2 + c_{01}|0\rangle_1|1\rangle_2 + c_{10}|1\rangle_1|0\rangle_2 + c_{11}|1\rangle_1|1\rangle_2 \tag{25.4}$$

which can be written in the following shortened form:

$$|\psi\rangle = c_{00}|00\rangle + c_{01}|01\rangle + c_{10}|10\rangle + c_{11}|11\rangle \tag{25.5}$$

In terms of polarization we can have two photons whose general state can be described by Eq. (25.5) with 0 representing horizontal polarization and 1 representing vertical polarization.

It is to be noted that although the information is stored in the four complex numbers c_{00}, c_{01}, c_{10}, and c_{11}, any measurement carried out on the system will then result in the collapse of the quantum system into one of the eigenstates with the respective probabilities. At the same time, as long as the qubits interact with each other coherently, the system evolves predictably and information is completely preserved.

Similarly the general state of a three qubit register will be a linear combination of eight states. The general state of a register with N qubits will be a linear superposition of 2^N states. The quantum information is stored in the 2^N complex amplitudes which of course satisfy the normalization condition $\sum|c_{nml\ldots}|^2 = 1$.

A quantum computer works with the qubits, and for this many quantum logic gates such as NOT, CNOT, and NAND have been developed.

Thus, in a NOT gate an input $|0\rangle$ gets converted to $|1\rangle$, while an input $|1\rangle$ gets converted to $|0\rangle$. Hence, when a NOT gate operates on an input state given by Eq. (25.3), the output state would be

$$|\psi\rangle = c_0|1\rangle + c_1|0\rangle \tag{25.6}$$

In a Hadamard gate, an input $|0\rangle$ gets converted to

$$|\psi\rangle_{out} = \frac{1}{\sqrt{2}}(|0\rangle + |1\rangle) \tag{25.7}$$

while an input $|1\rangle$ gets converted to

$$|\psi\rangle_{out} = \frac{1}{\sqrt{2}}(|0\rangle - |1\rangle) \tag{25.8}$$

The states of the two qubits can be represented by two column vectors:

$$|0\rangle = \begin{pmatrix} 1 \\ 0 \end{pmatrix}; \quad |1\rangle = \begin{pmatrix} 0 \\ 1 \end{pmatrix}$$

Thus, the state given by Eq. (25.3) can be expressed as

$$|\psi\rangle = c_0|0\rangle + c_1|1\rangle = c_0\begin{pmatrix} 1 \\ 0 \end{pmatrix} + c_1\begin{pmatrix} 0 \\ 1 \end{pmatrix} = \begin{pmatrix} c_0 \\ c_1 \end{pmatrix} \tag{25.9}$$

Single qubit gates can be represented by 2×2 matrices operating on input column vectors representing the quantum state.

Thus, a NOT gate (also referred to as the Pauli X-gate) will correspond to

$$X = \begin{pmatrix} 0 & 1 \\ 1 & 0 \end{pmatrix} \tag{25.10}$$

For an input given by Eq. (25.9), the output will be

$$|\psi\rangle_{out} = \begin{pmatrix} 0 & 1 \\ 1 & 0 \end{pmatrix}\begin{pmatrix} c_0 \\ c_1 \end{pmatrix} = \begin{pmatrix} c_1 \\ c_0 \end{pmatrix} = c_1|0\rangle + c_0|1\rangle$$

Similarly a Hadamard gate will correspond to

$$H = \frac{1}{\sqrt{2}}\begin{pmatrix} 1 & 1 \\ 1 & -1 \end{pmatrix} \tag{25.11}$$

For an input given by Eq. (25.9), the output will be

$$|\psi\rangle_{out} = \frac{1}{\sqrt{2}}\begin{pmatrix} 1 & 1 \\ 1 & -1 \end{pmatrix}\begin{pmatrix} c_0 \\ c_1 \end{pmatrix} = \frac{1}{\sqrt{2}}\begin{pmatrix} c_0 + c_1 \\ c_0 - c_1 \end{pmatrix} = \frac{1}{\sqrt{2}}(c_0 + c_1)|0\rangle + \frac{1}{\sqrt{2}}(c_0 - c_1)|1\rangle$$

Similarly a Z gate is represented by

$$Z = \begin{pmatrix} 1 & 0 \\ 0 & -1 \end{pmatrix} \tag{25.12}$$

For an input given by Eq. (25.9), the output will be

$$|\psi\rangle_{out} = \begin{pmatrix} 1 & 0 \\ 0 & -1 \end{pmatrix}\begin{pmatrix} c_0 \\ c_1 \end{pmatrix} = \begin{pmatrix} c_0 \\ -c_1 \end{pmatrix} = c_0|0\rangle - c_1|1\rangle$$

A two qubit NAND gate has the operating table given in Table 25.3.

Another very important two qubit quantum gate is the C-NOT gate or the controlled NOT gate consisting of a control qubit and a target qubit. If the control qubit is 0, then the target qubit does not change, while if the control qubit is 1, then the target qubit gets interchanged between 0 and 1. The operation table for C-NOT gate is given in Table 25.4.

Table 25.3 Operating table for a two qubit NAND gate

Input bits	Output bit
0 0	1
1 0	1
0 1	1
1 1	0

Table 25.4 Operating table for a two qubit C-NOT gate

Input qubit	Input qubit	Output qubit	Output qubit
Control qubit	Target qubit	Control qubit	Target qubit
0	0	0	0
0	1	0	1
1	0	1	1
1	1	1	0

If the state is written as $|\alpha\beta\rangle$ with the first symbol representing the control qubit and the second symbol the target qubit, then operating with C-NOT gate an input state given by Eq. (25.5) is transformed to the following:

$$|\psi\rangle = c_{00}|00\rangle + c_{01}|01\rangle + c_{11}|10\rangle + c_{10}|11\rangle \tag{25.13}$$

An interesting feature of the C-NOT gate is its ability to entangle or disentangle qubits. For example, if the input state corresponds to

$$|\psi\rangle_{in} = \frac{1}{\sqrt{2}}(|0\rangle \pm |1\rangle)_C|0\rangle_T \tag{25.14}$$

where the subscripts C and T correspond to control and target qubits, then as per the operation of the C-NOT gate, the output state would be

$$|\psi\rangle_{out} = \frac{1}{\sqrt{2}}\left(|0\rangle_C|0\rangle_T \pm |1\rangle_C|1\rangle_T\right) = |\phi^{\pm}\rangle \tag{25.15}$$

Similarly if the input is given by

$$|\psi\rangle_{in} = \frac{1}{\sqrt{2}}(|0\rangle \pm |1\rangle)_C|1\rangle_T \tag{25.16}$$

then the output will be given by

$$|\psi\rangle_{out} = \frac{1}{\sqrt{2}}\left(|0\rangle_C|1\rangle_T \pm |1\rangle_C|0\rangle_T\right) = |\psi^{\pm}\rangle \tag{25.17}$$

These output states correspond to the Bell states.

In order to perform universal quantum computation, single qubit transformations together with a two-qubit gate suffice.

Exercise 25.1 Consider a single photon state given by

$$|\psi\rangle_{in} = (\alpha|H\rangle + \beta|V\rangle)$$

The photon passes through a half wave plate oriented at 45° to the horizontal. Obtain the output state.

Solution When the photon passes through a half wave plate oriented at 45° to the horizontal, then an input state $|H\rangle$ gets converted to $|V\rangle$ and an input state $|V\rangle$ gets converted to $|H\rangle$. Hence, the output state will be given by

$$|\psi\rangle_{out} = (\beta|H\rangle + \alpha|V\rangle)$$

Hence, represented by the following matrix written in the $H - V$ basis:

$$\widehat{O}_{HWP} = \begin{pmatrix} 0 & 1 \\ 1 & 0 \end{pmatrix}$$

which is nothing but the Pauli X-gate.

Exercise 25.2 Consider the following Bell state:

$$|\psi\rangle_{in} = |\phi^+\rangle = \frac{1}{\sqrt{2}}(|H\rangle_1|H\rangle_2 + |V\rangle_1|V\rangle_2)$$

Obtain the output state if the photon identified by the subscript 2 passes through a half wave plate oriented at 45° to the horizontal.

Solution When the photon identified by subscript 2 is passed through the given half wave plate, then the half wave plate transforms $|H\rangle_2$ to $|V\rangle_2$ and $|V\rangle_2$ to $|H\rangle_2$. Thus, the output will be given by

$$|\psi\rangle_{out} = \frac{1}{\sqrt{2}}(|H\rangle_1|V\rangle_2 + |V\rangle_1|H\rangle_2)$$

which is nothing but the $|\psi^+\rangle$ Bell state.

25.3.1 Photonic Qubits and Photonic Quantum Computation

Quantum computing using linear quantum optics wherein the physical qubits are photons is very interesting from a practical implementation point of view as photons are much less subjected to decoherence with the result that the quantum information that is stored in the photon is maintained. One of the major problems in using photons for quantum processing is the requirement of nonlinear interaction especially to implement the C-NOT gate. Since nonlinear optical effects depend on the power of the interacting waves, this effect is almost ineffective at the single photon level. Interestingly in 2001 Knill, Laflamme, and Milburn (Knill et al. 2001) proposed a protocol in which using linear optics and projective measurements it is possible to implement probabilistically scalable quantum computing. For further discussions on this, readers are referred to some of the excellent reviews and research: Kok et al. (2007), Djordjevic (2010), Thompson (2011), Carolan et al. (2015), Bell and Walmsley (2021), Bogaerts et al. (2020), Blanco et al. (2022), Walmsley (2023) Yard et al. (2024), Romero and Milburn (2024).

In photonic quantum computers, physical qubits are the photons and various waveguide devices such as directional couplers, interferometers, and interactions between photons and a single atom are used for different operations on the various degrees of freedom such as polarization, path, spatial mode, time bin, etc. The primary advantages of photonic quantum computers are use of simple optical components which can be in integrated form and can operate at room temperature unlike other schemes and also their interaction with the environment is minimal.

In photonic quantum computing architectures, qubits are represented by photons using their various degrees of freedom such as polarization, spatial mode, and frequency.

A typical photonic computing chip would consist of a laser for generating single photons or entangled photon pairs, micro ring resonators, optical waveguides, interferometers and directional couplers for different operations on the photons, etc. (see, e.g., Assumpcao et al. 2024). In polarization encoding technique, a generic qubit corresponds to a state such as given in Eq. (25.3) as a superposition of two orthogonal polarization states.

In integrated quantum photonic computing architectures, the spatial degree of freedom is often used as such single qubit states can be easily manipulated using the various integrated photonic components.

We first look at integrated quantum computing employing the spatial degree of freedom of photons. A qubit has two distinct and discrete states possessing the fundamental quantum property of superposition.

As an example, we consider a pair of noninteracting identical waveguides (referred to as W_1 and W_2) which is excited by a *single photon*. If we refer to the qubit $|0\rangle$ as the photon occupying waveguide W_1 and the qubit $|1\rangle$ to refer to the photon occupying waveguide W_2, then the general state of the photon in the two-waveguide device would be a superposition as given by

$$|\psi\rangle = \alpha|0\rangle + \beta|1\rangle = \alpha\begin{pmatrix}1\\0\end{pmatrix} + \beta\begin{pmatrix}0\\1\end{pmatrix} = \begin{pmatrix}\alpha\\\beta\end{pmatrix} \tag{25.18}$$

with $|\alpha|^2 + |\beta|^2 = 1$. This represents a single qubit. For another qubit, we need to choose another waveguide pair. Thus, an n-qubit configuration will consist of n waveguide pairs, each pair representing one qubit. This spatial mode representation is also referred to as dual rail representation.

A single qubit operation operates on the waveguide pair through different devices such as the directional coupler representing the beam splitter, Mach-Zehnder interferometer, phase modulator (see Chaps. 5 and 7).

The unitary matrix connecting the input and output of a 3 dB directional coupler made up of two identical waveguides is given by (see Chap. 19)

$$\widehat{B}=\frac{1}{\sqrt{2}}\begin{pmatrix}1 & i\\ i & 1\end{pmatrix} \tag{25.19}$$

The beam splitter operates on the qubit given by Eq. (25.18) and gives an output

$$|\psi\rangle_{out}=\widehat{B}|\psi\rangle_{in}=\frac{1}{\sqrt{2}}\begin{pmatrix}1 & i\\ i & 1\end{pmatrix}\begin{pmatrix}\alpha\\ \beta\end{pmatrix}=\frac{1}{\sqrt{2}}\begin{pmatrix}\alpha+i\beta\\ i\alpha+\beta\end{pmatrix}$$

which can be written as

$$|\psi\rangle_{out}=\frac{1}{\sqrt{2}}(\alpha+i\beta)|0\rangle+\frac{1}{\sqrt{2}}(i\alpha+\beta)|1\rangle \tag{25.20}$$

As another component in photonic quantum computing, we introduce a phase shifting device that changes the phase of the propagating photon in either of the two waveguides (qubit) by ϕ_1 and ϕ_2. Such a phase shifter is represented by the following matrix:

$$\widehat{P}(\phi_1,\phi_2)=\begin{pmatrix}e^{i\phi_1} & 0\\ 0 & e^{i\phi_2}\end{pmatrix} \tag{25.21}$$

Such phase changes can be easily brought about by different effects such as the electro optic effect (see Chap. 6) or thermo-optic effect.

Figure 25.1 shows an integrated photonic Hadamard gate using a 3 dB directional coupler, followed by a phase shifter with $\phi_1=0$ and $\phi_2=\frac{\pi}{2}$ followed by another 3 dB directional coupler which is represented by the following transformation (note the order of the operators which sequentially operate from right to left):

$$\widehat{U}=\widehat{B}\widehat{P}\left(0,-\frac{\pi}{2}\right)\widehat{B}=\frac{1}{\sqrt{2}}\begin{pmatrix}1 & i\\ i & 1\end{pmatrix}\begin{pmatrix}1 & 0\\ 0 & -i\end{pmatrix}\frac{1}{\sqrt{2}}\begin{pmatrix}1 & i\\ i & 1\end{pmatrix}=\frac{1}{\sqrt{2}}\begin{pmatrix}1 & 1\\ 1 & -1\end{pmatrix}e^{i\pi/4} \tag{25.22}$$

which is the Hadamard gate (see Eq. 25.11) apart from a global phase $e^{i\pi/4}$.

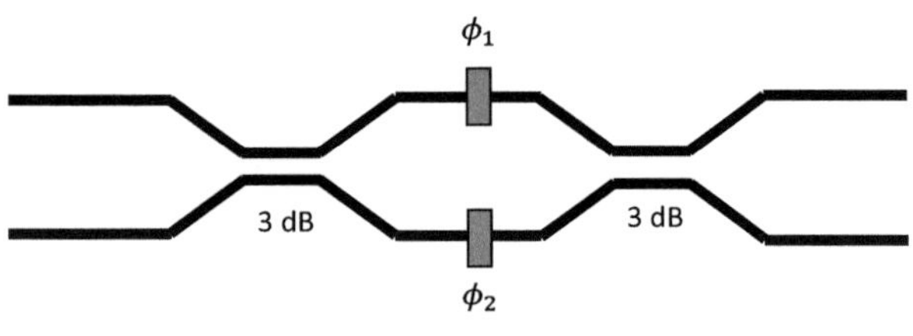

Fig. 25.1 A qubit gate consisting of two 3 dB directional couplers (or beam splitters) with phase shifters on both paths

If the input is in the state $|0\rangle = \begin{pmatrix} 1 \\ 0 \end{pmatrix}$, then the output would be in the state given by (apart from the global phase)

$$|\psi\rangle_{out} = \frac{1}{\sqrt{2}}\begin{pmatrix} 1 & 1 \\ 1 & -1 \end{pmatrix}\begin{pmatrix} 1 \\ 0 \end{pmatrix} = \frac{1}{\sqrt{2}}\begin{pmatrix} 1 \\ 1 \end{pmatrix} = \frac{1}{\sqrt{2}}\begin{pmatrix} 1 \\ 0 \end{pmatrix} + \frac{1}{\sqrt{2}}\begin{pmatrix} 0 \\ 1 \end{pmatrix} = \frac{1}{\sqrt{2}}(|0\rangle + |1\rangle)$$

Similarly it can be shown that an input state given by $|1\rangle = \begin{pmatrix} 0 \\ 1 \end{pmatrix}$ will get converted to $\frac{1}{\sqrt{2}}(|0\rangle - |1\rangle)$.

Similarly a Pauli X-gate represented by Eq. (25.10) can be realized by using two beam splitters in tandem (apart from a global phase factor):

$$\widehat{U} = \widehat{B}\widehat{B} = \frac{1}{\sqrt{2}}\begin{pmatrix} 1 & i \\ i & 1 \end{pmatrix}\frac{1}{\sqrt{2}}\begin{pmatrix} 1 & i \\ i & 1 \end{pmatrix} = \begin{pmatrix} 0 & 1 \\ 1 & 0 \end{pmatrix} \tag{25.23}$$

Similarly a Pauli Z-gate can be realized by a 3 dB beam splitter followed by a phase shifter with $\phi_1 = 0$ and $\phi_2 = \pi$ followed by another 3 dB beam splitter:

$$\widehat{U} = \widehat{B}\widehat{P}(0, \pi)\widehat{B} = \frac{1}{\sqrt{2}}\begin{pmatrix} 1 & i \\ i & 1 \end{pmatrix}\begin{pmatrix} 1 & 0 \\ 0 & -1 \end{pmatrix}\frac{1}{\sqrt{2}}\begin{pmatrix} 1 & i \\ i & 1 \end{pmatrix} = \begin{pmatrix} 1 & 0 \\ 0 & -1 \end{pmatrix} \tag{25.24}$$

The Pauli Y-gate can be realized using a phase shifter followed by a pair of 3 dB beam splitters:

$$\widehat{U} = \widehat{P}(\pi, 0)\widehat{B}\widehat{B} = \begin{pmatrix} -1 & 0 \\ 0 & 1 \end{pmatrix}\frac{1}{\sqrt{2}}\begin{pmatrix} 1 & i \\ i & 1 \end{pmatrix}\frac{1}{\sqrt{2}}\begin{pmatrix} 1 & i \\ i & 1 \end{pmatrix} = \begin{pmatrix} 0 & -i \\ i & 0 \end{pmatrix} \tag{25.25}$$

Using such devices, it is also possible to implement a C-NOT gate. Here we will discuss some examples of implementation of C-NOT gate in photonic quantum computing architectures. For a recent review, readers are referred to Kwon et al. (2024).

Exercise 25.3 What will be the operator corresponding to a device with two 3 dB directional couplers.

Solution A device consisting of two 3 dB directional couplers placed one after the other will be represented by the following transfer matrix:

$$\widehat{U} = \frac{1}{\sqrt{2}}\begin{pmatrix} 1 & i \\ i & 1 \end{pmatrix}\frac{1}{\sqrt{2}}\begin{pmatrix} 1 & i \\ i & 1 \end{pmatrix} = e^{i\pi/2}\begin{pmatrix} 0 & 1 \\ 1 & 0 \end{pmatrix}$$

25.3.2 Photonic C-NOT Gate

Here we present two photonic implementations of the C-NOT gate, one using dual rail qubits demonstrated by O'Brien et al. (2003) and the other using polarization qubits demonstrated by Crespi et al. (2011).

25.3.2.1 C-NOT Gate with Dual Rail Qubits

Figure 25.2 shows the device configuration consisting of five beam splitters or directional couplers in waveguide form, two of them with 50% reflectivity and three of them with 1/3 reflectivity. The photons also suffer a phase change of π when they enter through ports marked in the figure with the symbol π. The control qubits enter through the ports marked C_0 and C_1, while the target qubits enter through the ports marked T_0 and T_1. We represent the photon annihilation operators corresponding to the input ports marked C_0, C_1, T_0, and T_1 as $\hat{a}_{c0,i}$, $\hat{a}_{c1,i}$, $\hat{a}_{t0,i}$, and $\hat{a}_{t1,i}$, respectively. The corresponding annihilation operators corresponding to the output ports are written as $\hat{a}_{c0,o}$, $\hat{a}_{c1,o}$, $\hat{a}_{t0,o}$, and $\hat{a}_{t1,o}$, respectively. In addition to these ports, the figure also shows two ports with annihilation operators $\hat{a}_{s0,i}$, $\hat{a}_{s1,i}$, $\hat{a}_{s0,o}$, $\hat{a}_{s1,o}$ corresponding to the input and output ports. Here, the subscripts i and o represent input and output.

From the reflection coefficients of the various beam splitters and taking account of the phase shifts, identifying the paths taken by the control and target photons, we can write the following equation connecting the photon annihilation operators at the output port and the annihilation operators at the input ports (see Problem 25.4):

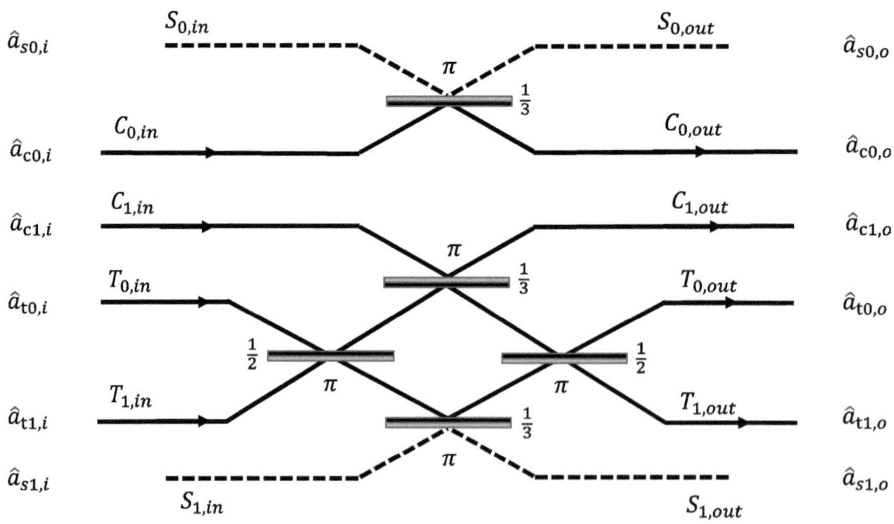

Fig. 25.2 An optical arrangement to achieve C-NOT gate operation using linear optical components

$$\begin{pmatrix} \hat{a}_{s0,o} \\ \hat{a}_{c0,o} \\ \hat{a}_{c1,o} \\ \hat{a}_{t0,o} \\ \hat{a}_{t1,o} \\ \hat{a}_{s1,o} \end{pmatrix} = \frac{1}{\sqrt{3}} \begin{pmatrix} -1 & \sqrt{2} & 0 & 0 & 0 & 0 \\ \sqrt{2} & 1 & 0 & 0 & 0 & 0 \\ 0 & 0 & -1 & 1 & 1 & 0 \\ 0 & 0 & 1 & 1 & 0 & 1 \\ 0 & 0 & 1 & 0 & 1 & -1 \\ 0 & 0 & 0 & 1 & -1 & -1 \end{pmatrix} \begin{pmatrix} \hat{a}_{s0,i} \\ \hat{a}_{c0,i} \\ \hat{a}_{c1,i} \\ \hat{a}_{t0,i} \\ \hat{a}_{t1,i} \\ \hat{a}_{s1,i} \end{pmatrix} \tag{25.26}$$

From the above equation, we can obtain the inverse of the 6x6 matrix and express the input annihilation operators in terms of the output annihilation operators as given by the following equation:

$$\begin{pmatrix} \hat{a}_{s0,i} \\ \hat{a}_{c0,i} \\ \hat{a}_{c1,i} \\ \hat{a}_{t0,i} \\ \hat{a}_{t1,i} \\ \hat{a}_{s1,i} \end{pmatrix} = \frac{1}{\sqrt{3}} \begin{pmatrix} -1 & \sqrt{2} & 0 & 0 & 0 & 0 \\ \sqrt{2} & 1 & 0 & 0 & 0 & 0 \\ 0 & 0 & -1 & 1 & 1 & 0 \\ 0 & 0 & 1 & 1 & 0 & 1 \\ 0 & 0 & 1 & 0 & 1 & -1 \\ 0 & 0 & 0 & 1 & -1 & -1 \end{pmatrix} \begin{pmatrix} \hat{a}_{s0,o} \\ \hat{a}_{c0,o} \\ \hat{a}_{c1,o} \\ \hat{a}_{t0,o} \\ \hat{a}_{t1,o} \\ \hat{a}_{s1,o} \end{pmatrix} \tag{25.27}$$

It is interesting to note that the transfer matrix given in Eq. (25.26) is unitary and satisfies the following relationship:

$$\widehat{U}^{\dagger}\widehat{U} = \widehat{U}\widehat{U}^{\dagger} = \widehat{I}$$

where $\widehat{U}^{\dagger}$ is the Hermitian conjugate of the matrix $\widehat{U}$. From the above equation, we conclude that

$$\widehat{U}^{-1} = \widehat{U}^{\dagger}$$

Using this relationship, it is easy to evaluate the inverse of the unitary matrix. Since the 6x6 matrix given in Eq. (25.26) is real and symmetric, the inverse is the same as the matrix itself as is apparent from Eq. (25.27).

Exercise 25.4 Obtain Eq. (25.26) connecting the output annihilation operators and the input annihilation operators for the quantum C-NOT gate shown in Fig. 25.2.

Solution Figure 25.3 shows the same device as shown in Fig. 25.2 with four planes P_j; $j = 1, 2, 3, 4$, marked.

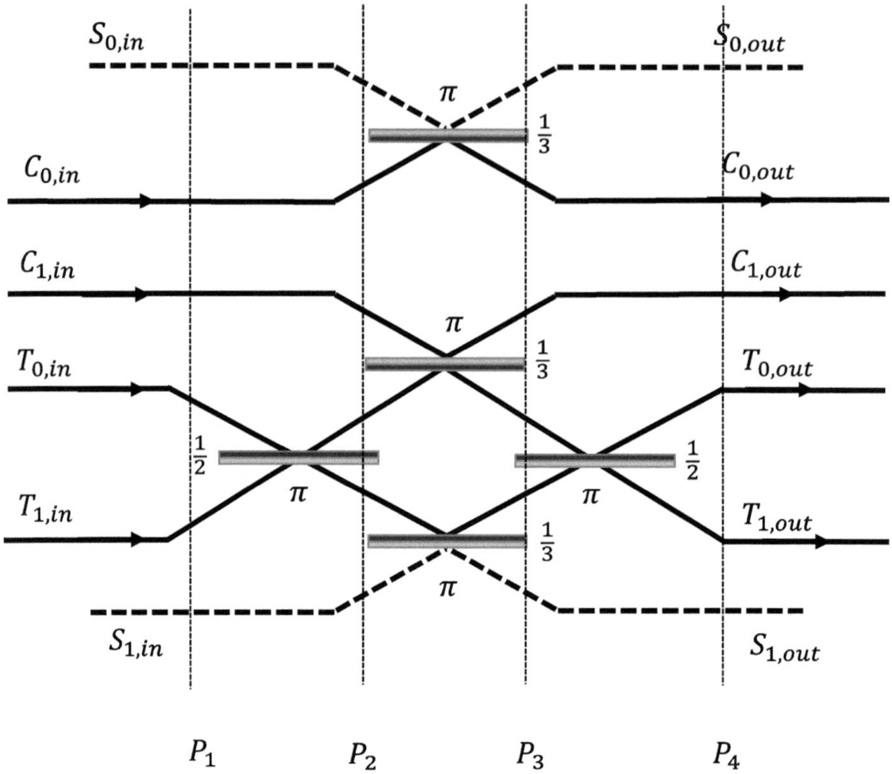

Fig. 25.3 Same as Fig. 25.2 with planes marked for evaluation of the transfer function

Plane P_1 is the input plane and plane P_4 is the output plane. Planes P_2 and P_3 are intermediate planes. At the input plane, we can write the annihilation operators at each path in terms of a column vector as below:

$$\begin{pmatrix} \hat{a}_{s0,i} \\ \hat{a}_{c0,i} \\ \hat{a}_{c1,i} \\ \hat{a}_{t0,i} \\ \hat{a}_{t1,i} \\ \hat{a}_{s1,i} \end{pmatrix}$$

From the transmission and reflection coefficients and the phase shifts of each beam splitter, the transformation from plane P_1 to P_4 is expressed in terms of matrices representing transformation from P_1 to P_2, then P_2 to P_3 and finally from P_3 to P_4 (note the order of the matrices which are from right to left:

right most matrix is for the transformation from P_1 to P_2, the middle matrix is for transformation from P_2 to P_3 and finally the left most matrix is for transformation from P_3 to P_4):

$$\begin{pmatrix} \hat{a}_{s0,o} \\ \hat{a}_{c0,o} \\ \hat{a}_{c1,o} \\ \hat{a}_{t0,o} \\ \hat{a}_{t1,o} \\ \hat{a}_{s1,o} \end{pmatrix} = \begin{pmatrix} 1 & 0 & 0 & 0 & 0 & 0 \\ 0 & 1 & 0 & 0 & 0 & 0 \\ 0 & 0 & 1 & 0 & 0 & 0 \\ 0 & 0 & 0 & 1/\sqrt{2} & 1/\sqrt{2} & 0 \\ 0 & 0 & 0 & 1/\sqrt{2} & -1/\sqrt{2} & 0 \\ 0 & 0 & 0 & 0 & 0 & 1 \end{pmatrix}$$
$$\begin{pmatrix} -1/\sqrt{3} & \sqrt{2/3} & 0 & 0 & 0 & 0 \\ \sqrt{2/3} & 1/\sqrt{3} & 0 & 0 & 0 & 0 \\ 0 & 0 & -1/\sqrt{3} & \sqrt{2/3} & 0 & 0 \\ 0 & 0 & \sqrt{2/3} & 1/\sqrt{3} & 0 & 0 \\ 0 & 0 & 0 & 0 & 1/\sqrt{3} & \sqrt{2/3} \\ 0 & 0 & 0 & 0 & \sqrt{2/3} & -1/\sqrt{3} \end{pmatrix}$$
$$\begin{pmatrix} 1 & 0 & 0 & 0 & 0 & 0 \\ 0 & 1 & 0 & 0 & 0 & 0 \\ 0 & 0 & 1 & 0 & 0 & 0 \\ 0 & 0 & 0 & 1/\sqrt{2} & 1/\sqrt{2} & 0 \\ 0 & 0 & 0 & 1/\sqrt{2} & -1/\sqrt{2} & 0 \\ 0 & 0 & 0 & 0 & 0 & 1 \end{pmatrix} \begin{pmatrix} \hat{a}_{s0,i} \\ \hat{a}_{c0,i} \\ \hat{a}_{c1,i} \\ \hat{a}_{t0,i} \\ \hat{a}_{t1,i} \\ \hat{a}_{s1,i} \end{pmatrix}$$

Multiplying the three 6x6 matrices, we can obtain Eq. (25.26).

Recalling our discussions from Chap. 19 where, in the case of a simple 2 × 2 beam splitter, we obtained the output state from the input state using the transformation representing the beam splitter, we follow the same procedure to obtain the output quantum state for a given input quantum state. Thus, the output quantum state will be given by replacing the input creation operators in terms of the output creation operators obtained from Eq. (25.27).

Let us assume an input state in which the control bit is in the state $|10\rangle$ corresponding to the logical bit $|0\rangle$ and the target bit to be in the state $|10\rangle$ corresponding to the logical state $|0\rangle$. In this case, the input state will be described by the following equation:

$$|\psi\rangle_{in} = |10\rangle_c|10\rangle_t \tag{25.28a}$$

which can be written in terms of input creation operators as

$$|\psi\rangle_{in} = \hat{a}^{\dagger}_{c0,i}\hat{a}^{\dagger}_{t0,i}|00\rangle_c|00\rangle_t = \hat{a}^{\dagger}_{c0,i}\hat{a}^{\dagger}_{t0,i}|0000\rangle \tag{25.28b}$$

Note that we are not writing the quantum state corresponding to the input ports S_0 and S_1.

In order to obtain the output state, we replace the input operators in terms of the output operators and obtain

$$|\psi\rangle_{out} = \frac{1}{\sqrt{3}}\left(\sqrt{2}\hat{a}^{\dagger}_{s0,o} + \hat{a}^{\dagger}_{c0,o}\right)\frac{1}{\sqrt{3}}\left(\hat{a}^{\dagger}_{c1,o} + \hat{a}^{\dagger}_{t0,o} + \hat{a}^{\dagger}_{s1,o}\right)|0000\rangle$$

or

$$|\psi\rangle_{out} = \frac{1}{3}\left(\sqrt{2}\hat{a}^{\dagger}_{s0,o}\hat{a}^{\dagger}_{c1,o} + \sqrt{2}\hat{a}^{\dagger}_{s0,o}\hat{a}^{\dagger}_{t0,o} + \sqrt{2}\hat{a}^{\dagger}_{s0,o}\hat{a}^{\dagger}_{s1,o} + \hat{a}^{\dagger}_{c0,o}\hat{a}^{\dagger}_{c1,o} + \hat{a}^{\dagger}_{c0,o}\hat{a}^{\dagger}_{t0,o} + \hat{a}^{\dagger}_{c0,o}\hat{a}^{\dagger}_{s1,o}\right)|0000\rangle$$

which can be written as

$$\begin{aligned}|\psi\rangle_{out} = &\frac{\sqrt{2}}{3}\hat{a}^{\dagger}_{s0,o}|0100\rangle + \frac{\sqrt{2}}{3}\hat{a}^{\dagger}_{s0,o}|0010\rangle + \frac{\sqrt{2}}{3}\hat{a}^{\dagger}_{s0,o}\hat{a}^{\dagger}_{s1,o}|0000\rangle + \frac{1}{3}|1100\rangle\\ &+ \frac{1}{3}|1010\rangle + \frac{1}{3}\hat{a}^{\dagger}_{s1,o}|1000\rangle\end{aligned} \tag{25.29}$$

Here we have used the fact that $\hat{a}^{\dagger}_{c1,o}|0000\rangle = |0100\rangle$, $\hat{a}^{\dagger}_{t0,o}|0000\rangle = |0010\rangle$, etc. Also note that the term $\hat{a}^{\dagger}_{s0,o}|0100\rangle$ implies creation of a photon in the output port S_0, etc. Thus, each individual term in Eq. (25.29) represents a state with two photons appearing in two output ports.

The only term that leads to photons in the control *and* target qubits is

$$|\psi\rangle_{out} = \frac{1}{3}|1010\rangle \tag{25.30}$$

Thus, when the control bit is in the logical state $|0\rangle = |10\rangle$, the input target bit $|0\rangle = |10\rangle$ remains unchanged, the probability of which is $\frac{1}{9}$.

We now consider the case when the control bit is in the logical state $|1\rangle = |01\rangle$ and the target bit is still in the logical state $|0\rangle = |10\rangle$. In this case, the input state is given by

$$|\psi\rangle_{in} = |0110\rangle = \hat{a}^{\dagger}_{c1,i}\hat{a}^{\dagger}_{t0,i}|00\rangle_c|00\rangle_t = \hat{a}^{\dagger}_{c1,i}\hat{a}^{\dagger}_{t0,i}|0000\rangle \tag{25.31}$$

In order to obtain the output state, we replace the input operators $\hat{a}^{\dagger}_{c1,i}$ and $\hat{a}^{\dagger}_{t0,i}$ in terms of the output operators using Eq. (25.26) and obtain

$$|\psi\rangle_{out} = \frac{1}{\sqrt{3}}\left(-\hat{a}^{\dagger}_{c1,o} + \hat{a}^{\dagger}_{t0,o} + \hat{a}^{\dagger}_{t1,o}\right)\frac{1}{\sqrt{3}}\left(\hat{a}^{\dagger}_{c1,o} + \hat{a}^{\dagger}_{t0,o} + \hat{a}^{\dagger}_{s1,o}\right)|0000\rangle$$

Opening the brackets, we obtain

$$|\psi\rangle_{out} = \frac{1}{3}\Big(-\hat{a}^{\dagger}_{c1,o}\hat{a}^{\dagger}_{c1,o} - \hat{a}^{\dagger}_{c1,o}\hat{a}^{\dagger}_{t0,o} - \hat{a}^{\dagger}_{c1,o}\hat{a}^{\dagger}_{s1,o} + \hat{a}^{\dagger}_{t0,o}\hat{a}^{\dagger}_{c1,o} + \hat{a}^{\dagger}_{t0,o}\hat{a}^{\dagger}_{t0,o} + \hat{a}^{\dagger}_{t0,o}\hat{a}^{\dagger}_{s1,o} + \hat{a}^{\dagger}_{t1,o}\hat{a}^{\dagger}_{c1,o} + \hat{a}^{\dagger}_{t1,o}\hat{a}^{\dagger}_{t0,o} + \hat{a}^{\dagger}_{t1,o}\hat{a}^{\dagger}_{s1,o}\Big)|0000\rangle$$

The second and fourth terms in the above equation cancel each other and we obtain

$$|\psi\rangle_{out} = \frac{1}{3}\Big(-\hat{a}^{\dagger}_{c1,o}\hat{a}^{\dagger}_{c1,o} - \hat{a}^{\dagger}_{c1,o}\hat{a}^{\dagger}_{s1,o} + \hat{a}^{\dagger}_{t0,o}\hat{a}^{\dagger}_{t0,o} + \hat{a}^{\dagger}_{t0,o}\hat{a}^{\dagger}_{s1,o} + \hat{a}^{\dagger}_{t1,o}\hat{a}^{\dagger}_{c1,o} + \hat{a}^{\dagger}_{t1,o}\hat{a}^{\dagger}_{t0,o} + \hat{a}^{\dagger}_{t1,o}\hat{a}^{\dagger}_{s1,o}\Big)|0000\rangle$$

which becomes

$$|\psi\rangle_{out} = -\frac{\sqrt{2}}{3}|0200\rangle - \frac{1}{3}\hat{a}^{\dagger}_{s1,o}|0100\rangle + \frac{\sqrt{2}}{3}|0020\rangle + \frac{1}{3}\hat{a}^{\dagger}_{s1,o}|0010\rangle + \frac{1}{3}|0101\rangle + \frac{1}{3}|0011\rangle + \frac{1}{3}\hat{a}^{\dagger}_{s1,o}|0001\rangle$$

where we have used the fact that $\hat{a}^{\dagger}_{t0,o}\hat{a}^{\dagger}_{t0,o}|0000\rangle = \hat{a}^{\dagger}_{t0,o}|0010\rangle = \sqrt{2}|0020\rangle$ and similarly for the term $\hat{a}^{\dagger}_{c1,o}\hat{a}^{\dagger}_{c1,o}|0000\rangle$.

The term that corresponds to having one photon each in the control and target qubits is

$$|\psi\rangle_{out} = \frac{1}{3}\hat{a}^{\dagger}_{t1,o}\hat{a}^{\dagger}_{c1,o}|0000\rangle = \frac{1}{3}|0101\rangle \tag{25.32}$$

Thus, when the control bit is in the logical state $|1\rangle = |01\rangle$, the target qubit changes from the logical state $|0\rangle = |10\rangle$ to the state $|1\rangle = |01\rangle$ showing the C-NOT gate operation. The control qubit remains in the same state as at the input. The probability of this process is again $\frac{1}{9}$.

It can be shown that if we consider the general state

$$\begin{aligned}|\psi\rangle_{in} &= \alpha|0\rangle_c|0\rangle_t + \beta|0\rangle_c|1\rangle_t + \gamma|1\rangle_c|0\rangle_t + \delta|1\rangle_c|1\rangle_t \\ &= \alpha|10\rangle_c|10\rangle_t + \beta|10\rangle_c|01\rangle_t + \gamma|01\rangle_c|10\rangle_t + \delta|01\rangle_c|01\rangle_t\end{aligned}$$

which can be written in terms of input creation operators as

$$|\psi\rangle_{in} = \Big(\alpha\hat{a}^{\dagger}_{c0,i}\hat{a}^{\dagger}_{t0,i} + \beta\hat{a}^{\dagger}_{c0,i}\hat{a}^{\dagger}_{t1,i} + \gamma\hat{a}^{\dagger}_{c1,i}\hat{a}^{\dagger}_{t0,i} + \delta\hat{a}^{\dagger}_{c1,i}\hat{a}^{\dagger}_{t1,i}\Big)|00\rangle_c|00\rangle_t \tag{25.33}$$

Replacing the input operators in terms of the output operators given by Eq. (25.27) and collecting terms which have one photon in each of the two qubits, we obtain the output state as

$$|\psi\rangle_{out} = \frac{\alpha}{3}|10\rangle_c|10\rangle_t + \frac{\beta}{3}|10\rangle_c|01\rangle_t + \frac{\gamma}{3}|01\rangle_c|01\rangle_t + \frac{\delta}{3}|01\rangle_c|10\rangle_t \tag{25.34}$$

which is nothing but

$$|\psi\rangle_{out} = \frac{1}{3}\left(\alpha|0\rangle_c|0\rangle_t + \beta|0\rangle_c|1\rangle_t + \gamma|1\rangle_c|1\rangle_t + \delta|1\rangle_c|0\rangle_t\right) \tag{25.35}$$

This is exactly the operation of the C-NOT gate. Thus, when photons are detected in both the control output and the target outputs, then it is confirmed that a C-NOT operation has taken place.

In the experimental demonstration of the C-NOT gate polarization qubits are converted to dual rail qubits and after the operation, they are converted back into polarization qubits (O'Brien et al. 2003). See also Politi et al. (2008) and Politi et al. (2009) for a waveguide implementation of the C-NOT gate operation.

25.3.2.2 C-NOT Operation with Polarization Qubits

Another scheme using integrated quantum photonic network involving polarization qubits has been demonstrated by Crespi et al. (2011). A schematic of the device is shown in Fig. 25.4. In the figure, PPDC represents partially polarizing directional coupler. Such a coupler has different coupling lengths for the two orthogonal polarizations and hence as the two polarizations propagate through the directional coupler, they exchange energy between the two waveguides at different rates. Hence, at any given length of interaction, the power coupled for the two orthogonal polarizations will be different and hence it is termed partially polarizing directional coupler.

The logical qubits for the control qubit are $|0\rangle_c = |H\rangle_c$, $|1\rangle_c = |V\rangle_c$ and the logical qubits for the target qubits are $|0\rangle_T = |D\rangle_T = \frac{1}{\sqrt{2}}\left(|H\rangle_T + |V\rangle_T\right)$ and $|1\rangle_T = |A\rangle_T = \frac{1}{\sqrt{2}}\left(|H\rangle_T - |V\rangle_T\right)$. Here H and V correspond to horizontal and vertical polarization states, respectively.

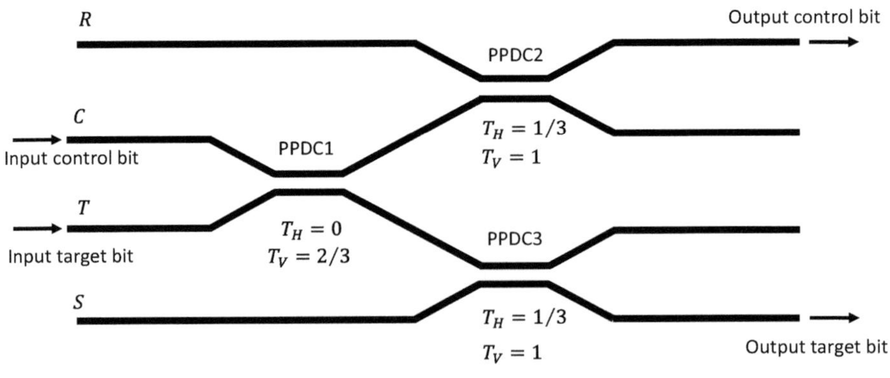

Fig. 25.4 C-NOT gate realization using polarization qubits

The device has four inputs and four outputs which we will refer to as ports *R*, *C*, *T* and *S*. The input control port is *C* and the input target port is *T*. The output control port is *R* and the output target port is *S*. Note that the output control and target waveguide ports are different from those of the control and target waveguides at the input. These ports are connected through partially polarizing directional couplers with the following transmission probabilities (represented by *T*) corresponding to horizontal (H) and vertical (V) polarization states:

PPDC1:

$$T_H = 0, T_V = \frac{2}{3} \tag{25.36a}$$

PPDC2 and PPDC3:

$$T_H = \frac{1}{3}, T_V = 1 \tag{25.36b}$$

Referring to Chap. 5 for a directional coupler made from identical waveguides, for an interaction length *L*, the amplitude transmission coefficient is $i \sin \kappa L$ (from the input port to the cross port) and the amplitude reflection coefficient (from the input port to the same port at the output) is $\cos\kappa L$. Thus, for $\kappa L \approx 0.6155$, the transmission probability will be $\sin^2\kappa L \approx \frac{1}{3}$ and $\cos^2\kappa L \approx \frac{2}{3}$. If the waveguide has a slight birefringence through either the geometry of the waveguide or the material of the waveguide, the value of the coupling coefficient for the *H* and *V* polarizations will be different. By an appropriate choice of the waveguide parameters, it is possible to have the transmission coefficients defined in Eqs. (25.36a) and (25.36b) for the *H* and *V* polarizations (Crespi et al. 2011).

We now need to relate the input and output from the device for both the horizontal and vertical polarization states. For this we first note that for the vertical polarization state the matrix connecting the input plane to just after PPDC1 (before PPDC2 and PPDC3) is given by

$$\widehat{M}_1 = \begin{pmatrix} 1 & 0 & 0 & 0 \\ 0 & 1/\sqrt{3} & i\sqrt{2/3} & 0 \\ 0 & i\sqrt{2/3} & 1/\sqrt{3} & 0 \\ 0 & 0 & 0 & 1 \end{pmatrix}$$

Similarly the transfer matrix between the plane just after PPDC1 and the output plane is given by

$$\widehat{M}_2 = \begin{pmatrix} 0 & i & 0 & 0 \\ i & 0 & 0 & 0 \\ 0 & 0 & 0 & i \\ 0 & 0 & i & 0 \end{pmatrix}$$

Hence, the transfer matrix connecting the input and outputs is given by

$$\widehat{M} = \widehat{M}_2\widehat{M}_1$$

Substituting for $\widehat{M}_1$ and $\widehat{M}_2$, we obtain the following relationship between the output and input annihilation operators for the vertical polarization state:

$$\begin{pmatrix} \widehat{a}_{RV,o} \\ \widehat{a}_{CV,o} \\ \widehat{a}_{TV,o} \\ \widehat{a}_{SV,o} \end{pmatrix} = \begin{pmatrix} 0 & i/\sqrt{3} & -\sqrt{2/3} & 0 \\ i & 0 & 0 & 0 \\ 0 & 0 & 0 & i \\ 0 & -\sqrt{2/3} & i/\sqrt{3} & 0 \end{pmatrix} \begin{pmatrix} \widehat{a}_{RV,i} \\ \widehat{a}_{CV,i} \\ \widehat{a}_{TV,i} \\ \widehat{a}_{SV,i} \end{pmatrix} \tag{25.37}$$

Here the subscripts refer to the port (R, C, T, S), the polarization state (V), and input or output (i, o).

We can use this equation and express the input operators in terms of the output annihilation operators and obtain

$$\begin{pmatrix} \widehat{a}_{RV,i} \\ \widehat{a}_{CV,i} \\ \widehat{a}_{TV,i} \\ \widehat{a}_{SV,i} \end{pmatrix} = \begin{pmatrix} 0 & -i & 0 & 0 \\ -i/\sqrt{3} & 0 & 0 & -\sqrt{2/3} \\ -\sqrt{2/3} & 0 & 0 & -i/\sqrt{3} \\ 0 & 0 & -i & 0 \end{pmatrix} \begin{pmatrix} \widehat{a}_{RV,o} \\ \widehat{a}_{CV,o} \\ \widehat{a}_{TV,o} \\ \widehat{a}_{SV,o} \end{pmatrix} \tag{25.38}$$

Note that the 4 × 4 matrix in Eq. (25.38) is the inverse of the 4 × 4 matrix in Eq. (25.37) and satisfies the relationship $\widehat{U}^{-1} = \widehat{U}^{\dagger}$.

We follow the same procedure for the horizontal polarization and we obtain the following expressions for the transfer matrices $\widehat{M}_1$ and $\widehat{M}_2$:

$$\widehat{M}_1 = \begin{pmatrix} 1 & 0 & 0 & 0 \\ 0 & 1 & 0 & 0 \\ 0 & 0 & 1 & 0 \\ 0 & 0 & 0 & 1 \end{pmatrix}$$

and

$$\widehat{M}_2 = \begin{pmatrix} \sqrt{2/3} & i/\sqrt{3} & 0 & 0 \\ i/\sqrt{3} & \sqrt{2/3} & 0 & 0 \\ 0 & 0 & \sqrt{2/3} & i/\sqrt{3} \\ 0 & 0 & i/\sqrt{3} & \sqrt{2/3} \end{pmatrix}$$

and obtain the following equation ($\widehat{M} = \widehat{M}_2\widehat{M}_1$) connecting the output annihilation operators and the input annihilation operators:

$$\begin{pmatrix} \widehat{a}_{RH,o} \\ \widehat{a}_{CH,o} \\ \widehat{a}_{TH,o} \\ \widehat{a}_{SH,o} \end{pmatrix} = \begin{pmatrix} \sqrt{2/3} & i\sqrt{1/3} & 0 & 0 \\ i\sqrt{1/3} & \sqrt{2/3} & 0 & 0 \\ 0 & 0 & \sqrt{2/3} & i\sqrt{1/3} \\ 0 & 0 & i\sqrt{1/3} & \sqrt{2/3} \end{pmatrix} \begin{pmatrix} \widehat{a}_{RH,i} \\ \widehat{a}_{CH,i} \\ \widehat{a}_{TH,i} \\ \widehat{a}_{SH,i} \end{pmatrix} \tag{25.39}$$

which on inversion gives us

$$\begin{pmatrix} \widehat{a}_{RH,i} \\ \widehat{a}_{CH,i} \\ \widehat{a}_{TH,i} \\ \widehat{a}_{SH,i} \end{pmatrix} = \begin{pmatrix} \sqrt{2/3} & -i/\sqrt{3} & 0 & 0 \\ -i/\sqrt{3} & \sqrt{2/3} & 0 & 0 \\ 0 & 0 & \sqrt{2/3} & -i/\sqrt{3} \\ 0 & 0 & -i/\sqrt{3} & \sqrt{2/3} \end{pmatrix} \begin{pmatrix} \widehat{a}_{RH,o} \\ \widehat{a}_{CH,o} \\ \widehat{a}_{TH,o} \\ \widehat{a}_{SH,o} \end{pmatrix} \tag{25.40}$$

In order to show the C-NOT gate operation, we first consider the following input in which both the control qubit and the target qubit are in logical $|0\rangle$ state:

$$\begin{aligned} |\psi\rangle_{in} &= |0\rangle_c|0\rangle_T = |H\rangle_c|D\rangle_T = \frac{1}{\sqrt{2}}|H\rangle_c(|H\rangle_T + |V\rangle_T) \\ &= \frac{1}{\sqrt{2}}(|H\rangle_c|H\rangle_T + |H\rangle_C|V\rangle_T) \end{aligned} \tag{25.41}$$

Now,

$$|H\rangle_c|H\rangle_T = \widehat{a}^\dagger_{CH,i}\widehat{a}^\dagger_{TH,i}|0\rangle_c|0\rangle_T$$

and

$$|H\rangle_c|V\rangle_T = \widehat{a}^\dagger_{CH,i}\widehat{a}^\dagger_{TV,i}|0\rangle_c|0\rangle_T$$

Hence,

$$|\psi\rangle_{in} = \frac{1}{\sqrt{2}}\left(\widehat{a}^\dagger_{CH,i}\widehat{a}^\dagger_{TH,i} + \widehat{a}^\dagger_{CH,i}\widehat{a}^\dagger_{TV,i}\right)|0\rangle_c|0\rangle_T \tag{25.42}$$

As before we obtain the output by using the relationship between the input operators and the output operators from Eq. (25.40). We work out term by term and have for the first term in Eq. (25.42)

$$\begin{aligned} \widehat{a}^\dagger_{CH,i}\widehat{a}^\dagger_{TH,i} &= \left(i\sqrt{1/3}\widehat{a}^\dagger_{RH,o} + \sqrt{2/3}\widehat{a}^\dagger_{CH,o}\right)\left(\sqrt{2/3}\widehat{a}^\dagger_{TH,o} + i\sqrt{1/3}\widehat{a}^\dagger_{SH,o}\right) \\ &= i\frac{\sqrt{2}}{3}\widehat{a}^\dagger_{RH,o}\widehat{a}^\dagger_{TH,o} - \frac{1}{3}\widehat{a}^\dagger_{RH,o}\widehat{a}^\dagger_{SH,o} + \frac{2}{3}\widehat{a}^\dagger_{CH,o}\widehat{a}^\dagger_{TH,o} + i\frac{\sqrt{2}}{3}\widehat{a}^\dagger_{CH,o}\widehat{a}^\dagger_{SH,o} \end{aligned}$$

Similarly, we obtain for the second term in Eq. (25.42)

$$\hat{a}^{\dagger}_{CH,i}\hat{a}^{\dagger}_{TV,i} = \left(i\sqrt{1/3}\hat{a}^{\dagger}_{RH,o} + \sqrt{2/3}\hat{a}^{\dagger}_{CH,o}\right)\left(-\sqrt{2/3}\hat{a}^{\dagger}_{RV,o} + i/\sqrt{3}\hat{a}^{\dagger}_{SV,o}\right)$$
$$= -i\frac{\sqrt{2}}{3}\hat{a}^{\dagger}_{RH,o}\hat{a}^{\dagger}_{RV,o} - \frac{1}{3}\hat{a}^{\dagger}_{RH,o}\hat{a}^{\dagger}_{SV,o} - \frac{2}{3}\hat{a}^{\dagger}_{CH,o}\hat{a}^{\dagger}_{RV,o} + i\frac{\sqrt{2}}{3}\hat{a}^{\dagger}_{CH,o}\hat{a}^{\dagger}_{SV,o}$$

Using these expressions, we will obtain the output state.

Now the only terms in the two expressions giving one photon each in the output arm R (which corresponds to the output control qubit) and output arm S (which corresponds to the output target qubit) are the second terms in the expressions. Hence, we write for the output state satisfying this condition

$$|\psi\rangle_{out} = -\frac{1}{3}\frac{1}{\sqrt{2}}\left(\hat{a}^{\dagger}_{RH,o}\hat{a}^{\dagger}_{SH,o} + \hat{a}^{\dagger}_{RH,o}\hat{a}^{\dagger}_{SV,o}\right)|0\rangle_R|0\rangle_S$$
$$= -\frac{1}{3}\hat{a}^{\dagger}_{RH,o}\left[\frac{1}{\sqrt{2}}\left(\hat{a}^{\dagger}_{SH,o} + \hat{a}^{\dagger}_{SV,o}\right)\right]|0\rangle_R|0\rangle_S = -\frac{1}{3}|H\rangle_R\frac{1}{\sqrt{2}}\left(|H\rangle_S + |V\rangle_S\right)$$

Since the ports R and S are now the control and target qubit outputs, we can rewrite the output in terms of the control and target qubits as

$$|\psi\rangle_{out} = -\frac{1}{3}|H\rangle_C\frac{1}{\sqrt{2}}\left(|H\rangle_T + |V\rangle_T\right) = -\frac{1}{3}|H\rangle_C|D\rangle_T = -\frac{1}{3}|0\rangle_C|0\rangle_T \quad (25.43)$$

Thus, when the input control bit is in state $|0\rangle_C = |H\rangle_C$, the output target bit remains unchanged as $|0\rangle_T = |D\rangle_T$ and the probability of this process is $\frac{1}{9}$.

We can similarly consider the following input state:

$$|\psi\rangle_{in} = |1\rangle_c|0\rangle_T = |V\rangle_c|D\rangle_T = \frac{1}{\sqrt{2}}|V\rangle_c\left(|H\rangle_T + |V\rangle_T\right) = \frac{1}{\sqrt{2}}\left(|V\rangle_c|H\rangle_T + |V\rangle_C|V\rangle_T\right) \quad (25.44)$$

and perform exactly similar calculations to obtain the output state with one photon each in the output ports R and S

$$|\psi\rangle_{out} = -\frac{1}{3}\frac{1}{\sqrt{2}}\left(\hat{a}^{\dagger}_{RV,o}\hat{a}^{\dagger}_{SH,o} - \hat{a}^{\dagger}_{RV,o}\hat{a}^{\dagger}_{SV,o}\right)|0\rangle_R|0\rangle_S = -\frac{1}{3}\hat{a}^{\dagger}_{RV,o}\left[\frac{1}{\sqrt{2}}\left(\hat{a}^{\dagger}_{SH,o} - \hat{a}^{\dagger}_{SV,o}\right)\right]|0\rangle_R|0\rangle_S$$
$$= -\frac{1}{3}|V\rangle_R\frac{1}{\sqrt{2}}\left(|H\rangle_S - |V\rangle_S\right) \quad (25.45)$$

Since the ports R and S are now the control and target qubit outputs, we have

$$|\psi\rangle_{out} = -\frac{1}{3}|V\rangle_C \frac{1}{\sqrt{2}}\left(|H\rangle_T - |V\rangle_T\right) = -\frac{1}{3}|H\rangle_C|A\rangle_T = -\frac{1}{3}|1\rangle_c|1\rangle_T \quad (25.46)$$

Thus, when the input control bit is in state $|1\rangle_C = |V\rangle_C$, the output target changes from $|0\rangle_T = |D\rangle_T$ to $|1\rangle_T = |A\rangle_T$ as expected for a C-NOT gate operation. The probability of this process is again $\frac{1}{9}$.

One can extend the analysis for a general input and show the C-NOT gate operation.

This shows the operation of the C-NOT gate using polarization qubits rather than spatial qubits (Crespi et al. 2011).

Exercise 25.5 Consider a symmetric Y-junction (see Chap. 7) made of single mode channel waveguide with the pair of waveguides (W1 and W2) of the Y-junction to be separated by a large separation to have no evanescent interaction between them. The input to the device is excited by a single photon. Assuming no loss in the system, what is the qubit state generated at the output of the device?

Solution Assuming adiabatic transition, the input photon will excite the symmetric fundamental mode of the device which will be a superposition state given by

$$|\psi\rangle = \frac{1}{\sqrt{2}}(|0\rangle + |1\rangle)$$

with $|0\rangle$ representing photon occupying waveguide W1 and $|1\rangle$ representing the photon occupying waveguide W2.

Exercise 25.6 Consider an asymmetric Y-junction (see Chap. 7) with single mode waveguides and the output waveguides separated by a large separation to have no evanescent interaction. What is the condition to be satisfied so that the output of the device generates a qubit state with amplitude $\frac{\sqrt{3}}{2}$ of the state $|0\rangle$ and $\frac{1}{2}$ of the state $|1\rangle$?

Solution Since all the waveguides are single mode, the input fundamental mode excitation will evolve to the fundamental mode of the two-waveguide region of the Y-junction. For nonidentical waveguides characterized by a coupling coefficient κ and waveguides with propagation constant difference $\Delta\beta$, the ratio of the amplitudes of the field in the two waveguides is given by (see Chap. 5) $\frac{S_0}{R_0} = \sqrt{\frac{\Delta\beta^2}{4\kappa^2} + 1} - \frac{\Delta\beta}{2\kappa}$. To achieve the ratio as $\frac{1}{2} / \frac{\sqrt{3}}{2} = \frac{1}{\sqrt{3}}$, we must have $\frac{\Delta\beta}{2\kappa} = \frac{1}{\sqrt{3}}$. The quantum state will then be given by $|\psi\rangle = \frac{\sqrt{3}}{2}|0\rangle + \frac{1}{2}|1\rangle$.

Exercise 25.7 Consider the device shown in Fig. 25.5. It consists of two phase shifters and two 3 dB beam splitters. Obtain the unitary operator representing the device.

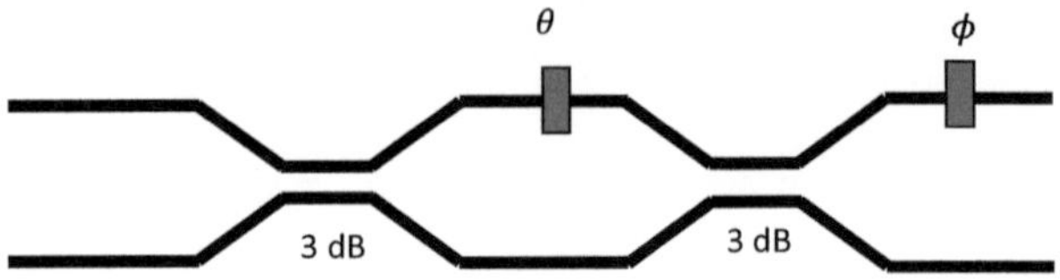

Fig. 25.5 Device configuration for Exercise 25.7

Solution We represent the phase shifter by Eq. (25.21) and the beam splitter by Eq. (25.19). Hence, the device is characterized by the following unitary operator:

$$\widehat{U} = \widehat{P}(\phi,0)\widehat{B}\widehat{P}(\theta,0)\widehat{B} = \begin{pmatrix} e^{i\phi} & 0 \\ 0 & 1 \end{pmatrix} \frac{1}{\sqrt{2}} \begin{pmatrix} 1 & i \\ i & 1 \end{pmatrix} \begin{pmatrix} e^{i\theta} & 0 \\ 0 & 1 \end{pmatrix} \frac{1}{\sqrt{2}} \begin{pmatrix} 1 & i \\ i & 1 \end{pmatrix}$$

Multiplying the matrices we obtain

$$\widehat{U} = ie^{i\theta/2} \begin{pmatrix} e^{i\phi} \sin\frac{\theta}{2} & e^{i\phi} \cos\frac{\theta}{2} \\ \cos\frac{\theta}{2} & -\sin\frac{\theta}{2} \end{pmatrix} \tag{25.47}$$

This represents a generic beam splitter with adjustable parameters of phase shifters. Check whether the matrix $\widehat{U}$ obtained in Eq. (25.47) is a unitary matrix.

Qubits based on polarization state are exploited to realize quantum optical processing. For example, the two basis states are a photon in horizontal polarization $|0\rangle = |H\rangle$ and vertical polarizations $|1\rangle = |V\rangle$. A general qubit is represented by a photon that is in a coherent superposition of the basis states and is given by $|\psi\rangle = \alpha|H\rangle + \beta|V\rangle = \alpha|0\rangle + \beta|1\rangle$, with $|\alpha|^2 + |\beta|^2 = 1$. In channel waveguides, the two basis states $|0\rangle$ and $|1\rangle$ would correspond to quasi TE and quasi TM modes, respectively (see Chap. 4). One-qubit logic gates can be implemented on a photon propagating in a single mode waveguide using birefringence of the waveguide. We refer to excellent research papers for more details on polarization encoded qubit operations using waveguides (Crespi et al. 2011; Morozko et al. 2023; Heilmann et al. 2014; Wang et al. 2016; Zeuner et al. 2018; Slussarenko and Pryde 2019). Linear optical quantum computation using frequency as a degree of freedom has also been proposed (Yamazaki et al. 2023). A nice review of programmable photonic circuits is given in Bogaerts et al. (2020).

25.3.2.3 Bell State Measurement

As discussed in Chap. 17, the four Bell states form an orthogonal set of two-qubit states. In a measurement termed Bell state measurement, it is desired to identify which Bell state is input to the device. In principle, we should be able to distinguish with 100% probability the specific Bell state that is incident, but many schemes work with only a lower probability. For optical qubits, Bell state measurement can be performed by sending the quantum state on a balanced beam splitter.

As an example let us consider the following Bell state in polarization degree of freedom (see Chap. 17):

$$|\psi^+\rangle = \frac{1}{\sqrt{2}}\left(|H\rangle_1|V\rangle_2 + |V\rangle_1|H\rangle_2\right) \tag{25.48}$$

where subscripts 1 and 2 refer to the two photons which are entangled. We let such a quantum state incident on a polarization independent balanced beam splitter with the states corresponding to 1 and 2 incident on the two input arms of the beam splitter.

We recall the relationship between the input ports (1 and 2) and output ports (3 and 4) of the annihilation operators of a beam splitter (see Eq. 19.43 and Fig. 19.1). Writing the relationships for the horizontal H and the vertical V polarizations, we have

$$\widehat{a}_{1H} = \frac{1}{\sqrt{2}}(-i\widehat{a}_{3H} + \widehat{a}_{4H}) \tag{25.49a}$$

$$\widehat{a}_{2H} = \frac{1}{\sqrt{2}}(\widehat{a}_{3H} - i\widehat{a}_{4H}) \tag{25.49b}$$

$$\widehat{a}_{1V} = \frac{1}{\sqrt{2}}(-i\widehat{a}_{3V} + \widehat{a}_{4V}) \tag{25.49c}$$

$$\widehat{a}_{2V} = \frac{1}{\sqrt{2}}(\widehat{a}_{3V} - i\widehat{a}_{4V}) \tag{25.49d}$$

We write the input state as

$$|\psi\rangle_{in} = |\psi^+\rangle = \frac{1}{\sqrt{2}}\left(\widehat{a}_{1H}^\dagger\widehat{a}_{2V}^\dagger + \widehat{a}_{1V}^\dagger\widehat{a}_{2H}^\dagger\right)|0\rangle_1|0\rangle_2 \tag{25.50}$$

As discussed in Chap. 19, we obtain the output state by replacing $\widehat{a}_{1H}^\dagger$, $\widehat{a}_{1V}^\dagger$, $\widehat{a}_{2H}^\dagger$ and $\widehat{a}_{2V}^\dagger$ in terms of $\widehat{a}_{3H}^\dagger$, $\widehat{a}_{3V}^\dagger$, $\widehat{a}_{4H}^\dagger$ and $\widehat{a}_{4V}^\dagger$ to obtain

$$|\psi\rangle_{out} = \frac{1}{2\sqrt{2}}\left[\left(\widehat{a}_{4H}^\dagger + i\widehat{a}_{3H}^\dagger\right)\left(\widehat{a}_{3V}^\dagger + i\widehat{a}_{4V}^\dagger\right) + \left(\widehat{a}_{4V}^\dagger + i\widehat{a}_{3V}^\dagger\right)\left(\widehat{a}_{3H}^\dagger + i\widehat{a}_{4H}^\dagger\right)\right]|0\rangle_3|0\rangle_4$$

which opens to

$$\begin{aligned}|\psi\rangle_{out} = \frac{1}{2\sqrt{2}}\Big[&\left(\widehat{a}_{4H}^\dagger\widehat{a}_{3V}^\dagger + i\widehat{a}_{4H}^\dagger\widehat{a}_{4V}^\dagger + i\widehat{a}_{3H}^\dagger\widehat{a}_{3V}^\dagger - \widehat{a}_{3H}^\dagger\widehat{a}_{4V}^\dagger\right)\\ &+\left(\widehat{a}_{4V}^\dagger\widehat{a}_{3H}^\dagger + i\widehat{a}_{4V}^\dagger\widehat{a}_{4H}^\dagger + i\widehat{a}_{3V}^\dagger\widehat{a}_{3H}^\dagger - \widehat{a}_{3V}^\dagger\widehat{a}_{4H}^\dagger\right)\Big]|0\rangle_3|0\rangle_4\end{aligned}$$

which gives

$$|\psi\rangle_{out} = \frac{1}{\sqrt{2}}\left(\hat{a}^{\dagger}_{3H}\hat{a}^{\dagger}_{3V} + \hat{a}^{\dagger}_{4H}\hat{a}^{\dagger}_{4V}\right)|0\rangle_3|0\rangle_4 = \frac{1}{\sqrt{2}}\left(|1\rangle_{3H}|1\rangle_{3V} + |1\rangle_{4H}|1\rangle_{4V}\right) \tag{25.51}$$

where we have ignored a global phase factor. Hence, in this case both the photons of the entangled state exit from the same port, either port 3 or port 4, of the beam splitter.

In an identical fashion, it is possible to obtain the following outputs when the other three Bell states are incident on the beam splitter:

$$|\psi\rangle_{in} = |\psi^-\rangle \rightarrow |\psi\rangle_{out} = \frac{1}{\sqrt{2}}\left(\hat{a}^{\dagger}_{3H}\hat{a}^{\dagger}_{4V} - \hat{a}^{\dagger}_{3V}\hat{a}^{\dagger}_{4H}\right)|0\rangle_3|0\rangle_4 \tag{25.52}$$

$$|\psi\rangle_{in} = |\phi^+\rangle \rightarrow |\psi\rangle_{out} = \frac{1}{2\sqrt{2}}\left(\hat{a}^{\dagger}_{3H}\hat{a}^{\dagger}_{3H} + \hat{a}^{\dagger}_{4H}\hat{a}^{\dagger}_{4H} + \hat{a}^{\dagger}_{3V}\hat{a}^{\dagger}_{3V} + \hat{a}^{\dagger}_{4V}\hat{a}^{\dagger}_{4V}\right)|0\rangle_3|0\rangle_4 \tag{25.53}$$

$$|\psi\rangle_{in} = |\phi^-\rangle \rightarrow |\psi\rangle_{out} = \frac{1}{2\sqrt{2}}\left(\hat{a}^{\dagger}_{3H}\hat{a}^{\dagger}_{3H} + \hat{a}^{\dagger}_{4H}\hat{a}^{\dagger}_{4H} - \hat{a}^{\dagger}_{3V}\hat{a}^{\dagger}_{3V} - \hat{a}^{\dagger}_{4V}\hat{a}^{\dagger}_{4V}\right)|0\rangle_3|0\rangle_4 \tag{25.54}$$

where we have ignored the global phase factor.

From Eq. (25.51), we can see that for an input state $|\psi^+\rangle$, both photons end up at the same detector but having orthogonal polarizations while for an input state $|\psi^-\rangle$ one photon reaches each of the two output detectors. Thus, these can be unambiguously discriminated.

However, when the input state is either $|\phi^+\rangle$ or $|\phi^-\rangle$, the output is ambiguous. Hence, only two of the Bell states can be precisely identified leading to an overall success probability of 50%.

There have been many techniques that have been proposed to have a higher probability of identification using ancilla photons (see, e.g., Kwiat and Weinfurter 1998; Kim et al. 2001; Grice 2011; Ewert and Loock 2014; Bayerbach et al. 2023).

25.3.3 Boson Sampling

One interesting photonic quantum computational problem is the Boson sampling problem (Aaronson and Arkhipov 2011) which tries to show quantum computational advantage of linear optical computing.

We start by looking at a 3 dB waveguide directional coupler (50/50 beam splitter) with identical photons entering each of the input arms. The probability that at the output of the beam splitter, we will find one photon in each of the output arms is zero (see Chap. 19). We recall that the directional coupler is represented by the following unitary evolution matrix (see Chap. 19):

$$U = \frac{1}{\sqrt{2}}\begin{pmatrix} 1 & i \\ i & 1 \end{pmatrix} \tag{25.55}$$

Now, the permanent of a 2 × 2 matrix

$$M = \begin{pmatrix} \alpha_1 & \alpha_2 \\ \beta_1 & \beta_2 \end{pmatrix} \tag{25.56}$$

is defined by

$$Per(M) = \alpha_1\beta_2 + \alpha_2\beta_1 \tag{25.57}$$

Note that the determinant of the matrix is $(\alpha_1\beta_2 - \alpha_2\beta_1)$ in contrast to the permanent. In obtaining the permanent, the sign changes in the determinant are not implemented. Thus, the permanent of a 3 × 3 matrix

$$M = \begin{pmatrix} \alpha_1 & \alpha_2 & \alpha_3 \\ \beta_1 & \beta_2 & \beta_3 \\ \gamma_1 & \gamma_2 & \gamma_3 \end{pmatrix} \tag{25.58}$$

is

$$Per(M) = \alpha_1\beta_2\gamma_3 + \alpha_1\beta_3\gamma_2 + \alpha_2\beta_3\gamma_1 + \alpha_2\beta_1\gamma_3 + \alpha_3\beta_1\gamma_2 + \alpha_3\beta_2\gamma_1 \tag{25.59}$$

To show the relationship between the permanent and the probability of detecting photons at the outputs, we consider the incidence of identical photons in each of the input arms of a 2 × 2 directional coupler.

We recall the relationship between the output and input amplitudes of a 3 dB directional coupler with identical waveguides (see Eq. 5.60):

$$\begin{pmatrix} R_{out} \\ S_{out} \end{pmatrix} = \frac{1}{\sqrt{2}}\begin{pmatrix} 1 & i \\ i & 1 \end{pmatrix}\begin{pmatrix} R_{in} \\ S_{in} \end{pmatrix} \tag{25.60}$$

The quantum equivalent of the directional coupler relates the output and input annihilation operators (see Chap. 19):

$$\begin{pmatrix} \hat{b}_1 \\ \hat{b}_2 \end{pmatrix} = \frac{1}{\sqrt{2}}\begin{pmatrix} 1 & i \\ i & 1 \end{pmatrix}\begin{pmatrix} \hat{a}_1 \\ \hat{a}_2 \end{pmatrix} \tag{25.61}$$

where $\widehat{a}_1$ and $\widehat{a}_2$ represent the annihilation operators of the two input arms and $\widehat{b}_1$ and $\widehat{b}_2$ represent the annihilation operators of the two output arms.

We can invert Eq. (25.61) and obtain the following equation:

$$\begin{pmatrix} \widehat{a}_1 \\ \widehat{a}_2 \end{pmatrix} = \frac{1}{\sqrt{2}} \begin{pmatrix} 1 & -i \\ -i & 1 \end{pmatrix} \begin{pmatrix} \widehat{b}_1 \\ \widehat{b}_2 \end{pmatrix} \tag{25.62}$$

which gives us

$$\begin{pmatrix} \widehat{a}_1^\dagger \\ \widehat{a}_2^\dagger \end{pmatrix} = \frac{1}{\sqrt{2}} \begin{pmatrix} 1 & i \\ i & 1 \end{pmatrix} \begin{pmatrix} \widehat{b}_1^\dagger \\ \widehat{b}_2^\dagger \end{pmatrix} = \begin{pmatrix} \alpha_1 & \alpha_2 \\ \beta_1 & \beta_2 \end{pmatrix} \begin{pmatrix} \widehat{b}_1^\dagger \\ \widehat{b}_2^\dagger \end{pmatrix} \tag{25.63}$$

We consider an input state with one photon entering each input arm of the directional coupler. Thus, the input state will be given by

$$|\Psi_{in}\rangle = |1_{a1}, 1_{a2}\rangle = \widehat{a}_1^\dagger \widehat{a}_2^\dagger |0_1, 0_2\rangle \tag{25.64}$$

and the output state will be obtained by replacing $\widehat{a}_1^\dagger$ and $\widehat{a}_2^\dagger$ from Eq. (25.63) and is given by (see Chap. 19)

$$|\Psi_{out}\rangle = \left(\alpha_1 \widehat{b}_1^\dagger + \alpha_2 \widehat{b}_2^\dagger\right) \left(\beta_1 \widehat{b}_1^\dagger + \beta_2 \widehat{b}_2^\dagger\right) |0_3, 0_4\rangle$$

which can be simplified to

$$|\Psi_{out}\rangle = \alpha_1 \beta_1 \widehat{b}_1^{\dagger 2} + \alpha_2 \beta_2 \widehat{b}_2^{\dagger 2} + (\alpha_1 \beta_2 + \alpha_2 \beta_2) \widehat{b}_1^\dagger \widehat{b}_2^\dagger \tag{25.65}$$

Thus, the probability of finding one photon in each of the output arms of the directional coupler will be

$$P = |\langle 1_{b1}, 1_{b2} || \Psi_{out}\rangle|^2 = |\alpha_1 \beta_2 + \alpha_2 \beta_1|^2 \tag{25.66}$$

The bracketed term in Eq. (25.66) is nothing but the square of the permanent of the transfer matrix. Hence, the probability of finding one photon in each of the output arms of the 3 dB directional coupler is just the permanent of the matrix.

For a 50/50 directional coupler, $\alpha_1\beta_2 + \alpha_2\beta_1 = 0$ and hence $P = Per(U) = 0$, that is, the probability of finding one photon in each of the output arm is zero.

We can see the same relationship for a three-waveguide directional coupler too. For a three-waveguide directional coupler with identical equally spaced waveguides, the coupled mode equations describing the evolution of the power are given by the following three coupled mode equations (see Eqs. 5.116–5.118):

$$\frac{dS_1}{dz} = i\kappa S_2 \tag{25.67a}$$

$$\frac{dS_2}{dz} = i\kappa S_1 + i\kappa S_3 \tag{25.67b}$$

$$\frac{dS_3}{dz} = i\kappa S_2 \tag{25.67c}$$

with $\kappa = \kappa_1 = \kappa_2$. Solving the above set of three equations (see Exercise 25.8), we obtain the following relation between the amplitudes of the fields in the three waveguides at a value $z = L$ and at $z = 0$:

$$\begin{pmatrix} S_1(L) \\ S_2(L) \\ S_3(L) \end{pmatrix} = \begin{pmatrix} \frac{1}{2}(1+C) & \frac{i}{\sqrt{2}}S & -\frac{1}{2}(1-C) \\ \frac{i}{\sqrt{2}}S & C & \frac{i}{\sqrt{2}}S \\ -\frac{1}{2}(1-C) & \frac{i}{\sqrt{2}}S & \frac{1}{2}(1+C) \end{pmatrix} \begin{pmatrix} S_1(0) \\ S_2(0) \\ S_3(0) \end{pmatrix} \tag{25.68}$$

where $C = \cos\sqrt{2}\kappa L$ and $S = \sin\sqrt{2}\kappa L$.

Exercise 25.8 Solve Eqs. (25.67a)–(25.67c) and obtain the relationship given in Eq. (25.68).

Solution Differentiating Eq. (25.67b) with respect to z and using Eqs. (25.67a) and (25.67c), we obtain

$$\frac{d^2S_2}{dz^2} = -2\kappa^2 S_2$$

whose general solution is given by

$$S_2(z) = a\cos\sqrt{2}\kappa z + b\sin\sqrt{2}\kappa z \tag{25.69a}$$

where a and b are constants of integration. Substituting the solution given by Eq. (25.69a) in Eqs. (25.67a) and (25.67c), we obtain the following two solutions:

$$S_1(z) = i\frac{a}{\sqrt{2}}\sin\sqrt{2}\kappa z - i\frac{b}{\sqrt{2}}\cos\sqrt{2}\kappa z + c \tag{25.69b}$$

$$S_3(z) = i\frac{a}{\sqrt{2}}\sin\sqrt{2}\kappa z - i\frac{b}{\sqrt{2}}\cos\sqrt{2}\kappa z + d \tag{25.69c}$$

where again c and d are constants of integration. Substituting the solutions given by Eqs. (25.69b) and (25.69c) in Eq. (25.67b), we obtain $d = -c$.

Using the initial conditions $S_1(z = 0) = S_1(0)$, $S_2(z = 0) = S_2(0)$ and $S_3(z = 0) = S_3(0)$, we obtain the following:

$$a = S_2(0) \tag{25.70}$$

$$b = \frac{i}{\sqrt{2}}[S_1(0) + S_3(0)] \tag{25.71}$$

$$c = -d = \frac{1}{2}[S_1(0) - S_3(0)] \tag{25.72}$$

Substituting the values of the constants of integration in Eqs. (25.69a)–(25.69c) and simplifying, we obtain Eq. (25.68) relating the amplitudes $S_1(z)$, $S_2(z)$ and $S_3(z)$ in terms of $S_1(0)$, $S_2(0)$ and $S_3(0)$.

Exercise 25.9 In Chap. 5, we had discussed the normal modes of a three waveguide directional coupler. A normal mode by definition does not change in the transverse distribution and only the phase changes. Consider the following three input conditions and show that in all the three cases the output amplitudes are independent of the length L of propagation and suffer only phase changes thus form modes: (a) $\begin{pmatrix} 1 \\ \sqrt{2} \\ 1 \end{pmatrix}$, (b) $\begin{pmatrix} 1 \\ 0 \\ -1 \end{pmatrix}$ and (c) $\begin{pmatrix} 1 \\ -\sqrt{2} \\ 1 \end{pmatrix}$. The first and last correspond to the two symmetric normal modes, while the central one is the antisymmetric normal mode.

Solution For an input given by (a), we have

$$\begin{pmatrix} S_1(L) \\ S_2(L) \\ S_3(L) \end{pmatrix} = \begin{pmatrix} \frac{1}{2}(1+C) & \frac{i}{\sqrt{2}}S & -\frac{1}{2}(1-C) \\ \frac{i}{\sqrt{2}}S & C & \frac{i}{\sqrt{2}}S \\ -\frac{1}{2}(1-C) & \frac{i}{\sqrt{2}}S & \frac{1}{2}(1+C) \end{pmatrix} \begin{pmatrix} 1 \\ \sqrt{2} \\ 1 \end{pmatrix} = e^{i\sqrt{2}\kappa L} \begin{pmatrix} 1 \\ \sqrt{2} \\ 1 \end{pmatrix}$$

which is thus a mode of the system. Similarly for the input given by (b), we have

$$\begin{pmatrix} S_1(L) \\ S_2(L) \\ S_3(L) \end{pmatrix} = \begin{pmatrix} \frac{1}{2}(1+C) & \frac{i}{\sqrt{2}}S & -\frac{1}{2}(1-C) \\ \frac{i}{\sqrt{2}}S & C & \frac{i}{\sqrt{2}}S \\ -\frac{1}{2}(1-C) & \frac{i}{\sqrt{2}}S & \frac{1}{2}(1+C) \end{pmatrix} \begin{pmatrix} 1 \\ 0 \\ -1 \end{pmatrix} = \begin{pmatrix} 1 \\ 0 \\ -1 \end{pmatrix}$$

and for the input given by (c), we have

$$\begin{pmatrix} S_1(L) \\ S_2(L) \\ S_3(L) \end{pmatrix} = \begin{pmatrix} \frac{1}{2}(1+C) & \frac{i}{\sqrt{2}}S & -\frac{1}{2}(1-C) \\ \frac{i}{\sqrt{2}}S & C & \frac{i}{\sqrt{2}}S \\ -\frac{1}{2}(1-C) & \frac{i}{\sqrt{2}}S & \frac{1}{2}(1+C) \end{pmatrix} \begin{pmatrix} 1 \\ -\sqrt{2} \\ 1 \end{pmatrix}$$

$$= e^{-i\sqrt{2}\kappa L} \begin{pmatrix} 1 \\ \sqrt{2} \\ 1 \end{pmatrix}$$

Note that the distribution among the amplitudes in the three waveguides does not change with L and thus the chosen inputs form normal modes of the three waveguide directional coupler. The exponential factors give us the propagation constant of the three modes as $\beta_0 + \sqrt{2}\kappa$, β_0 and $\beta_0 - \sqrt{2}\kappa$ (see Chap. 5). Here β_0 is the propagation constant of the individual mode of each waveguide.

As with the directional coupler with two waveguides, here too we can relate the annihilation operators of the output modes (represented by $\widehat{b}_j; j = 1, 2, 3$) in the three waveguides with the annihilation operators of the input modes (represented by $\widehat{a}_j; j = 1, 2, 3$) of the three waveguides through the following relation:

$$\begin{pmatrix} \widehat{b}_1 \\ \widehat{b}_2 \\ \widehat{b}_3 \end{pmatrix} = \begin{pmatrix} \frac{1}{2}(1+C) & \frac{i}{\sqrt{2}}S & -\frac{1}{2}(1-C) \\ \frac{i}{\sqrt{2}}S & C & \frac{i}{\sqrt{2}}S \\ -\frac{1}{2}(1-C) & \frac{i}{\sqrt{2}}S & \frac{1}{2}(1+C) \end{pmatrix} \begin{pmatrix} \widehat{a}_1 \\ \widehat{a}_2 \\ \widehat{a}_3 \end{pmatrix} \qquad (25.73)$$

We can invert Eq. (25.73) and obtain the relationship between the creation operators at the input with the creation operators at the output (remember that the transfer matrix is unitary):

$$\begin{pmatrix} \widehat{a}_1^\dagger \\ \widehat{a}_2^\dagger \\ \widehat{a}_3^\dagger \end{pmatrix} = \begin{pmatrix} \frac{1}{2}(1+C) & \frac{i}{\sqrt{2}}S & -\frac{1}{2}(1-C) \\ \frac{i}{\sqrt{2}}S & C & \frac{i}{\sqrt{2}}S \\ -\frac{1}{2}(1-C) & \frac{i}{\sqrt{2}}S & \frac{1}{2}(1+C) \end{pmatrix} \begin{pmatrix} \widehat{b}_1^\dagger \\ \widehat{b}_2^\dagger \\ \widehat{b}_3^\dagger \end{pmatrix} \qquad (25.74)$$

Now consider an input state with one photon in each of the input waveguides:

$$|\psi_{in}\rangle = |1_{a1}, 1_{a2}, 1_{a3}\rangle = \widehat{a}_1^\dagger \widehat{a}_2^\dagger \widehat{a}_3^\dagger |0_1, 0_2, 0_3\rangle \qquad (25.75)$$

Writing the 3 × 3 matrix in Eq. (25.74) as $\begin{pmatrix} \alpha_1 & \alpha_2 & \alpha_3 \\ \beta_1 & \beta_2 & \beta_3 \\ \gamma_1 & \gamma_2 & \gamma_3 \end{pmatrix}$, we can again calculate the probability of finding one photon in each of the output waveguides. These terms will have the product $\hat{b}_1^\dagger \hat{b}_2^\dagger \hat{b}_3^\dagger$. Collecting these terms, we obtain the probability of detecting one photon in each of the output waveguides as

$$\begin{aligned} P &= |\langle 1_{b1}, 1_{b2}, 1_{b3} || \psi_{out} \rangle|^2 \\ &= (\alpha_1\beta_2\gamma_3 + \alpha_1\beta_3\gamma_2 + \alpha_2\beta_3\gamma_1 + \alpha_2\beta_1\gamma_3 + \alpha_3\beta_1\gamma_2 + \alpha_3\beta_2\gamma_1)^2 \end{aligned} \tag{25.76}$$

which is nothing but the square of the permanent of the transfer matrix.

The result which we have shown for 2 × 2 and 3 × 3 directional couplers can be generalized to an *nxn* coupler consisting of *n* parallel waveguides. The probability amplitude of obtaining a certain output distribution of photons is directly proportional to the permanent of the matrix and obtaining the permanent of a matrix by conducting experiments with photons propagating through linear optical networks is an interesting photonic quantum computing problem.

Now let us consider the problem of finding the permanent of an *nxn* matrix. Using classical computation, this problem is known to be very hard if *n* is large. Photonic quantum computation comes in handy here. In order to obtain the permanent of the matrix, we need to design a directional coupler with *n* parallel identical and equally spaced waveguides, launch identical photons in each of the input waveguides, and perform the experiment to obtain the probability distribution which can lead us to the permanent of the matrix. For a randomly designed circuit, classically computing the permanent scales up exponentially in time with the size of the matrix and thus photonic quantum computing can help in obtaining the permanent in a much shorter time. Interesting experimental demonstrations of the Boson sampling problem can be found in Tillmann et al. (2013), Crespi et al. (2013), Spring et al. (2013), Spagnolo et al. (2014), Wang et al. (2019) and Brod et al. (2019), and Hoch et al. (2022).

25.4 Quantum Teleportation

Quantum teleportation is a technique to teleport information about a system from one location to another using entangled states (Bennett et al. 1993). We will very briefly discuss the basic principle behind the quantum teleportation protocol using an example of teleporting an unknown polarization state of a photon to another photon at a different location. For a recent review on quantum teleportation, see Hu et al. (2023).

For specificity, let us assume that we have a photon having an unknown polarization state written in the $H - V$ basis as

$$|\chi\rangle = \alpha|H_1\rangle + \beta|V_1\rangle \tag{25.77}$$

with the normalization condition $|\alpha|^2 + |\beta|^2 = 1$ which Alice wishes to teleport to Bob who is located in another far off location.

For the teleportation protocol, first Alice and Bob share a polarization entangled state. Let us assume that the polarization entangled state that Alice and Bob share is given by

$$|\phi^+\rangle = \frac{1}{\sqrt{2}}(|H_A H_B\rangle + |V_A V_B\rangle) \tag{25.78}$$

with the entangled photon with subscript A being with Alice and the photon with subscript B being with Bob. In general, the entangled photon pair can be in any one of the four Bell states.

The combined initial state with Alice

$$|\chi_{1A}\rangle = |\chi\rangle|\phi^+\rangle = \frac{1}{\sqrt{2}}(\alpha|H_1\rangle + \beta|V_1\rangle)(|H_A H_B\rangle + |V_A V_B\rangle) \tag{25.79}$$

which can be written as

$$|\chi_{1AB}\rangle = \frac{1}{\sqrt{2}}(\alpha|H_1 H_A H_B\rangle + \alpha|H_1 V_A V_B\rangle + \beta|V_1 H_A H_B\rangle + \beta|V_1 V_A V_B\rangle) \tag{25.80}$$

Alice now performs a Bell state measurement on the photon pair that she possesses. In this we first apply the C-NOT gate on the first two qubits (which are in Alice's possession) in the state. As discussed in Sect. 25.3.2, a C-NOT gate does nothing to the target qubit if the control qubit is H and changes the target qubit to the orthogonal state if the control qubit is V. Here we are assuming $|H\rangle = |0\rangle$ and $|V\rangle = |1\rangle$.

Hence, a C-NOT operation transforms $|\chi_{1AB}\rangle$ to the following state of all the three photons:

$$\frac{1}{\sqrt{2}}(\alpha|H_1 H_A H_B\rangle + \alpha|H_1 V_A V_B\rangle + \beta|V_1 V_A H_B\rangle + \beta|V_1 H_A V_B\rangle)$$

Alice now applies the Hadamard gate on the first qubit in the state which has the following operation:

$$|H_1\rangle \rightarrow \frac{1}{\sqrt{2}}(|H_1\rangle + |V_1\rangle)$$

$$|V_1\rangle \rightarrow \frac{1}{\sqrt{2}}(|H_1\rangle - |V_1\rangle)$$

Hence, after the Hadamard gate the state becomes

$$\frac{1}{2}\alpha(|H_1\rangle + |V_1\rangle)|H_AH_B\rangle + \frac{1}{2}\alpha(|H_1\rangle + |V_1\rangle)|V_AV_B\rangle + \frac{1}{2}\beta(|H_1\rangle - |V_1\rangle)|V_AH_B\rangle + \frac{1}{2}\beta(|H_1\rangle - |V_1\rangle)|H_AV_B\rangle$$

which can be written as

$$\frac{1}{2}\alpha|H_1H_AH_B\rangle + \frac{1}{2}\alpha|V_1H_AH_B\rangle + \frac{1}{2}\alpha|H_1V_AV_B\rangle + \frac{1}{2}\alpha|V_1V_AV_B\rangle + \frac{1}{2}\beta|H_1V_AH_B\rangle - \frac{1}{2}\beta|V_1V_AH_B\rangle + \frac{1}{2}\beta|H_1H_AV_B\rangle - \frac{1}{2}\beta|V_1H_AV_B\rangle$$

The above state can also be written in the following form:

$$\frac{1}{2}|H_1H_A\rangle(\alpha|H_B\rangle + \beta|V_B\rangle) + \frac{1}{2}|V_1H_A\rangle(\alpha|H_B\rangle - \beta|V_B\rangle) + \frac{1}{2}|H_1V_A\rangle(\alpha|V_B\rangle + \beta|H_B\rangle) + \frac{1}{2}|V_1V_A\rangle(\alpha|V_B\rangle - \beta|H_B\rangle) \tag{25.81}$$

Note that Alice has in possession photons marked with subscripts 1 and A on which she has been carrying out qubit operations.

Now Alice performs measurement on the pair of photons that she possesses. There is equal probability of 0.25 of getting any of the following combinations: $|H_1H_A\rangle$, $|V_1H_A\rangle$, $|H_1V_A\rangle$ or $|V_1V_A\rangle$. Depending on the result obtained by Alice, Bob needs to perform a unitary operation on his qubit to complete the teleportation protocol. Alice communicates through a *classical communication channel* as to which result Alice has obtained during her measurement and depending on the result, Bob needs to perform an operation on the qubit that is in his possession to transform the photon in his possession to the unknown photon polarization state.

If Alice finds her photon pairs to be in the state $|H_1H_A\rangle$, then the photon in possession of Bob gets transformed to the unknown state completing the teleportation protocol.

If Alice gets the state as $|V_1H_A\rangle$, then Bob has the state $(\alpha|H_B\rangle - \beta|V_B\rangle)$ and to get the unknown state has to do a transformation on his qubit which converts $(\alpha|H_B\rangle - \beta|V_B\rangle)$ to $(\alpha|H_B\rangle + \beta|V_B\rangle)$.

If Alice finds $|H_1V_A\rangle$, then Bob performs an operation to transform from the state $(\alpha|V_B\rangle + \beta|H_B\rangle)$ to $(\alpha|H_B\rangle + \beta|V_B\rangle)$.

And finally if Alice finds $|V_1V_A\rangle$, then Bob needs to carry out a transformation to convert the state $(\alpha|V_B\rangle - \beta|H_B\rangle)$ to the state $(\alpha|H_B\rangle + \beta|V_B\rangle)$.

It is important to note that until Alice transmits the result of her measurement to Bob through a classical communication channel, the protocol is incomplete. Hence, there is no information transmitted at superluminal speed. Also note that once the protocol is completed, the unknown state that was in possession of Alice has disappeared.

It is interesting to note that since the entire teleportation operation is a linear operation applied to the unknown state, the same process can be used for other states which are superpositions.

The first experimental demonstration of the teleportation protocol was carried by Anton Zeilinger's group (Bouwmeester et al. 1997). Quantum teleportation and entanglement distribution have been achieved over long distances via free space (see, e.g., Yin et al. 2012) as well as through fibers (Shen et al. 2023; Labonte et al. 2024).

Exercise 25.10 Express $|\chi_{1AB}\rangle$ given in Eq. (25.80) in terms of the four Bell states for the pair of photons with subscripts 1 and A.

Solution The four Bell states consisting of photons labeled 1 and A are given by (see Chap. 17)

$$|\psi^+\rangle = \frac{1}{\sqrt{2}}(|H_1V_A\rangle + |V_1H_A\rangle)$$

$$|\psi^-\rangle = \frac{1}{\sqrt{2}}(|H_1V_A\rangle - |V_1H_A\rangle)$$

$$|\phi^+\rangle = \frac{1}{\sqrt{2}}(|H_1H_A\rangle + |V_1V_A\rangle)$$

$$|\phi^-\rangle = \frac{1}{\sqrt{2}}(|H_1H_A\rangle - |V_1V_A\rangle)$$

From these equations, we obtain

$$|H_1H_A\rangle = \frac{1}{\sqrt{2}}(|\phi^+\rangle + |\phi^-\rangle)$$

$$|V_1V_A\rangle = \frac{1}{\sqrt{2}}(|\phi^+\rangle - |\phi^-\rangle)$$

$$|H_1V_A\rangle = \frac{1}{\sqrt{2}}(|\psi^+\rangle + |\psi^-\rangle)$$

$$|V_1H_A\rangle = \frac{1}{\sqrt{2}}(|\psi^+\rangle - |\psi^-\rangle)$$

Using these expressions, we can write $|\chi_{1AB}\rangle$ as

$$\begin{aligned}|\chi_{1AB}\rangle = &\frac{1}{2}|\phi^+\rangle(\alpha|H_B\rangle + \beta|V_B\rangle) + \frac{1}{2}|\phi^-\rangle(\alpha|H_B\rangle - \beta|V_B\rangle)\\ &+\frac{1}{2}|\psi^+\rangle(\alpha|V_B\rangle + \beta|H_B\rangle) + \frac{1}{2}|\psi^-\rangle(\alpha|V_B\rangle - \beta|H_B\rangle)\end{aligned}$$

Thus, if Alice performs a Bell state measurement, the probability of obtaining any of the four Bell states is 0.25. Depending on the result that

Alice obtains from her measurement, she can communicate the result to Bob who can then perform a unitary operation on his qubit to complete the teleportation protocol.

Problems

25.1. Show that if the photon is known to be linearly polarized but the polarization basis is not known, it is not possible to determine the state of polarization of the photon through a single measurement.

25.2. Representing a photon having horizontal polarization by the state $|0\rangle$, if an input photon in the state $|0\rangle$ passes through a quarter wave plate oriented with one of its axis oriented at 45° to the horizontal, what will be the output photon state written in the (H, V) basis and in the (D, A) basis?

25.3. How would you convert a photon in a state given by $|\psi\rangle = \cos\phi|0\rangle + \sin\phi|1\rangle$ to the state $|0\rangle$ with unit probability? Here $|0\rangle = |H\rangle$ and $|1\rangle = |V\rangle$.

25.4. Make a table similar to Table 25.2 that one will obtain by using the entangled state $|\psi^-\rangle$.

25.5. Consider the dual rail C-NOT gate discussed in Sect. 25.3.2. Obtain the output states with one photon in each qubit for the following input states: (a) $|\psi\rangle_{in} = |10\rangle_c|01\rangle_t$ and (b) $|\psi\rangle_{in} = |01\rangle_c|01\rangle_t$.

25.6. Assuming that the input annihilation and creation operators satisfy the necessary commutation relations, show that the output annihilation and creation operators as obtained from Eq. (25.27) also satisfy the necessary commutation relations.

25.7. Consider the C-NOT gate using polarization qubits as shown in Fig. 25.3. Assuming the input state to be (a) $|0\rangle_c|1\rangle_T$ and (b) $|1\rangle_c|1\rangle_T$, obtain the outputs and show that the operation corresponds to the C-NOT gate.

25.8. Refer to the C-NOT gate shown in Fig. 25.2. Consider an input state given by (a) $|\psi\rangle_{in} = |0\rangle_t = |10\rangle_t$ and (b) $|\psi\rangle_{in} = |1\rangle_t = |01\rangle_t$. The input state contains only one photon in the target bit and no control bit photon. Calculate the output state and show that at the output the target qubit exits in the same state as that in which it entered the device. What is the probability of finding the target bit in the two cases?

25.9. Derive Eqs. (25.52)–(25.54).

25.10. Carry out the analysis for the teleportation protocol if the other three Bell states are used.

25.11. Consider a directional coupler with input ports a_i and b_i and output ports a_o and b_o. In the analysis given in Sect. 25.3.2 discussing the polarization qubit operation of the C-NOT gate, the transfer function of the coupler between the input and output was assumed to be given by

$$\begin{pmatrix} a_o \\ b_o \end{pmatrix} = \begin{pmatrix} \sqrt{R} & i\sqrt{T} \\ i\sqrt{T} & \sqrt{R} \end{pmatrix} \begin{pmatrix} a_i \\ b_i \end{pmatrix}$$

where R and T represent the energy reflection and transmission coefficients and $R + T = 1$. Instead consider the directional coupler to be given by the following equivalent transfer matrix:

$$\begin{pmatrix} a_o \\ b_o \end{pmatrix} = \begin{pmatrix} -\sqrt{R} & \sqrt{T} \\ \sqrt{T} & \sqrt{R} \end{pmatrix} \begin{pmatrix} a_i \\ b_i \end{pmatrix}$$

Show that with this transfer function also the C-NOT gate performs exactly the same as given in Sect. 25.3.2.

25.12. Show that the transfer matrix contained in Eq. (25.68) is unitary.

25.13. Referring to Eq. (25.73) verify that the output creation and annihilation operators satisfy the requisite commutation relations.

25.14. Carry out the calculations and obtain Eq. (25.45) from Eq. (25.44) retaining only terms which give a photon each in the control and target qubits.

25.15. Obtain the value of $\langle\psi|\psi\rangle$ of the state $|\psi\rangle_{out}$ given in Eq. (25.29) and interpret the result.

References

Aaronson, S., and A. Arkhipov. 2011. *The computational complexity of linear optics.* In Proceedings of the 43rd annual ACM symposium on Theory of computing (STOC '11). New York, NY: ACM. pp. 333–342.

Assumpcao, D., D. Renaud, A. Baradari, B. Zeng, C. De-Eknamkul, C. J. Xin, A. Shams-Ansari, D. Barton, B. Machielse, and M. Loncar. 2024. A thin film lithium niobate near-infrared platform for multiplexing quantum nodes. *arXiv*:2405.0391212v1 [physics.optics].

Azuma, K., S. E. Economou, D. Elkouss, P. Hilaire, L. Jiang, H.-K. Lo, and I. Tzitrin. 2023. Quantum repeaters: from quantum networks to the quantum internet. *Rev Mod Phys* 95:045006.

Bayerbach, M. J., S. E. D'Aurelio, P. V. Loock, and S. Barz. 2023. Bell-state measurement exceeding 50% success probability with linear optics. *Sci Adv* 9:eadf4080.

Bell, B. A., and I. A. Walmsley. 2021. Further compactifying linear optical unitaries. *APL Photonics* 6:070804.

Bennett, C. H., G. Brassard, C. Crepeau, R. Jozsa, A. Peres, and W. K. Wootters. 1993. Teleporting an unknown quantum state via dual classical and Einstein-Podolsky-Rosen channels. *Phys Rev Lett* 70:1895–1899.

Blanco, X. P., C. M. Orille, J. Linares, H. G. Nunez, and D. Balado. 2022. Quantum projectors implemented with optical directional couplers in ion-exchanged glasses. *J Lightwave Tech* 40: 7676–7684.

Bogaerts, W., D. Perez, J. Capmany, D. A. B. Miller, J. Poon, D. Englund, F. Morichetti, and A. Melloni. 2020. Programmable photonic circuits. *Nature* 586:207–216.

Bouwmeester, D., J.-W. Pan, K. Mattle, M. Eibi, H. Weinfurter, and A. Zeilinger. 1997. Experimental quantum teleportation. *Nature* 390:575–579.

Brod, D. J., E. F. Galvão, A. Crespi, R. Osellame, N. Spagnolo, and F. Sciarrino. 2019. Photonic implementation of boson sampling: a review. *Adv Photonics* 1:034001.

Carolan, J., C. Harrold, C. Sparrow, E. Martín-López, N. J. Russel, J. W. Silverstone, P. J. Shadbolt, N. Matsuda, M. Oguma, M. Itoh, G. D. Marshall, M. G. Thompson, J. C. F. Mathews, T. Hashimoto, J. L. O'Brian, and A. Laing. 2015. Universal linear optics. *Science* 349:711–716.

Crespi, A., R. Ramponi, R. Osellame, L. Sansoni, I. Bongioanni, F. Sciarrino, G. Vallone, and P. Mataloni. 2011. Integrated photonic quantum gates for polarization qubits. *Nat Commun* 2: 566. https://doi.org/10.1038/ncomms1570.

Crespi, A., R. Osellame, R. Ramponi, D. J. Brod, E. F. Galvão, N. Spagnolo, C. Vitelli, E. Maiorino, P. Mataloni, and F. Sciarrino. 2013. Integrated multimode interferometers with arbitrary designs for photonic boson sampling. *Nat Photonics* 7:545–549.

Djordjevic, I. B. 2010. On the photonic implementation of universal quantum gates, Bell states preparation circuit, quantum relay and quantum LDPC encoders and decoders. *IEEE Photonics J* 2:81–91.

Ewert, F., and P. V. Loock. 2014. ¾ efficient Bell measurement with passive linear optics and unentangled ancillae. *Phys Rev Lett* 113:140403.

Grice, W. P. 2011. Arbitrary complete Bell-state measurement using only linear optical elements. *Phys Rev A* 84:042331.

Heilmann, R., M. Grafe, S. Nolte, and A. Szameit. 2014. Arbitrary photonic wave plate operations on chip: Realizing Hadamard, Pauli X and rotation gates for polarization qubits. *Sci Rep* 4:4118.

Hoch, F., S. Piacentini, T. Giordani, Z.-N. Tian, M. Iuliano, C. Esposito, A. Camillini, G. Carvacho, F. Ceccarelli, N. Spagnolo, A. Crespi, F. Sciarrino, and R. Osellame. 2022. Reconfigurable continuously-coupled 3D photonic circuit for boson sampling experiments. *npj Quant Inform* 8: 55. https://doi.org/10.1038/s41534-022-00568-6.

Hu, X.-M., Y. Guo, B.-H. Liu, C.-F. Li, and G.-C. Guo. 2023. Progress in quantum teleportation. *Nat Rev Phys* 5:339–353.

Kim, Y.-H., S. P. Kulik, and Y. Shih. 2001. Quantum teleportation of a polarization state with a complete Bell state measurement. *Phys Rev Lett* 86:1370–1373.

Knill, E., R. Laflamme, and G. J. Milburn. 2001. A scheme for efficient quantum computation with linear optics. *Nature* 409:46–52.

Kok, P., W. J. Munro, K. Nemoto, T. C. Ralph, J. P. Dowling, and G. J. Milburn. 2007. Linear optical quantum computing with photons. *Rev Mod Phys* 79:135–174.

Kwiat, P. G., and H. Weinfurter. 1998. Embedded Bell-state analysis. *Phys Rev A* 58:R2623–R2626.

Kwon, Y., A. Baldazzi, L. Pavesi, B.-S. Choi. 2024. Quantum circuit mapping for universal and scalable computing in MZI-based integrated photonics. https://arxiv.org/abs/2401.16875 (accessed March 16, 2024).

Labonté, L., O. Alibart, V. D'Auria, F. Doutre, J. Etesse, G. Sauder, A. Martin, É. Picholle, and S. Tanzilli. 2024. Integrated photonics for quantum communications and metrology. *Phys Rev X Quan* 5:010101.

Morozko, F., A. Novitsky, A. Mikhalychev, and A. Karabchevsky. 2023. On chip polarization encoded single qubit gates with twisted waveguides. *Phys Rev Res* 5:043155.

O'Brien, J. L., G. J. Pryde, A. G. White, T. C. Ralph, and D. Branning. 2003. Demonstration of an all-optical quantum controlled-NOT gate. *Nature* 426:264–267.

Politi, A., M. J. Cryan, J. Rarity, S. Yu, and J. L. O'Brien. 2008. Silica on silicon waveguide quantum circuits. *Science* 320:646–649.

Politi, A., J. C. F. Mathews, M. G. Thompson, and J. L. O'Brien. 2009. Integrated quantum photonics. *IEEE J Selected Topics Quant Electron* 15:1673–1684.

Romero, J., and G. Milburn. 2024. Photonic quantum computing. *arXiv*:2404:03367v1 [quant-ph].

Sax, R., A. Boaron, G. Boso, S. Atzeni, A. Crespi, F. Grünenfelder, D. Rusca, A. Al-Saadi, D. Bronzi, S. Kupijai, H. Rhee, R. Osellame, and H. Zbinden. 2023. High speed integrated QKD system. *Photonics Res* 11:1007–1014.

Shen, S., C. Yuan, Z. Zhang, H. Yu, R. Zhang, C. Yang, H. Li, Z. Wang, Y. Wang, G. Deng, H. Song, L. You, Y. Fan, G. Guo, and Q. Zhou. 2023. Herz-rate metropolitan quantum teleportation. *Light Sci Appl* 12:115. https://doi.org/10.1038/s41377-023-01158-7.

Sibson, P., J. E. Kennard, S. Stanisic, C. Erven, J. L. O'Brien, and M. G. Thompson. 2017. Integrated silicon photonics for high-speed quantum key distribution. *Optica* 4:172–177.

Slussarenko, S., and G. J. Pryde. 2019. Photonic quantum information processing: A concise review. *Appl Phys Rev* 6:041303.

Spagnolo, N., C. Vitelli, M. Bentivegna, D. J. Brod, A. Crespi, F. Flamini, S. Giacomini, G. Milani, R. Ramponi, P. Mataloni, R. Osellame, E. F. Galvão, and F. Sciarrino. 2014. Experimental validation of photonic boson sampling. *Nat Photonics* 8:615–620.

Spring, J. B., B. J. Metcalf, P. C. Humphreys, W. S. Kolthammer, X.-M. Jin, M. Barbieri, A. Datta, N. Thomas-Peter, N. K. Langford, D. Kundys, J. C. Gates, B. J. Smith, P. G. R. Smith, and I. A. Walmsley. 2013. Boson sampling on a photonic chip. *Science* 339:798–801.

Thompson, M. G., A. Politi, J. C. F. Mathews, and J. L. O'Brien. 2011. Integrated waveguide circuits for optical quantum computing. *IET Circuits Devices Syst* 5:94–102.

Tillmann, M., B. Dakić, R. Heilmann, S. Nolte, A. Szameit, and P. Walther. 2013. Experimental boson sampling. *Nat Photonics* 7:540–544.

Walmsley, I. 2023. Light in quantum computing and simulation: perspective. *Opt Quant* 1:35–40.

Wang, S. M., Q. Q. Cheng, Y. X. Gong, P. Xu, C. Sun, L. Li, T. Li, and S. N. Zhu. 2016. A 14 x 14 mm^2 footprint polarization-encoded quantum controlled NOT gate based on hybrid waveguide. *Nat Commun* 7:11490. https://doi.org/10.1038/ncomms11490.

Wang, H., J. Qin, X. Ding, M.-C. Chen, S. Chen, X. You, Y.-M. He, X. Jiang, L. You, Z. Wang, C. Schneider, J. J. Renema, S. Höfling, C.-Y. Lu, and J.-W. Pan. 2019. Boson sampling with 20 input photons and a 60-mode interferometer in a 10^{14} dimensional Hilbert space. *Phys Rev Lett* 123:250503.

Wei, K., X. Hu, Y. Du, X. Hua, Z. Zhao, Y. Chen, C. Huang, and X. Xiao. 2023. Resource-efficient quantum key distribution with integrated silicon photonics. *Photonics Res* 11:1364–1372.

Yamazaki, T., T. Arizono, T. Kobayashi, R. Ikuta, and T. Yamamoto. 2023. Linear optical quantum computation with frequency-comb qubits and passive devices. *Phys Rev Lett* 130:200602.

Yard, P., A. E. Jones, S. Paesani, A. Maïnos, J. F. F. Bulmer, and A. Laing. 2024. On-chip quantum information processing with distinguishable photons. *Phys Rev Lett* 132:150602.

Yin, J., J.-G. Ren, H. Lu, Y. Cao, H.-L. Yong, Y.-P. Wu, C. Liu, S.-K. Liao, F. Zhou, Y. Jiang, X.-D. Cai, P. Xu, G.-S. Pan, J.-J. Jia, Y.-M. Huang, H. Yin, J.-Y. Wang, Y.-A. Chen, C.-Z. Peng, and J.-W. Pan. 2012. Quantum teleportation and entanglement distribution over 100 km free-space channels. *Nature* 488:185–188.

Zeuner, J., A. N. Sharma, M. Tillmann, R. Heilman, M. Grafe, A. Moquanaki, A. Szameit, and P. Walther. 2018. Integrated-optics heralded controlled NOT gate for polarization-encoded qubits. *npj Quant Inform* 4:13. https://doi.org/10.1038/s41534-018-0068-0.

Index

K. Thyagarajan, *Integrated Quantum Photonics*, Graduate Texts in Physics,
https://doi.org/10.1007/978-3-031-85728-7

Q

R

S

T

U

V

MIX
Papier aus verantwortungsvollen Quellen
Paper from responsible sources
FSC® C105338

If you have any concerns about our products, you can contact us on
ProductSafety@springernature.com

In case Publisher is established outside the EU, the EU authorized representative is:
Springer Nature Customer Service Center GmbH
Europaplatz 3, 69115 Heidelberg, Germany

Printed by Libri Plureos GmbH
in Hamburg, Germany